OTHER TITLES OF INTEREST FROM ST. LUCIE PRESS

Development, Environment, and Global Dysfunction: Toward Sustainable Recovery

Resolving Environmental Conflict: Towards Sustainable Community Development

Economic Theory for Environmentalists

Sustainable Forestry: Philosophy, Science, and Economics

Everglades: The Ecosystem and Its Restoration

The Everglades Handbook: Understanding the Ecosystem

The Everglades: A Threatened Wilderness (Video)

From the Forest to the Sea: The Ecology of Wood in Streams, Rivers, Estuaries, and Oceans

Environmental Effects of Mining

Handbook of Trace Elements in the Environment

Contamination of Groundwaters

Lead in Soil: Recommended Guidelines

Metal Compounds in Environment and Life: Interrelation Between Chemistry and Biology

For more information about these titles call, fax or write:

St. Lucie Press
100 E. Linton Blvd., Suite 403B
Delray Beach, FL 33483
TEL (407) 274-9906 • FAX (407) 274-9927

S$_L^t$

ENVIRONMENTAL FATE and EFFECTS of PULP and PAPER MILL EFFLUENTS

Edited by

MARK R. SERVOS, Ph.D.
Great Lakes Laboratory for Fisheries and Aquatic Sciences,
Canada Department of Fisheries and Oceans

KELLY R. MUNKITTRICK, Ph.D.
Great Lakes Laboratory for Fisheries and Aquatic Sciences,
Canada Department of Fisheries and Oceans

JOHN H. CAREY, Ph.D.
Aquatic Ecosystems Conservation Branch,
National Water Research Institute,
Environment Canada

GLEN J. VAN DER KRAAK, Ph.D.
Department of Zoology, University of Guelph

St. Lucie Press
Delray Beach, Florida

Copyright ©1996 by St. Lucie Press

All rights reserved. No part of this publication may be reproduced, stored in a retrieval system or transmitted in any form or by any means, electronic, mechanical, photocopying, recording or otherwise, without the prior written permission of the publisher.

Printed and bound in the U.S.A. Printed on acid-free paper.
10 9 8 7 6 5 4 3 2 1

ISBN 1-884015-71-9

    All rights reserved. Authorization to photocopy items for internal or personal use, or the personal or internal use of specific clients, is granted by St. Lucie Press, provided that $.50 per page photocopied is paid directly to Copyright Clearance Center, 222 Rosewood Drive, Danvers, MA 01923 USA. The fee code for users of the Transactional Reporting Service is ISBN 1-884015-71-9 1/96/$100/$.50. The fee is subject to change without notice. For organizations that have been granted a photocopy license by the CCC, a separate system of payment has been arranged.
    The copyright owner's consent does not extend to copying for general distribution, for promotion, for creating new works, or for resale. Specific permission must be obtained from St. Lucie Press for such copying.
    Direct all inquiries to St. Lucie Press, Inc., 100 E. Linton Blvd., Suite 403B, Delray Beach, Florida 33483.

Phone: (407) 274-9906
Fax: (407) 274-9927

S^t_L

Published by
St. Lucie Press
100 E. Linton Blvd., Suite 403B
Delray Beach, FL 33483

Table of Contents

Preface xiii

Foreword: Where were we in 1991? *Anders Södergren (Sweden)* xv

Section I: Identification and Origin of Chemicals and Toxicity

Invited Overviews

Effects of internal process changes and external treatment on effluent chemistry. *Lars Strömberg, Roland Mörck, Filipe de Sousa and Olof Dahlman (Sweden)* 3

Sources of pulping and bleaching derived chemicals in effluents. *Larry E. LaFleur (U.S.A.)* .. 21

Contributed Research

Toxicity of TCF and ECF pulp bleaching effluents assessed by biological toxicity tests. *J. Ahtiainen, T. Nakari and J. Silvonen (Finland)* 33

The effect of waste constituents on the toxicity of TCF and ECF pulp bleaching effluents. *M. Verta, J. Ahtiainen, T. Nakari, A. Langi and E. Talka (Finland)* 41

The use of chronic bioassays in characterizing effluent quality changes for two bleached kraft mills undergoing process changes to increased chlorine dioxide substitution and oxygen delignification. *Timothy J. Hall, Richard K. Haley, Dennis L. Borton and Theresa M. Bousquet (U.S.A.)* 53

Reproduction effects on *Monoporeia affinis* of HPLC-fractionated extracts of sediments from a pulp mill recipient. *Ann-Kristin Eriksson, Brita Sundelin, Dag Broman and Carina Näf (Sweden)* 69

Examination of bleached kraft mill effluent fractions for potential inducers of mixed function oxygenase activity in rainbow trout. *L. Mark Hewitt, John H. Carey, D. George Dixon and Kelly R. Munkittrick (Canada)* 79

Comparative assessment of the toxic effects from pulp mill effluents to marine and brackish water organisms. *Britta Eklund, M. Linde and M. Tarkpea (Sweden)* 95

An assessment of the significance of discharge of chlorinated phenolic compounds from bleached kraft pulp mills. *Robert P. Fisher, Douglas A. Barton and Paul S. Weigand (U.S.A.)* 107

A comparison of gas chromatographic and immunochemical methods for quantifying resin acids. *Kai Li, Tao Chen, Paul Bicho, Colette Breuil and John N. Saddler (Canada)* 119

A method for detection of chlorinated, etherified lignin structures in high molecular weight materials of industrial and natural origin. *Olof Dahlman, Anders Reimann, Pierre Ljungquist and Roland Mörck (Sweden)* 129

Surface tension changes related to the biotransformation of dehydroabietic acid by *Mortierella isabellina*. *A.G. Werker, P.A. Bicho, J.N. Saddler and E.R. Hall (Canada)* 139

Correlations among contaminant profiles in mill process streams and effluents. *Thomas E. Kemeny and Sujit Banerjee (U.S.A.)* 151

Recent Research and Future Directions

Origins of effluent chemicals and toxicity: recent research and future directions. *Mark R. Servos (Canada)* 159

Section II: Environmental Fate

Invited Overviews

Environmental fate and distribution of substances. *G.E. Carlberg and T.R. Stuthridge (Norway and New Zealand)*. 169

Biodegradability of different size classes of bleached kraft pulp mill effluent organic halogens during wastewater treatment and in lake environments. *E.K. Saski, J.K. Jokela and M.S. Salkinoja-Salonen (Finland)* 179

Contributed Research

Analysis of pulp mill black liquor for organosulfur compounds using GC/atomic emission detection (AED). *I. Ya. Rod'ko, B.F. Scott and J.H. Carey (Russia and Canada)* 159

Turnover of chloroorganic substances in a Bothnian sea recipient receiving bleached kraft pulp mill effluents. *Per Nilsson, Lars Brydsten, Magnus Enell and Mats Jansson (Sweden)* . 203

Modeling the fate of 2,4,6-trichlorophenol in pulp and paper mill effluent in Lake Saimaa, Finland. *Donald Mackay, Jeanette M. Southwood, Jussi Kukkonen, Wan Ying Shiu, Debbie D. Tam, Dana Varhaníčkova, and Rebecca Lun (Canada and Finland)* 219

Load from the Swedish pulp and paper industry (nutrients, metals and AOX) quantities and shares of the total load on the Baltic Sea. *Magnus Enell (Sweden)* 229

Degradation of the high molecular weight fraction of bleached kraft mill effluent by biological and photochemical processes. *Kelly R. Millar, John H. Carey, B. Kent Burnison, Hung Lee and Jack T. Trevors (Canada)* 239

Concentrations of chlorinated organic compounds in various receiving environment compartments following implementation of technological changes at a pulp mill. *S.M. Swanson, K. Holley, G.R. Bourree, D.C. Pryke and J.W. Owens (Canada and U.S.A.)* . 253

Mesocosm simulation on sediment formation induced by biologically treated bleached kraft pulp mill wastewater in freshwater recipients. *E.K. Saski, A. Vähätalo, K. Salonen and M.S. Salkinoja-Salonen (Finland)* 261

Recent Research and Future Directions

Recent advances in environmental fate of chemicals from pulp mills. *John S. Gifford (New Zealand)* 271

Section III: Bioaccumulation of Substances from Pulp and Paper Mills to Fish and Wildlife

Invited Overviews

Bioaccumulation of bleached kraft pulp mill related organic chemicals by fish. *Derek C.G. Muir and Mark R. Servos (Canada)* 283

Persistent pulp mill pollutants in wildlife. *John E. Elliott, Phil E. Whitehead, P.A. Martin, G.D. Bellward and Ross J. Norstrom (Canada)* 297

Contributed Research

Dioxins and furans in crab hepatopancreas: use of principal components analysis to classify congener patterns and determine linkages to contamination sources. *Mark B. Yunker and Walter J. Cretney (Canada)* 315

Studies on eulachon tainting problem: analyses of tainting and toxic aromatic pollutants. *P. Mikkelson, J. Paasivirta, I.H. Rogers and M. Ikonomou (Finland and Canada)* 327

Monitoring trends in chlorophenolics in Finnish pulp mill recipient watercourses by bioaccumulation in incubated mussels. *Sirpa Herve, Pertti Heinonen and Jaakko Paasivirta (Finland)* 335

Bioaccumulation of pulp chlorobleaching-originated aromatic chlorohydrocarbons in recipient watercourses. *Tiina Rantio, Jaana Koistinen and Jaakko Paasivirta (Finland)* 341

Section IV: Field and Laboratory Studies of Biochemical Responses Associated with Pulp and Paper Mill Effluents

Invited Overviews

Mixed function oxygenase induction by pulp mill effluents: advances since 1991. *Peter V. Hodson (Canada)* 349

Biochemical responses in organisms exposed to effluents from pulp production: are they related to bleaching? *Karl-Johan Lehtinen (Finland)* 359

Contributed Research

Evaluation of biomarkers of exposure to 2-chlorosyringaldehyde and simulated ECF bleached eucalypt pulp effluent. *C.M. Brumley, V.S. Haritos, J.T. Ahokas and D.A. Holdway (Australia)* 369

Rainbow trout (*Oncorhynchus mykiss*) mixed function oxygenase responses caused by unbleached and bleached pulp mill effluents: a laboratory-based study. *Todd G. Williams, John H. Carey, B. Kent Burnison, D. George Dixon and Hing-Biu Lee (Canada)* 379

Lethal and sublethal effects of chlorinated resin acids and chloroguaiacols in rainbow trout. *C.J. Kennedy, R.M. Sweeting, J.A. Johansen, A.P. Farrell and B.A. McKeown (Canada)* ... 391

Effluents from Canadian pulp and paper mills: a recent investigation of their potential to induce mixed function oxygenase activity in fish. *P.H. Martel, T.G. Kovacs and R.H. Voss (Canada)* 401

The use of stable isotopic analyses to identify pulp mill effluent signatures in riverine food webs. *Leonard I. Wassenaar and Joseph M. Culp (Canada)* 413

Detection of steroid hormone disruptions associated with pulp mill effluent using artificial exposures of goldfish. *M.E. McMaster, K.R. Munkittrick, G.J. Van Der Kraak, P.A. Flett and M.R. Servos (Canada)* 425

Recent Research and Future Directions

Field and laboratory studies of biochemical responses associated with pulp mill effluents: status in 1991, 1994 and beyond. *P.J. Kloepper-Sams (Belgium)* 439

Section V: Field and Laboratory Studies of Whole Organism Responses Associated with Pulp and Paper Mill Effluents

Invited Overviews

In situ assessments of the impact of pulp mill effluent on life-history variables in fish. *Olof Sandström (Sweden)* 449

Laboratory responses of whole organisms exposed to pulp and paper mill effluents: 1991-1994. *Tibor G. Kovacs and Stanley R. Megraw (Canada)* 459

Contributed Research

Survival, growth, production and biomarker responses of fish exposed to high-substitution bleached kraft mill effluent in experimental streams. *Dennis L. Borton, William R. Streblow, W. Kenneth Bradley, Terry Bousquet, Peter A. Van Veld, Richard E. Wolke and Alexander H. Walsh (U.S.A.)* 473

Effect monitoring in pulp mill areas: response of the meiofauna community to altered process technique. *Brita Sundelin and Ann-Kristin Eriksson (Sweden)* 483

The western mosquitofish as an environmental sentinel: parasites and histological lesions. *Robin M. Overstreet, William E. Hawkins and Thomas L. Deardorff (U.S.A.)* 495

Abnormalities in winter flounder (*Pleuronectes americanus*) living near a paper mill in the Humber Arm, Newfoundland. *R.A. Khan, D.E. Barker, K. Ryan, B. Murphy and R.G. Hooper (Canada)* 511

Effects of pulp mill effluent on benthic freshwater invertebrates: food availability and stimulation of increased growth and development. *Richard B. Lowell, Joseph M. Culp, Frederick J. Wrona and Max L. Bothwell (Canada)* 525

Comparative evaluation of macrobenthic assemblages from the Sulphur River Arkansas in relation to pulp mill effluent. *Chet F. Rakocinski, Michael R. Milligan, Richard W. Heard and Thomas L. Deardorff (U.S.A.)* 533

Design and application of a novel stream microcosm system for assessing effluent impacts to large rivers. *Joseph M. Culp and Cheryl L. Podemski (Canada)* 549

Whole organism and population response of sentinel fish species in a large river receiving pulp mill effluents. *Stella M. Swanson, Richard Schryer and Barry Firth (Canada and U.S.A.)* 557

Nutrient and contaminant effects of bleached kraft mill effluent on benthic algae and insects of the Athabasca River. *Cheryl L. Podemski and Joseph M. Culp (Canada)* 571

Recent Research and Future Directions

Field and laboratory responses of whole organisms associated with pulp mills. *Stella Swanson (Canada)* 581

Section VI: Integrated Monitoring

Invited Overviews

The role of biological measurements in monitoring pulp and paper mills in Australasia and S.E. Asia. *Bruce M. Allender, John S. Gifford and Paul N. McFarlane (Australia and New Zealand)* 591

A review of recent studies in North America using integrated monitoring approaches to assessing bleached kraft mill effluent effects on receiving waters. *Timothy J. Hall (U.S.A.)* .. 599

Ecotoxicological effects of process changes implemented in a pulp and paper mill: a Nordic case study. *Aimo Oikari and Bjarne Holmbom (Finland)* 613

Contributed Research

Monitoring environmental effects and regulating pulp and paper discharges: Bay of Plenty, New Zealand. *Paul Dell, Fergus Power, Robert Donald, John McIntosh, Stephen Park and Liping Pang (New Zealand)* 627

Ecotoxicological impacts of pulp mill effluents in Finland. *Maarit H. Priha (Finland)* 637

Recent Research and Future Directions

Recent perspectives in integrated monitoring. *Peter M. Chapman (Canada)* 651

Section VII: Future Directions

Invited Overviews

Regulation of pulp mill aquatic discharges: current status and needs from an international perspective. *J.W. Owens (U.S.A.)* 661

Swedish environmental regulations for bleached kraft pulp mills. *Staffan Lagergren (Sweden)* 673

Contributed Research

Environmental aspects of ECF vs. TCF pulp bleaching. *Jens Folke, Lars Renberg and Neil McCubbin (Denmark, Sweden, and Canada)* 681

Recent Research and Future Directions

Future directions for environmental harmonization of pulp mills. *Jens Folke (Denmark)* 693

Keyword Index 701

Acknowledgments

We would like to acknowledge the assistance of the International Scientific Steering Committee for their guidance in planning the 2nd International Conference on Environmental Fate and Effects of Bleached Kraft Mill Effluents, Vancouver, British Columbia, Canada, November 6-10, 1994:

John H. Carey, National Water Research Institute, Canada (Co-Chair)
Kelly R. Munkittrick, GLLFAS, Department of Fisheries and Oceans, Canada (Co-Chair)
Peter M. Chapman, EVS Consultants, Canada
Anders Södergren, Lund University, Sweden
John Gifford, New Zealand Forest Research Institute, New Zealand
J. William Owens, Procter & Gamble Co., Cincinnati, Ohio, U.S.A.
Mirja Salkinoja-Salonen, University of Helsinki, Finland
Mark R. Servos, GLLFAS, Department of Fisheries and Oceans, Canada
Glen J. Van Der Kraak, University of Guelph, Canada

The international scientific community that participated in the conference by presenting their work, joining in panel discussions and reviewing submitted manuscripts played an essential role in the production of this book. Ruth Burr made a major contribution to this work by coordinating and processing the abstracts and manuscripts. The dedication of Sandy Pearlman and the St. Lucie Press staff greatly enhanced the presentation and quality of the entire publication. The administrative assistance by Jill Parker and the financial coordination by Linda Gysbers were gratefully appreciated. Artistic and graphics support were provided by Joanne Wotherspoon and Randy Bovaird. The photograph of the white sucker was provided by Monique Gagnon. The remaining photographs were provided by Mark Servos, Mike Keir and Kelly Munkittrick.

Primary financial support for this project was provided by the following departments of the Canadian federal government:

Environment Canada
Fisheries and Oceans
Industry Canada

Additional financial and other support was provided by:

American Forest & Paper Association
BC Environment
Canadian Occidental Petroleum, Ltd.
National Council of the Paper Industry for Air and Stream Improvement, Inc.
Pulp and Paper Research Institute of Canada
Sterling Pulp Chemicals, Ltd.
The Technical Association of the Australian & New Zealand Pulp & Paper Industry Inc.
University of Guelph

This book was published on paper generously donated by Georgia-Pacific Corporation, Atlanta, GA.

The Editors: Mark R. Servos (lower left), Kelly R. Munkittrick (lower right), John H. Carey (upper left) and Glen J. Van Der Kraak (upper right).

Mark R. Servos, Ph.D., is a research scientist with the Great Lakes Laboratory for Fisheries and Aquatic Sciences, Canada Department of Fisheries and Oceans in Burlington, Ontario, Canada. His research interests are primarily directed at understanding the environmental chemistry, distribution, and effects of organic contaminants in aquatic ecosystems.

Kelly R. Munkittrick, Ph.D., is a research scientist with the Great Lakes Laboratory for Fisheries and Aquatic Sciences. His research interests are directed at determining the responses of fish to industrial effluents and discriminating factors influencing their responses.

John H. Carey, Ph.D., is director of the Aquatic Ecosystems Conservation Branch, National Water Research Institute of Environment Canada in Burlington, Ontario, Canada. His research interests are in the processes controlling the distribution, fate and effects of contaminants in aquatic ecosystems.

Glen J. Van Der Kraak, Ph.D., is an associate professor in the Department of Zoology, University of Guelph, Guelph, Ontario, Canada. His research is centered on understanding how a variety of factors, including contaminants, influence endocrine function in fish.

Preface

In the late 1980s worldwide attention was focused on the potential environmental impact of bleached pulp mill effluents after scientists in Sweden identified that fish collected near bleached kraft pulp mill effluent discharges demonstrated changes in growth, carbohydrate metabolism, maturation, recruitment, mortality and community structure (reviewed in Södergren 1989). Many of the same responses were subsequently observed at a number of North American pulp mills (McMaster et al. 1991; Munkittrick et al. 1992; Hodson et al. 1992; Servos et al. 1992). Regulatory initiatives to address these concerns were initially focused on the reduction or elimination of chlorinated organic compounds which were believed to be associated with the effects. However the general applicability of these studies and assumptions was question and led to a meeting in Saltsjobaden in November of 1991 to review the scientific evidence (Södergren 1992).

Since the 1991 meeting, the environmental fate and effects of pulp and paper mill effluents has been one of the most active areas of research in the environmental sciences. Implementation of new regulations and monitoring programs in several countries has focused additional attention on this industrial sector and the potential impact of pulp and paper mill effluents. Industry has responded rapidly by implementing new process and treatment technologies, including elemental chlorine-free or totally chlorine-free bleaching, which were designed to minimize or eliminate potential impacts on the environment. The complexity of the process changes, the resulting effluents, and the receiving environments have made it difficult to evaluate the success of these efforts.

In November 1994, over 280 scientists from more than a dozen countries gathered in Vancouver, Canada to examine and compare new developments regarding the environmental fate and effects of pulp mill effluents. This book is a direct result of that meeting and explores the most recent and critical areas of research and experimentation during the 1990s. These contributions from industry, government and academia from around the world provide a balanced global perspective of the most recent scientific findings in this very dynamic and active area of research. The book was designed to present the current status of our understanding of the environmental fate and effects of contaminants originating from pulp and paper mill effluents and to identify the challenges faced by scientists in industry and government in the immediate future.

This book contains 61 peer reviewed manuscripts contributed by scientists from ten countries. It is organized into seven major themes which emphasize the active areas of research or policy development:

- identity and origin of chemicals in pulp mill effluents,
- environmental fate of chemicals from pulp and paper mills,
- bioaccumulation of substances from pulp mills to fish and wildlife,
- field and laboratory studies of biochemical responses associated with pulp and paper effluents,
- field and laboratory studies of whole organism responses associated with pulp and paper effluents,
- integrated monitoring, and
- future research and policy directions.

Each section begins with one or more invited reviews of the major developments in the specific area of research or policy over the last three to five years. These are followed by original contributed research papers which have significantly increased our understanding of the fate and effects of pulp and paper mill effluents. Each section ends with a review of the most recent developments, future directions and research needs which were presented, discussed or identified during the 2nd International Conference on Environmental Fate and Effects of Bleached Pulp Mill Effluents, held in Vancouver, British Columbia, Canada, November 6-10, 1994.

REFERENCES

Hodson, P.V. M. McWhirther, K. Ralph, B. Gray, D. Thivierge, J. Carey and G. Van Der Kraak. 1992. Effects of bleached kraft mill effluent on fish in the St. Maurice River, Quebec. Environ. Toxicol. Chem. 11:1635-1651.

McMaster, M.E., G.J. Van Der Kraak, C.B. Portt, K.R. Munkittrick, P.K. Sibley, I.R. Smith and D.G. Dixon. 1991. Changes in hepatic mixed function oxygenase (MFO) activity, plasma steroid levels and age at maturity of a white sucker (*Catostomus commersoni*) population exposed to bleached kraft pulp mill effluent. Aquat. Toxicol. 21:199-218.

Munkittrick, K.R., G.J. Van Der Kraak, M.E. McMaster and C.B. Portt. 1992. Response of hepatic mixed function oxygenase (MFO) activity and plasma sex steroids to secondary treatment and mill shutdown. Environ. Toxicol. Chem. 11:1427-1439.

Servos, M.R., J. Carey, M. Ferguson, G. Van Der Kraak, H. Ferguson, J. Parrott, K. Gorman, and R. Cowling. 1992. Impact of a modernized bleached kraft mill on white sucker populations in the Spanish River, Ontario. Water Pollut. Res. J. Can. 27:423-437.

Södergren, A. 1989. Biological effects of bleached pulp mill effluents. National Swedish Environmental Protection Board, Report 3558, 139 p.

Södergren, A. 1992. Environmental fate and effects of bleached pulp mill effluents. Proceedings of a SEPA Conference held at Grand Hotel Saltsjobaden, Stockholm, Sweden, 19-21 November 1991. Swedish Environmental Protection Agency, Report 4031, 394 p.

WHERE WERE WE IN 1991?

Anders Södergren

Department of Ecology, Lund University, Lund, Sweden

Three years ago Dr. Munkittrick opened the first conference in this series in Stockholm, Sweden, and stated that "Virtually all findings reported in the early Scandinavian investigations (1982-1985) of the impact of bleached kraft pulp mill effluent (BKME) on fish, including effects on mixed function oxygenases (MFO), conjugation enzymes, energetics, haematology, immune responses, non-specific biomarkers, growth and reproduction, have been found in North America, but changes are inconsistent and vary from site to site" (Munkittrick 1992).

Thus, at that time, there was a consensus that similar disturbances appeared in the receiving water of bleached pulp mills both in the new and old world; however, there was no clear and solid evidence of which compounds were the culprits behind the damages, although many of us, due to our previous knowledge about the properties of organochlorine compounds, considered them main suspects.

Today, in waters that no longer receive chlorinated compounds (due to a dramatic and rapid development of bleaching technology), we notice considerable improvements. Still, residual effects linger on. To make things more complicated, in receiving waters of unbleached pulp mills, we observe effects that we formerly believed only were connected to bleached pulp mills. In 1991, I think we were a little bit too optimistic and assumed that we were at last going to find Pandora's box around the corner and finally identify the culprit, or culprits, behind the observed environmental disturbances. Today it is safe to say that there was no box - instead we found several new corners!

In 1991, we paid little attention to chlorinated organic chemicals produced by natural processes. Today we know that there are approximately 1500 of these substances which are released into the environment by a number of organisms, ranging from bacteria to higher plants, and by various natural combustion processes (Willes *et al.* 1993). The substances represent a broad range of chemicals and are of the same classes of chemicals as those produced by human activities. However, the available information is inadequate to estimate total environmental releases for most of these chemicals. The distribution of the naturally produced organochlorine compounds in the environment, and the presence in many animals of an enzyme system capable of detoxifying these compounds, suggest that organisms have evolved in the presence of such chemicals. Thus, many organisms are capable of handling chlorinated organic chemicals at a certain level. But what is the threshold level and how do the anthropogenic compounds interact with the natural ones? A vast area of research is in front of us, where the significance of expressions like synergistic, additive and antagonistic has to be identified and evaluated.

Considering this, and adding the complicated nature of the wastewater from the pulp and bleaching processes which the ecosystem is exposed to, we have to realize that it is impossible to predict with complete confidence the ecological effects of a specific chemical in the mixture. Whether predictions can ever attain complete certainty is indeed an open question.

Methods to arrive at conclusions are not missing; I believe, however, that we must place more importance on integrating the observations in models or structures than collecting more and more data. Data alone and in huge volumes do not solve any problems: the challenge is to improve the interpretation of what seem to be unrelated data. For example, for effluent sources, we should consider consistency of the exposure, strength, specificity, temporal relationships and coherence of the associations. Models to integrate and evaluate these factors are debated and a long time will pass before consensus is established.

In addition, we are fighting a tough enemy from an ecotoxicological standpoint: technological development. In ecotoxicological studies it is of great importance that the exposure does not change. In receiving waters of pulp mills, effect studies will always be hampered by changing conditions due to technological improvements or process changes. The time since the last conference has seen great changes in this respect and today an intensive development characterizes the technology of bleaching chemical pulps. The need for new and more environmentally acceptable bleaching practices was realized in the 1960s and in the 1970s new processes were introduced. The development work was intensified in the middle of the 1980s. However, the changes that have taken place during the last 10 years, and especially during the last few years, are more dramatic and unexpected. In 1991, words like ECF (elemental chlorine free), TCF (totally chlorine free) or TEF (totally effluent free) did not exist. Today they are common language; moreover, TEF seems to be within grasp.

The use of non-chlorine-containing chemicals in the bleaching processes includes oxygen delignification and/or peroxide. For kraft pulps, oxygen and peroxide are not sufficient to produce TCF pulps. This has resulted in an interest in ozone as a chlorine-free bleaching agent. Ozone is a very strong electrophilic and oxidative agent but unfortunately is not very selective, which means that it also attacks cellulose.

In all, it is hoped that the development will drastically change and reduce the environmental impact of the pulp mills. A complete closure will, of course, result in no environmental impacts whatsoever. However, the use of all the new techniques and processes that look very harmless on paper may result in environmentally unacceptable compounds when the effluent encounters complex ecosystems. For example, the oxidative power of ozone may very well result in compounds that the ecosystem of the receiving waters may have some difficulties in handling. A comparison of the contributions at the conference in Stockholm in 1991 to those of this conference may not be appropriate but may nevertheless show in which direction the environmental research is heading.

It is safe to say that the Stockholm conference saw the breakthrough of the use of biochemical and physiological methods to estimate responses from fish to pulp mill effluents, both locally and regionally. These techniques to biologically confirm the distributional pattern of the effluent are now firmly established and routinely used, which is indicated by the number of contributions to this conference. There were more than four times the papers submitted to our conference in Vancouver within this field of research than at the Stockholm meeting. But it should be observed that in 1991, we regarded more or less all of the biochemical and physiological responses as effects, which would develop into population disturbances or ecosystem effects. Today we talk more about responses, and in many cases, the connection to population-level damage is still to be defined. We obviously need more knowledge on how to relate the first, primary responses at the molecular and biochemical level to toxic effects such as functional disorders at the organ level, reproductive failure or other survival disturbances that may affect a population. Will the contributions here advance our knowledge?

Again, compared to the Stockholm conference, we see today a lot more, about three times more, contributions that deal with pulp mill contaminants that are bioavailable and accumulate in organisms. I believe this reflects the development of the analytical techniques rather than requests from ecotoxicologists.

Finally, if we take a look at the industrial concepts, production costs have been reconsidered and processes that were not profitable to introduce some 10 years ago are now a necessary part of the fiber line. The oxygen stage is the most significant example. It is more likely that environmental concerns will continue to be a decisive factor for the pulp industry but the focus will shift from chlorine to other issues and the TEF pulp mill will be prominent on the agenda. When this last stage is completed, we have put ourselves in an awkward position - we have lost our jobs. In principle, is that not the ultimate goal for an ecotoxicologist? However, history tells us that the sum of mankind's sins is constant and, therefore, new challenges will appear. For us today, unknown environmental problems are awaiting to be discovered, defined and treated.

Before arriving at that point, I am sure that there will have been a 3rd, a 4th and a 5th conference on environmental effects of bleaching. This is especially obvious when looking at the distribution of global pulp bleaching technology: despite powerful marketing of the new processes, less than 5% of the global production of bleached chemical pulp is TCF (Vannerberg 1994).

REFERENCES

Munkittrick, K.R. 1992. Recent North American studies of bleached kraft mill impacts on wild fish. pp. 347-356 *In* (ed. A. Södergren) Environmental Fate and Effects of Bleached Pulp Mill Effluents, Proceedings November 19-21, 1991, Stockholm, Sweden. Swedish Environmental Protection Agency 4031.

Vannerberg, N.-G. 1994. ECF en rakare väg till den slutna fabriken. Svensk Papperstidning/Nordisk Cellulosa, 33-34.

Willes, R.F., E.R. Nestmann, P.A. Miller, J.C. Orr and I.C. Munro. 1993. Scientific principles for evaluating the potential for adverse effects from chlorinated organic chemicals in the environment. Regul. Toxicol. Pharmacol. 18:313-356.

SECTION I

IDENTIFICATION AND ORIGIN OF CHEMICALS AND TOXICITY

Pulp and paper mill effluents are complex mixtures. The characteristics of each effluent are dependent on numerous factors including wood furnish and process technology (including washing, cooking, bleaching, prebleaching, etc.), as well as final effluent treatment. The chemical composition of each effluent varies drastically, as does the toxicity. In an effort to remain competitive, as well as meet new environmental regulations, mills have been rapidly modifying process, bleaching and treatment technologies. These changes have led to dramatic changes in the chemical composition and toxicity of the final effluent. Recent studies have focused on understanding the chemistry, sources and fate of specific chemicals and toxicity within the mills. Considerable progress has been made toward understanding and controlling the release of contaminants, which has greatly reduced the potential threat to the environment.

EFFECTS OF INTERNAL PROCESS CHANGES AND EXTERNAL TREATMENT ON EFFLUENT CHEMISTRY

Lars Strömberg[1,2], Roland Mörck, Filipe de Sousa and Olof Dahlman

[1]STFI, Box 5604, S-11486 Stockholm, Sweden
[2]Present Address: AssiDomän, S-10522 Stockholm, Sweden

During the last decade, a rapid development and introduction to mill-scale of new and modified techniques for production of bleached kraft pulp has occurred in the kraft pulping industry. The main objectives of this development have been to reduce the amount of residual lignin entering the bleach plant and to reduce and subsequently eliminate the use of chlorine in the bleaching process, thereby minimizing the discharge of chlorinated organic matter. This development will also facilitate the closure of the water system in the bleach plant. Results from chemical characterization of effluents from Swedish bleached kraft mills show that the combination of extended delignification in the cooking stage, oxygen delignification and ECF bleaching is capable of bringing the discharge of AOX down to such a low level as 0.2 kg ptp before secondary effluent treatment and to 0.1 kg ptp after secondary treatment. The reduction in the quantity of chlorinated phenolic compounds, resulting from the introduction of ECF bleaching, is even more pronounced than the AOX reduction. TCF bleaching practically eliminates the discharges of chlorinated compounds. Thus, the chlorine content of high molecular weight effluent materials from TCF bleaching of softwood and hardwood kraft pulps was extremely low and fully comparable to the chlorine content found in naturally occurring humic materials. The contents of extractives such as fatty acids, resin acids and sterols in untreated kraft mill effluents were found to vary considerably between different mills. Other factors than the type of bleaching process used may be of large importance for these variations. An effective removal of extractives was observed both in aerated lagoons and in activated sludge plants.

INTRODUCTION

In response to environmental concerns as well as to government regulations on emission of chlorinated organic matter and to market demands, the pulp and paper industry has acted worldwide since the mid-1970s by developing and introducing a number of new processes and process modifications in order to minimize the effluent load of organic matter. Special interest has been directed towards the formation and discharge of chlorinated organic matter. The main strategy behind the development of internal process changes has been to remove as much lignin as possible before the pulp enters the bleach plant and in the last few years also to replace chlorine in the bleaching process by substituting it for other bleaching agents. These advancements have recently been reviewed from an environmental point of view (Axegård et al. 1993).

In order to evaluate the environmental impact of effluents from the production of bleached softwood and hardwood kraft pulps according to the pulping and bleaching technology applied in Swedish mills today, comprehensive chemical and biological characterization studies have been carried out (Haglind et al. 1993). This paper presents the results of studies on the chemical composition of bleached kraft mill effluents (BKMEs) from mills using either conventional or modified cooking in combination with oxygen delignification and elemental chlorine-free (ECF) or totally chlorine-free (TCF) bleaching. The studies have been carried out within the SSVL Environment 93-project – a research project conducted by the Swedish pulp and paper industry.

In order to illustrate the effects of recent process development on the chemical composition of the organic effluent material, the results obtained in the present investigation are compared with corresponding data from somewhat older processes using chlorine as a bleaching agent. In this project, effluents from mills with different types of process conditions were studied both before and after secondary effluent treatment.

MATERIALS AND METHODS

BKMEs were sampled at five kraft mills producing bleached softwood pulp and at two mills producing hardwood kraft pulp. All effluents except those labeled SW-1 and SW-5 were studied both before and after secondary effluent treatment. Process data relevant for the sampling occasions are given in Tables 1 and 2. All effluent samples studied were 24 h composite samples.

Table 1. Cooking, bleaching and effluent treatment processes studied.

Effluent	Type of Pulp	Cooking Process	Bleaching Process	Effluent Treatment
SW-1	Softwood kraft	Conventional	O(C92+D8)(EO)DED	—
SW-2	Softwood kraft	Conventional	O(D25,C70+D5)(EPO)D(EP)D	Aerated lagoon
SW-3	Softwood kraft	Conventional	OD(EPO)D(EP)D*	Activated sludge (pilot plant)
SW-4	Softwood kraft	Mod. (batch)	OD(EO)D(EP)D*	Aerated lagoon
SW-5	Softwood kraft	Mod. (batch)	OQP**	—
HW-1	Hardwood kraft	Mod. (batch)	(D80+C20)(EPO)DED	Activated sludge
HW-2	Hardwood kraft	Conventional	O(D27,C68+D5)(EPO)D(EP)D	Aerated lagoon (pilot plant)
HW-3	Hardwood kraft	Conventional	OQP**	Activated sludge (pilot plant)
HW-4	Hardwood kraft	Conventional	OQPZP**	Activated sludge (pilot plant)

* ECF bleaching; ** TCF bleaching.

The effluent SW-1 originated from a softwood kraft mill using conventional kraft cooking combined with oxygen delignification and chlorine bleaching. The mill producing the effluent SW-2 used conventional cooking combined with oxygen delignification and low multiple chlorination in the first bleaching stage. The effluent labeled SW-3 originated from a softwood kraft mill using conventional kraft cooking combined with oxygen delignification and ECF bleaching. The mill producing effluent SW-4 used modified kraft pulping to low kappa number according to the Super Batch method (extended delignification in the cooking stage) combined with oxygen delignification and ECF bleaching.

The effluent SW-5 was sampled at a softwood kraft mill using TCF bleaching (hydrogen peroxide) preceded by modified kraft pulping to low kappa number according to the Super Batch method and by oxygen delignification. The brightness of the bleached pulp was comparatively low (75% ISO). The TCF bleaching technique applied in this mill has, however, recently been improved, and the mill is now producing TCF bleached (hydrogen peroxide) softwood kraft pulp with a brightness of 85-88% ISO.

The effluent HW-1 originated from a mill with production of hardwood kraft pulp using modified continuous cooking and bleaching with a combination of chlorine dioxide and chlorine in the first bleaching stage. Oxygen delignification was not applied at this mill.

The hardwood effluents HW-2, HW-3 and HW-4 were all sampled at the same mill, although at different occasions. In 1989, when the effluent HW-2 was sampled, this mill used conventional cooking combined with oxygen delignification and low multiple chlorination in the first bleaching stage. TCF bleaching (hydrogen peroxide) was used in 1992, when effluent HW-3 was sampled. TCF bleaching was also applied in 1993 (effluent HW-4), this time using a combination of hydrogen peroxide and ozone. In 1993 the mill had also installed modified continuous cooking which gave the unbleached pulp a somewhat lower kappa number.

Table 2. Process parameters used during effluent sampling.

	SW-1	SW-2	SW-3	SW-4	SW-5	HW-1	HW-2	HW-3	HW-4
Kappa number after O-stage	19	21	20	8.5	10	15	13	12	10
Chlorine charge, first bleaching stage (kg ptp)	30	19	—	—	—	8.8	11.1	—	—
Total chlorine charge (kg ptp)	30	19	—	—	—	10.6	11.1	—	—
Chlorine multiple, first bleaching stage	0.16	0.09	—	—	—	0.06	0.08	—	—
Total charge of chlorine dioxide (kg a. Cl ptp)	27	30	56	30	—	56	24	—	—
Oxygen charge in E-stages (kg ptp)	4.6	6.1	3	5	—	4.5	6	—	—
Total charge of hydrogen peroxide (kg ptp)	—	2	3	2	36	1	2	35	30
Ozone charge (kg ptp)	—	—	—	—	—	—	—	—	3
EDTA charge (kg ptp)	—	—	—	—	2	—	—	2	2
Brightness, bleached pulp (% ISO)	89.9	90.5	90.3	89.4	75	90.4	89.5	82	89
Retention time, aerated lagoon (d)	—	8-9	—	4	—	—	5	—	—
Retention time, activated sludge (h)	—	—	12	—	—	8-10	—	16	18
Effluent volume (m^3 ptp)	147	50	45	96	110	40	54	56	45

The general principles for the chemical characterization of the BKMEs studied are shown in Fig. 1. The effluents were characterized with respect to AOX, COD, BOD$_7$ and color. In addition, the chemical characterization also included the quantification of specific compounds such as chlorinated phenolic compounds, chlorinated acetic acids, chloroform, polychlorinated dibenzodioxins and dibenzofurans (PCDDs and PCDFs), fatty acids, resin acids and sterols. Chlorinated compounds were not determined in effluents from mills with TCF bleaching. All data are presented as amount per tonne of air-dried bleached pulp.

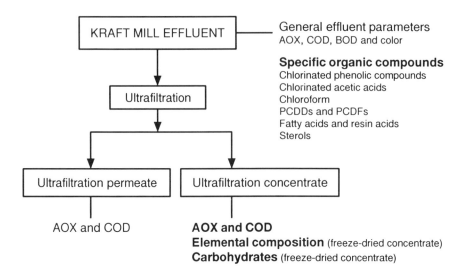

Figure 1. Analytical scheme showing the parameters studied in the chemical characterization of BKMEs.

COD was determined according to Dr. Lange's cuvette test (LCK 414). AOX and BOD_7 were determined according to Swedish Standards SS 028112 (AOX) and SS 028143 (BOD_7). Color was determined as absorption at 465 nm (pH 7.6 after centrifugation). Chlorinated phenolic compounds were determined by GC/ECD according to the method described by Starck *et al.* (1985). In the case of the effluents SW-3 and SW-4, the content of chlorinated phenolic compounds was determined using a sensitive GC/high-resolution MS technique. Chlorinated acetic acids were determined according to the method described by Lindström and Österberg (1986). Chloroform, fatty acids, resin acids and sterols were determined using sensitive GC/high-resolution MS techniques. PCDDs and PCDFs were determined according to the technique described by Swanson *et al.* (1988).

The effluents studied were also fractionated by ultrafiltration (nominal cut-off 1000 Da) followed by sequential batch diafiltration into a low molecular weight fraction (the ultrafiltration permeate) and a high molecular weight fraction (the ultrafiltration concentrate). The ultrafiltration procedure used has been described in detail elsewhere (Dahlman *et al.* 1996). The ultrafiltration permeates and concentrates obtained were characterized with respect to their contents of COD and AOX. Freeze-dried ultrafiltration concentrates were characterized with respect to elemental composition and carbohydrate content. The elemental analysis was carried out by Mikro Kemi AB, Uppsala, Sweden. The carbohydrate content was determined according to the procedure published by Theander and Westerlund (1986).

RESULTS

Effects of Process Modifications

The results in Table 3 show, as expected, that the BKME load of AOX depends strongly upon the bleaching process used. "Traditional" chlorine bleaching (SW-1) results in a higher AOX level than low multiple chlorination (SW-2 and HW-2). The introduction of ECF bleaching processes (SW-3 and SW-4) is favorable in order to reach AOX levels well below 1 kg ptp. As shown in Table 3, a very small amount of AOX ($\leq$0.01 kg ptp) was also detected in BKMEs from production of TCF bleached softwood (SW-5) and hardwood (HW-3, HW-4) kraft pulps.

The COD and BOD levels found in the softwood BKMEs studied do not show any clear correlation to the bleaching process used or to the kappa number of the oxygen-delignified pulp (Table 3). It can, however, be noted that the hardwood BKMEs HW-2, HW-3 and HW-4 all exhibited a considerably lower COD than the softwood BKMEs studied.

Upon ultrafiltration (nominal cut-off 1000 Da) of the two BKMEs from production of ECF bleached softwood pulp (SW-3 and SW-4), roughly half the total quantities of AOX were found in the ultrafiltration permeates (Table 3) and the other half in the concentrates. The results in Table 3 also indicate that the organic material in the softwood TCF effluent SW-5 is of lower molecular weight than the organic materials in the softwood ECF effluents. The data in Table 3 further indicate that a major part of the organic material in the effluents from production of TCF bleached hardwood pulp (HW-3 and HW-4) was of low molecular weight.

Table 3. Levels of AOX, COD, BOD_7 and color in the BKMEs studied (before secondary effluent treatment).

	AOX (kg ptp)	AOX <1000 Da (%)	COD (kg ptp)	COD <1000 Da (%)	BOD_7 (kg ptp)	Color (kg ptp)
SW-1	2.9	NA	64	NA	22	77
SW-2	1.9	54	52	56	7	80
SW-3	0.53	52	51	48	13	36
SW-4	0.19	53	57	57	19	52
SW-5	0.01	NA	41	67	19	12
HW-1	2.2	NA	82	NA	30	41
HW-2	1.2	58	33	64	10	27
HW-3	<0.01	NA	30	69	19	16
HW-4	0.01	NA	30	77	14	9

NA = Not analyzed.

The results in Table 3 clearly show that ECF bleaching of softwood kraft pulp produces BKMEs with lower values for color than BKMEs from chlorine bleaching. TCF bleaching reduces the BKME color even more than ECF bleaching.

The presence of chlorinated phenolic compounds in effluents from production of chlorine-bleached pulp has attracted much interest from an environmental point of view (Kovacs et al. 1993). Recent studies on mill and laboratory effluents from ECF bleaching have shown that ECF bleaching of softwood kraft pulp produces only very small quantities of chlorinated phenolic compounds (Wilson et al. 1992; Dahlman et al. 1993a; Anon. 1994; Pryke et al. 1994). The results from the present investigation confirm that the content of chlorinated phenolic compounds in BKMEs has decreased dramatically as a result of the introduction of ECF bleaching (Table 4 and Fig. 2). Also the degree of chlorination in these compounds has decreased significantly. Only monochlorinated and dichlorinated compounds were thus detected in the ECF effluents studied (SW-3 and SW-4). No tri- or tetrachlorinated phenolic compounds were detected (detection limit 0.005 g ptp).

A striking difference in the total content of chlorinated phenolic compounds was observed between the softwood and hardwood BKMEs from mills using chlorine in the bleaching process. Thus, the hardwood effluents HW-1 and HW-2 contained much lower quantities of these compounds than the softwood effluents SW-1 and SW-2 (Table 4 and Fig. 2).

Table 4. Levels of chlorinated phenolic compounds, chlorinated acetic acids, chloroform and PCDDs and PCDFs in the BKMEs SW-1, SW-2, SW-3, SW-4, HW-1 and HW-2 (before secondary effluent treatment).

	Chlorinated phenolic compounds				Chlorinated acetic acids			Chloroform (g ptp)	PCDDs PCDFs (µg ptp)*
	Monochloro (g ptp)	Dichloro (g ptp)	Trichloro (g ptp)	Tetrachloro (g ptp)	Monochloro (g ptp)	Dichloro (g ptp)	Trichloro (g ptp)		
SW-1	2.4	17.9	23.0	2.8	ND	88	71	33	0.4
SW-2	7.4	24.5	10.6	0.3	ND	43	24	8	0.03
SW-3	0.09	0.95	ND**	ND**	0.4	19	14	4	0.04
SW-4	0.16	0.03	ND	ND	0.2	8	2	4	ND***
HW-1	0.3	1.5	0.7	0.1	16	52	18	4	0.04
HW-2	0.7	1.4	0.5	0.2	14	77	27	8	0.07

*Nordic TCDD equivalents. **Detection limit ≈ 0.1 µg L^{-1}. *** Detection limit for 2,3,7,8-TCDD ≈ 0.05 pg L^{-1}. ND = not detected.

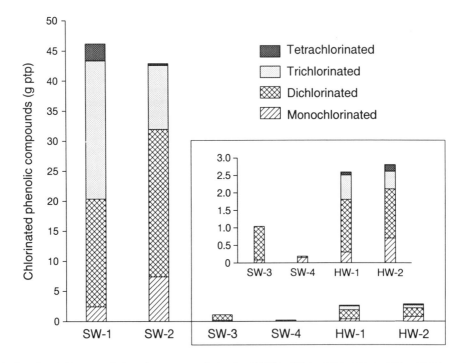

Figure 2. Levels of chlorinated phenolic compounds in the BKMEs SW-1, SW-2, SW-3, SW-4, HW-1 and HW-2 (before secondary effluent treatment).

Table 4 further shows that the loads of chlorinated acetic acids and chloroform also decrease markedly with the introduction of ECF bleaching. It is also worth noting that the effluent SW-4, representing the combination of modified cooking, oxygen delignification and ECF bleaching, contained considerably smaller amounts of chlorinated phenolic compounds and chlorinated acetic acids than the effluent SW-3, representing conventional cooking combined with oxygen delignification and ECF bleaching.

The discharge of PCDDs and PCDFs to water recipients from older kraft mills with chlorine bleaching attracted much attention from an environmental point of view during the later part of the 1980s. In the present investigation, PCDDs and PCDFs were detected in all effluents originating from mills with chlorine bleaching or ECF bleaching, except in the effluent SW-4 (Table 4). The quantities found in the effluents SW-2, SW-3, HW-1 and HW-2 were, however, extremely small and close to the detection limit (detection limit for 2,3,7,8-TCDD = 0.05 ppq).

Fatty acids and resin acids, which are natural constituents of wood, belong to those types of BKME components which are known to be toxic to fish. Some of the BKMEs studied in the present investigation exhibited rather high content of resin acids (Table 5) whereas others showed quite high content of fatty acids. The largest total quantities of fatty acids and resin acids were found in the effluents SW-4, SW-5 and HW-1.

It has been suggested that sterols may cause physiological and biochemical responses in fish by acting like hormone substances (Denton *et al.* 1985). Sterols occur both in softwood and hardwood species, often with ß-sitosterol as the dominating component. In the present study, a considerable variation in the content of sterols was observed between the BKMEs studied (Table 5). As expected, ß-sitosterol was the dominating compound in the sterol fraction in all effluents investigated (not shown in Table 5).

Table 5. Levels of resin acids, fatty acids and sterols in the BKMEs studied (before secondary effluent treatment).

	Resin Acids (g ptp)	Fatty Acids (g ptp)		Sterols (g ptp)
		Unsaturated	Saturated	
SW-1	11	2	23	NA
SW-2	14	1	8	NA
SW-3	78	3	13	3
SW-4	100	20	13	16
SW-5	24	93	17	14
HW-1	97	63	45	24
HW-2	9	ND	16	2
HW-3	8	6	23	19
HW-4	3	6	47	34

NA = not analyzed. ND = not detected.

The amount of high molecular weight material isolated varied considerably between the different BKMEs studied (Fig. 3), partly due to differences in the kappa number of the unbleached pulp and partly due to differences in the molecular weight of the organic effluent material. Fig. 3 also shows that the high molecular weight BKME materials originating from mills with TCF-bleaching (SW-5, HW-3 and HW-4) to a relatively large extent consisted of carbohydrates. The high molecular weight materials originating from mills with ECF and TCF bleaching exhibited, as expected, very low contents of chlorine compared to the materials originating from chlorine bleaching (Table 6).

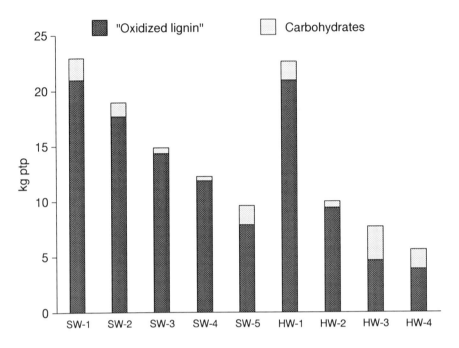

Figure 3. Yields of high molecular weight material (<1000 Da) obtained upon ultrafiltration of the BKMEs studied (before secondary effluent treatment).

Table 6. Elemental composition, chlorine content and C/Cl ratio in high molecular weight materials (>1000 Da) isolated from the BKMEs studied (before secondary effluent treatment).

Sample	Elemental Composition	Cl (%)	C/Cl
SW-1	$C_{100}H_{117}N_{0.7}O_{75}S_{4.8}Cl_{6.6}$	7.2	15
SW-2	$C_{100}H_{115}N_{0.4}O_{63}S_{1.4}Cl_{3.6}$	4.7	28
SW-3	$C_{100}H_{129}N_{0.3}O_{62}S_{1.6}Cl_{1.1}$	1.4	94
SW-4	$C_{100}H_{125}N_{1.2}O_{68}S_{2.3}Cl_{0.38}$	0.51	260
SW-5	$C_{100}H_{131}N_{0.6}O_{77}S_{1.9}Cl_{<0.08}$	<0.1	>1200
HW-1	$C_{100}H_{146}N_{0.8}O_{60}S_{2.1}Cl_{2.6}$	3.4	38
HW-2	$C_{100}H_{143}N_{1.6}O_{87}S_{4.8}Cl_{4.0}$	4.1	25
HW-3	$C_{100}H_{145}N_{1.1}O_{89}S_{2.3}Cl_{0.069}$	0.080	1400
HW-4	$C_{100}H_{129}N_{1.7}O_{73}S_{1.9}Cl_{0.084}$	0.10	1200

Effects of Secondary Effluent Treatment

In the present investigation, the effluents SW-3, HW-1, HW-3 and HW-4 were studied before and after treatment in an activated sludge plant (pilot plant with anoxic pre-zone), whereas the effluents SW-2, SW-4 and HW-2 were studied before and after treatment in aerated lagoons with anoxic zones. All treatment plants except the one used for treatment of the effluent HW-4 worked well when the treated BKMEs were sampled. In the case of HW-4, disturbances in the operation of the treatment plant made the effluent treatment less efficient than expected.

A COD reduction on the order of 45-55% was observed upon secondary treatment of the softwood BKMEs studied (Table 7). The COD reduction was considerably higher after treatment of the hardwood effluents HW-1 and HW-3 (exceptionally high in the case of HW-3). The COD reduction after treatment of the effluent HW-4 should also be considered as quite high, knowing that the treatment plant did not work at an optimum level.

An effective decrease of BOD_7 was observed upon secondary treatment in all treatment plants studied (Table 7). Secondary treatment was, however, not effective in decreasing the effluent color. Only a moderate decrease in color was observed after treatment of the effluents SW-2, SW-4, HW-2 and HW-3, whereas an increase in effluent color could be noted after treatment of the effluents SW-3, HW-1 and HW-4.

The reduction of AOX upon secondary treatment was around 50% in the case of the BKMEs from mills with ECF bleaching of softwood kraft pulp (SW-3 and SW-4). This is a comparatively large reduction in the case of aerated lagoons, but can be considered as a more normal reduction level in the case of activated sludge plants (Gergov et al. 1988).

A dominating part of the organically bound chlorine remaining in the ECF effluents SW-3 and SW-4 after secondary treatment was found to be incorporated into material of high molecular weight (>1000 Da). The AOX reduction obtained upon secondary treatment decreased the AOX load in the effluent SW-4 down to the very low value 0.09 kg ptp (Fig. 4). After secondary treatment of this effluent, 0.06 kg ptp of AOX was found in the high molecular weight fraction, whereas only 0.03 kg ptp of AOX was found in the low molecular weight fraction (<1000 Da). A similar distribution was observed also for COD (not shown in Fig. 4).

It has previously been observed that treatment of softwood bleached kraft mill effluents in aerated lagoons results in an effective removal of mono- and dichlorinated phenolic compounds, whereas tri- and tetrachlorinated phenolic compounds are more resistant towards degradation (Dahlman et al. 1991). The data obtained in the present study (Table 7) confirm that mono- and dichlorinated phenolic compounds are quite easily degradable in aerobic treatment plants and that polychlorinated phenolic compounds (especially the tetrachlorinated ones) often are more difficult to remove.

Chloroform was quite effectively removed in most of the treatment plants studied although large variations were observed between the treatment plants. The removal of chloroform is most likely due to evaporation rather than to biological degradation (Leuenberger et al. 1985).

An almost complete removal of resin acids was observed in all treatment plants studied, except in the one used for treatment of the effluent HW-4 (Table 7). The removal of fatty acids and sterols was also quite effective, but varied considerably between different treatment plants. The removal of unsaturated fatty acids seems to be somewhat more effective than the removal of saturated fatty acids, probably due to oxidation of the olefinic bonds in the unsaturated compounds. It is at present unclear whether the decrease of the amounts of resin acids, saturated fatty acids and sterols upon secondary treatment is due to biodegradation, chemical transformation or adsorption to the sludge.

The chelating agent ethylenediaminetetraacetic acid (EDTA) was charged in the bleaching process by the mills using TCF bleaching processes (Table 2). No major reduction of the EDTA load could be observed upon secondary treatment of the BKMEs studied (not shown in Table 7).

During chlorine dioxide bleaching of pulp, part of the chlorine dioxide is converted to chlorate (Germgård et al. 1981), which in aquatic environments is toxic towards bladder-wrack and other brown algae (Rosemarin et al. 1986). The occurrence of chlorate in the effluents from kraft mills equipped with secondary effluent treatment does not in most cases constitute an environmental problem since chlorate very effectively can be converted to chloride in the anoxic zones which often precede aerobic effluent treatments (Malmqvist et al. 1991). In the present investigation, no chlorate was detected in the biologically treated effluents.

Table 7. Effects of secondary effluent treatment on various effluent parameters in the BKMEs SW-2, SW-3, SW-4, HW-1, HW-2, HW-3 and HW-4.

	Removal (%) in Secondary Effluent Treatment						
	SW-2	SW-3	SW-4	HW-1	HW-2	HW-3	HW-4
COD	45	48	54	63	39	71	54
BOD_7	81	94	93	93	90	99	81
Color	28	0	28	0	19	32	0
AOX	34	47	53	45	56	NA	NA
Monochlorinated phenolic compounds	89	45	75	57	83	NA	NA
Dichlorinated phenolic compounds	66	93	73	24	79	NA	NA
Trichlorinated phenolic compounds	21	ND	ND	36	67	NA	NA
Tetrachlorinated phenolic compounds	0	ND	ND	40	0	NA	NA
Chloroform	100	67	88	63	54	NA	NA
Monochloroacetic acid	ND	100	70	100	100	NA	NA
Dichloroacetic acid	99	99	93	96	99	NA	NA
Trichloroacetic acid	96	99	35	0	92	NA	NA
Fatty acids (unsaturated)	98	67	99	100	ND	100	42
Fatty acids (saturated)	43	43	67	69	90	94	64
Resin acids	97	98	96	97	98	100	68
Sterols	NA	85	67	53	62	99	70

NA = not analyzed. ND = not detected.

The carbohydrates (mainly hemicelluloses) which constitute part of the high molecular weight BKME material are to a large degree (in some cases quantitatively) removed in aerobic effluent treatment plants (Table 8). Activated sludge plants seem to be somewhat more effective than aerated lagoons. The other constituents of the high molecular weight material (in this paper described as "oxidized lignin") are, however, only influenced to a minor extent by aerobic treatment. This indicates that those parts of the high molecular weight BKME material which originate from the residual lignin in the pulp are more or less resistant towards degradation during secondary effluent treatment.

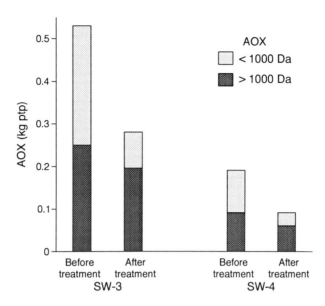

Figure 4. Distribution of AOX between concentrates (>1000 Da) and permeates (<1000 Da) upon ultrafiltration of the BKMEs SW-3 and SW-4 (before and after secondary effluent treatment).

Table 8. Effects of secondary effluent treatment on components of the high molecular weight BKME material.

	High Molecular Weight Material	
	"Oxidized lignin" (% removal)	Carbohydrates (% removal)
SW-2	11	77
SW-3	14	100
SW-4	14	50
HW-1	6	76
HW-2	0	50
IIW-3	17	97
HW-4	0	100

DISCUSSION

Process Development

The major advancements in the development of new and modified delignification processes in the closed part of the pulp mill are schematically illustrated in Fig. 5 for the production of bleached softwood kraft pulp. Conventional kraft cooking of softwood species will normally result in a kraft pulp with a kappa number of 30-35 (process 1 in Fig. 5). With the introduction of oxygen delignification, the kappa number of the pulp entering the bleach plant was reduced to around 20 (process 2 in Fig. 5). Recently, the combination of extended delignification in the cooking stage (modified cooking) and oxygen delignification has further reduced the kappa number of the softwood pulp entering the bleach plant to around 10 (process 3 in Fig. 5).

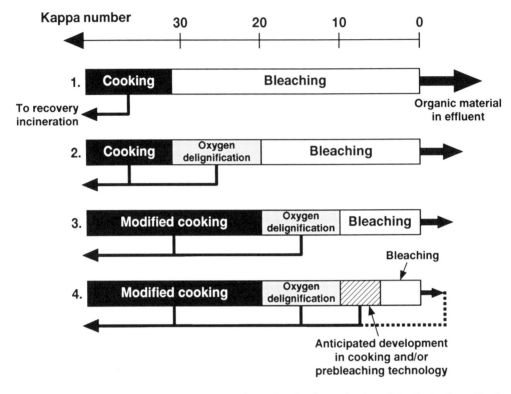

Figure 5. Schematic presentation of different process alternatives for the production of bleached softwood kraft pulp.

It can be anticipated that the process development will continue in the near future with further improvements of cooking and pre-bleaching delignification processes which probably will make it possible to produce pulps with kappa numbers well below 10 before bleaching (process 4 in Fig. 5).

Modified cooking and oxygen delignification constitute parts of the closed process system in modern kraft mills. Hence, the organic material dissolved in the waste liquors from these processes is returned to the recovery furnace for combustion. These processes, which have led to a substantial reduction of the quantities of residual lignin entering the bleach plant, are also favorable from the point of view that they reduce the demand (and thereby also the costs) for bleaching agents.

Recently, Call and Mücke (1994) reported on a possible breakthrough in enzyme bleaching technology. The enzyme bleaching process proposed by these authors is suggested as an alternative to existing bleaching stages and uses a combination of enzyme (laccase) and a low molecular weight redox mediator. According to Call and Mücke (1994), this system effectively reduces the kappa number of the pulp. The system may also be suitable for combination with established technologies to improve the flexibility of bleaching sequences with respect to consumption of chemicals and pulp quality. Effluents from mills using enzymatic bleaching stages have not been studied in the present investigation. No investigation on such effluents has to our knowledge been published.

In the development of the modern bleaching technologies, the charge of chlorine in the first bleaching stage has gradually been reduced by the introduction of low multiple chlorination and by a gradual substitution of chlorine dioxide for chlorine. This development has now reached the point where many kraft mills no longer use chlorine as a bleaching agent. Instead, bleaching is carried out using chlorine dioxide exclusively in the first bleaching stage (ECF bleaching). Reinforcement of the alkaline extraction stages with oxygen and/or hydrogen peroxide has facilitated this development by shifting parts of the delignification work from the first bleaching stage to the alkaline extraction stages.

Kraft bleaching processes based on hydrogen peroxide or ozone or combinations of these bleaching agents (TCF bleaching processes) have recently been introduced in full scale operation in several kraft mills worldwide, with Swedish and Finnish mills acting as forerunners. Also, other oxidative bleaching agents such as peracetic acid are now used in some mills as a complement to other bleaching alternatives. The main objective behind the introduction of TCF bleaching has in a short perspective been to minimize the discharge of chlorinated organic material to water recipients (Anon. 1992). In a somewhat longer perspective, the introduction of TCF bleaching as well as ECF bleaching with a minimum use of chlorine dioxide will facilitate the closure of the water system in the bleach plant (Anon. 1992).

Effects of Internal Process Development on Effluent Characteristics

The results presented in Table 3 show that production of softwood kraft pulp using conventional kraft cooking combined with oxygen delignification and chlorine-based bleaching results in an AOX level of approximately 2-3 kg ptp. The chlorine charge is, however, critical for the AOX level, and higher charges of chlorine than those applied here (Table 2) will result in a higher load of AOX in the effluent (Gergov et al. 1988). The results from the present investigation further show that oxygen delignification of conventionally cooked softwood kraft pulp followed by ECF bleaching (Table 3; effluent SW-3) may reduce the discharge of AOX to about 0.5 kg ptp. The combination of extended delignification in the cooking stage (modified kraft cooking), oxygen delignification and ECF bleaching (Table 3; effluent SW-4) is an even more powerful process alternative, capable of bringing the discharge of AOX from production of softwood kraft pulp down to such a low level as ≤ 0.2 kg ptp (before secondary effluent treatment). TCF bleaching, on the other hand, will practically eliminate the discharge of chlorinated compounds from the bleach plant.

It is difficult to draw any conclusions from the data presented in Table 3 concerning the effects of the process changes studied on BOD_7 and COD. It is well known that the COD load in bleach plant effluents is strongly related to the kappa number of the pulp prior to the bleach plant. However, the COD of the total mill effluent will, apart from the contribution from the bleach plant effluent, also have contributions from other sources such as evaporation condensates, black liquor residues, "carry-over" from the oxygen delignification stage and in some cases also from woodroom effluents. It should also be pointed out that some kraft mills are integrated with a paper mill. In such cases, the paper mill will also contribute to the total COD discharge. The importance of these "other" COD sources (and of variations in these sources) may increase when the contribution from the bleach plant effluent to the total COD discharge becomes smaller as a result of process changes. The mill producing the effluent SW-4 may serve as an example. In view of the low kappa number of the pulp entering the bleach plant (Table 2), the COD level in the effluent SW-4 was considerably higher than expected, and it is obvious that a large part of the COD in this BKME sample must have originated from sources other than the bleach plant.

The data presented in this investigation show that the reduction in the quantity of chlorinated phenolic compounds, resulting from the introduction of ECF bleaching, is far greater than suggested by the overall AOX reduction and that the small amounts of chlorinated phenolic compounds detected in effluents from mills with ECF bleaching exhibit a low degree of chlorination (mono- and dichlorinated). The generally lower degree of chlorination in the chlorinated phenolic compounds found in ECF effluents (as compared to effluents from chlorine bleaching) reduces the persistence and toxicity of these compounds as well as their potential for bioaccumulation.

The large difference observed in the content of chlorinated phenolic compounds between softwood and hardwood BKMEs from mills using chlorine in the first bleaching stage is interesting and indicates that (C+D) bleaching of hardwood kraft pulp using a low chlorine multiple produces only small quantities of chlorinated phenolic compounds.

During chlorine-based bleaching of kraft pulp, formation of chloroform can occur both in the acidic and in the alkaline bleaching stages. The largest quantities of chloroform have been found in effluents from older mills having a hypochlorite stage (H-stage) in the bleaching sequence (Kringstad and Lindström 1984). The removal of the H-stage from the bleaching process has therefore drastically reduced the amount of chloroform found in the mill effluent. The introduction of ECF bleaching has further decreased the chloroform level in the mill effluent. Thus, less than 5 g ptp of chloroform was detected in the ECF effluents SW-3 and SW-4.

Investigations published at the end of the 1980s showed that the formation of PCDDs and PCDFs in the kraft bleaching process can be avoided by the use of low multiple chlorination (Kringstad *et al.* 1989), by bleaching with a high substitution of chlorine dioxide in the first bleaching stage or by bleaching with chlorine dioxide exclusively (ECF bleaching) in the first bleaching stage (Berry *et al.* 1989). The extremely small amounts of PCDDs and PCDFs detected in the present investigation in the effluents SW-2, HW-1, HW-2 (low multiple chlorination) and SW-3/SW-4 (ECF bleaching) are close to the detection limit and can be considered as more or less equivalent to the background level.

As shown in Table 5, the discharges of fatty acids, resin acids and sterols from kraft mills may vary considerably from mill to mill. The causes of these variations are not fully clarified but factors such as the quality of the wood material used, the storage time of the wood chips before pulping, how the different process stages in the mill are designed and whether or not woodroom effluent is included in the total mill effluent may be of large importance. At present, it is therefore uncertain whether the bleaching process as such has any significant influence on the quantities of extractives like fatty acids, resin acids and sterols in BKMEs.

Resin acids were also detected in the hardwood BKMEs studied in spite of the fact that resin acids do not occur in common hardwood species used for kraft pulping. Tall oil (containing both fatty and resin acids) is, however, often added during kraft cooking of hardwood species in order to facilitate the removal of neutral extractives. BKMEs from production of hardwood kraft pulp may therefore contain certain amounts of resin acids.

In the present study, the high molecular weight effluent material was defined as the fraction of the organic BKME material which upon ultrafiltration (nominal cut-off 1000 Da) followed by sequential batch diafiltration was unable to penetrate the ultrafiltration membrane used, and thus was retained in the concentrate. The high molecular weight BKME material is mainly composed of strongly oxidized and hydrophilized degradation products originating from the residual lignin in the unbleached pulp and to a varying degree also of carbohydrates. In total mill effluents, such as those studied here, the high molecular weight material may also to some degree originate from sources other than the bleach plant effluent.

In effluents from older kraft mills with chlorine bleaching and no oxygen delignification, the high molecular weight material was quantitatively dominating and carried a very large part of the organically bound chlorine (Kringstad and Lindström 1984). The high molecular weight material seems, however, to be quantitatively less dominating in effluents from modern mills, especially in effluents from mills with TCF bleaching preceded by oxygen delignification (Table 3).

The very low content of chlorine (0.5%) and the correspondingly high C/Cl ratio (260/1) found in the present study in the high molecular weight material from the effluent SW-4 is very interesting and indicates that extended delignification in the kraft cooking stage combined with oxygen delignification and ECF bleaching may result in a very low degree of chlorination in the organic material discharged to water recipients.

Trace amounts of chlorine also were detected in the high molecular weight materials isolated from the TCF effluents studied (Table 5). The Cl contents of these samples were, however, extremely low and fully comparable to the chlorine contents found in naturally chlorinated humic substances (Grøn 1991; Dahlman et al. 1993b).

Effluent Treatment Technologies

Secondary effluent treatment in aerated lagoons or activated sludge plants is an important and technologically well-tried technique for reducing the load of organic material in the mill effluent before discharging it to the recipient. Although initially designed for BOD reduction, it is now recognized that biological treatment plants also are effective for controlling discharges of organic and inorganic matter measured as AOX, COD or specific compounds. These biological processes are generally also effective in reducing the acute toxicity of kraft mill effluents (Gergov et al. 1988).

Other effluent treatment methods applied in full or pilot scale operation include ultrafiltration, chemical precipitation and adsorption methods, all of which have proven more or less efficient in the removal of organic material from BKMEs (Amoth et al. 1992). These methods have, however, not been studied in the present investigation.

Effects of Secondary Effluent Treatment on Effluent Characteristics

The reduction of COD after secondary treatment of some of the hardwood effluents (HW-1 and HW-3) studied in the present investigation was unusually large (65-70%). This high reduction level may, at least to some degree, be due to the relatively large portion of low molecular weight organic material (<1000 Da) present in the hardwood BKMEs.

Ultrafiltration at nominal cut-off 1000 Da showed that the distribution of AOX and COD between permeates and concentrates changed significantly after secondary effluent treatment (AOX results are shown in Fig. 4). Dominating parts of COD and AOX were thus found in the ultrafiltration concentrates after secondary treatment. This indicates that the bacteria in aerobic effluent treatment plants preferably remove organic material of low molecular weight (Yin et al. 1989).

The data in Table 7 show that monochloroacetic acid and dichloroacetic acid are very effectively removed both in the activated sludge plants and in the aerated lagoons studied. An efficient removal of trichloroacetic acid in aerobic treatment plants seems to require quite long retention time (Lindström and Mohamed 1988). The retention times in the case of the effluents SW-4 (aerated lagoon) and HW-1 (activated sludge) may therefore have been too short to allow for complete degradation of trichloroacetic acid. The efficient removal of this compound in the activated sludge plant used for treatment of the effluent SW-3 is unexpected in view of the relatively short retention time.

The part of the high molecular weight BKME material which mainly originates from the residual lignin in the pulp ("oxidized lignin") is a strongly oxidized, hydrophilic material rich in carboxylic acid groups (Mörck et al. 1991; Dahlman et al. 1996). The results obtained in the present investigation show that this material is removed only to a minor extent in aerobic treatment plants (Table 8). The major part of the "oxidized lignin" in BKMEs from chlorine bleaching has been found to be mainly non-aromatic (Mörck et al. 1991), whereas aromatic lignin structures (mainly non-phenolic) are retained to a significant degree in high molecular weight BKME materials from ECF bleaching and hydrogen peroxide-based TCF bleaching of softwood kraft pulp (Dahlman et al. 1996). Degradation studies have shown that the major part of the phenolic structural elements in "oxidized lignins" from ECF bleaching is non-chlorinated (Dahlman et al. 1996). A minor part is monochlorinated (Dahlman et al. 1996). This means that there is no reason to fear that degradation of high molecular weight BKME materials from ECF bleaching in receiving waters should lead to the formation of highly chlorinated monomeric phenolic compounds.

ACKNOWLEDGMENTS

The work presented in this paper has been carried out within the SSVL project Environment 93. The authors are most grateful to the Swedish Forest Industries Water and Air Pollution Research Foundation (SSVL) for financial support for the investigation.

REFERENCES

Amoth, A., G. Hickman and J. Miller. 1992. Treatment technologies for reduction of color, AOX and resin & fatty acids. Proceedings of the Environmental Conference, Richmond, Virginia, U.S.A., April 12-15, Book 1, 339-346.

Anon. 1992. Miljöstörande anläggningar. 10-årigt program för översyn av utsläppsvillkoren. Report to the Swedish Government from the Swedish Environmental Protection Board, December 15, 1992.

Anon. 1994. NCASI Technical Workshop – Effects of alternative pulping and bleaching processes on production and biotreatability of chlorinated organics. NCASI Special Report No. 94-01.

Axegård, P., O. Dahlman, I. Haglind, B. Jacobson, R. Mörck and L. Strömberg. 1993. Pulp bleaching and the environment – the situation 1993. Nordic Pulp Pap. Res. J. 8:365-378.

Berry, R.M., B.I. Fleming, R.H. Voss, C.E. Luthe and P.E. Wrist. 1989. Toward preventing the formation of dioxins during chemical pulp bleaching. Pulp Pap. Can. 90(8):48-58.

Call, H.P., and I. Mücke. 1994. State of the art of enzyme bleaching and disclosure of a breakthrough process. Proceedings of the International Non-Chlorine Bleaching Conference, Amelia Island, Florida, March 6-10:Sect. 10-2.

Dahlman, O., R. Mörck, F. de Sousa and L. Johansson. 1991. Chemical composition of modern bleached kraft mill effluents. Proceedings of the International Conference on Environmental Fate and Effects of Bleached Pulp Mill Effluents (Swedish Environmental Protection Agency Report 4031), Stockholm, Sweden, Nov. 19-21:47-56.

Dahlman, O., I. Haglind, R. Mörck, F. de Sousa and L. Strömberg. 1993a. Chemical composition of effluents from chlorine dioxide-bleaching of kraft pulps before and after secondary effluent treatment. Proceedings of the EUCEPA, Symposium Pulp and Paper Technologies for a Cleaner World, Paris, France, April 27-29, Vol. 1, 193-215.

Dahlman, O., R. Mörck, P. Ljungquist, A. Reimann, C. Johansson, H. Borén and A. Grimvall. 1993b. Chlorinated structural elements in high molecular weight organic matter from unpolluted waters and bleached kraft mill effluents. Environ. Sci. Technol. 27:1616-1620.

Dahlman, O., A. Reimann, R. Mörck and P. Ljungquist. 1996. A method for the detection of chlorinated, etherified lignin structures in high molecular weight materials of industrial and natural origin. In Environmental Fate and Effects of Pulp and Paper Mill Effluents, M.R. Servos, K.R. Munkittrick, J.H. Carey and G. Van Der Kraak (ed.), St. Lucie Press, Delray Beach, FL.

Denton, T.E., W.M. Howell, J.J. Allison, J. McCollum and B. Marks. 1985. Masculinization of female mosquitofish by exposure to plant sterols and *Mycobacterium smegmatis*. Bull. Environ. Contam. Toxicol. 35:627-635.

Gergov, M., M. Priha, E. Talka, O. Valttila, A. Kangas and K. Kukkonen. 1988. Chlorinated organic compounds in effluent treatment at kraft mills. TAPPI J. 71(12):175-184.

Germgård, U., A. Teder and D. Tormund. 1981. Chlorate formation during chlorine dioxide bleaching of softwood kraft pulp. Pap. Puu 63:127-133.

Grøn, C. 1991. Organic Halogens in Danish Groundwaters, p. 495-506. In B. Allard, H. Borén and A. Grimvall (ed.). Humic Substances in the Aquatic and Terrestrial Environment. Lecture Notes in Earth Sciences 33. Springer-Verlag, Heidelberg.

Haglind, I., L. Strömberg and B. Hultman. 1993. SSVL Environment 93 – Results from a research project conducted by the Swedish pulp and paper industry. Proceedings of the 1993 Environmental Conference, Boston, Massachusetts, March 28-31, Book 1, 245-251.

Kovacs, T.G., P.H. Martel, R.H. Voss, P.E. Wrist and R.F. Willes. 1993. Aquatic toxicity equivalency factors for chlorinated phenolic compounds present in pulp mill effluents. Environ. Toxicol. Chem. 12:281-289.

Kringstad, K.P., and K. Lindström. 1984. Spent liquors from pulp bleaching. Environ. Sci. Technol. 18:236A-247A.

Kringstad, K.P., L. Johansson, M.C. Kolar, F. de Sousa, S.E. Swanson, B. Glas and C. Rappe. 1989. The influence of chlorine ratio and oxygen bleaching on the formation of polychlorinated dibenzofurans (PCDFs) and polychlorinated dibenzo-*p*-dioxins (PCDDs) in pulp bleaching. Part 2. A full mill study. TAPPI J. 72(6):163-170.

Lindström, K., and M. Mohamed. 1988. Selective removal of chlorinated organics from kraft mill total effluents in aerated lagoons. Nordic Pulp Pap. Res. J. 3(1):26-33.

Lindström, K., and F. Österberg. 1986. Chlorinated carboxylic acids in softwood kraft pulp spent bleach liquors. Environ. Sci. Technol. 20:133-138.

Malmqvist, Å., T. Welander and L. Gunnarsson. 1991. Anaerobic growth of microorganisms with chlorate as an electron acceptor. Appl. Environ. Microbiol. 57:2229-2232.

Mörck, R., A. Reimann and O. Dahlman. 1991. Characterization of high molecular weight organic materials in modern softwood and hardwood bleached kraft mill effluents. Proceedings of the International Conference on Environmental Fate and Effects of Bleached Pulp Mill Effluents (Swedish Environmental Protection Agency Report 4031), Stockholm, Sweden, Nov. 19-21:155-163.

Pryke, D.C., P. Winter, G.R. Bouree and C. Mickowski. 1994. The impact of chlorine dioxide delignification on pulp manufacture and effluent characteristics at Grande Praire, AB. Pulp Pap. Can. 95:T230-T236.

Rosemarin, A., J. Mattsson, K.-J. Lehtinen, M. Notini and E. Nylén. 1986. Effects of pulp mill chlorate on *Fucus vesiculosus* – A summary of projects. OPHELIA 4:219-224.

Starck, B., P.O. Bethge, M. Gergov and E. Talka. 1985. Determination of chlorinated phenols in pulp mill effluents – An intercalibration study. Pap. Puu 67:745-749.

Swanson, S.E., C. Rappe, J. Malmström and K.P. Kringstad. 1988. Emissions of PCDDs and PCDFs from the pulp industry. Chemosphere 17:681-691.

Theander, O., and E.A. Westerlund. 1986. Studies on dietary fiber. 3. Improved procedure for analysis of dietary fiber. J. Agric. Food Chem. 34:330-336.

Wilson, R., J. Swaney, D.C. Pryke, C. Luthe and B. O'Connor. 1992. Mill experience with chlorine dioxide delignification. Pulp Pap. Can. 93(10):35-43.

Yin, C.-F., T.W. Joyce and H.-M. Chang. 1989. Characterization and biological treatment of bleach plant effluent. Proceedings of the 44th Industrial Waste Conference, W. Lafayette, Indiana, May 8-11:747-754.

SOURCES OF PULPING AND BLEACHING DERIVED CHEMICALS IN EFFLUENTS

Larry E. LaFleur

National Council of the Paper Industry for Air and Stream Improvement,
P.O. Box 458, Corvallis, OR, 97339 U.S.A.

A large number of chemicals are formed in the course of converting wood to bleached pulp. Given the complexity and diversity of modern kraft mills and the transition state of the industry in response to environmental concerns, it is impossible to completely predict the chemical composition of effluents. This review uses selected examples of what has been learned about the formation of chemicals in kraft pulping and bleaching to illustrate factors to be considered in understanding sources of chemicals and in identifying new chemicals potentially related to emerging environmental concerns. The basic characteristics of wood, including extractives, lignin and polysaccharides, are the starting materials for most compounds formed in the process. Kraft pulping removes substantial portions of lignin, forming a wide array of by-products. Polysaccharides and extractives are also modified by the process. Basic differences in the chemistry of chlorine and chlorine dioxide result in significant differences in the types of by-products formed. Although many compounds are formed in the pulping and bleaching process, not all end up in the effluent. The chemical composition of the major discharges to the sewers (including debarking, black liquor, condensates, causticizing bleach plant wastewaters) varies significantly. Biological treatment further modifies the composition. Although removal is generally determined by comparing influent and effluent concentrations, it does not account for potential transformations. One example is the transformation of resin acids to either oxidized forms or to aromatized structures such as retene, depending on the conditions in the treatment system.

INTRODUCTION

The production of bleached pulp using the kraft process results in the concurrent production of a large array of chemical by-products. The composition and quantities produced depend on a number of factors. The complexity and diversity of modern bleached kraft mills preclude a complete summary of how these various factors interact, and thus accurate prediction of the discharge of chemicals from any given process is not feasible. Furthermore, the industry has been undergoing significant changes in recent years in response to environmental concerns and consumer demands. However, an understanding of the sources of chemicals in the process and the mechanisms by which they are formed is possible and may provide a useful background which can aid in the identification of additional compounds or in the evaluation of the fate and effects of kraft mill discharges.

The objective of this review is to illustrate through selected examples how the basic components of wood react in the various stages of the bleached kraft manufacturing process and to highlight current knowledge and hypotheses concerning the formation and discharge of organic compounds. Decades of research into the chemistry of pulping and bleaching provide a wealth of information, but a comprehensive review of all the literature within the space allowed would be impossible. Thus, this discussion focuses on modern bleaching processes and the chemicals that are formed in these processes (as opposed to process additives).

Wood Chemistry

Any understanding of the types and sources of chemicals produced in the pulping process must necessarily begin with some knowledge of the composition of the raw material. Table 1 summarizes the basic compositions of hardwoods and of softwoods.

Table 1. The basic composition of wood.

	Percent Composition	
	Hardwoods	Softwoods
Polysaccharides		
Alpha-cellulose	42-47%	51-55%
Hemicellulose	20-30%	15-20%
Lignin	16-25%	23-33%
p-Hydroxyl propyl	Trace	Trace
guaiacol propyl	29-35%	>95%
syringyl propyl	40-46%	1%
Extractives	0.2-3.5%	0.5-7%

The polysaccharides in wood include cellulose and hemicellulose. Cellulose is the main component of wood fiber and is the most important for pulp production (Browning 1975). Cellulose consists of a linear polymer of ß-D-glucopyranose units coupled through (1-4) glycosidic bonds. The hemicellulose is branched and made up of different subunits, the composition of which is species dependent. The main components of softwood hemicellulose are galactoglucomannans, arabinoglucuronoxylan and arabinogalactan. Hardwood hemicellulose is largely glucuoxylan and glucomannan. Hemicellulose is also characterized by its 4-*O*-methyl glucuronic acid side chains. Again, there is a difference between the softwoods and hardwoods, with the higher frequency of 4-*O*-methyl glucuronic acid side chains associated with hardwoods. Hydrolysis of the 4-*O*-methyl groups is the main source of methanol formation in the digester, and the higher frequency of these structures in hardwoods results in formation of larger quantities (Wilson and Hrutfiord 1971).

Native lignin is generally considered a heterogenous polymer of high aromatic content and essentially infinite molecular weight (Adler 1977). The main components of the polymer are arylpropyl groups linked in the *para* position through ß-aryl ether bonds. Guaiacyl propyl groups are the predominant component of softwood lignin. Hardwood lignin is composed of both guaiacyl propyl and syringyl propyl groups with the relative ratio varying between species. Both hardwoods and softwoods contain small amounts of benzyl propyl ethers, particularly in compression wood. Although the ß-aryl ether linkages predominate in the polymer, α-aryl ether linkages, aryl-aryl linkages, and aryl-alkyl linkages are also known to be present. In native lignin, only limited numbers of free phenolic groups exist.

Chemicals which can be extracted with organic solvents are termed extractives. Both wood and bark can contain significant quantities of these materials. Softwoods contain from 0.5 to 7% extractives depending on the species. Even within a species, the quantity and composition will vary with the age of the tree, the location where it was grown and the season when the tree was felled. The major monoterpenes are α-pinene and ß-pinene (Drew *et al.* 1971). Although neutral diterpene hydrocarbons, alcohols and aldehydes are present in wood, the major form of diterpenes is the resin acids, and the relative composition is species dependent (Holmbom and Ekman 1978; Zinkel and Foster 1980). As a group of compounds, softwood extractives have been long recognized as biologically active. This is hardly surprising, as they form part of the tree's natural insect and microbial defense mechanism. Extractives acutely toxic to fish include juvabione (and related compounds) (Leach *et al.* 1975), resin and fatty acids (Leach and Thakore 1977) and diterpene alcohols (Leach and Thakore 1977; McKague *et al.* 1977). In addition, resin acids have been found to cause sublethal effects (Tana 1988). Plant sterols such as ß-sitosterol are found both in the free form and as fatty acid esters. The triglycerides are predominantly made up of C_{14} to C_{24} fatty acids with stearic, oleic and linoleic acids most common. Hardwood extractives include plant sterols (both free and as fatty acid esters), waxes, triglycerol esters and higher terpenoids such as betulin (Mutton 1958).

The extractives are most susceptible to changes during storage prior to being pulped. Monoterpenes volatilize and/or metabolize in chip storage piles, thereby reducing turpentine yields (Drew *et al.* 1971). Air oxidation and/or microbial degradation increase the levels of free fatty acids and cause unsaturated fatty acids and resin acids to become oxidized in the form of alcohols, ketones, epoxides or peroxides (Lawrence 1959; Levitin 1967; Assarsson and Croon 1963; Mutton 1958). These changes influence both overall yields as well as the formation of precursors of compounds formed in subsequent pulping. Under anaerobic conditions, ethanol is formed in the chips through glycolysis. This ethanol is then released in the pulping process and is found in black liquor and condensates (Wilson and Hrutfiord 1971).

Kraft Pulping Formation Reactions

The main objective of kraft pulping is removing the bulk of lignin while minimizing the degradation of cellulose. Hundreds of compounds have been identified in black liquor and it is beyond the scope of this review to discuss this work in detail. The following is intended to provide a brief overview of some of the kinds of pulping by-products that can be formed.

The chemicals formed from the reaction of lignin with the pulping liquors are generally polar in nature. These chemicals often contain ionizable functional groups such as phenols or carboxylic acids which aid in their dissolution in the highly caustic cooking liquor. Compounds which have been identified include simple phenols, aromatic and aliphatic carboxylic acids and diacids, and reduced sulfur compounds. The aromatics can be formed from the basic aromatic structures found in native lignin; thus the guaiacyl and syringyl content will be species dependent. Aromatics include phenol, catechol, guaiacol and syringyl structures, as well as compounds with formyl, acetyl, carboxyl and more complex hydroxy and carboxylic acid groups *para* to the hydroxy group. In addition to phenol formed from kraft delignification, phenol can be formed through decarboxylation of *p*-hydroxybenzoic acid hydrolyzed from hardwoods during the cook (Shariff *et al.* 1989).

Although the intent is to minimize degradation of cellulose, some degradation reactions do occur. The well-known peeling reaction generates low molecular weight acids such as isosaccharinic acid, as well as a wide variety of other hydroxy acids and diacids. It has also been shown that carbohydrates will react with base to form a variety of acetyl and formyl thiophenes (Lunde *et al.* 1991), thiophene carboxylic acids (Niemelä 1989b) and alkyl cyclopentenones (Voss 1984). 3-Methyl catechol and a dimethyl catechol have been identified in black liquor, and it is hypothesized that their formation is due to alkaline degradation and aromatization of carbohydrates (Niemelä 1989a).

The conditions in the digester can also give rise to reactions between lignin (or lignin fragments) and carbohydrates (or carbohydrate fragments). A C-C bond formation mechanism involving aldol or ketol carbohydrate breakdown fragments and aldehydic lignin fragments through an aldol condensation has been proposed (Gierer and Wännström 1984). Formation of possible C-O linkage between lignin and carbohydrates through oxirane-substituted lignin fragments was suggested by Iverson and Wännström (1986). These examples serve to demonstrate the complexity and variety of reactions possible in the kraft pulping process.

Triglyceride and other esters are saponified in the digester, giving rise to free fatty acids. Resin acids undergo both rearrangement of double bonds to more stable structures and disproportionation. Other reactions generally change the relative concentrations or composition of resin acids during kraft pulping. Fatty acids which were hydroxylated or oxidized while being stored are thought to be converted, in part to α,β-unsaturated ketones. Through retro-aldol condensations, these ketones can yield both acetone and methylethyl ketone (Wilson and Hrutfiord 1971).

Compounds Formed in Bleaching

Bleaching by-products can result from reaction of bleaching chemicals with the residual lignin, with cellulose or with chemicals carried over from the pulping process. A basic knowledge of the main reactions of the bleaching chemicals with these basic components should help explain the composition of the resulting bleaching filtrates. In 1994, the most widely used bleaching chemicals were and chlorine dioxide. The significant differences in the chemistry of chlorine and chlorine dioxide provide a mechanistic basis for understanding the formation and character of chlorinated bleaching by-products. Chlorine not only oxidizes lignin to form quinone or ring- opened structures; it also reacts through an electrophilic substitution mechanism to chlorinate aromatic rings and some aliphatic side chain groups. Chlorine dioxide reacts with lignin only as an oxidant, forming ring-opened muconic acid (ester) structures. The small quantities of chlorinated organics formed in chlorine dioxide bleaching are thought to be due to formation of hypochlorous acid when chlorine dioxide oxidizes lignin (Kolar et al. 1983; Ni et al. 1994). The hypochlorous acid is in equilibrium with chlorine (depending on the pH), which is probably the actual chlorinating agent. Since hypochlorous acid/chlorine react extremely rapidly with lignin soon after they are formed, there is never a high concentration of chlorine present. Therefore, unlike full chlorine bleaching where the initial concentration of elemental chlorine is very high and polychlorination of a given compound can occur, the degree of multiple chlorination in chlorine dioxide bleaching is much lower.

Pre-bleaching with chlorine (and/or chlorine dioxide) and caustic continues the lignin removal process and produces the greatest quantity of by-products. The most well-studied group of compounds resulting from the reaction of bleaching chemicals with lignin are the chlorinated phenolics (see Fig. 1). Chlorine multiple, brownstock kappa number and substitution of chlorine dioxide for elemental chlorine and order of addition have all been shown to influence not only the quantity but also the degree of chlorination of the chlorinated phenolics. Some of these findings are summarized by Berry et al. (1991) and were further confirmed and reinforced with new data by Axegård et al. (1993). NCASI (1994) has also summarized these data in numerous studies. It is interesting to note, however, that despite the extensive attention given to this group of compounds, recent laboratory studies have identified several minor new compounds (Wallis et al. 1993) that are clearly related to lignin reactions.

Ultra-trace levels of furanones, hydroxy furanones and butenedioic acids (both chlorinated and non-chlorinated) similar or related to the potent mutagens MX (3-chloro-(dichloromethyl)-5-hydroxy-2-(5H)-furanone) and EMX (E2-chloro-3-dichloromethyl-4-oxobutenoic acid) have recently been identified in chlorine bleaching filtrates (Kronberg and Franzen 1993). Similar studies have not been reported for ECF (elemental chlorine free) or TCF (totally chlorine free) filtrates. The mechanism of formation of these compounds has not been confirmed but the formation of MX and EMX from model compounds structurally similar to components of lignin (Långvik and Hormi 1994) suggests that these compounds are by-products of delignification. Earlier studies indicated that MX and EMX were effectively reduced in biological treatment (Smeds et al. 1990). Thus, there is a strong possibility that the compounds identified by Kronberg and Franzen might similarly be removed in treatment systems.

The effects of chlorine multiple, brownstock kappa number and increased substitution of chlorine dioxide in reducing the formation of other chlorinated organics (e.g., AOX and polychlorinated dibenzo-p-dioxins and dibenzofurans [PCDD/F]) were also addressed by Berry et al. (1991). ECF bleaching has also been shown to reduce the formation of chlorinated acetic acids (O'Connor et al. 1993). The trends of decreased chlorine content and reduced polychlorination are all consistent with and readily explained by the basic reaction of chlorine dioxide discussed above. Driven by regulatory and consumer pressures, there has been an increasing trend towards complete substitution of chlorine dioxide for elemental chlorine because of the reduced formation of chlorinated organics.

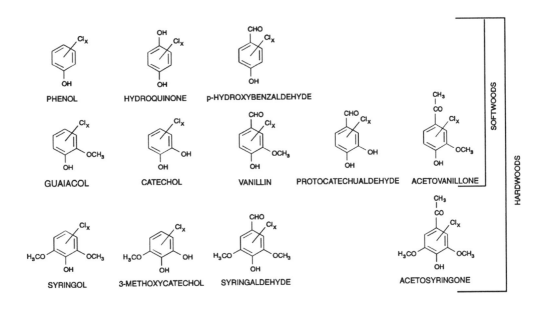

Figure 1. Chlorinated phenolics identified in bleaching filtrates.

Questions about the potential environmental significance of the high molecular weight material (HMWM) formed in the bleaching process have driven research into characterizing the structure of these materials. Increasing the level of substitution of ClO_2 for Cl_2 increases the carbon to chlorine ratio, indicating a lower degree of chlorination (Mörck et al. 1991; Dahlman et al. 1991, 1993a,b, 1996; O'Connor et al. 1993). These studies show that the HMWM has a higher carboxylic acid content, decreased methoxyl content and higher phenolic content than residual lignin. The aromatic content is low (0.5-2%) and only a small portion of the aromatic rings identified have chlorine substitution. Berry et al. (1991) and Mörck et al. (1991) correctly point out that this should allay concerns over potential formation of highly chlorinated, persistent aromatic breakdown products when the HMWM is discharged.

Recent model compound studies are providing insight into the chemical character of HMWM. McKague et al. (1993) have shown that a ß-aryl ether dimer model of lignin is only partially dearomatized by chlorine dioxide. Further reaction resulted in cleavage of the ß-aryl ether bond. This would suggest that ß-aryl ether linkages in residual lignin are largely broken during chlorine dioxide bleaching and thus contribute little to the structure of HMWM. Further studies, using dimer models of diaryl methane and biphenyl moieties known to be present in residual lignin, have shown that the resulting degradation products include some structures where the aromatic rings of the dimer are converted to muconic lactone structures (McKague et al. 1994a,b). The lactones open to muconic acids upon treatment with base and are apparently in equilibrium with the lactone form depending on pH. The compounds identified in these studies have many of the general characteristics of HMWM (low degree of chlorination, high carboxyl content, largely aliphatic, etc.) and thus provide some valuable insight into likely components of HMWM.

Carry-over of extractives, pulping by-products and process additives due to imperfect brownstock washing may result in formation of new compounds through reaction (mainly chlorination) with bleaching chemicals. As discussed earlier, the degree of chlorination will greatly depend on the bleaching chemicals in use. Chlorination of extractives carried over with the brownstock is the source of chlorinated resin and fatty acids and monoterpenes (Stuthridge *et al.* 1990). Chlorinated thiophenes (Lunde *et al.* 1991; McKague *et al.* 1989) and chlorinated sulfones (Voss 1983) are examples of compounds that arise from chlorination of pulping by-product carry-over. Carry-over of precursors found in certain oil-based defoamers has been recognized as a significant source of PCDD/Fs (Voss *et al.* 1988). Recently, ultra-trace levels of chlorinated polycyclic aromatics (chlorophenanthrenes, alkylphenanthrenes and alkylnapthalenes) have been reported (Koistinen *et al.* 1992, 1994). Although the mechanism of formation was not identified, the authors suggest that these compounds may also be formed from chlorination of precursors in oil-based defoamers. Further work will be required to determine how significant these findings are.

Although numerous non-chlorinated compounds arise from reaction of bleaching agents with polysaccharides, chlorinated compounds are not major by-products. Characterization of laboratory bleaching filtrates recently revealed the presence of 4-chloro-3-hydroxy-2H-pyran-2-one (Smith *et al.* 1994). The chlorinated pyrone was identified in bleaching liquors from both hardwood and softwood kraft pulps, but not from any bisulfite pulps. The authors hypothesized that the compound was formed from polysaccharide precursors. Although the chlorinated pyrone was stable in strong sodium hydroxide solutions, the compound quickly disappeared when C filtrates were neutralized. Thus, it is unlikely that the chlorinated pyrone will ever be detected in effluents discharged into the environment.

Today, some mills use TCF bleaching. Since the objective is to recycle TCF filtrates back to recovery, there has been very little attention given to characterization of by-products. Sonnenberg *et al.* (1992) conducted some ozone filtrate characterization. This work focused on low molecular weight compounds and identified numerous aliphatic acids and glyoxal, glyoxylic acid and methyl glyoxyloate. Studies by Dahlman *et al.* (1996) and O'Connor *et al.* (1993) have investigated the HMWM formed in TCF bleaching. As expected, the material has very little chlorine associated with it. The most notable observation was that the material had a very high carbohydrate content: 48% versus 3-18% associated with conventional or ECF HMWM.

Process Sources

The previous discussion focused on how compounds are formed but did not identify whether or not they were sewered; formation does not necessarily imply discharge into the receiving environment. Since not all sources of chemicals are sewered, the partitioning of compounds in the process determines to a large extent which compounds end up being discharged. The main process sources of wastewater discharged to biological treatment systems, or in some cases directly into the receiving environment, include debarking wastewater, sewered condensates (both digester and evaporator), brownstock washing/screening wastewater and bleach plant filtrates. Causticizing wastewater can also contribute small amounts of organics. Essentially all bleach plant filtrates are discharged to the treatment system, and thus, the previous section serves to characterize the types of compounds discharged. General chemical characteristics of various woodroom and pulping sources are summarized in Table 2.

Wastewaters from the wet debarking process contain high levels of wood extractives such as resin acids, fatty acids, sterols, monoterpenes, etc. (Holmbom and Lehtinen 1980; Talka and Priha 1987). In mills pulping softwoods, this can represent a significant source of resin and fatty acids. The present trend is towards dry debarking, which produces minimal discharge to the wastewater system and thus will minimize this resin/fatty acid discharge.

Black liquor can enter into wastewater sewers through spills or leaking seals in pumps. Generally, the amount of black liquor lost in this manner is minor, and thus only a small amount of pulping by-products will be discharged directly into the sewer. Another source of pulping by-products is brownstock washing. The extent to which the brownstock washing and screening is closed will determine the amounts of pulping by-products that will enter into the sewer.

Table 2. Summary of composition of various pulping process and wastewaters streams.

Compound Class	Debarking	Black Liquor	Condensates	Causticizing
Extractives				
Monoterpenes (hydrocarbons, alcohols and ketones)	x	x	x	x
Sesquiterpenes	x	x	x	
Diterpene aldehydes/alcohols	x	x	x	
Diterpene acids	x	x	x	
Sterols	x	x	x	
Misc. extractives (e.g., juvabione, *epi*-juvabione)	x	x	x	
Fatty acids	x	x	x	
Pulping by-products				
Aliphatic alcohols (e.g., methanol, ethanol)		x	x	
Aromatic alcohols (e.g., phenol, guaiacol, *p*-cresol)		x	x	
Aromatic acids		x		
Aliphatic acids (e.g., acetic, propionic acids)	x	x	x	x
Aliphatic aldehydes (e.g., acetaldehyde, furfural)		x	x	
Aliphatic ketones (e.g., acetone, methyl ethyl ketone, alkyl cyclopentenones)		x	x	x
Aromatic aldehydes (e.g., vanillin, syringaldehyde)		x	x	
Aromatic ketones (e.g., acetovanillone, acetosyringone, acetophenone)		x	x	
Thiophene		x	x	
Acetyl and formyl thiophenes		x	x	x
Misc. other sulfur compounds (e.g., dimethylsulfide, dimethyldisulfide)		x	x	

Many compounds formed in the digester are prone to steam distillation and thus will partition into digester relief gas, blow gas or evaporator condensates. Low molecular weight alcohols, aldehydes and ketones, terpenes, sulfur-bearing compounds and phenolics have been found in condensates (Blackwell *et al.* 1979; Holmbom and Lehtinen 1980; Talka and Priha 1987). The quantities, quality and composition depend on wood species, digester type and type of evaporators used. If these condensates are used for washing in the bleach plant, they may represent a source of compounds which may become chlorinated and discharged with bleach plant filtrates. Steam stripping of condensates substantially reduces the BOD load to the treatment system, largely through removal of methanol, but its effects on other components such as phenols and terpenes have not been systematically documented.

Biological Treatment

The beneficial effects of biological treatment in the removal of BOD and acute toxicants have long been recognized. Many of the compounds that have been identified in pulping and bleaching sewers may never be discharged into the environment if biological treatment is utilized. Removal of resin and fatty acids (Easty *et al.* 1978; Holmbom and Lehtinen 1980; Voss and Rapsomatiotis 1985) and chlorinated phenolics (Dahlman *et al.* 1996; NCASI 1994) is well documented in the literature. It has also been shown that chlorinated acetones (Gergov et al. 1988; Smeds et al. 1990), chlorinated acetic acids (Lindström and Mohamed 1988; Dahlman et al. 1991, 1993a), MX/EMX (Smeds et al. 1990) and sterols (Holmbom and Lehtinen 1980; Dahlman et al. 1993a) are removed or substantially reduced in biological treatment systems. Estimated removal efficiencies of various groups of compounds are summarized in Table 3. Recent studies on the effects of biological treatment on HMWM indicate that the guaiacol and/or syringyl end groups are somewhat reduced, the overall molecular weight shifts toward higher masses and the carbohydrate content is reduced (Dahlman et al. 1993a, 1996).

Table 3. Summary of range of treatment efficiencies for various groups on compounds.

	Treatment Efficiency Range
Non-Chlorinated Compounds	
Monoterpenes	80-100%
Diterpene aldehydes	57-98%
Diterpene resin acids	47-100%
Sterols	62-98%
Fatty acids	80-100%
Phenolics	90-100%
Chlorinated Compounds	
Mono/dichlorophenolics	70-95%
Tri/tetrachlorophenolics	20-80%
Chlorosulfones	8-17%
Chloroacetic acids	54-100%
Chloroacetones	64-100%
MX/EMX	0-98%
Chloro-resin/fatty acids	38-100%

Removal in biological treatment systems can be due to microbial degradation, physio-chemical processes such as adsorption, or air stripping. In some cases, microbial degradation may not be complete, resulting in compounds being partially metabolized. This results in the formation and discharge of new compounds. For example, early studies showed that some monoterpenes discharged to biological treatment systems can be converted to camphor or fenchone (Wilson and Hrutfiord 1975). Recent studies report that resin acids can be hydroxylated (Wilkins et al. 1988; Zender et al. 1993) or, under anaerobic conditions, undergo reduction (McFarlane and Clark 1988; Zender et al. 1993). Surveys of aerobically treated effluents in the United States showed detectable levels of 13-abieten-18-oic acid (NCASI 1989; NCASI unpublished data) and 7-oxodehydroabietic acid (NCASI unpublished data). NCASI has also surveyed several additional mill effluents for 7-isopimarenic acid and found the compound present only once at just above the detection limit (NCASI 1993; NCASI unpublished data). Additional studies would be required to determine how significant these findings are, but the low levels detected indicate these reduced and oxidized resin acids may not be a problem in well-operated aerobic treatment systems.

Evidence of anaerobic decarboxylation and reduction or aromatization of resin acids has also been reported (Wilkins and Panadam 1987). In this work, fichtelite, dehydroabietin, 1,2,3,4-tetrahydroretene and retene were identified in mill discharges. The source was identified as a small overflow stream from a sludge lagoon.

SUMMARY

The origins of a large number of the compounds present in pulping and bleaching wastewaters have been identified. The by-products result from the reaction of pulping and bleaching chemicals with various components of wood, including extractives, lignin, cellulose and hemicellulose. The amounts and types of these by-products entering the aquatic receiving environment will be dependent on the species being pulped, the conditions of the pulping, the techniques used for bleaching, how the bleaching sewers are mixed prior to discharge and the presence (or absence) of biological treatment.

Much work has been done in identifying compounds in process sewers, but comparatively less on biologically treated effluents. The reasons for this include the higher levels found in process sewers (i.e. no dilution with other wastewaters and/or no biological removal) and less complexity since the wastewaters have not been mixed with other wastewaters. Also, most of this work was done many years ago and focused on elemental chlorine bleaching. Modern bleaching practices should substantially minimize the formation of chlorinated organics, and consequently one should not make the mistake of assuming bleached kraft mill effluents contain all of the 300-plus compounds previously identified in wastewaters. Finally, little is known about the composition of TCF wastewaters. Although the intent is to return TCF filtrates to recovery, the likely use of small purge streams and the potential for chemicals to remain with the pulp suggest it would be prudent to determine if any deleterious materials are present.

REFERENCES

Adler, E. 1977. Lignin chemistry - past, present and future. Wood Sci. Technol. 11:169-218.

Assarsson, A. and I. Croon. 1963. Studies on wood resin, especially the change in chemical composition during seasoning of the wood, part 1. Changes in the composition of the ethyl ether soluble part of the extractives from birch wood during log seasoning. Svensk Papperstidn. 66(21):876-883.

Axegård, P., O. Dahlman, I. Haglind, B. Jacobson, R. Mörk and L. Strömberg. 1993. Pulp bleaching and the environment - the situation in 1993. Nordic Pulp and Pap. Res. J. 4:365-378.

Berry, R., C. Luthe, R. Voss, P. Wrist, P. Axegård, G. Gellerstedt, P-O. Lindbald and I. Pöpke. 1991. The effects of recent changes in bleached softwood kraft mill technology on organochlorine emissions: an international perspective. Pulp Pap. Can. 92(6):T155-T165.

Blackwell, B., W. MacKay, F. Murray and W. Oldham. 1979. Review of kraft foul condensates. TAPPI J. 62(10):33-37.

Browning, B.L. Ed. 1975. The Chemistry of Wood. Robert E. Krieger Publishing Co., Huntington, NY, 689 p.

Dahlman, O., R. Mörck, I. Johansson and F. de Sousa. 1991. Chemical composition of modern bleached kraft mill effluents. Environmental Fate and Effects of Bleached Pulp Mill Effluents, SEPA Conference, Stockholm, Sweden, November.

Dahlman, O., I. Haglind, R. Mörck, F. de Sousa and L. Strömberg. 1993a. Chemical composition of effluents from chlorine-dioxide bleaching of kraft pulps before and after secondary treatment. International Environmental Symposium, EUCEPA, Paris, France, April.

Dahlman, O., R. Mörck, P. Ljungquist, A. Reimann, C. Johansson, H. Boren and A. Grimvall. 1993b. Chlorinated structural elements in high molecular weight organic matter from unpolluted waters and bleached-kraft mill effluents. Environ. Sci. and Tech. 27(8):1616-1620.

Dahlman, O., A. Reimann, R. Mörck and P. Ljungquist. 1996. On the nature of high molecular weight effluent materials from modern ECF- and TCF-bleaching. In Environmental Fate and Effects of Pulp and Paper Mill Effluents, M.R. Servos, K.R. Munkittrick, J.H. Carey and G. Van Der Kraak (ed.), St. Lucie Press, Delray Beach, FL.

Drew, J., J. Russell and H. Bajak. Eds. 1971. Sulfate Turpentine Recovery. Pulp Chemicals Association, New York, NY, 147 p.

Easty, D., L. Borchardt and B. Wabers. 1978. Wood derived toxic compounds. Removal from mill effluents by waste treatment processes. TAPPI J. 61(10):57-60.

Gergov, M., M. Priha, E. Talka, O. Välttilä, A. Kangas and K. Kukkonen. 1988. Chlorinated organic compounds in effluent treatment at kraft mills. TAPPI J. 71(12):175-184.

Gierer, J. and S. Wännström. 1984. Formation of alkali-stable C-C-bonds between lignin and carbohydrate fragments during kraft pulping. Holzforschung 38(4):181-184.

Holmbom, B. and R. Ekman. 1978. Tall oil precursors of scots pine and common spruce and their change during sulfate pulping. Acta Acad. Abo. Ser. B 38(3):1-11.

Holmbom, B. and K-J. Lehtinen. 1980. Acute toxicity to fish of kraft pulp mill waste waters. Paperi ja Puu 62(11):673-684.

Iverson, T. and S. Wännström. 1986. Lignin-carbohydrate bonds in a residual lignin isolated from pine kraft pulp. Holzforschung 40(1):19-22.

Koistinen, J., T. Nevalainen and J. Tarhanen. 1992. Identification and level estimation of aromatic coeluates of polychlorinated dibenzo-p-dioxins and dibenzofurans in pulp mill products and wastes. Environ. Sci. and Tech. 26(12):2499-2507.

Koistinen, J., J. Paasivirta, T. Nevalainen and M. Lahtiperä. 1994. Chloro phenanthrenes, alkylchlorophenanthrenes and alkyl-chloronaphthalenes in kraft pulp mill products and discharges. Chemosphere 28(7):1261-1277.

Kolar, J., B. Lindgren and B. Pettersson. 1983. Chemical reactions in chlorine dioxide stages of pulp bleaching. Wood Sci. Technl. 17:117-128.

Kronberg, L. and R. Franzen. 1993. Determination of chlorinated furanones, hydroxy furanones and butenedioic acids in chlorine-treated water and in pulp bleaching liquor. Environ. Sci. and Tech. 27(9):1811-1818.

Långvik, V-A. and O. Hormi. 1994. Possible reaction pathways for the formation of 3-chloro-4-(dichloromethyl)-5-hydroxy-2(5H)-furanone (MX). Chemosphere 28(6):1111-1117.

Lawrence, R. 1959. Oxidation of resin acids in wood chips. TAPPI J. 42(10):867-869.

Leach, J. and A. Thakore. 1977. Compounds toxic to fish in pulp mill waste streams. Prog. Wat. Tech. 9:787-798.

Leach, J., A. Thakore and J. Manville. 1975. Acute toxicity to rainbow trout (*Salmo gairdneri*) of naturally occurring insect juvenile hormone analogues. J. Fish. Res. Board Can. 32(12):2556-2559.

Levitin, N. 1967. Review of effect of chip storage on wood resins and pulps. Pulp and Paper Mag. Can. 68(9):T454-460.

Lindström, K. and M. Mohamed. 1988. Selective removal of chlorinated organics from kraft mill total effluents in aerated lagoons. Nordic Pulp and Pap. Res. J. 1:26-33.

Lunde, A., J. Skramstad and G. Carlberg. 1991. Identification, mutagenicity and origin of chlorinated thiophenes in kraft bleaching effluent. Paperi ja Puu 73(6):522-526.

McFarlane, P. and T. Clark. 1988. Metabolism of resin acids in anaerobic systems. Nat. Sci. Tech. 20(1):273-276.

McKague, B., J. Leach, R. Soniassy and A. Thakore. 1977. Toxic constituents in woodroom effluents. Transactions of the Technical Section (CPPA) 3(3):75-81.

McKague, B., M-C. Kolar and K. Kringstad. 1989. Nature and properties of some chlorinated, lipophilic, organic compounds in spent liquors from pulp bleaching, 2. Environ. Sci. and Tech. 23(9):1126-1129.

McKague, B., G. Kang and D. Reeve. 1993. Reaction of a lignin model dimer with chlorine and chlorine dioxide. Holzforschung 47:497-500.

McKague, B., F. Xi, G. Kang and D. Reeve. 1994a. The chemistry of high molecular weight chlorinated organic matter from chlorine dioxide bleaching. 1994 International Pulp Bleaching Conference, Vancouver BC, Canada, June 13-16.

McKague, B., G. Kang and D. Reeve. 1994b. Reactions of lignin model dimers with chlorine dioxide. Nordic Pulp and Pap. Res. J. 2:84-87, 128.

Mörck, R., A. Reimann and O. Dahlman. 1991. Characterization of high molecular weight organic materials in modern softwood and hardwood bleached kraft mill effluents. Environmental Fate and Effects of Bleached Pulp Mill Effluents, SEPA Conference, Stockholm, Sweden, September 19-21, 155-163.

Mutton, D. 1958. Hardwood resin. TAPPI J. 41(11):632-643.

NCASI. 1989. Effects of biologically treated bleached kraft mill effluent on cold water stream productivity in experimental streams - fifth progress report. NCASI Technical Bulletin No. 566, NCASI, 260 Madison Ave., New York, NY 10016.

NCASI. 1993. Aquatic community effects of biologically treated bleached kraft mill effluent before and after conversion to increased chlorine dioxide substitution: results from an experimental streams study. NCASI Technical Bulletin No. 653, NCASI, 260 Madison Ave., New York, NY 10016.

NCASI. 1994. NCASI Technical Workshop - Effects of alternative pulping and bleaching processes on production and biotreatment of chlorinated organics. NCASI Special Report No. 94-01, NCASI, 260 Madison Ave. New York, NY 10016.

Ni, Y., X. Shen and A. van Heiningen. 1994. Studies on the reaction of phenolic and non-phenolic lignin model compounds with chlorine dioxide. J. Wood Chem. Tech. 14(2):243-262.

Niemelä, K. 1989a. GLC-MS studies on pine kraft black liquors part V. Identification of catechol compounds. Holzforschung 43(2):99-103.

Niemelä, K. 1989b. GLC-MS studies on pine kraft black liquors part VI. Identification of thiophenecarboxylic acids. Holzforschung 43(3):169-171.

O'Connor, B., T. Kovacs, R. Voss, P. Martel and B. Van Lierop. 1993. A laboratory assessment of the environmental quality of alternative pulp bleaching effluents. International Environmental Symposium, EUCEPA, Paris, France, April.

Shariff, A., R. Lowe, D. Berthiaume, J. Bryce and R. McLean. 1989. Unexpected source of phenol in the sulfur-free semichemical pulping of hardwood. TAPPI J. 72(3):177-183.

Smeds, A., B. Holmbom and L. Tikkanen. 1990. Formation and degradation of mutagens in kraft pulp mill sewers. Nordic Pulp and Pap. Res. J. 3:142-147.

Smith, T., R. Wearne and A. Wallis. 1994. An ubiquitous chlorohydroxypyrone in C stage filtrates from bleaching alkaline pulps. 1994 International Pulp Bleaching Conference, Vancouver, B.C., Canada, June 13-16.

Sonnenberg, L., K. Poll, R. Le Lacheur and R. Murphy. 1992. Characterization of pulp ozonolysis products. 1992 TAPPI Environmental Conference, Richmond, VA, April 12-15.

Stuthridge, T., A. Wilkins and A. Langdon. 1990. Identification of novel chlorinated monoterpenes formed during kraft pulp bleaching of *Pinus radiata*. Environ. Sci. and Tech. 24(6):903-908.

Talka, E. and M. Priha. 1987. Fractionation and identification of some biologically active compounds in bleached kraft mill effluents. Paperi ja Puu 3:220-228.

Tana, J. 1988. Sublethal effects of chlorinated phenols and resin acids on rainbow trout (*Salmo gairdneri*). Water Sci. Tech. 20(2):77-85.

Voss, R. 1983. Chlorinated neutral organics in biologically treated bleached kraft mill effluents. Environ. Sci. and Tech. 17(9):530-537.

Voss, R. 1984. Neutral organic compounds in biologically treated bleached kraft mill effluents. Environ. Sci. and Tech. 18(12):938-946.

Voss, R. and A. Rapsomatiotis. 1985. An improved solvent-extraction based procedure for the gas chromatographic analysis of resin and fatty acids in pulp mill effluents. J. Chromatography 346:205-214.

Voss, R., C. Luthe, B. Flemming, R. Berry and L. Allen. 1988. Some new insights into the origins of dioxins formed during chemical pulp bleaching. Pulp Paper Can. 89(12):151-162.

Wallis, A., T. Smith and R. Wearne. 1993. Determination of chlorinated phenols in effluent from bleached eucalypt kraft mills. National Pulp Mills Research Program, Tech. Report No. 2, INRE Project Office, PO Box 225, Dickson, ACT 2602, Australia.

Wilkins, A. and S. Panadam. 1987. Extractable organic substances from the discharges of a New Zealand pulp and paper mill. Appita 40(3):208-212.

Wilkins, A., A. Langdon, G. Mills, S. Panadam and T. Stuthridge. 1988. Kinleithic acid: a new hydroxylated resin acid from the biological treatment system of a New Zealand kraft pulp and paper mill. Aust. J. Chem. 42:983-986.

Wilson, D. and B. Hrutfiord. 1971. Sekor IV. Formation of volatile organic compounds in the kraft pulping process. TAPPI J. 54(7):1094-1098.

Wilson, D. and B. Hrutfiord. 1975. The fate of turpentine in aerated lagoons. TAPPI J. 76(6):91-93.

Zender, J., T. Stuthridge, A. Langdon, A. Wilkins, K. Mackie and P. McFarlane. 1993. Removal and transformation of resin acids during secondary treatment at a New Zealand bleached kraft pulp and paper mill. 4th IAWQ Symposium on Forest Industry Wastewaters, Tampere, Finland.

Zinkel, D. and D. Foster. 1980. Tall oil precursors in the sapwood of four southern pines. TAPPI J. 63(5):137-139.

TOXICITY OF TCF AND ECF PULP BLEACHING EFFLUENTS ASSESSED BY BIOLOGICAL TOXICITY TESTS

J. Ahtiainen, T. Nakari and J. Silvonen

National Board of Waters and the Environment,
P.O. Box 250, 00101 Helsinki, Finland

The toxicity of eighteen different untreated or secondary treated TCF (total chlorine free), ECF (elemental chlorine free) and conventional (Cl_2) bleaching effluents was assessed by a battery of biological tests. The toxicity tests used were: *Pseudomonas putida* growth inhibition test, *Vibrio fisheri* luminescence bacteria test, *Selenastrum capricornutum* algal growth inhibition test, *Daphnia magna* mobility inhibition (24 h) test and *Brachydanio rerio* zebra fish hatching and survival test. In the *P. putida* growth inhibition test only conventional bleaching effluents and ECF birch pulp effluent gave a slightly toxic response. The *V. fisheri* test was more sensitive. The EC50 values of most untreated ECF and TCF effluents were under 10% effluent concentrations (conventional effluent 15%) and secondary treated effluents were not toxic. All untreated bleaching effluents gave a toxic response in *S. capricornutum* algal test. EC50 values varied between 12 and 46% effluent concentrations. Treated effluents were not toxic and had a stimulative impact on algal growth. The results obtained by the *D. magna* test showed that effluents of TCF, ECF (birch pulp) and conventional (pine pulp) bleaching were equally toxic (LC50 values about 40%). Untreated TCF and ECF (pine and mixed pine and birch pulp) like secondary and pilot treated effluents were all nontoxic. The results obtained by the egg/larvae test of zebra fish showed that the lowest effect concentration (LOEC) values differed between hatching and mortality of the larvae. Some effluents were more harmful to hatching while others had a greater effect on mortality. Secondary treated effluents did not have any significant effects. Of the untreated effluents, TCF and conventional (both pine pulp) were the most toxic (LOEC values of hatching between 1 and 3.2% and mortality between 12 and 6.8%). The effects of all the other samples were nearly the same (LOEC values varying between 5 and 25%). There was no significant difference in toxicity between untreated conventional, ECF and TFC bleaching effluents. The natural constituents of wood are probably responsible for the toxicity observed in ECF and TCF effluents.

INTRODUCTION

The discharge of different organic compounds in bleach effluents from pulp mills has been implicated in a variety of effects observed in aquatic ecosystems. The research regarding toxic impacts of wastewaters has focused mainly on chlorinated organic compounds but also on phenolics and resin acids. The current improvements in pulp bleaching technology (ECF and TCF) will reduce the concentrations of chlorinated organic compounds to very low levels. These changes may have effects on the toxicity of untreated and treated wastewaters. In recent studies it has been concluded that non-chlorinated compounds in mill effluents probably contribute the harmful effects observed in laboratory and field studies. Most of these toxic compounds are natural constituents of wood (Axegård et al. 1993).

Single species toxicity tests can be used to estimate the potential harmful impact of wastewaters on aquatic ecosystems. In order to simulate the aquatic food chain, representatives of different trophic level organisms such as algae, daphnids and various fish often are used as tests organisms. One major advantage of biological toxicity tests, over chemical analysis, is their direct assessment of potential biotic impact without extrapolation from chemical analysis of uncertain completeness of substances analyzed.

In recent years various bacterial bioassays have been developed for screening of wastewater toxicity. Most of them are based on the measurements of growth inhibition, respiration and viability of bacterial cells. Microorganisms, bacteria in particular, have several attributes which make them attractive for use in wastewater toxicity testing. Microbial tests are simple, rapid, sensitive and inexpensive toxicity assays (Bitton and Dutka 1986). With these tests it is possible to estimate the harmful effects of wastewaters on natural aquatic microbes. By using bacterial toxicity tests it is also possible to estimate the toxic impacts of different, often fluctuating, untreated wastewaters on activated sludge wastewater treatment processes. Most modern wastewater treatment relies on microorganisms, so their activity is of primary importance in proper wastewater treatment.

The bacterium *Pseudomonas putida* represents a common aquatic heterotrophic microorganism. When *P. putida* cells are cultured under specified conditions, in a defined medium with different concentrations of wastewater over several generations, toxic substances present in the wastewater sample can inhibit the cell multiplication of the bacteria (Brinkmann and Kühn 1977).

The inhibition of light production by *Vibrio fisheri* luminescent bacteria indicates disturbance of the energy metabolism of this heterotrophic bacterium. Because this luminescence pathway is a direct branch of the electron transport chain, the luminescent measurement assesses the metabolic status of this bacterium (Hastings 1978). Hence, the change in bacterial luminescence when these bacteria are exposed to wastewater samples can be used as an indicator of potential toxicity (Bulich *et al.* 1981).

Aquatic single cell algae are sensitive test organisms used to assess the toxicity of wastewaters. Algal toxicity tests are based on measurement of cell growth inhibition or photosynthetic activity. When exponentially growing cultures of green alga *Selenastrum capricornutum* are exposed to various dilutions of wastewater sample over several generations, the substances in the wastewater can either inhibit or stimulate the algal growth.

The sensitivity of organisms to the toxic properties of a substance may vary considerably from one species to another, owing to differences in their metabolism and the nature of their habitats. The *Daphnia* test (*Daphnia magna* Straus) is one of the most commonly used toxicity tests for the determination of the acute toxicity of water soluble chemicals, industrial effluents, sewage effluents, and surface and ground waters. The test is based on the determination of the sample concentration which immobilizes 50% of exposed *D. magna*.

Fish are particularly susceptible to the influence of toxic substances during the reproductive (gametogenesis) and early developmental stages (embryo and fry stages) (McKim *et al.* 1978; Hodson and Blunt 1981). Determination of the influence on early developmental stages is a more sensitive index of fish tolerance than obtained by determination of acute toxicity on adults. The embryo-fry test is a test method for the determination of toxicity of chemicals, products or sewage water using embryos and early fry stages of a species of freshwater fish *Brachydanio rerio* (Hamilton-Buchanan zebra fish).

The aim of the study was to compare the toxicity of the wastewaters of conventional chlorine (Cl_2) and novel ECF and TCF bleaching with a battery of the above mentioned biotests.

MATERIALS AND METHODS

Sampling

The two pulp mills chosen for the study produce both bleached hardwood kraft pulp and softwood kraft pulp. Mill A (Wisaforest) produces nonbleached board and bleached and nonbleached kraft pulp. Mill B (Enocell) produces bleached kraft pulp on two lines, hardwood and softwood on separate lines. More precise descriptions of these pulp mills are presented in a following paper (Verta *et al.* 1995). Both treated and untreated effluents were sampled and tested from these mills. In Mill A, samples after secondary treatment were taken at parallel occasions as bleached effluents were sampled. The secondary treated samples from Mill A represent the mixtures of two bleaching lines (conventional and ECF or TCF) plus effluents from a paper mill. The ECF effluent (22.05) and the TCF effluent (02.06) were secondary treated in a pilot scale sewage plant. From Mill B, the bleaching effluents are mixtures of a softwood and a hardwood line. Parallel samples were also taken after secondary treatment with municipal sewage and after sewage treatment pond.

Samples were taken as daily pooled samples (Mill A) or as grab samples (Mill B). Samples were divided into 500 or 1000 mL polyethylene bottles and frozen at -18°C. For algal and bacterial toxicity tests the pH of the thawed samples was measured and adjusted at 7.0.

Chemical Characterization

The samples were analyzed for general characteristics, BOD_7, COD_{Cr}, TOC, loss on ignition, suspended solids, color, total nitrogen and total phosphorus by standard methods. Specific compounds, DCM extract, AOX, 25 major and trace elements, phenolic compounds, fatty acids, resin acids, molecular weight distribution, terpenes and sterols, EDTA and DTPA were also determined (Verta et al. 1995).

Toxicity Tests

The toxicity of effluent samples was tested by automated modification of the standard *Pseudomonas putida* test (DIN 38 412 Teil 8 1991) with Bioscreen C® analyzer (Labsystems). In this test *P. putida* MIGULA (DSM 50026) bacteria are grown in a liquid medium in special cuvettes and turbidity due to bacterial growth is measured by vertical photometry. The EC20 values (effective concentration which inhibited 20% of the growth) for the wastewater samples were then estimated by growth inhibition (%) in different dilutions compared with deionized water as presented in the above mentioned standard procedure.

The toxicity of the effluents was tested according to the standardized luminescence bacteria test procedure (DIN 38 412 Teil 34 1991). The luminescence inhibition test was accomplished by combining different dilutions of the wastewater with the luminescent bacteria *Vibrio fisheri* NRRL B-11177. These test cuvettes were then incubated in a 15+ -1°C^{-1} waterbath. Luminescence was measured after 30 min incubation with a luminometer (BioOrbit 1253 model). The EC50 values of the extracts were estimated as in the above mentioned standard procedure by luminescence inhibition (H%) in different dilutions compared with deionized water.

The algal growth inhibition testing of the sterile filtered effluent samples was performed by a standard method (ISO 8692 1989). In this test, exponentially growing *Selenastrum capricornutum* Printz ATCC 22662 cells are cultured for several generations (72 h) in a defined medium. The algal growth was measured by whole cell fluorometry [Sequoia-Turner 450 Digital Fluorometer, filters: NB440 (excitation) and SC665 (emission)]. The toxicity of the effluents was estimated as presented in this standard (ISO 8692) by growth inhibition in effluent dilutions compared with the control sample.

The toxicity of the water samples to *Daphnia magna* was tested according to a standardized method (ISO 6341 1989). The standard describes a method for the determination of the acute toxicity of the initial concentration of the samples which, in 24 h, immobilizes 50% of exposed animals, under the conditions defined in the standard. This concentration is known as the effective initial inhibitory concentration (24 h EC50). Since the sensitivity of *D. magna* to toxicants is age dependent, the animals used in the test shall be of the same age, less than 24 h old. The samples were neutralized if their pH values differed greatly from pH 7 (under pH 5.5 or over pH 8.5). According to Priha (pers. comm.) pH values under 4 and over 10 are lethal to *D. magna*.

The toxicity of the samples on zebra fish fertilized eggs and hatched larvae was tested according to the standardized method (SFS 5501 1991). Using a series of concentrations of the sample and controls, determination is made of the percentage of hatching and survival of eggs and early fry stages of zebra fish. The test continues with daily observations until at least 90% of the eggs or fry have died in all test solutions, which are renewed daily. This takes about two weeks as no food is supplied. The data obtained are used to calculate the mean time for hatching and survival in the concentrations tested and in the control series. The results are evaluated with regard to the highest concentration without significant effects and the lowest concentration with significant effects in relation to the controls. The tests are made on newly fertilized eggs, at most 4 h old. As distinct from the standard, the dilution water used in these tests was pure Lake Päijänne water. One sample was tested using both Lake Päijänne water and the standard water as the dilution water. There were no differences between the results. All samples were neutralized after thawing if the pH values differed greatly from pH 7.

RESULTS

Chemical Analysis

The chemical composition of metals and organic constituents of ECF and TCF bleach effluents did not show large differences. Color, instead, was clearly lower in TCF than in ECF bleached effluents. Secondary treatment was effective in the elimination of organic compounds, but not color or metals. Total nitrogen concentrations were clearly higher in the TCF bleached than in other effluents. The amount of absorbed organic chlorine was proportional to the use of chlorine in bleaching. All the results of the chemical analysis are presented in a following paper (Verta *et al.* 1995).

Toxicity

P. putida and V. fisheri Tests
In the *P. putida* growth inhibition test only untreated conventional bleaching and one ECF (birch) bleaching effluent gave slight toxic responses (Table 1). Untreated TCF effluents and all secondary treated effluents stimulated the growth of this bacteria. The strongest stimulation was often achieved in 10% dilution of these effluents.

The ECF and TCF bleaching effluents before secondary treatment from Mill A (Wisaforest) inhibited quite strongly the bacterial light production in the *V. fisheri* test. These effluents were even more toxic (EC50 dilutions between 2-12%) than conventional bleaching effluent (EC50 dilution 15%). Untreated effluents from Mill B (Enocell) were less toxic. ECF effluent gave a stronger response in light production inhibition than TCF bleaching effluent. The secondary treated effluents did not inhibit light production.

S. capricornutum Algae Tests
Only the untreated bleaching effluents inhibited the algal growth. There were no significant differences in the effluent toxicity of conventional bleaching or ECF and TCF bleaching. One Mill A ECF (birch) effluent was even more toxic than conventional bleaching effluent. Most of the treated effluents stimulated the algal growth.

D. magna Tests
According to the results all the effluents of Mill B (Enocell) were nontoxic to *D. magna*. The samples were tested unneutralized, because the pH values were within the limits of the animals' physiology.

The effluents from Mill A (Wisaforest) with pH under 3 were all toxic when tested unneutralized, so these samples were neutralized before the tests. From the effluents of Mill A, TCF and ECF bleached birch pulp and conventionally bleached pine pulp were equally toxic. Untreated wastewaters of TCF (pine and birch) and ECF (pine) bleaching as well as secondary and pilot treated effluents were all nontoxic.

B. rerio Egg-Larvae Tests
The results obtained from the egg and larvae test of zebra fish showed that the effects of the wastewaters, presented as the lowest effective concentration (LOEC), differed when considering the survival and hatching of the eggs and the mortality of the larvae. The pilot treatment decreased the toxicity

Table 1. The toxicity values (EC20, EC50 or LOEC) expressed as percentage concentrations (%) of the tested pulp mill effluents. (S = pine pulp, H = birch pulp, P = pilot treated effluent, $Effl_{St}$= secondary treated effluent, St = secondary treatment, Lt = lagoon treatment, - = nontoxic).

Effluent	Date	P. putida EC20	V. fisheri EC50	S. capricornutum EC50	D. magna EC50	B. rerio LOEC	Hatch/Surv. LOEC
Mill A							
$CONV_S$	Apr. 30	50	15	20	42	3	7
$Effl_{St}$	Apr. 30	50	—	—	—	100	—
ECF_H	May 22	40	2	18	40	25	5
ECF_H	Jul. 13	—	6	30	37	10	10
$Effl_{St}$	Jul. 13	—	—	—	—	—	25
ECF_S	Aug. 6	—	7	42	—	34	25
$Effl_{St}$	Aug. 6	—	—	—	—	—	—
ECF_P	May 25	—	60	—	—	60	50
TCF_H	Jun. 2	—	12	26	—	10	19
TCF_P	Jun. 2	—	—	—	—	50	50
TCF_S	Aug. 16	—	5	—	—	1	12
$Effl_{St}$	Aug. 16	—	—	—	—	—	—
Mill B							
$ECF_{S,H}$	Dec. 16	—	15	32	—	20	17
ECF_{St}	Dec. 16	—	—	—	—	—	—
ECF_{Lt}	Dec. 16	—	—	—	—	—	—
$TCF_{S,H}$	Aug. 31	—	45	42	—	22	5
TCF_{St}	Aug. 31	—	—	—	—	—	—
TCF_{Lt}	Aug. 31	—	—	—	—	—	—

of the wastewaters, but did not eliminate it completely. Secondary treated effluents did not have any significant effects. Of the untreated wastewaters, TCF and conventionally bleached pine pulp effluents were the most toxic. The effects of all other samples were nearly the same. Wastewaters from the pilot treatment and conventional bleaching delayed the hatching time of the roe, while all other samples advanced it (from 7 h to 24 h depending on the toxicity of the sample), in relation to the controls.

In order to compare the toxicity with effluent water quality, a toxicity class was formed based on the results of the *V. fisheri, S. capricornutum* (EC50) and *B. rerio* tests (LOEC). The original results of each test were classified into 5 toxicity classes. At first, a value 0 was given to all those effluents that were not toxic according to the test in question. Second, values 1-4 were given to all toxic effluents according to their position in the frequency distribution of the toxic results. If the effluent's toxicity lay within the first fractile it was placed in class 1, if in the second fractile class 2, etc. (Verta *et al.* 1995). The toxicity class values for each effluent and these test are shown in Fig. 1.

Also, the toxicity unit (TU, Equation 1) and toxicity emission factor (TEF, Equation 2) were calculated for untreated effluents. The TU assessed by *V. fisheri, S. capricornutum* and *D. magna* tests are presented in Figure 2 and TEF in Figure 3 accordingly.

$$TU = 100\ EC50^{-1} \quad [1]$$

$$TEF = TU \times Q\ P^{-1} \quad [2]$$

where Q = effluent flow ($m^3\ d^{-1}$) and P = pulp production ($t\ d^{-1}$)

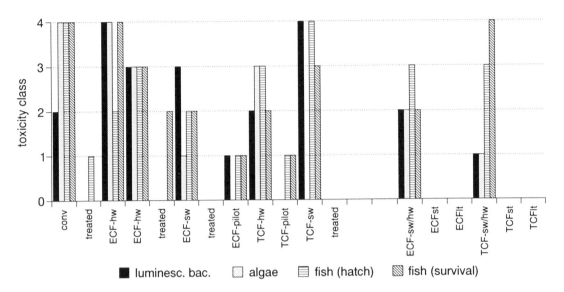

Figure 1. Toxicity class values (0-4, 0 = not toxic, 4 = most toxic) of the effluents assessed by most sensitive biotests (*V. fisheri* luminescent bacteria), *S. capricornutum* algae, *B. rerio* fish hatching and survival).

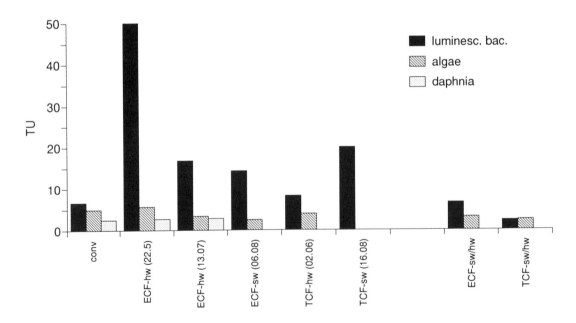

Figure 2. Toxicity units (TU) of untreated bleaching effluents assessed by *V. fisheri* luminescent bacteria, *S. capricornutum* algae and *D. magna* tests.

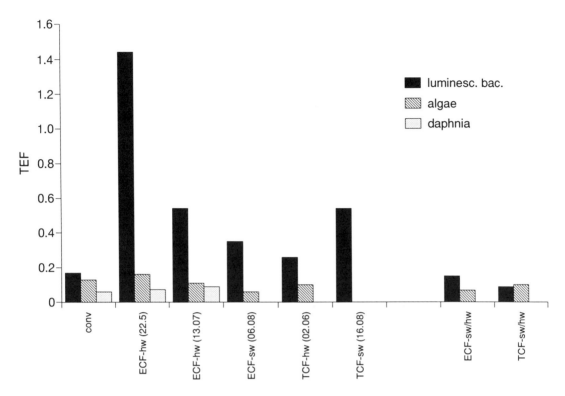

Figure 3. Toxicity emission factors (TEF, related to effluent flow and pulp production) of untreated bleaching effluents assessed by V. fisheri luminescent bacteria, S. capricornutum algae and D. magna tests.

DISCUSSION

The results indicate that natural constituents in wood may be responsible for most of the acute toxicity of pulp bleach effluents. These bioactive compounds seem to be eliminated by the secondary treatment of effluents and this will lead to a reduction in effluent toxicity. The toxicities of ECF and TCF bleached untreated effluents do not differ remarkably.

The results between the effluents of the two mills were slightly contradictory. According to the chemical analyses (Verta et al. 1995) there were high levels of resin acids in the effluent of the conventional bleaching of Mill A (Wisaforest) in the treated effluent twice the amount as in the untreated sample. The secondary treated effluent represents a mixture of different bleaching and other wastewaters of the mill, which might explain this. In the untreated effluents of Mill B (Enocell, only one bleaching line at the time) the concentration of resin acids was even higher than in the effluent of the conventional bleaching of Wisaforest mill. Also the concentration of fatty acids was high. The concentrations of phenolic compounds in these samples were almost the same. All these compounds are known to be harmful to aquatic animals (Lindström-Seppä 1990). However, the effluents of conventionally and TCF bleached pine pulps and ECF bleached birch pulps of Wisaforest mill were the most harmful and the most toxic of all tested samples.

Because the luminescent bacteria test measures the hazards to bacterial energy metabolism and the *P. putida* growth inhibition test measures cell multiplication of the heterotrophic microbes, these two very simple tests form an ecotoxically relevant pair of tests measuring two important, but different, impacts of wastewaters. However, in the assessing of environmental impacts of industrial effluents a set of different trophic level biotests should be used.

Toxicity tests that are conducted in the laboratory under arbitrarily defined conditions are incapable of perfectly simulating environmental conditions. They have only a limited value in assessing the effects of a substance or effluent in actual environmental conditions, in which many other factors may have an influence, for example the presence of organic and inorganic materials. Therefore, for an accurate prediction of the environmental impacts in a specific environmental situation, the results obtained by strict application of standardized methods should be complemented by data obtained under conditions which better simulate the effects of the environment and by data under field conditions.

ACKNOWLEDGMENTS

We would like to thank M. Aalto, K. Heinonen and S. Paattakainen for excellent technical assistance.

REFERENCES

Axegård, P., O. Dahlman, I. Haglind, B. Jacobson, R. Mörck and L. Strömberg. 1993. Pulp bleaching and the environment - the situation 1993. Nordic Pulp and Paper Research Journal 4:365-378.

Bitton, G. and B. J. Dutka (eds.). 1986. Introduction and review of microbial and biochemical toxicity screening problems. In Toxicity Testing Using Microorganisms (Vol. 1) CRC Press, Boca Raton, Florida, pp. 1-8.

Brinkmann, G. and R. Kühn. 1977. Limiting values for the damaging action of water pollutants to bacteria *Pseudomonas putida* and green algae *Scenedesmus quadricauda* in cell multiplication inhibition test. Z. f. Wasser und Abwasser-Forschung 10:87-98.

Bulich, A.A., M.W. Greene and D.L. Isenberg. 1981. The reliability of the bacterial luminescence assay for the determination of toxicity of pure compounds and complex effluents. In. D.R. Branson and K.L. Dickson (eds.). Aquatic Toxicology and Hazard Assessment. STP 737. American Society for Testing and Materials. pp. 338-347.

DIN 38 412 Teil 8. 1991. Bestimmung der Hemmwirkung von Wasserinhaltstoffen auf Bakterien Pseudomonas-Zellvermehrungs-Hemmtest. German Institute for Standardization.

DIN 38 412 Teil 34. 1991. Bestimmung der Hemmwirkung von Abwasser auf die Lichtemission von *Photobacterium phosphoreum*. German Institute for Standardization.

Hastings, J.W. 1978. Bacterial luminescence: An overview. In: M. Deluca (ed.). Methods in Enzymology (Vol. 57). Academic Press, New York.

Hodson, P.V. and B.R. Blunt. 1981. Temperature-induced changes in pentachlorophenol chronic toxicity to early life stages of rainbow trout. Aquat. Toxicol. 1:113-127.

ISO 6341. 1989. Water quality. Determination of the inhibition of the mobility of *Daphnia magna* Straus (*Cladocera, Crustacea*). International Organization for Standardization.

ISO 8692. 1989. Water quality. Fresh water algal growth inhibition test with *Scenedesmus subspicatus* and *Selenastrum capricornutum*. International Organization for Standardization.

Lindström-Seppä, P. 1990. Biotransformation in fish: Monitoring inland water pollution causes by pulp and paper mill effluents. Publications of University of Kuopio, 8/1990.

McKim, J.M., J.G. Eaton and G.W. Holcombe. 1978. Metal toxicity to embryos and larvae of eight species of freshwater fish. Bull. Environ. Contam. Toxicol. 19:608-616.

SFS 5501. 1991. Water quality. Determination of embryo-larval toxicity to freshwater fish. Semistatic method. Finnish Standardization Organization.

Verta, M., J. Ahtiainen, T. Nakari, A. Langi and E. Talka. 1995. The effect of waste constituents on the toxicity of TCF and ECF pulp bleaching effluents. (this volume).

THE EFFECT OF WASTE CONSTITUENTS ON THE TOXICITY OF TCF AND ECF PULP BLEACHING EFFLUENTS

M. Verta[1], J. Ahtiainen[1], T. Nakari[1], A. Langi[2] and E. Talka[2]

[1]Finnish Environment Agency, P.O. Box 140, FIN-00251 Helsinki, Finland
[2]The Finnish Pulp and Paper Research Institute, P.O. Box 70, FIN-02151 Espoo, Finland

Eighteen untreated, pilot treated and secondary treated bleach effluents from two pulp mills were studied in order to compare toxicity to a large number of elements and organic compounds. The toxicity tests used were: *Pseudomonas putida* growth inhibition, *Vibrio fisheri* luminescence bacteria test, *Daphnia magna* mobility inhibition, zebra fish (*Brachydanio rerio*) hatching and survival and *Selenastrum capricornutum* algae growth. General characteristics, BOD_7, COD_{Cr}, TOC, loss on ignition, suspended solids, color, DCM extract, as well as specific compounds, AOX, totN, totP, 25 major and trace elements, phenolic compounds, fatty acids, resin acids, molecular weight (MW) distribution, terpenes and sterols, EDTA and DTPA were determined. A toxicity index (TI) based on EC50 and lowest effective concentration (LOEC) values of the three most sensitive tests was calculated for every effluent. COD_{Cr}, BOD_7 and TOC showed the strongest positive correlation with TI ($r = 0.928-0.774$). Significant positive correlation with TI was observed for certain specific compounds as well: phenols, fatty acids, MW>10,000, AOX, totP, and MW<1,000, and Mn ($r = 0.755-0.591$). From the specific compounds, stepwise regression analysis selected phenols, fatty acids and resin acids as the best independent variables that explained 79% of the variance of TI. The correlation of AOX with toxicity was interpreted due to its correlation with other constituents in ECF and conventional effluents. The sum concentrations of phenols, fatty acids and resin acids in toxic dilutions of effluents exhibited levels comparable to EC50 values in the literature. Natural wood constituents were concluded to be responsible for most of the toxicity of the effluents studied.

INTRODUCTION

The discharge of different organic compounds in the bleach effluents from pulp mills has been implicated in a variety of effects observed in aquatic ecosystems. The research regarding toxic impacts of wastewaters has focused mainly on chlorinated organic compounds (e.g., Södergren *et al.* 1988). The recent improvements in pulp bleaching technology (elemental chlorine free, ECF, and total chlorine free, TCF) have reduced the concentrations of chlorinated organic compounds to very low levels. No relationship between absorbed organic chlorine (AOX) and environmental effects has been found at the current low AOX levels in effluents from chlorine dioxide bleaching, however (e.g., Axegård *et al.* 1993; Tana *et al.* 1994). Based on several recent studies Axegård *et al.* (1993) further concluded that non-chlorinated compounds in mill effluents can contribute to the biological responses, observed in laboratory and field studies, and some of these compounds are natural constituents of wood. However, data on occurrence and exposure level of nonchlorinated substances, except for resin acids (Oikari and Holmbom 1986) are almost totally lacking.

Single species toxicity tests can be used to estimate the potential harmful impact of wastewaters on aquatic ecosystems. In order to simulate the aquatic food chain, often representatives of different trophic level organisms such as algae, the water flea and various fish are used as tests organisms. In recent years various bacterial bioassays have also been developed for the screening of wastewater toxicity.

The objective of this study was to compare the toxicity of the effluents, from conventional chlorine (low Cl_2 charge) bleaching, ECF bleaching and TCF bleaching procedures, to a large number of constituents measured from the effluents. In order to get a comprehensive picture of toxicity, a battery of five different tests (Ahtiainen et al. 1996) was used. The tests used reflect both acute and subacute responses on several organisms. The effect of effluent treatment on toxicity and waste components was also studied.

MATERIAL AND METHODS

Sampling

The two pulp mills chosen for the study produce both bleached hardwood pulp and softwood pulp. Mill A produces bleached and nonbleached pulp and nonbleached board with a capacity of 560,000 air dried metric tonnes (ADt) d^{-1} of pulp and 130,000 ADt d^{-1} of board. Both hardwood and softwood are used. Mill B produces bleached pulp on two lines, hardwood and softwood on separate lines. The capacity is 515,000 ADt d^{-1}.

The samples collected from both mills represented both untreated and treated effluents. The production of pulp and the process data during sampling are presented in Table 1. During each sampling conventional bleaching (softwood) was run on the other line of Mill A, and the secondary treated samples represent the mixtures of the two lines plus effluents from an adjacent paper mill. One ECF effluent (may 22) and one TCF effluent (June 2) were secondary treated in a pilot scale sewage plant and included as well. From Mill B the bleached effluents are mixtures of a softwood and a hardwood line. Parallel samples were taken after secondary treatment with municipal sewage (1,000 m^3 d^{-1}) and after sewage pond treatment (retention 3-5 d). Samples were taken as daily pooled samples (Mill A) or as grab samples (Mill B). Samples were divided into 500 or 1,000 mL polyethylene bottles and frozen in -18°C.

Table 1. Production of pulp d^{-1}, effluent flow and some running data of the two mills during the sampling (sw=softwood, hw=hardwood).

Effluent	Sampling Date (1993)	Pulp Production ADt d^{-1}	Bleaching Sequence	Effluent Flow (bleached) m^3 t^{-1}	Effluent Flow (total) m^3 t^{-1}
Mill A					
$Conv_{sw}$	Apr. 30	800	D/C_{40}-E_O-D_1-E_2-D_2	20	78
ECF_{sw}	Aug. 6	900	D_O-E_O-D_1-E_2-D_2	22	81
ECF_{hw}	May 22	900	D_O-E_O-D_1-E_2-D_2	56	100
ECF_{hw}	Jul. 13	900	D_O-E_O-D_1-E_2-D_2	28	64
TCF_{sw}	Aug. 8	700	O-Q-Z-E_{OP}-A_Z-E_P-E	19	71
TCF_{hw}	Jun. 2	600	O-Q-Z-E_{OP}-A_Z-E_P-E	16	68
Mill B					
$ECF_{sw,hw}$	Dec. 16	1326	D-E-D-D	30	45
$TCF_{sw,hw}$	Aug. 31	980	Q-P-P-P	40	58

Chemical Characterization

The samples were analyzed for general characteristics, BOD_7, COD_{Cr}, TOC, loss on ignition, suspended solids, color, total nitrogen and total phosphorus after standardized methods. Specific compounds, DCM extract, AOX, 25 major and trace elements, phenolic compounds, fatty acids, resin acids, molecular weight (MW) distribution, terpenes and sterols, EDTA and DTPA were also determined. Briefly, the analytical methods were as follows:

DCM Extract

The sample was extracted with dichloromethane from acid solution for 24 h. The solvent was distilled and the extract dried at 105°C. The residue was weighed and the amount of extract calculated.

Total Phenols

The method is based on the shift of the UV absorption maximum of free phenols as a function of pH (Tamminen *et al.* 1993). The sample is diluted with NaOH (0.2 mol L^{-1}) and with a buffered (pH=12, NaOH 0.1 mol L^{-1}) solution. UV absorption spectra were run at wavelengths between 230 and 400 nm with a reference (pH=6) solution. The peaks were measured at wavelengths 300 and 350 nm. Both total phenols and conjugated phenols were determined.

AOX

The sample was acidified with nitric acid and its organic constituents absorbed on activated carbon in suspension. Inorganic ions containing chlorine were displaced by nitrate ions. The carbon was combusted with oxygen in a quartz tube at 1,000 °C. The hydrogen chloride formed was absorbed in an electrolyte solution and determined by microcoulometric titration (SCAN_W 9 1989).

MW Distribution

A MW distribution of UV-absorbent compounds was determined using gel chromatography (SEPHADEX G-50) and NaOH solution (0.5 mol L^{-1}) as an eluent (Forss and Sågfors 1984). Cytochrome C, glucagon and bacitracin were used as model compounds.

Fatty and Resin Acids

The acidified sample was extracted with petroleum ether-acetone-methanol. The acids (myristine, palmitine, stearine, oleine, linole, linoleine, arakine, behene, lignoserine, pimare, isopimare, levopimare, palustrine, abietine, neoabietine, dehydroabietine) were determined with GC using capillary column and FID detector. (KCL method 218:86).

Sterols

The acidified sample was extracted with dichloromethane. The solvent was evaporated and the extract dried at <40°C. Addition of BSTFA (*N,O-bis*(trimethylsilyl)trifluoroacetamide) was made and the sample was heated for one hour at 70°C. If necessary the sample was diluted with tetrahydrofuran. GC-MS determination was carried out using a silica capillary column (SE-54). Dotriacontane (C32) was used as an inner standard. Compounds quantified were campesterol, stigmasterol, β-sitosterol, β-sitostanol and betulinol.

EDTA and DTPA

The sample was dried and EDTA and DTPA derivated with boron trifluoride-methanol complex (Lennart 1990). Complexing agents were extracted and analyzed with CG using mass selective detector and pure model compounds. The recoveries from distilled and natural lake water after addition ranged from 25 to 80%.

Metals

The samples were analyzed by ICP-MS (Perkin-Elmer Sciex Elan 5000) for Na, Ca, K, Mg, Si, Fe, Mn, Al, Zn, Sr, Ba, Ti, Ni, Cu, Cr, Rb, V, Pb, As, Mo, Co, Cd and U. Rhodium was used as an internal standard. SLRS-2 (Riverine Water Reference Material for Trace Metals, National Research Council, Canada) was used as a reference sample. The validation of the analysis was performed using EPA's method 200.8 (Long and Martin 1991).

Biological Characterization

A full description of the methods used is given by Ahtiainen et al. (1996). Briefly, the methods were as follows:

Pseudomonas putida Test

An automated modification of the standard *P. putida* test (DIN 38 412 Teil 8 1991) with Bioscreen C® analyzer (Labsystems) was used. The EC50 values for the effluents were estimated by growth inhibition (%) in different dilutions compared to deionized water.

Luminescence Bacteria Test

A standardized luminescence bacteria test procedure (DIN 38 412 Teil 34 1991) was used. The luminescence inhibition test was accomplished by combining different dilutions of the effluent with the luminescent bacteria *Vibrio fisheri* NRRL B-11177. The EC50 values of the extracts were estimated by luminescence inhibition (H%) in different dilutions compared to deionized water.

Daphnia Test

A standardized method (ISO 6341 1989) was used. The standard describes a method for the determination of the acute toxicity of the initial concentration of the samples which, in 24 h, immobilizes 50% of exposed animals, under the conditions defined in the standard.

Zebra Fish Test

The toxicity of the samples on zebra fish fertilized eggs and hatched larvae was tested according to a standardized method (SFS 5501 1991). Using a series of concentrations of the sample and controls, determination is made of the percentage of hatching and the survival of eggs and early fry stages of zebra fish. The results presented here are evaluated with regard to the lowest concentration with significant effects in relation to the controls. As distinct from the standard, the dilution water used in the tests was lake Päijänne water.

Algae Growth Test

The algal growth inhibition testing of the sterile filtered effluent samples was performed by a standard method (ISO 8692 1989). Exponentially growing *Selenastrum capricornutum* Printz ATCC 22662 cells are cultured for several generations (72 h) and the growth is measured by whole cell fluorometry. The toxicity was estimated by growth inhibition in effluent dilutions compared to the control sample.

RESULTS AND DISCUSSION

Chemistry

The chemical composition of metals and organic constituents of ECF and TCF bleach effluents did not show large differences (Tables 2 and 3). Color, instead, was clearly lower in TCF- than in ECF-bleached effluents. Secondary treatment was effective in the elimination of organic compounds (Table 2), but not color or metals (Table 3). Total nitrogen concentrations were clearly higher in the TCF- bleached than in the other effluents due to the use of EDTA and DTPA. The amount of absorbed organic chlorine was proportional to the use of chlorine in bleaching (Tables 1 and 2). As indicated by AOX concentrations, the conventional Cl_2 bleaching in the other line of Mill A, and the long retention during effluent treatment in Mill B, resulted in mixtures of several bleaching sequences in the treated effluents sampled.

Table 2. Some chemical characteristics of the studied pulp mill effluents (sw = softwood, hw = hardwood, P = pilot treated, $Effl_{St}$ = secondary treated effluent, $Effl_{La}$ = lagoon treated effluent, - = not analysed).

Effluent	Date	COD (mg L^{-1})	BOD (mg L^{-1})	TOC (mg L^{-1})	AOX (mg L^{-1})	Tot. N (mg L^{-1})	Phenols (total) (mmol L^{-1})	Fatty acids (mg L^{-1})	Resin acids (mg L^{-1})	Sterols (mg L^{-1})	EDTA+ DTPA (mg L^{-1})
Mill A											
$Conv_{sw}$	Apr. 30	1950	400	750	95	2.6	1.3	1.3	1.4	0.04	<0.005
$Effl_{St}$	Apr. 30	460	9.5	150	10	3.0	0.24	1.3	2.4	0.22	<0.005
ECF_{hw}	May 22	1720	360	250	52	3.4	0.75	10.1	0.8	0.30	<0.005
ECF_{hw}	Jul. 13	1220	320	420	41	2.3	0.95	13.2	0.5	0.61	<0.005
$Effl_{St}$	Jul. 13	340	4.0	100	14	2.5	0.28	0.7	0.1	0.40	-
ECF_{sw}	Aug. 6	1300	270	490	37	2.0	1.0	1.8	0.4	0.02	<0.005
$Effl_{St}$	Aug. 6	310	8.5	88	15	2.8	0.31	0.7	0.2	0.02	<0.005
ECF_P	May 25	970	55	380	22	2.0	0.23	4.3	0.5	-	5.5
TCF_{hw}	Jun. 2	1850	900	710	0.12	8.0	0.42	6.5	0.1	0.68	11
TCF_P	Jun. 2	540	61	220	0.59	3.8	0.14	0.3	0.1	-	11
TCF_{sw}	Aug. 16	1250	380	380	0.07	7.9	0.40	1.2	0.2	0.03	43
$Effl_{St}$	Aug. 16	210	8.2	69	7.1	2.5	0.16	0.1	0.1	<0.01	1.4
Mill B											
$ECF_{sw,hw}$	Dec. 16	970	300	290	8.9	4.2	0.29	5.3	3.4	1.06	<0.005
ECF_{St}	Dec. 16	300	4.3	110	3.0	1.5	0.10	0.2	<0.1	0.05	0.1
ECF_{La}	Dec. 16	250	3.4	88	2.3	1.8	0.08	0.2	<0.1	<0.01	16
$TCF_{sw,hw}$	Aug. 31	820	320	250	0.13	9.7	0.08	7.8	5.5	3.42	185
TCF_{St}	Aug. 31	170	4.5	64	0.30	5.7	0.01	0.2	<0.1	<0.01	150
TCF_{La}	Aug. 31	220	2.1	82	1.3	4.8	0.06	0.4	<0.1	0.21	45

Table 3. Mean concentration of metals in the studied effluents classified according to effluent type ([1] denotes high concentrations due to pilot plant steel).

		Raw Water (Mill A)	Conv.	ECF	TCF	Pilot	Treated
Na	(mg L^{-1})	3.3	680	460	550	760	330
Ca	(mg L^{-1})	4.5	32	62	75	43	80
K	(mg L^{-1})	2.0	10	6.7	2.9	3.9	11
Mg	(mg L^{-1})	1.7	5.5	6.7	20	10	5.4
Si	(mg L^{-1})	4.6	8.7	4.8	6.6	3.4	3.5
Fe	(mg L^{-1})	2.5	2.3	1.3	1.4	2.5;20[1]	1.9
Mn	(mg L^{-1})	0.17	1.7	1.3	1.4	1.2	0.47
Al	(mg L^{-1})	0.74	1.1	0.89	0.98	0.71	0.99
Zn	(μg L^{-1})	12.2	220	240	380	470	140
Sr	(μg L^{-1})	36	150	270	280	260	230
Ba	(μg L^{-1})	13	180	240	350	160	130
Ti	(μg L-1)	26	96	89	160	120	100
Ni	(μg L-1)	3.9	23	15	16	31;100[1]	8.6
Cu	(μg L-1)	3.1	21	16	16	30	9.8
Cr	(μg L-1)	1.5	26	13	16	16;125[1]	7.7
Rb	(μg L-1)	4.0	43	20	7.6	13	40
V	(μg L-1)	2.6	30	10	4.9	4.7	15
Pb	(μg L-1)	0.43	2.4	5.0	6.9	7.2	4.9
As	(μg L-1)	0.84	4.1	2.8	1.3	4.0	2.2
Mo	(μg L-1)	0.7	3.1	1.3	1.2	0.29;14[1]	0.73
Co	(μg L-1)	1.8	1.9	1.3	1.2	6.8	1.6
Cd	(μg L-1)	<0.01	1.7	1.6	1.5	0.83	1.6
U	(μg L-1)	0.13	1.7	0.39	0.71	0.35	1.1

Toxicity

Three of the toxicity tests performed gave measurable responses for about 50% or more of the effluents tested. These were *Vibrio fisheri* luminescence bacteria test, zebra fish (*Brachydania rerio*) hatching and survival test and *Selenastrum capricornutum* algae growth test. *Pseudomonas putida* growth inhibition test and *Daphnia magna* mobility inhibition, instead, gave responses only to a limited number of effluents (Ahtiainen et al. 1996).

The toxicities of all the untreated effluents were comparatively high (Table 4). Generally EC50 or LOEC values were less than 30% and in several cases (every third) less than 10%. Secondary treatment eliminated toxicity almost totally. No significant differences were found between different bleaching procedures.

In order to compare toxicity with effluent water quality, a toxicity index (TI) was formed on the basis of the results of the most sensitive tests, *V. fisheri, S. capricornutum* and *B. rerio*. The original results of each test were classified into 5 toxicity classes. At first, a value 0 was given to all those effluents that were not toxic according to the test in question. Second, values from 1 to 4 were given to all toxic effluents according to their position in the frequency distribution of the test in question. If the effluent's toxicity lay within the first fractile it was placed in class 1, if in the second fractile, class 2, etc. The toxicity class values for each effluent and each test are given in Table 4.

Table 4. The toxicity (EC50 and LOEC values as %) of the studied pulp mill effluents, the toxicity class values (TC_i) and computed toxicity index (TI) values for each effluent. (sw = softwood, hw = hardwood, P = pilot treated effluent, $Effl_{St}$ = secondary treated effluent, $Effl_{Lt}$ = lagoon treated effluent, - = nontoxic.)

Effluent	Date	Vibrio fisheri		Selenastrum capricornutum		Brachydanio rerio				Toxicity Index
						Hatching		Survival		
		EC50	TC_{Vf}	EC50	TC_{Sc}	LOEC	TC_{Brh}	LOEC	TC_{Brs}	TI
Mill A										
$Conv_{sw}$	Apr. 30	15	2	20	4	3	4	7	4	0.88
$Effl_{St}$	Apr. 30	-	0	-	0	100	1	-	0	0.06
ECF_{hw}	May 22	2	4	18	4	25	2	5	4	0.88
ECF_{hw}	Jul. 13	6	3	30	3	10	3	10	3	0.75
$Effl_{St}$	Jul. 13	-	0	-	0	-	0	25	2	0.13
ECF_{sw}	Aug. 6	7	3	42	1	34	2	25	2	0.50
$Effl_{St}$	Aug. 6	-	0	-	0	-	0	-	0	0.00
ECF_P	May 25	60	1	-	0	60	1	50	1	0.19
TCF_{hw}	Jun. 2	12	2	26	3	10	3	19	2	0.63
TCF_P	Jun. 2	-	0	-	0	50	1	50	1	0.13
TCF_{sw}	Aug. 16	5	4	-	0	1	4	12	3	0.69
$Effl_{St}$	Aug. 16	-	0	-	0	-	0	-	0	0.00
Mill B										
$ECF_{sw,hw}$	Dec. 16	15	2	32	2	20	3	17	2	0.56
ECF_{St}	Dec. 16	-	0	-	0	-	0	-	0	0.00
ECF_{Lt}	Dec. 16	-	0	-	0	-	0	-	0	0.00
$TCF_{sw,hw}$	Aug. 31	45	1	42	1	22	3	5	4	0.56
TCF_{St}	Aug. 31	-	0	-	0	-	0	-	0	0.00
TCF_{Lt}	Aug. 31	-	0	-	0	-	0	-	0	0.00

Finally, TI was computed according to the formula:
$$TI = (TC_{Vf} + TC_{Sc} + TC_{Brh} + TC_{Brg})/16 \tag{1}$$

where TI = toxicity index; TC_{Vf} = toxicity class according to *V. fisheri*; TC_{Sc} = toxicity class according to *S. capricornutum*; TC_{Brh} = toxicity class according to *B. rerio*/hatching; TC_{Brg} = toxicity class according to *B. rerio*/survival.

The computed TI values are given in Table 4 and in Figure 1. The mean TI values for different types of effluent are shown in Figure 1 and were as follows (range in parenthesis):

Conventional	0.88
ECF	0.67 (0.50-0.88)
TCF	0.63 (0.56-0.69)
Pilot treated	0.16 (0.13-0.19)
Secondary treated	0.02 (0.00-0.13)

Based on the TI values the only conventionally (Cl_2, low charge) bleached effluent was the most toxic. Both ECF- and TCF-bleached effluents were only 25-30% lower, indicating only minor reduction in toxicity compared to the conventional effluent. A major reduction in TI was achieved with secondary treatment either at pilot scale or at pulp mill treatment plants.

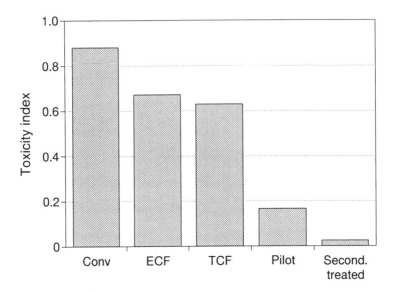

Figure 1. The calculated mean toxicity index values (TI) for different type of untreated, pilot treated and full scale treated pulp mill effluents.

Correlation and Regression Analysis

A correlation analysis was carried out between TI and the effluent chemistry. The variables that showed significant correlation with TI were (correlation coefficient in parenthesis): COD (0.928), BOD (0.803), TOC (0.774), total phenols (0.755), Mn (0.751), DCM extract (0.743), the sum of fatty acids (0.692), loss on ignition (0.686), MW >10,000 (0.617), AOX (0.605), Abs_280 (0.602), total phosphorus (0.598), and MW <1000 (0.591).

The toxicity of effluents was strongly related to organic material. The correlation was not only seen in differences between untreated and treated effluents but was also evident between similarly bleached and untreated effluents (Fig. 2a,b,c). Of the specific compounds, known to have toxic properties, phenols and fatty acids gave a fairly good correlation with toxicity (Fig. 2d,e). Typically, a large variation in concentrations was, however, found within a narrow range in toxicity. Even more clearly, AOX, although having fairly good overall correlation with toxicity, showed no correlation within untreated effluents (Fig. 2f). Equally high toxicity was recorded in TCF and ECF effluents regardless of practically any AOX in the former. Furthermore, the fairly good correlation of toxicity with phenols (Fig. 2d) cannot be explained by chlorophenols, since AOX concentrations reveal that these substances were not present in TCF-bleached effluents (organic absorbed chlorine < 0.01 mmol L^{-1}). It seems obvious that the correlation of AOX with toxicity is due to its high intercorrelation with organic material (AOX/COD, r=0.688; AOX/phenols r=0.907).

Stepwise multiple regression was used to model the relationship between toxicity (TI) and effluent water quality (PROC REG, SAS Institute Inc. 1989; Belsley *et al.* 1980). In the first phase, all measured variables were introduced to the model run. From these only COD was selected as an explaining variable. In the second phase, only the variables which are related more closely to specific compounds were introduced. These were AOX, resin acids, fatty acids, phenols, sterols, EDTA, DTPA, DCM extract,

A_280, MW >10,000, MW 5,000-10,000, MW 3,000-5,000, MW 1,500-3,000, and MW <1,000. From these variables the model selected phenols, fatty acids and resin acids in an equation, which explained 79% of the variation in TI (Table 5).

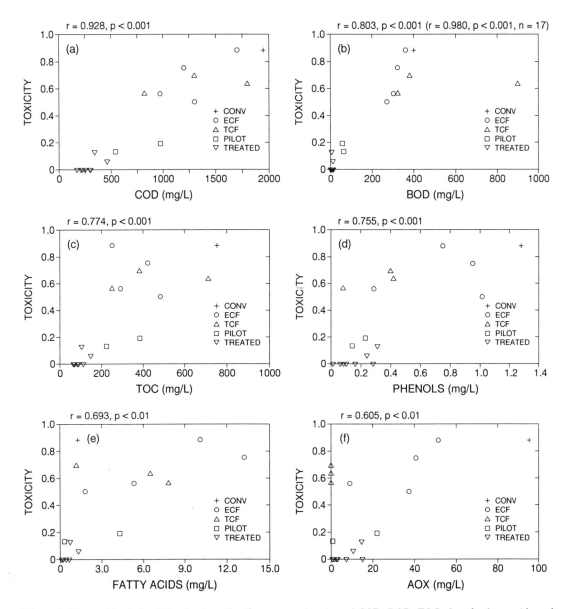

Figure 2. The toxicity index (TI) of pulp mill effluents as a function of COD, BOD, TOC phenols, fatty acids and AOX. TI=1.0 is given to the effluent, which according to every toxicity test is placed in the most toxic group. TI=0 is given to the effluent, which showed no toxicity in any of the tests.

Table 5. Stepwise multiple regression model predicting toxicity (TI) of treated and untreated mill effluents from phenols, fatty acids and resin acids.

	Variable	Variance Explained (%)	F	P
Step 1	Phenols	57	21.26	<0.001
Step 2	Phenols + Fatty acids	75	11.23	<0.01
Step 3	Ph + Fa + Resin acids	79	2.47	0.14

The regression equation was:

$$TI = 0.006 \cdot Phenols + 0.032 \cdot Fatty\ acids + 0.048 \cdot Resin\ acids - 0.02 \qquad [2]$$

where concentration of each group of compounds is expressed as mg L^{-1}.

Each of the groups selected by the model consists of several compounds that have toxic properties. An attempt was made to compare the concentrations of these compounds in the toxic dilutions of the effluents with the LC50 values reported in the literature. A sum concentration of phenols, fatty acids and resin acids was computed for each toxic effluent using the largest toxic dilution (Table 4). This concentration varied from 2.3 to 21 mg L^{-1}. For Mill A always a major fraction (82-98%) of the sum consisted of phenols. For the two toxic effluents of Mill B the phenolic fraction was smaller (36-76%). The LC50 values for rainbow trout (*Salmo gairdneri*) reported in the literature (Verschueren 1983; McLeay 1987; Nikunen *et al.* 1990) for several specific compounds in the phenolic, fatty acid and resin acid groups were:

Phenols   0.15-10.5 mg L^{-1}
Fatty acids   0.32-8.2 mg L^{-1}
Resin acids   0.32-1.7 mg L^{-1}

Comparison of the measured concentrations of these substances (2.3-21 mg L^{-1}) with these LC50 values implies that the toxic responses measured can at least partially be caused by these three groups originated from wood material.

CONCLUSIONS

The results indicate that the natural constituents in wood may be responsible for most of the acute toxicity of the studied pulp bleach effluents. The phenolic compounds and some of the acids are degradation products originating mainly from lignin and can be found both in brownstock washings and in bleach plant effluents. Resin acids and fatty acids occur in wood-debarking waters, brownstock washings, condensates and bleach plant effluents (Axegård *et al.* 1993). Elimination of these substances and other bioactive compounds by secondary treatment of effluents is suggested as the primary reason for the reduction in effluent toxicity. The toxicity of ECF- and TCF-bleached effluents, under the process technology and raw materials used in this study, may not differ much, probably because different bleaching effluents do not differ in their total amount of these substances. The toxicity of pulp mill effluents seems to be a result of numerous bioactive substances rather than a few specific compounds. AOX cannot explain toxicity of pulp bleach effluents at the current low AOX levels.

ACKNOWLEDGMENTS

The study was funded by the Finnish National Board of Waters and the Environment, A. Ahlström Corporation, Enso-Gutzeit Oy, Oy Finnish Peroxides Ab, The Finnish Pulp and Paper Research Institute,

Kymi Paper Mills Ltd., Oy Metsä-Botnia Ab, Sunila Oy, Veitsiluoto Oy, United Paper Mills Ltd., and Aracruz Celulose S.A. Mr. Jukka Puustinen and Mr. Jussi Uotila were responsible for the EDTA and DTPA analysis and are gratefully acknowledged.

REFERENCES

Ahtiainen, J., T. Nakari and J. Silvonen. 1996. Toxicity of TCF and ECF pulp bleaching effluents assessed by biological toxicity tests. In Environmental Fate and Effects of Pulp and Paper Mill Effluents, M.R. Servos, K.R. Munkittrick, J.H. Carey and G. Van Der Kraak (ed.), St. Lucie Press, Delray Beach, FL.

Axegård, P., O. Dahlman, I. Haglind, B. Jacobson, R. Mörck and L. Strömberg. 1993. Pulp bleaching and the environment - the situation 1993. Nordic Pulp and Paper Research Journal 4:365-378.

Belsley, D.A., E. Kuh and R.E. Welsh. 1980. Regression diagnostics: identifying influential data and sources of collinearity. John Wiley & Sons, New York, N.Y., 292 p.

DIN 38 412 Teil 8. 1991. Bestimmung der Hemmwirkung von Wasserinhaltstoffen auf Bakterien Pseudomonas-Zellvermehrungs-Hemmtest. German Institute for Standardization.

DIN 38 412 Teil 34. 1991. Bestimmung der Hemmwirkung von Abwasser auf die Lichtemission von Photobacterium phosphoreum. German Institute for Standardization.

Forss, K. and P.-E. Sågfors. 1984. Ligninet i kokvätskan - hur ser det ut. Nordisk Cellulosa, No. 4.

ISO 6341. 1989. Water quality. Determination of the inhibition of the mobility of Daphnia magna Strauss (Cladocera, Crustacea). International Organization for Standardization.

ISO 8692. 1989. Water quality. Fresh water algal growth inhibition test with Scenedesmus subspicatus and Selenastrum capricornutum. International Organization for Standardization.

Lennart, K. 1990. Bestämmningsmetod för komplexbildare i vatten. Swedish EPA, Stockholm, 2.11.1990. Mimeographed report. 3 p.

Long, S.E. and T.D. Martin. 1991. Method 200.8. Determination of trace elements in waters and wastes by inductively coupled plasma-mass spectrometry. ICP Information Newsletter 18(3):170-171.

McLeay, D. 1987. Aquatic toxicity of pulp and paper mill effluent: A review. Environment Canada, Ottawa, Report EPS 4/PF/1, 191 p.

Nikunen, E., R. Leinonen and A. Kultamaa. 1990. Environmental properties of chemicals. Ministry of the Environment, Environment Protection Department, Helsinki, Research Report 91, 1084 p.

Oikari, A. and B. Holmbom. 1986. Assessment of water contamination by chlorophenolics and resin acids with the aid of fish bile metabolites. Aq. Toxicol. Env. Fate. 9th vol. ASTM STP 921, T.M. Poston and R. Purdy (Eds.), Philadelphia, 252-267.

SAS Institute Inc. 1989. SAS/STAT user's guide, version 6, edition 4. Vol. 2. SAS institute Inc., Cary, N.C., 1989.

SCAN_W 9. 1989. Organically bound chlorine by the AOX method. Scandinavian Pulp, Paper and Board Testing Committee.

SFS 5501. 1991. Water quality. Determination of embryo-larval toxicity to freshwater fish. Semistatic procedure. Finnish Standardization Organization.

Södergren, A., B.-E. Bengtsson, P. Jonsson, S. Lagergren, Å. Larsson, M. Olsson and L. Renberg. 1988. Summary of results from the Swedish project "Environment-Cellulose". Wat. Sci. Technol. 20:49-60.

Tamminen, T., B. Hortling, K. Poppius-Levlin and J. Sundqvist. 1993. Connections between bleachability and residual lignin structure. Int. Symp. Wood Pulp Chem., CTAPI, Beijing, P.R. China, May 25-28, poster presentation.

Tana, J., A. Rosemarin, K.-J. Lehtinen, J. Härdig, O. Grahn and L. Landner. 1994. Assessing impacts on Baltic coastal ecosystems with mesocosms and fish biomarker tests: A comparison of new and old wood pulp bleaching technologies. Sci. Total Env.: In press.

Verschueren, K. 1983. Handbook of environmental data of organic chemicals. Van Nostrand Reinhold Co. Inc., New York, 1310 p.

THE USE OF CHRONIC BIOASSAYS IN CHARACTERIZING EFFLUENT QUALITY CHANGES FOR TWO BLEACHED KRAFT MILLS UNDERGOING PROCESS CHANGES TO INCREASED CHLORINE DIOXIDE SUBSTITUTION AND OXYGEN DELIGNIFICATION

Timothy J. Hall[1], Richard K. Haley[1], Dennis L. Borton[2] and Theresa M. Bousquet[3]

[1]NCASI, 1900 Shannon Point Road, Anacortes, WA, 98221 U.S.A.
[2]NCASI, P.O. Box 28561, New Bern, NC, 28561-2868 U.S.A.
[3]NCASI, P.O. Box 458, Corvallis, OR, 97339 U.S.A.

A suite of marine and freshwater chronic bioassays was used to characterize effluent from two pulp and paper mills using high chlorine dioxide substitution before and after conversion to oxygen delignification. One of these mills was located in the northwest U.S. and the other in the Southern U.S. Both incorporated effluent secondary treatment. For one of the mills this data set is also provided for comparison with effluent quality before the initial conversion to increased chlorine dioxide substitution. Marine bioassays included those with echinoderm sperm and eggs and larval fish. Freshwater testing included those with fathead minnows and *Ceriodaphnia dubia*. Concurrent chemical analysis included AOX, individual chlorinated organic compounds, resin/fatty acids, and a large number of other chemical analytes. Data are presented to characterize both the biological and chemical relevance of mill process conversions designed to reduce the usage of elemental chlorine.

INTRODUCTION

Research into the ecological impacts of pulp bleaching in the U.S. has spanned several decades during which the industry's wastewater treatment and bleaching practices have continued to evolve. The hazard assessment review by Owens (1991) indicated that, especially in the case of secondary treated effluents, environmental effects are minimal and if they occur are limited in most cases to eutrophication effects in the area immediately adjacent to the effluent discharge. There has, however, remained a great concern internationally about the potential for adverse effects from chlorinated organics in receiving water. Within the U.S., and elsewhere, an increasing number of mills have modified the bleaching or delignification processes to replace some or all of the elemental chlorine with ClO_2. This report describes a study of the corresponding benefits of such mill process changes on the chemical and physical characteristics of effluent as well as an effort to determine whether there are associated benefits as measured by chronic bioassay procedures.

A suite of marine and freshwater bioassays was used to characterize effluent quality changes for a bleached kraft mill (Mill A) undergoing conversion from conventional bleaching to bleaching using increased ClO_2 substitution and subsequently after the addition of oxygen delignification. Results for an additional bleached kraft mill (Mill B) are described before and after the addition of oxygen delignification to a process that was already at 100% ClO_2 substitution. The bioassay methods are those sanctioned by the U.S. EPA to estimate chronic toxicity (U.S. EPA 1988, 1989). Marine bioassays included the echinoid sperm/egg fertilization test using the sand dollar (*Dendraster excentricus*) and fish larval growth tests using either *Cyprinodon variegatus* or *Menidia beryllina*. Freshwater bioassays included those based on reproduction using *Ceriodaphnia dubia* and larval growth with the fathead minnow (*Pimephales promelas*).

Corresponding physical and chemical measurements were also made to characterize changes in effluent quality and to determine to what extent these changes were associated with changes in the bioassay responses.

These bioassay studies were one part of a broader investigative program carried out at both mill locations that included the exposure of outdoor experimental streams to the effluents from these same mills during periods before and after process conversions (Hall *et al.* 1991; NCASI 1993, 1994a,b). The experimental streams results indicated an absence of lethal or sublethal effects on rainbow trout or alterations in the supporting aquatic food web at effluent test concentrations of 1.5 and 5.0% v/v with effluent from Mill A (the highest test concentrations tested). Similar studies with Mill B indicated an absence of deleterious effects on several warm-water fish species at concentrations >10% v/v effluent. The integration of both experimental stream and laboratory bioassay approaches was considered important in accurately characterizing effluent quality changes that might be associated with the mill's conversion to increased ClO_2 substitution.

METHODS AND EXPERIMENTAL DESIGN

Description of Mill A

Mill A is a bleached kraft mill located in the northwest U.S., producing bleached kraft pulp from a variety of softwood species. Production from the mill is approximately 1200 air dried metric tonnes per day (ADMT d^{-1}) of bleached kraft pulp, with water use about 132,500 m^3 d^{-1} (about 111 m^3 of water/ADMT of pulp). Pulp is converted on site to tissue and paper. The original bleaching sequence was CEHHD for pulp from wood chips and a CEHD sequence for pulp from sawdust. For the chip line, up to 17% ClO_2 substitution was made for molecular chlorine in the initial bleaching stage. The chip line incorporated peroxide enhanced extraction to reduce ClO_2 usage. After conversion to increased ClO_2 substitution, the level of substitution increased to 60-70% for both the chip and sawdust lines. Effluent used in the study was taken from the outlet of the mill's 14-d retention time aerated stabilization basin. The wastewater treatment plant routinely achieves >90% BOD_5 reduction.

The time frame of the study included a 7-mo period (October 1990 to May 1991) prior to conversion to increased ClO_2 substitution, a 10-mo period (May 1991 through March 1992) following conversion, and a 16-mo period (January 1993 through April 1994) after the initiation of oxygen delignification.

Description of Mill B

Mill B is located in the southeast U.S. and produces approximately 700 ADMT d^{-1} of bleached kraft pulp from a furnish of 75% softwood and 25% hardwood species. Water use by this mill averaged about 90,840 m^3 d^{-1}, representing on a production basis about 130 m^3 $ADMT^{-1}$ of pulp produced. The original bleaching sequence of CEHDED was modified to O/C (E_{op}) wash DED which reduced the use of ClO_2 by 70%. The bleaching sequence became OD (E_{op}) wash DED after oxygen delignification was installed. The effluent used for the bioassays was pumped from the final effluent canal after approximately 14 d of treatment in an aeration stabilization basin which reduced 90-95% of the BOD_5. The time frame of the study included an 8-mo period before the addition of oxygen delignification (October 1990 through May 1991) and a 14-mo period after oxygen delignification was added (January 1992 to February 1994).

Effluent Sampling Procedures

Effluent grab samples were collected from the respective NCASI Experimental Streams site located adjacent to the two mills. The effluent sample sites and sampling procedures were chosen to provide effluent of similar quality to that which was discharged from each mill. Samples were collected and shipped in glass containers with effluent shipment by overnight courier, in cooled, insulated containers.

Effluent Physical/Chemical Measurements

Effluent grab samples for chemical analysis were carried out in conjunction with each bioassay. BOD_5 and total suspended solids (TSS) analyses were provided by mill personnel at the two locations. Chlorophenol and resin and fatty acid analyses were conducted with GC-MS procedures described by NCASI (1986a,b). AOX concentrations were determined by non-purgable organic halide (NPOX) analysis (American Public Health Association et al. 1985). Methods for effluent color, conductivity, hardness, pH, and turbidity parameters are described by NCASI (1993) but generally followed the methods of the American Public Health Association et al. (1985). Tannin/lignin concentrations were determined with the tyrosine test method (Hach Company 1989). Additional effluent chemical characterization carried out in conjunction with the experimental streams studies at the two locations are provided by NCASI (1993, 1994a,b).

Bioassay Procedures

Marine bioassays were carried out by the NCASI West Coast Aquatic Biology Research Station in Anacortes, Washington. The echinoderm sperm/egg bioassay method was based on Cherr et al. (1987) with modifications as described by Hall et al. (1993) and NCASI (1992). Sperm and eggs from the sand dollar, *Dendraster excentricus*, were used with sperm exposed to effluent for a 10-min period followed by a 10-min-period egg exposure. The bioassay was scored based on fertilization success determined with a light microscope. Tests with larval marine fish included the sheepshead minnow, *Cyprinodon variegatus*, and the inland silverside, *Menidia beryllina*. These tests were based on 7-d exposure times with survival and growth as endpoints. For the marine tests a dilution series of 0.0, 0.3, 1.0, 3.0, 10.0, 30.0, and 70.0% v/v effluent was used. The effluents tested had <1 g kg^{-1} salinity and required a salinity adjustment to achieve full strength seawater salinity (30 g kg^{-1}) requirements for these organisms. Blends of hypersaline brine and seawater were used to achieve a constant 30 g kg^{-1} salinity of the test solutions.

Freshwater bioassays were carried out at the NCASI Southern Streams Site, in Vanceboro, North Carolina. These were 7-d tests with *Ceriodaphnia dubia* and the fathead minnow, *Pimephales promelas*. The *C. dubia* bioassay included survival and reproduction as endpoints and the fathead minnow bioassay endpoints were survival and growth. For the freshwater bioassays a dilution series of 0.0, 1.0, 10.0, 30.0, and 100.0% v/v effluent was used.

Statistical Procedures

Bioassay endpoints were calculated using the methods suggested by the U.S. EPA (1989), including the Dunnett's test for determining the NOEC (no observable effect concentration) and the Bootstrap method for determining the 25% inhibition response level (IC25). For the echinoderm test, Abbott's correction was used to standardize responses to 100% fertilization in the controls. Echinoderm NOEC and IC25 response endpoints were calculated using the computer program ToxCalc (Tidepool Scientific Software, McKinleyville, CA). The student's t-test was used to determine whether there were significant differences (two-tail, $P < 0.05$) between the IC25 bioassay responses or effluent chemical parameters before or after mill process changes. The non-parametric Mann-Whitney test was used to test for significant differences (two-tail, $P < 0.05$) in NOEC values before or after mill process changes.

RESULTS AND DISCUSSION

Results for Mill A

Effluent Quality

Effluent chemical characteristics before and after mill conversions are provided in Table 1 and Fig. 1. Although these comparisons, and others herein, are based on concentrations rather than mass loadings, the effluent flow and production rates were comparable during the periods of interest, allowing meaningful comparisons of concentration based parameters. Other modifications may have taken place during the time frame of the process changes. Thus, not all of the reported changes in effluent quality are necessarily attributable to mill conversion.

Table 1. Mean values for chemical/physical parameters for the effluent from Mill A before and after process conversion from <20% ClO_2 to >70% ClO_2 substitution and oxygen delignification.

Parameter	Unit	Before ClO_2	Sample Size	After ClO_2	Sample Size	After O.D.	Sample Size
Alkalinity	mg L^{-1}	252	7	272	5	294	9
BOD	mg L^{-1}	24	7	26	8	31	13
Color	CPU	1785	7	1700*	7	1268*	16
Conductivity	µmoh cm^{-1}	2736	7	2567*	7	2097*	16
Hardness	mg L^{-1}	111	7	123	5	105	9
pH		7.7	7	7.6	8	7.8	
Tannin/lignin	mg L^{-1}	65	7	66*	7	54*	16
Turbidity	NTU	36.4	7	61.0	7	72.0	16
Non-chlorinated phenolics	µg L^{-1}	2.3	7	2.9	8	2.0	12
AOX	mg L^{-1}	22.4*	7	18.4*	8	10.0*	10
Chlorinated phenolics	µg L^{-1}	63.3	7	66.2*	8	10.4*	12
Chlorocatechol	µg L^{-1}	23.5*	7	51.3*	8	11.0*	12
Chloroguaiacol	µg L^{-1}	26.9*	7	11.8*	8	1.7*	12
Chlorophenol	µg L^{-1}	12.9*	7	3.1*	8	1.2*	12
Aromatic acids	µg L^{-1}	1.4	7	4.7	7	1.3	12
Fatty/resin acids	µg L^{-1}	38.7	7	124.3	7	52.5	12
Chlorinated fatty/resin acids	µg L^{-1}	2.8	7	8.4	7	3.7	12
Chloroaldehydes	µg L^{-1}	7.9	7	4.2*	7	1.3*	12

* = Significantly different ($p < 0.05$).

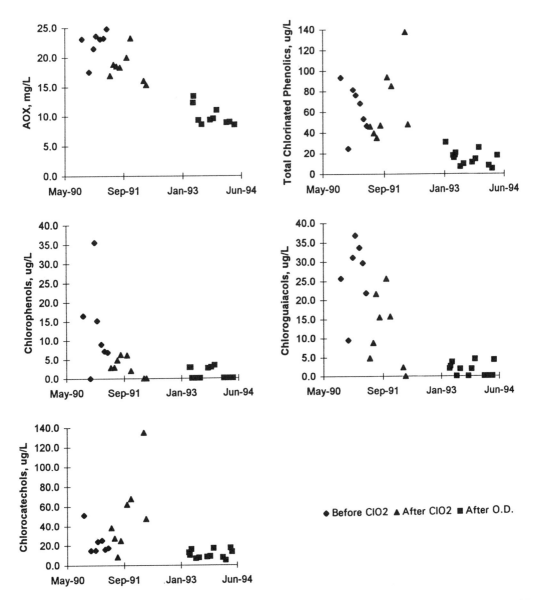

Figure 1. Mill A effluent chemical characteristics before and after process conversion from <20% ClO_2 to >70% ClO_2 substitution and oxygen delignification.

For the period before/after the conversion to increased ClO_2 substitution there were few significant differences in physical/chemical parameters (Table 1). None of the differences for conventional effluent parameters such as BOD, TSS, or pH were significant. Effluent color declined significantly following the addition of oxygen delignification but appeared to be unaffected by the initial conversion to increased ClO_2 substitution. A similar corresponding pattern of significant reduction in tannin/lignin was also observed.

There were significant reductions in AOX, following conversions to increased ClO_2 substitution and oxygen delignification (Table 1). Before conversion, effluent AOX averaged 22.4 mg L^{-1}, with a decline to 18.4 mg L^{-1} after conversion to increased ClO_2 substitution and further reduction to 10.0 mg L^{-1} after the addition of oxygen delignification. The combination of both mill modifications reduced AOX by more than 50% from pre-conversion levels. Measurements for chlorinated phenolic compounds indicated no significant change following the conversion to increased ClO_2 substitution but a significant decline occurred following the addition of oxygen delignification (Table 1, Fig. 1). Mean total chlorinated phenolics declined to 10.4 ug L^{-1}, a reduction to <20% of the levels of the pre-conversion mill. Similar significant reductions occurred for the chlorophenolic subgroups, including chlorophenols, chloroguaiacols, and chlorocatechols. Greatest reductions occurred for chlorophenols and chloroguaiacols, with mean values after the addition of oxygen delignification <10% of those before mill conversion. Both of these chlorinated phenolic subgroups also indicated significant reductions after the initial conversion to increased ClO_2 substitution, with values reduced to 24 and 44% of pre-conversion mill operation values for the chlorophenols and chloroguaiacols, respectively. Chlorocatechols were less affected by conversion to oxygen delignification with a reduction to 47% relative to conventional mill operation. The chlorocatechols indicated an unexpected significant increase following the conversion to increased ClO_2 substitution. There is no clear explanation for the increase in chlorocatechols although, as noted for other non-chlorinated chemical parameters, other mill changes not related to the process change could have been responsible. Changes in furnish, for example, during the various study periods could have influenced effluent chemical properties.

Marine Bioassays

Bioassay data for both freshwater and marine tests indicated process change related significant differences for only one of the test endpoints, the echinoderm IC25 (Table 2). In this case, the IC25 following conversion to increased ClO_2 substitution declined (i.e., increased toxicity) significantly below those before conversion or after the addition of oxygen delignification. The reason for the decreased IC25 is unclear and may relate to other parameters within the mill other than the process change. The echinoderm bioassay NOEC endpoint did not change significantly before or after the mill process changes.

The echinoderm bioassay IC25 ranged from 0.7 to 4.3% v/v effluent during pre-conversion, from 0.1 to 1.8% v/v effluent following conversion to ClO_2 substitution, and from 0.0 to 11.3% v/v effluent following the addition of oxygen delignification. Substantial variability in response was evident in each of the study periods (Fig. 2). The majority of the IC25 values below 1.0% effluent occurred following conversion to increased ClO_2 and prior to the addition of oxygen delignification.

The echinoderm bioassay has on other occasions indicated sensitive responses to pulp and paper mill effluents. The nature of this bioassay response has not, however, been clearly defined. Pinza *et al.* (1991) carried out sperm/egg tests with *D. excentricus* using individual chlorinated phenolics with EC50s ranging from 2.1 mg L^{-1} for pentachlorophenol to 26 mg L^{-1} for tetrachloroguaiacol. A comparison with literature values indicated *D. excentricus* to be relatively insensitive to chlorinated phenolic compounds. The total chlorinated phenolic concentration of whole effluent for the mill in this study was nearly two orders of magnitude lower than the EC50 of the most sensitive individual chlorinated phenolic measured by Pinza *et al.* (1991). Considering that most echinoderm NOEC and IC25 endpoints in the present study were <3.0% v/v effluent it seems unlikely that the *D. excentricus* responses were due to chlorinated phenolic compounds either before or after the mill process conversions.

Similar individual chemical tests were carried out with resin and fatty acids by Cherr *et al.* (1987) using sperm/egg from the sea urchin, *Stongylocentroetus purpuratus*. For the chemicals tested EC50s ranged from 0.28 mg L^{-1} for linoleic acid to >20 mg L^{-1} for abietic acid. EC50s for these effluent constituents were noted as being substantially greater (less toxic) in seawater for echinoderms as compared to literature values for rainbow trout in freshwater for similar chlorinated phenolics. The EC50s reported by Cherr *et al.* (1987) for individual resin/fatty acids are much greater than the combined concentrations of these chemical groups for undiluted effluent in the present study. Considering, again, that echinoderm IC25s for the effluent in the present study were <3.0% v/v effluent, this group of chemical constituents is probably not responsible for test responses either before or after mill conversions.

Table 2. Bioassay data for Mill A effluent before and after process conversion from <20% ClO_2 to >70% ClO_2 substitution and oxygen delignification. Response values (NOEC = no observable effect concentration and IC25 = inhibition concentration at 25% reduction from control) are in % v/v effluent with significant differences ($p<0.05$) indicated in bold. The number of samples with no response at highest effluent concentration is indicated in ( ) after the NOEC and IC25 with *** indicating there were no responses at the highest effluent concentration.

		Before ClO_2	Sample Size	After ClO_2	Sample Size	After O.D.	Sample Size
Marine Chronic Bioassays							
Dendraster excentricus	NOEC	0.5	6	0.2	8	1.3	23
Fertilization	IC25	**2.5**	6	**0.6**	8	**3.5**	23
Menidia beryllina	NOEC	50.0 (1)	4	***	1	60.0 (1)	4
Survival	IC25	57.6 (2)	4	***	1	***	4
Growth	NOEC	***	4	***	1	60.0 (1)	4
	IC25	66.9 (3)	4	***	1	***	4
Cyprinodon variegatus	NOEC	***	6	***	3		
Survival	IC25	***	6	***	3		
Growth	NOEC	***	6	***	3		
	IC25	***	6	***	3		
Freshwater Chronic Bioassays							
Ceriodaphnia dubia	NOEC	***	7	***	5	88.9 (8)	9
Survival	IC25	***	7	***	5	84.1 (8)	9
Reproduction	NOEC	33.4	7	34.4	5	34.4	9
	IC25	41.8	7	456.5	5	44.8	9
Pimephales promelas	NOEC	***	7	***	5	88.9	9
Survival	IC25	***	7	***	5	***	9
Growth	NOEC	82.0 (5)	7	***	5	84.0 (7)	9
	IC25	83.3 (5)	7	95.4 (4)	5	96.6 (7)	9

There were no significant differences for bioassays with the two marine fish species in the period before or after mill conversions (Table 2 and Fig. 2). No responses were measured for either species at the highest test concentration (70% v/v effluent), either as IC25 or NOEC, following ClO_2 substitution. This was also true for tests with *C. variegatus* for effluent prior to mill conversion. These data indicate *M. beryllina* to be somewhat more sensitive to effluent since there were some occasions when effects on both survival and growth endpoints were measured at high effluent concentrations (>50% v/v) in the pre-conversion period. Following the addition of oxygen delignification, three additional tests with *M. beryllina* were conducted. The postoxygen delignification effluent did not generate an IC25 endpoint for either growth or survival. Although there is a suggestion of improved effluent quality following mill conversion, the ability to quantify this improvement is limited since most bioassay responses were at or near the upper limit of detection (i.e., no response at 70% v/v effluent, the highest concentration which could be tested).

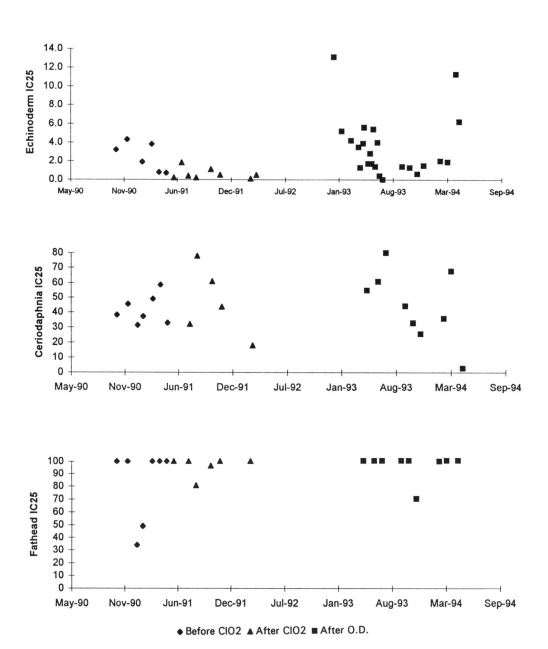

Figure 2. Mill A bioassay response before and after process conversion from <20% ClO_2 to >70% ClO_2 substitution and oxygen delignification. Response values are IC25s as % v/v effluent. Echinoderm responses are for sperm exposure and egg fertilization with Dendraster excentricus, Ceriodaphnia dubia responses for reproduction, and fathead minnow responses for growth.

Freshwater Bioassays

Comparisons of *C. dubia* and *P. promelas* bioassay endpoints indicated no significant change following mill conversions (Table 2 and Fig. 2). Similar to marine fish responses, improvements in effluent quality were difficult to quantify because many of the bioassays did not produce effects even at the highest effluent concentration tested (100% v/v effluent). In computing mean responses these values were considered as being 100% v/v effluent. This produces an artificial bias downward (i.e., increased bioassay response) in the mean since a response wasn't actually measured at 100% v/v effluent.

The reproduction endpoint for *C. dubia* was the most sensitive response measured for the freshwater 7-d tests (Table 2 and Fig. 2), with only one of the tests indicating a survival effect even at 100% v/v effluent. The IC25 for reproduction ranged from 31 to 59% v/v effluent in the pre-conversion period to 18 to 78% v/v effluent after conversion to increased ClO_2 substitution, and from 2.6 to 70.8% effluent after the addition of oxygen delignification. The mean IC25 (41.8% v/v effluent) before conversion was not significantly different from the mean response (46.5% v/v effluent) following conversion to increased ClO_2 substitution or after the addition of oxygen delignification (44.8% v/v effluent).

Similar to *C. dubia* bioassay results, there was an absence of significant response differences for fathead minnows, before or after mill changes (Table 2 and Fig. 2). Responses were typically at or near the upper limit of detection for the method. For example, with the survival endpoint, a response was measured for only 1 of the 21 tests conducted. For the IC25 growth endpoint, 4 of 7 bioassays in the pre-conversion period, 4 of 5 bioassays in the ClO_2 substitution period, and 7 of 9 bioassays after the addition of oxygen delignification failed to exhibit a bioassay response even at 100% v/v effluent. These non-detect values were counted as a response at 100% effluent in calculating the mean for process change comparisons. The resulting mean IC25 of 83.3% v/v effluent before conversion was not significantly different from the IC25 of 95.4% v/v effluent and 96.7% v/v after conversions to increased ClO_2 substitution and oxygen delignification, respectively.

Results from Mill B

Effluent Quality

Effluent quality comparisons for Mill B before and after process changes are provided in Table 3 and Fig. 3. Effluent flow and production rates were similar in the period of time before and after the process change to oxygen delignification, allowing comparisons for effluent chemical quality and bioassay responses without adjustments for water use. A detailed multi-species bioassay characterization just prior to conversion high ClO_2 substitution was unavailable for this effluent.

Following the addition of oxygen delignification, effluent color and tannin/lignin content were significantly reduced. Mean color declined from 1034 color units prior to the addition of oxygen delignification to 678 color units after. A significant increase in hardness was measured, with a mean of 78 mg L^{-1} before oxygen delignification and 106 mg L^{-1} after. No other non-chlorinated chemical effluent parameters indicated significant differences before or after the mill process change.

Mean effluent AOX declined from 8.2 mg L^{-1} prior to the addition of oxygen delignification to 1.3 mg L^{-1} after (Table 3 and Fig. 3). This 84% reduction in AOX is substantially greater than that which occurred with Mill A effluent (55% AOX reduction for the combined effects of conversion to 70% ClO_2 substitution and oxygen delignification). The concentrations of total chlorinated phenolics and the chlorocatechol, chloroguaiacol, and chlorophenol subgroups for Mill B effluent all declined to non-detect levels after the addition of oxygen delignification (Table 3 and Fig. 3). Significant reductions also occurred for chloroaldehydes and chlorinated fatty and resin acids. Although not statistically significant, the mean concentration of non-chlorinated fatty and resin acids increased, after the addition of oxygen delignification. The mean concentration of aromatic acids also increased, in this case the difference was significant. Changes in some of the non-chlorinated effluent parameters may relate to other mill activities and be unrelated to the conversion to oxygen delignification.

Table 3. Mean values for chemical/physical parameters for the effluent from Mill B at 100% ClO_2, before and after the addition of oxygen delignification. Significant differences (p<0.05) indicated in bold. Non-detect (N.D.) values for chlorinated phenolic compounds generally ranged from 0.2-2.2 µg L^{-1} and were counted as 0.0 in tests for statistical differences.

Parameter	Unit	Before O.D.	Sample Size	After O.D.	Sample Size
Alkalinity	mg L^{-1}	134	8	155	15
BOD	mg L^{-1}	13	8	17	14
Color	CPU	**1034**	8	**678**	14
Conductivity	µmoh cm^{-1}	1552	8	1477	15
Hardness	mg L^{-1}	**78**	8	**106**	15
pH		7.5	8	7.5	14
Tannin/lignin	mg L^{-1}	**47**	8	**34**	14
Turbidity	NTU	11.6	8	13.7	14
Non-chlorinated phenolics	µg L^{-1}	0.1	9	0.3	15
AOX	mg L^{-1}	**8.3**	9	**1.3**	13
Chlorinated phenolics	µg L^{-1}	**20.7**	9	**N.D.**	15
Chlorocatechol	µg L^{-1}	**16.4**	9	**N.D.**	15
Chloroguaiacol	µg L^{-1}	**3.6**	9	**N.D.**	15
Chlorophenol	µg L^{-1}	**0.7**	9	**N.D.**	15
Aromatic acids	µg L^{-1}	**0.2**	9	**4.2**	15
Fatty/resin acids	µg L^{-1}	64.0	9	125.3	15
Chlorinated Fatty/resin acids	µg L^{-1}	**11.2**	9	**0.4**	15
Chloroaldehydes	µg L^{-1}	**36.9**	9	**0.6**	15

Marine Bioassays

A summary of bioassay data for Mill B before and after the addition of oxygen delignification is provided in Table 4 and Fig. 4. Similar to results for Mill A, the echinoderm response was the most sensitive of the tests used. Mean echinoderm IC25 following the addition of oxygen delignification was significantly greater (i.e., reduced toxicity) than before mill change (4.1% v/v vs 12.5% v/v after). Echinoderm IC25 values were generally higher than for Mill A both before and after the addition of oxygen delignification, with no IC25 values falling below 1.0% v/v effluent (Fig. 2. and Fig. 4).

Larval fish bioassays with both *C. variegatus* and *M. beryllina* were used prior to the addition of oxygen delignification, and *M. beryllina*, only, after mill conversion. There were no significant response differences before or after the addition of oxygen delignification by the mill. None of the *C. variegatus* tests produced a response even at the highest concentration (70% v/v effluent). This result is similar to testing with Mill A and indicates that *C. variegatus* is somewhat less sensitive to effluent than *M. beryllina*. Only *M. beryllina* was used in effluent testing after the mill added oxygen delignification. Before oxygen delignification none of the three tests indicated a response at the highest effluent concentration and two of three tests after oxygen delignification did not indicate an effluent response.

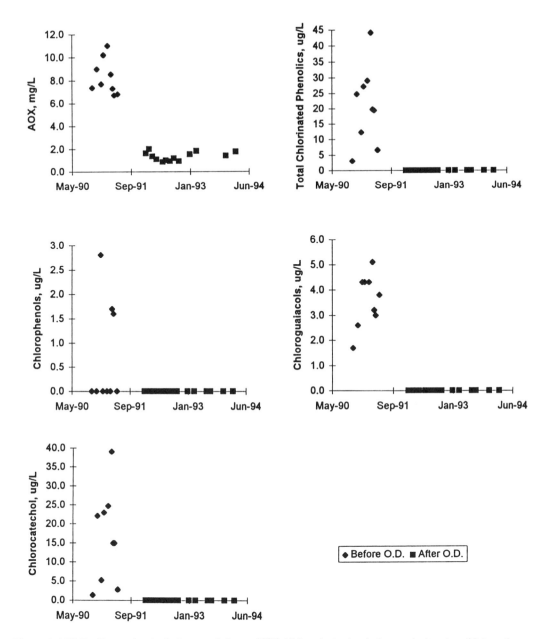

Figure 3. Mill B effluent chemical characteristics at 100% ClO_2 substitution before and after the addition of oxygen delignification.

Freshwater Bioassays

No significant response differences were indicated for tests with *C. dubia* with either survival or reproduction endpoints (Table 4). There was an absence of a survival response, even at the 100% v/v effluent, for the 23 tests carried out during the period before and after the addition of oxygen delignification. The mean IC25 for reproduction was similar before and after the addition with respective values of 50.1 and 51.7% v/v effluent. The reproduction IC25 endpoint ranged from 29 to 70.5% v/v effluent prior to the addition of oxygen delignification to between 21.8 and 100% v/v effluent after the process change (Fig. 4).

Table 4. Bioassay data for Mill B effluent at 100% ClO$_2$ substitution before and after the addition of oxygen delignification. Response values (NOEC = no observable effect concentration and IC25 = inhibition concentration at 25% reduction from control) are in % v/v effluent with significant differences (p<0.05) indicated in bold. The number of samples with no response at highest effluent concentration is indicated in ( ) after the NOEC and IC25 with *** indicating there were no responses at the highest effluent concentration.

		Before O.D.	Sample Size	After O.D.	Sample Size
Marine Chronic Bioassays					
Dendraster excentricus	NOEC	2.3	7	4.0	14
Fertilization	IC25	**4.1**	7	**12.5**	14
Menidia beryllina	NOEC	56.7 (2)	3	***	3
Survival	IC25	***	3	65.7 (2)	3
Growth	NOEC	***	3	***	3
	IC25	***	3	***	3
Cyprinodon variegatus	NOEC	***	4		
Survival	IC25	***	4		
Growth	NOEC	***	4		
	IC25	***	4		
Freshwater Chronic Bioassays					
Ceriodaphnia dubia	NOEC	***	8	***	15
Survival	IC25	***	8	***	15
Reproduction	NOEC	42.3	8	34.7	15
	IC25	50.1	8	51.7	15
Pimephales promelas	NOEC	***	8	***	15
Survival	IC25	***	8	***	15
Growth	NOEC	**72.5 (3)**	8	**94.1(14)**	15
	IC25	**81.9 (1)**	8	**99.7(14)**	15

The mean fathead minnow IC25 growth response increased significantly from 81.9% v/v before the addition of oxygen delignification to 99.7% v/v effluent after oxygen delignification (Table 4). The mean IC25 for growth increased. Prior to the addition of oxygen delignification, 1 test out of 8 failed to yield a response at 100% v/v effluent, as contrasted to the period after process change when 14 of 15 tests yielded no response at 100% v/v effluent (Fig. 4).

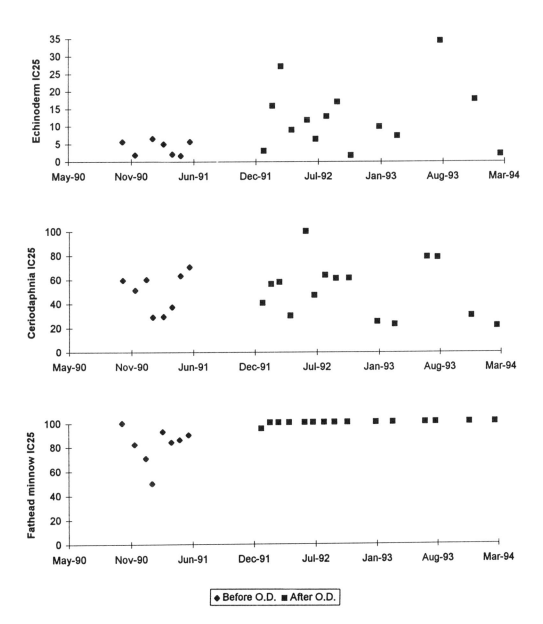

Figure 4. Mill B bioassay response at 100% ClO_2 substitution before and after the addition of oxygen delignification. Response values are IC25s as % v/v effluent. Echinoderm responses are for sperm exposure and egg fertilization with Dendraster excentricus, Ceriodaphnia dubia responses for reproduction, and fathead minnow responses for growth.

SUMMARY AND CONCLUSIONS

Characterization data for Mill A indicated that although the conversion to increased ClO_2 substitution resulted in significant reductions in AOX and reductions in some other effluent chlorinated organic chemical constituents, it did not lead to significant improvements in bioassay responses. This was in part due to the difficulty in making comparisons when many responses were at or near the upper limit of detection for the test methods. The only significant difference in bioassay response was for the echinoderm sperm/egg IC25 fertilization endpoint. In this case, the response following conversion to increased ClO_2 substitution was significantly worse than either the period during conventional mill operation or after the addition of oxygen delignification. Although echinoderm responses were sensitive (i.e., IC25 values <5% v/v effluent), comparisons of effluent chlorinated phenolic compounds with those for *D. excentricus* in the literature indicated that the responses were probably not related to these chemical groups. Similar literature value comparisons for resin/fatty acid constituents of Mill A effluent also suggest an absence of concentrations sufficient to account for bioassay responses.

The addition of oxygen delignification at Mill A resulted in significant reductions in effluent chlorinated phenolics and AOX compared to concentrations during pre-conversion and increased ClO_2 substitution periods. With the exception of the echinoderm IC25 response, none of the bioassay responses showed significant changes before or after the addition of oxygen delignification. The echinoderm response indicated a reduction in toxicity when compared to the tests conducted in the period after increased ClO_2 substitution. The response levels following the addition of oxygen delignification were similar to those during the period of pre-conversion mill operation.

Bioassay comparisons for Mill B were made for the period before/after the addition of oxygen delignifcation. Similar to Mill A, the addition of oxygen delignification resulted in significant reductions in chlorinated phenolics and AOX. Concentrations for all of the chlorophenolic subgroups fell to non-detect levels following the addition of oxygen delignification. Reductions in effluent chlorinated organics content also coincided with significant improvements in echinoderm and fathead minnow bioassay responses (i.e., increased IC25). None of the other bioassays, including those with *C. dubia*, indicated changes after the addition of oxygen delignification.

The ability to make direct comparisons between Mill A and Mill B are limited due to the likelihood of a number of operational parameter differences, including furnish (Mill A = 100% softwood, Mill B = 75% and 25% hardwood). For both effluents, the *C. dubia* growth response and echinoderm fertilization responses were the most sensitive. The *C. dubia* response did not change significantly, however, for either mill before or after process changes. Mean IC25 growth response for both mills was similar, falling within the range of 41.8-44.8% v/v and 50.1-51.7% v/v for the various study periods at Mill A and Mill B, respectively. There was, however, a substantial difference in echinoderm response for the two effluents. Following the addition of oxygen delignification, Mill B echinoderm IC25 was 3.6 times greater (i.e., less toxic) than for Mill A. Although literature values would not indicate echinoderm bioassay responses at the chlorophenolic levels found in the effluent from Mill A, these levels were substantially higher than the non-detect levels found in Mill B. In addition, the AOX content in Mill A effluent was 7 times greater than that measured in effluent from Mill B. The possibility of echinoderm responses being ameliorated by reductions in chlorinated organic compounds is further suggested by the significant improvement in response for Mill B following the addition of oxygen delignification and corresponding reductions in AOX and specific chlorinated organic chemical groups. Effluent tannin/lignin and color, parameters not necessarily associated with chlorinated organics, also indicate a similar pattern of differences between Mill A and Mill B and point to an alternate cause/effect association with echinoderm response differences between the mills and before/after process change.

For both Mill A and Mill B, bioassay responses for most tests with marine and freshwater organisms were at or near the limit of detection. The effectiveness of effluent secondary treatment in reducing acute and chronic bioassay responses in receiving water and effluent has been demonstrated by others (Robinson *et al.* 1994; Firth and Backman 1990) and may have been the dominant factor in the present study rather than the effect of process changes on chlorinated organics. The experimental streams studies carried out with effluent from Mill A and Mill B provide additional evidence of the non-harmful effects of these effluents, before/after the addition of process changes to reduce chlorinated organics, on fish and supporting food web components (NCASI 1993, 1994a,b).

REFERENCES

American Public Health Association, American Water Works Association and Water Pollution Control Federation. 1985. Standard Methods for the Examination of Water and Waste Water, 16th ed. American Public Health Association, Washington D.C.

Cherr, G., C. Shenker, C. Lundmark and K. Turner. 1987. Toxic effects of selected bleached kraft mill effluent constituents on the sea urchin sperm cell. Environ. Toxicol. Chem. 6:561-569.

Firth, B.K., and C.J. Backman. 1990. Comparison of Microtox testing with rainbow trout (acute) and *Ceriodaphnia* (chronic) bioassays in mill wastewaters. TAPPI J. 169-174.

Hach Company. 1989. Water Analysis Handbook. Loveland, CO.

Hall, T., R. Haley and L. Lafleur. 1991. Effects of biologically treated bleached kraft mill effluent on cold water stream productivity in experimental stream channels. Environ. Toxicol. Chem. 10:1051-1060.

Hall, T.J., R.K. Haley and K.J. Battan. 1993. Turbidity as a method of preparing sperm dilutions in the echinoid sperm/egg bioassay, Environ. Toxicol. Chem. 12:2133-2137.

NCASI. 1986a. NCASI methods for the analysis of chlorinated phenolics in pulp industry wastewaters," Technical Bulletin No. 498. National Council of the Paper Industry for Air and Stream Improvement, New York, NY.

NCASI. 1986b. Procedures for the analysis of resin and fatty acids in pulp mill effluents. Technical Bulletin No. 501. National Council of the Paper Industry for Air and Stream Improvement, New York, NY.

NCASI. 1992. Marine chronic bioassays - Technical evaluations and effluent responses with the echinoderm sperm/egg test. Technical Bulletin No. 627. National Council of the Paper Industry for Air and Stream Improvement, New York, NY.

NCASI. 1993. Aquatic community effects of biologically treated bleached kraft mill effluent before and after conversion to increased chlorine dioxide substitution: Results from an experimental streams study. Technical Bulletin No. 653. National Council of the Paper Industry for Air and Stream Improvement, New York, NY.

NCASI. 1994a. Characterization of a bleach kraft mill effluent with high substitution of chlorine dioxide for chlorine using laboratory bioassays and experimental streams. Part 1. Comparison of effects of a bleached kraft pulp mill effluent on laboratory bioassay responses and fish growth and production in experimental streams before and after high substitution with ClO_2. Tech. Bull. National Council of the Paper Industry for Air and Stream Improvement, New York, NY.

NCASI. 1994b. Characterization of a bleach kraft mill effluent with high substitution of chlorine dioxide for chlorine using laboratory bioassays and experimental streams. Part 2. Comparison of biochemical, physiological, and histopathological biomarkers to the survival growth and production of fish exposed to high substitution bleached kraft mill effluent in experimental streams. Tech. Bull. National Council of the Paper Industry for Air and Stream Improvement, New York, NY.

Owens, W.J. 1991. The hazard assessment of pulp and paper effluents in the aquatic environment: A review. Environ. Toxicol. Chem. 10:1511-1540.

Pinza, R.M., R.A. Matthews and T.J. Hall. 1991. Measuring the toxicities of selected chlorinated phenolics to *Dendraster excentricus* using the echinoderm sperm cell toxicity test. Environ. Sciences 1:89-97.

Robinson, R.D., J.H. Carey, K.R. Solomon, I.R. Smith, M.R. Servos and K.R. Munkittrick. 1994. Survey of receiving-water environmental impacts associated with discharges from pulp mills. 1. Mill characteristics, receiving water chemical profiles and lab toxicity tests. Environ. Toxicol. Chem. 13:1075-1088.

U.S. EPA (U.S. Environmental Protection Agency). 1988. Short-term methods for estimating the chronic toxicity of effluents and receiving waters to marine and estuarine organisms. EPA/600/4-87/028. Washington D.C.

U.S. EPA (U.S. Environmental Protection Agency). 1989. Short-term methods for estimating the chronic toxicity of effluents and receiving waters to freshwater organisms, Second Edition, EPA/600/4-89/001. Washington D.C.

REPRODUCTION EFFECTS ON *MONOPOREIA AFFINIS* OF HPLC-FRACTIONATED EXTRACTS OF SEDIMENTS FROM A PULP MILL RECIPIENT

Ann-Kristin Eriksson[1], Brita Sundelin[1], Dag Broman[2] and Carina Näf[2]

[1]Institute of Applied Environmental Research, Laboratory for Aquatic Ecotoxicology,
Stockholm University, S-611 82 Nyköping, Sweden
[2]Aquatic Chemical Ecotoxicology, Department of Zoology, Stockholm University,
S-106 91 Stockholm, Sweden

The deposit-feeding amphipod *Monoporeia affinis* was exposed in soft bottom microcosms to fractionated organic extracts of bottom sediments collected outside a pulp and paper mill. Changes in the biological effects of sediment extracts were tested before and after altering the bleaching process at the mill, i.e., a change from chlorine gas to elemental chlorine free (ECF) bleaching in combination with substantial decreases in total wastewater and COD discharges. Sediment samples representing material deposited before (10-12 cm) and after (0-2 cm) this event, as well as sediments (0-2 cm) from the central Bothnian Sea, were Soxhlet extracted with toluene. The extracts were fractionated by HPLC into three fractions containing aliphatic/monoaromatic, diaromatic and polyaromatic compounds. Control sediments were dosed with the extracts and placed in soft bottom microcosm test systems. After an exposure period of 5 months, reproduction success, fecundity, embryonic development and different types of embryonic malformation were studied in amphipods. All fractions of the sediment from the pulp mill were toxic, except the polyaromatic fraction from surface sediment (0-2 cm) and the aliphatic/monoaromatic fraction from 10-12 cm depth. However, the effects of the fractions were lower than for corresponding non-fractionated sediment extracts in both depths. The polyaromatic fraction from the 10-12 cm depth exceeded the surface sediment fraction and the others in toxicity, indicating improved conditions as a result of the reduced discharge from the mill.

INTRODUCTION

A great number of pulp and paper mills are located along the coast of the northern Baltic, particularly the Gulf of Bothnia. Discharges from these industries have been reported as one of the main sources of organic pollutants found in the Bothnian Sea (Thorman 1987). Negative biological effects in combination with increased concentrations of various organochlorine compounds have been observed especially in areas close to the pulp mills (Södergren 1989, 1993; Sundelin 1989). Since the early 1980s, Swedish environment programs have focused on the discharge and environmental effects of persistent organic compounds, such as organochlorines, emitted from the pulp and paper industry.

In 1991, Sweden issued a new environmental policy which sets a goal that discharges from the pulp and paper industry should be reduced by the beginning of next century to such levels that the environment is not noticeably affected. The emphasis will, in the future, be put on environmental fate from both bleached and unbleached kraft production. Efforts to reduce the discharges have resulted in new technologies and processes at the mills as well as demands for effect variables which give a quick response to the decreased discharges. Great effort has been made to characterize mill effluents in terms of identification and quantification of more than 300 specific organic compounds, chlorinated as well as non-chlorinated (Dahlman *et al.* 1992). However, these are most likely still a minor part of the effluent composition. Also, since the production processes at pulp and paper mills continuously undergo changes, a significant number of unknown compounds might be found in the future. Therefore, in this paper we have made an attempt to define the fraction in which the most negative biological effects can be found. The fractionation method used for this is a previously described HPLC technique which produces one aliphatic/monoaromatic, one diaromatic and one polyaromatic fraction (Zebühr *et al.* 1989). The advantage of using this technique is that it gives an indication of the aromaticity of the compounds which are responsible for the demonstrated effect. It is also possible to go further with toxic fractions and subfractionate them for renewed toxicity testing.

The test system used in this paper is based on the Baltic soft bottom benthic community. The ecological relevance is obvious since soft bottoms represent 70% of the Baltic bottom areas. The sediment also plays an important role in the recycling of nutrients and toxic substances such as metals and organic compounds. A high proportion of pollutants and nutrients is associated with the solid phase in water, and after deposition they accumulate in the bottom sediments. Once sorbed in the sediment, noxious substances are slowly released into the water column. Conclusively, sediments often represent a sink for these pollutants and can cause long-lasting or chronic contamination of the benthos (Giere 1993). Additionally, sediment dwelling invertebrates are often exposed to two sources of contamination, the substrate, which is often their food, and the interstitial and overlying water (Moore *et al.* 1979).

Monoporeia affinis is by far the most productive macrofaunal species in the Baltic. By its burrowing and feeding activities together with intraspecies interactions (Hill *et al.* 1992) in the sediment, the amphipod probably regulates both oxygenation of the sediments as well as populations of bacteria, meiofauna (Ölafsson and Elmgren 1991; Sundelin and Elmgren 1991) and other species such as *Macoma balthica* (Segerstråle 1962, 1973; Elmgren *et al.* 1986). The abundance, body size and high lipid content (Hill *et al.* 1992) of this deposit-feeding amphipod make it very important in the transfer of carbon as well as chlorinated organic contaminants from the sediment to the fish community (Dermott and Corning 1988) and call for careful investigations regarding its susceptibility to anthropogenic compounds.

To enable relevant measurements of biological effects of sediment-bound pollutants, a soft bottom microcosm system, comprised of the natural meiofauna community and the dominating macrofauna species *Monoporeia affinis*, was designed to simulate the sublittoral bottoms dominating the Baltic (Sundelin 1983). Continuous water flow through the microcosms contributes to the long-term stability and the similarity with the parent system *in situ*. In spite of manipulation of the original fauna association, the microcosms develop very much in accordance with the parent system *in situ*, at least over a little more than one annual cycle (Sundelin 1983; Sundelin and Elmgren 1991).

Reproduction has been demonstrated as the most sensitive period in the life cycle when considering ecotoxicological effects on benthic invertebrates. Harpactcoid copepods were shown to be resistant to the insecticide fenvalerate when population variables were examined, whereas fecundity was affected at considerably lower concentrations (Chandler 1990). For *Nitocra spinipes* (Harpacticoida), Bengtsson and Bergström (1987) found a 96-h LC50 of 0.4 to 0.8 mg Cd L^{-1}, while fecundity was reduced at 100 µg Cd L^{-1}. Samoiloff (1980) and Samoiloff *et al.* (1980) reported reduced survival for a nematode at cadmium concentrations of 11.2 mg L^{-1}, while fecundity was affected at 11.2 µg L^{-1}.

Variables describing the reproduction of *Monoporeia affinis,* such as embryonic malformation, were the most sensitive variable when soft bottom microcosms were exposed to metals such as cadmium and lead (Sundelin 1983, 1984), arsenic (Blanck *et al.* 1989), organic compounds such as 4,5,6-trichloroguaiacol, heavy metals, contaminated sediment and pulp mill effluents (Sundelin 1988, 1989). The microcosm test system is sensitive and suitable for analyzing individual and population effects from xenobiotics, associated with the sediment or the incoming water. This microcosm test system has high ecological realism and could offer a possibility to translate laboratory results to the natural environments.

The reproduction cycle of *Monoporeia affinis* starts at the beginning of August and is triggered by the reduction of light (Segerstråle 1971). The amphipod assimilates and stores food resources from the diatom bloom, and the amphipod's lipid content rises to a maximum in late summer and declines during gonad maturation. The maturation continues until the beginning of November when mating starts. The mature males leave their sperm in the marsupium (brood pouch) of the females where fertilization occurs. The males die within a few weeks after mating. Embryonic development continues for about 3.5 months, and the first juveniles hatch in the middle of February (Sundelin and Eriksson, unpubl. data). The high lipid resources of *Monoporeia* prior to reproduction (Hill *et al.* 1992) are utilized during maturation of the gonads, indicating increasing levels of lipophilic compounds in the gonads and comparatively higher levels than for the rest of the body. Albaigés *et al.* (1987) have demonstrated gonad concentrations of PCB one order of magnitude higher than in muscle and liver in both benthic (*Mullus* sp.) and pelagic (*Trachurus* sp.) fishes. When lipid resources are utilized during embryo development, the associated lipophilic compounds are released and may exert their toxic effects either metabolized or unmetabolized. Obvious malformations are detectable during early embryogenesis, while minor malformations are easier to distinguish after embryo differentiation, which means that the examination of juveniles ought to be carried out during the latter part of embryogenis.

The purpose of the present study was 1) to examine whether the reproductive effects of HPLC-fractionated sediment extracts could be related to a specific group of compounds (aliphatics/monoaromatics, diaromatics or polyaromatics), 2) to investigate if previously identified pulp and paper mill related compounds could explain the possible effect or if it is related to unidentified compounds, and 3) to determine whether the usefulness and sensitivity of the effect variables of *Monoporeia* reproduction should be examined when changing the bleaching process from chlorine to elemental chlorine free (ECF) bleaching.

MATERIALS AND METHODS

Test Organisms and Sediment

At the beginning of August, the sediment, including the natural populations of meio- and macrofauna, was collected with a modified Ockelman dredge (Blomqvist and Lundgren 1995) (mesh size 450 μm). Collections were made at 40 m depth at Hållsfjärden in the Askö area in the northern Baltic proper. The sediment was sieved through 1.0 and 0.5 mm nets to separate macro- and meiofauna. Of the macrofauna (1.0 mm fraction), only *Monoporeia affinis* was used (for details see Sundelin 1983).

Microcosm System

The microcosms used were two-liter Erlenmeyer flasks (Fig. 1) containing 4 cm of sediment with its natural meiofauna (227 cm^2). Sea water from 40 m depth entered the microcosms at a rate of 2.2 L h^{-1}. The pH varied between 7.3 and 7.7, salinity between 6.8 and 7.0‰, and the temperature ranged from 4 to 6°C during the experiment. Further details are given in Sundelin (1983).

Experimental Design

The experiment was started in the middle of August, when maximum lipid values were reached in amphipods and the situation is optimal for bioaccumulation of lipophilic organic substances. *Monoporeia affinis* was exposed to sediment extracts from two stations, one situated in the recipient of a pulp and paper mill (stn. 3, depth 18 m) and the other in the open Bothnian Sea (stn. 2, depth 120 m). Reference sediment from station 1 (depth 37 m), an uncontaminated area in the Baltic proper, was used as a control. Both surface (0-2 cm) and 10-12 cm depth sediments were used from the pulp mill site, whereas only the surface layer was used from the open sea sediment (stn. 2). Sedimentation rate at this open sea station is low relative to the pulp mill station, which means that 10-12 cm sediment depth corresponds to preindustrial years and should not be compared with the 10-12 cm depth at the near shore station 3. Carbon content in extracted sediment from the near mill station was 5.6% (of dry weight) and 5.0% for

0-2 cm and 10-12 cm, respectively, while sediment from the open sea station 2 showed carbon content of 6.5%.

Monoporeia in the microcosms were exposed to 4 different series of extracts (unfractionated and three fractionated extracts containing aliphatic/monoaromatic, diaromatic and polyaromatic compounds) of surface sediment and deeper sediment (10-12 cm) from station 3 and open sea surface sediment (0-2 cm) (4 x 3 series in total). Five replicates of each extract treatment and a blank extract control were used, which means an experimental size of 65 microcosms. The different fractionated and unfractionated extracts from 69 g sediment (dry weight, dwt) (available sediments from 13 core samples) were dissolved in 25 mL acetone before being mixed and adsorbed in 550 mL of control sediment (67% water content) from an uncontaminated area in the Baltic proper (stn. 1). The extracts were slowly blended into the uncontaminated sediment while stirring continuously. The sediment extracts were then stirred for an additional 20 min. Before addition of 100 mL contaminated sediment to each microcosm, 300 mL control sediment was allowed to settle in the aquaria for 2 d. This revealed a sediment depth of 3.5 cm in the microcosms and corresponds to the natural habitat depth of the amphipods. Water flow was allowed to flow at a rate of 2.2 L h^{-1} after 4 d. To avoid effects of acetone the test organisms (50 preadults, sex unknown) were added after one more day. Examination of 200 individuals showed no deviation from the expected 1:1 sex ratio.

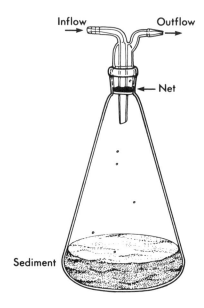

Figure 1. The microcosm system. To prevent surface contact and escape of amphipods, the outlet was covered with a 0.25 mm mesh net (from Sundelin 1983).

Chemical Fractionation

The wet sediment samples were placed in preextracted cellulose thimbles and Soxhlet extracted with toluene for 24 h. The apparatus was equipped with a Dean-Stark trap for collecting water. After evaporation of the toluene (<10 torr; 35-40°C), the residue was cleaned on a silica column with *n*-hexane as a mobile phase. To fractionate the extracts, we used an HPLC system (Zebühr *et al.* 1989) consisting of a MERCK-Hitachi L-6200 pump, Rheodyne 7125 valve injector, Rheodyne 7067005 automatic valve station and switching valves, and MERCK-Hitachi 1-4200 UV-VIS detector. A semipreparative µBondpak® amino column (300 x 7.8 mm, Waters) was used with *n*-hexane as the mobile phase. Three fractions were collected: aliphatics/monoaromatics, diaromatics and polyaromatics. At a rate of 3 mL min^{-1}, the aliphatics/monoaromatics were eluted first, followed by the diaromatics in a forward direction (Zebühr et al. 1989). The column was then back flushed (5 mL min^{-1}) and the polyaromatics eluted as one narrow peak. The compounds were monitored at 254 nm. Low-boiling aliphatics/monoaromatics were probably lost during the evaporation steps.

Monoporeia Analyses

After an exposure period of 5 months, females were examined in terms of fertilization frequency, fecundity (eggs/female), completely or partly dead brood, infested parasites and other visible deviations from normal appearance. Eggs and embryos were removed from the female and analyzed, using a stereomicroscope (50x magnification), in terms of embryo development, undeveloped/unfertilized eggs, enlarged embryos, two degrees of malformation (grade I: moderately malformed, assumed to live until hatching; grade II: severely malformed, assumed to die before hatching) and dead eggs.

Statistical Evaluation

The statistical evaluations were performed by logit analyses (malformations and dead brood) (Aldrich and Nelson 1984; Demaris 1992) and Holm's procedure, a simple sequentially rejective multiple test procedure (Holm 1979). One-way analysis of variance was used to test fecundity and egg development. The conventional significance level of $\alpha < 0.05$ was applied.

RESULTS

The experiment was terminated after 5 months exposure, a few weeks before the first juvenile will hatch *in situ*. A comparison with the natural system *in situ* showed no deviation from expected embryo development. There was no difference between microcosms treated with extracts and the control according to reproductive success, fecundity and females infested with parasites (Table 1). Extracts from the open sea surface sediment (stn. 2) did not show any toxicity in comparison with controls (Fig. 2). In extracts of the surface sediment from the station close to the mill (stn. 3), several characteristic malformations were identified, such as weakened membranes, fat abnormalities, malformed extremities and light greyish colored eggs (Fig. 3). However, when separated to different types of malformation, undifferentiated and dead eggs, there were no significant differences in comparison with controls. The total amount of abnormal eggs and embryos was however higher in microcosms exposed to all sediment extracts except the polyaromatic fraction from the surface sediment and the aliphatic/monoaromatic fraction from 10-12 cm depth (both from the pulp and paper mill station 3). These two extracts showed borderline significance when compared with the controls (Fig. 2). The toxicity will henceforth be referred to as the total sum of abnormal eggs and embryos.

Besides different types of damage identified to eggs and embryos, females exposed to the non-fractionated extracts from station 3 (0-2 cm as well as 10-12 cm) showed a higher degree of partly or entirely dead brood than the control (Table 1). Several eggs or embryos from the brood had died and remained as a fat residue. Though insignificant, the trend was an increased frequency of females with dead brood in all extracts with the exception of the aliphatic fraction from 10-12 cm depth at station 3 (Table 1).

Toxicity of the non-fractionated extracts was higher from both depths at station 3 than each of the separate fractions. The various fractionated extracts from the surface sediment at station 3 did not differ in toxicity, while the pattern was different in the 10-12 cm depth, where compounds in the polyaromatic fraction caused more serious effects than those in the diaromatic and aliphatic fractions (Fig. 2). The polyaromatic fraction from the 10-12 cm depth also caused a higher degree of abnormal eggs and embryos

Table 1. Effects on the reproduction of Monoporeia affinis in microcosms exposed to unfractionated and fractionated sediment extracts for 5 months. Sediments (0-2 cm and 10-12 cm) near a pulp mill (station 3) were Soxhlet extracted and HPLC fractionated. From an open sea station (2) in the Bothnian Sea, the 0-2 cm depth was used. The control sediment originated from an uncontaminated area in the northern Baltic proper. Five replicates of each treatment were used. Data are given as mean ± S.D.

Station	Sediment Extract	Females n	Eggs/Female n	Females with Dead Brood %	Females with Nematode Parasite %
1	Control	14.2 ± 5.3	42.8 ± 11.5	16.0 ± 14.2	1.5 ± 1.4
2 (0-2 cm)	Diaromatic	10.2 ± 3.8	42.4 ± 12.5	16.9 ± 5.2	3.9 ± 2.7
"	Polyaromatic	9.0 ± 3.0	40.3 ± 12.5	27.0 ± 6.4	2.2 ± 2.2
"	Aliph./monoar.	14.6 ± 3.2	40.6 ± 9.9	12.3 ± 3.9	4.1 ± 2.3
"	Non-fraction.	8.8 ± 6.6	40.0 ± 10.9	15.9 ± 5.5	2.3 ± 2.3
3 (0-2 cm)	Diaromatic	16.2 ± 1.0	42.3 ± 11.6	16.0 ± 4.0	3.7 ± 2.1
"	Polyaromatic	14.0 ± 1.9	40.2 ± 12.4	17.1 ± 4.7	2.9 ± 2.0
"	Aliph./monoar.	16.0 ± 4.4	42.8 ± 11.2	19.4 ± 4.0	2.6 ± 1.8
"	Non-fraction.	12.2 ± 2.9	39.7 ± 9.4	32.7 ± 5.6	1.6 ± 1.6
3 (10-12 cm)	Diaromatic	15.4 ± 6.4	41.6 ± 11.9	20.2 ± 4.5	5.2 ± 2.5
"	Polyaromatic	14.0 ± 6.5	42.2 ± 10.4	23.8 ± 6.3	7.1 ± 4.0
"	Aliph./monoar.	15 ± 3.5	42.4 ± 10.1	19.5 ± 4.9	0
"	Non-fraction.	10.2 ± 5.6	40.4 ± 11.0	32.7 ± 6.6	3.9 ± 2.7

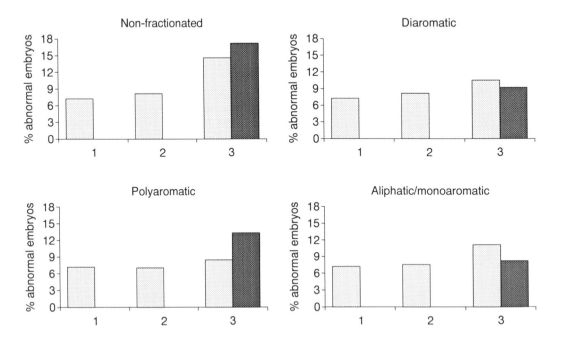

Figure 2. Frequency of abnormal eggs and embryos of the fractionated and non-fractionated extracts in the experiment. 1 = control, 2 = open sea station, 3 = station close to the pulp and paper mill. Light dotted bars = surface sediment (0-2 cm), heavy dotted bars = deeper sediment (10-12 cm).

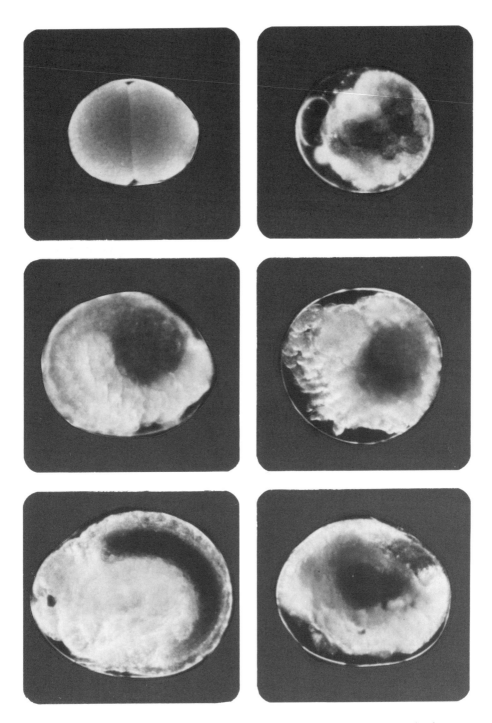

Figure 3. Normal and abnormal eggs and embryos of Monoporeia affinis during embryonic development.

than corresponding extracts from the surface sediment (Fig. 2). This indicates improved conditions in the more recently settled sediment as a possible result of less toxic discharges from the mill, when the bleaching process at the mill was changed from chlorine to ECF bleaching. There were no differences between other extracts, non-fractionated as well as diaromatic and aliphatic/monoaromatic, but borderline significance to higher toxicity of the non-fractionated extract from 10-12 cm depth compared with the 0-2 cm depth sediment was observed.

DISCUSSION

In this microcosm experiment, as well as in earlier experiments with *Monoporeia affinis,* the reproduction success, fecundity and rate of embryo development were unaffected at concentrations where frequency of abnormal eggs and embryos increased significantly (Sundelin 1983, 1989). Thus, embryogenesis seems to be even more sensitive than other variables of reproduction.

The toxicity of the non-fractionated extract was higher when compared with each fraction. When control values were subtracted from toxicity levels of each fraction, the total sum of the different fractions was of the same order as the unfractionated, indicating compounds associated with the fractionated extracts were responsible for a major part of the total effect. However, the polyaromatic fraction of the surface sediment exerts no measurable toxicity in comparison with the control, indicating the minor importance of this fraction to the observed total effect. The results indicate that a major portion of the attained toxicity was associated with compounds in fractionated extracts, in particular the aliphatic/monoaromatic and the diaromatic extracts. Knowledge about compounds in the aliphatic/monoaromatic fraction is limited but PCBs, polychlorinated dibenzodioxins and dibenzofurans (PCDDs and PCDFs) will be retained in the diaromatic fraction (Zebühr *et al.* 1989). Näf *et al.* (1992) demonstrated comparatively high concentrations of total PCDDs and PCDFs in sediment trap material from station 3 close to the mill, which suggests that these compounds can, at least partly, contribute to the demonstrated effects of the diaromatic fraction. Effects of fractions and extracts from station 3 near the pulp mill were comparatively small, when compared with sediment extracts from station 2 in the open Bothnian Sea.

Also in the deeper (10-12 cm) sediments from station 3, the major portion of harmful compounds seems to be related to the fractionated extracts once the control has been subtracted. No measurable toxicity was exerted by the aliphatic fraction, indicating low persistence of these compounds, since this fraction was toxic in the extract of the surface sediment which represents more recently deposited substances. A comparison between the two depths, 0-2 cm and 10-12 cm, demonstrated significant differences between the polyaromatic fraction of the surface sediment extract and the polyaromatic fraction of the deeper sediment extract which exerts the most serious effect of all fractionated extracts. Since there is a decrease in toxicity of the polyaromatic fraction from previously deposited sediment (10-12 cm) to the more recently deposited (0-2 cm) there are reasons to believe that the change in bleaching process at the mill has resulted in a decrease in toxic substances in the effluents from the mill. A study by Näf *et al.* (1992) has shown that levels of polycyclic aromatic hydrocarbons (PAHs) in sediment trap material from the area are just moderately increased, when compared with background levels. Furthermore, it is unlikely that discharges of PAH from the pulp mill during the time period corresponding to the 10-12 cm depth differ from recent effluents. This suggests that polyaromatic compounds other than the parent PAHs analyzed by Näf *et al.* (1992) may be responsible for the demonstrated effects.

Effects from the present study were lower in comparison with previously performed experiments where microcosms were exposed to naturally contaminated sediments from an area 2 km from a bleaching pulp and paper mill located about 100 km south of the present study (Sundelin 1989). More serious effects in terms of malformed eggs and embryos in combination with reduced fecundity were demonstrated on *Monoporeia affinis* (Sundelin 1989). Also *in situ*, comparatively higher frequencies of abnormal eggs and embryos were demonstrated in areas outside the pulp and paper mill station 3 sampled in this study (Sundelin 1992).

A calculation to compare the *in situ* levels with those obtained in micrcosms suggests higher concentrations of sediment-bound contaminants *in situ*. Each microcosm was exposed to extracts (non-fractionated or fractionated) originated from 13.6 g sediment (dwt). The total amount of sediment transferred to each microcosm was 132 g (dwt). Clearly, to simulate *in situ* concentrations, extracts from 132 g sediment (dwt) should have been used. However, compounds in extracts are concentrated to the top centimeter of the microcosm sediment layer, which means 25% of the total sediment depth. *Monoporeia affinis* feeds almost exclusively on the top surface centimeter of the sediment (Lopez and Elmgren 1989). This feeding strategy would theoretically result in exposure to concentrations of less than half of the concentrations *in situ*.

Another explanation for lower effects of extracted organic compounds compared with effects from naturally contaminated sediments might be a loss of more volatile and labile compounds during the extraction procedure. Since the level of extractable organic chlorine (EOCL) (Sundelin 1989) was substantially higher in sediment from the present study than in previously performed experiments with contaminated sediments, it can also be suggested that other sediment characteristics than extractable organic pollutants are responsible for the difference in toxicity.

The present study is the first in a series of subsequent experiments and the main purpose of this study was to identify toxic fractions and to examine the relative toxicity of fractionated extracts with various aromaticies. The extraction technique is highly efficient and is likely to increase the bioavailability of the toxic compounds in comparison with a natural sediment. However, it is not likely that this procedure has increased the relative bioavailability of the various fractions since they ought to be reabsorbed in a similar manner once *in situ*. Also, the comparison between the various stations ought to be valid since the extraction method has the same degree of efficiency on all sediments included in the present study. Further studies on the comparison between sediments spiked with toluene extracted fractions and total extracts as well as naturally contaminated sediments need to be carried out in the future. The method of using soft bottom microcosms where the sediment has been dosed with fractionated extracts has proved to be a useful tool. In order to narrow down the problem even more, concerning what kind of compounds are responsible for most of the effects of sediments from contaminated locations, the extracts can be further subfractionated in combination with chemical characterization of the compounds in the most toxic fractions

ACKNOWLEDGMENTS

This work was supported by grants from the National Swedish Environmental Protection Agency and from the Swedish Pulp and Paper Research Institute. Stig Johan Wiklund gave valuable comments on the statistical evaluation.

REFERENCES

Albaigés, J., A. Farran, M. Soler, A. Gallifa and P. Martin. 1987. Accumulation and distribution of biogenic and pollutant hydrocarbons. PCBs and DDT in tissues of Western Mediterranean fishes. Mar. Environ. Res. 22:1-18.

Aldrich, J.H., and F.D. Nelson. 1984. Linear probability, logit, and probit models. Series: Quantitative applications in the social sciences. Sage Publications, Inc., California.

Bengtsson, B.-E., and B. Bergström. 1987. A flowthrough fecundity test with *Nitocra spinipes* (Harpacticoidea Crustacea) for aquatic toxicity. Ecotox. and Environ. Safety 14:260-268.

Blanck, H., K. Holmgren, L. Landner, H. Norin, M. Notini, A. Rosemarin and B. Sundelin. 1989. Advanced hazard assessment of arsenic in the Swedish environment, p. 85-122. In L. Landner [ed.] Chemicals in the aquatic environment. Advanced Hazard Assessment. Springer Ser. Environ. Quality Comm. CM E 58.

Blomqvist, S., and L. Lundgren. 1995. A benthic sled for sampling soft bottoms. Submitted.

Chandler, G.T. 1990. Effects of sediment-bound residues of the pyrethroid insecticide fenvalerate on survival and reproduction of meiobenthic copepods. Mar. Environ. Res. 29:65-76.

Dahlman, O., R. Mörck, L. Johansson and F. de Sousa. 1992. Chemical composition of modern bleached kraft mill effluents, p. 47-56. In A. Södergren [ed.] Environmental fate and effects of bleached pulp mill effluents. Sweden Env. Protect Agency Report 4031.

Demaris, A. 1992. Logit modelling practical applications. Series: Quantitative applications int the social sciences. Sage Publications, Inc., California.

Dermott, R.M., and K. Corning. 1988. Seasonal ingestion rates of *Pontoporeia hoyi* (Amphipoda) in Lake Ontario. Can. J. Fish. Aquat. Sci. 45:1886-1895.

Elmgren, R., S. Ankar, B. Marteleur and G. Ejdung. 1986. Adult interference with postlarvae in soft sediments: The *Pontoporeia-Macoma* example. Ecology 67:827-836.

Giere, O.G. 1993. Meio-Benthology, the microscopic fauna in aquatic sediments, 328 pp. Springer-Verlag, Berlin.

Hill, C., M.A. Quigley, J.F. Cavaletto and W. Gordon. 1992. Seasonal changes in lipid content and composition in the benthic amphipods *Monoporeia affinis* and *Pontoporeia femorata*. Limnol. Oceanogr. 37(6):1280-1289.

Holm, S. 1979. A simple sequentially rejective multiple test procedure. Scand. J. Statist. 6:65-70.

Lopez, G. and R. Elmgren. 1989. Feeding depths and organic absorption for the deposit-feeding benthic amphipods *Pontoporeia affinis* and *Pontoporeia femorata*. Limnol. Oceanogr. 34(6):982-991.

Moore, J.W., V.A. Beaubien and D.J. Sutherland. 1979. Comparative effects of sediment and water contamination on benthic invertebrates in four lakes. Bull. Environ. Contam. Toxicol. 23:840-847.

Näf, C., D. Broman, H. Pettersen, C. Rolff and Y. Zebühr. 1992. Flux estimates and pattern recognition of particulate polycyclic aromatic hydrocarbons, polychlorinated dibenzo-*p*-dioxins, and dibenzofurans in the waters outside various emission sources on the Swedish Baltic coast. Environ. Sci. Tech. 26(7):1444-1457.

Samoiloff, M.R. 1980. Action of chemical and physical agents on freeliving nematodes, p. 81-98. In B.M. Zuckerman [ed.] Nematodes as biological models, Vol. 2. Aging and other model systems. Academic Press, London.

Samoiloff, M.R., S. Schulz, J. Jordan, K. Denich and E. Arnolt. 1980. A rapid simple long-term toxicity assay for aquatic contaminants using the nematode *Panagrellus redivivus*. Can. J. Fish. Aquat. Sci. 37:1167-1174.

Segerstråle, S.G. 1962. Investigations on Baltic populations of the bivalve *Macoma baltica* (L.) Part II. What are the reasons for the periodic failure of recruitment in the deeper waters of the inner Baltic. Comm. Biol. Soc. Sci. Fenn. 25:1-26.

Segerstråle, S.G. 1971. Light and gonad development in *Pontoporeia affinis*, p. 573-581. In D. J. Crisp [ed.] Fourth European Marine Biology Symposium, Cambridge University.

Segerstråle, S.G. 1973. Results of bottom fauna sampling in certain localities in the Tvärminne area (inner Baltic) with special reference to the so-called *Macoma-Pontoporeia* theory. Comm. Biol. Soc. Sci. Fenn. 67:3-12.

Sundelin, B. 1983. Effects of cadmium on *Pontoporeia affinis* (Crustacea: Amphipoda) in laboratory soft-bottom microcosms. Mar. Biol. 74:203-212.

Sundelin, B. 1984. Single and combined effects of lead and cadmium on *Pontoporeia affinis* (Crustacea, Amphipoda) in laboratory soft-bottom microcosms, p. 237-258. In G. Persoone, E. Jaspers and C. Claus [ed.] State Univ. Ghent and Inst. Mar. Scient. Res., Bredene, Belgium. Vol. 2. 588 p. Ecotox. Testing for the Mar. Env.

Sundelin, B. 1988. Effects of sulphate pulp mill effluents on soft bottom organisms - a microcosm study. Wat. Sci. Tech. 20 (2):175-177.

Sundelin, B. 1989. Ecological effect assessment of pollutants using Baltic benthic organisms. Ph.D thesis, University of Stockholm.

Sundelin, B. 1992. Effect monitoring in pulp mill areas using benthic macro- and meiofauna. In A. Södergren [ed.] Environmental fate and effects of bleached pulp mill effluents. Sweden Env. Protect. Agency Report 4031.

Sundelin, B., and R. Elmgren. 1991. Meiofauna of an experimental soft bottom ecosystem - effects of macrofauna and cadmium exposure. Mar. Ecol. Prog. Ser. 70:245-255.

Södergren, A. [ed.]. 1989. Biological effects of bleached pulp mill effluents. Nat. Swedish Environ. Protect. Bd. Report 3558.

Södergren, A. [ed.]. 1993. Bleached pulp mill effluents. Composition, fate and effects in the Baltic Sea. Sweden Environ. Protect. Agency Report 4047.

Thorman, S. 1987. Miljökvalitetsbeskrivning av Bottniska viken och dess kustområden. Nat. Swedish Environ. Protect. Bd. Report 3363.

Zebühr, Y., C. Näf, D. Broman, K. Lexén, A. Colmsjö and C. Östman. 1989. Sampling techniques and clean up procedures for some complex environmental samples with respect to PCDDs and PCDFs and other organic contaminants. Chemosphere 19:39-44.

Ölafsson, E., and R. Elmgren. 1991. Effects of biological disturbance by benthic amphipods *Monoporeia affinis* on meiobenthic community structure: a laboratory approach. Mar. Ecol. Prog. Ser. 74:99-107.

EXAMINATION OF BLEACHED KRAFT MILL EFFLUENT FRACTIONS FOR POTENTIAL INDUCERS OF MIXED FUNCTION OXYGENASE ACTIVITY IN RAINBOW TROUT

L. Mark Hewitt[1*], John H. Carey[2], D. George Dixon[1] and Kelly R. Munkittrick[3]

[1]Department of Biology, University of Waterloo, Waterloo, Ontario N2L 3G1 Canada
[2]National Water Research Institute, Environment Canada, 867 Lakeshore Road, Burlington, Ontario L7R 4A6 Canada
[3]Great Lakes Laboratory for Fisheries and Aquatic Sciences, Department of Fisheries and Oceans, 867 Lakeshore Road, Burlington, Ontario L7R 4A6 Canada
*Present Address: Great Lakes Laboratory for Fisheries and Aquatic Sciences, Department of Fisheries and Oceans, 867 Lakeshore Road, Burlington, Ontario L7R 4A6 Canada

The induction of mixed function oxygenase (MFO) enzymes in fish exposed to pulping effluents is well documented but the responsible compounds are unidentified. Liver ethoxyresorufin-O-deethylase (EROD) activity was determined in rainbow trout after exposures to fractions generated from the effluent of a modernized bleached kraft pulp and paper mill. Exposures to whole and filtered (<1 µm) effluent, resuspended solids, and two fractions generated from nanofiltration were conducted for primary effluent, effluent after secondary treatment (aerated lagoon), and effluent collected during a scheduled pulping shutdown. For each fraction exposed, effluent constituents were detected using GC with ECD, and FID as well as GC-MS. Exposure concentrations were compared with EROD activities to evaluate their potential to cause induction. Resin acids, fatty acids, bacterial fatty acids, terpenes, chlorophenolics, aliphatic alkanes, plant sterols, and chlorinated dimethylsulfones for which authentic standards were available were eliminated as potential EROD inducers. Several chlorophenolics, including tetrachloroguaiacol, exhibited a correlation with observed induction patterns but subsequent exposures to pure tetrachloroguaiacol failed to cause MFO induction. The correlations exhibited by these compounds may indicate the potential source of the chemicals responsible for MFO induction.

INTRODUCTION

Physiological responses in wild fish populations exposed to pulping discharges have been well documented (Förlin et al. 1985; Andersson et al. 1988; Rogers et al. 1989; Munkittrick et al. 1991; Hodson et al. 1992). These responses include changes in the ability of fish to control their production of reproductive hormones (Van Der Kraak et al. 1992), which has been linked to impaired reproductive performance (McMaster et al. 1991). These hormone effects have generally occurred in parallel with increased activity (induction) of the hepatic mixed function oxygenase (MFO) enzymes that are involved in phase I metabolism of exogenous compounds. Induction is known to occur following exposure to several environmental contaminants such as polynuclear aromatic hydrocarbons (PAHs) (Klotz et al. 1983), polychlorinated biphenyls (Addison et al. 1978), and polychlorinated dibenzo-p-dioxins and dibenzofurans (Vodicnik et al. 1981). Since these compounds are also highly toxic, increased MFO activity is interpreted as both chemical exposure and as potential toxicity. Because MFO induction occurs rapidly after exposure to pulp mill effluent, it may provide an early indicator of potential sublethal toxicity. Previous investigations have determined that MFO induction is not associated with a particular pulping or bleaching process or degree of effluent treatment (Munkittrick et al. 1992; Martel et al. 1993; Munkittrick et al. 1994). Furthermore, recent changes in pulp bleaching and effluent treatment processes, instituted to meet increasingly stringent environmental regulations, have not eliminated induction and reproductive effects at modernized facilities (Munkittrick et al. 1992; Servos et al. 1992). The chemical(s) responsible for the MFO induction associated with pulping effluents is currently unknown.

Final effluents discharged by pulp mills are a combination of wastewaters from many internal chemical processes and are highly complex. Effluents contain both inorganic and organic low molecular weight extractives as well as a large proportion of high molecular weight (>1 kDa) material (Brownlee and Strachan 1977). Several investigations have utilized gas chromatography (GC) to characterize the major extractive families in effluents and receiving waters of different pulping processes (Voss 1983, 1984; Leuenberger et al. 1985; Wilkins and Panadam 1987; McKague et al. 1988; Stuthridge et al. 1990). In addition to the lignin-derived phenols, guaiacols, catechols, and vanillins, other chemical classes such as fatty acids, resin acids, terpenes, dimethylsulfones, and sterols are frequently encountered. These materials can also be chlorinated to various degrees if chlorine-based bleaching is used to brighten pulp (Kringstad and Lindström 1984).

Since known MFO inducers are extractable, monomeric, low molecular weight (<1 kDa) substances, it is likely that the inducer(s) associated with pulp mill effluent possesses these characteristics. The aim of this study was to determine whether constituents of the major extractive families could be correlated with MFO induction by exposing fish to chemically characterized effluent fractions. Previous experience using nanofiltration in our laboratory (Martin et al. 1995) indicated that this would be an appropriate technique for removing high molecular weight interferences and efficiently fractionating the large volumes of effluent required for bioassays.

EXPERIMENTAL SECTION

Materials

All gases were ultra high purity carrier grade (Canox, Mississauga, ON), solvents were pesticide residue grade distilled in glass (Baxter Diagnostics Corp., Canlab Division, Mississauga, ON), and concentrated acids were analytical grade (BDH Chemical Co., Toronto, ON). Bioassay holding water (hardness, 130 mg L^{-1} $CaCO_3$; pH 8) was from the local water treatment plant and was dechlorinated by charcoal filters and a sodium sulfite drip system. Sources of other equipment and reagents are cited upon initial mention.

A total of three effluent samples obtained from different stages of the treatment system of a bleached kraft pulp and paper mill were subjected to filtration and fractionation by nanofiltration (NF) as summarized in Fig. 1. Each fraction generated, including whole effluent, was exposed to rainbow trout (*Oncorhynchus mykiss*) in 96 h static bioassays, after which MFO activity was determined using a hepatic ethoxyresorufin-*O*-deethylase (EROD) assay. Subsamples of effluent fractions were extracted and analyzed by GC using authentic standards of expected chemical families (Fig. 2). Comparisons between exposure concentrations and EROD activity were used to assess the induction potential of confirmed constituents.

Effluent samples were obtained from the treatment system of a bleached kraft pulp and paper mill located in Espanola, ON, Canada. The mill produces approximately 1000 air dried tonnes d^{-1} of pulp from two lines. The first line produces softwood pulp (jackpine) in a continuous digester and, at the time of this study, bleached with an OD_CE_OHD (O, oxygen delignification; D_C, chlorination with 60% chlorine dioxide substitution for chlorine; E_O, caustic extraction with oxygen addition; H, reaction with sodium or calcium hypochlorite; D, reaction with aqueous chlorine dioxide) sequence. The second line processes a blend of dense hardwood (maple and birch), softwood, and poplar in batch digesters and bleached with an OC_DEHD sequence (C_D: 30% chlorine dioxide substitution). Independent of the pulping process, two paper machines produce 170 tonnes d^{-1} of specialty and fine papers. The process streams from the pulp and paper lines are combined and enter a primary and secondary treatment system. Primary treatment utilizes a settling basin with a 12 h hydraulic retention time (HRT). Secondary treatment consists of a 760,000 m^3 lagoon equipped with 22 aerators and a HRT of 5.5-6.0 d (Munro et al. 1990). The mill discharges approximately 100,000 m³ d⁻¹ of effluent. Grab samples of 400 L were taken in April, 1992 at the exit of each effluent treatment stage; primary effluent, secondary effluent (corresponding to the primary sample, based on lagoon HRT), and secondary effluent at the end of a scheduled 10 d pulping shutdown (paper mill still operational). Effluent samples were kept at 2°C and transported to the laboratory in sealed 200 L polyethylene barrels and were received within 2 d.

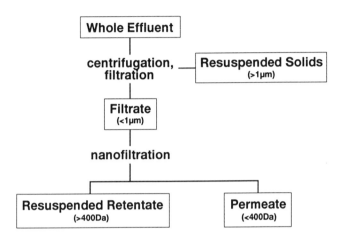

Figure 1. Fractions generated for each effluent sample collected. Each fraction was subjected to extraction and analysis as depicted in Fig. 2. The MFO induction potentials for each fraction were determined using rainbow trout waterborne bioassays.

Effluent Fractionation

Effluent fractionation was initiated immediately upon sample arrival using the NF method described in Martin et al. (1995). Briefly, suspended solids were removed through a two-step process prior to NF. Effluent was pumped through a continuous flow centrifuge (Westfalia Separator, Firing Industries, Niagara-on-the-Lake, ON) operating at 10,000 rpm with a product flow rate of 2 L min⁻¹. Solids were recovered and weighed for each sample. Using nitrogen, the centrifugate was subsequently pressure filtered through 1 µm glass fiber filters (Gelman Sciences, Rexdale, ON) held in a Millipore 316 Stainless Sanitary Filter Holder (Millipore Corporation, Minneapolis, MN). Filters were saved and filtering requirements were recorded for each sample. A maple sap concentrator (Les Equipments Lapierre Inc., St. George de Beauce, PQ) modified with a greaseless ceramic-piston high-pressure pump was used for NF. The membrane (NF-40, 400 Da nominal molecular weight cut-off, Filmtec Corporation, Minneapolis, MN) was cleaned, and the integrity checked, before each fractionation. Cleaning used a 30°C 10 g L⁻¹ solution of P3 Ultrasil 10 (Cleans Aid, Toronto, ON) followed by a copious flush with 25°C potable water under benchmark conditions (product flow rates, 2 L min⁻¹; membrane pressure, 225 psi). The membrane integrity was checked with NaCl, $MgSO_4 \cdot 7H_2O$ and d-raffinose pentahydrate solutions. Two modes of NF were used to process each 200 L effluent sample under benchmark conditions. Filtrate was initially passed through the membrane and both the retentate (>400 Da) and permeate (<400 Da) were collected separately (transfer mode). The retentate was then continually recycled while permeate collection was maintained (concentrate mode) until the retentate volume was reduced to the internal apparatus volume (ca. 10 L). Permeate collection was then stopped while retentates were dialyzed eight times with 100 L of distilled water.

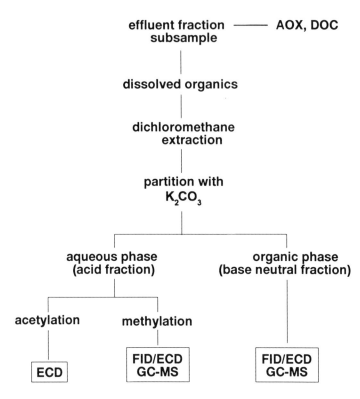

Figure 2. Extraction and analytical subfractionation performed on each effluent fraction used in rainbow trout exposures.

Bioassay Procedures

Immature rainbow trout (5 to 10 g) were obtained from Rainbow Springs Hatchery, Thamesford, ON and acclimated to 15°C for >10 d. Fish were exposed in static, 96 h non-renewed regimes to undiluted effluent fractions at a loading density of approximately 10 g L^{-1}. Fish were fed *ad libitum* until 48 h prior to exposures; fish were not fed during exposures. Triplicate aerated exposures were conducted in 12 to 16 L covered glass aquaria held in temperature regulated water baths using a randomized block design. Dissolved oxygen, pH, and temperature were measured daily. Over the course of the exposures, mean ± SE tank temperatures were 14.7 ± 0.8°C and the dissolved oxygen content was 8.8 ± 1.1 mg L^{-1}. pH values ranged from 7.1 to 8.4 and were not adjusted. For fish exposures, dialysed retentate and recovered solids were resuspended to their estimated effluent proportions in holding water. Dialyzed retentate was diluted so that the absorbance of the resulting solution at 600 nm equalled that of the corresponding filtrate. Centrifuged solids were combined with 2 cm^2 sections of fouled filters and were resuspended according to recoveries from centrifugation and filtration.

Fish were exposed for 96 h to tetrachloroguaiacol (>99% purity, Helix Biotech, Vancouver, BC) by waterborne and intraperitoneal (i.p.) injection. Tetrachloroguaiacol was dissolved in methanol and spiked into holding water for waterborne exposures of 5 and 500 μg L^{-1}. For i.p. injection, tetrachloroguaiacol was dissolved in corn oil, and fish were anaesthetized with a solution of tricaine methane sulfonate (Syndel Laboratories, Vancouver, BC) and dosed at 10 and 100 mg kg^{-1}. For each series of exposures, three replicates of dilution water, with one replicate containing an unused filter, were employed as reference treatments. One tank of fish was anaesthetized and injected i.p. with 50 µL of a 1 mg mL^{-1} solution of β-napthoflavone (BNF) (Sigma Chemical Co., St. Louis, MO) in corn oil and sampled after 2 d to ensure induction capability. During sampling, fish were killed by concussion and spinal severance. Livers were removed and placed in cryovials, immediately homogenized using a microhomogenizer (Kontes; Baxter Diagnostics Corp.) in 1 mL of cold HEPES-KCl buffer (0.02 M HEPES, 0.15 M KCl, pH 7.5), and centrifuged at 11,500 rpm for 20 min at 3°C. Supernatants containing microsomes were removed and stored in cryovials at -80°C until assayed for EROD activity following the spectrofluorometric assay of Hodson et al. (1991).

Statistics

EROD activities were log transformed and treatment effects on EROD activity were determined by one-way analysis of variance followed by Tukey's test and post hoc contrast comparisons (SYSTAT 1992).

Fraction Chemical Characterizations

Samples for dissolved organic carbon (DOC), adsorbable organic halogen (AOX), and dissolved organics were taken from thoroughly mixed effluent fractions and reconstituted exposure solutions prior to fish addition. AOX and DOC samples were acidified with respective additions of concentrated nitric or sulfuric acid, sealed in 125 mL polyethylene bottles and frozen (-20°C) until analysis. AOX was quantified with a Mitsubishi TOX-10 organohalogen analyzer, while DOC was determined with a Beckman 915-B total carbon analyzer (Tokats 1987). Samples for dissolved organics were collected in 1 L amber glass bottles, acidified to pH 2 with concentrated sulfuric acid and spiked with 2,4-dibromophenol in methanol. Then 100 mL dichloromethane was added and the solutions stirred. Bottles were sealed with caps lined with hexane-rinsed foil and stored at 4°C in darkness until extraction (Fig. 2). Upon removal from storage, 500 mg of L-ascorbic acid was added to each solution to serve as a phenolic preservative (Allard et al. 1985). Samples were extracted three times with 100 mL dichloromethane and emulsions were broken by centrifugation. The combined extracts were reduced in volume under vacuum at room temperature and partitioned with 0.1 M potassium carbonate to separate acidic components. The organic phases (base neutral fraction) were dried through anhydrous sodium sulfate. The aqueous extracts (acid fraction) were collected in 100 mL graduated cylinders and were gently sparged with nitrogen for 10 min to ensure complete mixing. One tenth of each acid fraction was removed for acetylation and chlorophenolic analysis (Chau and Coburn 1974). The remainder was acidified, back extracted with dichloromethane and methyl esterified using freshly prepared diazomethane (Lindström and Nordin 1976). All analytical fractions were reduced in volume and solvent exchanged to iso-octane before final evaporation under nitrogen to 1 mL and analysis by GC.

Gas Chromatography

Each analytical fraction was profiled by GC with flame ionization detection (FID), dual column electron capture detection (ECD), and gas chromatography-mass spectrometry (GC-MS); chlorophenolics were characterized separately by dual column ECD only. All GC fused silica capillary columns (Supelco Canada, Oakville, ON) were bonded to a 0.25 µm phase thickness, were 30 m X 0.25 mm i.d. and injections were made through split/splitless injectors operating in the splitless mode. FID analyses used a Hewlett Packard 5890 gas chromatograph with a J&W DB-5 column; acid and base neutral ECD analyses used a Hewlett Packard 5890 II gas chromatograph with injection split between a DB-5 and an OV-1 column; chlorophenolics were analyzed on a Hewlett Packard 5890 II gas chromatograph with injection split between a DB-5 and a J&W DB-17 column. GC-MS analyses were performed on a Hewlett

Packard 5989A gas chromatograph and a DB-5 column interfaced to a quadruple MS Engine using electron impact (EI) ionization.

All injections were 2 µL and purge delays were 0.5 min. Acidic component determinations were conducted using FID, ECD, and GC-MS under identical conditions. Oven temperatures were set at 90°C for 2 min, then programmed at 4°C min^{-1} to 280°C and maintained for 15 min. Injection temperatures were 280°C and the FID and ECD temperatures were 280°C and 350°C, respectively. Hydrogen and helium were used as the FID and ECD carriers, respectively; head pressures were 15 psi and both used a nitrogen make-up. Helium was used as the GC-MS carrier with a 9 psi head pressure. For chlorophenolic analyses, oven temperatures were held at 90°C for 2 min, increased at 3°C min^{-1} to 220°C, then 10°C min^{-1} to 280°C for a 10 min hold. Full scan EI mass spectra were obtained by scanning from 40 to 550 m/z at a rate of 0.84 scans s^{-1} and a scan threshold of 100. The electron energy and electron multiplier voltages were 70 eV and 2480 V, respectively; source and quadrupole temperatures were 250°C and 100°C, respectively. Autotuning was performed by twice optimizing the m/z 69, 219, and 502 abundances of perfluorotributylamide before each chromatographic sequence. With the exception of oven temperatures, base-neutral analyses used the same chromatographic conditions as described above. Base-neutral oven temperatures were initially set at 90°C for 2 min, then programmed at 2°C min^{-1} to 280°C, and held for 25 min. Authentic standards of expected effluent constituents were prepared and analyzed against sample extracts for confirmation and quantification of individual constituents (Hewitt 1993). A summary of the compound classes that were quantified and used for confirmation is given in Fig. 3.

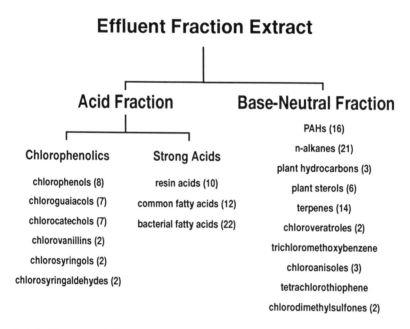

Figure 3. Chemical families quantified in exposed effluent fractions using GC. Values in parentheses represent the number of isomers for which there were authentic standards.

Internal recoveries of 2,4-dibromophenol were 105 ± 2% (mean ± SE). In addition to internal recoveries, compound class recoveries were determined with method spikes (Table 1). Reported concentrations have not been adjusted for recovery. Method detection limits (MDLs), determined for several compound classes and for each detector (Hewitt 1993), were determined by the method of Keith et al. (1983) using derivitization blanks. MDLs ranged from 0.02 µg L^{-1} for tetrachloroguaiacol using ECD to 9.9 µg L^{-1} for β-sitosterol using full scan GC-MS.

Table 1. Mean (± SE) percent method recoveries representative of the compound classes quantified.

Compound or Class	Number in Mixture	Recovery (%)
resin acid mixture	8	66 ± 3
tetradecanoic acid	-	69
hexadecanoic acid	-	63
chlorophenol mixture	8	122 ± 3
chloroguaiacol mixture	7	95 ± 3
chlorocatechol mixture	7	120 ± 25
chlorosyringol and chlorosyringaldehyde mixture	4	86 ± 3
chlorovanillin mixture	2	127 ± 19
1,1,3-trichlorodimethylsulfone	-	61
β-sitosterol	-	118

RESULTS

Rainbow Trout Hepatic EROD Activity

Fish injected with BNF demonstrated high levels of induction in all experiments, indicating that fish exposed to effluent fractions were capable of induction. No significant effects were noted between reference exposures containing unused filters and those in which filters were absent ($p = 0.17$). Respective mortalities of 67 and 28% occurred in fish exposed to whole and filtrate fractions of primary effluent after 2 d; surviving fish were sampled immediately for determination of EROD activity. Field trials have shown that 2 d EROD activities are generally equivalent to 4 d activity (van den Heuvel et al. 1995), which would indicate that these values are not underestimating 4 d induction values.

With primary effluent, high EROD induction was confined to whole and filtrate fractions; primary permeate exhibited slight induction relative to reference values ($p = 0.004$; Fig. 4). With secondary effluent all fractions showed induction, but whole and filtrate fractions showed the highest induction ($p < 0.001$; Fig. 4). For the sample obtained during the pulping shutdown, only whole ($p < 0.001$) and filtered ($p = 0.03$) effluent showed induction relative to reference activities. The mean EROD activity in fish exposed to whole shutdown effluent was 25% of the induction associated with the secondary effluent sample obtained during normal mill operation.

Chemical Characterization of Exposure Fractions

The relative distribution of AOX and DOC among the fractions of each effluent was similar, with the exception of the AOX concentration in the secondary solids exposures (Fig. 5). The AOX and DOC concentrations were reduced by approximately 50% in the secondary effluent sample relative to the corresponding primary sample; a further 10% reduction was apparent with the shutdown sample. NF consistently resulted in high rejections of both AOX and DOC. With the exception of the secondary effluent solids fraction, AOX levels indicated that fish were exposed to greater effluent equivalents in the solids and retentate exposures than in the whole effluent. This was also evident with retentate DOC concentrations.

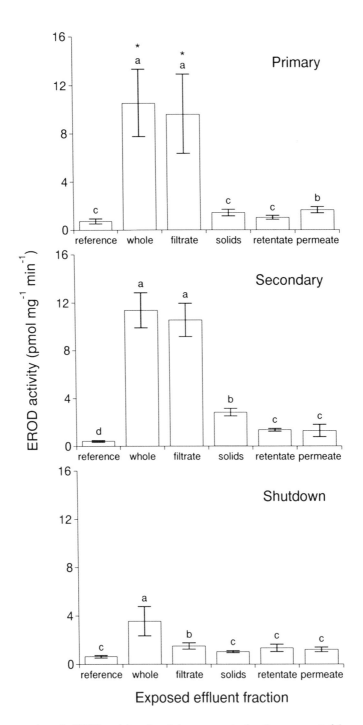

*Figure 4. Rainbow trout hepatic EROD activity after 4 d exposures to fractions generated from samples of primary, corresponding secondary, and shutdown effluents. Values are replicate means ± SE. * denotes 2 d activity. Bars showing the same alphabetical superscripts are not significantly different ($p<0.05$).*

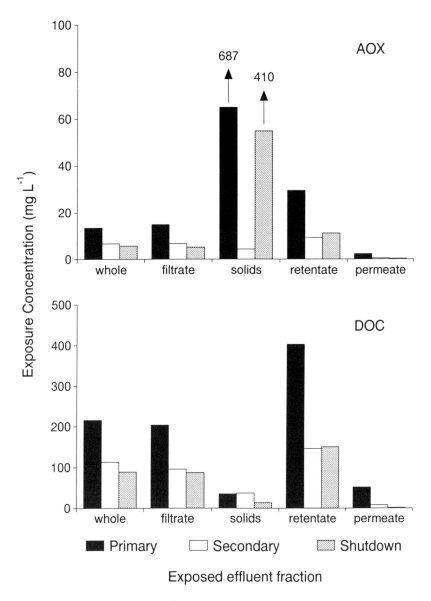

Figure 5. AOX and DOC concentrations (mg L^{-1}) determined for exposed effluent fractions.

Several compound classes were confirmed and quantified in the exposure solutions by GC (Tables 2 to 5). Generally, the major constituents evident in the profiles of each effluent fraction were confirmed with authentic standards. Resin acids, fatty acids, bacterial fatty acids, chlorophenolics, unsaturated straight chain alkanes, a plant sterol, and several chlorinated neutrals were the most consistent materials detected in the effluent samples collected. Sixteen priority PAHs (detection limit 0.3 µg L^{-1}), Arochlors 1232, 1248, and 1262 (detetection limit 0.02 µg L^{-1}), and selected common plant hydrocarbons (detection limit 0.6 µg L^{-1}) were not detected in any of the effluent samples. The sole terpenoid detected was α-terpineol; trace levels were found in primary effluent only. The distinct behavior of each chemical family during effluent treatment and NF facilitated the assessments for their induction potential.

Table 2. Resin acid concentrations in effluent exposures. Concentrations are in µg L^{-1} and have not been adjusted for recovery. Non-detectable levels designated by (-). Resin acids were eliminated as potential EROD inducers after comparisons to secondary effluent (high induction) and primary solids (no induction). Shaded fractions showed highest EROD induction.

Resin Acid	Primary			Secondary			Shutdown	
	whole	filtrate	solids	whole	filtrate	solids	whole	filtrate
pimaric	58.2	50.1	42.7	-	-	-	-	-
sandaracopimaric	18.2	16.5	10.9	-	-	-	-	-
isopimaric	166.6	132.9	181.5	-	-	-	-	-
dehydroabietic	312.5	352.7	71.8	1.8	2.6	-	1.8	3.8
abietic	363.8	216.7	181.6	-	-	-	-	-
14-chlorodehydroabietic	-	-	645.6	-	-	-	-	-
12,14-dichlorodehydroabietic	787.1	611.0	417.8	0.5	-	-	-	-

Table 3. Fatty and bacterial acid concentrations in effluent exposures. Concentrations are in µg L^{-1} and have not been adjusted for recovery. Shaded fractions showed highest EROD induction. Non-detectable levels designated by (-). Fatty and bacterial acids were eliminated as candidates for EROD induction based on losses during secondary treatment (no loss in EROD induction) or high concentrations in solids exposures (after loss of EROD induction).

Fatty or Bacterial Acid	Primary			Secondary			Shutdown		
	whole	filtrate	solids	whole	filtrate	solids	whole	filtrate	solids
cis-9-hexadecenoic	54.7	21.9	33.5	25.0	3.9	17.9	14.8	-	22.4
hexadecanoic	106.7	51.1	85.2	15.3	10.3	7.9	-	-	-
octadecenoic	38.2	12.3	35.1	15.3	3.3	13.9	2.2	-	7.8
9,10-epoxystearic	214.2	165.4	-	-	-	-	-	-	-
eicosanoic	41.3	22.2	42.4	-	-	-	-	-	-
2-hydroxydecanoic	4.4	2.6	-	-	-	-	-	-	-
dodecanoic	7.3	4.2	0.5	1.9	1.9	0.5	-	2.0	0.5
13-methyltetradecanoic	-	-	-	4.3	1.6	2.7	-	-	-
pentadecanoic	3.6	2.0	2.1	-	-	-	-	-	-
heptadecanoic	6.6	3.5	5.4	-	-	-	-	-	-
cis-9,12-octadecadienoic	31.5	18.4	58.5	-	-	-	-	-	-
cis-9-octadecenoic	38.2	7.8	35.1	-	-	-	-	-	-
nonadecanoic	7.7	4.7	7.5	-	-	-	-	-	-

Table 4. Chlorophenolic concentrations in effluent exposures. Shaded fractions showed highest EROD induction. Concentrations are in µg L^{-1} and have not been adjusted for recovery. Non-detectable levels designated by (-). Chlorophenolics (except tetrachloroguaiacol) were eliminated as potential inducers based on reductions after secondary treatment or high concentrations in permeate of primary treated effluent.

Compound	Primary			Secondary		Shutdown	
	whole	filtrate	perm.	whole	filtrate	whole	filtrate
2,4-dichlorophenol	2.3	1.6	1.7	0.8	0.8	0.5	0.1
2,4,6-trichlorophenol	8.5	7.6	7.1	1.0	1.2	1.7	2.0
4,6-dichloroguaiacol	1.1	1.0	0.9	1.0	1.0	0.7	0.8
3,4-dichloroguaiacol	1.6	1.6	1.3	0.9	1.0	0.6	0.8
4,5-dichloroguaiacol	11.5	10.5	8.0	1.0	1.6	0.7	1.1
2,3,4,6-tetrachlorophenol	0.8	0.9	0.7	0.2	0.2	0.2	0.2
6-chlorovanillin	16.9	13.8	11.8	1.7	2.0	1.3	2.3
3,4,5-trichloroguaiacol	6.7	6.6	3.5	4.0	4.1	2.8	2.7
4,5,6-trichloroguaiacol	4.7	4.5	3.0	1.7	1.6	1.0	0.9
5,6-dichlorovanillin	13.7	13.7	8.9	0.7	1.0	0.5	1.4
pentachlorophenol	0.2	0.3	0.1	0.1	0.1	0.1	0.1
tetrachloroguaiacol	4.0	4.2	1.6	2.4	2.3	2.0	2.0
2,6-dichlorosyringaldehyde	2.0	2.7	0.4	0.6	0.6	0.3	0.3

Table 5. Concentrations of neutral constituents in effluent exposures. Concentrations are in µg L^{-1} and have not been adjusted for recovery. Non-detectable levels designated by (-). Shaded sections showed highest EROD induction. Most neutrals were eliminated as potential inducers based on declines after secondary treatment or high concentrations in solids exposures. Dichlorodimethylsulfone was eliminated based on high concentrations in primary and secondary permeate (11.4, 7.1 µg L^{-1}, respectively).

Compound	Primary			Secondary			Shutdown		
	whole	filtrate	solids	whole	filtrate	solids	whole	filtrate	solids
1,1-dichlorodimethylsulfone	15.2	8.7	-	14.7	8.3	-	10.4	5.8	-
1,1,3-trichlorodimethylsulfone	0.9	0.1	-	0.1	0.1	-	-	-	-
3,4,5-trichloroveratrole	-	-	-	0.4	0.4	-	0.4	0.3	-
4,5,6-trichlorotrimethoxybenzene	0.3	0.3	0.1	0.1	0.1	-	0.1	0.1	-
tetrachloroveratrole	0.1	0.1	-	0.1	0.1	-	0.1	-	-
tetrachlorothiophene	0.2	0.2	0.1	-	-	-	-	-	-
2,4,6-trichloroanisole	0.1	0.1	-	0.05	0.04	-	-	-	-
octadecane (C_{18})	12.8	-	8.6	4.3	-	-	1.6	1.2	-
nonadecane (C_{19})	2.6	-	1.8	1.1	-	-	-	-	-
cosane (C_{20})	11.7	-	9.3	3.2	1.6	-	-	-	-
heneicosane (C_{21})	3.5	-	3.1	-	-	-	-	-	-
eicosane (C_{22})	31.6	-	29.1	0.5	5.3	4.3	-	-	-
tricosane (C_{23})	5.8	0.7	13.8	2.7	-	1.7	-	-	-
tetracosane (C_{24})	46.7	13.1	36.0	7.9	7.8	6.9	3.1	-	6.7
pentacosane (C_{25})	16.6	4.1	11.0	-	-	-	-	-	2.2
β-sitosterol	1100.0	619.7	674.9	279.7	150.6	25.3	206.7	145.9	157.3

Evaluation of Constituents for Induction Potential

The majority of extractives quantified in the effluent samples exhibited no correlation with the induction patterns observed in exposed fish. Within each effluent type, many individual constituents were present in equivalent or greater concentration in solutions which exhibited low induction potential (solids, retentate, permeate) compared to concentrations in solutions which caused induction in exposed fish (whole, filtrate). Additionally, many compounds became either non-detectable (resin and fatty acids) or were significantly reduced in concentration in the corresponding sample of secondary effluent (n-alkanes, β-sitosterol) which caused comparable levels of induction. Effluent constituents common to both primary and secondary effluent were not considered to be responsible for induction if: i) concentrations were significantly reduced after secondary treatment since induction was not reduced; ii) concentrations in solids, retentate, and permeate exposures, which had low induction potential, were equivalent to or greater than those in whole and filtrate; and iii) concentrations in any fractions of shutdown effluent were equivalent to or greater than those in whole and filtrate. Several assumptions were made in the comparisons conducted. It was assumed that inducing compounds were identical in all effluent samples. Synergistic or antagonistic interactions and multiple compound toxicities were not considered. Whole and filtrate solutions of primary and secondary effluent were considered bioactive. Resuspended solids, retentate, and permeate solutions were considered inactive. Although induction associated with some of these exposures was statistically significant, the magnitude of this induction was low and substantially lower than the induction for active fractions.

The presence and concentrations of tetrachloroguaiacol (Table 4), 4,5,6-trichlorotrimethoxybenzene, tetrachloroveratrole, and 2,4,6-trichloroanisole (Table 5) correlated with induction patterns. With the exception of tetrachloroguaiacol, the concentrations of these compounds were near the detection limits determined for this study. The reliability of quantitations decreases near detection limit values (Keith *et al.* 1983). Tetrachloroguaiacol was present in concentrations well above detection limits and more confidence is therefore associated with the levels measured and the correlation with hepatic MFO activity in exposed fish. Subsequent bioassays using waterborne exposures at and exceeding effluent concentrations as well as i.p. injection of tetrachloroguaiacol demonstrated no induction of MFO activity after 4 d (Fig. 6).

DISCUSSION

Similar induction patterns were observed among the fractions of primary, secondary, and shutdown effluent samples and induction was primarily associated with whole and filtrate exposures. Low levels of induction were observed in fish exposed to the resuspended solids and filters and both fractions generated from NF. Dialyzed retentate resuspensions attempted to attain a solution which was comparable in dissolved organic matter to the corresponding filtrate. Color was chosen as a crude means of approximating this. The absorbance at 600 nm was arbitrarily chosen as a wavelength in the visible spectrum which would enable these estimations. It was expected that these estimations and the estimations made for resuspended solids would overestimate the actual proportion in filtrate and whole effluent, but this was felt preferable to foregoing any induction that could be associated with these fractions. The higher proportions in retentate and solids exposures are reflected in AOX and DOC values (Fig. 5).

Low levels of induction observed in fish exposed to retentate and permeate imply that the material responsible was either i) removed from product streams, ii) inactivated during NF, or iii) the result of interactions between two or more compounds which became separated during NF. The third possibility was considered and in a separate experiment. Fractions from the sample of secondary effluent were recombined at their estimated effluent proportions; however the results were impeded by a rapid loss of induction potential in all fractions during short term storage (Hewitt 1993).

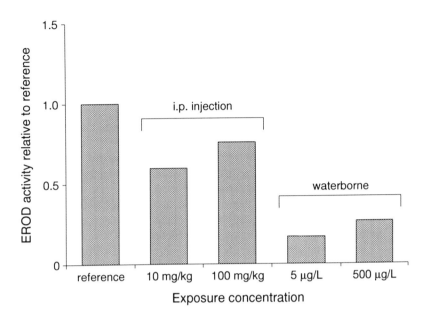

Figure 6. Relative rainbow trout hepatic EROD induction after intraperitoneal injection and waterborne exposures to tetrachloroguaiacol.

There has been some speculation about the potential of plant sterols to induce MFO activity. Denton et al. (1985) showed that female mosquitofish (*Gambusia affinis affinis*) were masculinized by exposure to microbial degradation products of the plant sterol β-sitosterol. More recently, Howell and Denton (1989) showed that microbial degradation products of stigmastanol caused similar effects. Since the sterol degradation products exhibit steroid-like activity and since MFOs are involved with the catabolism of steroids (Lu and West 1980; Lee 1988), it has been suggested that exposure to plant sterols and/or their degradation products may induce MFO activity. β-Sitosterol was the only sterol detected in the effluents of this study and comparisons of exposed concentrations with rainbow trout MFO activity showed no correlation with EROD induction.

There is little information regarding the MFO induction potential of effluent extractives. In a study comparing EROD activity in fish exposed to effluents and their components, Oikari and Lindström-Seppä (1990) exposed juvenile rainbow trout to a mixture of resin acids for 8 d at a total concentration of 350 μg L^{-1}, to dehydroabietic acid at 50 and 200 μg L^{-1}, and to tetrachloroguaiacol at 5 and 25 μg L^{-1}. No induction was associated with the resin acid mixture, dehydroabietic acid, or tetrachloroguaiacol at the concentrations exposed. Ferguson et al. (1992) also reported no induction in rainbow trout with waterborne and i.p. exposure to resin acid mixtures. These results are supported by this study, but it should be noted that previous investigations employed exposures to pure compounds and the present study measured concentrations of these materials exposed under effluent conditions.

Tetra- and higher polychlorinated debenzo-*p*-dioxin and dibenzofurans (PCDD/Fs) are potent inducers of fish EROD activity (Vodicnik et al. 1981; Parrott et al. 1995) and were initially thought to be responsible for pulp mill induction (Rogers et al. 1989; Hodson et al. 1992; Lindström-Seppä and Oikari 1990); however, there is some uncertainty regarding the exact role of these compounds. Dietary exposure to 2,3,4,7,8-penta-chlorodibenzofuran has resulted in persistent induction (Muir et al. 1990), while pulp mill induction has been shown to be relatively short lived, as evidenced during temporary mill shutdowns (Munkittrick et al. 1992) and in caging exposures (Munkittrick et al. 1995). Effluent levels of 2,3,7,8-

substituted PCDD/Fs from this mill were non-detectable (detection limit 3-10 pg L^{-1}) during the time of this study (Ministry of Environment and Energy 1992, Sudbury, Ontario, unpublished data). Although recent studies at several mills have shown that liver 2,3,7,8-TCDD toxic equivalents (TEQs) in wild fish correlate with EROD activity between mills, no relationship was found between EROD activity and TEQs in fish compared with bleached kraft mill sites (Servos et al. 1994). There were also fish below other mills which had low TEQs and elevated EROD activity. This suggests that other compounds are at least partially responsible for induction at some mills, but the contribution of dioxins at other mills remains unresolved.

The compounds displaying a correlation with the induction patterns in this study are chlorophenolic derivatives, whose structures are unlike those of known inducers. EROD induction in fish has been associated with materials which possess a molecular shape which conforms to the Ah receptor (Okey 1990). In particular, inducers have been characterized by a planar multi-ringed structure with a high degree of symmetry, whereas chlorophenolics are single-ring structures. Although the chlorophenolics implicated are likely not the compounds responsible for the MFO induction observed in this study, their correlation with induction may be suggestive regarding the source of the actual inducers. Phenolics are among the major products of the oxidation of residual lignin during bleaching. When chlorine-based bleaching is used, some of these phenolics may become chlorinated. The observed correlation with lignin-derived phenolics suggests that the inducers may also be lignin-derived. If it can be assumed that the inducer(s) in pulp mill effluents have the planar, multi-ringed structures of known inducers, this implies that compounds possessing these structural characteristics are either liberated from lignin during bleaching or they are formed through subsequent reactions of lignin-derived single-ring compounds. The observed correlations also imply that the inducers possess physical-chemical properties similar to the chlorophenolics implicated in this study.

In summary, exposures of rainbow trout to treated kraft mill effluent demonstrated that MFO induction potential, measured as EROD activity, was present after secondary treatment. During effluent fractionations, the majority of induction potential was not recovered after NF. Comparisons of the EROD activity of exposure solutions with the chemical content detectable by GC demonstrated that nearly all dissolved organics exhibited no correlation with induction, with the exception of several chlorophenolics including tetrachloroguaiacol. Subsequent testing showed no induction in fish subjected to either i.p. injection or waterborne tetrachloroguaiacol exposure. All of the resin acids, fatty acids, bacterial fatty acids, terpenes, chlorinated phenols, vanillins, catechols, syringols, syringaldehydes, aliphatic alkanes, plant sterols, and chlorodimethylsulfones detected and verified with authentic standards exhibited no correlation with induction.

ACKNOWLEDGMENTS

The authors acknowledge the assistance of Vince Martin, Bev Blunt, Joan Yaromich, Sean Backus, and Colleen O'Neill. E.B. Eddy Forest Products Ltd., Espanola Ontario, generously provided the effluent samples and details on mill operating conditions. Financial support was provided through the National Water Research Institute, Environment Canada, and the Forest Industries Program of Industry, Science and Technology Canada. Additional support to L.M. Hewitt was provided by an IAGLR/Mott Foundation Fellowship.

REFERENCES

Addison, F., M.E. Zinck and D.E. Willis. 1978. Induction of mixed-function oxidase (MFO) enzymes in trout (*Salvelinus fontinalis*) by feeding Arochlor or 3-methylcholanthrene. Comp. Biochem. Physiol. 61C:323-325.

Allard, A-N., M. Remberger and A.H. Neilson. 1985. Bacterial O-methylation of chloroguaiacols: Effect of substrate concentration, cell density, and growth conditions. Appl. Environ. Microbiol. 49:279-288.

Andersson, T., L. Förlin, J. Hardig and Å. Larsson. 1988. Physiological disturbances in fish living in coastal water polluted with bleached kraft pulp mill effluents. Can. J. Fish. Aquat. Sci. 45:1525-1536.

Brownlee, B., and W.M.J. Strachan. 1977. Distribution of some organic compounds in the receiving waters of a kraft pulp and paper mill. J. Fish. Res. Bd. Can. 34:830-837.

Chau, F.S., and J.A. Coburn. 1974. Determination of pentachlorophenol in natural and waste waters. J. AOAC 57:389-393.

Denton, T.E., W.M. Howell, J.J. Allison, J. McCollum and B. Marks. 1985. Masculinization of female mosquitofish by exposure to plant sterols and *Mycobacterium smegmatis*. Bull. Environ. Contam. Toxicol. 35:627-632.

Ferguson, M., M. Servos and K. Munkittrick. 1992. Inability of resin acid exposure to elevate EROD activity in rainbow trout (*Oncorhynchuss mykiss*). Wat. Poll. Res. J. Can. 27(3):561-573.

Förlin, L., T. Andersson, B.E. Bengtsson, J. Hardig and Å. Larsson. 1985. Effects of pulp bleach plant effluents on hepatic xenobiotic biotransformation enzymes in fish: Laboratory and field studies. Marine Environ. Res. 17:109-112.

Hewitt, L.M. 1993. A gas chromatographic evaluation of treated kraft mill effluent constituents as sources of mixed function oxygenase induction in rainbow trout. M.Sc. thesis, University of Waterloo, Waterloo, ON, 174 p.

Hodson, P.V., P.J. Kloeper-Sams, K.R. Munkittrick, W.L. Lockhart, D.A. Metner, P.L. Luxon, I.R. Smith, M.M. Gagnon, M. Servos and J.F. Payne. 1991. Protocols for measuring mixed function oxidases in fish liver. Can. Tech. Rept. Fish. Aquat. Sci. No. 1829, 51 p.

Hodson, P.V., M. McWhirther, K. Ralph, B. Gray, D. Thivierge, J. Carey, G. Van Der Kraak, D.M. Whittle and M.C. Levesque. 1992. Effects of bleached kraft mill effluent on fish in the St. Maurice River, Quebec. Environ. Toxicol. Chem. 11:1635-1651.

Howell, W.M., and T.E. Denton. 1989. Gonopodial morphogenesis in female mosquitofish, *Gambusia affinis affinis*, masculinized by exposure to degradation products from plant sterols. Environ. Biol. Fish. 24:43-51.

Keith, L.H., W. Crummet, J. Deegan, R.A. Libby, J.K. Taylor and G. Wentler. 1983. Principles of environmental analysis. Anal. Chem. 55:2210-2218.

Klotz, A.V., J.J. Stegeman and C. Walsh. 1983. An aryl hydrocarbon hydroxylating hepatic cytochrome P-450 from the marine fish *Stenotomus chrysops*. Arch. Biochem. Biophys. 226:578-592.

Kringstad, K.P., and K. Lindström. 1984. Spent liquors from pulp bleaching. Environ. Sci. Technol. 18:236A-266A.

Lee, R.F. 1988. Possible linkages between mixed function oxygenase systems, steroid metabolism, reproduction, molting, and pollution in aquatic animals. *In* M.S. Evans, Ed., Toxic Contaminants and Ecosystem Health: A Great Lakes Focus. John Wiley & Sons, New York, pp. 201-213.

Leuenberger, C., W. Giger, R. Coney, J.W. Graydon and E. Molnar-Kubica. 1985. Persistent chemicals in pulp mill effluents: Occurrence and behaviour in an activated sludge treatment plant. Water Res. 19:885-894.

Lindström, K., and J. Nordin. 1976. Gas chromatography-mass spectrometry of chlorophenols in spent bleach liquors. J. Chromogr. 128:13-26.

Lindström-Seppä, P., and A. Oikari. 1990. Biotransformation and other toxicological physiological responses in rainbow trout (*Salmo gairdneri* Richardson) caged in a lake receiving effluents of pulp and paper industry. Aquatic Toxicol. 16:187-204.

Lu, A.H.Y., and S.B. West. 1980. Multiplicity of mammalian microsomal cytochromes P-450. Pharm. Rev. 31:277-295.

Martel, P.H., T.G. Kovacs, B.I. O'Connor and R.H. Voss. 1993. A survey of pulp and paper mill effluents and their potential to induce mixed function oxidase enzyme activity in fish. Proc. 79th Annual Meeting, Technical Section, Can. Pulp & Paper Assoc., Montreal Quebec, p. A165-A177.

Martin, V.J.J., B.K. Burnison, H. Lee and L.M. Hewitt. 1995. Chlorophenolics from high molecular weight chlorinated organics isolated from bleached kraft mill effluents. Holzforschung: In press.

McKague, A.B., M-C. Kolar and K.P. Kringstad. 1988. Nature and properties of some chlorinated, lipophillic, organic compounds in spent liquors from pulp bleaching. 1. Liquors from conventional bleaching of softwood kraft pulp. Environ. Sci. Technol. 22:523-526.

McMaster, M.E., G.J. Van Der Kraak, C.B. Portt, K.R. Munkittrick, P.K. Sibley, I.R. Smith and D.G. Dixon. 1991. Changes in hepatic mixed function oxygenase (MFO) activity, plasma steroid levels and age at maturity of a white sucker (*Catostomus commersoni*) population exposed to bleached kraft pulp mill effluent. Aquatic Toxicol. 21:199-218.

Muir, D.C.G., A.L. Yarechewski, D.A. Metner, W.L. Lockhart, G.R. Barrie Webster and K.J. Friesen. 1990. Dietary accumulation and sustained hepatic mixed function oxidase enzyme induction by 2,3,4,7,8-pentachlorodibenzofuran in rainbow trout. Environ. Toxicol. Chem. 9:1463-1472.

Munkittrick, K.R., C. Portt, G.J. Van Der Kraak, I. Smith and D. Rokosh. 1991. Impact of bleached kraft mill effluent on population characteristics, liver MFO activity and serum steroid levels of a Lake Superior white sucker (*Catostomus commersoni*) population. Can. J. Fish. Aquat. Sci. 48:1371-1380.

Munkittrick, K.R., G.J. Van Der Kraak, M.E. McMaster and C.B. Portt. 1992. Response of hepatic MFO activity and plasma sex steroids to secondary treatment of bleached kraft pulp mill effluent and mill shutdown. Environ. Toxicol. Chem. 11:1427-1439.

Munkittrick, K.R., G.J. Van Der Kraak, M.E. McMaster, C. Portt, M.R. van den Heuvel and M.R. Servos. 1994. Survey of receiving water environmental impacts associated with discharges from pulp mills. II. Gonad size, liver size, hepatic MFO activity, and plasma sex steroid levels in white sucker. Environ. Toxocol. Chem. 13:1089-1101.

Munkittrick, K.R., M.R. Servos, K. Gorman, B. Blunt, M.E. McMaster and G. Van Der Kraak. 1995. *In* Environmental Toxicology and Risk Assessment: Fourth Volume, T.W. La Pointe, F.T. Price and E.E., Little, Eds., ASTM STP 1262, American Society for Testing and Materials, Philadelphia: In press.

Munro, F.C., S. Chandrasekaran, C.R. Cook and D.C. Pryke. 1990. Impact of high chlorine dioxide substitution on oxygen delignified pulp. TAPPI J. 73:123-130.

Oikari, A., and P. Lindström-Seppä. 1990. Responses of biotransformation enzymes in fish liver: Experiments with pulp mill effluents and their components. Chemosphere 20:1079-1085.

Okey, A.B. 1990. Enzyme induction in the cytochrome P-450 system. Pharmac. Ther. 45:241-298.

Parrott, J.L., P.V. Hodson, M.R. Servos, S.L. Huestis and D.G. Dixon. 1995. Relative potency of polychlorinated dibenzo-p-dioxins and dibenzofurans for inducing mixed-function oxygenase activity in rainbow trout. Environ. Toxicol. Chem. 14:1041-1050.

Rogers, I.H., C.D. Levings, W.L. Lockhart and R.J. Norstrom. 1989. Observations on overwintering juvenile chinook salmon *Oncorhynchus tshawytscha* exposed to bleached kraft mill effluent in the upper Fraser River, British Columbia. Chemosphere 19:1853-1868.

Servos, M., J. Carey, M. Ferguson, G.J. Van Der Kraak, H. Ferguson, J. Parrott, K. Gorman and R. Cowling. 1992. Impact of a modern bleached kraft mill with secondary treatment on white suckers. Water Poll. Res. J. Can. 273:423-437.

Servos, M.R., S.Y. Huestis, D.M. Whittle, G.J. Van Der Kraak and K.R. Munkittrick. 1994. Survey of receiving-water environmental impacts associated with discharges from pulp mills. 3. Polychlorinated dioxins and furans in muscle and liver of white sucker (*Catostomus commersoni*). Environ. Toxicol. Chem. 9:1103-1115.

Stuthridge, T.R., A.L. Wilkins and A.G. Langdon. 1990. Identification of novel chlorinated monoterpenes formed during kraft pulp bleaching of *Pinus radiata*. Environ. Sci. Technol. 24:903-908.

SYSTAT for Windows: Graphics, Version 5 Edition. 1992. SYSTAT Inc., Evanston, IL, 636 pp.

Tokats, P. 1987. Operation of the Beckman industrial 915-B total organic carbon analyzer. National Water Research Institute Instruction Manual. 10 p.

van den Heuvel, M.R., K.R. Munkittrick, G.J. Van Der Kraak, M.R. Servos and D.G. Dixon. 1995. Hepatic-7-ethoxyresorufin-O-deethylase activity, plasma steroid hormone concentration and liver bioassay-derived 2,3,7,8-TCDD toxic equivalent concentrations in wild white sucker (*Catostomus commersoni*) caged in bleached kraft pulp mill effluent. Can. J. Fish. Aquat. Sci.: In press.

Van Der Kraak, G.J., K.R. Munkittrick, M.E. McMaster, C.B. Portt and J.P. Chang. 1992. Exposure to bleached kraft pulp mill effluent disrupts the pituitary gonadal axis of white sucker at multiple sites. Toxicol. Appl. Pharmacol. 115:224-233.

Vodicnik, M.J., D.R. Elcombe and J.J. Lech. 1981. The effect of various types of inducing agents of hepatic microsomal monooxygenase activity in rainbow trout. Toxicol. Appl. Pharmacol. 56:364-374.

Voss, R.H. 1983. Chlorinated neutral organics in biologically treated bleached kraft mill effluents. Environ. Sci. Technol. 17:530-537.

Voss, R.H. 1984. Neutral organic compounds in biologically treated bleached kraft mill effluents. Environ. Sci. Technol. 18:938-946.

Wilkins, A.L., and S. Panadam. 1987. Extractable organic substances from the discharges of a New Zealand pulp and paper mill. Appita 40:208-212.

COMPARATIVE ASSESSMENT OF THE TOXIC EFFECTS FROM PULP MILL EFFLUENTS TO MARINE AND BRACKISH WATER ORGANISMS

B. Eklund, M. Linde and M. Tarkpea

Institute of Applied Environmental Research, Laboratory for Aquatic Ecotoxicology
Stockholm University, Studsvik, S-611 82 Nyköping SWEDEN

The acute and chronic toxicities of whole mill effluents from a totally chlorine free (TCF) and elemental chlorine free (ECF) process, before and after secondary treatment, and an effluent from an unbleached process were assessed and compared. The tests performed were acute tests with Microtox® and the brackish water harpacticoid copepod *Nitocra spinipes* and two reproduction tests, one with *N. spinipes* and one with the marine red alga *Ceramium strictum*. All tests ranked the TCF effluent as the least toxic. The unbleached water was the most toxic according to all organisms except *C. strictum*. Both reproduction tests were 2-4 times more sensitive than Microtox and the acute test using *N. spinipes*. Secondary treatment of TCF and ECF effluents reduced the toxicity according to these tests by 2-5 times. The five effluents were separated into a lipophilic and a hydrophilic fraction using a filter with an impregnated reversed phase material (R_{18}). The hydrophilic fractions were assessed in all tests, except the reproduction test with *N. spinipes*. This fraction did not cause any harmful responses to Microtox or the acute test with *N. spinipes*. In all effluents this fraction, however, contained a large amount of substances harmful to the reproduction of *C. strictum*.

INTRODUCTION

In Sweden, research activities concerning chemical and biological characterization of pulp mill effluents have been of great interest since the middle of the 1970s. Within the Swedish Environmental Protection Agency's project, Environment/Cellulose I and II (Södergren *et al.* 1989; Södergren 1993), a large amount of data has been produced about groups of chemicals having an adverse effect on biota in the environment. The focus has been mainly on the bleaching process. These results, among others, have led to a considerable reduction in chlorine compounds in the bleaching process. One way has been to substitute chlorine gas with chlorine dioxide in the bleaching process, which reduces the discharge of chlorinated compounds considerably. The elemental chlorine free (ECF) process is in common use in Sweden today. The Swedish industry has also developed a method to bleach pulp that is totally chlorine free (TCF).

All pulp mill plants differ from each other in some way such as raw material used, cooking length, the prebleaching procedure, etc. In the present investigation the pulp industry's organization in Sweden together with the Swedish Protection Agency have chosen the factories where the samples were taken so that the samples were as representative as possible of the processes.

The aim of this study was to use short-term biological tests to investigate the impact of the above mentioned processes and to estimate the degree of reduction of observed acute and chronic effects by secondary treatment. The test organisms used were brackish and marine species which are ecologically relevant for the receiving waters around Sweden. These include an acute test with the brackish water harpacticoid copepod *Nitocra spinipes* (Bengtsson and Tarkpea 1995) and reproduction tests with *N. spinipes* (Bengtsson and Bergström 1987) and with the marine red alga *Ceramium strictum* (Eklund 1993). Acute tests using the Microtox method (Microbics 1992) have also been included.

The Swedish debate has strongly focused on the effects caused by the bleaching process. Less has been published about the effects from other parts of the pulp producing process. In this investigation we have made an attempt to study in which fraction and to what degree the samples contain substances harmful to the organisms mentioned above. This has been done by performing tests on samples filtered through discs impregnated with a reversed phase material (C_{18}) which separates the sample into lipophilic compounds, that are adsorbed on the filter, from the hydrophilic ones which pass through. The test results from the hydrophilic fractions are compared to the toxicity in the total water samples.

MATERIALS AND METHODS

Effluents

Individual (grab) samples of whole mill effluents were taken from one plant using the TCF process, one using the ECF process and also an effluent from an unbleached (UB) process. The latter is from the same plant as the ECF taken during a period of time when the bleaching process was not running. For the TCF and ECF processes, tests have also been carried out on samples taken after secondary treatment. The TCF plant used a pilot plant with activated sludge for biological treatment and the ECF plant used an aerated lagoon. All effluent samples were taken when the processes were running well. The samples were deep frozen in 200-1000 mL polyethylene containers and thawed in 15°C running water immediately before use. A description of the processes and some of the characteristics of the effluents are given in Table 1.

Table 1. Description of the whole mill effluents.

	TCF Process		ECF Process		Unbleached
	Before biological treatment activated sludge (pilot plant)	After biological treatment activated sludge (pilot plant)	Before biological treatment aerated lagoon	After biological treatment aerated lagoon	Before aerated lagoon
Bleaching sequence	Q Q P Z P	Q Q P Z P	D(EO)D(EP)D	D(EO)D(EP)D	
Raw material	hardwood	hardwood	softwood	softwood	softwood
Kappa number after O_2 stage	10	10	10.8	10.8	
Chlorine dioxide substitution %			100	100	
Total charge of chlorine dioxide (kg a.Cl ptp)			ca 40	ca 40	
Waste water volume (m^3 t $pulp^{-1}$)	50	50	99	99	134
Pulp produced (t d^{-1})	1000	1000	674	674	426
pH of waste water	7.14	7.62	9.56	7.45	8.79
Duration of secondary treatment		18 h		4 d	
BOD_7 (mg L^{-1})	300	58	310	78	310
COD_{cr} (mg L^{-1})	548	257	749	459	807
Sampling date	9-Oct-93	9-Oct-93	15-Nov-93	15-Nov-93	12-Apr-94

Q = EDTA; P = H2O2; Z = O3; D = ClO2; E = NaOH

Hydrophilic Fractions

All effluents were separated into a lipophilic and a hydrophilic phase using a C_{18} solid-phase filter (SPEC-47-C18AR, ANSYS). The retention capability of these filters for 47 substances is reported elsewhere (EPA 1991). Biological tests were run only on the hydrophilic fraction.

The C_{18} filter was placed in a membrane filter system and activated with 5 mL methanol (analytical grade). The filter was then rinsed with 5 mL deionized and 5 mL effluent sample under low pressure; 150 mL effluent was then immediately filtered through the freshly prepared C_{18} filter for use in the Microtox®, the *Nitocra* acute and the *Ceramium* reproduction test.

Biological Tests

Because of the complex character of bleached and unbleached kraft mill effluents, a complete chemical characterization is impossible. Accordingly, the need for biological tests of different complexity to identify and assess the possible environmental hazards of these effluents is evident. The objective in the present study was to use a battery of four acute and chronic tests with three species, two of which reflect different trophic levels and important functions in the brackish and marine water systems. In the case of marine recipients, it is important to use marine species since the salinity itself may alter the state in which different compounds are found and this surely influences the degree to which organisms are affected.

Inhibition of Bacteria Luminescence: The Microtox® Method

The Microtox method was performed using a Microtox instrument (Model 500) and following the "Basic Test" and the "100% Test" protocols described in the Microtox manual (Microbics 1992). Freeze-dried marine luminescent bacteria *Vibrio fisheri* (Bulich 1982; ISO 1994) were used as test organisms. The toxicity of the sample was measured after 15 and 30 min by registration of the reduction of light emission from the bacteria. The sensitivity of the Microtox method has been proven to be comparable to acute toxicity tests with fish and crustaceans in several comparative investigations (Bulich *et al.* 1981; Qureshi *et al.* 1982; Tarkpea *et al.* 1986).

All effluent samples were adjusted to 2 ± 0.2% NaCl to provide osmotically favorable conditions for the bacteria. The test temperature was 15 ± 0.1°C. pH adjustments were made to 7 on samples having a pH outside the range 6-8.

Calculations of 15 and 30 min EC50 and EC20 values were made automatically using Microtox software. EC50 is expressed as the concentration of a water sample that causes a 50% reduction of light output from *V. fisheri*. EC20 is regarded as the lowest EC value in the Microtox test indicating a toxic effect (Microbics 1992).

Acute Toxicity Test with *Nitocra spinipes*

The acute lethal toxicity test with *N. spinipes* Boeck (Crustacea) was performed according to the Swedish standard (SIS 028106 1991). A technical method document is also in press (Bengtsson and Tarkpea 1995). The harpacticoid *N. spinipes* is the brackish water counterpart to the freshwater *Daphnia* and *Ceriodaphnia* (ISO 1989; Environment Canada 1992; EPA 1993a; OECD 1993) which are frequently used in standardized tests. *N. spinipes* is a typical brackish water organism, playing an important role in the littoral food chain of the Baltic Sea. It is also found in coastal regions of the North Sea, the Mediterranean, Asia, Africa and North America (Lang 1948). The length of the adult animal is 0.6-0.8 mm. Culturing in the laboratory is easy to carry out and development from egg to the adult stage takes about 2 wk.

The test was carried out as a static test for 96 h in tubes containing 10 mL of sample, with a salinity of 7 ‰. The test temperature was 20 ± 1°C. The mean lethal concentration when 50% and 10% of the animals died was expressed as 96 h LC50 and 96 h LC10. The pH of the sample was adjusted to 7 if outside the range 6-9.

Calculations of the 96 h LC50 and LC10 were made using a computer program for probit analysis developed at the Swedish EPA (Finney 1971).

Reproduction Test with *Nitocra spinipes*

The reproduction test with *N. spinipes* has been developed by Bengtsson and Bergström (1987) as a complement to the acute toxicity test. It was shown to be a sensitive test for chlorine bleached pulp mill effluents (Tarkpea, ITM Stockholm University, pers. comm.).

In the present study a semi-static procedure was used, changing the test water every second day for 7 d. Newly fertilized females (20 for each concentration) with ovigerous bands from the laboratory culture were put in separate test vessels at the start of the experiment. Only the offspring (metanauplia and copepodids) from day 3 to 7 were recorded. The females and offspring were fed each day with a fish-food suspension with a growing culture of bacteria.

For experimental conditions such as salinity, temperature and pH see the acute *N. spinipes* test above.

The concentration which gave 50% (EC50) and 10% (EC10) reduction of live offspring at the end of the test compared to the control was determined. The calculations were done with an EPA program (ICp vers. 2.0) for linear interpolation for sublethal toxicity (EPA 1993b). EC10 is regarded as the lowest concentration with significant effect in this test.

Reproduction Test with *Ceramium strictum*

The marine red alga *C. strictum* is used in the *Ceramium* reproduction test and is described in Eklund (1993). The test is based on that with *Champia parvula* (Steele and Thursby 1983; EPA 1988). *Ceramium strictum* is a cosmopolitan species in temperate waters. In Sweden it grows along the west coast. *C. tenuicorne* is one of the most common red algae in the Baltic Sea and is found throughout the Bothnian Bay (Kautsky and Kautsky 1994; H. Kautsky pers. comm.). According to Rueness (1978), *C. tenuicorne* is a subspecies of *C. strictum*. The results from tests using *C. strictum* are thus expected to be valid for *C. tenuicorne* and the Baltic environment.

The end point of the test is the number of successful fertilizations, the cystocarps, found in the different concentrations compared to the control. In this investigation eight concentrations were used with at least four replicates. Female (10-15) and male (1-2) plants were exposed to different concentrations of the samples for 24 h. The light regime was 16:8 h L:D with an intensity of 35 µmol $m^{-2}s^{-1}$. The temperature was 22 ± 2°C. The pH of the sample was adjusted to 8 if outside the range of 7-9. NaCl was added to all samples to get a salinity of 20 ‰, which is the optimum salinity for the species (Rueness and Kornfeldt 1992). After exposure, female plants were transferred to a fresh enriched medium. The algae were allowed to grow for another 6 d, after which the cystocarps have grown to a size where they are easily detected.

The concentration which gave 50% (EC50) and 10% (EC10) reduction in the number of cystocarps compared to the control was determined. The calculations, with 95% confidence interval, were done by using the ToxCalc ver. 4.0 program (Tidepool Scientific Software 1992-1993).

Presentation of Results: Toxic Emission Factor (TEF) Values

The calculated LC and EC values are based on the concentration of effluent causing a toxic effect. To estimate the potential environmental impact of the polluting discharges, the amount of waste water generated per tonne pulp produced must also be considered. To get comparable data the results were transformed into TEF values (Toxic Emission Factor).

TEF = (100 LC50^{-1}) x (m^3 waste water tonne^{-1} pulp)
Instead of LC50, other LC or EC values can be put into the formula

Unlike the LC or EC values where a smaller number corresponds to a more toxic effluent, a smaller TEF value corresponds to a less toxic effluent. LC, EC and TEF values will be presented in this report.

RESULTS

Toxicity of the Total Effluent

The LC50, EC50, LC10, EC10 and EC20 values are shown in Table 2. All three pulp producing processes produce toxic effluents, before and after secondary treatment, according to at least one of the four test methods used. When comparing the processes before secondary treatment, the most sensitive organism, *C. strictum*, gives an EC50 value of 14.9% for the TCF process, 6.18% for the ECF process and 2.02% for the unbleached process.

All four test methods show a reduction in toxicity for the TCF and ECF processes after secondary treatment (Table 2). The EC50 value for the most sensitive organism, *C. strictum*, increased to 26.2% for the TCF process and 12.7% for the ECF process. This means a 50% reduction in the toxicity for both processes.

According to the Microtox 15 min and the *Nitocra* acute tests, no toxicity could be detected after biological treatment of the effluents from the TCF process (Table 2). However, after the 30 min Microtox test a 20% reduction of the luminescence was noticed at 78.7% effluent. In the acute test with *N. spinipes* 10% of the animals died at 42.7% TCF effluent. The reproduction test with *C. strictum* was the most sensitive test for TCF effluent. A 50% effect on reproduction was already observed at 14.9%, before biological treatment. After treatment, toxicity was reduced to almost half and the EC50 value was 26.2%.

All the tested organisms showed toxic responses to effluents from the ECF process (Table 2). The most sensitive tests were the reproduction tests with *N. spinipes* and *C. strictum*, where 50% effect was observed for the effluent before secondary treatment at only 6% of the waste water concentration. According to all five test methods, toxicity was reduced two to five times after secondary treatment.

All tests, except the *Ceramium* reproduction test of unbleached effluent, show this effluent to be the most toxic of the tested effluents before biological treatment (Table 2). The EC50 values were Microtox 15 min 2.02%, Microtox 30 min 2.67%, *Nitocra* acute 8.86% and *Nitocra* reproduction 3.55%. The *Ceramium* test ranked this water as a little less toxic than the ECF effluent and had an EC50 of 8.2%.

Toxicity of the Hydrophilic Fraction

Results from tests on the hydrophilic fraction of the five effluent samples are shown in Table 2.

No toxicity was found in the hydrophilic part of the TCF effluent according to the Microtox test after 15 or 30 min (Table 2). LC50 for the acute test with *Nitocra* was 52.3% before the secondary treatment and no toxicity was detected after it. The hydrophilic part of the effluent consisted of a large amount of harmful substances towards the reproduction of *C. strictum*. EC50 before the secondary treatment was 35.3% for this test and the toxicity was only reduced to 40.2% after the aerated lagoon.

The hydrophilic fraction of the effluent from the ECF process did not give any toxic response, before or after the secondary treatment, according to the Microtox test or the acute test with *Nitocra*. However, the *Ceramium* reproduction test showed responses both before and after the aerated lagoon (Table 2). The EC50 value before the lagoon was 14.7% and 30.5% after.

Table 2. Acute toxicity and reproduction disturbances from TCF (total chlorine free), ECF (elemental chlorine free) and unbleached (UB) pulp mill processes. 95% fiducial limits are shown in parentheses.

	Acute Tests							Reproduction Tests				
	Microtox, 15 min		Microtox, 30 min		Nitocra, 96 h			Nitocra, 7 d			Ceramium, 24 h	
	EC50 vol%	EC20 vol%	EC50 vol%	EC20 vol%	LC50 vol%	LC10 vol%		EC50 vol%	EC10 vol%		EC50 vol%	EC10 vol%
Total Effluent												
TCF before biological treatment	24.0 (21.2-27.2)	6.36 (5.36-7.56)	23.6 (21.2-26.2)	6.27 (5.40-7.29)	40.9 (33.7-50.2)	7.02 (-7.02-14.8)		20.0 (15.0-26.4)	5.48 (1.66-13.7)		14.9 (9.95-25.1)	6.31 (0.49-6.64)
TCF after biological treatment	>99	>99	>99	78.7 (26.4-235)	>100	42.7 (-7.4-∞)		>97	>97		26.2 (23.0-29.2)	13.0 (10.4-14.7)
ECF before aerated lagoon	31.3 (28.2-34.8)	10.2 (8.42-12.3)	29.4 (26.1-33.2)	9.48 (7.58-11.8)	14.3 (10.6-19.5)	2.31 (-3.75-6.15)		6.22 (2.88-12.7)	2.33 (0.44-4.23)		6.18 (4.10-7.29)	2.76 (2.42-3.14)
ECF after aerated lagoon	60.7 (44.9-91.9)	12.8 (8.95-18.3)	46.7 (37.3-58.5)	8.76 (5.89-13.1)	34.2 (25.2-43.7)	-11.5 (-32.8-1.11)		30.3 (20.3-36.4)	13.1 (4.90-26.7)		12.7 (5.94-14.6)	2.45 (0.0-3.06)
UB before aerated lagoon	2.02 (1.62-2.54)	0.66 (0.56-0.77)	2.67 (2.42-2.94)	0.93 (0.87-1.00)	8.86 (5.86-21.3)	-4.13 (-22.8-0.52)		3.55 (1.12-5.56)	0.15 (0.11-0.88)		8.20 (4.55-8.58)	3.12 (2.52-3.42)
Hydrophilic Fraction												
TCF before biological treatment	>99	26.0 (20.1-33.6)	99.7 (70.0-142)	20.4 (15.8-26.3)	52.3 (42.4-65.8)	9.02 (-8.19-20.1)		not tested	not tested		35.3 (30.7-39.8)	13.2 (4.90-24.3)
TCF after biological treatment	>99	36.8 (4.53-299)	>99	71.2 (32.9-154)	>100	49.0 (- - -)		not tested	not tested		40.2 (30.6-43.6)	12.9 (0.0-15.5)
ECF before aerated lagoon	>90	36.3 (23.8-55.3)	>90	32.0 (21.8-46.9)	>100	18.5 (- - -)		not tested	not tested		14.7 (13.6-15.6)	6.86 (0.30-12.0)
ECF after aerated lagoon	>90	78.8 (22.4-278)	>90	75.0 (12.0-469)	>100	29.3 (-43.6-63.0)		not tested	not tested		30.5 (29.6-32.3)	14.8 (13.7-16.6)
UB before aerated lagoon	>99		>99		63.0 (55.6-73.2)	32.3 (18.2-40.8)		not tested	not tested		28.8 (26.2-32.0)	6.73 (0.51-28.2)

The hydrophilic part of the unbleached effluent did not contain any harmful substances towards the bacteria used in the Microtox method (Table 2). Some toxicity was measured in the *Nitocra* acute test, giving an EC50 value of 63%. The *Ceramium* reproduction test showed a 50% response at 28.8% effluent water, which means that 28% of the toxicity was derived from water that had passed through the solid phase filter.

TEF

The same trend as observed for the EC50 values, but accentuated, is seen when looking at the figures for TEF per tonne pulp. The results from total effluents before and after secondary treatment and toxicity in the hydrophilic fraction are shown in Table 3.

Table 3. TEF (total emission factor) per tonne pulp produced from plants using different processes; TCF (total chlorine free), ECF (elemental chlorine free) and unbleached (UB) pulp mill effluents. The corresponding EC50 values are shown in parentheses.

	Acute Tests			Reproduction Tests	
	Microtox, 15 min TEF t^{-1}	Microtox, 30 min TEF t^{-1}	*Nitocra*, 96 h TEF t^{-1}	*Nitocra*, 7 d TEF t^{-1}	*Ceramium*, 24 h TEF t^{-1}
Total Effluent					
TCF before biological treatment	208 (24.0)	212 (23.6)	122 (40.9)	250 (20.0)	336 (14.9)
TCF after biological treatment	50 (>99)	50 (>99)	50 (>100)	50 (>97)	191 (26.2)
ECF before aerated lagoon	318 (31.3)	338 (29.4)	695 (14.3)	1596 (6.22)	1609 (6.18)
ECF after aerated lagoon	164 (60.7)	213 (46.7)	291 (34.2)	328 (30.3)	783 (12.7)
UB before aerated lagoon	6624 (2.02)	5011 (2.67)	1510 (8.86)	3769 (3.55)	1632 (8.20)
Hydrophilic Fraction					
TCF before biological treatment	50 (>99)	50 (99.7)	96 (52.3)	not tested	142 (35.3)
TCF after biological treatment	50 (>99)	50 (>99)	50 (>100)	not tested	125 (40.2)
ECF before aerated lagoon	110 (>90)	110 (>90)	99 (>100)	not tested	676 (14.7)
ECF after aerated lagoon	110 (>90)	110 (>90)	99 (>100)	not tested	326 (30.5)
UB before aerated lagoon	134 (>99)	134 (>99)	212 (63.0)	not tested	465 (28.8)

DISCUSSION

Internationally there are few published investigations concerning short term tests with algae and crustaceans to predict effects from pulp mill effluents.

A combination of chemical analyses and biological short term tests has been used by Renberg (1992) to evaluate the effects of different chlorine dioxide substitutes for chlorine as a bleaching component. The tests used were the Microtox test, the acute toxicity test with *Ceriodaphnia dubia* and a growth inhibition test using the green alga *Selenastrum capricornutum*. EC50 results from one pulp mill with 100% chlorine dioxide bleaching were 85% for Microtox, 100% for the *Ceriodaphnia* test and 11% for the *Selenastrum* test. These results, like ours, demonstrate that the algae test is the most sensitive for this type of effluent.

Härdig et al. (1988) reported 96 h LC50 values for *N. spinipes* of 7.4-28.0 vol% and 5 min EC50 values for Microtox of 8.7-18.0 vol% for six waste waters from two pulp mills with chlorine bleaching, one using pine and the other birch. No attempt was made to interpret the short term effects. Borton et al. (1991) showed effects of effluents, treated in aerated lagoons, from three mills practicing oxygen bleaching, on *C. dubia* (7 d reproduction) and fathead minnow, *Pimephales promelas* (7 d growth of early life stage). They found the *Ceriodaphnia* test to be more sensitive than the fathead minnow test. Their conclusion was that oxygen bleaching was less toxic than conventional bleaching. O'Connor et al. (1993) used chemical analyses and acute/chronic aquatic toxicity tests and preliminary screening for the potential to induce mixed function oxygenase (MFO) enzyme activity to evaluate effluents from various bleaching technologies. Chronic toxicity of *C. dubia* showed IC25 values for reproduction of 0.7 and 4% for TCF and 40-60% for ECF softwood bleached effluents.

Both bleached and unbleached effluents have a considerable impact on living organisms. This has been shown by Leach and Thakore (1977), who studied the unbleached effluents from Canadian softwood kraft, sulphite and mechanical pulping operations, wood debarking and effluents from chlorine bleached mills. The samples were whole mill effluents which were not secondary treated. Although the results differed between mills, their overall results showed about the same level of toxicity at all types of pulp producing plants. However, the compounds responsible for the toxicity from the unbleached procedure were found to be mainly from seven resin acids, and the toxicity from the bleached mills was mainly caused by chlorolignins.

Toxicity According to the Biological Tests

All tests ranked the TCF effluent as the least toxic and, excluding the *Ceramium* test, the effluents from the unbleached process as the most toxic. The toxicity for the ECF was a little higher than for the unbleached effluent according to the *Ceramium* test. The raw material in this study was hardwood in the TCF process and softwood in the two other processes. Preliminary results from our laboratory using the same biological test methods on TCF softwood show even less toxicity than from TCF hardwood. This indicates that the difference in toxicity seen in our results can probably not entirely be related to raw material.

The bacteria used in the Microtox test showed high sensitivity to all types of effluent but especially to the unbleached. After secondary treatment acute toxicity disappeared in the TCF water but some remained in the ECF water. The slightly higher toxicity seen in the ECF effluent after secondary treatment by the 30 min test (Table 2) could indicate the presence of metals in toxic concentrations (Microbics 1991). However, no toxicity was found in the hydrophilic fraction (Table 2), where the metals would be expected to be found. They could, however, have been adsorbed on particles and thus caught in the filter. Since toxicity is found in the total water after secondary treatment for the ECF process but not in the hydrophilic fraction, lipophilic compounds are probably responsible for this effect.

According to *N. spinipes*, the TCF effluent was also the least and unbleached the most acutely toxic. After secondary treatment the effluent from the TCF process was non-toxic to this organism, but some acutely toxic substances were still left in the ECF water. Since no toxicity was noticed in the hydrophilic fraction after secondary treatment (Table 2), the substances toxic towards *N. spinipes* are probably of a lipophilic character. Before secondary treatment some toxicity was also noticed in the hydrophilic fraction. Since this toxicity disappeared after secondary treatment, the compounds responsible must be readily degradable.

The reproduction test using *N. spinipes* ranked the water samples in the same way as the Microtox and *Nitocra* acute tests. By using the reproduction test, sensitivity increased twice compared to the acute test on the same organism. No reproduction tests using *N. spinipes* were carried out on the hydrophilic fractions. Tarkpea (ITM Stockholm University, pers. comm.) conducted short-term acute toxicity and chronic tests on zebrafish (*Brachydanio rerio*), *C. dubia*, *N. spinipes*, *S. capricornutum*, duckweed (*Lemna minor*) and Microtox on three effluents from industries using softwood and conventional bleaching (O(C70

+ D30)). The reproduction test with *N. spinipes* and the growth inhibition test with *S. capricornutum* were found to be the most sensitive tests. EC50 values for reproduction before and after the aerated lagoon were 8.0 and 39.7% respectively, which is in the same order of magnitude as for the softwood ECF effluents in the present study.

The toxicity seen with the *Ceramium* reproduction test again ranked the TCF process as the least toxic. However, the differences are not large between the three processes (Table 2). All effluents before and after secondary treatment, as well as the corresponding hydrophilic fractions, contained substances harmful to the reproduction of *C. strictum*. In the TCF and the ECF effluent, before and after secondary treatment, this toxicity may be partly due to the content of Cu and Zn. Quantifications of heavy metals, made by C. Andrén (ITM Stockholm University, pers. comm.), reveal concentrations of Cu and Zn high enough to be toxic to *C. strictum*. The EC50 values for the reproduction of *C. strictum* were 16 µg L^{-1} for Cu and 125 µg L^{-1} for Zn (Eklund in prep).

The slightly more toxic response for the ECF relative to the unbleached water may be due to chloric derivatives from the bleaching procedure. Chlorate, which is a product of chlorine dioxide, is known to be harmful to in particular brown algae (Rosemarin *et al.* 1986). Lindblad (1994) found that a chlorate concentration of 80 µg L^{-1} led to sublethal stress reactions in *Fucus vesiculosus* community metabolism. Measurements were made the week before and the week after the samples for this project were taken. The mean from these two measurements was 24 mg L^{-1} before and 0.7 mg L^{-1} after secondary treatment (Michel, Swedish EPA, pers. comm.). The reproduction test with *C. strictum* on chlorate gave an EC50 value of 1.5 mg L^{-1} and an EC10 value of 0.63 mg L^{-1}. This means that part of the toxicity seen may be due to the chlorate formed during the ECF bleaching process.

For all five effluents, a considerable part of the toxicity towards the reproduction of *C. strictum* was found in the hydrophilic fraction. Compared to the total toxicity, 42% was observed in the hydrophilic fraction before and 65% after secondary treatment for the TCF process. For the ECF process, 42% of the toxicity was found in the hydrophilic fraction both before and after secondary treatment. For the unbleached water it was 28.5%.

Reduction in Toxicity After Secondary Treatment

The toxicity after secondary treatment, according to all the biological test methods, was reduced to half for the TCF and ECF processes. This is much less than was observed by O'Connor *et al.* (1993), who found the toxicity to be typically ten times less after secondary treatment. They based their conclusions on chronic tests with fathead minnow (*P. promelas*) and *Ceriodaphnia dubia*.

Chemical analyses of resin acids within this project show an almost complete disappearance of these compounds in the TCF water after the secondary treatment (Adolpsson-Erici, ITM Stockholm University, pers. comm.). Hardly any difference was seen in the total amount of resin acids for the ECF process. This means that harmful compounds other than resin acids might be toxic to the actual organisms.

Hydrophilic Fraction

In this study an attempt has been made to find out if the toxic responses of the test organisms were due mainly to lipophilic or hydrophilic substances. We found the filtering method easy to perform. The test procedures on the total water sample and on the hydrophilic fraction give an indication as to whether the most acute and chronically toxic substances have a lipophilic or hydrophilic character (see discussion above for the single tests). This may be useful in further chemical characterizations of the water samples. To fully evaluate this filtering method, chemical analyses on compounds adsorbed on the filter and in the filtrate, along with biological tests, should be more thoroughly characterized. The retention of typical substances for pulp mill effluents should also be further investigated.

TEF per Tonne Pulp

The TEF per tonne pulp produced for the three investigated plants is shown in Table 3. The same trend as observed for the EC50 values is seen when looking at the figures for TEF per tonne pulp, but accentuated. 50 m^3 water per tonne pulp produced was required for the TCF producing plant, 99 m^3 for the ECF producing plant and 134 m^3 for the plant producing the unbleached pulp. This means that the toxicity per tonne pulp produced is about doubled for the ECF and the unbleached process compared to the TCF process.

SUMMARY

- The short term tests used were found to be suitable to evaluate toxicity in pulp mill effluents.
- Both reproduction tests were two to four times more sensitive than the Microtox test and the acute test with *N. spinipes*.
- The effluent from the plant using TCF process was ranked the least toxic by all four test methods.
- The effluent from the plant using the unbleached process was the most toxic according to all tests except the *Ceramium* reproduction test.
- Secondary treatment reduced the toxicity two to five times to all test organisms.
- The hydrophilic fraction did not produce any toxic responses according to the Microtox test in any of the tested effluents.
- The hydrophilic fractions of the effluent waters from the TCF process before secondary treatment and the unbleached process resulted in some toxic responses according to the *Nitocra* acute test.
- The hydrophilic fraction of all effluents contained substances harmful to the reproduction of the red alga *Ceramium strictum*.

REFERENCES

Bengtsson, B.-E. and B. Bergström. 1987. A flow through fecundity test with *Nitocra spinipes* (Harpacticoidea Crustacea) for aquatic toxicity. Ecotoxicol. Environ. Saf. 14:260-268.

Bengtsson, B.-E. and M. Tarkpea. 1995. A 96-h acute toxicity test with the brackish water crustacean *Nitocra spinipes*. Assessment of chemicals, products and effluents. Environ. Tox. Wat. Qual.: In press.

Borton, D.L., W.R. Streblow and W.K. Bradley. 1991. Biological characterization studies of oxygen delignification effluents using short-term chronic toxicity tests. Proceedings of the 1992 TAPPI Environmental Conference, Austin, Texas:135-145.

Bulich, A.A. 1982. A practical and reliable method for monitoring the toxicity of aquatic samples. Process Biochemistry. March/April.

Bulich, A.A., M.W. Green and D.L. Isenberg. 1981. Reliability of the bacterial luminescence assay for determination of the toxicity of pure compounds and complex effluents. Amer. Soc. Test. Mater. STP 737:338-347.

Eklund, B. 1993. A 7-day reproduction test with the marine red alga *Ceramium strictum*. The Science of the Total Environment, Supplement. Elsevier Science Publishers B.V., Amsterdam. 749-759.

Environment Canada. 1992. Biological test method: Test of reproduction and survival using the cladoceran *Ceriodaphnia dubia*. Environmental Protection Series Report EPS 1/RM/21

EPA. 1988. Short-term methods for estimating the chronic toxicity of effluents and receiving waters to marine and estuarine organisms. U.S. Environmental Protection Agency. Report EPA/600/4-87/028.

EPA. 1991. Determination of organic compounds in drinking water by liquid-solid extraction and capillary column gas chromatography/mass spectrometry. U.S. Environmental Protection Agency Method 525.1.

EPA. 1993a. Methods for measuring the acute toxicity of effluents and receiving waters to freshwater marine organisms. U.S. Environmental Protection Agency. Report EPA/600/4-90/027F.

EPA. 1993b. A linear interpolation method for sublethal toxicity. The inhibition concentration (ICp) approach. Version 2.0. Environmental Protection Agency. Environmental Research Laboratory, Duluth, Minnesota.

Finney, D. 1971. Probit Analysis. Cambridge University Press.

Härdig, J., T. Andersson, B.-E. Bengtsson, L. Förlin and Å. Larsson. 1988. Long-term effects of bleached kraft mill effluents on red and white blood cell status, ion balance, and vertebral structure in fish. Ecotoxicol. Environ. Saf. 15:96-106.

ISO. 1989. Water quality - Determination of the inhibition of the mobility of *Daphnia magna* Straus (*Cladocera, Crustacea*). International Standard 6341.

ISO. 1994. Water quality - Determination of the inhibitory effect of water samples on the light emission of *Vibrio fischeri* (luminiscent bacteria test). ISO/CD 11348.

Kautsky, H. and U. Kautsky. 1994. Coastal productivity in the Baltic Sea. The 28th European Marine Biology Symposium, EMBS, Crete, Sept 23rd-28th 1993.: In press.

Lang, K. 1948. Monographie der Harpacticiden, Vols. I, II. Ohlsson, Lund.

Leach, J.M. and A.N. Thakore. 1977. Compounds toxic to fish in pulp mill waste streams. Prog. Wat. Tech. 9:787-798.

Lindblad, C. 1994. Functional disturbances in Baltic Sea *Fucus vesiculosus* communities after chlorate exposure. *In* C. Lindblad (ed) Perturbation of Functions in Shallow Water Benthic Ecosystems. Thesis at Stockholm University.

Microbics Corporation. 1991. Carlsbad, California No M102.

Microbics Corporation. 1992. Microtox Manual, A Toxicity Testing Handbook, Vol. 1-5.

O'Connor, B.I., T.G. Kovacs, R.H. Voss, P.H. Martel and B. Van Lierop. 1993. A laboratory assessment of the environmental quality of alternative pulp bleaching effluents. International Environmental Symposium, EU CE PA, Paris.

OECD. 1993. OECD test no 202. *Daphnia* sp. Acute immobilization test and reproduction test. Guidelines for Testing of Chemicals. Section 2.

Qureshi, A.A., K.W. Flood, S.R. Thompson, S.M. Janhurst, C.S. Inniss and D.A. Rokosh. 1982. Comparison of a luminescent bacterial test with other bioassays for determining toxicity of pure compounds and complex effluents. Amer. Soc. Test. Mater. STP 766:179-195.

Renberg, L. 1992. The use of cost-effective chemical and biological tests for the estimation of the environmental impact of bleaching plant effluents. Proceedings of the 1992 TAPPI Environmental Conference, Richmond, Virginia:317-329.

Rosemarin, A., J. Mattsson, K.-J. Lehtinen, M. Notini and E. Nylén. 1986. Effects of pulp mill chlorate (ClO_3^-) on *Fucus vesiculosus*. Ophelia, Suppl. 4:219-224.

Rueness, J. 1978. Hybridization in red algae. *In* D.E.G. Irvine and J.H. Price (ed.) Modern Approaches to the Taxonomy of Red and Brown Algae. Academic Press, London and New York.

Rueness, J. and R.-A. Kornfeldt. 1992. Ecotypic differentiation in salinity responses of *Ceramium strictum* (Rhodophyta) from Scandinavian waters. Sarsia 77:207-212.

SIS (Standardiseringsorganisationen I Sverige). 1991. Determination of acute lethal toxicity of chemical substances and effluents to *Nitocra spinipes* Boeck - Static procedure. Svensk Standard SS 02 81 06. In Swedish.

Steele, R.L., and G.B. Thursby. 1983. A toxicity test using life stages of *Champia parvula* (Rhodophyta). p. 73-89. *In* W.E. Bishop, R.D. Cardwell and B.B. Heidolph. (Ed.) Aquatic Toxicology and Hazard Assessment: Sixth Symposium. ASTM STP 802. American Society for Testing and Materials, Philadelphia, Pennsylvania.

Södergren, A. 1993. Bleached pulp mill effluents. Composition, fate and effects in the Baltic Sea. Swedish Environmental Protection Agency Report 4047.

Södergren, A., P. Jonsson, B.-E. Bengtsson, K. Kringstad, S. Lagergren, M. Olsson and L. Renberg. 1989. Biological effects of bleached pulp mill effluents. Swedish Environmental Protection Agency Report 3558.

Tarkpea, M., M. Hansson and B. Samuelsson. 1986. Comparison of the Microtox test with the 96-hr LC50 test for the harpacticoid *Nitocra spinipes*. Ecotoxicol. Environ. Saf. 11:127-143.

Tidepool scientific software. 1992-1993. ToxCalc users guide. Comprehensive toxicity data analysis and database software. Version 4.0. Mc Kinleyville, California.

AN ASSESSMENT OF THE SIGNIFICANCE OF DISCHARGE OF CHLORINATED PHENOLIC COMPOUNDS FROM BLEACHED KRAFT PULP MILLS

Robert P. Fisher, Douglas A. Barton and Paul S. Wiegand

National Council of the Paper Industry for Air and Stream Improvement, Inc.,
P.O. Box 141020, Gainesville, FL 32614 U.S.A.

Laboratory and field studies conducted over the past several decades have confirmed earlier findings that pulp and paper mill effluents which have been well treated in conventional biological treatment processes in general do not constitute adverse aquatic life or human health impacts. Nonetheless, questions continue to arise regarding the effects of various components of bleached mill effluents on aquatic and human health. To address these concerns, the concentrations of chlorinated phenolic compounds in waters impacted by bleached kraft mills were examined relative to aquatic life and human health water quality protection values. Water quality protection values were assembled, discharge rates of chlorinated phenolics were estimated, and low flow stream conditions were determined. These data were combined to calculate frequency distribution plots of in-stream concentrations for each of the compounds downstream of each mill. The plots were used to determine whether differences exist in exceedance of water quality protection values for any of the compounds of concern under baseline conditions and under either of two currently proposed technology options. Results indicate that a minority of mills show some exceedances of water quality protection values under baseline conditions, but that no facilities will produce exceedances after implementation of either complete chlorine dioxide substitution or this option plus oxygen delignification.

INTRODUCTION

Chlorinated phenolic compounds contained in bleached kraft mill effluent have been the subject of a number of studies and reviews (e.g., Suntio *et al.* 1988; Voss *et al.* 1980), and the toxicity and environmental fate of chlorinated phenols have been investigated and reported (e.g., Shiu *et al.* 1994; Saarikoski and Viluksela 1982). As a component of its clustered rule-making proposal for the pulp and paper industry (EPA 1993a), the U.S. Environmental Protection Agency proposed limitations on the release from certain categories of pulp bleaching of twelve chlorinated phenolic compounds. Implicit in this proposal and in the agency's analysis of postulated benefits which will result therefrom (EPA 1993b) are the concepts that (a) water quality values for these compounds may be exceeded downstream of bleached mill discharges in the absence of new regulations and (b) the regulatory option proposed by the Agency, viz., oxygen delignification and complete substitution of chlorine dioxide for chlorine in the bleach plant, is the option incorporating the minimum of technology and process changes which will result in avoidance of exceedances of water quality values for chlorinated phenolic compounds. The purpose of this paper is to address these concepts.

MATERIALS AND METHODS

Overview of Strategy

To determine the frequency with which water quality protection values for the chlorinated phenolic compounds are approached or exceeded under baseline and two option conditions, the following activities were carried out:

(a) Water quality values for the chlorinated phenolic compounds of concern were assembled. These include EPA criteria, water quality protection values derived by EPA and its contractor, and water quality screening values derived by NCASI and its contractor.
(b) Discharge rates of 28 chlorinated phenolic compounds from individual bleached kraft mills were calculated. Data are available for 69 mills, which were organized into six categories depending upon the degree of or use of oxygen delignification, extended delignification, and chlorine dioxide substitution for chlorine.
(c) Low flow stream conditions (harmonic mean, 7Q10, 1Q10, 30Q5) were determined for each of the streams receiving effluent from each of 86 bleached kraft mills.
(d) The mill effluent pollutant concentrations, the mill effluent flow rate, and low flow stream conditions were combined to calculate frequency distribution plots of in-stream concentrations for each of the compounds of concern downstream of each mill under four design flow conditions: 1Q10, 7Q10, 30Q5, and harmonic mean. The water quality protection bases corresponding to these design flow conditions are acute aquatic toxicity, chronic aquatic toxicity, human health - non-cancer, and human heath - cancer, respectively.
(e) The frequency distribution plots were used to determine whether exceedances of water quality values exist for any of the compounds of concern under baseline conditions and under either of the two currently proposed technology options.

Development of Water Quality Screening Values

The twelve chlorinated phenolic compounds for which EPA has proposed regulation are designated higher chlorinated phenolic derivatives of lignin (HCPDs). The HCPDs include four chlorinated phenols (2,4,5-tri-, 2,4,6-tri-, 2,3,4,6-tetra-, and pentachlorophenol), four chlorinated guaiacols (3,4,5-tri-, 3,4,6-tri-, 4,5,6-tri-, and tetrachloroguaiacol), three chlorinated catechols (3,4,5-tri-, 3,4,6-tri-, and tetrachlorocatechol), and 3,4,5-trichlorosyringol. U.S. EPA has developed aquatic life water quality criteria only for pentachlorophenol and for 2,4,6-trichlorophenol and has proposed draft aquatic life criteria for 2,4,5-trichlorophenol. EPA's contractor, Versar, Inc., has developed guidance water quality protection values for these and other compounds (Versar 1993), and these values have been used by the agency in its benefits analysis (EPA 1993b). EPA and/or Versar have also developed human health based water quality protection values or criteria for four of these compounds: pentachlorophenol, 2,3,4,6-tetrachlorophenol, 2,4,5-trichlorophenol, and 2,4,6-trichlorophenol.

NCASI and its contractor, TERRA, Inc., prepared chronic aquatic life water quality screening values for ten of the HCPDs (NCASI 1994a), based upon new data and the use of structure activity relationships and sensitive multi-life cycle assays. These values are defensible and conservative. They were not derived by the use of arbitrary application factors applied to acute toxicity values, as is the case for many of the values used by the agency (EPA 1993b) and derived by Versar (1993). In some of the material which follows, the different water quality values are referred to by the names of the contract toxicology laboratories which participated in their development (i.e., Versar, TERRA).

The term *criterion* is used here to denote a water quality value which has been derived by EPA according to standard protocol (EPA 1985) and which is a component of existing EPA guidelines, or state regulations. The term *water quality protection value* is used to describe the water quality values derived by Versar for EPA. The term *water quality screening value* is used to indicate the water quality values derived by NCASI and TERRA. Neither the water quality protection values nor the water quality screening values are to be considered true criteria for chlorinated phenolic compounds.

A second category of compounds of interest in this activity is designated the lower chlorinated phenolic derivatives of lignin (LCPDs). These include two catechols (4-chloro- and 4,5-dichlorocatechol), three phenols (4-chloro-, 2,4-dichloro-, and 2,6-dichlorophenol), two vanillins (6-chloro- and 5,6-dichlorovanillin), and 2,6-dichlorosyringealdehyde. EPA did not address regulation of·the LCPDs in its 1993 proposed regulations, but benefits of reducing LCPDs were considered in the benefits discussion of the Regulatory Impact Assessment (EPA 1993). NCASI/TERRA also developed water quality screening values for the LCPDs (NCASI 1994a).

Numerous aspects of the development of water quality screening values for the HCPDs and the LCPDs, including development methodology, are discussed in a previously published NCASI technical bulletin (NCASI 1994a). The information contained in Table 1 includes a summary of the human health and aquatic life water quality values developed by EPA, Versar, and NCASI. Both the EPA/Versar values and the NCASI/TERRA values are used in the assessment described elsewhere in this paper.

Table 1. Summary of water quality screening and protection values for chlorinated phenolic compounds (all values are freshwater, µg L^{-1}).

Compounds	Human Health (Ingestion Wtr & Org)			Aquatic Toxicity Acute Effects	Aquatic Toxicity Chronic Effects	
	Versar	TERRA	Endpt[1]	Versar	Versar	TERRA
LCPDs						
4CC	NC(1)	NC		1580	79	650
4CP	24	24	N	3800	1100	650
6CV	NC	NC		2130	107	650
4,5 DCC	NC	NC		890	44.5	376
2,4 DCP	93	NC	N	2020	70	376
2,6 DCP	0.2	NC	O	3240	162	376
2,6 DCS	NC	NC		2950	147.5	376
5,6 DCV	NC	NC		1720	87.5	376
HCPDs						
PCP	0.28	0.4	C	20	13	NC
TeCC	NC	NC		726	7.3	200
TeCG	NC	NC		320	3.2	240
2,3,4,6 TeCP	810	NC	N	85	10	NC
3,4,5 TCC	NC	NC		1800	18	200
3,4,6 TCC	NC	NC		3820	38.2	200
3,4,5 TCG	NC	NC		750	7.5	240
3,4,6 TCG	NC	NC		3020	30.2	240
4,5,6 TCG	NC	NC		307	3.1	240
2,4,5 TCP	490	2200	N	450	4.5	150
2,4,6 TCP	1.6	4	C	320	3.2	500
TCS	NC	NC		5280	52.8	200

(1) NC = no criterion (or value) developed, N = non-cancer endpoint, C = cancer endpoint, O = organoleptic properties.

Calculation of Rates of Release of Chlorinated Phenolic Compounds

The data used to derive treated effluent phenolics loading estimates from bleached kraft mills were obtained from several sources. These include (1) data generated as part of the EPA and Paper Industry Effluent Variability Studies, (2) data generated during EPA's effluent guidelines-related Short Term Study, (3) mill-supplied data, and (4) NCASI file data. From these sources, data for 28 chlorinated phenolic compounds were consistently available. Phenolics data were available for samples of final treated effluent and bleach plant effluent (both acid and alkaline effluent) although data for bleach plant effluents were much more limited in number.

In total, data were available for more than 60 mills. However, the vast majority of these data were gathered from mills with low levels of chlorine dioxide substitution and without oxygen or extended delignification. Data for only 20 mills using high levels of chlorine dioxide substitution and mills using oxygen or extended delignification were available. The amount of data available from each source varied from a single 72-h composite sample to a series of eighteen 24-h composite samples. Further, virtually all data sets contained a high frequency of values reported as "not detected" (ND) above the analytical method minimum level (ML). To facilitate comparison between mills with different bleaching technologies, the mean value for each analyte in each matrix (final effluent or bleach plant effluent) was calculated.

Data were categorized by mill bleaching process type. Category divisions were based on the use of oxygen or extended delignification and the level of substitution of chlorine dioxide for chlorine. Six process categories were created as follows:

Complete substitution
Greater than 50% substitution
Less than 50% substitution
Oxygen/extended delignification and complete substitution
Oxygen/extended delignification and greater than 50% substitution
Oxygen/extended delignification and less than 50% substitution.

These categories were selected because they represent a range spanning the conventional bleaching technology used during the past decade and modern bleaching technology which is expected to become common during the next decade. It should be noted, however, that data from only two mills employing extended delignification were available. Therefore, treated final effluent discharge loading estimates for categories identified as "oxygen/extended delignification" are based almost entirely on mills employing oxygen delignification.

The high frequency of ND values in the data sets greatly complicated the process of making treated effluent loading estimates. If an analyte was detected in any of the treated effluent samples, only data from final effluents were used to derive the loading estimate. If an analyte was not detected in any of the final effluents tested, a loading estimate was calculated using bleach plant effluent data and known effluent treatment efficiencies. In this case, ND values were assumed to be equal to one-half of the analytical method minimum level. (Only minimum level values, as opposed to detection limit values, were available. Using one-half the minimum level value is nonetheless environmentally conservative.) Details of the procedures used to develop final effluent loading estimates are provided in a separate NCASI technical bulletin (NCASI 1994b).

The resulting final effluent loading estimates are intended to be used as a means of comparing the relative characteristics of treated final effluents derived from the use of different kraft pulp bleaching operations. No distinction has been made concerning wood species (i.e., hardwood vs. softwood). In addition, some loading estimates are based entirely on concentration data for which no values were above the ML. Such loading estimates are particularly common to the two complete substitution production categories. As such, it is recommended that these estimates not be used for purposes other than the stated purpose in the current analysis.

Determination of Receiving Water Low Flow Conditions

Low flow stream conditions for each of the 79 bleached kraft mills which discharge to rivers were obtained from the EPA STORET on-line data base using the DFLOW procedure available on the mainframe computer at the EPA National Computing Center. The following flow conditions were obtained: 7Q10, 1Q10, 30Q5, and harmonic mean.

An additional ten mills discharge to open waters such as oceans, estuaries, lakes, or ship channels. For these mills, the zone of initial dilution factor provided in the EPA TCDD Risk Assessment Document (EPA 1990) was used to estimate the low flow stream conditions. The calculated flow was used to estimate the in-stream concentrations for all four of the toxicity values.

For various reasons pertaining to discharge conditions, four mills were excluded from this analysis. The total number of discharges included in this analysis was therefore reduced from 90 to 86.

Various other special conditions existing at other mills were not accounted for in this analysis. These conditions include factors such as seasonally intermittent discharges and tidally controlled discharges. In the former case, the actual in-stream concentration is likely overstated since the mill probably does not discharge during low flow stream conditions. In the latter case, the receiving water concentrations are also likely to be overstated since discharge only during ebb tide enhances the flushing rate of the compound from the point of discharge.

Determination of In-Stream Concentrations Under Low Flow Conditions

In-stream concentration estimates of the twenty chlorinated phenolic compounds were made for each of three conditions: (1) baseline conditions representing pulping and bleaching equipment in place as of January 1993, (2) conditions following implementation of oxygen delignification (or extended delignification) and complete ClO_2 substitution at all bleached kraft mills, and (3) conditions following implementation of complete ClO_2 substitution but not necessarily including oxygen or extended delignification. The resulting estimates of in-stream concentrations are compared with the water quality criteria/values presented above.

Mill-specific estimates of mass effluent loadings (g d^{-1}) were developed for each of the twenty chlorinated phenolic compounds based on the unit mass effluent loadings (gm $tonne^{-1}$) described above and on the bleaching configuration of the mill. Information contained in the NCASI data base of existing (as of January 1993) bleaching conditions for each bleached kraft mill was used to assign each bleach line to one of the six groupings used in the development of the final effluent mass loadings.

The effluent mass loading for a specific compound, not a specific mill, under the baseline conditions was calculated in the following manner. For each bleach line at a given mill, the unit mass effluent loading for the bleaching group corresponding to the bleaching conditions practiced at that bleach line was multiplied by the bleach line production capacity to yield a mass effluent loading. Then, the mass effluent loadings for all bleach lines at that mill were summed to arrive at the mill total mass effluent loading.

Effluent mass loadings projected under the conditions of complete ClO_2 substitution with and without oxygen delignification were estimated in a manner similar to that used for the baseline conditions except that all bleach lines were assigned unit effluent loadings derived for the pulping/bleaching configuration under consideration.

Table 2 lists the estimated industry total discharge loading under each of the three pulping/bleaching configurations for which mill-specific effluent loading estimates were made. Also listed in Table 2 are similar effluent loading estimates prepared by EPA and published in the Cost Effectiveness Document (EPA 1993c).

Recognizing that there is some uncertainty in developing these estimates of effluent mass loadings for the chlorinated phenolic compounds, close agreement between the NCASI estimate and the EPA estimate would not necessarily be expected. However, the EPA estimates of loadings for several compounds, including two of the more highly substituted compounds, exceed the NCASI effluent loading estimates by more than an order of magnitude. Comparison of the EPA estimates with actual effluent data indicates that EPA grossly overstated the final effluent loadings for the following four compounds: 4-chlorocatechol, 4-chlorophenol, 2,4,5-trichlorophenol, and pentachlorophenol (NCASI 1994b.)

Table 2. Estimated annual discharge of chlorinated phenolic compounds under baseline, BAT, and ClO_2 substitution only conditions.

	Annual Discharge (lb yr^{-1}) (a)				
	NCASI Estimate (b)			EPA Estimate (c)	
Compound	Baseline	BAT Options	ClO_2 Sub. Only	Baseline	BAT Options
4-Chlorocatechol	3421	276	1113	343117	2325
4-Chlorophenol	2910	269	749	2325597	1984
6-Chlorovanillin	26370	15669	6693	19622	3262
4,5-Dichlorocatechol	58447	327	327	22987	2502
2,4-Dichlorophenol	20390	327	327	35631	2169
2,6-Dichlorophenol	1769	327	327	7716	2338
2,6-Dichlorosyringealdehyde	12648	1440	1440	13309	4006
5,6-Dichlorovanillin	20493	1055	793	14683	4201
Pentachlorophenol	3489	2604	2604	108298	3996
Tetrachlorocatechol	71442	2604	2604	20091	4099
Tetrachloroguaiacol	41489	2604	2604	32455	3834
2,3,4,6-Tetrachlorophenol	7210	1302	1302	6379	1985
3,4,5-Trichlorocatechol	89268	2233	2233	43749	4120
3,4,6-Trichlorocatechol	12235	2604	2604	11121	3991
3,4,5-Trichloroguaiacol	69555	1258	1258	127607	2200
3,4,6-Trichloroguaiacol	2049	1302	1302	5230	1954
4,5,6-Trichloroguaiacol	19900	1542	1302	23615	2316
2,4,5-Trichlorophenol	1972	1302	1302	171466	1876
2,4,6-Trichlorophenol	40171	1557	1557	54756	2908
3,4,5-Trichlorosyringol	20125	1302	1302	17460	4467

(a) Based on estimated unit mass effluent loading and bleached pulp production; (b) papergrade and dissolving kraft only; (c) from Tables C-1 & C-3 of EPA-821-R-93-016.

For each of the twenty chlorinated phenolic compounds, in-stream concentrations were estimated at up to three different low flow stream conditions for each of the 86 dischargers. The resulting in-stream concentrations were sorted into an ascending order and plotted on a probability scale. A total of 47 combinations of chlorinated phenolic compounds and water quality criteria/values were examined. Representative examples of these frequency distributions are provided in Figure 1. Note that for some compounds (e.g., 2,3,4,6-tetrachlorophenol and pentachlorophenol) unit mass effluent loadings are identical for the cases of complete ClO_2 substitution with oxygen delignification and without oxygen delignification due to the widespread occurrences of non-detectable concentrations in bleach plant filtrates and final effluents under both of these conditions.

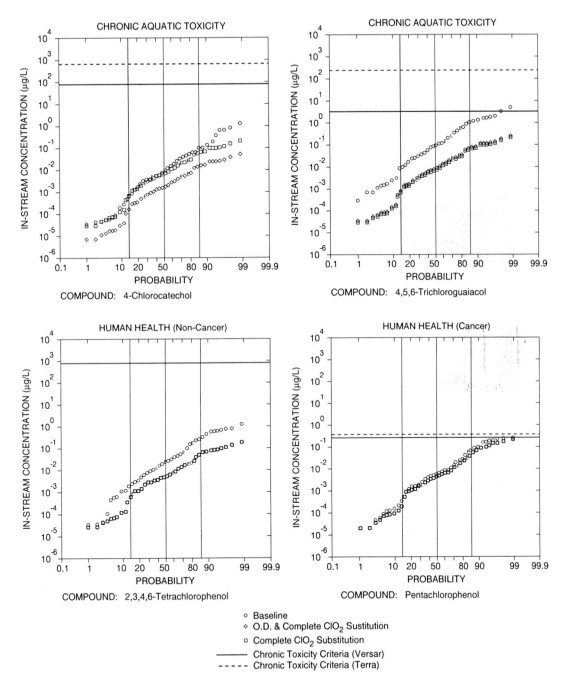

Figure 1. Example probability distribution plots of in-stream chlorinated phenolic concentrations estimated under three conditions.

RESULTS AND DISCUSSION

Comparison of In-Stream Concentrations with Water Quality Values

The four types of water quality values examined in this analysis apply to different low flow conditions, as discussed above. Table 3 presents a matrix of the water quality values available for each of the twenty chlorinated phenolic compounds. Where available, both the EPA/Versar water quality protection value and the NCASI/TERRA water quality screening value were used for comparison with the projected distribution of in-stream concentrations.

Table 3. Matrix of water quality values availability for twenty chlorinated phenolic compounds.

Compound	Acute Aquatic	Chronic Aquatic	Human Non-Cancer	Human Cancer
4-Chlorocatechol	X	X		
4-Chlorophenol	X	X	X	
6-Chlorovanillin	X	X		
4,5-Dichlorocatechol	X	X		
2,4-Dichlorophenol	X	X	X	
2,6-Dichlorophenol	X	X	X	
2,6-Dichlorosyringealdehyde	X	X		
5,6-Dichlorovanillin	X	X		
Pentachlorophenol	X	X		X
Tetrachlorocatechol	X	X		
Tetrachloroguaiacol	X	X		
2,3,4,6-Tetrachlorophenol	X	X	X	
3,4,5-Trichlorocatechol	X	X		
3,4,6-Trichlorocatechol	X	X		
3,4,5-Trichloroguaiacol	X	X		
3,4,6-Trichloroguaiacol	X	X		
4,5,6-Trichloroguaiacol	X	X		
2,4,5-Trichlorophenol	X	X	X	
2,4,6-Trichlorophenol	X	X		X
Trichlorosyringol	X	X		

Note: X indicates water quality value available.

Based on the effluent mass loadings developed in this analysis, the estimated in-stream concentrations of chlorinated phenolic compounds were projected to exceed the EPA/Versar water quality protection values under baseline conditions in a total of 29 instances (7 compounds at 10 different mills). In only one case was an estimated in-stream concentration projected to exceed a water quality screening value derived by NCASI/TERRA, and that was the human health value of 4.0 µg L^{-1} for 2,4,6-trichlorophenol, which was projected to be exceeded downstream of one mill at a value of 4.05 µg L^{-1}. In three other cases, EPA/Versar values for 2,6-dichlorophenol are exceeded slightly at baseline, and since NCASI/TERRA did not develop an alternative water quality screening value for this compound, the NCASI analysis assumes exceedance for this compound at baseline. These results are summarized in Table 4.

Table 4. Number of water quality value exceedances estimated in this study under baseline conditions* and following implementation of BAT.

Condition and WQ Protection Values	Number of Exceedances Under Toxicity/Flow Regimes				
	Aquatic Acute	Aquatic Chronic	Human Non-Cancer	Human Cancer	Total
Baseline					
EPA/Versar	0	20	3	6	29
NCASI/TERRA	0	3	0	1	4
Oxygen Delig. & Complete Sub.					
EPA/Versar	0	0	0	0	0
NCASI/TERRA	0	0	0	0	0
Complete Sub. Only					
EPA/Versar	0	0	0	0	0
NCASI/TERRA	0	0	0	0	0

* Based on mill discharge estimates developed by NCASI.

In comparison, EPA's Water Quality Assessment Document estimates that 80 instances of water quality exceedances occur under baseline conditions. The number of water quality value exceedances projected by the EPA analysis and the NCASI analysis are compared in Table 5. The largest number of discrepancies occur for two compounds, pentachlorophenol and 2,4,5-trichlorophenol, which alone account for 38 instances of water quality value exceedance in the EPA analysis. As indicated above, it appears that EPA has overestimated the mass effluent loadings for these compounds by one to two orders of magnitude. (Note that these compounds are rarely detected in analyses of current bleach plant effluents. They are not thought to be produced in pulp bleaching, rather their presence in effluents historically was due to use of slimicide formulations which are no longer employed by the industry.) The projected number of water quality exceedances under baseline conditions for each of the twenty compounds examined in this analysis is available elsewhere (NCASI 1994b).

Data provided in the Water Quality Assessment Document indicate that implementation of the EPA Proposed BAT Technologies (oxygen or extended delignification and complete ClO_2 substitution) would reduce the number of water quality exceedances of chlorinated phenolic compounds to five: two for 2,6-dichlorophenol and three for pentachlorophenol. This information is also summarized in Table 5. In contrast, based on the effluent mass loadings developed in this analysis, in-stream concentrations projected by NCASI under EPA's proposed BAT technology indicate no exceedances of water quality values will occur (Table 4). This applies both to water quality protection values developed by EPA/Versar and to water quality screening values developed by NCASI/TERRA.

NCASI also used the effluent loadings information developed in this analysis to examine the number of water quality value exceedances which would be expected to occur if the BAT technology were defined as complete ClO_2 substitution, without the requirement for extended delignification or oxygen delignification. Using mass effluent loadings estimated for this bleaching configuration, and the water quality values described above, the resulting in-stream concentrations were found not to exceed any of the four water quality value types evaluated (Table 4). Again, this applies both to water quality protection values developed by EPA/Versar and to water quality screening values developed by NCASI/TERRA.

Table 5. Comparison of EPA and NCASI analyses of numbers of estimated water quality protection value exceedances under baseline conditions and following implementation of BAT.

Condition and WQ Protection Values	Number of Exceedances Under Toxicity/Flow Regimes				
	Aquatic Acute	Aquatic Chronic	Human Non-Cancer	Human Cancer	Total
Baseline					
EPA Analysis	2	36	7	35	80
NCASI Analysis	0	20	0	9	29
Oxygen Delig. & Complete Sub.					
EPA Analysis	0	0	0	5	5
NCASI Analysis	0	0	0	0	0
Complete Sub. Only					
NCASI Analysis	0	0	0	0	0

* The analyses summarized here considered exceedances of the EPA/Versar water quality protection values only (Versar 1993).

CONCLUSIONS

Based on estimated mass effluent loadings for various pulping/bleaching configurations, in combination with appropriate low flow stream conditions, this study showed that under baseline (January, 1993), low flow conditions, only a minority of bleached mills (ten) are anticipated to show any exceedances of even the highly conservative water quality protection values developed by EPA and Versar for chlorinated phenolic compounds, and these exceedances are not large. Considering the more realistic and scientifically sound water quality screening values developed by NCASI and TERRA, only four mills are estimated to show exceedances at baseline under low flow conditions.

The frequency distribution plots also show that after implementation of either the BAT option of complete chlorine dioxide substitution plus oxygen delignification or the option of complete chlorine dioxide substitution alone, it is estimated that no exceedances of water quality values will occur. This important finding is restated below.

As a result of this activity, several conclusions regarding U.S. EPA's proposed effluent guideline regulations may be made:

(1) Regarding proposed benefits of the newly proposed rule, EPA overestimated the industry's biologically treated effluent mass loadings at baseline by a factor of at least 30 for four chlorinated phenolic compounds, including two of the more highly substituted compounds.

(2) EPA overestimated by a factor of three the number of chlorinated phenolic compound related water quality exceedances under baseline conditions. At least two-thirds of the margin of overestimation is believed to result directly from the overestimation of effluent mass loadings for the same four compounds indicated in point (1), above.

(3) EPA overestimated the number of water quality exceedances following implementation of proposed BAT technologies, in part due to the overestimation of the effluent mass loading for pentachlorophenol. This study has shown that no occurrences of water quality value exceedance should take place following implementation of the BAT option.

(4) The results of this study also indicate that biologically treated effluent discharges of chlorinated phenolic compounds from mills which employ complete chlorine dioxide substitution without oxygen delignification or extended delignification will not result in exceedances of water quality values. This demonstrates that, considering chlorinated phenolic compounds and water quality value exceedances, complete chlorine dioxide substitution is equivalent to oxygen delignification plus complete chlorine dioxide substitution.

ACKNOWLEDGMENTS

In-stream concentration calculations and frequency distribution plots were prepared by HydroQual, Inc., Mahwah, NJ.

REFERENCES

EPA. 1985. Guidelines for Deriving Numerical National Water Quality Criteria for the Protection of Aquatic Organisms and Their Uses, NTIS Publication PB85-227049.

EPA. 1990. Risk Assessment for 2378-TCDD and 2378-TCDF Contaminated Receiving Waters From U.S. Chlorine-Bleaching Pulp and Paper Mills, NTIS Publication PB90-272-873.

EPA. 1993a. Effluent Limitations Guidelines, Pretreatment Standards, and New Source Performance Standards: Pulp, Paper, and Paperboard Category; National Emission Standards for Hazardous Air Pollutants for Source Category: Pulp and Paper Production; Proposed Rule, FR 58 (241), 66078-66216.

EPA. 1993b. Regulatory Impact Assessment of Proposed Effluent Guidelines and NESHAP for the Pulp, Paper, and Paperboard Industry, EPA-821-R-93-020.

EPA. 1993c. Cost Effectiveness Analysis of Proposed Effluent Guidelines for the Pulp, Paper, and Paperboard Industry, EPA-821-R-93-016.

NCASI. 1994a. Aquatic Life Water Quality Screening Values for Chlorinated Phenolic Compounds, NCASI Technical Bulletin No. 659, NCASI, 260 Madison Avenue, New York, NY 10016.

NCASI. 1994b. An Assessment of the Significance of Discharge of Chlorinated Phenolic Compounds From Bleached Kraft Pulp Mills, NCASI Technical Bulletin No. 658, NCASI, 260 Madison Avenue, New York, NY 10016.

Saarikoski, J., and M. Viluksela. 1982. Relation between physicochemical properties of phenols and their toxicity and accumulation in fish. Ecotoxicol. Environ. Safety. 6:501-512.

Shiu, W.Y., K.-C. Ma, D. Varhanickova and D. McKay. 1994. Chlorophenols and alkylphenols: A review and correlation of environmentally relevant properties and fate in an evaluative environment. Chemosphere 29:1155-1224.

Suntio, L.R., W.Y. Shiu and D. McKay. 1988. A review of the nature and properties of chemicals present in pulp mill effluents. Chemosphere. 17:1249-1290.

Versar, Inc. 1993. Human Health and Aquatic Life Toxicity Data and References for Pulp, Paper, and Paperboard Industry Proposed Effluent Guidelines Environmental Assessment, prepared for U.S. EPA, Office of Science and Technology, Standards and Applied Science Division, Washington, D.C.

Voss, R.H., J.T. Wearing, R.D. Mortimer, T. Kovacs and A. Wong. 1980. Chlorinated organics in kraft bleachery effluents. Pap. Puu 62:809-814.

A COMPARISON OF GAS CHROMATOGRAPHIC AND IMMUNOCHEMICAL METHODS FOR QUANTIFYING RESIN ACIDS

Kai Li, Tao Chen, Paul Bicho, Colette Breuil and John N. Saddler

Department of Wood Science, University of British Columbia, Vancouver, BC, V6T 1Z4, Canada

Resin acids are major sources of fish toxicity in many Canadian softwood pulp mill effluents. Therefore, their accurate detection is an important aspect in the evaluation of the environmental impact of these effluents. Among the various methods used to detect and quantify resin acids, gas chromatography (GC) is usually the method of choice due to its high resolution and accuracy. However, major disadvantages such as tedious sample pretreatment protocols and low sample output reduce the suitability of this method for analyzing large numbers of samples in a short period of time. In this paper, a traditional gas chromatographic method and an immunological method were compared. In the GC method, liquid-liquid extraction was further evaluated using four different solvents (diethyl ether, dichloromethane, methyl *t*-butyl ether and ethyl acetate) for their extraction efficiency. An enzyme-linked immunosorbent assay (ELISA) based on polyclonal antibodies was developed for detecting dehydroabietic acid (DHA), one of the most abundant resin acids found in softwood pulp mill effluents. The direct ELISA can detect DHA in the range between 10 and 500 µg L^{-1} with 50% inhibition concentration of 261.6 µg L^{-1}. Both GC and the direct ELISA were used to quantify DHA in a spiked effluent and the results were compared. Since the immunoassay requires no pretreatment, a small volume of sample and results in a larger sample output, this alternative method of resin acid analysis shows promise.

INTRODUCTION

Resin acids are diterpenoid carboxylic acids present in most softwoods and are usually released from wood chips during pulping processes (Taylor *et al.* 1988; Rogers *et al.* 1979; McKague *et al.* 1977). They are of concern to the pulp and paper industry because of their acute toxicity towards fish and other aquatic life. For example, the 96 h LC50 of most resin acids for rainbow trout falls between 0.2 and 1.7 ppm (Rogers 1973; Walden and Howard 1981; Priha and Talka 1986; Taylor *et al.* 1988; McCarthy *et al.* 1990). It has been reported that resin acids may contribute to as much as 70% of the toxicity of whole effluents (Leach and Thakore 1977). The eight common resin acids usually detected in Canadian softwood pulp mill effluents are listed in Fig. 1. Dehydroabietic acid (DHA) is one of the most abundant resin acids found in pulp mill effluents and can form chlorinated derivatives by electrophilic substitution of chlorine on the aromatic ring during pulp bleaching when chlorine or chlorine dioxide is used. These chlorinated resin acids are reported to be more persistent and toxic than their parents (McLeay and Associates 1987; Taylor *et al.* 1988).

The accurate detection of resin acids in pulp mill effluents is crucial for the evaluation of the environmental impact of these compounds. Even though several methods of resin acid analysis have been reported, the primary method of choice is gas chromatography (GC), which requires extraction of analytes from a sample matrix, derivatization to increase analyte volatility, and separation and quantification by capillary GC (Zinkel and Engler 1977; Foster and Zinkel 1982). Extraction is usually performed by either liquid-liquid extraction or solid phase extraction. Derivatization methods for GC analysis include ethylation (NCASI 1986) or methylation (Voss and Rapsomatiotis 1985) for flame ionization detector (FID) or pentafluorbenzylation when using an electron capture detector (ECD) (Lee *et al.* 1990).

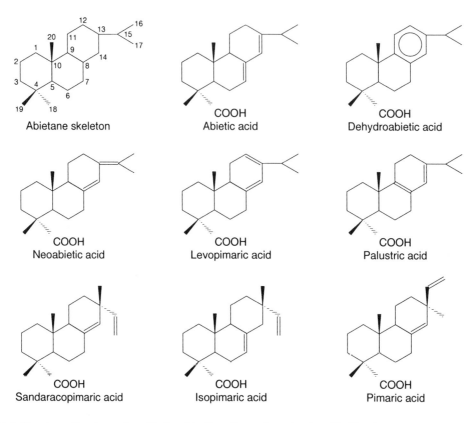

Figure 1. Structure of major resin acids found in Canadian softwood pulp mill effluents.

During the past two decades, various efforts have been made to extract and quantify resin acids in pulp mill effluents. Different pHs (2-10) have been used in the extraction step to ensure good recovery (NCASI 1975; Richardson and Bloom 1982; NCASI 1978; Wilkins and Panadam 1987; NCASI 1986; Voss and Rapsomatiotis 1985; Morales *et al.* 1992). Different organic solvents such as diethyl ether, methylene chloride and chloroform have also been investigated for their extraction efficiency. In order to minimize emulsion problems, specific solvents such as methyl *tert*-butyl ether (MTBE) (Voss and Rapsomatiotis 1985), petroleum ether-acetone-methanol (McMahon 1980), hexane-acetone-methanol (Wearing *et al.* 1984) and MTBE-dichloromethane (Morales *et al.* 1992) have been suggested. In order to solve the current contradictory recommendations in the literature, Chen *et al.* (unpublished data, Department of Wood Science, University of British Columbia, Vancouver, BC, Canada) studied the effects of extraction efficiency using a broad range of pHs and different solvents. It was found that the total resin acids were extracted more effectively at pH 2-3 than at higher pHs. Previously Voss and Rapsomatiotis (1985) had found higher extraction efficiencies at pH 9.

In this paper, we further investigate the extraction efficiency using four common solvents, diethyl ether, dichloromethane, MTBE and ethyl acetate. Since GC analysis requires tedious sample pretreatment, sophisticated instrumentation and specially trained personnel, this method is not easily available to most pulp mills. Together with the disadvantage of low sample throughput, a GC method is not suitable for applications where large numbers of samples are analyzed and results are required within a short period of time. To overcome these problems, an enzyme-linked immunosorbent assay (ELISA) was developed as an alternative method for an easy and fast quantification of DHA one of the major resin acids found in softwood pulp mill effluents. A direct ELISA format was used for the quantification of DHA. Finally, a DHA spiked effluent was analyzed by both GC and direct ELISA and the results were compared. Some advantages of an ELISA method over a conventional GC method are also discussed.

MATERIALS AND METHODS

Resin Acids Analysis by GC

Authentic standards of resin acids were obtained from Helix Biotech Corporation (Richmond, BC, Canada). Diethyl ether, dichloromethane, MTBE and ethyl acetate (Ominosol) were HPLC grade and used without further purification. Carbonate-bicarbonate buffer capsules, phosphate-citrate buffer with sodium perborate capsules, and o-phenylenediamine (OPD) were purchased from Sigma (St. Louis, MO). Bleached kraft effluents were obtained from a British Columbia interior mill. Fresh effluent was divided into volumes large enough to accommodate all of the analysis and kept at 4°C until ready for extraction. The pH of the effluent was adjusted to 2 and 25 mL of pH adjusted effluent was spiked with 25 µL of surrogate spiking solution (1.0 mg mL^{-1} of o-methyl podocarpic acid in methanol) and saturated with about 10 g of NaCl for 10 min before extraction. The saturated samples were then transferred to a separatory funnel and extracted two times with an equal volume of either ethyl acetate, diethyl ether, MTBE or dichloromethane. The organic phases from each extraction were pooled and dried over anhydrous Mg_2SO_4. The solvent was then removed by evaporation under reduced pressure. The dried extracts containing resin acids were dissolved in one drop of methanol and internal standards (25 µL of methyl henecosinoate acid and tricosanoic acid, 1.0 mg mL^{-1} of each) were added. The etheral solution of diazomethane was added until a persistent yellow color was obtained. The mixture was then evaporated to dryness under nitrogen and re-dissolved in 1.0 mL of ethyl acetate. GC was performed on a Hewlett Parkard HP 5890 series II gas chromatograph equipped with an HP 7673 auto injector and a FID. Individual resin acids were identified and quantified using a DB-5 fused silica capillary column (ID, 0.25 mm, length 30 m). Helium and nitrogen were used as the carrier and make-up gases, respectively. Injector and detector temperatures were 260 and 290°C. The temperature program used was as follows: 150°C for 3 min, 1.2°C/min to 170°C, then 0.6°C/min to 190°C, finally 5.0°C/min to 275°C.

ELISA

The 96-well microtitration plate (IMMULON® 4 flat bottom) was purchased from Dynatech Laboratories Inc. (Chantilly, VA) and a THERMOmax™ microplate reader was obtained from Molecular Devices Corp. (Menlo Park, CA). The polyclonal antibodies against DHA were obtained by immunization of rabbits with DHAM-SUC-KLH (Li et al. 1994). Enzyme conjugate, DHAM-SUC-HRP, was prepared by the carbodiimide method using DHAM-SUC and HRP. A microtiter plate was coated with polyclonal antibodies (1:8,000, 100 µL/well) in carbonate-bicarbonate buffer (pH 9.6) at 37°C overnight. The plate was washed four times with PBS buffer and blocked with 2% milk in PBS buffer (200 µL/well) by incubation at 37°C for 1 h. After washing as described above, diluted DHAM-SUC-HRP (1:10,000) was mixed with different amounts of DHA, added to the microtiter plate (100 µL/well) and incubated at 37°C for 2 h. After washing, o-phenylenediamine (1.0 mg mL^{-1} in citrate-phosphate buffer with sodium perborate, 100 µL/well) was added and the reaction was stopped after 5 minutes by adding sulfuric acid (2.5 M, 50 µL/well). Optical density was measured at 490 nm on a THERMOmax™ microplate reader.

RESULTS

GC Analysis

Different pHs have been used by various groups for the liquid-liquid extraction of resin acids from pulp mill effluent. Based on our experience, significantly improved extraction of the resin acids from the pulp mill effluents was observed at lower pHs. This agreed with the observations reported by other workers (NCASI 1978, 1986) who showed that the extraction efficiency of resin acids was inversely proportional to the pH of the aqueous medium. However, Voss *et al.* (1985) reported that recoveries of 95% or greater for resin and fatty acids were obtained when the pH of the aqueous phase was adjusted to between 6 and 10. At pH 2, recoveries were less than 70%. It is recognized that the un-ionized form of resin acids is extracted more efficiently than their ionized forms. As the dissociation constants (pK_as) of resin acids are approximately 7, at pHs higher than 7, most of the resin acids were in the ionized form (Equation 1) and were therefore less extractable by organic solvents. At pH 9, as suggested by Voss and Rapsomatiotis (1985), more than 95% of the resin acids were present in the ionized form. Only under acidic conditions (pH <4) should all of the resin acids be in their neutral state (McLeay and Associates 1979).

$$RCOOH \iff RCOO^- + H^+ \qquad [1]$$

Based on this argument, we performed the extraction at pH 2-3 with four different solvents, diethyl ether, dichloromethane, MTBE and ethyl acetate. It was found that ethyl acetate gave the best resin acids recoveries from the pulp mill effluent sample under our experimental conditions. The good recoveries obtained with the two most abundant resin acids, DHA and abietic acid (Fig. 2), confirmed that ethyl acetate should be used in all subsequent work. An effluent free of resin acids was subsequently spiked with DHA and extracted in exactly the same way using ethyl acetate and the extract was subjected to GC analysis after derivatization with ethereal diazomethane. Part of the DHA spiked effluent was used for the ELISA procedure.

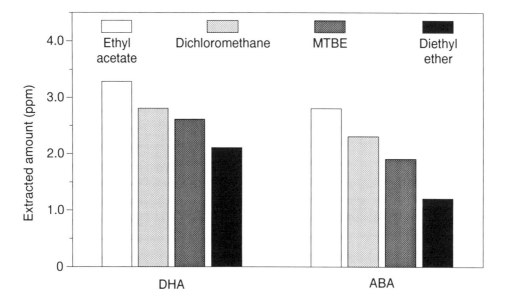

Figure 2. Solvent effect on the extraction of resin acids from a bleached kraft mill effluent.

ELISA

Recently, immunoassays have been introduced into the area of environmental analysis for the detection and quantification of toxic chemicals, such as pesticide residues and industrial pollutants (Vanderlaan et al. 1990; Van-Emon and Lopez-Avila 1992). ELISA is one of the most popular assay formats. It is based on the interaction between an antibody and antigen (Equation 2). The antibodies will react with the antigen, or the analyte of interest, to form an Ag-Ab complex.

$$Ab + Ag \iff Ab\text{-}Ag \qquad [2]$$

Antibodies are immunoglobulins produced by vertebrates upon immunization with an immunogen (antigen). Since resin acids are low molecular weight compounds (haptens) that are not immunogenic, they can not be used to immunize the animal directly. One way to make resin acids immunogenic is to covalently link them to a larger carrier such as protein. The hapten-protein conjugate can then be used to immunize the animal. The antibodies generated will recognize not only the hapten portion in the hapten-protein conjugate, but also the free hapten (Robins 1986).

In order to develop an ELISA for resin acids, DHA was chosen as the target compound because of its high concentration in pulp mill effluents and its stable chemical structure. Polyclonal antibodies against DHA were raised by immunization of rabbits with dehydroabietylamine conjugated to keyhole limpet hemocyanin (DHAM-SUC-KLH). Without optimization, the indirect ELISA using the biotin-streptavidin system had an I_{50} of 20.2 µg L^{-1} and a detection limit of 1.9 µg L^{-1} (Fig. 3). It was found that the polyclonal antibodies were not specific to DHA and cross-reacted with mono- and dichlorodehydroabietic acids and two abietic type resin acids, abietic and palustric acids (Li et al. 1994).

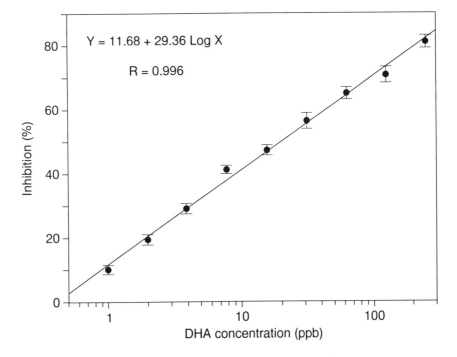

Figure 3. Standard curve for DHA in the indirect ELISA using biotin-streptavidin system.

Of the various formats of ELISA, we favored the indirect ELISA for determining the titration of the polyclonal serum and direct competitive ELISA for detecting DHA in effluent. Competitive ELISA using an enzyme tracer greatly reduces the amount of time needed to perform the traditional ELISA. There are still several ways to perform the direct competitive ELISA (Schneider and Hammock 1992). Fig. 4 shows the principle of the direct competitive ELISA format used in this study. The polyclonal antibodies against DHA were immobilized on the plate and reacted competitively with a mixture of analyte and hapten-enzyme conjugate to form a complex. When the substrate was added, the enzyme will catalyze the reaction and lead to the formation of a colored product. The intensity of the color will be inversely proportional to the concentration of analyte present in the system and can be determined by photometric methods. However, the enzyme tracer, DHAM-SUC-HRP, is not available commercially and must be synthesized. It was found that with the direct competitive ELISA, the polyclonal antibodies could be used in a 1:8,000 dilution and the 50% inhibition concentration was 261.6 µg L^{-1} and the detection limit was 17.4 µg L^{-1}, which was defined as the low limit (10% inhibition or I_{10}) of the linear portion of the standard curve (Fig. 5). Compared with the indirect ELISA using the biotin-streptavidin system (Li et al. 1994), the direct ELISA has a lower sensitivity. However, less assay steps are required in the direct ELISA.

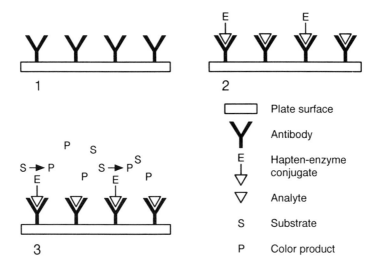

Figure 4. Principle of the direct competitive ELISA.

To determine if the developed ELISA was suitable for the detection of DHA in pulp mill effluents, a bio-treated, resin acid-free effluent was spiked with various amounts of DHA, and then analyzed by both GC and ELISA, following the procedures described previously. The effluent sample was extracted with ethyl acetate at pH 2-3, derivatized using etheral diazomethane and analyzed by GC. For the direct ELISA, the microtiter plate was coated with polyclonal antiserum (1:8,000) and blocked as described previously. Next, 200 µL of DHA spiked effluent was mixed with 800 µL of diluted enzyme conjugate, DHAM-SUC-HRP (1:8,000 in PBS with 0.1% milk). This maintained the final enzyme conjugate at 1:10,000 dilution. Then 100 µL of this solution was transferred to each well in the microtitration plate and, after a 2 h incubation, the substrate (OPD) was added and the OD was read as described previously. The recoveries determined by both the GC and the ELISA method are described in Table 1. It was found that the direct ELISA can recover more than 84.6% while GC gave recoveries greater than 93.6%.

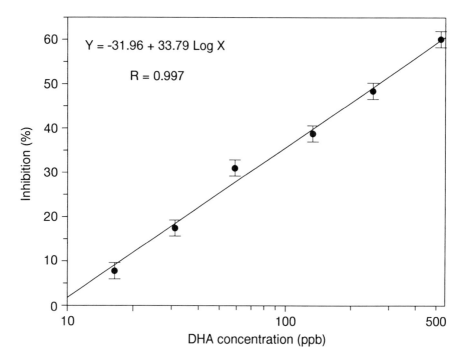

Figure 5. Standard curve for DHA in the direct ELISA.

Table 1. Recovery of DHA in spiked effluent using the direct ELISA and GC.

DHA Spiked ($\mu g\ L^{-1}$)	Recovery by ELISA		Recovery by GC	
	($\mu g\ L^{-1}$)	(%)	($\mu g\ L^{-1}$)	(%)
2500	3100 ± 385	124.0	2437	97.5
1250	1178 ± 131	94.2	1203	96.2
625	530 ± 67	84.8	592	94.7
312	293 ± 32	93.9	323	103.5
156	164 ± 25	105.1	146	93.6
78	66 ± 11	84.6	75	96.1

ELISA data were the average from 8 wells and GC was single determination.

DISCUSSION

The primary method for quantification of resin acids in pulp mill effluents is GC since all resin acid components can be resolved and quantified simultaneously. However, the major shortcomings of this method are the tedious sample pre-treatment such as extraction, purification and derivatization prior to GC analysis and the low sample output. Since the GC program for resin acid analysis requires around 70 minutes for each sample, less than eight samples can usually be processed daily. Although an ELISA method based on our polyclonal antibodies can not provide information on the concentration of individual resin acids, it should provide a fast way to screen samples for total abietic type resin acids, and 32 samples can be assayed simultaneously in triplicate using one plate. Processing of one plate using a direct ELISA format can usually be finished in 2.5 h if the plate has been precoated and blocked. This provides an enormous time saving over GC analysis. Also, ELISA can be adapted for automation, which should further increase the sample output. Therefore, ELISA was considered as a useful method for screening large numbers of samples such as effluent samples from different sources.

As the toxicities of the various individual resin acids are relatively similar (LC50 between 0.2-1.7 ppm), it is primarily the concentration of the total resin acids rather than the predominance of one type that will affect the toxicity of an effluent. As DHA is usually the most abundant resin acid found in most pulping processes when softwood species are used as the furnish (Taylor et al. 1988), the concentration of DHA in the effluents should usually reflect the total concentration of resin acids. If DHA concentrations are low, concentrations of other resin acids will be probably lower and of less environmental concern. Therefore, it is suggested that DHA concentration is a suitable parameter for approximating the concentration of total resin acids present in an effluent. The use of DHA as a measure of the total resin acids present in an effluent has also been suggested by Taylor et al. (1988) during the development of water quality guidelines. From available literature data and our own experience in resin acid analysis, abietic type resin acids, mainly DHA and abietic acid, usually constitute more than 70% of the total resin acid concentration present in a typical kraft mill effluent. To our knowledge, there is no circumstance where pimaric type resin acids predominate either in BKME or CTMP effluents.

Previously we had shown that the polyclonal antibodies used in this study recognized not only DHA, but also other abietic type resin acids (Li et al. 1994), This was probably due to similarities in the resin acid structures (Fig. 1) since all abietic type resin acids have an isopropyl group at the C13 position. Pimaric type resin acids, which have both methyl and vinyl groups at the C13 position, showed negligible cross-reactivity to the polyclonal antibodies. As most of the work reported in the literature indicated that the abietic type resin acids predominated in both treated and untreated effluents, an ELISA method which primarily detects the abietic type resin acids was thought to be of considerable use as it should approximate the total amount of resin acids that are present.

As an extension of this work, we are currently investigating the relationship between the toxicity and immunoreactivity to see whether the products released during the resin acid degradation and detoxification will be recognized by these polyclonal antibodies. It has been shown by numerous researchers that biological treatment of softwood pulp mill effluents greatly reduces toxicity, and that this is partially due to the removal of resin acids. We hope to use both GC and immunological methods to help identify the intermediates and metabolic pathways involved in the aerobic mineralization of resin acids. An ELISA could be used for monitoring resin acids present in the various pulp mill streams. If successful, this method will enable pulp mills to monitor resin acids more frequently. The rapid and routine availability of resin acid concentrations could perhaps enable mills to adjust various process streams or control effluent treatment parameters to ensure effective detoxification of the pulp mill effluents.

REFERENCES

Foster, D.O. and D.F. Zinkel. 1982. Qualitative and quantitative analysis of diterpene resin acids by glass capillary gas-liquid chromatography. J. Chromatogr. 248:89-98.

Leach, J.M. and A.N. Thakore. 1977. Compounds toxic to fish in pulp mill waste streams. Progress in Water Technology 9(4):787-798.

Lee, H.B., T.E. Peart and J. Carron. 1990. Gas chromatographic and mass spectrometric determination of some resin and fatty acids in pulp mill effluents as their pentafluorobenzyl ester derivatives. J. Chromatogr. 498:367-379.

Li, K., M. Chester, J.P. Kutney, J.N. Saddler and C. Breuil. 1994. Production of polyclonal antibodies for the detection of dehydroabietic acid in pulp mill effluents. Analytical Letters 27(9):1671-1688.

McCarthy, P., K.J. Kennedt and R.L. Droste. 1990. Role of resin acids in the anaerobic toxicity of chemithermomechanical pulp wastewater. Water Res. 24(11):1401-1405.

McKague, A., J.M. Leach, R.N. Soniassy and A.N. Thakore. 1977. Toxic constituents in woodroom effluents. Pulp Pap. Can. Trans. Tech. Sect. 3(3):75-81.

McLeay and Associates Ltd. 1987. Aquatic toxicity of pulp and paper mill effluents: A review. Prepared for Environment Canada, Fisheries and Oceans Canada, Canadian Pulp and Paper Assoc., and Ontario Ministry of the Environment, Report EPS 4/PF/1. 191 pp.

McLeay, D.J., C.C. Walden and J.R. Munro. 1979. Influence of dilution water on the toxicity of kraft pulp and paper mill effluent, including mechanisms of effect. Water Res. 13:151-158.

McMahon, D.H. 1980. Analysis of low levels of fatty and resin acids in kraft mill process streams. Tappi 63(9):101-103.

Morales, A., D.A. Birkholz and S.E. Hrudey. 1992. Analysis of pulp mill effluent contaminants in water, sediment, and fish bile - fatty and resin acids. Water Environ. Res. 64:660-666.

NCASI. 1975. Improved procedures for the gas chromatographic analysis of resin and fatty acids in kraft mill effluent. National Council of the Paper Industry for Air and Stream Improvement Inc., NCASI Tech. Bull. No. 281.

NCASI. 1978. An examination of porous polymer resins for analytical use and removal of compounds in kraft effluent responsible for short term bioassay response. National Council of the Paper Industry for Air and Stream Improvement Inc., NCASI Tech. Bull. No. 302.

NCASI. 1986. Procedures for the analysis of resin and fatty acids in pulp mill effluents. National Council of the Paper Industry for Air and Stream Improvement Inc., NCASI Technical Bulletin No. 501.

Priha, M.H. and E.T. Talka. 1986. Biological activity of bleached kraft mill effluent (BKME) fractions and process streams. Pulp Pap. Can. 87:143-147 (T447).

Richardson, D.E. and H. Bloom. 1982. Analysis of resin acids in untreated and biologically treated thermomechanical pulp effluent. Appita 35(6):477-482.

Richardson, D.E., J.B. Bremmer and B.V. O'Grady. 1992. Quantitative analysis of total resin acids by high performance liquid chromatography of their coumarine ester derivatives. J. Chromatogr. 595:155-162.

Robins, R.J. 1986. The measurement of low-molecular-weight, non-immunogenic compounds by immunoassay, *in* H.F. Linskens and J.F. Jackson (Eds.). Immunology in Plant Sciences. Springer-Verlag, pp. 86-141.

Rogers, I.H. 1973. Isolation and chemical identification of toxic components of kraft mill wasters. Pulp Pap. Mag. Can. 74(9):111-116 (T303).

Rogers, I., H. Mahood, J. Servizi and R. Gordon. 1979. Identifying extractives toxic to aquatic life. Pulp Pap. Can. 80(9):94-99 (T286).

Schneider, P. and B.D. Hammock. 1992. Influence of the ELISA format and the hapten-enzyme conjugates on the sensitivity of an immunoassay for s-triazine herbicides using monoclonal antibodies. J. Agric. Food Chem. 40:525-530.

Taylor, B.R., K.L. Yeager, S.G. Abernethy and G.F. Westlake. 1988. Scientific criteria document for development of provincial water quality objectives and guidelines: resin acids. Ontario Ministry of the Environment, Water Resources Branch, Toronto, Ontario.

Vanderlaan, M., L.H. Stanker, B.E. Wakkins and D.W. Robert (Eds.). 1990. Immunoassays for Trace Chemical Analysis, Monitoring Toxic Chemicals in Human, Food, and the Environment. ACS Symposium Series 451, American Chemical Society, Washington, DC.

Van Emon, J.M. and V. Lopez-Avila. 1992. Immunochemical methods for environmental analysis. Analyt. Chem. 64:79A-88A.

Voss, R.H. and A. Rapsomatiotis. 1985. An improved solvent extraction based procedure for the gas chromatographic analysis of resin and fatty acids in pulp mill effluents. J. Chromatog. 346:205-214.

Walden, C.C. and T.E. Howard. 1981. Toxicity of pulp and paper mill effluent - a review. Pulp Pap. Can. 82(4):115-124(T143).

Wearing, J.T., M.D. Ouchi, R.D. Mortimer, T.G. Kovacs and A. Wong. 1984. Factors controlling resin and fatty acid dissolution in sulfite cooking of black spruce. J. Pulp Pap. Sci. 10(6):J178-181.

Wilkins, A.L. and S. Panadam. 1987. Extractable organic substances from the discharges of a New Zealand pulp and paper mill. Appita 40(3):208-212.

Zinkel, D.F. and C.C. Engler. 1977. Gas-liquid chromatography of resin acid esters. J. Chromatogr. 136:245-252.

A METHOD FOR DETECTION OF CHLORINATED, ETHERIFIED LIGNIN STRUCTURES IN HIGH MOLECULAR WEIGHT MATERIALS OF INDUSTRIAL AND NATURAL ORIGIN

Olof Dahlman, Anders Reimann, Pierre Ljungquist and Roland Mörck

STFI, Box 5604, S-11486 Stockholm, Sweden

A method based on thioacidolysis and advanced gas chromatography/mass spectrometry (GC/HRMS) analysis of degradation products has been developed in order to assess whether trace amounts of chlorinated nonphenolic alkyl-aryl ether structures of the β-O-4 type occur in high molecular weight lignin-derived or lignin-containing materials of industrial or natural origin. The identification and determination of chlorinated thioacidolysis degradation products was focused on GC/HRMS analysis in the selected ion monitoring (SIM) mode of silyl esters of degradation products originating from etherified lignin structures of the β-O-4 type. The identification of the chlorinated degradation products was based on the comparison of GC retention times and accurate mass numbers for the base peaks in the mass spectra with corresponding data for reference compounds. A mixture of monochlorinated reference compounds (thioacidolysis degradation products originating from guaiacyl and syringyl structures) was obtained by chlorination of a hardwood steam explosion lignin followed by thioacidolysis and silylation of the degradation products. The method has been found to be useful for detection of trace amounts (>1 µg/g) of chlorinated etherified lignin structures of the β-O-4 type in lignin-containing samples.

INTRODUCTION

Several studies have shown that the high molecular weight (>1000 Da) fraction of the organic material in bleached kraft mill effluents (BKMEs) from mills using chlorine bleaching or elemental chlorine free (ECF) bleaching, accounts for a considerable part of the organically bound chlorine measured as AOX (Lindström et al. 1981; Kringstad and Lindström 1984; Dahlman et al. 1991, 1993a, 1994a; O'Connor et al. 1993). In addition, degradation studies (permanganate oxidation) have shown that a part of the organically bound chlorine in the high molecular weight BKME material is found in the residual lignin-derived phenolic structural elements (Lindström and Österberg 1984; Österberg and Lindström 1985; Mörck et al. 1991; Dahlman et al. 1993a, 1993b, 1994a, 1994b).

It has also been demonstrated that certain high molecular weight organic materials of natural origin (humic substances) may contain detectable amounts of organically bound halogens, measured as AOX (Wigilius et al. 1988; Asplund et al. 1989; Asplund and Grimvall 1991; Enell and Wennberg 1991; Grøn 1991). Recently, by combining oxidative degradation (permanganate oxidation) with gas chromatography and high resolution mass spectrometry (GC/HRMS), we have identified chlorinated phenolic structural elements in naturally occurring humic substances (Dahlman et al. 1993b, 1994b; Johansson et al. 1994). In these studies, we were also able to show that the chlorinated aromatic carboxylic acids identified as degradation products of the humic substances investigated were mainly of the same types as those found upon degradation of high molecular weight organic matter from BKMEs.

The results from our studies on high molecular weight organic materials of industrial and natural origin raised the question whether or not small amounts of chlorinated aromatic structures could also be present as structural elements in native lignin materials, e.g., in wood, in decaying wood or in annual plants. It is well known that the lignin in native wood to a large extent is composed of nonphenolic alkyl-aryl ether structural elements of the β-O-4 type (Adler 1977). A major drawback of the oxidative degradation technique (permanganate oxidation) used by us in previous investigations is that only aromatic structures carrying free phenolic hydroxyl groups can be analyzed (Gellerstedt 1992). Thus, no information on the major structural elements (nonphenolic elements) present in native wood lignin samples can be obtained by using permanganate oxidation.

The aim of the present study was therefore to develop an analytical procedure which would enable us to assess whether chlorinated nonphenolic aromatic structures of the β-O-4 type occur in lignins and lignin-derived materials of natural and industrial origin, such as native wood lignin, lignin in annual plants, lignin from decaying plant materials, high molecular weight BKME materials and humic substances. In search for such an analytical technique, we investigated the possibility of using the thioacidolysis reaction according to the method developed by Lapierre *et al.* (1985, 1986a). The thioacidolysis reaction, which proceeds via acidic cleavage of α-O-4 and β-O-4 linkages, results in depolymerization of the lignin sample under formation of trithioethyl-phenylpropane compounds. Thus, low molecular weight lignin-derived degradation products are formed which are amenable to analysis by gas chromatography/mass spectrometry (GC/MS). This paper describes an analytical technique for detection of chlorinated thioacidolysis degradation products and reports some preliminary application results.

MATERIALS AND METHODS

Samples

The origin and preparation of the hardwood (Aspen) steam-explosion lignin (SEL) used for chlorination experiments was described by Milne *et al.* (1994). It is a standard sample which has been used by a number of laboratories throughout the world in an international round robin on lignin analysis organized by the International Energy Agency. Chlorinated SELs with varying degree of chlorination were obtained by treatment of SEL samples (20 mg) suspended in 2 mL methanol with different amounts of chlorine added by drop-wise addition of chlorine-water at room temperature. The resulting chlorinated lignin suspensions were neutralized with sodium bicarbonate and thereafter freeze-dried.

The bleach-plant effluent (BPE) sample (originating from bleaching of softwood kraft pulp) was sampled at a mill producing bleached softwood and hardwood kraft pulp. The mill used conventional kraft cooking combined with oxygen delignification and ECF bleaching.

High molecular weight organic material was isolated from the effluent by ultrafiltration (nominal cut-off 1000 Da). The starting volume was 10 L whereas the volume of the concentrate was 0.5 L. The ultrafiltration concentrate was "washed" with 5 x 1.0 L ultrapure (Millipore Milli-Q Plus) water using a sequential batch diafiltration procedure (Beaton and Klinkowski 1983) and was thereafter freeze-dried.

The spruce wood sample and the pine wood sample were collected at industrially unaffected areas situated in the southern part of Sweden. The samples were air-dried and grounded in a mortar to fine particles.

Chemical Degradation

Thioacidolysis was carried out essentially according to the method previously developed by Lapierre *et al.* (1985, 1986a). The main principles of the thioacidolysis reaction are outlined in Fig. 1. About 20-50 mg of dry sample was suspended in 10 mL of a 9:1 mixture of dioxane and ethanethiol containing 0.2 M BF$_3$-etherate. The reaction mixture was deaerated (Argon), heated at 100°C for 4 h, cooled to room temperature, poured into a diluted (0.4 M) sodium carbonate solution and extracted with dichloromethane. The dichloromethane extract was dried (sodium sulfate), filtered and evaporated to about 1 mL and silylated using N,O-bis(trimethylsilyl)trifluoroacetamide. The degradation products were thereafter analysed as their trimethylsilyl ethers by GC/MS.

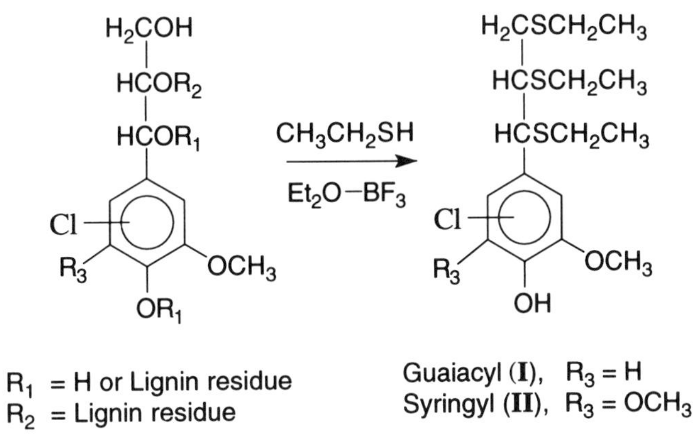

Figure 1. Main principles of the thioacidolysis reaction.

GC/MS Instrumentation

The GC/MS analyses were performed using a VG-250 SE high resolution mass spectrometer connected to a HP 5890 gas chromatograph equipped with a fused silica column (Supelco PTE-5, 60 m x 0.32 mm, phase thickness 0.25 µm). Temperature program: 50°C for 1 min, then 30°C/min to 100°C, 4°C/min to 230°C and finally 20°C/min to 310°C. Split less injection: 2 µl. Carrier gas: He.

RESULTS

Chemical Characterization of the Monochlorinated Thioacidolysis Products

In the initial part of the study, chlorinated steam-explosion lignin samples were subjected to thioacidolysis followed by identification of the monochlorinated lignin-derived thioacidolysis products formed. The chemical characterization of the monochlorinated thioacidolysis products of the guaiacyl (I) and syringyl (II) types (Fig. 1) was based on GC combined with full-scan MS of silylated samples. The total ion current (TIC) chromatogram obtained on GC/MS analysis of a degraded chlorinated SEL sample is shown in Fig. 2. In the TIC chromatogram two strong double peaks are found at retention times of approx. 18.6 and 20.2 minutes. These peaks correspond to the threo and erythro isomers of the nonchlorinated guaiacyl and syringyl thioacidolysis products. In addition to these peaks, two minor peaks, assigned to the chlorinated thioacidolysis products of the guaiacyl (I) and syringyl (II) type, were found at retention times 20.6 and 21.3 minutes, respectively.

The mass spectra obtained for the monochlorinated compounds I and II are shown in Fig. 3a, whereas the mass spectra of the corresponding nonchlorinated compounds, also previously described by Lapierre *et al.* (1985, 1993) and by Rolando *et al.* (1992), are shown in Fig. 3b. As can be seen in Fig. 3a, the mass spectra of the monochlorinated compounds I and II exhibit a strong base peak (fragment ion A^+) at m/e 303 and m/e 333, respectively. The formation of the fragment ion A^+ corresponds to a loss of 149 mass units from the molecular ions (not shown in Fig. 3a) of compounds I and II. A strong base peak, formed by loss of 149 mass units from the molecular ion, is also seen in the mass spectra of the nonchlorinated compounds in Fig. 3b. Fragmentation of the molecular ion of the nonchlorinated guaiacyl and syringyl thioacidolysis products, by loss of 149 mass units (probably loss of a diethylthioethyleneglycol radical), has also previously been reported by Lapierre *et al.* (1985, 1993) and by Rolando *et al.* (1992). A tentative chemical structure for the fragment ion A^+ is shown in Fig. 4.

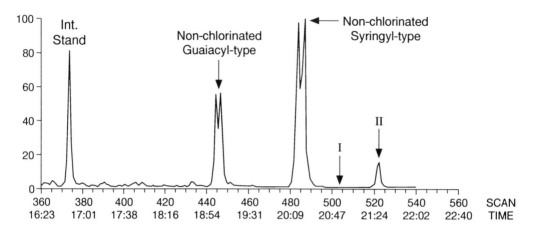

Figure 2. TIC chromatogram obtained on GC/MS analysis of a degraded (thioacidolysis) chlorinated SEL sample.

In the mass spectra of compounds I and II (Fig. 3a), the base peak exhibits a characteristic chlorine cluster, A^+ and $(A+2)^+$ ions, which strongly indicates the presence of a chlorine atom in these compounds. Thus, the chlorinated thioacidolysis products I and II were after silylation identified as the trimethylsilyl ethers of 1-(chloro-4-hydroxy-3-methoxyphenyl)-1,2,3-trithioethylpropane (I) and 1-(2-chloro-4-hydroxy-3,5-dimethoxyphenyl)-1,2,3-trithioethylpropane (II) by comparison of their mass spectra with those of the corresponding nonchlorinated products.

Method for Detection of Compounds I and II in Materials of Industrial and Natural Origin

In the second part of the study, an analytical method was developed to facilitate detection of trace amounts of the monochlorinated thioacidolysis products I and II in the samples investigated. The analytical method chosen was based on gas chromatography and high resolution mass spectrometry in the selected ion monitoring mode (GC/HRMS/SIM). Retention times and accurate mass numbers for the most abundant fragment ion A^+ and the $(A+2)^+$ ion of compounds I and II were used for monitoring. The mass numbers used in the GC/HRMS/SIM analyses are shown in Table 1. The GC/MS -instrumentation and gas chromatographic conditions used were essentially the same as in the full-scan GC/MS experiments, except that a somewhat faster temperature program was used in the GC/HRMS/SIM experiments.

Table 1. Accurate mass numbers used in GC/HRMS/SIM analyses of trimethylsilyl ethers of monochlorinated thioacidolysis products. (I) and (II) refers to the labels shown in Fig. 1.

Thioacidolysis Product (trimethylsilyl ether)	Accurate Mass Numbers of Fragment Ion A^+ and $(A+2)^+$
1-(chloro-4-hydroxy-3-methoxyphenyl)-1,2,3-trithioethylpropane (I)	303.0641, 305.0612
1-(2-chloro-4-hydroxy-3,5-dimethoxyphenyl)-1,2,3-trithioethylpropane (II)	333.0747, 335.0717

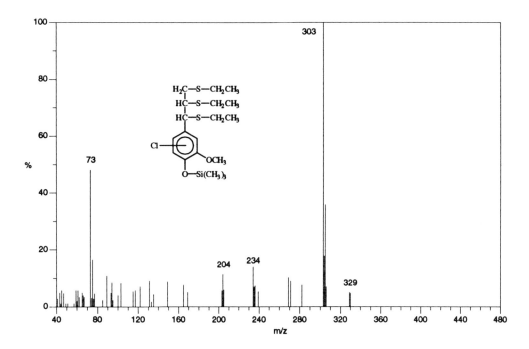

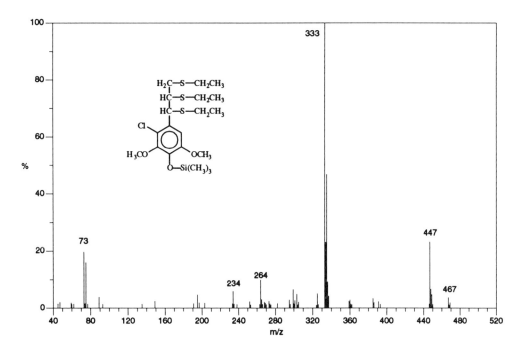

Figure 3a. Mass spectra of the monochlorinated compounds I (upper spectrum) and II (lower spectrum).

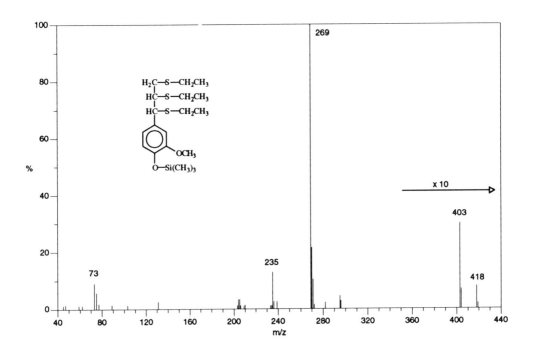

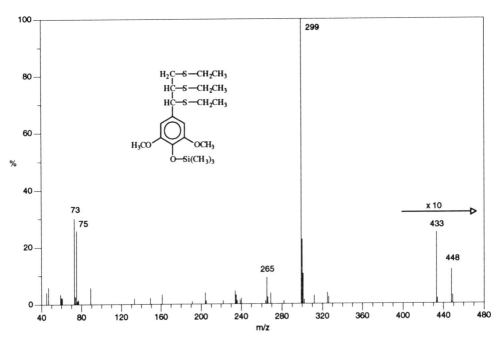

Figure 3b. Mass spectra of the nonchlorinated thioacidolysis products of the guaiacyl (upper spectrum) and syringyl (lower spectrum) types.

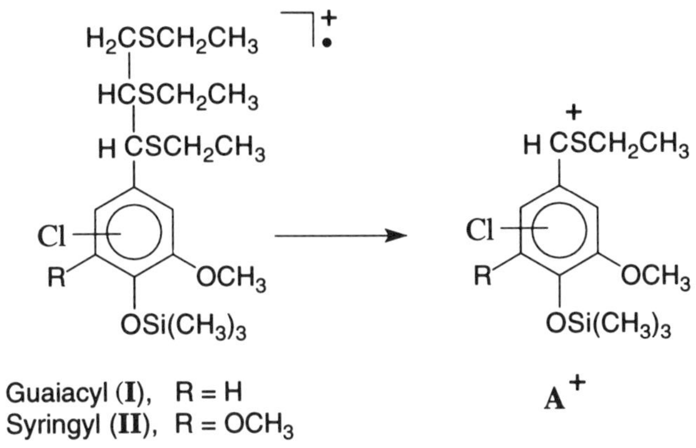

Figure 4. Tentative chemical structure for the fragment ion A$^+$.

It was found that operating the MS instrument at a resolution of 5000 and using an electron energy of about 28 eV during GC/HRMS/SIM analysis were optimal conditions in order to detect compounds I and II in the degraded high molecular weight (BPE) sample. Fig. 5 shows the GC/HRMS/SIM detection of compound I in the degraded BPE sample obtained by monitoring the accurate mass numbers of the fragment ion A$^+$ and the (A+2)$^+$ ion. In a similar way, trace amounts of compound II was also detected in the BPE sample investigated (not shown in Fig. 5). It should be observed that the retention time for compound I differs somewhat between Figs. 2 and 5, due to the slightly faster GC temperature program used in the latter case.

The amounts of compounds I and II in the degraded BPE sample were estimated to about 700 and 10 ppm, respectively. However, it should be noted that the quantification was somewhat uncertain due to the lack of pure reference compounds. The quantities of compounds I and II were thus estimated by assuming that the GC/MS response factors for the fragment ions monitored [A$^+$ and (A+2)$^+$] were the same as those found for the corresponding nonchlorinated thioacidolysis products.

The spruce wood sample and the pine wood sample were also analyzed using thioacidolysis in combination with GC/HRMS/SIM. However, in these two samples no detectable amounts of compounds I and II were found (estimated detection level about 1 ppm).

DISCUSSION

Chemical degradation based on thioacidolysis according to the method developed by Lapierre *et al.* (1985, 1986a) has been used in several studies in order to gain information on nonchlorinated alkyl-aryl ether structures in native and industrial lignins (Lapierre *et al.* 1986a, 1986b, 1987; Rolando *et al.* 1992; Lapierre *et al.* 1993, 1994). In the present study, it has also been shown that monochlorinated alkyl-aryl ether structures of the guaiacyl (I) and syringyl (II) types can be analyzed by the use of thioacidolysis in combination with GC/HRMS/SIM.

The GC/HRMS/SIM detection of compound I (Fig. 5) and compound II (not shown in Fig. 5) at the ppm level in the high molecular weight BPE sample investigated serves as an example of the usefulness of this technique. As can be seen in Fig. 5, a minor peak, probably an isomer of compound I, appears shortly before the main peak (compound I). The appearance of a minor isomer of compound I in the degraded BPE sample is quite reasonable. Several monochlorinated isomers of phenolic guaiacyl structures have previously been found in BPE samples by using oxidative degradation combined with GC/HRMS/SIM (Dahlman *et al.* 1994b).

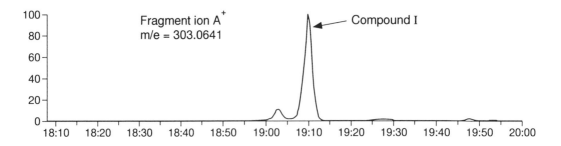

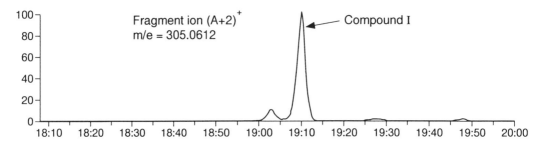

Figure 5. GC/HRMS/SIM detection of the trimethylsilyl ether of 1-(chloro-4-hydroxy-3-methoxyphenyl)-1,2,3-trithioethylpropane in the degraded high molecular weight BPE sample studied.

Using the analytical technique presented here, we were not able to detect (<1 ppm) the monochlorinated compounds I and II, in either the spruce wood sample or in the pine wood sample investigated. This strongly indicates that compounds I and II were not artifacts formed during the thioacidolysis reaction by chlorination involving chloride ions present in the samples.

No firm conclusions should be drawn from the limited number of samples investigated regarding the possibility of chlorinated etherified aromatic structures being present in native lignin materials. Thus, further studies are needed in order to clarify whether or not trace amounts of such structures can be found in native or naturally decaying lignin-containing materials.

REFERENCES

Adler, E. 1977. Lignin chemistry - past, present and future. Wood Sci. Technol. 11(3):169-218.
Asplund, G., and A. Grimvall. 1991. Organohalogens in nature. Environ. Sci. Technol. 25(8):1347-1350.
Asplund, G., A. Grimvall and C. Pettersson. 1989. Naturally produced adsorbable organic halogens (AOX) in humic substances from soil and water. Sci. Tot. Environ. 81/82:239-248.
Beaton, N.C., and P.R. Klinkowski. 1983. Industrial ultrafiltration design and application of diafiltration processes. J. Separ. Proc. Technol. 4(2):1-10.
Dahlman, O., R. Mörck, F. de Sousa and L. Johansson. 1991. Chemical composition of modern bleached kraft mill effluents. Proceedings of the International Conference on Environmental Fate and Effects of Bleached Pulp Mill Effluents, Stockholm, Sweden, Nov. 19-21:47-56.
Dahlman, O., I. Haglind, R. Mörck, F. de Sousa and L. Strömberg. 1993a. Chemical composition of effluents from chlorine dioxide-bleaching of kraft pulps before and after secondary effluent treatment. Pulp and Paper Technologies for a Cleaner World. Proceedings of the EUCEPA Symposium, Paris, France, April 27-29:Vol. 1, 193-215.

Dahlman, O., R. Mörck, P. Ljungquist, A. Reimann, C. Johansson, H. Borén and A. Grimvall. 1993b. Chlorinated structural elements in high molecular weight organic matter from unpolluted waters and bleached-kraft mill effluents. Environ. Sci. Technol. 27(8):1616-1620.

Dahlman, O., A. Reimann, L. Strömberg and R. Mörck. 1994a. On the nature of high molecular weight effluent materials from modern ECF and TCF-bleaching. Proceedings of the 1994 International Pulp Bleaching Conference, Vancouver, Canada, June 13-16:123-132.

Dahlman, O., A. Reimann, P. Ljungquist, R. Mörck, C. Johansson, H. Borén and A. Grimvall. 1994b. Characterization of chlorinated aromatic structures in high molecular weight BKME-materials and in fulvic acids from industrially unpolluted waters. Wat. Sci. Technol. 29(5-6):81-91.

Enell, M., and L. Wennberg. 1991. Distribution of halogenated organic compounds (AOX) - Swedish transport to surrounding sea areas and mass balance studies in five drainage systems. Wat. Sci. Technol. 24:385-395.

Gellerstedt, G. 1992. Chemical degradation methods: Permanganate oxidation, 322-333. In S.Y. Lin and C.W. Dence (eds.). Methods in Lignin Chemistry. Springer-Verlag, Heidelberg.

Grøn, C. 1991. Organic halogens in Danish groundwaters, 495-506. In B. Allard, H. Borén, and A. Grimvall (eds.). Humic Substances in the Aquatic and Terrestrial Environment. Lecture Notes in Earth Sciences 33. Springer-Verlag, Heidelberg.

Johansson, C., I. Pavasars, H. Borén, A. Grimvall, O. Dahlman, R. Mörck and A. Reimann. 1994. A degradation procedure for determination of halogenated structural elements in organic matter from marine sediments. Environ. Int. 20(1):103-111.

Kringstad, K.P., and K. Lindström. 1984. Spent liquors from pulp bleaching. Environ. Sci. Technol. 18(8):236A-247A.

Lapierre, C., B. Monties and C. Rolando. 1985. Thioacidolysis of lignin: Comparison with acidolysis. J. Wood Chem. Technol. 5(2):277-292.

Lapierre, C., B. Monties and C. Rolando. 1986a. Thioacidolysis of poplar lignins: Identification of monomeric syringyl products and characterization of guaiacyl-syringyl lignin fractions. Holzforschung 40(2):113-118.

Lapierre, C., B. Monties and C. Rolando. 1986b. Preparative thioacidolysis of spruce lignin: Isolation and identification of main monomeric products. Holzforschung 40(1):47-50.

Lapierre, C., B. Monties and C. Rolando. 1987. Degradation of various lignins and lignin model compounds by thioacidolysis. Proceedings of the Fourth International Symposium on Wood and Pulping Chemistry, Paris, France, April 27-30:Vol. 2, 431-435.

Lapierre, C., B. Pollet, M.-T. Tollier, B. Chabbert, B. Monties and C. Rolando. 1993. Molecular profiling of lignins by thioacidolysis. Proceedings of the Seventh International Symposium on Wood and Pulping Chemistry, Beijing, China, May 25-28:Vol. 2, 818-828.

Lapierre, C., B. Pollet and C. Rolando. 1994. Determination of the main bonding patterns in native and industrial softwood lignins. Proceedings of the Third European Workshop on Lignocellulosics and Pulp, Stockholm, Sweden, August 28-31:54-57.

Lindström, K., and F. Österberg. 1984. Characterization of the high molecular mass chlorinated matter in spent bleach liquors (SBL). Part 1. Alkaline SBL. Holzforschung 38(4):201-212.

Lindström, K., J. Nordin and F. Österberg. 1981. Chlorinated organics of low and high relative molecular mass in pulp mill bleachery effluents, 1039-1058. In L.H. Keith (ed.). Advances in the Identification & Analysis of Organic Pollutants in Water. Ann Arbor Science Publishers Inc., Ann Arbor, Michigan.

Milne, T.A., H.L. Chum, F. Agblevor and D.K. Johnson. 1994. Biomass conversion annex VII. Standardized analytical methods activity report. Biomass and Bioenergy: In press.

Mörck, R., A. Reimann and O. Dahlman. 1991. Characterization of high molecular weight organic materials in modern softwood and hardwood bleached kraft mill effluents. Proceedings of the International Conference on Environmental Rate and Effects of Bleached Pulp Mill Effluents, Stockholm, Sweden, Nov. 19-21:155-163.

O'Connor, B.I., T.G. Kovacs, R.H. Voss, P.H. Martel and B. van Lierop. 1993. A laboratory assessment of the environmental quality of alternative pulp bleaching effluents. Pulp and Paper Technologies for a Cleaner World, Proceedings of the EUCEPA Symposium, Paris, France, April 27-29:Vol. 1, 273-297.

Rolando, C., B. Monties and C. Lapierre. 1992. Thioacidolysis, 334-349. In S.Y. Lin and C.W. Dence (eds.). Methods in Lignin Chemistry. Springer-Verlag, Heidelberg.

Wigilius, B., H. Borén and A. Grimvall. 1988. Determination of adsorbable organic halogens (AOX) and their molecular weight distribution in surface water samples. Chemosphere 17:1985-1994.

Österberg, F., and K. Lindström. 1985. Characterization of the high molecular mass chlorinated matter in spent bleach liquors (SBL). Part II. Acidic SBL. Holzforschung 39(3):149-158.

SURFACE TENSION CHANGES RELATED TO THE BIOTRANSFORMATION OF DEHYDROABIETIC ACID BY *MORTIERELLA ISABELLINA*

A.G. Werker[1], P.A. Bicho[2], J.N. Saddler[2] and E.R. Hall[1]

[1]NSERC/COFI Industrial Research Chair in Forest Products Waste Management,
Department of Civil Engineering, 2324 Main Mall, University of British Columbia, Vancouver,
British Columbia, V6T 1Z4 Canada
[2]Forest Products Biotechnology Laboratory, Wood Science Department, Faculty of Forestry,
University of British Columbia, British Columbia, V6T 1Z4 Canada

Dynamic surface tension measurements can be used to follow changes in adsorption kinetics following biotransformation of dehydroabietic acid to metabolites of reduced toxicity. Since the hydrophobic fraction of kraft mill effluents contains the major toxicants, such surface tension measurements may have application in assessing effluent quality. In a controlled study, hydroxylation of dehydroabietic acid reduced the rate of adsorption at the gas-liquid interface. The maximum bubble pressure method was the technique for dynamic surface tension measurement. Adsorption kinetics are explained in terms of the polarity conferred by the hydroxyl group. Concentration of resin acids at liquid surfaces has bearing on the fate of these toxicants in the environment.

INTRODUCTION

More than seventy percent of the acute toxicity found in pulp mill effluent is attributable to the hydrophobic fraction (Ng *et al.* 1974). Within the hydrophobic fraction, resin and fatty acids are the primary toxicants frequently related to incidences of toxicity breakthrough (Taylor *et al.* 1988; McLeay 1987). Being natural wood extractives, resin and fatty acids are waste by-products endemic to the pulping process. They are of environmental concern primarily because they exhibit acute toxicities (LC50) towards aquatic life in the order of 1 mg L^{-1} (Taylor *et al.* 1988).

Resin acids are weak monocarboxylic acids of rather limited solubility. Dehydroabietic acid (DHA) has an ionization constant, pKa, of 5.7 and a total solubility of 6.6 mg L^{-1} (Nyrén and Back 1958). The solubility of the unionized acid is 4.9 mg L^{-1}. DHA is the most soluble of the diterpene resin acids because it possesses the highest number of double bonds (Bruun 1952). Salts of the resin acids can readily be formed with sodium, calcium, zinc, magnesium and aluminum (Soltes and Zinkel 1989). Salts of resin acids, in contrast to the acids themselves, are orders of magnitude more soluble in water and exhibit amphiphilic properties that make them useful as soaps. The critical micelle concentration of potassium dehydroabietate is approximately 7.5 to 9 g L^{-1} (Kolthoff and Stricks 1948).

Rosin soaps, or saponified refined diterpene resin acids, find application in paper sizing (Strazdins 1989) and as emulsifiers for polymerization reactions (Davis 1989). Rosin is readily refined from tall oil, being derived from the soap skimmings off black liquor. The kraft pulping process acts to solubilize natural resinous material in trees by saponification. Therefore, resin acids in the effluent stream are initially in their resinate or soap form. The soaps can be converted back to their acids in the presence of free hydrogen ions. Thus pH changes following the combination of acidic or alkaline sewer lines will influence the dissolved resinate. If these soaps are converted back into acids, they may be precipitated, but they can also be held in solution by association with remaining soaps (Nyrén and Back 1958) or with other organics in solution such as lignin.

These toxicants are surface active compounds that tend to gravitate towards interfaces. This property is manifested in conventional secondary biological treatment systems by resin acid accumulation in surface foams. Resin and fatty acid concentrations in surface foams above biobasins attain levels that are orders of magnitude higher than in the underlying liquid (Fein *et al.* 1992; Servizi *et al.* 1975; Zitko and Carson 1971). Collapse of the foam layer is a probable cause for some of the treatment detoxification failures that have been reported (Leach *et al.* 1977; Servizi *et al.* 1975). While aerobic secondary treatment systems can usually remove resin and fatty acids to sublethal concentrations, periodic toxicity breakthroughs continue to force mill shutdowns. Furthermore, even if sublethal concentrations at the outfall are achieved, these compounds tend to bioaccumulate and persist and can therefore pose a chronic environmental threat (Niimi and Lee 1992; Brownlee *et al.* 1977).

The fate of these chemicals in wastewater treatment systems mirrors their predisposition in the open environment. One often considers the partitioning of a pollutant from an effluent stream into biological, solid, aqueous or gaseous environmental compartments. Partitioning is dependent on a pollutant's physico-chemical properties. Fate of pulp mill effluent chemicals in treatment systems (Mackay and Southwood 1992) and in the environment (Kolset and Heiberg 1988) has been modelled using fugacity concepts. Fugacity measures the thermodynamic driving force for partitioning into the air, soil, water, biota, suspended solids and sediment. From such calculations, predictions are made about where most of the solute partitions and where the highest concentrations occur (Mackay and Paterson 1981). Each compartment has associated rate constants for physical or biological degradation. A chemical will persist in the environment, or in a treatment system, if it is partitioned away from dominant degradative mechanisms. For instance, Brownlee *et al.* (1977) calculated that the half-life for DHA was about 21 years in sediments and only 0.12 years in the water column. The half-life is important if the chemical remains readily accessible.

Interfaces, between so-called environmental compartments, can be an important bioaccessible, but overlooked, region where chemicals collect and concentrate. Resin acids are non-volatile and poorly soluble, and so naturally congregate at air-liquid interfaces. Soaps of resin acids are also strongly associated with interfaces since they are anchored in the aqueous phase by their hydrophilic end while their hydrophobic portion is repulsed into the gas phase. In this way, surfactancy can be an important determinant of the fate of toxicants from kraft pulp mill effluents.

As adsorption of solute at the air-liquid interface will act to lower the liquid surface tension, kraft mill effluent quality and the extent of biological detoxification should be inferrable from measurements of effluent surface tension. Berk *et al.* (1979) reported a strong correlation between *Daphnia magna* median survival time and surface tension for foam fractionated spent sulphite liquor. Keirstead (1978) reported the influence of fungal treatment on the surface tension of a 1% (w/v) sodium lignosulfonate solution.

The objective in the present study was to assess the utility of monitoring the dynamic surface tension in a controlled medium containing resin acids, during the course of biological detoxification. Dynamic surface tension measures both the extent and the rate of surfactant adsorption at a gas/liquid interface. Kutney *et al.* (1981ab, 1982ab) have demonstrated that the zygomycete *Mortierella isabellina* transforms resin acids through cell bound hydroxylation reactions. Substantial reductions in acute toxicities were achieved by the hydroxylation of dehydroabietic, isopimaric and abietic acids and of chlorinated forms of DHA (Servizi *et al.* 1986). In the present study, *M. isabellina* was grown in a dextrose yeast extract medium containing DHA (Kutney *et al.* 1981a). It was then of interest to observe changes in the dynamic surface tension related to biological activity and the transformation of DHA, presumably, to less toxic metabolites.

METHODS AND MATERIALS

Maintenance and Growth of *Mortierella isabellina*

M. isabellina was grown with dextrose yeast extract broth (DYE) containing 5.0 g dextrose, 0.5 g yeast extract, 1.0 g KH_2PO_4, 1.0 g $(NH_4)_2HPO_4$, 1.0 g NaCl and 0.5 g $MgSO_4 \cdot 7H_2O$ per liter of deionized water. Aliquots of technical grade (ICN) DHA, dissolved as a concentrate in HPLC grade methanol (MeOH), were introduced drop wise to this medium to form a colloidal suspension. For control cultures without DHA, an identical volume of MeOH was added to the medium. Stock cultures were maintained on agar plates of 1.5% pure agar mixed with the DYE. Shake flask medium was prepared as follows: (1) a 10-fold concentrate of dextrose was autoclaved and a 10-fold concentrate of yeast extract was filter sterilized, (2) shake flasks containing the remaining salts were brought to 80% of the final liquid volume, (3) MeOH with or without DHA was added to the 80% salt solution which was then autoclaved, and finally (4) the dextrose and yeast extract concentrates were added to the sterilized salt solutions by aseptic techniques to make the final volume. It was necessary to keep the ammonia and sugar separate during autoclave sterilization in order to prevent their chemical reaction.

Biodegradation of DHA by *Mortierella isabellina*

Shake flask cultures (100 mL in 250 mL Erlenmeyer flasks) of DYE with technical grade DHA were inoculated with a plug taken from *M. isabellina* stock culture on agar plates. The shake flasks were agitated by vigorous magnetic stirring in order to prevent mycelial balling and to provide aeration. Cultures were incubated at room temperature (25 ± 2°C). A 2% seed from these matured cultures was used as inoculum for a second growth cycle at the same volume. The 2% seed was found to be a minimum inoculum for growth of *M. isabellina* in the presence of DHA. The matured culture from this second growth cycle was used to inoculate the experimental shake flasks (500 mL in 1000 mL Erlenmeyer flasks). Measurement of the inoculum dry weight was made by passing 5 mL of culture and an additional 5 mL of deionized water through tarred 25 mm diameter, 0.45 μm cellulose acetate filter paper and by drying the biomass for one hour at 105°C.

For the degradation study, four experimental shake flasks were run in parallel. Three of the flasks were spiked with 1 mL MeOH containing DHA, while the fourth was spiked with only 1 mL MeOH. Of the three flasks with DHA, two were inoculated with a 2% seed and the third was kept as an abiotic control. The single flask without DHA was also inoculated with the same 2% seed for use as a control culture.

Sampling Protocol

At time intervals chosen to best follow the growth curve, 8 mL samples were aseptically removed from the shake flasks. From this volume, 1 mL was used for pH measurement and 1 mL, diluted as required, was used for optical density determination at 600 nm. Optical density at 600 nm was used as a direct indication of biological growth. The remainder was preserved by raising the pH to 9, through addition of $1N$ NaOH and refrigeration at 4°C. Samples were stored for less than one week before subsequent analysis. From the preserved sample, 1 mL was extracted for DHA evaluation and 0.5 mL was used in total organic carbon (TOC) analysis. Dynamic surface tension analysis required a 5 mL sample.

Analysis for DHA

For an extraction surrogate, 50 μL of 1345 mg L^{-1} 99% pure (Helix) dichloro-DHA (2Cl-DHA) in HPLC grade MeOH was added to 1 mL whole culture samples. Acidified (10% H_2SO_4) samples were extracted twice with 1 mL and 2 mL aliquots of HPLC grade ethyl acetate (EtAc), respectively, in 10 mL glass culture tubes. Emulsions were broken by centrifugation at 2000 rpm for 10 minutes. EtAc extractions were dried under nitrogen. For HPLC analysis, dried samples were redissolved in 1 mL MeOH to which 20 μL of 1000 mg L^{-1} 99% pure methyl-*o*-podocarporic acid (O-MPCA) in MeOH was added as an internal standard. Samples were filtered (0.45 μm) into GC vials. Standards of DHA were prepared from a 1000 mg L^{-1} 99% pure (Helix) DHA stock solution in MeOH.

Analysis for DHA was performed by HPLC (Hewlett Packard Series 1050), running with a 1 mL min^{-1} carrier flow of 80:20 MeOH:H$_2$O with the water acidified by 0.5% phosphoric acid. Draw and ejection rates were 200 μL min^{-1} with a 10 μL injection volume. Under these operating conditions and UV detection at 280 nm, the retention times were as follows: 3.1 min (O-MPCA), 6.3 min (DHA) and 13.1 min (2Cl-DHA).

Measurements of Total Organic Carbon

TOC was measured with a Shimadzu TOC 500 Analyzer. After it was verified that samples contained negligible inorganic carbon, only total carbon was assayed. Aliquots (0.5 mL) drawn from the supernatant of settled 6 mL culture samples in 10 mL vials were diluted to 7 mL, giving a working TOC range from 0 to 500 mg L^{-1} carbon. Calibration standards were diluted from a 1000 mg L^{-1} (as carbon) KHC$_8$H$_4$O$_4$ stock solution.

Dynamic Surface Tension

Dynamic surface tension measurements were made using the maximum bubble pressure method (Mysels 1990). In this method, gas is forced out of a capillary tube of diameter D immersed at a depth h (Fig. 1).

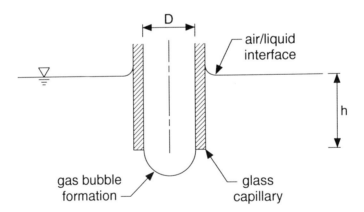

Figure 1. Schematic for the maximum bubble pressure method of surface tension measurement. Nitrogen gas is bubbled through a capillary of diameter D at a depth h in a liquid sample. The pressure required to produce each bubble is measured.

By the Laplace equation, the pressure, P, required to produce an assumed spherical bubble is a maximum (P$_{max}$) when the emerging bubble becomes a hemisphere with the capillary's radius:

$$P_{max} = P - P_o = \frac{4\sigma(t)}{D} + \rho gh \tag{1}$$

where P_o is the atmospheric pressure, ρ is the liquid density, g is the acceleration due to gravity and σ is the dynamic surface tension as a function of the bubble formation time, t. For a pure solvent, P_{max} is theoretically independent of the rate of bubble production. However, if surfactants are present in solution, then a longer time for bubble formation permits a greater extent of surfactant adsorption at the gas-liquid interface. Increased surfactant concentration at the gas-liquid interface lowers the surface tension as observed by a corresponding change in the measured P_{max}.

Bubbles were produced in a 5 mL sample contained in a 10 mL test tube immersed in a thermostated bath at 25 ± 1°C. Bubbles were formed at essentially constant pressure supplied by a nitrogen charged reservoir through a 10 µL micropipet having a capillary diameter in the order of 0.04 cm. The nitrogen reservoir was pressurized by displacement with deionized water. Bubbling frequency decreased as the reservoir pressure decreased with gas lost from the system in each bubble. Bubbling frequencies from about 10 Hz down to 0.04 Hz were recorded. The effective capillary diameter, D, was assessed by measurement and calibration against the surface tension of deionized water (72 dyn cm^{-1}). Capillary immersion depth, h, was measured to the closest 0.025 mm with a needle gauge micrometer. Pressure values during bubble formation were logged at 50 Hz directly onto a computer (12 bit A/D) through a calibrated differential pressure transducer (0-20 cm H_2O) producing a 1 volt cm^{-1} H_2O signal. All glassware was thoroughly rinsed with deionized water and fired at 450°C in a muffle furnace for over an hour to remove any organic impurities. Acquired data were numerically analyzed for maximum pressures and bubble life span by applying Savitzky and Golay (1964) least squares smoothing and differentiation. Figure 2 illustrates a typical segment from a pressure-time recording showing the release of four bubbles during the course of a measurement. A slight decrease in the (plateau) reservoir pressure level can be seen with the release of each bubble. Figure 3 is a close-up view of the pressure data during one bubble release. First and second time derivatives of the raw pressure data help to define the distinct phases of bubble production permitting the discretization of the pressure-time data into single bubble events. For each bubble produced, the formation time (or bubble life) and the corresponding maximum pressure were ascertained. Bubble life is defined by the period from the point of bubble release to the point of maximum pressure (P_{max}). The bubble life represents the time available for surfactant adsorption at the bubble's gas-liquid interface. A longer bubble life allows more surfactant to be adsorbed at the bubble surface with a correlated lower P_{max} and surface tension. The incremental extent of adsorption for a longer bubble life depends on the surfactant's bulk concentration, interfacial concentration, molecular diffusivity and chemical structure.

Figure 4 provides an example of a reduced pressure-time data set from one sample measurement. The data were fit by non-linear least squares approximation to an exponential decay function of the form:

$$\sigma(t) = A_o + A_1 \exp\left(\frac{-t}{\tau_1}\right) + A_2 \exp\left(\frac{-t}{\tau_2}\right) \qquad (2)$$

where σ is the dynamic surface tension, t is the bubble formation time, A_i are scaler constants and τ_i are decay constants. This function was used for data interpolation in order to align and compare results from different samples.

RESULTS

The biomass concentration of the fungal seed was 1.5 g L^{-1} dry weight. General trends for the growth curves of *M. isabellina* with and without the presence of DHA are the same in form as those observed by Kutney *et al.* (1981a). With reference to optical density changes at 600 nm (Fig. 5), presence of DHA in the growth medium resulted in a longer lag period before exponential growth.

Fungal activity acidified the culture environment (Fig. 5). Definite pH minima at 64 h and 30 h for the cultures with and without DHA, respectively, occurred at the height of the exponential growth phase.

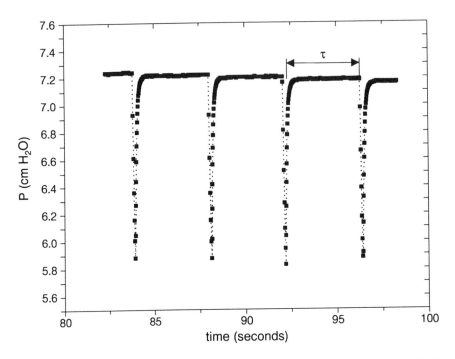

Figure 2. Segment of typical pressure-time raw data acquired at 50 Hz. Three complete cycles of bubble formation and release are shown. Bubble life, τ, for one cycle is shown. Plateau pressures correspond to the current nitrogen reservoir pressure and sudden drops in pressure relate sequences of bubble formation and release.

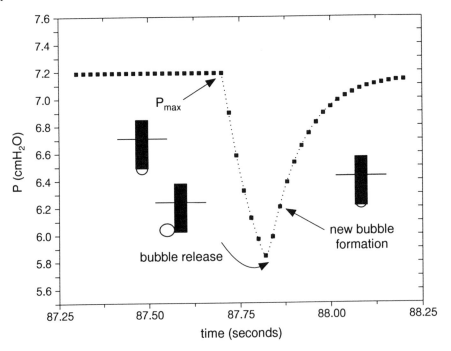

Figure 3. Detailed view of capillary pressure during bubble release and new bubble formation.

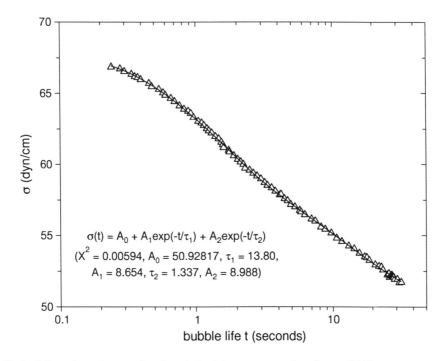

Figure 4. Typical dynamic surface tension data derived from pressure-time data at 50 Hz.

Subsequent to these minima, the media pH recovered but not to their initial levels over the period monitored. *M. isabellina* is found mainly in forest, heath or sandy soils with pH between 3 and 6 (Turner 1963). The increase in pH as the culture entered stationary phase is likely a result of the consumption of acidic metabolic intermediates from dextrose degradation.

Biological growth as marked by the depression in pH and the increase in optical density is paralleled by the consumption of TOC (Fig. 5). An offset is maintained between the TOC for the culture with DHA versus the one without DHA. Based on previous reports (Kutney 1981a), this offset, deemed to be related to the additional organic carbon added as DHA, was taken as an indication that the resin acid was not mineralized but only transformed by the fungus. *M. isabellina* is known to detoxify DHA through membrane bound (Servizi *et al.* 1986) hydroxylation reactions at C-2 with subsequent hydroxylation at C-15 or C-16 (Fig. 6). Secondary hydroxylation occurs only for resting cell cultures. Hydroxylation will increase the polarity of the diterpene molecule and thereby increase its water solubility. The improved water solubility is likely linked to the corresponding decrease in toxicity.

Drop in medium DHA concentrations (Fig. 5) closely followed the kinetics of pH acidification, biological growth and TOC consumption. The reduction in the measured DHA concentration was taken as the production of hydroxylated metabolites (Fig. 6).

The kinetics of surfactant adsorption were judged qualitatively by comparing surface tensions for short (0.2 seconds) and long (20 seconds) bubble lives between cultures with and without DHA (Fig. 7). Consider first the dynamic surface tensions at 20 seconds (σ_{20}) between the cultures with and without DHA. For a fixed bubble life, a reduction in the surface tension with incubation time indicates a greater bulk surfactant concentration. For the control culture, surface tension lowering with incubation time was found to be closely correlated to growth. A minimum surface tension nearly coincident in time to the minimum pH was observed. The molecules producing the medium acidity were probably also adsorbing and causing the observed surface tension depression. Therefore the acidic metabolite, associated with M. isabellina growth on DYE medium, apparently also had surfactant properties.

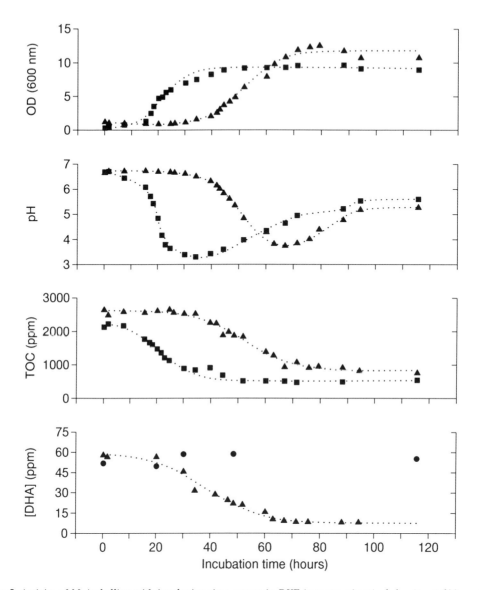

Figure 5. Activity of M. isabellina with incubation time grown in DYE in terms of optical density at 600 nm (OD), pH, TOC and DHA concentrations for: (■) control culture without DHA, (▲) culture with DHA and (•) abiotic control.

Resin acid surface activity is evidenced by the markedly lower surface tensions for cultures with DHA (Fig. 7). However in this case, the observed σ_{20} minimum at about 45 h is not in correspondence with the media acidification minimum at 64 h. Since the production of the surfactant acidic metabolite is concurrent with DHA hydroxylation, it is not surprising that these minima do not align. Surfactancy should increase with more of the acidic metabolite and decrease with DHA hydroxylation. Therefore the

σ_{20} did not seem to give any indication for loss of surfactancy through DHA hydroxylation until well into the stationary phase (incubation time >64 h) where secondary hydroxylation was assumed to be occurring.

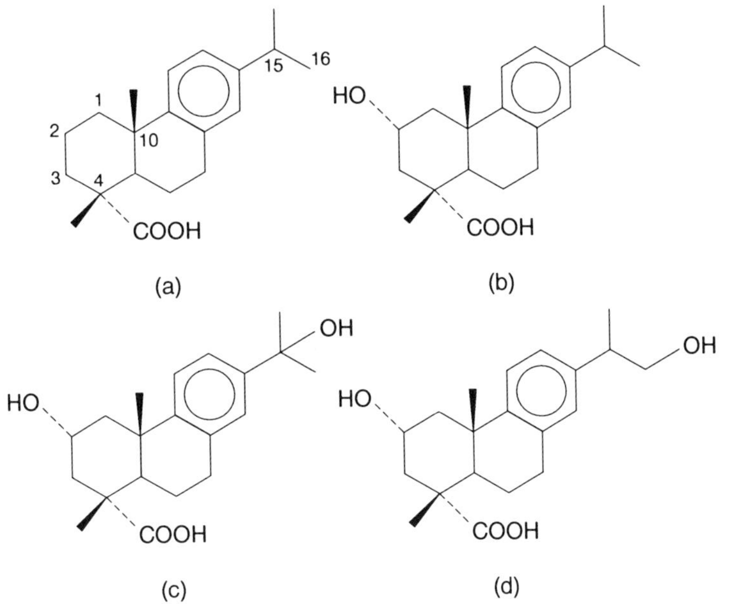

Figure 6. Chemical structure of DHA (a) and the primary (b) and secondary (c & d) hydroxylates produced by M. isabellina during growth on DYE.

The dynamic surface tension at 0.2 seconds ($\sigma_{0.2}$) proved to be more sensitive to resin acid biotransformation (Fig. 7). With the control culture, $\sigma_{0.2}$ remained relatively constant at a level comparable to pure water. The $\sigma_{0.2}$ for cultures with DHA showed a steady upward trend corresponding to the resin acid transformation. The subsequent decrease in σ to the consistently lower levels at σ_{20} suggests that the hydroxylated intermediates were also surface active. Therefore, although the hydroxylated resin acid still became adsorbed at the gas-liquid interface, the kinetics of adsorption were affected by the biotransformation.

DISCUSSION

Given that organic toxicants in a pulp effluent stream have hydrophobic properties, they are likely to gravitate towards interfaces where their non-polar carbon backbones can escape from their polar aqueous environment. Adsorption at an air-liquid interface can be observed by the reduction in the gas-liquid surface tension. The more strongly hydrophobic a molecule is, the greater will be the driving force to remove it from an aqueous phase. This driving force can be visualized as an increased probability for adsorption at a gas-liquid interface for every solute molecule of random orientation impinging on the surface. A solute molecule, diffusing to the air-liquid interface with its polar end up, will tend to be rejected from adsorption. The net effect for a hydrophobic molecule with slightly polar sites would be a decrease in the rate of adsorption.

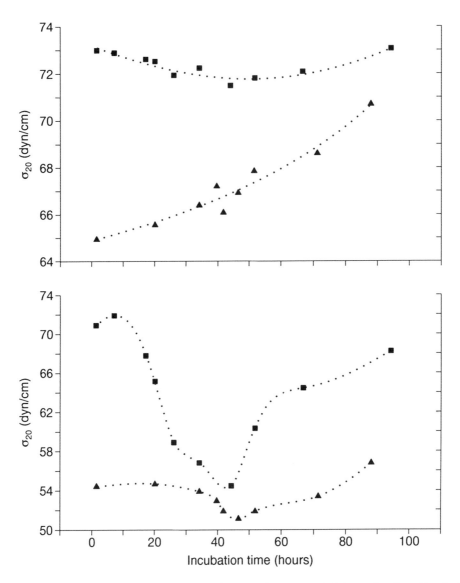

Figure 7. Media surface tension after 0.2 and 20 seconds ($\sigma_{0.2}$ and σ_{20} respectively) as a function of incubation time during growth of M. isabellina for: (■) control culture without DHA and (▲) culture with DHA. Media samples were measured at pH 9 and 25 ± 1°C.

From the dynamic surface tension measurements in this work, changes in the kinetics of resin acid interfacial adsorption due to hydroxylation are believed to have been observed. These changes were observed with short term surface tension measurements. Hydroxylation of resin acids will confer additional polarity to the diterpene molecule. However, the non-polar segment of the molecule still maintains some degree of affinity for gas-liquid interfaces. Since the hydroxylated metabolites are less toxic, adsorption kinetics may be a useful indicator of effluent quality. Dynamic surface tension measurements on whole or evapo-concentrated effluent samples could yield useful information on the extent to which secondary biological treatment removes a significant portion of the effluent toxicity. The method is attractive because it is simplistic and without elaborate sample preparation.

In the consideration of surface tension, this work has also served to emphasize the point that resin acids are adsorbed and therefore concentrate at air-liquid interfaces. This should become an important consideration in an assessment of the fate of these chemicals in the receiving environment. Scum or foam downstream of outfalls is likely rich in these toxicants with potential threat to aquatic life.

ACKNOWLEDGMENT

The authors would like to thank Dr. Bill Mohn for laboratory support. In addition, technical assistance from Ron Dolling, Susan Harper and Paula Parkinson was greatly appreciated. A. Werker was a grateful recipient of a Natural Sciences and Engineering Council post-graduate scholarship.

REFERENCES

Berk, D., J.E., Zajic and L.A. Behie. 1979. Foam fractionation of spent sulphite liquor. Part II: Separation of toxic components. Can. J. Chem. Eng. 57:327-332.

Brownlee, B., M.E. Fox, M.J. Strachan and S.R. Joshi. 1977. Distribution of dehydroabietic acid in sediments adjacent to a kraft pulp and paper mill. J. Fish. Res. Board Can. 34:838-843.

Bruun, H. 1952. Properties of monolayers of rosin acids. Acta Chem. Scand. 6:494-501.

Davis, C.B. 1989. Rosin soaps as polymerization emulsifiers. p. 625-642. In D.F. Zinkel and J. Russell [ed.] Navel Stores. Pulp Chemicals Association, New York.

Fein, J., M. Beavan, A. Effio, N. Gray, N. Moubayed and P. Cline. 1992. Comprehensive study of a bleached kraft mill aerobic secondary treatment system. Water Poll. Res. J. Canada 27(3):575-599.

Keirstead, K.F. 1978. Foam fractionation and surface tension method of characterizing soluble lignins. Pulp Paper Mag. Can. 4:T6-9.

Kolset, K., and A. Heiberg. 1988. Evaluation of the "FUGACITY" (FEQUM) and the "EXAMS" chemical fate and transport models: A case study on the pollution of the Norrsundet Bay (Sweden). Wat. Sci. Tech. 20(2):1-12.

Kolthoff, I.M., and W. Stricks. 1948. Solubilization of dimethylaminoazobenzene in solutions of detergents. I. The effect of temperature on the solubilization and upon the critical concentration. J. Phys. Colloid Chem. 52:915-941.

Kutney, J.P., M. Singh, G. Hewitt, P.J. Salisbury, B.R. Worth, J.A. Servizi, D.W. Martens and R.W. Gordon. 1981a. Studies related to biological detoxification of kraft pulp mill effluent. I. The biodegradation of dehydroabietic acid with *Mortierella isabellina*. Can. J. Chem. 59:2334-2341.

Kutney, J.P., M. Singh, G. Hewitt, P.J. Salisbury, E. Dimitriadis, B.R. Worth, J.A. Servizi, D.W. Martens and R.W. Gordon. 1981b. Studies related to biological detoxification of kraft pulp mill effluent. II. The biotransformation of isopimaric acid with *Mortierella isabellina*. Can. J. Chem. 59:3350-3355.

Kutney, J.P., M. Singh, G. Hewitt, P.J. Salisbury, E. Dimitriadis and B.R. Worth. 1982a. Studies related to biological detoxification of kraft pulp mill effluent. III. The biodegradation of abietic acid with *Mortierella isabellina*. Helv. Chim. Acta, 65(3):661-670.

Kutney, J.P., M. Singh, G. Hewitt, P.J. Salisbury, E. Dimitriadis, J.A. Servizi, D.W. Martens and R.W. Gordon. 1982b. Studies related to biological detoxification of kraft pulp mill effluent. IV. The biodegradation of 14-chlorodehydroabietic acid with *Mortierella isabellina*. Helv. Chim. Acta 65(5):1343-1350.

Leach, J.M., J.C. Mueller and C.C. Walden. 1977. Biodegradability of toxic compounds in pulp mill effluents. Trans. Tech. Section CPPA 3(4):TR126-TR130.

Mackay, D., and S. Paterson. 1981. Calculating fugacity. Environ. Sci. Technol. 15(9):1006-1014.

Mackay, D., and J.M. Southwood. 1992. Modelling the fate of organochlorine chemicals in pulp mill effluents. Water Poll. Res. J. Canada 27(3):509-537.

McLeay, D., and Assoc., Ltd. 1987. Aquatic Toxicity of Pulp and Paper Mill Effluents: A Review. EPS Report 4/PF/1, Environment Canada.

Mysels, K.J. 1990. The maximum bubble pressure method of measuring surface tension, revisited. Colloids Surfaces 43:241-262.

Niimi, A.J., and H.B. Lee. 1992. Free and conjugated concentrations of nine resin acids in rainbow trout (*Oncorhynchus mykiss*) following waterborne exposure. Env. Tox. Chem. 11:1403-1407.

Ng, K.S., J.C. Mueller and C.C. Walden. 1974. Study of Foam Separation as a Means of Detoxifying Bleached Kraft Mill Effluents, Removing Suspended Solids and Enhancing Biotreatability. CPAR Project Report 233-1, Canadian Forestry Service, Ottawa.

Nyrén, V., and E. Back. 1958. The ionization constant, solubility product and solubility of abietic and dehydroabietic acid. Acta Chem. Scand. 12(7):1516-1520.

Savitzky, A., and M.J.E. Golay. 1964. Smoothing and differentiation of data by simplified least squares procedures. Anal. Chem. 36(8):1627-1639.

Servizi, J.A., R.W. Gordon, I.H. Rogers and H.W. Mahood. 1975. Chemical characteristics, acute toxicity and detoxification of foam on two aerated lagoons. CPPA Technical Section, Environment Improvement Conference. p. 45-52.

Servizi, J.A., D.W. Martens, R.W. Gordon, J.P. Kutney, M. Singh and E. Dimitriadis. 1986. Microbiological detoxification of resin acids. Water Poll. Res. J. Canada 21(1):119-129.

Soltes, E.J., and D.F. Zinkel. 1989. Chemistry of rosin. p. 261-345. In D.F. Zinkel and J. Russell [ed.] Navel Stores. Pulp Chemicals Association, New York.

Strazdins, E. 1989. Paper sizes and sizing. p. 575-624. In D.F. Zinkel and J. Russell [ed.] Navel Stores. Pulp Chemicals Association, New York.

Taylor, B.R., K.L. Yeager, S.G. Abernethy and G.F. Westlake. 1988. Resin Acids. Scientific Criteria Document for Development of Provincial Water Quality Objectives and Guidelines, Queen's Printer for Ontario.

Turner, M. 1963. Studies in the genus *Mortierella*. I. *Mortierella isabellina* and related species. Trans. Brit. Mycol. Soc. 46(2):262-272.

Zitko, V., and W.V. Carson. 1971. Resin Acids and Other Organic Compounds in Groundwood and Sulfate Mill Effluents and Foams. Fish. Res. Board Can. Manuscript Report Series No. 1134, 28 p.

CORRELATIONS AMONG CONTAMINANT PROFILES IN MILL PROCESS STREAMS AND EFFLUENTS

Thomas E. Kemeny[1] and Sujit Banerjee[2]

[1]Georgia-Pacific Corporation, 133 Peachtree Street N.E., Atlanta, GA, 30348-5605 U.S.A.
[2]Institute of Paper Science and Technology, 500 Tenth Street NW, Atlanta, GA, 30318 U.S.A.

Two intensive monitoring studies conducted at the Georgia-Pacific facilities at Leaf River, MS and Brunswick, GA showed the absence (within the limits of detection) of TCDD/F, trisubstituted and higher chlorophenols, and acrolein in any of the process sewers. The eighteen sampling episodes conducted at Leaf River allowed correlations to be drawn among the various contaminant profiles. For example, there was a strong BOD-COD relationship in the pulp mill sewer, suggesting that the carbon in the pulp mill sewer (or a fixed fraction thereof) is biodegradable. Also, AOX and COD in the final effluent were strongly correlated, suggesting that variations in each parameter were governed by process changes.

INTRODUCTION

In 1993/1994, Georgia-Pacific conducted two studies to assess the environmental status of its Leaf River and Brunswick operations. Both of these are elemental chlorine-free (complete ClO_2 substituted) bleach mills. The Leaf River study covered a five week period in August-September 1993 and a four week interval in January-February 1994. The study was designed to capture variability resulting from both seasonal and day-to-day process changes. Process lines were sampled along with the final effluent, which led to data on the impact of various unit operations on final effluent quality. The Brunswick effort was shorter and occurred over a three day period in January 1994. A principal finding in both mills was the absence of dioxins, chlorophenols or acrolein in any of the samples taken. Additionally, the large body of data acquired for Leaf River allowed interrelationships to be developed among the concentration-time profiles of the various constituents.

DESCRIPTION OF SITE ACTIVITIES

Leaf River

The Leaf River mill is located in New Augusta, Mississippi. The mill began operations in 1984 and is in the market bleached kraft subcategory. During the study, daily production approximated 1650 tonnes of both hardwood and softwood bleached kraft market pulp. Approximately 60% of the wood received is softwood and 40% is hardwood. The mill operates a single pulp line that swings between hardwood and softwood. The digester is a continuous Kamyr dual-vessel hydraulic unit.

During this study, the bleaching sequence for both hardwood and softwood consisted of (1) ClO_2 bleaching, (2) extraction with sodium hydroxide reinforced with oxygen and peroxide, (3) ClO_2 bleaching, (4) extraction with sodium hydroxide and peroxide and (5) final bleaching with ClO_2 (D-E_{op}-D-E_p-D). Pulp dryer whitewater is used as wash water for the final bleaching (D_2) stage, while D_2 stage filtrate is used as make up for the E_2 washer. Filtrate is normally recycled to the previous stage of the bleach plant all the way back to the first stage.

The nominal kappa number of the pulp entering the bleach plant was 30 for softwood and 14 for hardwood. The pulp is considered elemental chlorine free and is designated as "ECF" in the industry. The kappa factors for softwood and hardwood were at 0.12. Both softwood and hardwood are bleached to a final ISO brightness of 90. The chemical recovery plant includes a Kamyr falling film evaporator with a capacity of 0.5 million kg water h^{-1} and a Gotaverken recovery boiler rebuilt to handle over 2.7 million kg of liquor solids d^{-1}.

Approximately 75 million L of water are taken daily from the Leaf River, treated by coagulation in a clarifier and passed through gravity filters before process use. Only the first and second stage filtrates are sewered; the others are used as wash water in the preceding stage. The treatment system consists of a bar screen, two primary clarifiers, equalization, cooling, activated sludge treatment in an aeration basin, two secondary clarifiers and a final holding pond.

Wastewater from the chemical recovery area, the pulping area, the pulp dryer, the woodyard and the power boiler, as well as stormwater, enters the wastewater treatment system through the bar screen. The wastewater is then pumped to two 52-meter-diameter primary clarifiers in parallel. Detention time in each of the primary clarifiers is four hours. Effluent from the primary clarifiers is discharged to an equalization basin after introduction of the bleach plant acid and alkaline sewers and addition of lime for pH adjustment.

The equalization basin, with ten floating aerators, each driven by a 40 horsepower motor, has a surface area of 26,000 m^2 and a detention time of 16.5 h. Following equalization, nutrients (ammonia and phosphoric acid) are added, and the wastewater flows to a cooling tower which discharges to a 180 million L aeration basin. The aeration basin has a detention time of 45 h and is aerated by four 700 horsepower aerators. After the aeration basin, the flow is divided between two 18-m-diameter secondary clarifiers, each with a detention time of 9.4 h. Approximately 30 million gallons d^{-1} (40%) of the secondary sludge is recycled to the aeration basin. The secondary clarifiers discharge to a holding pond which discharges continuously through diffusers to the Leaf River. The typical detention time for the holding pond is approximately 60 h. The mill discharges approximately 70 million L d^{-1} of treated wastewater to the Leaf River. A schematic for wastewater treatment is provided in Fig. 1.

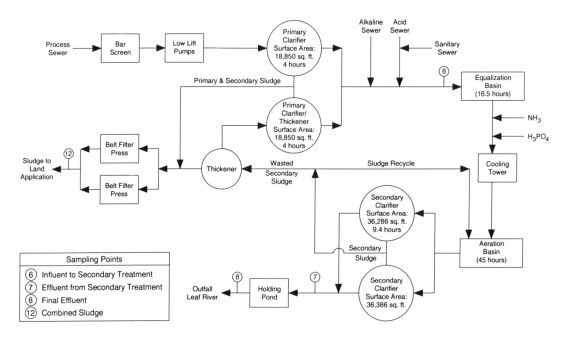

Figure 1. Leaf River wastewater treatment schematic.

Samples were taken over a five week period in August-September 1993 (10 episodes) and a four week interval in January-February 1994 (8 episodes). Six grab samples were taken over a 24 h period and composited where appropriate. Every 4 h the sampler (a) recorded pH, conductivity and temperature of the sample and also (where necessary) the temperature of the cooled sample from the sampling train, (b) added about 0.5 L to the compositing container and, if required, (c) collected volatile and methanol samples. For samples taken from the chlorination stage, the sampler additionally measured residual chlorine in the sample and added sodium thiosulfate to the first sample added to the compositing container.

The sampling plan is provided in Table 1. Some of the summer analyses were discontinued in the winter event since the analytes were not detected in any of the summer episodes. Since the ACF and ACS locations gave very similar results for all the analytes considered, ACF sampling was discontinued in the winter. Chlorophenols were dropped from the WTI and WTE, and acrolein analysis was discontinued completely, since they were not observed in the summer event. Useful resin acid data could not be obtained for the summer sampling due to analytical difficulties; these were resolved before the winter event.

Table 1. Sampling plan at Leaf River.

	MPW	ACF[1]	ACS	ALF	WTI	PMS	RCS	EVS	WTE	CSI	CSO
color	x	x	x	x	x	x	x	x	x		
pH	x	x	x	x	x	x	x	x	x		
cond.	x	x	x	x	x	x	x	x	x		
TSS	x				x	x	x		x		
COD	x	x	x	x	x	x	x	x	x	x	x
DOC	x	x	x	x	x	x	x	x	x		
TOC	x				x	x	x	x	x		
AOX	x	x	x	x	x				x		
BOD	x				x	x		x	x	x	x
resin acids	x	x	x	x	x	x	x	x	x		
methanol	x	x	x	x	x	x	x	x	x	x	x
chlorophenols	x	x	x	x	x				x		
volatiles	x	x	x	x	x	x		x	x		
TCDD/F[2]		x		x							

[1]Dropped for the winter sampling; [2]episodes 1 and 11 only.
MPW: mill process water; ACF: acid filtrate; ACS: acid sewer; ALF: alkaline filtrate; WTI: influent to waste treatment; PMS: pulp mill sewer; RCS: recovery sewer; EVS: evaporator sewer; WTE: waste treatment effluent; CSI: condensate stripper inlet; CSO: condensate stripper outlet.

Chlorophenol analysis was limited to the following congeners: 2,4,6-trichlorophenol, 2,4,5-trichlorophenol, 2,3,4,6-tetrachlorophenol, pentachlorophenol, 3,4,6-trichloroguaiacol, 3,4,5-trichloroguaiacol, 4,5,6-trichloroguaiacol, tetrachloroguaiacol, 3,4,6-trichlorocatechol, 3,4,5-trichlorocatechol, tetrachlorocatechol and trichlorosyringol. A typical detection limit for each congener was 2.5-5.0 µg L^{-1}.

Analysis of volatiles was restricted to methylene chloride, acetone, chloroform, 2-butanone and acrolein (summer only). The principal analytical methods used are summarized in Table 2.

Table 2. Analytical methods used.

Analyte	Method	Reference	Basis
methanol	8015A	EPA 1986	GC
chlorophenolics	EPA 1653	U.S. EPA 1991	acetylation, GC-MS
volatiles	624	U.S. EPA 1990	purge & trap GC-MS
TCDD/F	NCASI 551	NCASI 1989a	HRGC-MS
AOX	1650A	U.S. EPA 1992	combustion/coulometry
resin acids	NCASI 501	NCASI 1989b	ethylation, GC-FID

Brunswick

The facility pulps mostly softwood (loblolly and slash pine) and some hardwood (mainly gum, oak, hickory and ash). The mill operates 19 digesters: 11 for softwood and 8 for both softwood and hardwood. Pulp yields are 11.6 tons per small digester and 17.9 tons per large digester.

Three bleach lines are run under the following specifications:

	Species	Production (tonnes h^{-1})	
		Avg	Maximum
No. 1 plant	pine	24	26
No. 2 plant	hwd/pine	35	44
No. 3 plant	pine	40	44
Total: 2375 tonnes d^{-1}			

The sequence used is DE$_{op}$DE$_p$D. Peroxide is used for high brightness EF-100 and for preventing brightness reversion. The towers are all upflow in the No. 2 and No. 3 plants. The No. 1 plant uses both upflow and downflow towers.

The D$_{100}$ stage (for pine) has a retention time of 15-25 min and is controlled to a CEK-number of 3.5 determined by a correlation of wet brightness to CEK-number. The consistency is 4%, and the temperature is 32-35°C. The D$_1$ stage has a retention time of 1.5 h and is controlled to 82 Elrepho brightness. The consistency is 11%, and the temperature is 57-63°C. The E$_{op}$ stage has a retention time of 30 min, an oxygen dosage of 5 kg tonne^{-1} and a peroxide dosage of 1-4 kg tonne^{-1}. The extraction stage is peroxide enhanced with 0.75% NaOH. Temperature is 74°C, and pH is controlled to a 10.8 target with in-line probes. The D$_2$ stage is basically the same as the third stage but with 0.4% ClO$_2$ added, and is run at a temperature of 91°C.

Washing is counter-current with machine white water (or fresh water) to 5th stage, 5th to 4th stage, 4th to 3rd stage, 3rd to 2nd stage, 2nd to top showers 1st stage, sewer 2nd stage remainder. Fresh water is used on the bottom of the 1st stage showers. Of the 83,000 L tonne^{-1} of total water use, the bleach plant discharges 36,000 L tonne^{-1}.

The recovery boilers used are (i) a Babcox and Wilcox unit installed in 1971 with a capacity of 1.5 million kg solids d^{-1}, and (ii) a Tampella boiler installed in 1990 capable of processing 2 million kg solids d^{-1}.

Primary treatment occurs in a Dorr-Oliver clarifier operated at a rate of 76-91 million L d^{-1}. Secondary treatment is conducted in a 2.4 ha presettling basin (8.6 hr retention), a 35 ha aeration lagoon (5.14 d retention) and a 2.4 ha settling lagoon (8.6 h retention). The aeration lagoon has 39 Ashbrook (75 or 100 horsepower) and two Aire-O_2 100 horsepower surface aerators. Total flow to the river is 150 million L d^{-1}. BOD removal is 40,000 kg d^{-1} which represents a 90% removal rate. Water is drawn from 8 wells at an average depth of 24 m. The mill uses 150 million L d^{-1} of fresh water.

Samples were taken from the acid and alkaline filtrates from each of the three bleach lines, the acid sewer, the process water intake, the clarifier effluent and the final effluent. The analytes determined at Brunswick were identical to those examined at Leaf River except that DOC, TOC and resin acids were not run.

RESULTS

The following analytes were not detected at the listed locations in any of the samples taken in the Leaf River: chlorophenols (all locations), acrolein (all locations), dichlorodehydroabietic acid (all locations), neoabietic acid (ALF, ACS); 14-chlorodehydroabietic acid (ACS, PMS, RCS, EVS, WTI, WTE), 12-chlorodehydroabietic acid (ACS, PMS, RCS, EVS).

A full scale dioxin analysis was run on ALF and ACF samples taken during episode 1 (Leaf River). No 2,3,7,8-TCDD/F (detection limit: 0.01 ng kg^{-1}) was detected, but OCDD was found at 0.06 ng kg^{-1} in the ALF sample (detection limit 0.02 ng kg^{-1}). A single 2,3,7,8-TCDD/F scan was run on the ALF sample taken during episode 11; the analytes were non-detectable.

No chlorophenols or dioxins were found within the limits of detection in any of the samples collected at Brunswick.

Correlations of Contaminant Profiles at Leaf River

Concentration profiles of all the constituents were regressed against one another. Only a few of these relationships will be discussed here to illustrate the utility of the approach; a fuller account will be published later.

Regressions involving COD with r >0.8 are presented in Table 3. The COD-TOC relationships are not surprising since they probably reflect similar constituents. The PMS-COD vs PMS-BOD relationships illustrated in Fig. 2 are interesting since they are the only strong BOD-COD correlations observed in any of the regressions. It appears that the carbon in the pulp mill sewer (or a fixed fraction thereof) is biodegradable. The PMS contains a substantial amount of resin acids which are biodegradable, although at a slower rate than a smaller compound such as methanol. The pond efficiency at Leaf River was about 98% for resin acids. One conclusion from this relationship is that the PMS-COD will not be a heavy contributor to the COD of the final effluent since much of it will biodegrade. Similar reasoning applies to the evaporator sewer (EVS) BOD-COD relationship.

The WTE-COD and WTE-AOX profiles are illustrated in Fig. 3. Since the range of COD concentration is much greater than that of AOX, the two constituents are not related directly; rather they must both reflect a common cause. Since end-of-pipe AOX does not biodegrade in the treatment system, it should correlate with other constituents in the process that are also not readily biodegradable such as WTE-COD. Both parameters are, therefore, related to the amount of lignin removed, and the profiles reflect changes in operations such as hardwood/softwood swings. Most of the COD at Leaf River originates from the bleaching process, e.g., half of the COD entering the influent to waste treatment in the first episode originated from the acid and alkaline sewers. This situation reflects a clean well-run mill; presumably, the relationship would not be as good if COD entered the system from sources other than the bleach plant. Based on this *limited site-specific data* it appears that in the absence of upsets, end-of-pipe AOX and COD are surrogates for each other. More importantly, it implies that efforts to reduce COD through treatability will also reduce AOX.

Table 3. Relationships between analyte profiles involving COD[1].

	r		r
PMS-COD vs PMS-BOD	.997		
RCS-COD vs RCS-TOC	.995	WTE-COD vs WTI-TSS	.87
RCS-COD vs RCS-TSS	.99	WTI-COD vs ALF-DOC	.86
PMS-COD vs PMS-color	.99	PMS-COD vs RCS-methanol	.85
RCS-COD vs RCS-conductivity	.98	EVS-COD vs EVS-BOD	.84
PMS-COD vs PMS-MEK	.98	EVS-COD vs EVS-conductivity	.84
RCS-COD vs RCS-color	.96	ALF-COD vs ALF-color	.83
ALF-COD vs ALF-DOC	.93	ALF-COD vs ALF-methanol	.82
WTE-COD vs WTE-color	.92	WTE-COD vs WTE-AOX	.82
RCS-COD vs RCS-DOC	.90	WTE-COD vs WTE-TOC	.80

[1]n=18

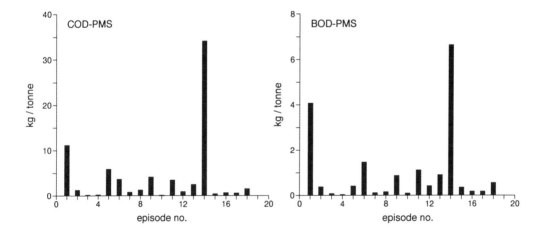

Figure 2. COD and BOD profiles for the pulp mill sewer (PMS).

The RCS-COD and RCS-conductivity correlations are somewhat misleading. COD was less than 900 mg L^{-1} for all but episodes 11 and 12 where it shot up to 32,000 mg L^{-1} and 100,000 mg L^{-1}, respectively. The conductivity was correspondingly high on these occasions. Clearly, the relationship reflects a spill rather than steady-state situation and confirms that conductivity is a good way to monitor black liquor spills.

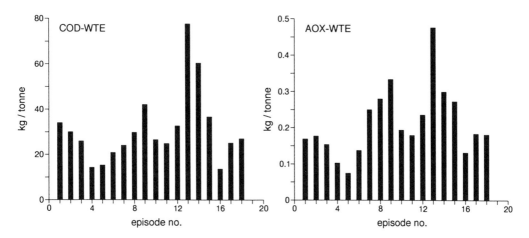

Figure 3. COD and AOX profiles for the final effluent (WTE).

Some correlations were expected but not found. For instance, a relationship between ALF-AOX and ACS-AOX was expected on the grounds that they should both proportionately vary with hardwood/softwood swings and other process variables. Surprisingly, this was not the case as illustrated in Fig. 4. It appears that the AOX split between acid and alkaline filtrates is different for hardwood and softwood. Also, there is much more variability in the ALF-AOX than in the ACF-AOX, suggesting that hardwood/softwood differences have a stronger influence on base-extractable AOX.

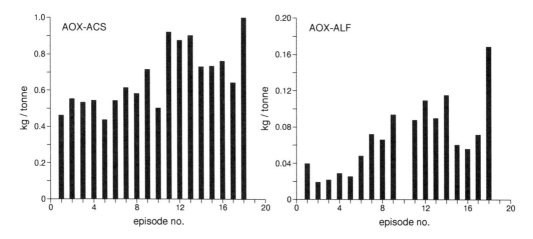

Figure 4. AOX profiles for the acid (ACS) and alkaline (ALF) sewer.

In summary, we have demonstrated the absence of measurable quantities of dioxins, chlorophenols (containing three or more chlorines) or acrolein in 100% ClO_2 bleaching. We have shown the utility of developing relationships among contaminant profiles across various locations and analytes. These correlations will be extended and we expect them to establish source-sink relationships and to provide insight into the transport of these materials through the mill.

REFERENCES

NCASI (National Council for Air and Stream Improvement). 1989a. NCASI procedures for the preparation and isomer specific analysis of pulp and paper industry samples for 2,3,7,8-TCDD and 2,3,7,8-TCDF. Technical Bulletin No. 551.

NCASI (National Council for Air and Stream Improvement). 1989b. Procedures for the analysis of resin and fatty acids in pulp mill effluents. Technical Bulletin No. 501.

U.S. EPA. 1986. Test methods for evaluating solid waste, physical chemical methods. Method 8015, SW-846, 3rd. Edition.

U.S. EPA. 1990. Method 624, 40 CFR, Part 136.

U.S. EPA. 1991. Chlorinated phenols in wastewater by *in situ* acetylation and GC-MS. EPA Draft Method 1653.

U.S. EPA. 1992. Method 1650 Revision A.

ORIGINS OF EFFLUENT CHEMICALS AND TOXICITY: RECENT RESEARCH AND FUTURE DIRECTIONS

Mark R. Servos

Great Lakes Laboratory for Fisheries and Aquatic Sciences, Fisheries and Oceans, Box 5050, 867 Lakeshore Road, Burlington, Ontario, L7R 4A6 Canada

The application of a variety of chemical and biological tests has enhanced our understanding of the origins and fate of toxic chemicals within mill process and treatment systems. Recent implementation of a wide variety of new process and treatment strategies, particularly at bleached kraft mills, has greatly altered the chemical composition and toxicity of final effluents. Rapid developments and improvements in cooking and prebleaching technologies that reduce the kappa number are expected to further reduce the residual lignin entering the bleach plant. The substitution of elemental chlorine for chlorine dioxide in the bleaching sequence significantly reduces the release of organochlorines and alters the chemical composition of the effluents. It seems inevitable that the industry trend toward conversion to elemental chlorine-free (ECF) bleaching will continue and will reduce the discharge of organochlorines and toxicity of mill effluents. New bleaching technologies, including hydrogen peroxide, ozone and enzymes, which are totally chlorine free (TCF), are now entering production and will continue to be developed. Unfortunately, there is relatively little known about the chemical composition and toxicity of TCF effluents and this represents a much needed area of research.

INTRODUCTION

The major objective of pulping processes is to remove lignin and other wood constituents while maximizing the yield of cellulose. There are a number of commonly used processes, including thermomechanical, sulfite and sulfate (kraft) pulping, which differ greatly in both yield and pulp quality. Attention has recently been focused on bleached kraft mills because of the concern surrounding the discharge of organochlorine contaminants. Kraft pulp is produced by mixing wood chips under high temperature and pressure with caustic sodium hydroxide and sodium sulfide. After extended washing, the pulp is brightened or bleached to achieve the desired brightness. The reaction of pulp and bleaching chemicals with the components of wood, including extractives, lignin, cellulose and hemicellulose, results in a number of chemicals in the process streams. The wood furnish, pulping conditions, bleaching techniques and sequence, and biological treatment strongly influence the levels and composition of chemicals in the final effluent (Kringstad and Lindström 1984; Berry et al. 1991).

In recent years there has been tremendous change within the pulp and paper industry as it moves toward introducing new pulping and bleaching processes to meet new environmental regulations and concerns. The greatest emphasis has been on the reduction of the production and release of organochlorines from bleached kraft mills. A large number of organochlorine compounds have been identified in process streams and final effluents of mills. The primary goals of the process changes have been to reduce the residual lignin entering the bleach plant and to eliminate the use of elemental chlorine in the bleach plant. Ultimately complete closure of the water system in the bleach plant could be achieved, thereby eliminating any release of contaminants into the aquatic environment. Rather than an exhaustive review (see Strömberg et al. 1996; LaFleur 1996 and Ahtiainen et al. 1996) this work examines the most recent advances, trends and future directions of research on the origins and fate of toxic chemicals in the pulp and paper mill process streams and final effluents.

CHEMICAL COMPOSITION

Changes or modifications in mill process or effluent treatment can dramatically change the chemical composition and toxicity of the final effluent. Many different approaches to reducing or eliminating the discharged organochlorines in particular have been recently implemented by the industry. These approaches generally include some form of pretreatment of the pulp, changes in the bleaching sequences and technology, or optimization of effluent treatment strategies. Although many compounds are formed in the various processes, not all are released into the environment. Debarking wastewater, sewered condensates, brownstock washing/screening and wastewater and bleach plant filtrates are usually released into treatment ponds. Small amounts of causticizing wastewater may also be released. Black liquor can reach wastewater but usually only through leaks or spills (LaFleur 1996). Although it would be very rare in modernized mills, poorly operated mills, especially in developing countries, may operate under unstable conditions resulting in frequent accidental discharges and spills of fiber and cooking chemicals (Oanh et al. 1994). Implementation of adequate waste monitoring with process control may minimize these discharges. Care must be taken to consider the operation of mills in other jurisdictions where regulations and market demands differ.

Reducing the amount of residual lignin in the pulp entering the bleach plant significantly reduces and alters the amount and types of bleaching chemicals required to achieve the desired brightness. Numerous techniques have been employed by mills to reduce the kappa number of the pulp prior to bleaching, i.e., extended delignification in cooking stage and oxygen delignification. Reduction in kappa number from 30 to 35 using conventional bleaching sequences to 10 or less using extended/modified delignification and oxygen delignification has been achieved (Strömberg et al. 1996).

The chlorine multiple, kappa number, chlorine charge, substitution of chlorine dioxide for elemental chlorine and order of addition have been shown to influence the quantity and composition of chlorophenols in the effluent (Berry et al. 1991; Åxegard et al. 1993). The most commonly introduced process modification has been the substitution of elemental chlorine with chlorine dioxide. The chemistry of these two bleaching agents is very different and therefore leads to a very different effluent composition. The substitution of elemental chlorine with other bleaching agents results in dramatic decreases in the discharge of organochlorine contaminants including total adsorbable organohalide (AOX) (Berry et al. 1991; Graves et al. 1993). Increased substitution of chlorine dioxide in the bleaching process also results in lower multiple chlorination of compounds such as chlorophenols, and 100% substitution (ECF) generally leads to nondetectable levels of polychlorinated dioxins and furans as well as tri- and tetra-substituted chlorophenols. Chemicals such as chlorinated acetic acids and chloroform are dramatically reduced by ECF. Lower chlorinated compounds are more susceptible to degradation within the secondary treatment systems which further reduces the release of these compounds into the receiving environment. Kemeny and Banerjee (1996) reported nondetectable levels of polychlorinated dioxins, highly chlorinated phenols or acrolein in a mill using 100% chlorine dioxide substitution. Charlet and Claudio-da-Silva (1994) found that introduction of oxygen delignification decreased the concentration of contaminants and toxicity of the final effluent. Fisher et al. (1996) compared the water quality protection values for chlorophenolic compound to the concentrations expected in the receiving environment of 69 mills during low flow. Only a minority of mills (10) currently produce exceedances of these values, and these exceedances were not large. No mills were predicted to have exceeded the water quality values after the implementation of complete chlorine dioxide substitution with or without oxygen delignification.

The majority of the mass of AOX in bleached kraft effluents using elemental or ECF bleaching is high molecular weight (>1000 Da) materials (Dahlman et al. 1996). However, the structure of the high molecular weight material is different depending on the bleaching process e.g., ECF, TCF (Dahlman et al. 1994a). Approximately half of the AOX in a softwood ECF effluent was high molecular weight while a major part of TCF hardwood pulp was low molecular weight (Dahlman et al. 1996). Halogenated high molecular weight material in softwood TCF effluents also appears to be composed of lower molecular weight material than that in softwood ECF effluents. Dahlman et al. (1996) have explored new methods to further examine and compare the structure of lignin. A method using thioacidolysis and GC/HRMS to detect chlorinated nonphenolic alkyl-aryl ether structures has been developed. Previous work has shown that part of the AOX is found in the lignin-derived phenolic structural elements. Ristolainen (1994) examined high molecular weight lignin fractions dissolved during TCF bleaching of a hardwood kraft pulp which were characterized by analyzing the alkaline oxidation products. Ultrafiltration and HPLC indicated a variety of phenolic products including acetosyringone, acetovanillone, syringaldehyde, vanillin and syringic acid. Differences in the total amount and composition of the product fraction from different bleaching stages were detected.

Organically bound halogens have also been detected in naturally occurring humic substances (Dahlman *et al.* 1993, 1994b). The chlorinated aromatic carboxylic acids identified as degradation products of humic substances are mainly the same as those found in the degradation of high molecular weight organic matter from bleached kraft mill effluent. Banerjee *et al.* (1994) produced radioactive AOX by bleaching pulp with $^{36}ClO_2$ and made site-specific biodegradation measurements in treatment lagoons. Degradation of AOX in treatment lagoons was a function of both depth and location (distance to aerators). While the implementation of ECF bleaching can bring the discharge of AOX to as low as 0.2 kg ADMT^{-1} and secondary treatment can reduce these values by half to 0.1 kg ADMT^{-1}, TCF processes produce effluent with AOX values essentially the same as natural wood products, <0.001 kg ADMT^{-1} (Fisher *et al.* 1996). After secondary treatment the AOX shifted toward higher molecular weight compounds (Strömberg *et al.* 1996; Jokela *et al.* 1993). The treatment lagoons tend to remove the low chlorinated compounds while the tri- and tetra-substituted compounds (e.g., chlorophenols) are more resistant. There is a large reduction in chloroform and almost complete removal of the resin acids, fatty acids and sterols in most treatment systems (Strömberg *et al.* 1996).

Concentrations and extractives such as fatty acids, resin acids and sterols in untreated effluent varies considerably among mills. The move to dry debarking has greatly reduced the release of resin and fatty acids. The extractives are also susceptible to changes during storage (LaFleur 1996). Aerated lagoons and activated sludge plants are very effective at removing but not eliminating the extractives from final effluent. The wood furnish is also an important factor; coniferous trees (i.e., *Picea abies* bark) contain relatively high amounts of stilbenes including resveratrol, piceatannol, isohapontigenin, astingin and isorhapontin (Mannila and Talvitie 1994) which have been shown to have a number of toxicological effects. O'Connor *et al.* (1992) clearly demonstrated that the wood furnish significantly altered the toxicity of simulated mechanical pulping effluents. More attention needs to paid to the extractive material in effluents not only from bleached kraft mills but all pulping processes. Recent attention has focused on the plant sterols which have been shown to cause a number of biological effects including altered plasma sex steroid profiles and decreased reproductive fitness in fish (Maclatchy and Van Der Kraak 1994). Although there are a number of plant sterols detectable in effluents, β-sitosterol is the dominant compound in the sterol fraction and has received the most attention.

Several mills are already producing pulps using TCF processes. Hydrogen peroxide and/or ozone have replaced chlorine or chlorine dioxide as the bleaching agents. Biological processes using enzymes are currently being tested (Mao and Smith 1994; Nelson *et al.* 1994b) to decolorize pulp. Unfortunately there is very little information currently available on the chemistry and toxicology of these newly introduced technologies.

TOXICITY

Considerable attention has been directed at examining the chemicals and toxicological responses of effluents from various process and treatment streams. Recent work has focused on characterizing the biological responses resulting from exposure to effluents from both ECF and TCF processes. The results of these studies suggest that the toxicity and physiological responses observed in biota may be associated not only with organochlorines produced in the bleachery but with the wood extractives.

Ahtiainen *et al.* (1996) and Verta *et al.* (1996) examined 18 different untreated or secondary treated TCF, ECF and conventional bleaching effluents using a variety of biological tests: *Pseudomonas putida*, growth inhibition; *Vibrio fisheri*, luminescence bacteria; *Selenastrum capricornutum*, algal growth inhibition; *Daphnia magna*, mobility inhibition; and *Branchydanio rerio*, zebra fish hatching and survival test. There was no significant difference in toxicity between untreated conventional ECF and TCF bleaching effluents. However, none of the biologically treated effluents caused toxicity. Toxicity was strongly related to organic material in the effluents (Verta *et al.* 1996). Verta *et al.* (1996) suggested that natural constituents of wood are probably responsible for the toxicity observed in ECF and TCF effluents, and this toxicity appears to be eliminated by biological treatment of the effluents.

Eklund *et al.* (1996) determined that TCF was less toxic than an ECF process using Microtox, acute copepod toxicity, *Nitocra spinipes* and reproduction of *N. spinipes* and the marine red algae *Ceraminium strictum*. Effluent from the unbleached process (from the ECF mill) was the most toxic. Secondary treatment reduced toxicity by 2-5 times for both ECF and TCF process effluents. Middaugh *et al.* (1994) used a 7-8 d embryo toxic/teratogenic responses of early life stage inland silverside *Menidia beryllina* to determine the effectiveness of a 15 d aerated stabilization basin. Hall *et al.* (1994, 1996) used a suite of marine and freshwater chronic bioassays to characterize effluents from two mills before and after implementation of oxygen delignification and >70% chlorine dioxide substitution (with secondary treatment). Although conversion of the mills to chlorine dioxide and/or oxygen delignification reduced the AOX and some other chlorinated organics, it did lead to clear improvements in the bioassay responses. Priha (1994) examined the response of a number of organisms to 13 treated effluents and showed very little toxicity. Robinson *et al.* (1994) also report that receiving waters exposed to treated effluents did not cause toxicity in short term bioassays with fathead minnows or *Cerodaphnia*.

Several studies have attempted to isolate or characterize the chemicals responsible for toxicity in pulp mill effluents. Using the amphipod reproduction test with *Monoporeia affinis*, Eriksson *et al.* (1996) determined that the majority of toxicity in the HPLC fractionated extracts of sediment from the Baltic was in the aliphatic/ monoaromatic and the diaromatic extracts. The aliphatic fraction was not toxic in deeper sediments. The hydrophilic fraction (based on reverse phase C_{18} material adsorption) affected the reproduction of *C. strictum* (Eklund *et al.* 19965). Adolfsson-Erici *et al.* (1994) used a battery of biological and biochemical tests as well as chemical characterization to examine effluents from ECF and TCF bleaching. PUF and XAD2 resin extracts of samples taken before and after secondary treatment were extracted and injected into fish. Nelson *et al.* 1994a tested water and acetone extracts of lodgepole pine sapwood and heartwood for toxicity to *Daphnia* 48 h and Microtox. Higher toxicity of the heartwood extracts was attributed to a higher concentration of leachable extractives and not a change in chemical composition. Nelson *et al.* (1994b) reported that the toxicity of effluent from xylanase prebleaching of a peroxide-based chlorine-free bleaching sequence was strongly dependent on the wood furnish. Förlin *et al.* (1994) reported a wide variety of biochemical responses resulting from exposure of fish to both ECF or TCF effluents. After conducting a comprehensive review, Solomon *et al.* (1994) concluded that treated ECF effluents represent no significant risk to the environment.

MIXED FUNCTION OXYGENASE INDUCTION

Mixed function oxygenase (MFO) induction is a specific response that has been extensively studied in fish exposed to a variety of effluents and process streams. Numerous recent studies have developed a variety of procedures and approaches to identify or characterize the chemicals responsible for this response in fish. Hodson *et al.* (1994) examined black liquors for chemicals responsible for the induction of MFO activity in rainbow trout. They determined that black liquor from softwood production was more potent than derived from hardwoods. Induction resulted from exposure to 0.0032 to 0.2%; exposure above 0.2% was lethal. Burnison *et al.* (1994) demonstrated that methanol extracts of particulates from at least one bleached kraft mill resulted in MFO induction in rainbow trout. Munkittrick *et al.* (1995) have also shown that dichloromethane extracts of a bleached kraft mill effluent caused MFO induction in fish. Burnison et al. (1994) isolated an MFO active fraction using both methanol and/or solid phase C_{18} extractions of whole effluent or particulates. HPLC separation of the extracts isolated the ethoxyresorufin-o-deethylase (EROD) activity to the relatively polar region of the chromatograph with a predicted K_{ow} in the 2 to 4.5 range.

Parrott et al. (1994) took a single chemical approach to eliminate a variety of suspected chemicals as the inducers of MFO activity in fish including retene, potassium biphthalate, phthalic anhydride, phenanthroline, benzophenone, diphenyl ethane, dicoumarol, 2-methyl naphthalene, phenanthrene and chrysene. Ferguson et al. (1992) previously examined a number of resin acids which failed to induce MFO activity in rainbow trout. Hewitt et al. (1996) ruled out tetrachloroguaiacol using direct exposures to rainbow trout. Hewitt et al. (1996) fractionated effluents from a bleached kraft mill before and after secondary treatment and during a mill shutdown using a 1 µm filter followed by ultrafiltration. The concentration of a wide variety of resin acids, fatty acids, bacterial fatty acids, terpenes, chlophenolics, aliphatic alkanes, plant sterols and chlorinated dimethylsulfones in the effluent fractions failed to correlate with MFO activity in fish.

Although the chemical(s) responsible for MFO induction have not been identified, there is strong evidence that at least some of them are not chlorinated. Williams et al. (1996) observed EROD activity in a mill using an unbleached process. Hodson et al. (1994) reported that spent cooking liquors as well as C- and E-stage effluents were very effective at causing MFO induction. Förlin et al. (1994) reported that EROD was greater in fish exposed to ECF vs. TCF effluents. Martel et al. (1996) reported that after examination of a large number of bleached and unbleached kraft, thermomechanical and chemi-thermomechanical mills, the EROD inducing potential of effluents could not be linked to a particular bleaching or pulping process. Numerous field studies (reviewed or presented in later chapters of this book, e.g., Lehtinen 1996 and Hodson 1996) support these laboratory observations.

FUTURE DIRECTIONS

Our understanding of the origins and fate of toxic chemicals within mill process and treatment systems has been greatly enhanced by the application of a variety of chemical and biological tests. Recent implementation of a wide variety of process and treatment strategies, particularly at bleached kraft mills, has greatly altered the chemical composition and toxicity of final effluents. There will continue to be rapid developments and improvements in cooking and prebleaching technologies that will reduce the kappa number and further reduce the residual lignin entering the bleach plant. New bleaching technologies including hydrogen peroxide, ozone and enzymes (TCF) are now entering production and will continue to be developed. Most of the previously published work on effluent chemical composition and toxicity is based on elemental chlorine bleaching of kraft pulps. Many of the chemicals previously reported may not be detectable in modern mills. Great care must be taken in future studies to document the operating conditions under which experiments or observations are made.

New bleaching processes including hydrogen peroxide, ozone and enzymes (TCF) are now in production. There is little known about the toxicity and chemical composition of these TCF effluents. Although the intention is to eventually recover the bleachery effluent stream, there is potential for some release into the environment. The elimination of elemental chlorine will reduce the discharge of organochlorines but a major benefit will be that it may allow the eventual closure of the bleach plant. The move toward a closed water system in the bleach plant is approaching but will not become widespread for some time to come.

More attention needs to paid to the extractive material in effluents not only from bleached kraft mills but all pulping processes. Recent findings have suggested that some of the toxicity and physiological responses in biota may be associated not only with organochlorines produced in the bleachery, but with the wood extractives. The natural constituents of wood are possibly responsible for the toxicity observed in ECF and TCF effluents, and this toxicity appears to be greatly reduced or eliminated by biological treatment of the effluents.

Reeve (1994) questioned the overall environmental benefit of the many processes currently being implemented when factors such as slightly lower yields, energy costs, etc. are considered. Many of the changes may be driven greater by market demand and public perception rather than environmental improvements. Totally effluent-free processes still have numerous problems to overcome and implementation at older mills may be very complex and costly. However, it seems inevitable that conversion to ECF and TCF bleaching will continue and additional research will eventually determine the associated environmental benefits.

REFERENCES

Adolfsson-Erici, M., T. Alsberg and E. Gravenfors. 1994. Sampling and analysis of effluent water from pulp mills using elemental chlorine free (ECF) and total chlorine free (TCF) bleaching. Presented at the Second International Conference on Environmental Fate and Effects of Bleached Pulp Mill Effluents, November 6-10, 1994, Vancouver, B.C., Canada.

Ahtiainen, J., T. Nakari and J. Silvonen. 19965. Toxicity of TCF and ECF pulp bleaching effluents assessed by biological toxicity tests. In Environmental Fate and Effects of Pulp and Paper Mill Effluents, M.R. Servos, K.R. Munkittrick, J.H. Carey and G. Van Der Kraak (ed.), St. Lucie Press, Delray Beach, FL.

Åxegard, P., O. Dahlman, I. Haglind, B. Jacobson, R. Mörck and L. Strömberg. 1993. Pulp bleaching and the environment - the situation in 1993. Nordic Pulp Paper Res. J. 4:365-378.

Banerjee, S., C.L. Williams and S.J. Severtson. 1994. Distribution and fate of AOX and model BKME chloro-organics in secondary treatment systems. Presented at the Second International Conference on Environmental Fate and Effects of Bleached Pulp Mill Effluents, November 6-10, 1994, Vancouver, B.C., Canada.

Berry, R., C. Luthe, R. Voss, P. Wrist, P. Åxegard, G. Gellerstedt, L-O. Lindbald and I. Pöpke. 1991. The effects of recent changes in bleached softwood kraft mill technology on organochlorine emissions: an international perspective. Pulp Pap. Can. 92(6):T155-T165.

Burnison, B.K., P.V. Hodson, D.J. Nuttley and S. Efler. 1994. Isolation and characterization of a MFO-inducing fraction from BKME. Presented at the Second International Conference on Environmental Fate and Effects of Bleached Pulp Mill Effluents, November 6-10, 1994, Vancouver, B.C., Canada.

Charlet, P., and E. Claudio-da-Silva, Jr. 1994. Study of the effectiveness of aerated lagoons for removal of conventional pollutants, chlorophenolic compounds and acute toxicity. Presented at the Second International Conference on Environmental Fate and Effects of Bleached Pulp Mill Effluents, November 6-10, 1994, Vancouver, B.C., Canada.

Dahlman, O., R. Mörck, P. Ljungquist, A. Reimann, C. Johansson, H. Borén and A. Grimvall. 1993. Chlorinated structural elements in high molecular weight organic matter from unpolluted waters and bleached-kraft mill effluents. Environ. Sci. Technol. 27(8):1616-1620.

Dahlman, O., A. Reimann, P. Ljungquist and R. Mörck. 1994a. A method for detection of chlorinate etherified lignin structures in high molecular weight materials of industrial and natural origin. Presented at the Second International Conference on Environmental Fate and Effects of Bleached Pulp Mill Effluents, November 6-10, 1994, Vancouver, B.C., Canada.

Dahlman, O., A. Reimann, P. Ljungquist, R. Mörck, C. Johansson, H. Borén and A. Grimvall. 1994b. Characterization of chlorinated aromatic structures in high molecular weight BKME-materials and in fulvic acids from industrially unpolluted waters. Wat. Sci. Technol. 29(5-6):81-91.

Dahlman, O., A. Reimann, P. Ljungquist and R. Mörck. 1996. A method for detection of chlorinated, etherified lignin structures in high molecular weight materials of industrial and natural origin. In Environmental Fate and Effects of Pulp and Paper Mill Effluents, M.R. Servos, K.R. Munkittrick, J.H. Carey and G. Van Der Kraak (ed.), St. Lucie Press, Delray Beach, FL.

Eklund, B., M. Linde and M. Tarkpea. 1996. Comparative assessment of the toxic effects from pulp mill effluents to marine and brackish water organisms. In Environmental Fate and Effects of Pulp and Paper Mill Effluents, M.R. Servos, K.R. Munkittrick, J.H. Carey and G. Van Der Kraak (ed.), St. Lucie Press, Delray Beach, FL.

Eriksson, A-K., B. Sundelin, D. Broman and C. Näf. 1996. Reproduction effects on *Monoporeia affinis* of HPLC-fractionated extracts of sediments from a pulp mill recipient. In Environmental Fate and Effects of Pulp and Paper Mill Effluents, M.R. Servos, K.R. Munkittrick, J.H. Carey and G. Van Der Kraak (ed.), St. Lucie Press, Delray Beach, FL.

Ferguson, M.L., M.R. Servos, K.R. Munkittrick and J. Parrott. 1992. Inability of resin acid exposure to elevate EROD activity in rainbow trout (*Oncorhynchus mykiss*). Water Poll. Res. J. Can. 27:561-573.

Fisher, R.P., D.A. Barton and P.S. Weigand. 1996. An assessment of the significance of discharge of chlorinated phenolic compounds from bleached kraft pulp mills. *In* Environmental Fate and Effects of Pulp and Paper Mill Effluents, M.R. Servos, K.R. Munkittrick, J.H. Carey and G. Van Der Kraak (ed.), St. Lucie Press, Delray Beach, FL.

Förlin, L., E. Lindesjöö and G. Ericson. 1994. Biochemical, histological and genotoxic effects in rainbow trout treated with extracts of effluent waters from pulp mills. Presented at the Second International Conference on Environmental Fate and Effects of Bleached Pulp Mill Effluents, November 6-10, 1994, Vancouver, B.C., Canada.

Graves, J.W., T.W. Joyce and H. Jameel. 1993. Effect of chlorine dioxide substitution, oxygen delignification, and biological treatment on bleach-plant effluent. Tappi J. 76:153-158.

Hall, T.J., D.L. Borton and T.J. Bousquet. 1994. Chronic bioassay responses and chemical characteristics of effluent from two bleached kraft mills following conversions to increased chlorine dioxide substitution and oxygen delignification. Presented at the Second International Conference on Environmental Fate and Effects of Bleached Pulp Mill Effluents, November 6-10, 1994, Vancouver, B.C., Canada.

Hall, T.J., R.K. Haley, D.L. Borton and T.M. Bousquet. 1996. The use of chronic bioassays in characterizing effluent quality changes for two bleached kraft mills undergoing process changes to increased chlorine dioxide substitution and oxygen delignification. *In* Environmental Fate and Effects of Pulp and Paper Mill Effluents, M.R. Servos, K.R. Munkittrick, J.H. Carey and G. Van Der Kraak (ed.), St. Lucie Press, Delray Beach, FL.

Hewitt, L.M., J.H. Carey, D.G. Dixon and K.R. Munkittrick. 1996. Examination of bleached kraft mill effluent fractions for potential inducers of mixed function oxygenase activity in rainbow trout. *In* Environmental Fate and Effects of Pulp and Paper Mill Effluents, M.R. Servos, K.R. Munkittrick, J.H. Carey and G. Van Der Kraak (ed.), St. Lucie Press, Delray Beach, FL.

Hodson, P.V. 1996. MFO induction by pulp mill effluents - advances since 1991. *In* Environmental Fate and Effects of Pulp and Paper Mill Effluents, M.R. Servos, K.R. Munkittrick, J.H. Carey and G. Van Der Kraak (ed.), St. Lucie Press, Delray Beach, FL.

Hodson, P.V., M.M. Maj, S. Efler, A. Schnell and J. Carey. 1994. Kraft black liquor as a source of MFO inducers. Presented at the Second International Conference on Environmental Fate and Effects of Bleached Pulp Mill Effluents, November 6-10, 1994, Vancouver, B.C., Canada.

Jokela, J.K., M. Laine, M. Ek and M. Salkinoja-Salonen. 1993. Effect of biological treatment on halogenated organics in bleached kraft pulp mill effluents studied by molecular weight distribution analysis. Environ. Sci. Technol. 27:547-557.

Kemeny, T.E., and S. Banerjee. 1996. Correlations among contaminant profiles in mill process streams and effluents. *In* Environmental Fate and Effects of Pulp and Paper Mill Effluents, M.R. Servos, K.R. Munkittrick, J.H. Carey and G. Van Der Kraak (ed.), St. Lucie Press, Delray Beach, FL.

Kringstad, K.P., and K. Lindström. 1984. Spent liquors from pulp bleaching. Environ. Sci. Technol. 18:236-248.

LaFleur, L.E. 1996. Sources of pulping and bleaching derived chemicals in effluents. *In* Environmental Fate and Effects of Pulp and Paper Mill Effluents, M.R. Servos, K.R. Munkittrick, J.H. Carey and G. Van Der Kraak (ed.) St. Lucie Press, Delray Beach, FL.

Lehtinen, K-J. 1996. Biochemical responses in organisms exposed to effluents from pulp production - are they related to bleaching? *In* Environmental Fate and Effects of Pulp and Paper Mill Effluents, M.R. Servos, K.R. Munkittrick, J.H. Carey and G. Van Der Kraak (ed.), St. Lucie Press, Delray Beach, FL.

Maclatchy, D.L., and G. Van Der Kraak. 1994. The plant sterol β-sitosterol decreases reproductive fitness in goldfish. Presented at the Second International Conference on Environmental Fate and Effects of Bleached Pulp Mill Effluents, November 6-10, 1994, Vancouver, B.C., Canada.

Mannila, E., and A. Talvitie. 1994. Bioactive stilbenes in Norway spruce (*Picea abies*) bark; a group of potential ecotoxic compounds. Presented at the Second International Conference on Environmental Fate and Effects of Bleached Pulp Mill Effluents, November 6-10, 1994, Vancouver, B.C., Canada.

Mao, H., and D.W. Smith. 1994. Decolorization and dechlorination of pulp mill effluents using a new immobilized living cell system. Presented at the Second International Conference on Environmental Fate and Effects of Bleached Pulp Mill Effluents, November 6-10, 1994, Vancouver, B.C., Canada.

Martel, P.H., T.G. Kovacs and R.H. Voss. 1996. Effluents from Canadian pulp and paper mills: a recent investigation of their potential to induce MFO activity in fish. *In* Environmental Fate and Effects of Pulp and Paper Mill Effluents, M.R. Servos, K.R. Munkittrick, J.H. Carey and G. Van Der Kraak (ed.), St. Lucie Press, Delray Beach, FL.

Middaugh, D.P., N. Beckham, T.L. Deardorff and J.W. Fournie. 1994. Responses of fish embryos to bleached kraft mill process water. Presented at the Second International Conference on Environmental Fate and Effects of Bleached Pulp Mill Effluents, November 6-10, 1994, Vancouver, B.C., Canada.

Munkittrick, K.R., M.R. Servos, K. Gorman, B. Blunt, M.E. McMaster and G.J. Van Der Kraak. 1995. Characteristics of EROD induction associated with exposures to pulp and paper mill effluent. *In* Environmental Toxicology and Risk Assessment: Fourth Volume, ASTM STP 1262, T.W. LaPointe, F.T. Price and E.E. Little (eds.), American Society for Testing and Materials, Philadelphia: In press.

Nelson, S., S.A. Brueckner, T. Chen, P. Bicho, J.N. Saddler and C. Breuil. 1994a. Identification of leachable toxic fractions from lodgepole pine. Presented at the Second International Conference on Environmental Fate and Effects of Bleached Pulp Mill Effluents, November 6-10, 1994, Vancouver, B.C., Canada.

Nelson, S., K.K.Y. Wong, P.A. Bicho and J.N. Saddler. 1994b. Toxicity profile for peroxide bleaching of softwood kraft pulps that involves xylanase prebleaching. Presented at the Second International Conference on Environmental Fate and Effects of Bleached Pulp Mill Effluents, November 6-10, 1994, Vancouver, B.C., Canada.

Oanh, K., N. Thi and B-E. Bengtsson. 1994. Wastewater monitoring, a possible tool to optimize process control and reduce contamination from bleached kraft pulp and paper industry with an example from a Vietnamese mill. Presented at the Second International Conference on Environmental Fate and Effects of Bleached Pulp Mill Effluents, November 6-10, 1994, Vancouver, B.C., Canada.

O'Connor, B.I., T.G. Kovacs and R.H. Voss. 1992. The effect of wood species composition on the toxicity of simulated mechanical pulping effluents. Environ. Toxicol. Chem. 11:1259-1270.

Parrott, J.L., B.K. Burnison, P.V. Hodson, M.E. Comba and M.E. Fox. 1994. Retene-type compounds - inducers of hepatic mixed function oxygenase (MFO) in rainbow trout (*Oncorhynchus mykiss*)? Presented at the Second International Conference on Environmental Fate and Effects of Bleached Pulp Mill Effluents, November 6-10, 1994, Vancouver, B.C., Canada.

Priha, M. 1994. Ecotoxicological impacts of pulp mill effluents in Finland. Presented at the Second International Conference on Environmental Fate and Effects of Bleached Pulp Mill Effluents, November 6-10, 1994, Vancouver, B.C., Canada.

Reeve, D.W. 1994. The evolution of processes for production of bleached pulp. Presented at the Second International Conference on Environmental Fate and Effects of Bleached Pulp Mill Effluents, November 6-10, 1994, Vancouver, B.C., Canada.

Ristolainen, M. 1994. Elucidation of lignin-derived materials from TCF bleaching effluents. Presented at the Second International Conference on Environmental Fate and Effects of Bleached Pulp Mill Effluents, November 6-10, 1994, Vancouver, B.C., Canada.

Robinson, R.D., J.H. Carey, K.R. Solomon, I.R. Smith, M.R. Servos and K.R. Munkittrick. 1994. Survey of receiving water environmental impacts associated with discharges from pulp mills. I. Mill characteristics, receiving water chemical profiles and laboratory chronic toxicity tests. Environ. Toxicol. Chem. 13:1075-1088.

Solomon, K., H. Bergman, D. Mackay, B. McKauge and D. Pryke. 1994. Ecotoxicological risks from organochlorine compounds produced by the use of chlorine dioxide for the bleaching of pulp. Presented at the Second International Conference on Environmental Fate and Effects of Bleached Pulp Mill Effluents, November 6-10, 1994, Vancouver, B.C., Canada.

Strömberg, L., R. Mörck, Filipe de Sousa and Olof Dahlman. 1996. Effects of internal process changes and external treatment on effluent chemistry. *In* Environmental Fate and Effects of Pulp and Paper Mill Effluents, M.R. Servos, K.R. Munkittrick, J.H. Carey and G. Van Der Kraak (ed.), St. Lucie Press, Delray Beach, FL.

Verta, M., J. Ahtiainen, T. Nakari, A. Langi and E. Talka. 1996. The effect of waste constituents on the toxicity of TCF and ECF pulp bleaching effluents. *In* Environmental Fate and Effects of Pulp and Paper Mill Effluents, M.R. Servos, K.R. Munkittrick, J.H. Carey and G. Van Der Kraak (ed.), St. Lucie Press, Delray Beach, FL.

Williams, T.G., J.H. Carey, B.K. Burnison, D.G. Dixon and H-B. Lee. 1996. Rainbow trout MFO responses caused by unbleached and bleached pulp mill effluents: a laboratory based study. *In* Environmental Fate and Effects of Pulp and Paper Mill Effluents, M.R. Servos, K.R. Munkittrick, J.H. Carey and G. Van Der Kraak (ed.), St. Lucie Press, Delray Beach, FL.

SECTION II

ENVIRONMENTAL FATE

Once an effluent is discharged from an outfall, the dispersion and transportation of the various chemicals in the receiving environment are dependent on the physical/chemical properties of both the chemical itself and the environment into which it is released. Predicting the fate and ultimate exposure of aquatic biota to contaminants has been the focus of considerable research. The dynamic nature of the receiving environments as well as the wide diversity of chemical structures found in pulp and paper mill effluents make this a very difficult but critical task.

ENVIRONMENTAL FATE AND DISTRIBUTION OF SUBSTANCES

G.E. Carlberg[1] and T.R. Stuthridge[2]

[1]The Norwegian Pulp and Paper Research Institute, P.O. Box 24 Blindern, 0313 Oslo, Norway
[2]PAPRO, New Zealand Forest Research Institute, Private Bag 3020, Rotorua, New Zealand

The processes influencing the distribution and fate of bleached pulp mill effluent constituents in receiving waters are reviewed. Water column concentrations of organochlorine, chlorophenolics and resin acids rapidly diminish upon discharge to the recipient. Dilution and hydrodynamic transport appear to be the most important mechanisms for dispersal. However, a significant proportion of the observed decrease in recipient waters can be attributed to sedimentation of compounds adsorbed to colloidal or particulate matter or reversible adsorption onto high molecular mass organics, such as chlorolignin or humic acids. Sediment accumulation, transport and continued adsorption processes may enable effluent constituents to be detected at elevated concentrations over a significant region of the recipient.

Changes in pulping and bleaching technology and the trend towards closed-cycle processes will decrease the levels of bleached pulp mill effluent constituents discharged to the recipient. The increased hydrophilicity and biodegradability of the organic material formed during such bleaching processes is also expected to decrease sediment concentrations of these compounds. However, it appears that previously accumulated bleached pulp mill effluent constituents are likely to persist in sediments for a significant period of time.

INTRODUCTION

To assess the ecological consequences of an industrial discharge, a description of the exposure situation and potential biological effects is needed. In this context knowledge about the recipient behavior of the discharged compounds is essential. Bleached pulp mill effluents contain a complex mixture of organic constituents of various compound classes over a broad molecular mass range. More than 300 compounds have been identified to date (McKague *et al.* 1988). Wood room, pulping and bleaching processes and the paper production contribute significantly to the mill's discharge of organic material and the effluent toxicity. The distribution and fate of these compounds in effluent receiving waters is highly dependent upon their physical/chemical characteristics and the nature of the recipient.

Studies relating to the recipient distribution and fate of compounds from bleached pulp mill effluents have predominantly focused on two specific classes of environmentally significant compounds, namely chlorinated phenolics and resin acids, and on sum parameters, particularly AOX (adsorbable organically bound halogen), TOX (total organically bound halogen) and EOX (extractable organically bound halogen). This paper describes processes influencing the distribution and fate of pulp mill discharged compounds through results from recent investigations.

BLEACHED PULP MILL EFFLUENT CONSTITUENTS IN RECEIVING WATERS

One of the difficulties in assessing the fate of bleached pulp mill effluent constituents in mill recipients is the high degree of temporal variability of the discharges from mills. Concentrations of individual constituents may vary over orders of magnitude during mill operations and sampling (Robinson *et al.* 1994). Similarly, differences in mill configuration, feedstock, operating conditions and the nature of the effluent recipient restrict direct intercomparison of mill studies.

Effluents from pulp mills are discharged into rivers, lakes, estuaries, bays and the open ocean. These environments provide different mechanisms and rates of mixing, dispersion, flushing and degradation as well as different water qualities (Volkman et al. 1993). The effluent plume from a bleached pulp mill discharge may disperse both horizontally and vertically into the receiving waters. Differences in effluent temperature and density may lead to stratification and restrict mixing and dilution of the effluent with recipient waters. For example, undiluted "plugs" of effluent may remain in the marine water column for over 36 km (Environment Canada 1991).

A number of recent studies have determined the concentrations of bleached pulp mill effluent constituents in the aqueous phase of mill recipients (Lindström-Seppä and Oikari 1990; Grimvall et al. 1991; Holmbom et al. 1992; Kukkonen 1992; Södergren 1993; Volkman et al. 1993; Comba et al. 1994; Robinson et al. 1994). Their results generally correlate with those of previous studies reviewed earlier by McLeay (1987).

Concurrent reductions in sodium levels and concentrations of sum parameters (AOX, DOC) indicate that dilution and hydrodynamic transport are the most important mechanisms for the dispersal of bleached pulp mill effluent constituents (Kukkonen 1992). Holmbom et al. (1992) found that about 80% of the AOX material was transported over 10 km and, correcting for dilution, remained at a constant level after this point (Fig. 1). This indicated that the chlorinated organic material is mainly stable towards chemical, photochemical and biochemical degradation and is also not sedimented to any notable extent in the recipient area.

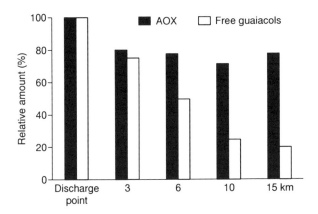

Figure 1. AOX and free guaiacol levels in downstream recipient waters (Holmbom et al. 1992).

Concentrations of chlorophenolic compounds in the water column rapidly decrease within 15 km of the point of discharge (McLeay 1987; Lindström-Seppä and Oikari 1990). However, the levels of these compounds exceed background levels over significantly greater distances (Grimvall et al. 1991; Lindström-Seppä and Oikari 1990). The proportion of free to total phenolics decreases in the recipient water samples relative to that in the effluent (15 and 30%, respectively; Holmbom et al. 1992). This difference could be due to adsorption and microbiological degradation of the free chlorophenolics. Aqueous phase concentrations of resin acids rapidly reach background levels within 15 km of the effluent discharge point (Lindström-Seppä and Oikari 1990; Holmbom et al. 1992; Volkman et al. 1993). The concentrations of these compounds in suspended and colloidal material in recipient waters and in fresh sediments collected at the same site indicate that sedimentation is the main process for the decrease of these compounds in recipient waters (Fig. 2). Binding of resin acids and associated extractives to suspended material may retard their microbiological degradation and bioavailability.

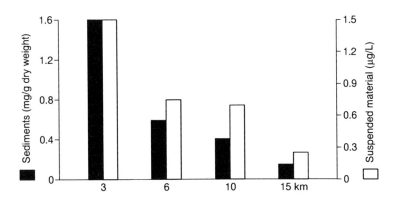

Figure 2. Resin acid levels in downstream sediments and suspended solids (Holmbom et al. 1992).

The high molecular mass chlorolignin present in bleached pulp mill effluents has previously been considered to be non-bioaccumulative, non-toxic and inert because of its large molecular size and water solubility. However, studies indicate that bioaccumulation of some of this material can occur in exposed organisms (Pellinen 1994). The chlorolignin is also implicated as the major toxicant in Echinoderm bioassays of bleached pulp mill effluents (Higashi et al. 1992).

Recent reports have shown that chlorolignin can adsorb lipophilic compounds in a similar manner to humic acids (Kukkonen 1992; Hinton et al. 1993). It has now been established that over 95% of the chlorocatechols and chloroguaiacols in the aqueous phase are reversibly adsorbed to the high molecular mass chlorolignin (O'Connor and Voss 1992). This association has important implications for the distribution and bioavailability of the chlorophenolic compounds. The high water solubility and stability of the chlorolignin allow concentrations of these compounds to remain above background levels and to be transported over distances in excess of 100 km (Grimvall et al. 1991). It has also been found that addition of high molecular weight lignin fractions reduced the bioaccumulation of chlorinated guaiacols in salmon by 30% compared to a solution without chlorolignin addition (Carlberg, G.E., The Norwegian Pulp and Paper Research Institute).

Biodegradation also contributes to the removal of bleached pulp mill effluent organic constituents from the aqueous phase. Taylor et al. (1988) found natural microbial populations in river water downstream of a mill discharge which were capable of rapidly degrading resin acids. Brownlee et al. (1977) determined the half-life of dehydroabietic acid in water and in sediments adjacent to a bleached kraft/groundwood mill. The half-life in water was found to be approximately 6 wks and in the sediment the half-life was estimated to be 21 yrs.

Some of the compounds in mill effluents may be biotransformed to more recalcitrant products in the receiving waters. Under aerobic conditions, chlorinated catechols and guaiacols can be metabolized to veratroles while chlorophenols can be biotransformed into anisoles (Neilson et al. 1990). These metabolites have been found to be stable towards further biotransformation and are more toxic and bioaccumulative than the parent compounds (Neilson et al. 1990). These compounds are also considered to be the primary source of odor tainting in water and fish taken downstream of bleached pulp mills. In a recent investigation Brownlee et al. (1993) identified chlorinated veratroles in receiving waters and fish downstream from a pulp mill. The compounds did not decrease to steady-state concentrations for over 180 km and were still detected 1100 km from the effluent discharge point.

BLEACHED PULP MILL EFFLUENT CONSTITUENTS IN RECIPIENT SEDIMENTS

The distribution and fate of hydrophobic bleached pulp mill effluent organic constituents in the aquatic recipient is principally determined by their sorption onto sediments. The rate at which these compounds accumulate in the sediment is highly site specific and is a function of the input concentrations of compound and adsorbent, the hydraulic loading and flow, the settling rate of the adsorbent, the sediment's organic carbon content and particle size distribution, and the compound's lipophilicity (Lijklema et al. 1993; Loonen et al. 1994).

In the short term, bleached pulp mill effluent recipient sediments act as organic sinks and reduce the uptake of bleached pulp mill effluent organic compounds by pelagic organisms. However, surficial sediments are an important habitat for benthic organisms and provide a vital food source for filter feeders and benthivorous fish. Accumulation in sediments may therefore substantially enhance toxicity effects relative to the overlying water and attenuate bioaccumulation through the food chain (Lijklema et al. 1993; Owens et al. 1994).

Several recent studies have determined the levels of organochlorine in downstream recipient sediments (Table 1). The sediment concentrations of EOX, the sum parameter which measures the concentration of lipophilic, slowly degradable chlorinated compounds, rapidly decrease with increasing distance from the source of bleached pulp mill effluents. Martinsen et al. (1994) found that the relative contribution of EOX to TOX also steadily diminished below the discharge point. EOX constituted 30% of the TOX in sediments near the mill discharge, but only 1.5% approximately 150 km downstream. Håkanson et al. (1988) showed that most of the EOX is bound in the upper 5 cm of the sediments. The precise nature of the EOX fraction is still not known. Sixty percent of the material has a molecular mass greater than 300 Da and only 5-8% of the EOX can be accounted for by known compounds (Kvernheim et al. 1993).

Table 1. Recent studies of sum organohalogen parameters in downstream recipient sediments.

Compound Class	Effluent	Treatment	Recipient	Country	Distance (km)	Concentration ($\mu g\ g^{-1}$ d.w.)	Reference
EOX	BKME	secondary	river	New Zealand	1.5	6	Stuthridge et al. 1992
					20	2.8	
					80	0.4	
	BKME	secondary	marine	Sweden	5	5270	Håkanson et al. 1988
					15	700	
					36	250	
	BKME	secondary	marine	Sweden	1	1100	Martinsen et al. 1994
					4	720	
					12	70	
					29	30	
TOX	BKME	secondary	lake	Finland	1	9600	Maatela et al. 1990
					1.5	1260	
	BKME	secondary	lake	Finland	3	4870	Paasivirta et al. 1988
					15	980	
					40	790	
					55	460	
					85	340	
	BKME	secondary	river	New Zealand	1.5	433	Stuthridge et al. 1992
					20	467	
					80	217	
	BKME	secondary	marine	Sweden	1	3600	Martinsen et al. 1994
					9	700	
					150	300	

TOX levels in recipient sediments show a similar decrease in concentrations downstream of the discharge (Table 1). Pellinen (1994) has shown that high molecular mass chlorolignin, the predominant constituent of TOX, can adsorb relatively rapidly to recipient sediments. Only a small fraction of this adsorbed material (<30% TOC, <20% AOX) can be subsequently desorbed. Relatively high concentrations of TOX remain in sediments sampled some distance from the mill discharge. It has now been established that naturally-occurring organohalogen constituents are ubiquitous in the environment. Asplund et al. (1989) found relatively high levels of AOX in sediments sampled from uncontaminated background sites. Stuthridge et al. (1992) measured no significant difference in TOX levels in sediments sampled upstream and downstream of the mill discharge.

Grimvall et al. (1991) investigated the long-term accumulation and degradation of bleached pulp mill effluent organochlorine constituents in Lake Vättern in Sweden. The 100 km long lake has a very long hydraulic residence time (56 yrs) and a considerable history (36 yrs) as a recipient of discharges from a bleached kraft mill. The levels of AOX and chlorinated guaiacols were elevated in the vicinity of the pulp mill, but decreased rapidly to levels found throughout the entire lake at all depths. These levels were 15 µg Cl L^{-1} for AOX and about 1 µg L^{-1} for 3,4,5-trichloroguaiacol. Core samples from the largest accumulation bottom of the lake showed that the ratio AOX/LOI (loss on ignition) was almost constant (about 1 mg Cl g^{-1} LOI) down to a depth of 15-30 cm despite the sediment accumulation being less than 10 cm since the introduction of chlorine bleaching in 1955. Furthermore, chlorinated guaiacols, indicator compounds of pulp mill effluents, were only found to a depth of 5-7 cm. Mass balance calculations showed that less than 15% of the AOX and 10% of the total chlorophenols discharged since 1955 could be accounted for. Degradation was thus concluded to be the major removal mechanism for AOX and chlorinated phenols. These results indicate that the half-life degradation rate of AOX is of the order of months.

Studies of chlorinated phenolics in recipient sediments indicate that concentrations of these compounds rapidly decrease downstream of the discharge point (Table 2). Over 90% of the chlorophenolic compounds are strongly associated with organic material and metal ions in the sediments (Allard et al. 1994). Therefore, it is generally considered that these compounds are not available to biodegrade or bioaccumulate. However, Remberger et al. (1993) have shown that chlorocatechols may be released from the sediments under recipient conditions, particularly where higher pH conditions occur, for example in marine or brackish waters. Release of chlorocatechols from the sediment material into the interstitial medium has been shown to cause adverse effect on the early life stages of fish.

Resin acid concentrations in downstream sediments have been found to reach near-background levels within 30 km of the discharge point (Table 2). This is expected given the strong association of these compounds with suspended and colloidal material. Movement of resuspended sediments is probably the predominant transport mechanism for these compounds (Owens et al. 1994).

Biodegradation processes in sediments exposed to bleached pulp mill effluent discharges have been investigated. Neilson et al. (1990) determined the degradation pathways of sediment-bound chlorophenolic compounds. Formation of chloroveratroles cannot occur due to the anoxic nature of sediments. Chloroguaiacols and accumulated chloroveratroles are demethylated to produce the equivalent catechol compounds. Partial dechlorination of the catechols also occurs. Use of elemental chlorine free (ECF) bleaching has led to the predominance of chlorovanillins in the chlorophenolic fraction of bleached pulp mill effluents. These compounds were also shown to biotransform in the sediments to produce chlorinated methyl-catechols.

Recent studies have shown that the resin acids in sediments can be biotransformed into a range of acidic and neutral biotransformation products (Stuthridge et al. 1992). These compounds have a higher lipophilicity than the parent compounds and have been detected in sediments and fish downstream of mill discharges, in some cases at concentrations approaching that of the parent resin acids (Hynning et al. 1993; Judd et al. 1994a,b).

Table 2. Recent studies of selected effluent constituents in downstream recipient sediments.

Compound Class	Effluent	Treatment	Recipient	Country	Distance (km)	Concentration (µg g^{-1} d.w.)	Reference
Chloro-phenolics	BKME	secondary	lake	Finland	3 15 40 55 85	0.8 (149.0)* 0.1 (44.8) 0.1 (5.3) 0.1 (3.5) nd (0.2)	Paasivirta et al. 1990
	BKME/TMP	secondary	lake	Finland	2.5 5.5 10 14	1.9 (54.4) 1.0 (32.3) 1.2 (31.7) 1.0 (27.5)	Holmbom et al. 1992
	BKME	secondary	river 1	New Zealand	1.5 11 20 80	0.159 0.043 0.001 0.002	Judd et al. 1994a
	BKME	secondary	river 2	New Zealand	2.5 14 16.5	0.017 0.016 0.007	Judd et al. 1994b
	BKME	secondary	marine	Sweden	1 2 6 11	1.0 (8.4) (7.2) (1.3) (0.1)	Kvernheim et al. 1993
Resin acids	newsprint	unspecified	estuarine	Australia	0 1.5 6 12	87 34 5 2	Volkman et al. 1993
	BKME	secondary	river	Canada	2 5	211 88	Lee and Peart 1991
	BKME/TMP	secondary	lake	Finland	2.5 5.5 10 14	1600 600 350 200	Holmbom et al. 1992
	BKME	secondary	river 1	New Zealand	1.5 11 20 80	85 56 37 1	Judd et al. 1994a
	BKME	secondary	river 2	New Zealand	2.5 14 16.5	23 7 10	Judd et al. 1994b

* Free (bound)

IMPLICATIONS OF PULP PROCESSING MODERNIZATION

The major process changes to bleached kraft pulp production have included the use of pre-bleaching techniques such as extended cooking and oxygen delignification and the introduction of ECF bleaching through the replacement of chlorine with chlorine dioxide in the first bleaching stage. In some cases, totally chlorine free (TCF) bleaching is also being implemented.

The implementation of ECF bleaching has led to reductions in AOX levels of over 90% and the virtual elimination of dioxins (O'Connor et al. 1994). Concentrations of bioaccumulative and persistent EOX, chlorophenolic compounds and other low molecular weight chlorinated organic compounds have also been substantially reduced. The remaining organic constituents in ECF bleaching effluents have a lower average molecular mass and are substantially less chlorinated (Axegård et al. 1993; O'Connor et al. 1994). These factors have increased the biodegradability and reduced the lipophilicity of these materials, leading to reduced bioconcentration and accumulation in recipient sediments (Mackay 1989; Axegård et al. 1993; O'Connor et al. 1994).

Utilization of modern bleaching technology will not significantly decrease the levels of non-chlorinated organic compounds such as resin acids, fatty acids and phytosterols. However, this can be achieved by changes in wood room and pulping processes and by accompanying measures to reduce total mill water usage and discharges from bleaching and non-bleaching process sources through improved washing and recirculation. The majority of mills are now also adopting external control measures such as secondary treatment systems. These measures should lead to lower recipient burdens and decreases in sediment concentrations of bleached pulp mill effluent organic constituents.

LONG-TERM FATE AND EFFECTS

A decrease in the discharges of organochlorine compounds from the pulp and paper industry will, in most cases, result in a significant reduction in the organochlorine levels in the recipient sediments. Särkkä et al. (1993) sampled sediments in Lake Ladoga close to the site where a sulfite mill had been operating until 1986. Figure 3 shows the concentrations of organically bound chlorine (OCl) and chlorinated phenols (>95% bound) in sediments dated between 1918 and 1990. Closing the mill resulted in a considerable reduction in the organochlorine content in the top sediment layer compared to those underneath. Similarly, cessation of chlorobleaching at the Jämsä Mill in 1981 caused sediment concentrations of bound chlorophenolics immediately downstream to decrease by 86% after 7 yrs (Paasivirta et al. 1990).

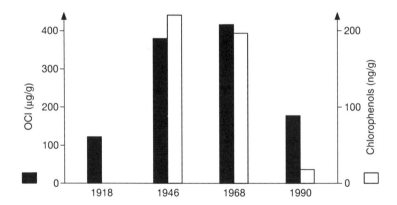

Figure 3. Organochlorine and chlorophenolic levels in Lake Ladoga sediments (Särkkä et al. 1993).

The continued bioavailability of the sedimented bleached pulp mill effluent organic compounds after reduction of discharges/shut down of mills is of major environmental concern. Carlberg et al. (1987) showed that fish in Lake Mjøsa in Norway contained similar organochlorine levels 16 months after mill closure as when the mill was in operation. Three years after the mill shut down, the organochlorine levels in the fish were down to background levels. In a recent investigation Martinsen et al. (1993) sampled fish in a Norwegian fjord one year after the shutdown of a sulfite plant discharging to the fjord. Eels were found to contain a very high concentration of OCl (2400 µg g^{-1} lipid). The OCl was mainly in the form of chlorinated fatty acids. Low concentrations of chlorinated phenols and resin acids were also detected.

To determine the long term degradation and depuration of sediment-bound chlorinated organic compounds, sediments were incubated for two years under different conditions (Granmo *et al.* 1992). Under aerobic conditions, sediment extractable organic chlorine (EOCl) concentrations decreased by 10-14%. The decrease could be attributed to degradation processes in the sediment or to diffusion into the aqueous phase. No significant changes in the EOCl levels were observed for sediments incubated under anaerobic conditions.

Based on measurements of EOCl inputs and sediment and water concentrations, large-scale models for the occupance and fate of EOCl in the Baltic Sea have been developed (Södergren 1993). Direct emissions from the pulp mills were the major source (500 tonne yr^{-1}), followed by atmospheric deposition (147 tonne yr^{-1}) and river discharges (9 tonne yr^{-1}). It has been estimated that pulp mills have discharged 21,000 tonnes of EOCl to the Baltic Sea since 1940. Approximately half of that amount (12,555 tonnes) has accumulated in the sediment. Since EOCl is associated with particulate material, today's decreased input of EOCl to the Baltic Sea compared to 1988-89 will result in a rapid removal of EOCl from the water by sedimentation processes. However, EOCl stored in the sediments will only decrease slowly. On this basis, the amounts of EOCl in the sediments in the year 2040 might be expected to be of a similar magnitude to those in the early 1970s.

Following a suitable period after cessation of bleached pulp mill effluent discharges and with anoxic conditions, sedimentation of unpolluted material may cover and immobilize the organochlorine compounds and make them biologically unavailable. The immobilization will, however, require a longer time in areas with oxic accumulation bottoms due to the effect of bioturbation.

CONCLUSIONS

A review of the current literature on the distribution and fate of bleached pulp mill derived substances in mill recipients suggests the following conclusions:

a) Water column concentrations of effluent constituents will decline downstream of mill discharge points. The decline rate will depend on receiving water conditions.
b) The major dispersal mechanisms are dilution (AOX), association with dissolved organic material (chlorophenolics) and adsorption/sedimentation with suspended particulates (resin acids).
c) Distribution trends in sediment matrices are similar to those observed in water column.
d) Binding of bleached pulp mill effluent constituents to sediments or dissolved organic matter may significantly reduce their biodegradability and bioaccumulation potentials. However, these same processes may significantly extend the distance and time over which they may cause potential environmental effects.
e) Implementation of modern bleaching processes, if accompanied by improvements in general mill water utilization, is likely to substantially decrease the levels and impacts of BKME constituents in the recipient.
f) Previously accumulated BKME constituents are likely to persist in the environment for a significant period of time and potentially cause long-term environmental effects.

REFERENCES

Allard, A.-S., P.-Å. Hynning, M. Remberger and A. Neilson. 1994. Bioavailability of chlorocatechols in naturally contaminated sediment samples and of chloroguaiacols covalently bound to C2-guaiacyl residues. Appl. Environ. Microbiol. 60(3):777-784.

Asplund, G., A. Grimvall and C. Pettersson. 1989. Naturally produced adsorbable organic halogens (AOX) in humic substances from soil and water. Sci. Tot. Environ. 81/82:239-248.

Axegård, P., O. Dahlman, I. Haglind, B. Jacobson, R. Mörck and L. Strömberg. 1993. Pulp bleaching and the environment - The situation 1993. Nord. Pulp Pap. Res. J. 8(4):365-378.

Brownlee, B.G., M.E. Fox, W.M.J. Strachan and S.R. Joshi. 1977. Distribution of dehydroabietic acid in sediments adjacent to a kraft pulp and paper mill. J. Fish. Res. Board. Can. 34:838-843.

Brownlee, B.G., G.A. MacInnis and L.R. Noton. 1993. Chlorinated anisoles and veratroles in a Canadian river receiving bleached kraft pulp mill effluent. Identification, distribution and olfactory evaluation. Environ. Sci. Technol. 27:2450-2455.

Carlberg, G.E., A. Kringstad, K. Martinsen and O. Nashaug. 1987. Environmental impact of organochlorine compounds discharged from the pulp and paper industry. Paperi ja Puu 69:337-341.

Comba, M.E., V.S. Palabrica and K.L.E. Kaiser. 1994. Volatile halocarbons as tracers of pulp mill effluent plumes. Environ. Toxicol. Chem. 13(7):1065-1074.

Environment Canada. 1991. Effluents from Pulp Mills Using Bleaching. Priority Substances List Assessment Report No. 2, Canadian Environmental Protection Act.

Granmo, Å., P. Jonsson, L. Brydsten, R. Ekelund, K. Magnusson and M. Berggren. 1992. Long term degradation test with coastal Baltic Sea sediments contaminated with extractable organic halogens. In A. Södergren [ed.] Environmental fate and effects of bleached pulp mill effluents. Swedish Environmental Protection Agency, Report 4031:68-73.

Grimvall, A., H. Borén, S. Jonsson, U. Lundström and R. Sävenhed. 1991. Organohalogens of natural and industrial origin in large recipients of bleach-plant effluents. Wat. Sci. Tech. 24(3/4):373-383.

Higashi, R.M., G.N. Cherr, J.M. Shenker, J.M. MacDonald and D.G. Crosby. 1992. A polar high molecular mass constituent of bleached kraft mill effluent is toxic to marine organisms. Environ. Sci. Technol. 26(12):2413-2420.

Hinton, S.W., R. Brunck and L. Walbridge. 1993. Mass transfer of TCDD/F from suspended sediment particles. Wat. Sci. Tech. 28(8/9):181-190.

Holmbom, B., J. Hemming and P. Mäki-Arvela. 1992. Environmental fate of effluent components from the Kaukas pulp and paper mill in South Saimaa Lake system 1992. In P. Hakamies [ed.] Saimaa-Seminar 1992, University of Joensuu, Finland publication, No. 103:39-52.

Hynning, P.-Å., M. Remberger and A.H. Neilson. 1993. Identification and quantification of 18-nor- and 19-norditerpenes and their chlorinated analogues in samples of sediment and fish. J. Chrom. 643:439-452.

Håkanson, L., B. Jonsson, P. Jonsson and K. Martinsen. 1988. Distribution of chlorinated organic substances from pulp mills. Wat. Sci. Tech. 20:25-37.

Judd, M.C., T.R. Stuthridge, M.H. Tavendale, P.N. McFarlane, K.L. Mackie, S.J. Buckland, C.J. Randall, C.W. Hickey, S.M. Anderson and D. Steward. 1994a. Bleached kraft pulp mill sourced organic chemicals in sediments from a New Zealand river. Part 1: Waikato River. Chemosphere: In press.

Judd, M.C., I.J. Bergman, P.N. McFarlane, S.M. Anderson and T.R. Stuthridge. 1994b. Bleached kraft pulp mill sourced organic chemicals in sediments from a New Zealand river. Part 2: Tarawera River. Chemosphere: In press.

Kukkonen, J. 1992. Effects of lignin and chlorolignin in pulp mill effluents on the binding and bioavailability of hydrophobic organic pollutants. Wat. Res. 26(11):1523-1532.

Kvernheim, A.L., K. Martinsen, G.E. Carlberg, B.E. Berg, M. Fresvig and T. Greibrokk. 1993. Characterization of organohalogen matter in sediments from the receiving water of a bleach plant. Chemosphere 27:733-745.

Lee, H.-B. and T.E. Peart. 1991. Determination of resin and fatty acids in sediments near pulp mill locations. J. Chrom. 547:315-323.

Lijklema, L., A.A. Koelmans and R. Portielje. 1993. Water quality impacts of sediment pollution and the role of early diagenesis. Wat. Sci. Tech. 28(8/9):1-12.

Lindström-Seppä, P. and A. Oikari. 1990. Biotransformation and other toxicological and physiological responses in rainbow trout (*Salmo gairdneri* Richardson) caged in a lake receiving effluents of pulp and paper industry. Aquatic Toxicol. 16:187-204.

Loonen, H., J.K. Parsons and H.A.J. Govers. 1994. Effect of sediment on the bioaccumulation of a complex mixture of polychlorinated dibenzo-p-dioxins (PCDDs) and polychlorinated dibenzofurans (PCDFs) by fish. Chemosphere 28:1433-1446.

Maatela, P., J. Passivirta, M.A. Grachev and E.B. Karabanov. 1990. Organic chlorine compounds in lake sediments. V. Bottom of Baikal near a pulp mill. Chemosphere 21(12):1381-1384.

Mackay, D. 1989. The environmental fate and effects of chemicals from the pulp and paper industry: A role for structure activity relationships. Proceedings, 1989 TAPPI Environ. Conf.:345-350.

Martinsen, K., K. Pedersen, A. Kringstad and A.L. Kvernheim. 1993. Investigation of Organic Contaminants in Sediments and Biota from Iddefjorden in 1992. Report No. SFT 27A 93032, SINTEF, Oslo, Norway.

Martinsen, K., A.L. Kvernheim and G.E. Carlberg. 1994. Distribution of organohalogen in sediments outside pulp mills using sum parameters. Sci. Total Environ. 144:47-57.

McKague, A.B., M. Jarl and K.P. Kringstad. 1988. An Up-to-date List of Compounds Identified in Bleaching Effluent as of January 1989. *SSVL Miljö 90*, Project Bleaching, Stockholm, 1988. Report available from ÅF-IPK, Box 8309, 104-20, Stockholm, Sweden.

McLeay, D. 1987. Aquatic Toxicity of Pulp and Paper Mill Effluent: A Review. Report EPS 4/PF/1, Environment Canada.

Neilson, A.H., A.-S. Allard, P.-Å. Hynning, M. Remberger and T. Viktor. 1990. The environmental fate of chlorophenolic constituents of bleachery effluents. TAPPI J. 73:239-247.

O'Connor, B.I. and R.H. Voss. 1992. A new perspective (sorption/desorption) on the question of chlorolignin degradation to chlorinated phenolics. Environ. Sci. Technol. 26(3):556-560.

O'Connor, B.I., T.G. Kovacs, R.H. Voss, P.H. Martel and B. van Lierop. 1994. A laboratory assessment of the environmental quality of alternative pulp bleaching effluents. Pulp Pap. Can. 95(3):47-56.

Owens, J.W., S.M. Swanson and D.A. Birkholz. 1994. Bioaccumulation of 2,3,7,8-tetrachlorodibenzo-p-dioxin, 2,3,7,8-tetrachlorodibenzofuran and extractable organic chlorine at a bleached-kraft mill site in a northern Canada river system. Environ. Toxicol. Chem. 13(2):343-354.

Paasivirta, J., J. Knuutinen, P. Maatela, R. Paukku, J. Soikkeli and J. Särkkä. 1988. Organic chlorine compounds in lake sediments and the role of chlorobleaching effluents. Chemosphere 178(1):137-146.

Paasivirta, J., H. Hakala, J. Knuutinen, T. Otollinen, J. Särkkä, L. Welling, R. Paukku and R. Lammi. 1990. Organic chlorine compounds in lake sediments. III. Chlorohydrocarbons, free and chemically bound chlorophenols. Chemosphere 21:1355-1370.

Pellinen, J. 1994. Sorption of high molecular weight chlorolignin on sediment. Chemosphere 28(10):1773-1789.

Remberger, M., P.-Å. Hynning and A.H. Neilson. 1993. Release of chlorocatechols from a contaminated sediment. Environ. Sci. Technol. 27(1):158-164.

Robinson, R.D., J.H. Carey, K.R. Solomon, I.R. Smith, M.R. Servos and K.R. Munkittrick. 1994. Survey of receiving-water environmental impacts associated with discharges from pulp mills. 1. Mill characteristics, receiving-water chemical profiles and lab toxicity tests. Environ. Toxicol. Chem. 13(7):1075-1088.

Stuthridge, T.R., M.H. Tavendale, M.C. Judd, K.L. Mackie and P.N. McFarlane. 1992. Determination of bleached kraft mill effluent organic constituents in recipient media. Proceedings, 1992 TAPPI Environ. Conf., Vol. 1:219-231.

Särkkä, J., J. Paasivirta, E. Häsänen, J. Joistinen, P. Manninen, K. Mäntykoski, T. Rantio and L. Welling. 1993. Organic chlorine compounds in lake sediments. VI. Two bottom sites of Lake Ladoga near pulp mills. Chemosphere 26:2147-2160.

Södergren, A. 1993. Bleached Pulp Mill Effluents. Composition, Fate and Effects in the Baltic Sea. Report 4047, Swedish Environmental Protection Agency.

Taylor, B.R., K.L. Yeager, S.G. Abernethy and G.F. Westlake. 1988. Scientific Criteria Document for Development of Provincial Water Quality Objectives and Guidelines: Resin Acids. Ontario Ministry of the Environment Report.

Volkman, J.K., D.G. Holdsworth and D.E. Richardson. 1993. Determination of resin acids by gas chromatography and high-performance liquid chromatography in paper mill effluent, river waters and sediments from the upper Derwent Estuary, Tasmania. J. Chrom. 643:209-219.

BIODEGRADABILITY OF DIFFERENT SIZE CLASSES OF BLEACHED KRAFT PULP MILL EFFLUENT ORGANIC HALOGENS DURING WASTEWATER TREATMENT AND IN LAKE ENVIRONMENTS

E.K. Saski, J.K. Jokela and M.S. Salkinoja-Salonen

Department of Applied Chemistry & Microbiology, P.O. Box 27, FIN-00014, University of Helsinki, Finland

Less than 5% of the organic halogen emissions from bleaching of pulp are molecules of known structures. The major part consists of mainly hydrophilic molecules of unknown structure, chlorohumus. This paper focuses on the impact of size of the halogenated molecules on the environmental fate. The tetrahydrofuran-soluble fraction of the wastewaters was studied because >90% of the wastewater contained organic halogen (AOX) dissolves in this solvent. We show that halogen emissions of modern kraft pulp mills range from 100 to 1000 g mol^{-1} in size. The molecular size had little limitation on biodegradation during secondary wastewater treatment. Thirty to 70% of the AOX was removed during wastewater treatment (full scale). Molecules of <500 g mol^{-1} were slightly more degraded than those >500. Fifty to 80% of the remaining halogens were biodegraded in the lake ecosystem, studied in 2 m^3 outdoor mesocosms over four seasons. The molecular size distribution of the tetrahydrofuran-soluble fraction of wastewater showed little change in clear lake water mesocosms, although over 50% of the AOX was removed. Sedimentation explained only a minor fraction of AOX removal from the water column. The tetrahydrofuran-soluble halogenated material which accumulated in the sediments showed a molecular weight distribution completely different from that in the water column or in the wastewater, indicating extensive metabolism.

INTRODUCTION

The size and the polarity of xenobiotic molecules are assumed to be the decisive factors for predicting the biomobility and the effects on biota in the environment. This assumption explains why most published studies deal with the small molecular (<500 g mol^{-1}) fraction, although this fraction comprises less than 5% of the total bleached kraft pulp mill discharged organic halogens. The main part of the pulp mill halogen discharge consists of extremely heterogenous, mainly hydrophilic, molecules of unknown structure and assumed of high molecular weight, 10^3 to 10^5 g mol^{-1}, often called chlorolignin. Chlorolignins are recognized as slowly biodegradable and considered humus-like in not being bioavailable, which would explain their recalcitrance.

Most pulp mills in Finland discharge into shallow inland waters or equally shallow Baltic bay estuaries. For this reason the mills have been required to build extensive secondary wastewater treatment. Since 1995 all mills have external secondary wastewater treatment. The data accumulating on the performance of biological secondary treatment in Finland and elsewhere show that treatment removes 30 to 60% of the mill effluent organic halogen (measured as AOX, active carbon adsorbable organic halogen), depending on the mill and the method of treatment (Jokela *et al.* 1993, also a review by Graves and Joyce 1994). Only a small part of the removal relies on conversion to small volatile molecules escaping AOX analyses (<1%) or adsorption to sludge and carry-away with excess sludge (<5%), indicating that the main part has truly become mineralized (Gergov *et al.* 1988; Jokela *et al.* 1993; Saski *et al.* 1995a). After being discharged into recipient waters, the organic halogen concentration (AOX) of the water phase decays faster than can be accounted for by dilution alone (Jokela *et al.* 1992).

The present paper deals with the environmental fate, with an emphasis on biodegradability, of bleached kraft pulp mill discharged organic halogens. The molecular size distribution of the halogenated compounds was studied during the biological purification of the wastewater and in the recipient lake (large enclosures placed in lakes) water column and sediment over four seasons.

EXPERIMENTAL

Molecular Weight Distribution Analysis

HPSEC. For molecular weight determination, nonaqueous high performance size exclusion chromatography (HPSEC) was used as described earlier (Jokela and Salkinoja-Salonen 1992). A comparative study and validation of the method used was performed with aqueous HPSEC, vapor pressure osmometry and ultrafiltration (Jokela and Salkinoja-Salonen 1992). Both synthetic lignin model compounds and commercial narrow molecular weight distribution polystyrenes (Pressure Chemicals, Pittsburgh, PA, U.S.A.), with average molelular weight of 498,000, 50,000, 17,500, 4000, 2200 and 800, were used. The lignin model compounds used for calibration contained 2, 3, 4 and 6 aromatic nuclei with 3C side chains, all of which were synthesized under the supervision of G. Brunow at the Dept. of Chemistry, University of Helsinki. The synthesis and structures of the 2- (320 g mol^{-1}) and 4-ring molecules (638 and 668 g mol^{-1}) were described by Jokela *et al.* (1985, 1987) and 6-ring (1080 g mol^{-1}) compound structures by Jokela (1995). The structures of the synthesized lignin models are shown Fig. 1. A total of 75 different chlorine compounds, obtained from commercial sources, were also used for calibration (for a list of the compounds used, see Jokela and Salkinoja-Salonen 1992). The ratio of C:Cl of the compounds varied from 0.5:1 (tetrachloroethene) to 20:1 (monochlorodehydroabietic acid) and molar weights ranged from 60 (acetic acid) to 369 (dichlorodehydroabietic acid).

Sample Preparation
Water samples were freeze dried and residues dissolved in tetrahydrofuran containing $\leq 2\%$ (v/v) of concentrated nitric acid in a bath sonicator. The solution was filtered through a 0.45 µm nylon filter and stored at -20°C under nitrogen gas until analyzed.

^{36}Cl-Chlorolignin
^{36}Cl-labeled bleaching effluent was a gift of Clas Wesen (University of Lund). Softwood kraft pulp (kappa 35, consistency 3%) was bleached with Cl$_2$ (70 kg tonne^{-1}), C-stage at room temperature, E-stage at 70°C and pH 12. The ^{36}Cl-labeled effluent was stored at -20°C in a polyethene bottle. Inorganic ^{36}Cl was removed using Sep-Pak C18 cartridges (Waters Associates, Milford, MA). The cartridges were conditioned with 10 mL of methanol, 10 mL of tetrahydrofuran and 20 mL of deionized water. Ten mL of the ^{36}Cl-labeled bleaching effluent was impregnated into the cartridge, then washed with 10 mL aqueous HCl (pH 1.2). The organic components were eluted from the cartridge with 5 mL of methanol. One mL of methanol was evaporated to dryness in a flow of N$_2$ and the residue dissolved in 100 µL of tetrahydrofuran.

Assay of Halogen
Halogen was assayed using microcoulometric assay, neutron activation analysis or scintillation counting (^{36}Cl). AOX (adsorbable organic halogen) was assayed microcoulometrically after adsorption onto active carbon (background <0.2 µg of Cl per 50 mg of C, from Euroglas, Delft, NL) according to the ISO 9562 protocol. HPSEC fractions were assayed after evaporating the solvent under nitrogen gas prior to inserting the sample into the oven of the microcoulometer. Some samples were parallel assayed by the neutron activation technique at the Technical Research Centre (VTT, Otaniemi, Finland by P. Manninen). The carbon used for adsorption was ENC activated granular carbon (grain size 0.15-0.35 mm, S-bet 1020 m^2 g^{-1}). The samples were irradiated in a Triga Mark II reactor, usually for 300 s, in a neutron flux of 4

· 10^{12} n cm^{-2} s^{-1}. Gamma rays were measured using a GeLi crystal detector (Ortec) and a multichannel analyzer (Canberra MCA 40) and a Rockwell AIM computer.

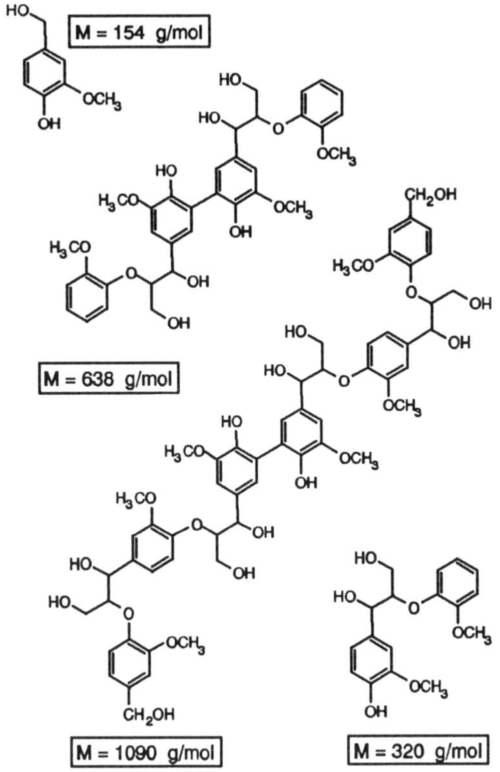

Figure 1. Synthetic lignin model compounds used for molecular weight calibration (HPSEC) along with the narrow molecular weight distribution polystyrenes and various halogenated compounds. Calibration was performed to fit both the lignin models and polystyrenes. Halogen compounds fitted reasonably well into the calibration curve thus obtained when the C:Cl ratio was 5:1 or higher.

Elemental Analyses

Organic carbon was measured (<0.18 mm particles) with a TOC-5000 total carbon analyzer (Shimadzu, Japan) according to ISO standard 8245. Calibration was done with sodium hydrogen phthalate (total carbon) and sodium hydrogen carbonate plus sodium carbonate (inorganic carbon) manufactured by Kanto Chemical Co. (3-chome Chuo-ku, Tokyo).

Mesocosm Set-up

Enclosures of 2 m^3, made of polyethene (translucent for daylight, black for darkened controls), were floating in the lake, aided by wood-styrofoam flotation frames. Wastewaters from two different mills, A and C, both bleaching hardwood and sofwood, approx. 1:1, with elemental chlorine free (ECF) bleaching (Mill A: D-EO-D-E-D) or with oxygen delignification (Mill C: O-C/D-EO-D-E-D in 1991 and O-D-EO-D-E-D in 1992), were mixed with lake water, 2%....13% (v/v, as indicated). Mill A wastewater contained 3.6 (1991) to 2.0 mg (1992) of total nitrogen, 0.9 to 0.3 mg of total phosphorus, 380 to 300 mg of total organic carbon (TOC) and 26 to 17 mg of Cl as organic halogen (AOX) per liter. Mill C wastewater contained 5.0 (1991) to 4.5 mg (1992) of total nitrogen, 1.0 to 0.9 mg of total phosphorus, 280 to 190 mg of TOC and 21 to 7 mg of Cl as AOX, respectively, per liter. Mill C discontinued the use of elementary chlorine after 1991 sampling and Mill A bleached ECF both years. Further details of the wastewaters are given by Saski et al. (1996).

One lake was clear (color 60 mg Pt L^{-1}) and oligotrophic, and the other highly humic (color 380 mg Pt L^{-1}). The lake water concentrations of nitrogen, phosphorus and organic carbon were 1/5 to 1/10 (N), 1/30 to 1/100 (P) and 1/60 to 1/10 (C) of that contained by the wastewaters. More details are described elsewhere (Saski et al. 1996).

RESULTS AND DISCUSSION

Molecular Size of Chlorinated Organic Molecules in Bleached Kraft Pulp Mill Effluents

The first thing that happens to wastewater leaving the pulp mill is dilution. At a wastewater treatment plant the bleaching wastewaters are diluted by other waste streams and sometimes cooling waters. In the receiving water, further dilution by a factor of 20 to 1000 occurs, depending on the recipient size and flow. It was found that while 70 to 80% of both the organic halogen (AOX) and carbon of undiluted wastewater (from which insolubles had been removed by G/FA filtration) was retained by the 1000 g mol^{-1} ultrafilter, the retained portion of the same wastewater decreased upon dilution in distilled water. Less than 50% was retained after 10-fold dilution, and after diluting 100-fold or more, the major part of the wastewater contained TOC and AOX permeated the 1000 g mol^{-1} filter (Jokela and Salkinoja-Salonen 1992). Since dilution with pure water may not be expected to cause chemical cleavage of covalent bonds, the phenomenon must be of a physical nature, i.e., dissolution of colloidal and/or micellar aggregates. One may therefore expect the wastewater contained organics behave like molecules of <1000 g mol^{-1} after entering the recipient water ecosystem.

Vapor pressure osmometry and different variants of HPSEC were used to analyze the molecular weight distribution organic compounds in wastewaters of six Finnish kraft pulp mills and one Swedish mill, with different pulping and bleaching protocols. The results showed that when measures were taken to counteract hydrophobic and ionic interactions between the solute molecules during analysis, the bleached kraft pulp mill organohalogens behaved as rather low molecular weight material, M_n (average number) varied from 200 to 300 g mol^{-1} and M_w (average weight) from 300 to 600 g mol^{-1} (Jokela and Salkinoja-Salonen 1992; Pellinen and Salkinoja-Salonen 1985a,b). In most wastewaters, 85 to 95% of the halogen was bound to molecules in the size range from 200 to 1000 g mol^{-1}.

The microcoulometric analyzer used as the chlorine detector of the HPLC chromatographic fractions in the above studies is halogen specific rather than chlorine specific. It also may give (although to our experience, rarely) false positive reponses if the studied sample contains nitrocompounds. We used neutron activation analysis (chlorine specific) to calibrate for the mass balance (solvent extracts of organohalogens), but the sensitivity of this method is too low to allow for measuring the chlorine contents of chromatographic fractions. In this paper we describe confirmatory results obtained using radiolabeled chlorine ($^{36}Cl_2$) as the chlorine indicator. This is a direct chlorine-specific method. Kraft pulp was bleached with $^{36}Cl_2$, inorganic chloride removed and the molecular weight distribution of the radiochlorinated organic fraction was analyzed by the same HPLC column and eluent system as before. Figure 2 shows the result: also in this case the major part of the molecules were of sizes ranging from 200 to 2000 g mol^{-1}. We therefore conclude that the molecular size of solvent soluble chlorolignin is in this range.

Ultrafiltration technology has been applied to remove the major part of chlorolignins from wastewater at pilot scale for the purpose of purification (reviewed by Frostell et al. 1994). This operates in practice, but the large apparent molecular size in concentrated aqueous solution probably depends on intermolecular associations, maybe similar in nature to the micelles formed by hydrophilic molecules with hydrophobic domains, e.g., (bio)surfactants. The surfactant nature of chlorolignins has in fact been indicated by studies where purified chlorolignin fractions ("HMM") were shown to inhibit sea urchin sperm acrosome reaction, an event prerequisite for succesful fertilization (Higashi et al. 1992).

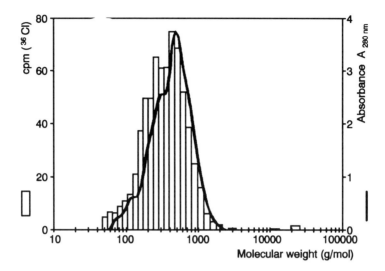

Figure 2. The distribution of ^{36}Cl among the different sizes of molecules in the spent liquor bleached with ^{36}Cl. The high performance size exclusion chromatogram shows the distribution of tetrahydrofuran-soluble halogens from softwood bleaching. The size calibration was done with synthetic lignin model compounds (Fig. 1) and narrow molecular weight range polystyrenes.

Impact of Molecular Weight on the AOX Degradability During Secondary Wastewater Treatment

The secondary wastewater treatment plants of bleached kraft pulp mills were operated to remove 90 to 97% of the BOD, but the concomitant removal of organic halogen (AOX) and carbon (TOC) was only 30 to 60% efficient (Gergov et al. 1988; Laine et al. 1991; Jokela et al. 1993). This was also true for the mills bleaching ECF (Mill A) and using oxygen delignification (Mill C) studied here (Jokela et al. 1993). This means that only a part of the AOX compounds contribute to BOD.

To answer the question whether molecular weight is of importance for biodegradation during wastewater treatment, we studied the molecular weight distributions of the organic bound halogens in the influent and the effluent of the secondary wastewater treatment units. Figure 3 shows examples of results from two mills, one of which had an oxygen delignification system and anaerobic-aerobic lagoon system for wastewater (Mill C), and the other conventional ECF bleaching with activated sludge treatment for wastewater (Mill A). Figure 3 shows that the removal of organic halogens (lower panel) was more efficient than removal of color (top panel). Figure 3 also shows that anaerobic-aerobic lagooning (AL) removed all size classes of the organic halogens and that the activated sludge plant (AS) removed efficiently up to 500 g mol^{-1} and less efficiently beyond this size.

The patterns obtained for the halogen removal according to molecular size distribution in the treatment plants, sampled at different seasons, during pulping of hardwood or softwood, and minor changes of bleaching sequence, were surprisingly similar to the examples shown in Figs. 2 and 3 and to those observed for other types of mills and biological treatments (Jokela et al. 1993). It is interesting to observe that anaerobic biotreatment was not inferior to aerobic in dehalogenating the chlorolignins (larger molecules). Our early work showed that anaerobic pretreatment at pilot and mill scale purification dehalogenated organohalogens in bleached kraft pulp mill wastewater (Salkinoja-Salonen et al. 1981 1984). More recently, Yu and Welander (1994) showed that a laboratory scale anaerobic fixed-film process removed 50 to 60% of pulp mill wastewater organohalogens (measured as AOX) with a hydraulic

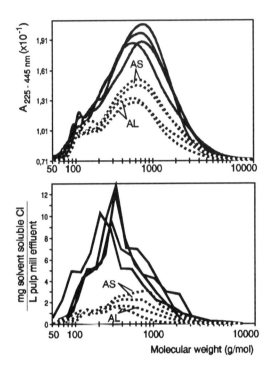

Figure 3. Changes of color (wide window absorbance, 225 to 445 nm, upper panel) and halogen content (lower panel) in different sizes of molecules in the wastewater of Mills A and C during wastewater purification. AL (Mill A) = anaerobic-aerobic lagooning; AS (Mill C) = activated sludge. Solid line = before treatment, broken lines = after treatment. The lines are drawn to scale and reflect the contents of the same amount of wastewater both before and after treatment. Both mills (A and C) bleached ECF at the time of the observation, either after oxygen delignification (Mill C) or hemicellulase prebleaching (Mill A). Mill A had an activated sludge plant (HRT 12 h+12 h), and Mill C an anaerobic (3d)-aerobic (7d) lagoon system. Both mills bleached alternatingly softwood and hardwood.

When the efficiencies of the different stages of wastewater treatment in halogen removal were compared over the molecular size distribution range, it was found that efficiency of the anaerobic lagoon remained constant or declined slightly with the molecular size from 100 to 2000 g mol^{-1}, while the tendency during aerobic treatment was the opposite. The efficiency of the activated sludge plant to remove halogens declined somewhat with an increasing molecular weight (Fig. 3, see also Laine *et al.* 1991; Jokela *et al.* 1993). It is interesting that approx. 50% of the total dehalogenation was already achieved during the equalization, prior to contact with biomass. This may reflect the hydrolysis of labile organohalogens (C-stage) upon pH neutralization observed in other studies (Smeds *et al.* 1994).

The overall carbon to chlorine ratio of the wastewater organic fraction increased from about 35:1 to 45:1 during anaerobic lagooning, but dropped back to the original during subsequent aerobic lagooning. Activated sludge treatment brought no significant change in the C to Cl ratio. Ninety percent of the pulp mill wastewater AOX was soluble in tetrahydrofuran before secondary treatment, but only about 60% of the AOX that resisted biotreament was soluble to tetrahydrofuran or any other organic solvent we tried. In fact, it seemed that the concentration of tetrahydrofuran-insoluble AOX remained constant in the wastewater during biological purification, i.e., either was not biodegraded or was both formed and degraded. The halogen content at this level (Cl:C <1:25) was less important for biodegradability than was the solubility in a hydrophilic solvent (tetrahydrofuran) (Laine *et al.* 1991; Jokela *et al.* 1993). The chemical nature of the tetrahydrofuran-insoluble, nonbiodegradable part of the wastewater containing AOX is not known.

Biodegradability studies using chlorinated model compounds larger than two aromatic rings are rarely published in the literature and to our knowledge have not been used to test biological wastewater purification. Such compounds are not commercially available. Sandermann *et al.* (1990, 1993) linked chloroaniline covalently to glucosides and lignin model compounds, thereby creating molecules of 400 to 650 g mol^{-1} and carbon to chlorine ratios similar to those in pulp mill chlorolignins. They found that the organohalogen parent compound was released in rat digestive tract and became bioavailable. They also found that linking of chloroaniline covalently to a synthetic lignin model improved its mineralization by a ligninolytic fungus, *Phanerochaete chrysosporium*. It may be that in some environments and for some microbes, a larger molecular size of the xenobiotic molecule may improve rather than hamper biodegradation.

Fate of Pulp Mill Organic Halogens in Finnish Fresh Water Recipient

Organic halogens become slowly removed from the water column after the wastewater is released into the lake ecosystem. We studied die-away of AOX in A Vuoksi river basin waterway, hosting 6 bleached kraft pulp mills over a distance of 350 km. The wastewater concentrations of AOX of the mills concerned ranged from 20,000 to 80,000 µg Cl L^{-1}, enabling the following of the wastewater plume over a large dilution. The natural background of AOX in these waters ranges from 10 (clear, oligotrophic) to 30 µg Cl L^{-1} (humic waters). Parallel halogen assays of the elevated AOX contents showed that the higher AOX figure found microcoulometrically in humic waters could not always be observed by neutron activation analysis and may therefore represent some compound other than an organohalogen. No such discrepancy was observed with any of the pulp mill wastewater containing samples.

In a study in 1989 we found that when the dilution was accounted for, the AOX was halved over a distance ($D_{1/2}$) of 47 km (Jokela *et al.* 1992). We repeated the study in 1990 and found a $D_{1/2}$ of 41 km (Fig. 4). In order to get more information on this environmental degradation, mass balance and the changes of molecular weights involved, we used mesocosmic enclosures holding a mixture of pulp mill wastewater and lake water (Saski *et al.* 1991). We observed the biodegradation of bleached kraft pulp mill wastewater organic fraction in the water column and the *de novo* formation of sediment in 22 separate mesocosm enclosures, incubated for four seasons in two different lakes, one clear, oligotrophic, and low in color and organic matter (60 mg Pt L^{-1}), and the other rich in humic matter and dissolved carbon (color 380 mg Pt L^{-1}). The mesocosms were not seeded with external sediment or any additive other than wastewater and lake water. Some of the mesocosm enclosures were kept out of daylight (dark) to be able to assess the impact of photochemistry.

Fate of Different Sizes of Organohalogens Studied in Fresh Water Recipient Mesocosms

The mesocosmic enclosures reflected a real lake environment in many respects: the extended incubation periods allowed for observing true seasonal variation including natural cycles of changes in light intensity, temperature, oxygen, and pH and allowing for growth of periphyton on the enclosure walls (mimicking natural biofilm surfaces on stones and shorelines) and sediment production.

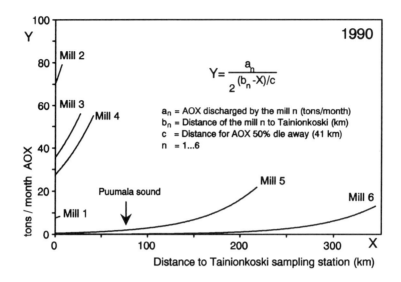

Figure 4. The calculated die-away of AOX discharged from 6 kraft pulp mills into the river basin Vuoksi in 1990. AOX data were collected for 12 months at the sampling points indicated in the figure. The discharge of AOX of each mill was known (as monthly averages). Correction was made for the background AOX contents of the recipient.

A clear stratification into winter and summer epi- and hypolimnion developed in the mesocosms where humic lake water was used as the diluent. The mesocosms with clear lake water diluent did not stratify; neither did the the lake where the water was taken from. Figure 5 shows the vertical gradients of temperature (5A) and oxygen (5B) in the mesocosms during one experiment over 500 d. Also, pH gradients were established in the mesocosms, vertical and temporal. The pH (not included in Fig. 5) varied from 6 (bottom of humic enclosures) up to 9...10.5 (springtime top layer of the epilimnion). The high epilimnion pH may promote alkaline dehalogenation of chlorolignins, as has been observed in wastewater treatment studies (Strehler and Welander 1994). The mesocosm enclosures mimicked a real receiving water except for water flow and plankton composition. The plankton were different in each mesocosm, so that neither phytoplankton nor zooplankton was quantitatively similar in species composition in any of the mesocosms, not even in the duplicate parallels (Salkinoja-Salonen *et al.* 1993).

Figure 6 shows examples of the time course of AOX concentration in some mesocosm enclosures started in June 1991. It shows that the organic halogen concentration (AOX) declined in all mesocosms, relatively rapidly in the summer, stagnating for winter and resuming the decline in the second summer. By the end of the second summer the total decline (corrected for storm water dilution on the basis of inorganic chloride concentration and conductivity) ranged from 50 to 80%. The half-life of AOX in the mesocosms thus was less than 500 d, as were those mesocosms that were kept dark. A detailed analysis of the mineralization kinetics of organic halogen and carbon in these mesocosms, and a similar set of mesocosms started in Nov. 1992, is presented elsewhere (Saski *et al.* 1995).

Figure 7 shows the accumulation of the organic halogens of pulp mill wastewater origin in the mesocosms into the sediments formed *de novo* during one year. Mass balance calculation of the sediment contribution to the organic halogen budget showed that only a minor fraction, 0.2 to 10%, of the AOX removed from the water column during the year accumulated into the sediment. Therefore, the disappearance of AOX from the water column (Fig. 6) was likely to be due to true biodegradation. These calculations are described in detail elsewhere (Saski *et al.* 1995, 1996).

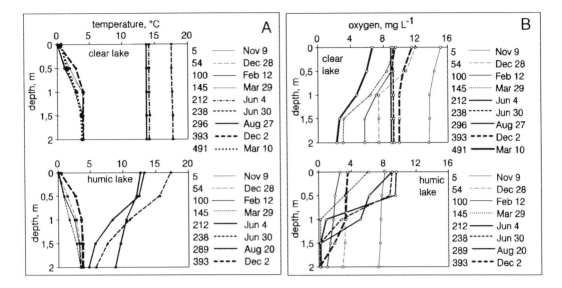

Figure 5. The vertical and temporal gradients of temperature (A) and dissolved oxygen (B) in the mesocosms in a clear water and humic water lake. The mesocosms shown in the figure were filled on Nov 5th. The data collected for mesocosms containing 2 to 13% of wastewater of Mill A or Mill C in lake water were close to those shown above. Similar results were obtained for mesocosms filled in June (data not shown).

Several interesting observations can be made from Fig. 7. Mill C wastewater induced more halogen accumulation into the mesocosms than did Mill A wastewater, although Mill C wastewater contained less AOX. It may be correlated to the larger amount of sediment formed in Mill C mesocosms, probably because Mill C wastewater contained more nutrients (N and P). When the wastewater was introduced into the mesocosm enclosures in November ("winter to winter") and incubated over one year, more sediment bound organic halogen accumulated than happened when the same operation was started in June ("summer to summer") and also incubated over one year. Moreover, the tetrahydrofuran-soluble fraction of the "winter to winter" sediment was larger than that of "summer to summer" sediment. Therefore, the long-term behavior of the organic halogen discharge from the mill seems to be affected by the season of discharge (for data and further discussion, see Saski et al. 1996).

We made tetrahydrofuran extracts of the water column and the sediment in order to study the changes in molecular weight distribution during the one-year incubation in the mesocosms. We used tetrahydrofuran extraction rather than cyclohexane or other less polar solvents to analyze sediments, because tetrahydrofuran dissolves 5 to ca. 100% (mean 34%, n = 29, see Fig. 7) instead of the 2 to 5% usually found to dissolve in nonpolar solvents used in other studies (Martinsen et al. 1988, 1994).

Figures 8A and B show examples of molecular size distributions of tetrahydrofuran-extractable organic halogens in the wastewaters of Mills A and C after incubation for one year in clear water lake. They show that the integrated peak area indicating the tetrahydrofuran-soluble halogens diminished about 50% (Mill A) or 20 to 30% (Mill C) in 393 d in the water column, but the molecular weight distributions remained essentially similar. The UV profile also showed no major change.

Some but usually not the major part of the AOX disappeared from the mesocosm water column during the year (Fig. 6) and accumulated in the sediment (Fig. 7). The molecular size distributions of tetrahydrofuran-soluble halogens extracted from the sediment formed *de novo* are shown in the lower panels of Fig. 8. To facilitate comparisons, the molecular size distribution of compounds of the wastewater origin is drawn in the same panels.

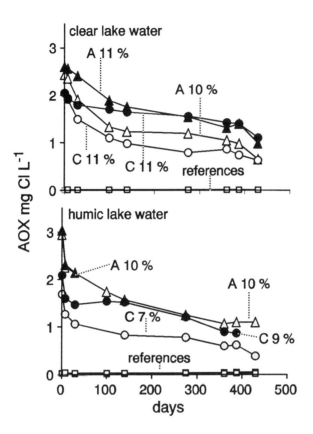

Figure 6. The time course of the AOX contents of the water column of some lake water containing enclosures with 0 to 11% (v/v) wastewater from Mill A or Mill C. The incubation was started June 1991. Some enclosures were built from black polyethene and covered with the same material to keep out daylight. Open symbols are daylight-exposed mesocosms; black symbols are mesocosms kept in the dark.

A striking difference is seen between the molecular sizes of halogenated compounds in the water column and those in the sediment of the same mesocosm: the average sizes of the halogenated molecules in the sediment were higher by almost an order of magnitude, with maxima appearing around and over 1000 g mol^{-1} (Fig. 8A, B). Some tetrahydrofuran-soluble halogenated material also sedimented in the 0% wastewater containing mesocosms, but the amount was negligible compared to those where wastewater was added (Fig. 7). Therefore the large molecular size halogen profiles (lower panels of Figs. 8A, B) of the sediment must have originated from the wastewater introduced 290 d before.

To elucidate the role of daylight on the molecular weight changes of organohalogens in the mesocosms, parallel mesocosms were studied, one open to daylight (Fig. 9, lower panel), and the other kept dark (Fig. 9, top panel). Figure 9 shows that large-sized halogenated molecules had accumulated in the sediment in 290 d, both in the light and in the dark.

The accumulation of large-sized halogenated molecules in the sediment was observed in all the 7 different mesocosms, both humic and clear lakes (Saski *et al.* 1996). At this moment we can only speculate on the mechanisms behind this change. The change was much larger than the slight shift towards higher molecular weight during activated sludge treatment of bleached kraft mill wastewater, while no shift was observed during aerobic-anaerobic treatment of the wastewater (Fig. 3, also Jokela *et al.* 1993).

It is possible that the chlorolignins may have become degraded into smaller molecules, then entered the intermediary metabolism of the mesocosm biota (Neilson 1994), and become incorporated into cellular macromolecules such as lipids and other tetrahydrofuran solubles, but also insolubles (Fig. 7).

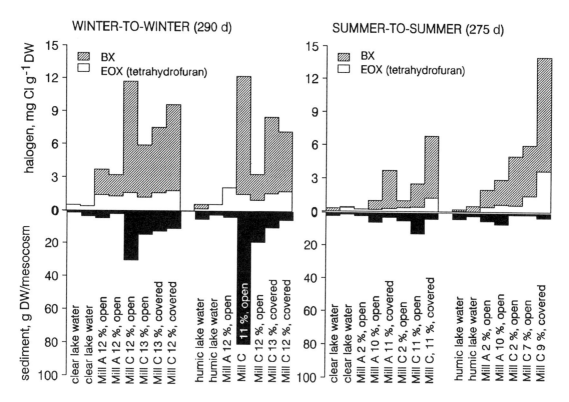

Figure 7. De novo produced sediment and sediment bound halogen in mesocosm enclosures filled with lake water and biologically treated bleached kraft mill wastewater. The mesocosms were filled with lake water (clear or humic) to which wastewaters of Mill A or Mill C were added to the percentage indicated in the figure. Fifteen enclosures were filled in June 1991, and another fifteen in November 1992. Some were covered to keep out daylight. All mesocosms were incubated floating in the lake for one year before the sediment was collected and analyzed. The black bars (lower panel) indicate sediment total dry weight, grey bars indicate the sediment-bound halogen (BX) and the white bars indicate the sediment contained tetrahydrofuran-soluble (EOX) halogen (upper panel).

We are not aware of other studies of the molecular weight of sediment-bound organohalogens, but in our laboratory a similar phenomenon, increase of molecular weight, was observed in micrososms of soil polluted with chlorophenols from wood preservation. Organic halogens occurred at old polluted sites (>10 yrs) as large molecules (similar in size to halogens in bleached pulp mill wastewater), especially if the soil was humic (high in organic matter). Earthworms exposed to such soils resorbed the smaller and the larger halogenated matter and converted it into even larger molecules, ranging from 10^3 to 10^4 g mol^{-1} (Laine et al. 1995). These data, and those presented in this paper, support the hypothesis that the biota may indeed extensively incorporate halogenated organic compounds into cellular materials from which they may further migrate into other forms of life, as molecules very different from the original structure.

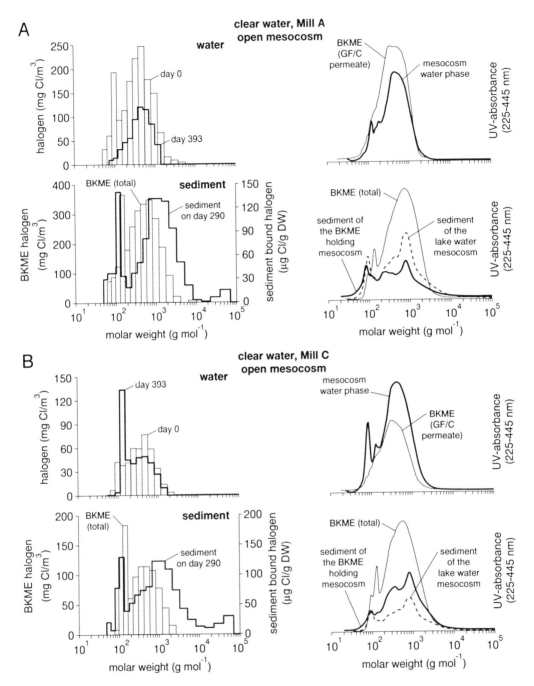

Figure 8. Changes in molecular weight distributions of the UV-absorbing materials and the organic halogens of wastewater (BKME) from two mills during weathering in daylight-exposed lake water mesocosms over four seasons. The vertical axis refers to tetrahydrofuran-soluble halogens calculated per m^3 of the input wastewater. The mesocosm water columns were sampled on day 0 and on day 393 (top panels). The mesocosm sediments formed de novo in either the absence (lake water mesocosm) or in the presence of wastewater (holding ca. 10% BKME) were collected on day 290 (lower panels). The mesocosms with Mill A wastewater (Fig. 8A); and Mill C wastewater (Fig. 8B) were simultaneously incubated in the same lake.

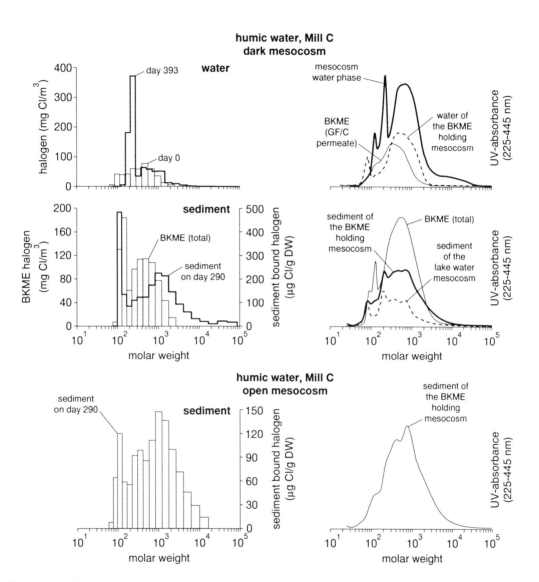

Figure 9. Daylight dependence of the weathering effect on molecular weight of bleached kraft mill wastewater halogens. Mill C wastewater was incubated in lake water holding mesocosms over four seasons exposed to (lower panel) or covered from (top panel) the daylight. The vertical axis refers to tetrahydrofuran-soluble halogens calculated per m^3 of the input wastewater. The mesocosm water columns were sampled on day 0 and on day 393. The mesocosm sediments formed de novo in the absence (lake water mesocosm) or in the presence of wastewater (holding ca. 10% BKME) were collected on day 290. All mesocosms were simultaneously incubated in the same lake. Size exclusion chromatograms of both UV-absorbing and the halogen-containing materials are shown.

CONCLUSIONS

In the present paper the molecular size distribution and solvent extractability (polarity) of chlorolignins were determined at various stages of exposure to treatment and water ecosystems, with the aim of increasing the understanding of the factors determining the ecological fate. We studied the fate of kraft mill-generated AOX in the mill secondary treatment and in outdoor mesocosms where the mill water was mixed with lake water and incubated under realistic conditions.

Molecular size does not limit bioremovability during biological wastewater treatment. The chlorolignins very likely behave in the water ecosystem as relatively small molecules from 200 to 1000 g mol^{-1}.

After discharge to recipient water, the kraft mill organic halogens, were removed from the water column: die-away distances of 40 to 50 km, averaged over one year in the Vuoksi river basin in southeast of Finland.

In the wastewater containing mesocosms (2 m^3, placed *in situ* outdoor), sediments with a high content of bound halogens were accumulated.

The sediment accumulated <10 % of the wastewater organohalogens in one year, while 50 to 80% of the wastewater organic halogens were removed from the water column. Accordingly, the sediment was not the main organohalogen sink; most likely the organic halogens mainly became mineralized.

The molecular weight distribution of kraft mill organic halogens remaining in the water column underwent no major change during weathering in mesocosms, in clear or humic water, over four seasons.

Organohalogens accumulated in the sediment showed a higher molecular size than those present in the water column. The molecular weight was significantly higher than in the wastewater from which the organic halogens originated, indicating extensive metabolism by the biota in the mesocosm.

ACKNOWLEDGMENTS

We acknowledge the financial support of The National Board of Waters and Environment funded SYTYKE 18 project, Maj and Tor Nessling Foundation (MSS), Academy of Finland (ES) and Foundation of Helsinki University (JJ & MSS). We thank Clas Wesen for the gift of ^{36}Cl$_2$ bleaching liquor, G. Brunow for the synthesis of multiring lignin model compounds, M. Herranen from Metsä-Sellu Oy, A.-C. Eriksson from Metsä-Botnia Ab, H. Jussila from Kymi-Kymmene Oy and R. Lammi from Visa-Forest for cooperation in permissions and practicalities in wastewater sampling, P. Manninen for neutron activation analyses, and Riitta Boeck for expert analytical help.

REFERENCES

Frostell, B., B. Boman, M. Ek and B. Palvall. 1994. Influence of bleaching conditions and membrane filtration on pilot scale biological treatment of kraft mill bleach plant effluent. Wat. Sci. Technol. 29:163-176.

Gergov, M., M. Priha, E. Talka, O. Välttilä, A. Kangas and K. Kukkonen. 1988. Chlorinated organic compounds in effluent treatment at kraft mills. TAPPI J. 71:175-184.

Graves, J.W., and T.W. Joyce. 1994. A critical review of the ability of biological treatment systems to remove chlorinated organics discharged by the paper industry. Water SA 20:155-160.

Higashi, R.M., G.N. Cherr, J.M. Shenkar, J.M. Macdonald and D.G. Crosby. 1992. A polar high molecular mass constituent of bleached kraft mill effluent is toxic to marine organisms. Environ. Sci. Technol. 26:2413-2420.

Jokela, J. 1995. Biodegradability and biodehalogenation synthetic lignin model compounds and bleached kraft pulp mill organic fraction. Tools for study. Ph.D. Thesis, Univ. Helsinki, Finland: In press.

Jokela, J.K., and M. Salkinoja-Salonen. 1992. Molecular weight distribution of organic halogens in bleached kraft pulp mill effluents. Environ. Sci. Technol. 26:1190-1197.

Jokela, J., J. Pellinen, M. Salkinoja-Salonen and G. Brunow. 1985. Biodegradation of two tetrameric lignin model compounds by a mixed bacterial culture. Appl. Microbiol. Biotechnol. 23:38-46.

Jokela, J., J. Pellinen and M. Salkinoja-Salonen. 1987. Initial steps in the pathway for bacterial degradation of two tetrameric lignin model compounds. Appl. Environ. Microbiol. 53:2642-2649.

Jokela, J., M.S. Salkinoja-Salonen and E. Elomaa. 1992. Adsorbable organic halogens (AOX) in drinking water and the aquatic environment in Finland. J Water SRT - Aqua 41(1):4-12.

Jokela, J.K., M. Laine, M. Ek and M.S. Salkinoja-Salonen. 1993. Effect of biological treatment on halogenated organics in bleached kraft pulp mill effluents studied by molecular weight distribution analysis. Environ. Sci. Technol. 27:547-557.

Laine, M., J. Jokela and M. Salkinoja-Salonen. 1991. Removal of different sizes of organic halogen compounds during biological treatment of bleached kraft pulp mill effluents. p. 113-120. In A. Södergren (ed.) Environmental fate and effects of bleached pulp mill effluents. SEPA Report 4031, Stockholm, Sweden.

Laine, M., J.K. Jokela and M. Salkinoja-Salonen. 1995. Biomobility of organic halogen compounds from contaminated soil - earthworms as a tool. In M. Munawar (ed.) Chemicals in the arctic-boreal environment. Ecovision, World Monograph Series, SPB, Academic Publishing (NL): In press.

Martinsen, K., A. Kringstad and G.E. Carlberg. 1988. Methods for determination of sum parameters and characterization of organochlorine compounds in spent bleach liquors from pulp mills and water, sediment and biological samples from receiving waters. Wat. Sci. Technol. 20:13-24.

Martinsen, K., A.L. Kvernheim and G.E. Carlberg. 1994. Distribution of organohalogen in sediments outside pulp mills using sum parameters. Sci. Tot. Environ. 144:47-57.

Neilson, A.H. 1994. Organic Chemicals in the Aquatic Environment. Distribution, Persistence and Toxicity. Lewis Publishers, CRC Press, Boca Raton, p. 183-310.

Pellinen, J., and M. Salkinoja-Salonen. 1985a. High performance size exclusion chromatography (HPSEC) of lignin and its derivatives. J. Chromatogr. 328:299-308.

Pellinen, J., and M. Salkinoja-Salonen. 1985b. Aqueous size-exclusion chromatography of industrial lignins. J. Chromatogr. 322:129-138.

Salkinoja-Salonen, M., M.-L. Saxelin, J. Pere, T. Jaakkola, J. Saarikoski, R. Hakulinen and O. Koistinen. 1981. Analysis of toxicity and biodegradability of organochlorine compounds released into the environment in bleaching effluents of kraft pulping. p. 1131-1164. In L. Keith (ed.) Advances in the identification and analysis of organic pollutants in water. Vol 2. Ann Arbor Science Publ., MI.

Salkinoja-Salonen, M.S., R. Valo, J. Apajalahti., R. Hakulinen, L. Silakoski and T. Jaakkola. 1984. Biodegradation of chlorophenolic compounds in wastes from woodprocessing industry. p. 668-676. In M.J. Klug and C.A. Reddy (ed.) Current perspectives in microbial ecology. American Society for Microbiology, Washington DC.

Salkinoja-Salonen, M.S., E. Saski, K. Salonen and A. Vähätalo. 1993. Decomposition and environmental impacts of organic substances in the wastewater of a mill producing bleached sulphate pulp. Part III. Decomposition. National Board of Waters and the Environment. SYTYKE project report, Ser. A 130:75-171 (In Finnish).

Sandermann, Jr., H., M. Arjmand, I. Gennity, R. Winkler, C.B. Struble and P.W. Aschbacher. 1990. Animal bioavailability of defined xenobiotic lignin metabolites. J. Agric. Food Chem. 38:1877-1880.

Sandermann, Jr., H., D.-H. Pieper and R. Winkler. 1993. Mineralization of lignin-bound and free xenobiotics by the white-rot fungus Phanerochaete chrysosporium. p. 499-503. In J.F. Kennedy, G.O. Phillips, and P.A. Williams (ed.) Cellulosics: pulp, fibre and environmental aspects. Ellis Horwood Series in Polymer Science and Technology, N.Y.

Saski, E., M. Salkinoja-Salonen, A. Vähätalo and K. Salonen. 1991. Degradation and environmental fate of bleached kraft pulp mill effluents studied in freshwater mesocosms. p. 183-189. In A. Södergren (ed.) Environmental fate and effects of bleached pulp mill effluents. SEPA Report 4031, Stockholm, Sweden.

Saski, E.K., K. Salonen, A. Vähätalo and M. Salkinoja-Salonen. 1995. Degradation of the halogenated organic fraction of biologically treated bleached kraft pulp mill effluents in Finnish lake water mesocosms. In M. Munawar (ed.) Chemicals in the arctic-boreal environment. Ecovision, World Monograph Series, SPB, Academic Publishing (NL): In press.

Saski, E.K., A. Vähätalo, K. Salonen and M.S. Salkinoja-Salonen. 1996. Mesocosm simulation on sediment formation induced by biologically treated bleached kraft pulp mill wastewater in freshwater recipient. In Environmental Fate and Effects of Pulp and Paper Mill Effluents, M.R. Servos, K.R. Munkittrick, J.H. Carey and G. Van Der Kraak (ed.), St. Lucie Press, Delray Beach, FL.

Smeds A., B. Holmbom and C. Pettersson. 1994. Chemical stability of chlorinated components in pulp bleaching liquors. Chemosphere 28:881-895.

Strehler, A., and T. Welander. 1994. A novel method for biological treatment of bleached kraft mill wastewaters. Wat. Sci. Technol. 29:295-301.

Yu, P., and T. Welander. 1994. Anaerobic treatment of kraft bleaching plant effluent. Appl. Microbiol. Biotechnol. 40:806-811.

ANALYSIS OF PULP MILL BLACK LIQUOR FOR ORGANOSULFUR COMPOUNDS USING GC/ATOMIC EMISSION DETECTION

I. Ya. Rod'ko[1], B.F. Scott[2] and J.H. Carey[2]

[1]Institute of Chemical Physics RAN, Moscow, Russia
[2]Aquatic Ecosystem Conservation Branch, National Water Research Institute, P.O. Box 5050, Burlington, Ontario, L7R 4A6 Canada

Solvent extracts of samples of waste cooking liquor (WBL, weak black liquor) from several Canadian kraft pulp mills were analyzed by gas chromatography/atomic emission detection (GC/AED). Of particular interest were the sulfur-containing compounds that might be useful chemical markers of WBL in final effluent. The WBL samples were extracted under four conditions: air excluded, moderate aeration, after oxidation by hydrogen peroxide (H_2O_2) and after treatment by alkaline zinc oxide (ZnO_2^{-2}) solutions. The types and amounts of compounds present in the extracts were dependent on the degree of oxidation of the black liquor prior to sampling, the treatment after sampling and the degree of oxidation during the extraction phase, and prior to extraction. One WBL was fractionated using silica gel and the chromatograms of the fractions were compared to chromatograms of similar fractions derived from treated final effluent from the same mill. One S-containing compound was found in both the black liquor and the effluent. The major sulfur compounds observed in the slightly oxidized black liquors were elemental sulfur (S_8), dimethyltrisulfide and dimethyltetrasulfide. The occurrence of these compounds in the extracts discussed.

INTRODUCTION

In the Canadian pulp and paper industry, the most widely used pulping process is the kraft or sulfate process. The first stage in this process involves cooking wood chips in a strongly basic solution containing sodium sulfide (Na_2S) to solubilize lignin and free the cellulose fibers. After cooling, the fibers are separated from the spent cooking liquor by filtration and the spent cooking liquors, known as weak black liquor (WBL), are sent to a chemical recovery system where, after evaporation and combustion, some of the cooking chemicals are recovered. In addition to economic benefits from recovery of cooking chemicals, the recovery process also reduces environmental impacts by diverting the highly toxic WBL from the pulp mill effluent system. Despite the emphasis on recovery of WBL, spills and incidental releases are known to occur. These spills, plus a small percentage of WBL that remains with the pulp and is carried into the washing and bleaching stages, are suggested to be contributors to the acute toxicity of untreated pulp mill effluent to fish (McLeay *et al.* 1979; Hodson *et al.* 1994a). Recently, it has been suggested that compounds present in WBL could also be responsible for sublethal effects on liver enzymes and steroid hormones in fish (Hodson *et al.* 1994b). It is therefore of interest to detect the presence of WBL in pulp mill effluent.

A number of previous studies have utilized gas chromatography/mass spectrometry (GC/MS) to identify compounds present in WBL (Niemela 1990; Ziobro 1984; Kringstad and Lindstrom 1984; McKague *et al.* 1989). The major components are derived from lignin, cellulose and hemicellulose, the major components of wood. Although GC/MS is a powerful tool, the multicomponent composition of WBL results in very complex chromatograms. Many of these compounds do not survive secondary treatment, and therefore unsuitable as chemical tracers of WBL in final effluent. In addition, the bleaching processes employed at many mills to reduce the lignin content of the pulp contribute substantial amounts of organic compounds to the effluent treatment system. The resulting increased complexity of the GC/MS chromatograms is a source of considerable uncertainty regarding the in-mill source of individual organic compounds.

The objectives of the current study were two-fold. The first objective was to compare the chemical profiles between WBL from several sources and identify similarities that could provide potential chemical tracers of WBL. We particularly wished to focus on sulfur-containing compounds as potential WBL tracers. We theorized that the use of Na_2S in the kraft process to solubilize lignin could result in the formation of organo-sulfur compounds that would be potential WBL markers. We wished to investigate whether the use of a gas chromatographic detector specific for sulfur would considerably decrease the complexity of the chromatograms and simplify the analyses since contributions from non-sulfur containing compounds should be absent. The atomic emission detector (AED) provides such element specificity. The second objective of our study was to determine if these potential WBL tracers survived secondary treatment and were present in treated final effluent from pulp mills. We report here the first results of this study.

MATERIAL AND METHODS

Six WBL samples and one final effluent sample from Canadian pulp mills were analyzed in this study. Five WBL samples (**A-E**) were collected in February 1993. Samples **A** and **B** were from the same Ontario mill. Sample **A** was obtained when the mill was using a poplar wood furnish and **B** from a mixed softwood (60% jack pine, 20% spruce and 20% balsam) furnish. Samples **C** and **D** were from a second Ontario mill. **C** was derived from a cook of 100% poplar and **D** from dense hardwood (maple/birch). Sample **E** was from a British Columbia mill employing a mixed pine (50%) and spruce (50%) wood furnish. In August 1993, a sixth WBL sample (**F**), consisting of a 50/50 mixture of WBL from softwood and dense hardwood pulping, was collected together with a final treated effluent sample from the same mill as samples **C** and **D**. The WBL samples were stored in closed containers at 4°C in the dark.

For extractions, 25 mL of the WBL was shaken twice with 5 mL of dichloromethane (DCM). The DCM was then reduced to a volume of 2 mL using a flow of dry N_2. Under these conditions the pH of the WBL remained about 13. When WBL was extracted in the air, the pH decreased to 8 or 9 and a precipitate resulted with an attendant change of color from black to brown. Two WBLs (**F** and **B**) were subjected to fractionation. For this, 25 mL of the WBL were mixed with 500 mL of aerated water. The solution was then extracted with 3 X 200 mL of DCM. The extract was dried using Na_2SO_4, and the solvent was evaporated. This residue was then dissolved in a minimum of 40% DCM and 60% n-hexane and placed on a silica gel column. The sample was eluted using 70 mL aliquots of 33% DCM + 67% *n*-hexane, 67% DCM + 33% n-hexane, followed with DCM and finally MEOH. These fractions were each reduced to a volume of 10 mL. The effluent sample was subjected to silica gel fractionation. For this experiment, 2 L of the effluent was extracted with 2 X 200 mL of DCM dried with Na_2SO_4 and then the solvent was reduced with a stream of dry N_2. This was reduced to dryness and weighed before dissolving in a minimum of 40% DCM and 60% *n*-hexane. The elution of fractions from a silica gel column was similar to that of the WBL.

Certain WBLs were oxidized using H_2O_2. A concentrated (30%) H_2O_2 solution was added in excess to a 100 mL sample of the black liquor. The reaction mixture was extracted with 100 mL of DCM. During the reaction the pH of the solution decreased from 13 to about 9, producing a precipitate which was retained in the aqueous phase. Several WBL solutions were treated with excess Zn^{+2} under basic conditions. This was prepared from $ZnSO_4$ dissolved in KOH solution.

All analyses were carried out using capillary column GC/AED. An HP 5890B GC, equipped with an HP 7673A automatic sampler, was connected to an HP 5921A AE detector. These were controlled by the HP AED Pascal Chemstation. An XE-52 XL capillary column, supplied by HIRESCO (Mississauga, Ont.), for all chromatograms. This column was 30 m in length, with an i.d. of 0.25 mm and a film thickness of 0.25 μ. The elements regularly monitored were C (193 nm), S (181 nm), H (478 nm) and O (777 nm). Reagent gases were added to the plasma to enhance detection; these were H_2 and O_2 for S and C, O_2 for H, and the analysis for oxygen used H_2 and a N/CH_4 mixture. Injections were made in the splitless mode with a 0.8 min purge. The cavity block was kept at a temperature of 260°C and the temperature of the transfer line was 257°C. The carrier gas was purified He with a column head pressure of 17 psi. All injections were 1 µL. For chromatographic analysis, an initial temperature of 85°C was held for 2 min;, then the temperature was increased at a rate of 10°C min^{-1} until 140°C, and the rate was then increased to 30°C min^{-1} to 255°C. This temperature was maintained for 9 min. When analyzing for volatiles, 2 µL vapor injections were made, but no solvent vent (1.3 min to 3.8 min) was incorporated into the analysis as no solvent was present.

Identifications were based on mass spectra obtained from an HP 5989A mass spectrometer coupled to an HP 5890 series II GC equipped with an HP 7673A automatic sampler. Spectral searches were carried out using a Wiley 138K mass spectral data base. The capillary column, supplied by Supelco, was 30 m in length with a 0.25 mm i.d. and a 0.25 µ film thickness. Injections were made in the splitless mode with a 0.75 min purge. The temperature of the transfer line was 250°C. The carrier gas was purified He with a column head pressure of 10 psi with injections of 2 µL. For chromatographic analysis, the initial temperature of 90°C was held for 2 min, and then the temperature was increased at a rate of 4°C min^{-1} until 280°C, which was maintained for 10 min. The solvent delay before the MS filament was turned on was 3 min.

RESULTS

Preliminary experiments indicated that oxygen was consumed during subsampling, extraction and workup of the WBL samples, suggesting that some WBL components were susceptible to oxidation. To avoid this complication, samples were transferred and extracted under N_2. Element-specific C-chromatograms of the WBL samples indicated the presence of five major and many minor organic components (Fig. 1A). These organic compounds all appeared to be stable in air since little difference was discernable between the C-chromatograms for samples extracted under N_2 or under air [Fig. 1A (i), (ii), (iii)]. The corresponding element- specific S-chromatograms are shown in Fig. 1B. In the sample extracted under N_2, there are several S-containing compounds present at low levels as denoted by the low instrument response. After the sample had been exposed to a minimal amount of air, the response and the complexity of the S-chromatogram increased [Fig. 1A (ii)]. In the S-chromatogram of the WBL sample extracted in open air [Fig. 1 (iii)], a peak due to elemental sulfur, S_8, was the dominant feature of the chromatogram, accompanied by major peaks eluting at approximately 4, 6 and 8 min and several minor peaks. These results suggested that the major S-containing compounds in WBL extracts resulted from oxidation reactions involving air.

To facilitate comparisons, WBL samples **C**, **D** and **F** were extracted under N_2 to inhibit oxidation. The S-chromatograms of the resulting extracts are shown in Fig. 1C. There are significant qualitative differences in response between the three samples. The most significant feature of these chromatograms is the lack of a significant peak common to all three samples. These samples came from the same mill but differed in either the type of wood being pulped or time of sampling. This suggests that it will likely not be possible to use S-chromatography of simple extracts to detect chemical markers for WBL. Moreover, in most mills, WBL contributions to final effluent will be exposed to air or oxidizing conditions both in the bleaching process or in the mill sewers and also in final effluent treatment. In view of the apparent susceptibility of the S-containing components in WBL to oxidation, our search for chemical markers of WBL focused on the readily detectable S-containing oxidation products.

The identity of these WBL components was investigated with an aliquot of WBL **F**, oxidized such that a minimal amount of S_8 was formed. An extract of this solution was analyzed by GC/AED and GC/MS. Element-specific S- and C-chromatograms are shown in Fig. 2. Two of the eleven compounds identified contain sulfur. These were identified by their mass spectra as dimethyltrisulfide (R = 4.2 min) and dimethyltetrasulfide (R = 8.02 min). The AED instrument responses for these two compounds confirm a 2:3 and 2:4 C:S ratio.

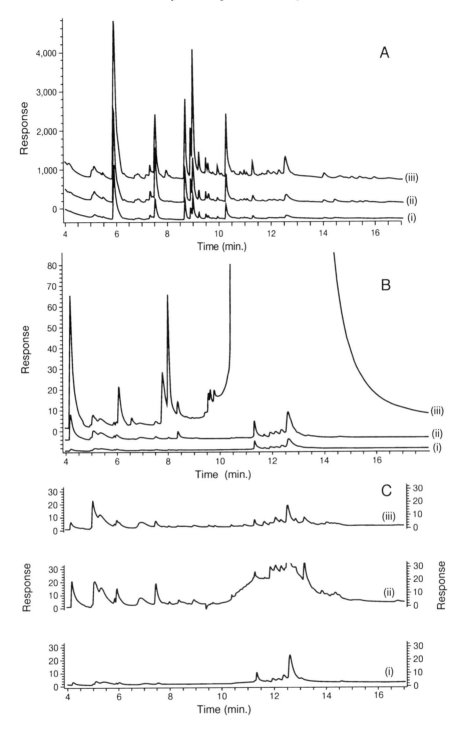

Figure 1. Element-specific chromatograms illustrating the degree of oxidation of black liquors: (A) carbon element-specific chromatograms for black liquor E with (i) no oxidation, (ii) minimal oxidation, (iii) air extracted; (B) sulfur element-specific chromatograms for black liquor E showing (i) no oxidation, (ii) minimal oxidation, (iii) air extracted; (C) sulfur-specific chromatograms for black liquor (i) F, (ii) D and (iii) C after extracting under N_2.

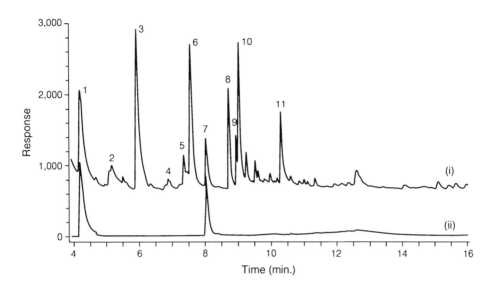

*Figure 2. Element-specific (i) C- and (ii) S-chromatograms for weak black liquor **F**. Peaks are ascribed the following designation: (1) dimethyltrisulfide, (2) benzenemethanol, (3) 2-methoxyphenol, (4) endo-boneol, (5) 4-methyl-1-(1-methylene)-3-cyclohexene-1-ol, (6) l-a-terpinol, (7) dimethyltetrasulfide, (8) 4-ethyl-2-methoxyphenol, (9) 10. 2-(2-hydroxy-2-propyl)-5-methylcyclohexanol, (11) 2-methoxy-4-(1-propenyl)-phenol.*

Further evidence that the observed S-containing compounds are oxidation products was obtained by studying the effect of added Zn^{+2}. The influence of added Zn^{+2} on the S- and C-chromatograms of extracts of WBL **B** is shown in Fig. 3A. Chromatogram (i) in Fig. 3A contains dimethyltrisulfide and dimethyltetrasulfide peaks as well as a substantial S_8 peak. In chromatogram (ii) of the same figure, obtained from an extract of the same WBL exposed to identical extraction conditions but to which Zn^{+2} had been added, all three compounds are absent. This latter chromatogram contains no large S-containing peaks, indicating that the added Zn^{+2} has inhibited the formation of the S-containing compounds.

Oxidation was enhanced by treating an aliquot of WBL **E** with H_2O_2. The element-specific S- and C-chromatograms resulting from this treatment are shown in Fig. 3B. These chromatograms can be compared with those in Figs. 1A and B that were obtained from aliquots of the same sample extracted under less oxidizing conditions. H_2O_2 treatment has significantly altered the C-profile and eliminated the two early eluting C-containing peaks at 4 and 8 min. A new S-containing compound eluting at 6 min appears to be produced but it has not been identified since it is obscured in the mass spectral analysis by 2-methoxyphenol that is present in higher concentrations and with which it coelutes.

To ascertain the reactivity of S_8, phenol was heated with sulfur in basic solution. The early eluting portion of the resulting S-chromatogram after extraction with DCM is shown in Fig. 3C. A large S_8 peak is present but not shown. In the early eluting region of the chromatogram there are two smaller peaks eluting at 6.1 0 and 7.87 min which were not present before reaction. Peaks eluting at the same times are also found in chromatograms derived from the oxidized WBL to which phenol had not been added.

The S-chromatograms derived from the methanol fractions for both the effluent as well as that from an aerated aqueous solution of WBL are shown in Fig. 4A. Since effluent treatment at this site involved aeration, considerable oxidation of the S-containing compounds had occurred prior to sampling. Therefore when fractionating the WBL, no special precautions were made to exclude air. From the S-chromatograms, there are many more S-containing compounds in the effluent than the WBL. The MEOH fractions of the

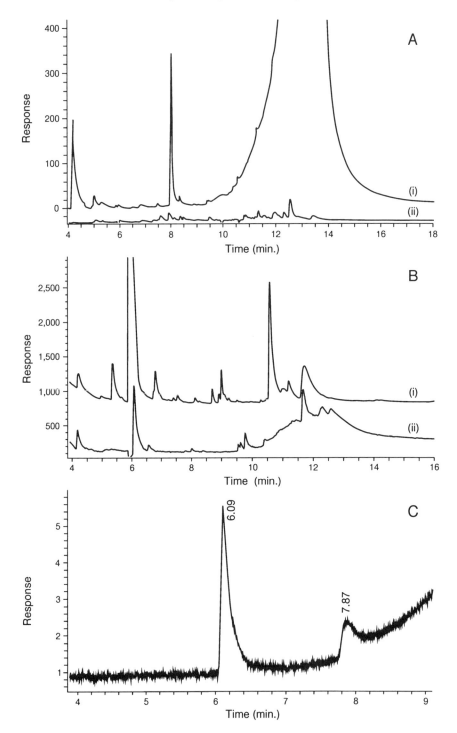

Figure 3. Element-specific chromatograms illustrating the influence of chemicals added to the black liquor: (A) element-specific S-chromatograms for (i) air oxidized DCM extract and (ii) similar extract but which had Zn^{+2} added before extraction; (B) element specific chromatograms for (i) sulfur and (ii) carbon of black liquor **E** after being oxidized with excess H_2O_2; (C) element-specific S-chromatogram of the basic reaction mixture containing phenol and sulfur.

fractionated WBLs are shown in Fig. 4B. There are several peaks that occur at the same time in these two chromatograms. One common peak in all three fractions elutes at 7.5 min.

Throughout this study, the reagents and solvents were extracted or concentrated and then analyzed to ensure that none of the peaks observed in the element-specific chromatograms originated from these possible sources of contaminants. In addition, white liquor, regenerated WBL, was extracted and no chromatographable sulfur or carbon compounds were detected.

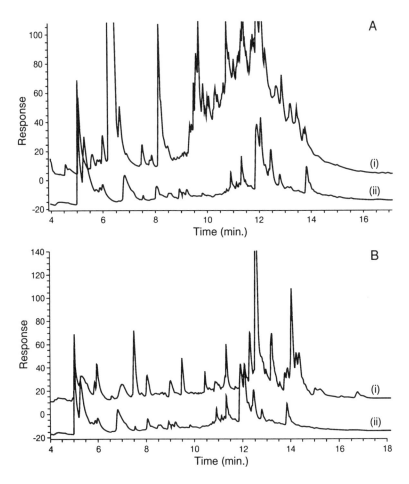

Figure 4. Sulfur element-specific chromatograms of MeOH extract of silical gel chromatography for A. (i) effluent, (ii) black liquor from the same mill (F); B (i) black liquor A compared to (ii) black liquor F.

DISCUSSION

The DCM extracts of WBLs from different mills produce similar element-specific S- and C-chromatograms. From examination of Fig. 1, the C-chromatogram contains a large number of peaks at moderate responses, but the S-chromatograms have few if any peaks and these few peaks have very low responses. On exposure to air, the C-chromatograms for a partiular WBL change little, but the S-chromatograms exhibit more and larger peaks. The WBL is the spent cooking liquid from the kraft process which was utilized to solubilize the wood lignin at high temperatures and increased pressures. When returned to ambient temperatures, the WBL is in thermodynamic equilibrium as the chemical processes are complete. However it is not in redox equilibrium.

Spills or other incidents in mill operations can cause the WBL to be directed into the effluent stream of the facility. There organic compounds in the WBL can be oxidized by air, peroxide solutions or other oxidants. Also oxidation products and the extent of oxidation would depend on the various mill processes such as action of air on the pulping mixture during opening of the digestor, washing efficiency and bleaching conditions. However, experiments in this work show that addition of Zn^{+2} ion will inhibit oxidation in the WBL.

As shown in this work, the action of air on the WBL will result in the formation of S_8 and dimethylpolysulfides. These compounds, which are usually not stable in the alkaline WBL, were removed from the basic solution by the extraction solvent in which they are stable. As oxidation proceeds, these compounds decrease in concentration and other unidentified compounds are formed. When the oxidation is complete as forced by addition of hydrogen peroxide, these dimethylpolysulfides are not detected. Minimal detectable amounts of S_8 are estimated to be 10 fg (as S_8), 100 fg for $CH_3S_3CH_3$ and 50 fg for $CH_3S_4CH_3$.

Some of the WBL may enter the effluent stream in a pulp mill, and its components will be oxidized. Comparison of the element-specific S-chromatograms from the MeOH eluants for both WBL **F** and the effluent from the same mill indicates there are many differences but there are similarities. One of these is the small peak at 7.55 min. A similar comparison of the S-chromatograms for the methanol eluant for different WBLs shows that although the two chromatograms appear different, many peaks occur in both chromatograms, including the peak at 7.55 min.

The mechanism of the formation of the oxidation products is largely unknown. Using only alkaline Na_2S solution with phenol present, at least one S-containing organic compound was formed. Studies are continuing on the mechanisms and products of the oxidation reactions.

ACKNOWLEDGMENTS

The authors wish to thank Dr. K. Burnison for collecting the weak black liquor (**F**) and effluent samples. The contributions of Mr. D. Bennie and R. Wilkinson, who analyzed selected samples with GC/MS and provided confirmation of the dimethylpolysulfides, are appreciated.

REFERENCES

Hodson, P.V., J.H. Carey, K.R. Munkittrick and M.R. Servos. 1994a. Canada and Sweden - contrasting regulations for chlorine discharge from pulp and paper industries. Proceedings of SETAC workshop on Risk Assessment of Organochlorine Compounds, Alliston, Ont., July 24-29, 1994.

Hodson, P.V., M.M. Maj, S. Efler, A. Schnell and J. Carey. 1994b. Kraft black liquor as a source MFO inducers. Presented at the Second International Conference on Environmental Fate and Effects of Bleached Pulp Mill Effluents, November 6-10, 1994, Vancouver, B.C., Canada.

Kringstad, K.P. and K. Lindstrom. 1984. Spent liquor from pulp bleaching. Environ. Sci. Technol. 18(8):236A-248A.

McKague, A.B., M.C. Kolar and K.P. Kringstad. 1989. Nature and properties of some chlorinated, lipophilicorganic compounds in spent liquor from pulp bleaching. Environ. Sci. Technol. 23(9):1126-1129.

McLeay, D.J., C.C. Walden and J.R. Munro. 1979. Effect of pH on toxicity of kraft pulp and paper mill effluent to salmonid fish in fresh and sea water. Water Res. 13:249-254.

Niemela, K. 1990. Low molecular weight organic compounds in birch kraft black liquor. Ann. Acad. Sci. Fenn., Ser. A, II. Chemica 229:1-142.

Scott, B.F. 1993. Investigation of tire leachates by gas chromatography with atomic emission detector (GC/AED). NRWI Contribution No. 93-63, pp. 1-25.

Ziobro, G.C. 1984. Origin and nature of Kraft color: 1. Role of aromatics. J. Wood Chem. Technol. 10(2):133-149.

TURNOVER OF CHLOROORGANIC SUBSTANCES IN A BOTHNIAN SEA RECIPIENT RECEIVING BLEACHED KRAFT PULP MILL EFFLUENTS

Per Nilsson[1], Lars Brydsten[1], Magnus Enell[2] and Mats Jansson[1]

[1]Department of Physical Geography, Umeå University, S 901 87 Umeå, Sweden
[2]Swedish Environmental Research Institute, Box 21060, S 100 31 Stockholm, Sweden

Cycling of chloroorganic substances from a pulp mill effluent has been studied with a model in the estuary serving as recipient for the discharge. The model is calibrated against emission of adsorbable organic halogens from the mill and the cycling of extractable organic chlorine. Outflow of tri- and tetrachlorinated substances from the recipient to the sea, related to emissions prior to reconstruction of the mill in 1986, was larger than the input to the recipient, due to mobilization of these substances from the sediment by resuspension. 6-Chlorovanillin and dichlorinated substances, considered to represent emissions of today, are exported directly to the open sea to a higher degree than the tri- and tetrachlorinated substances without being trapped within the recipient. There is a temporal variation in the retention, depending on seasonal variation in the hydraulic regime, with high export of tri- and tetrachlorinated substances during the spring and fall circulation. The export of 6-chlorovanillin and dichlorinated substances decreases during circulation due to increased sedimentation. The morphology of the basin and weather variations are shown to be the two most important factors controlling the cycling of chloroorganic substances in Gårdsfjärden.

INTRODUCTION

Bleached kraft pulp mill effluents (BKME) have been the focus of scientific interest during the last decade, particularly around the Baltic Sea, Bothnian Sea and the Bothnian Bay, where several pulp mills are located (Södergren 1988, 1992). Several investigations have been conducted on the ecological and biological effects of BKME (Landner et al. 1977; Förlin et al. 1985; Oikari et al. 1985; Oikari 1986; Bengtsson et al. 1988; Härdig et al. 1988; Söderström et al. 1994). The chemical and biological turnover, and degradation of BKME, and substances related to BKME, has been studied extensively by (Neilson et al. 1983; Remberger et al. 1986; Neilson et al. 1987; Allard et al. 1988; Häggblom et al. 1988; Neilson et al. 1989; Rosemarin et al. 1990; Remberger et al. 1993). The local, regional and large-scale distribution of BKME has been investigated (Paasivirta 1985; Xie et al. 1986; Paasivirta 1987; Håkansson et al. 1988; Paasivirta et al. 1988, 1990). A model for the turnover in the Baltic Sea was developed to predict the fate of BKME in the sea area (Södergren 1992). There are, however, few investigations (van Leeuwen et al. 1993) of the relationship between the sediment with its associated pollutants and the overlying water column in pulp mill recipients.

The pulp mills around the Bothnian Sea are situated at river mouths and estuaries. Bothnian Sea estuaries are characterized by the absence of tides, low salinity and an ice cover during winter. These characteristics make them different from high salinity estuaries, with tidal-induced mixing of the water column. The most important factor determining the retention capacity of the studied recipient, Gårdsfjärden, is the morphology of the basin. Gårdsfjärden is an almost enclosed bay of the Bothnian Sea and can be regarded as one extreme, where the other extreme is an open coastline, without any sheltering islands.

The pulp mill studied, located at Iggesund, went through a large-scale reconstruction from 1986 to 1991, including an altered bleaching technology, which reduced chlorine consumption by almost 90% (Södergren 1992), and improved handling of the BKME, including aerated lagoons. The concentration of adsorbable organic halogens in the BKME from the Iggesund mill has been reduced by about 90%, compared with the levels before 1986 (Södergren 1992), and the composition of the BKME has changed due to a new bleaching technology and waste water purification. Emissions of tri- and tetrachlorinated substances have decreased more than dichlorinated substances (Dahlman *et al.* 1991). 6-Chlorovanillin discharges increased after 1986 (Bengtsson, Iggesunds, pers. comm.). 12,14-Dichlorodehydroabietic acid is not formed in the new bleachery process (Remberger, Swedish Environmental Research Institute (IVL), Stockholm, pers. comm.).

The present study describes a detailed investigation of the seasonal variations in the turnover of chloroorganic substances in the primary recipient of the pulp mill at Iggesund. The purpose of the study was to assess the most important factors controlling the chloroorganic dynamics in the recipient and to quantify different pathways in the cycling of chloroorganic substances. The changed composition of the BKME, after reconstruction of the mill, made it possible to distinguish between recent and former emissions. Changes in the sedimentation, turnover and export from the recipient of chloroorganic substances to the Baltic Sea after the reconstruction are evaluated.

The following abbreviations are used in the text:

Bleached kraft pulp mill effluents	(BKME)
Adsorbable organic halogens	(AOX)
Extractable organic chlorine	(EOCl)
Suspended particulate matter	(SPM)
Characterized chloroorganic substances	(CCS)
4,5-Dichloroveratrol	(4,5-DCV)
4,5-Dichloroguaiachol	(4,5-DCG)
3,5-Dichlorocatechol	(3,5-DCC)
3,4,5-Trichloroveratrol	(3,4,5-TCV)
6-Chlorovanillin	(6-CVN)
3,4-Dichlorocatechol	(3,4-DCC)
4,5-Dichlorocatechol	(4,5-DCC)
Tetrachloroveratrol	(TeCV)
3,4,5-Trichloroguaiachol	(3,4,5-TCG)
3,4,6-Trichlorocatechol	(3,4,6-TCC)
3,4,5-Trichlorocatechol	(3,4,5-TCC)
Tetrachloroguaiachol	(TeCG)
Tetrachlorocatechol	(TeCC)
12,14-Dichlorodehydroabietic acid	(DCHAA)

MATERIALS AND METHODS

Study Area

The recipient, the Gårdsfjärden estuary, is situated in the middle of the western side of the Bothnian Sea (Fig. 1). The estuary is an almost enclosed basin, with an area of $3.92 \times 10^6 \, m^2$ and with a narrow passage, Dukarsundet, to the Bothnian Sea. The water volume is $22.4 \times 10^6 \, m^3$ and the mean depth 5.9 m. In nearshore areas the bottoms are relatively steep, and at 8-10 m depth there is a flat plain that covers a large part of the recipient. In the eastern part of the recipient the deepest part is 18 m. The salinity of the Bothnian Sea at this latitude is about 5 ppt, and there is no tide in the area. The natural fluvial input of suspended and particulate matter to Gårdsfjärden is through the regulated River Delångersån. The annual mean runoff is $10 \, m^3 \, s^{-1}$ and the annual maximum, which usually occurs during the spring flood, is about 20 m³ s⁻¹. The Iggesund pulp and paper mill is situated a few kilometers north of Gårdsfjärden (Fig. 1) and discharges about 1 m³ s⁻¹ via the aerated lagoons to the recipient.

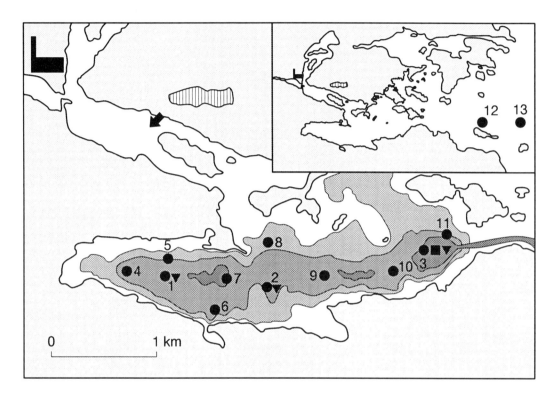

Figure 1. Gårdsfjärden with the mill in the upper left corner. The striped area is the aerated lagoons. The arrow indicates the outlet from the aerated lagoons. Circles show sediment sampling stations, triangles are sediment trap stations and the square shows the station for water sampling. The stippled areas show bottoms deeper than 4 m, 8 m and 11 m, respectively. Note the narrow dredged channel through Dukarsundet to the sea. The insert shows an overview of the area with sediment sampling stations 12 and 13.

Sampling

Salinity and temperature were measured every second week, from early May to late November 1992 and from late April until late June 1993, at three locations in Gårdsfjärden and at one location, (no. 13) about 4 km east of Dukarsundet, in the open sea (Fig. 1). Measurements were made at 1 m increments from the surface to the bottom. Wind speed and wind direction were obtained from the SMHI's (Swedish Meteorological and Hydrological Institute) weather stations. Sea level measurements were obtained from the closest SMHI station and from the local pilot station.

Three locations were used to trap suspended matter along an east-west gradient in the recipient (Fig. 1) and to investigate if an uneven distribution of the sedimentation occurred. At each location a sediment trap with two replicate tubes was placed at three levels: one surface trap 2.5 m below the surface, a bottom trap 1.5 m above the bottom and one sediment trap between the other two. The sediment trap tubes are made of plexiglass, with a length of 47 cm and a diameter of 5.4 cm. From early May to late November 1992, the sediment traps were emptied every second week to prevent algal growth and mineralization. No preservatives were added to the sediment traps. They were left during the winter and emptied in April 1993, and then deployed according to the same scheme as in the previous summer, until the field work was finished in late June 1993. The captured matter was kept refrigerated until freeze-drying and analysis.

Water samples were taken at two levels, above and below the halocline, at the three trap stations (Fig. 1) and on the same occasions as the sediment traps were emptied. About 50 L of water from each level was pumped through a 30 cm GFF filter, which was used for analysis of SPM. One liter of the filtrated water was taken for analysis of dissolved and colloidal matter.

Sediment samples were taken on August 11, 1992, with a core sampler at 13 locations (Fig. 1). Two samples were taken east of Dukarsundet to obtain data from the open sea. Three samples were taken at the same positions as the sediment traps. The rest were taken evenly over Gårdsfjärden to map the distribution of chloroorganic substances and the bottom dynamic conditions in the area. All samples were sliced before analysis. At stations 1, 2 and 3, the sediment cores were sliced in 1 cm intervals down to 5 cm depth and then in 2.5 cm slices to the end of the core. At the other stations, the cores were sliced in a 0-1 cm surface slice and then 5 cm slices. All samples were kept refrigerated before freeze-drying and analysis.

Analysis

The material caught in traps was analyzed for dry weight (DW), ^{137}Cs activity, C, N, Tot-P, PO_4^{3-}, Tot-N, NH_4^+, NO_2^-, NO_3^-, Kjeldahl-N, EOCl and the following CCS: 4,5-DCV, 4,5-DCG, 3,5-DCC, 3,4,5-TCV, 6-CVN, 3,4-DCC, 4,5-DCC, TeCV, 3,4,5,TCG, 3,4,6-TCC, 3,4,5-TCC, TeCG, TeCC, DCHAA.

The filtrated water and the particulate matter collected on filters were analyzed for the same substances as the suspended matter. Bleachery-specific substances were only analyzed in water samples from station 3. ^{137}Cs-activity and EOCl were not analyzed in water samples.

The sediment samples were analyzed exactly in the same way as the parameters for suspended matter from the sediment traps. No samples were taken of the BKME discharge from the mill.

The ^{137}Cs activity analyses were done at the Department of Limnology, Uppsala University, and the C, N, and P analyses were made at the Center for Marine Sciences at Umeå University. Analyses of the CCS, except for EOCl, were done at IVL in Stockholm. The EOCl analyses were made at SINTEF SI in Oslo.

Calculation of the Water Balance

The outflow of water from Gårdsfjärden is calculated as the sum of inflowing fresh water and inflowing seawater. The fresh water inflow is known from daily measurements in River Delångersån. With Knudsen's hydrographical theorem, using the dissolved salts in the seawater as a conservative tracer, it is possible to calculate the inflow of seawater to the recipient if the salinities of the two water bodies are known.

The salinity in Gårdsfjärden is calculated as the sum of salt in every one meter layer at stations 1-3 (Fig. 1), divided by the total volume of the recipient. The salinity in the inflowing seawater was taken directly from measurements from the station in the open sea, as there was no salinity gradient in the open sea.

Temperature and salinity data showed that Gårdsfjärden had a stable two-layer stratification during a large part of the year. In the upper part of the water mass, the warmer fresh water with a low density flows on top of the colder and more saline water at the bottom. The water below the combined halothermocline, the pycnocline, is relatively stagnant and anoxic conditions can develop. On a few occasions the stratification is broken and the water mass in the recipient is mixed (Table 1).

Therefore, it was necessary to separate the two different situations, when the water mass had a stable two-layer structure (Fig. 2) and when the water mass was mixed. When the stratified situation occurred, the water flow in and out of the recipient was calculated by using the salinities of the upper layer of water, as described above. When a mixed situation occurred the whole water volume was included in the calculations. Temperature, salinity, sea level, wind data and water chemistry data from the recipient were used to differentiate the two situations. The stratification situation in the recipient from the spring 1992 to the spring 1993 is presented in Table 1. Since no temperature and salinity data were available from the period immediately before the spring flood in 1993, the salinity data from the end of November 1992 were used. The water flow for this period was calculated as a mixed period. The winter period was calculated as a stratified period.

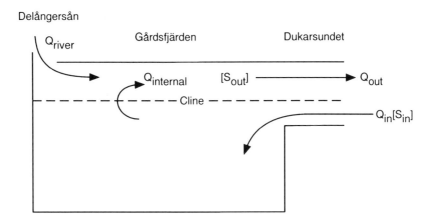

Figure 2. General view of the hydrodynamic circulation in Gårdsfjärden.

In order to obtain information on the stability of the water mass in Gårdsfjärden, the density of the water was calculated for one meter increments in depths at stations 1-3 in the recipient (Fig. 1). The position of the pycnocline was defined as the depth where the greatest difference in density occurred. The numerical value of the density difference, here called the cline index, also represents the stability of the stratification.

Calculation of Particle Dynamics and Turnover

The particle turnover model is based on measurements of sediment trap fluxes, analysis of surficial sediments from cores and from calculated water flows from the water dynamic model.

A generalized picture of the dynamics of particulate matter in Gårdsfjärden is given in Fig. 3. The total outflow of SPM (arrow 8, Fig. 3) through Dukarsundet to the open sea is calculated by multiplying the water flow with the concentration of suspended particles in the outflowing water.

The concentration of SPM above the pycnocline is calculated by using the concentration of TeCC from sediment trap 3 and the concentration of particulate TeCC collected on filters from the same station. This calculation gives an integrated value of the particle concentration in the water during the exposure period. TeCC in the particulate state is expressed as mg g^{-1} DW^{-1}, and in the dissolved state as mg L^{-1}. By dividing the former by the latter, the particle concentration is given in g L^{-1}. When there was a mixed water mass the average concentration of the surface sample and the bottom sample was used.

The gross sedimentation of particles (arrow 4, Fig. 3) is calculated by multiplying the trap yield by the ratio between the bottom area of the recipient and the sediment trap area.

Radioactive cesium (^{137}Cs) from the Chernobyl accident was used as a tracer to calculate the part of the total sedimentation that was primarily deposited (arrow 3, Fig. 3) and redeposited as resuspended particles, respectively (arrow 6, Fig. 3). There are high activities of radioactive ^{137}Cs in the surficial sediment in the recipient and therefore also in the resuspended matter. Primarily sedimenting matter has a lower activity, and the activity is estimated to be 1 Bq g^{-1} DW^{-1} in this area of Sweden (Carbol 1993); this is in the same range as the activity in sediment traps above the cline during calm and stratified periods.

Table 1. Seasonal variation of the stratification in Gårdsfjärden, May 1992 to May 1993.

Period	Dates	Duration, days	Situation	Remarks
1	920504-920617	45	Stratified	Rapidly rising surface temperatures and a high inflow of fresh water, leading to a stable stratification.
2	920618-920701	14	Mixed	An inflow of salt and warm seawater caused by strong winds and/or a sea level rise which mixed the water mass.
3	920702-920810	40	Stratified	The stratification was governed by high surface temperature and a decreasing sea level. An inflow of cold salt water under the cline increased the oxygen concentration in the bottom water.
4	920811-920813	3	Mixed	The seiche situation caused a rapid and complete mixing of the water in the recipient.
5	920814-920922	40	Stratified	Stable stratification developed by the salinity and a high fresh water inflow. A small inflow of cold seawater. Rising levels of ammonia in the bottom water indicated decreasing oxygen conditions.
6	920923-921017	25	Mixed	The fall turnover was caused by decreasing surface temperature and a strong wind from the east. The low oxygen conditions at the bottom disappeared as seawater replaced the bottom water.
7	921018-930412	165	Stratified	A rapid decrease of the surface temperature in the recipient gave a reversed temperature gradient with colder water in the surface layer than in the bottom. Large volumes of seawater were brought into the recipient during the fall turnover and determined the level of the cline. A smaller inflow of salt water occurred in late November. The recipient was ice covered from approximately the 10th of December to the 10th of April. It is likely that the water column was stratified during this period.
8	930413-930509	26	Mixed	Spring turnover.

The primarily deposited part (arrow 3, Fig. 3) of the total sedimentation is calculated by multiplying the gross sedimentation by the ratio between the ^{137}Cs activity in the bottom trap minus the primary ^{137}Cs activity input and the ^{137}Cs activity in the surface sediment, minus the primary ^{137}Cs activity input.

The resuspended, redeposited part (arrow 6, Fig. 3) is the difference between the gross sedimentation and the primary input.

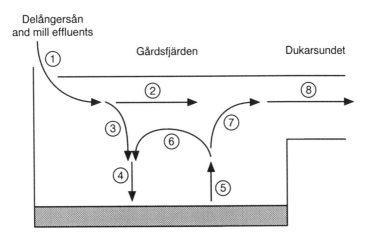

Figure 3. General view of the circulation of particulate matter in Gårdsfjärden.

In order to separate the total outflow of particulate matter (arrow 8, Fig. 3) in primary outflow (arrow 2, Fig. 3) and outflow of resuspended particles (arrow 7, Fig. 3), the same method as used to distinguish between primary and redepositing matter was used, but the ^{137}Cs activity in surface traps was used instead of bottom traps.

The total resuspension (arrow 5, Fig. 3) is the sum of resuspended, redeposited and resuspended exported matter.

Calculation of the Dynamics of Particulate and Dissolved Chloroorganic Substances

In order to quantify the different pathways of chloroorganic substances in the recipient (Fig. 3), the concentration of each substance was multiplied by the corresponding particle flow.

In the case of dissolved chloroorganic substances, the outflow of dissolved substances from the recipient during different periods was calculated by multiplying the concentration of the substance in the dissolved state by the calculated water flow for the period.

RESULTS AND DISCUSSION

A summary of the hydrodynamic situation in Gårdsfjärden during the different periods of the investigation is given in Table 1. The turnover of particulate matter in Gårdsfjärden during the different periods of the investigation is summarized in Fig. 4 and Fig. 5.

The sum of concentrations of CCS is less than 1% of the EOCl in Gårdsfjärden (Södergren 1992). The calculated particulate inflow of the CCS to Gårdsfjärden during the investigation was 120 kg, of which DCHAA makes up more than 80%. Tri- and tetrachlorinated substances make up over 80% of the rest of the CCS. Almost 80% of the export was DCHAA and most of the remaining part of the CCS was tri- and tetrachlorinated substances. In contrast to the situation for the particulate bound substances, over 90% of the dissolved substances was 6-CVN and dichlorinated substances. The total export of the dissolved CCS was 70 kg.

The two different groups of CCS, 6-CVN and dichlorinated substances, which are supposed to represent recent emissions, and tri- and tetrachlorinated substances and DCHAA, which are supposed to represent emissions before the reconstruction of the mill (Södergren 1992), have a similar vertical distribution in the sediment (Fig. 6). A slight increase of the former group in the surficial sediment supports the conclusion that they have increased in relative concentration in the BKME. On the other hand, 50% of the total inventory of the substances that were supposed to represent new emissions were found below 5 cm depth in the sediment, which was about the same as for tri- and tetrachlorinated substances and DCHAA. For example, 6-CVN showed almost exactly the same vertical distribution as TeCC (Fig. 6).

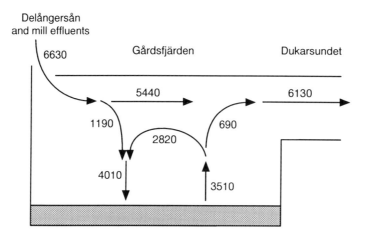

Figure 4. Turnover of particulate matter in Gårdsfjärden in tonnes per year.

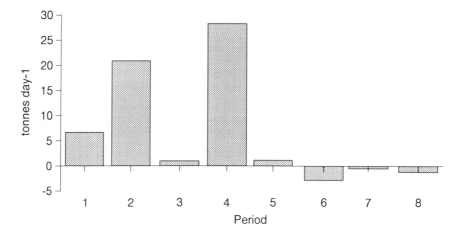

Figure 5. The retention of particulate matter in Gårdsfjärden in tonnes per day for different periods (Table 1).

The two groups of CCS had different dynamics in the recipient. 6-CVN and dichlorinated substances occurred mainly as dissolved forms, and in very low concentrations as particulate species. They were to a high degree transported through the recipient without settling (Fig. 7). The tri- and tetrachlorinated substances and DCHAA were mainly found in a particulate form (Fig. 8). These compounds thus have a complex turnover, including settling, resuspension, redeposition and export of resuspended matter (Fig. 8). The reconstruction of the mill is supposed to have almost totally eliminated emissions of tri- and tetrachlorinated substances and DCHAA. Nevertheless, the annual turnover of the latter was about one order of magnitude higher than the annual input of 6-CVN and the dichlorinated substances from the mill.

The concentration of particulate tri- and tetrachlorinated substances above the pycnocline increased when the stratification was disrupted or broken during a resuspension event. At stable stratification these substances increased in concentration below the pycnocline. The concentrations in the dissolved form were very low. Above the pycnocline there was no variation between the periods, and below the cline there was a slight increase, when resuspension took place.

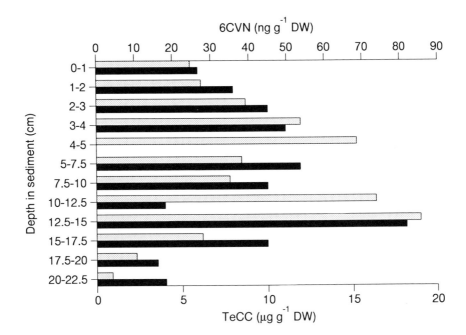

Figure 6. Concentration of 6-CVN and TeCC sediment from core 2.

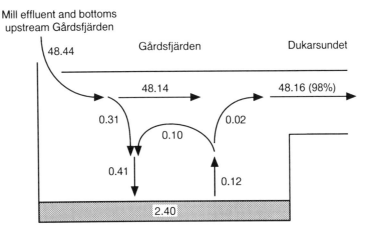

Figure 7. Calculated turnover of 6-CVN between 920504-930509. Value in shaded area represents the total inventory in the sediment in Gårdsfjärden. Value in brackets is the percentage of dissolved substance in the export to the sea.

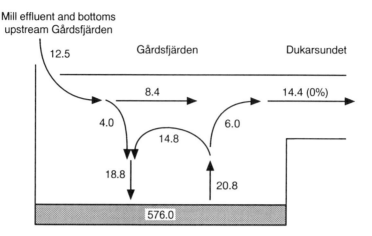

Figure 8. Calculated turnover of TeCC between 920504-930509. Value in shaded area represents the total inventory in the sediment in Gårdsfjärden. Value in brackets is the percentage of dissolved substance in the export to the sea.

Particulate 6-CVN and dichlorinated substances do not show the same variations. In the dissolved phase they occurred in higher concentrations, especially above the pycnocline. In deeper water layers they only occurred when the water column was mixed. The temporal variation of dissolved 6-CVN and dichlorinated substances did not show any correlation with resuspension events and periods of mixing.

To some extent, Gårdsfjärden acted as a sink for particulate 6-CVN and dichlorinated substances, while the dissolved fraction was exported directly out to sea. The total retention of 6-CVN and dichlorinated substances was about 1%. The tri- and tetrachlorinated substances had a negative retention, (i.e., Gårdsfjärden was a source of these substances), and the outflow to the Bothnian Sea was 17% larger than the input to the recipient.

The sediment had a much higher concentration of tri- and tetrachlorinated substances, compared to with 6-CVN and dichlorinated substances (Table 2). The calculated retention rate for the latter was nevertheless higher. The percentage of 6-CVN and dichlorinated substances was higher in the sediment than on SPM in the water mass. The vertical distribution was almost similar for both groups of substances, but there was a slight increase in 6-CVN and dichlorinated substances relative to tri- and tetrachlorinated substances in the uppermost centimeter of the sediment. Fifty percent of the CCS was found below 5 cm depth. The highest concentrations of tri- and tetrachlorinated substances in the sediment were found close to the inflow to Gårdsfjärden, sediment cores 7, 8 and 2 (Fig. 1). The concentrations decreased towards the mouth of the recipient and a few kilometers east of Dukarsundet they were at least one order of magnitude lower. 6-CVN and dichlorinated substances showed a wider distribution within the recipient.

The veratrols (4,5-DCV, 3,4,5-TCV and TeCV) occurred only in dissolved forms. The guaiacols, except TeCG and 3,4,5-TCG, were only found in a few samples and in very low concentrations in the dissolved phase. TeCG mainly occurred in the dissolved phase. The absence of veratrols and the low concentration of some of the guaiacols in the particulate state might be explained by microbial transformations (Neilson *et al.* 1987, 1989).

The total inventory of analyzed catechols in the sediment represents several hundred years of emissions with the calculated net accumulation. The reduction of these substances in the BKME by the reconstruction of the mill can explain the discrepancy between the accumulation rate and the total inventory. A microbial transformation of guaiacols to catechols in the sediment may be another explanation (Neilson *et al.* 1987). The low concentration of TeCG in the particulate state could be a result of such reactions.

Table 2. Total inventory of CCS, except for veratroles, which were not detected, in sediment cores. The average values (mg m^{-2}) are for cores 1-11, located in Gårdsfjärden (Fig. 1).

Station	4,5-DCG	3,5-DCC	6-CVN	3,4-DCC	4,5-DCC
1	1.00	23.78	0.75	5.57	14.86
2	2.09	28.57	1.61	6.61	10.51
3	0.74	19.04	0.97	4.87	7.73
4	1.27	22.18	0.30	5.60	8.98
5	0.21	10.66	0.32	2.23	4.01
6	1.40	27.58	1.97	5.50	8.47
7	1.54	24.15	1.62	4.73	9.16
8	2.07	23.19	1.43	5.48	9.43
9	0.41	15.20	0.17	3.44	5.12
10	1.04	27.00	1.20	8.55	8.81
11	0.00	31.07	0.93	8.60	14.06
12	1.21	10.14	0.00	4.38	3.27
13	0.53	11.88	0.33	7.46	2.84
Average	1.07	22.95	1.03	5.56	9.19

Station	3,4,5-TCG	3,4,6-TCC	3,4,5-TCC	TeCG	TeCC	DCHAA
1	2.49	76.89	86.74	1.05	214.68	1730
2	4.84	153.93	182.19	2.39	397.59	2800
3	1.85	51.16	115.87	0.79	204.29	1520
4	2.10	52.90	102.81	0.94	222.60	1470
5	2.48	26.93	123.57	1.11	250.44	960
6	2.73	85.07	514.46	1.63	387.41	1940
7	3.15	87.33	142.45	1.47	407.68	2040
8	4.20	143.36	212.81	2.28	555.87	3260
9	1.49	75.57	93.96	0.66	177.03	1380
10	2.59	106.61	166.96	1.32	228.85	1910
11	2.36	56.73	251.56	0.93	334.78	1550
12	0.52	16.72	23.18	0.13	33.30	380
13	0.33	12.65	22.06	0.12	30.69	230
Average	2.75	83.32	181.22	1.32	307.38	1870

There seems to be a primary inflow (arrow 3, Fig. 3) of substances that were supposed to be eliminated in the post-reconstruction emissions to the recipient. This might be an effect of reactivated substances from the area between Gårdsfjärden and the discharge point (Fig. 1). There could also be a leakage from earlier deposits north of the present discharge point. It is also possible that these substances are a component of the present emissions, since no analyses were made on the emissions during the investigation.

The different turnover of DCHAA and tri- and tetrachlorinated substances on one hand, and 6-CVN and dichlorinated substances on the other, depends on their different tendencies to be associated with particles and their different emission history. 6-CVN and dichlorinated substances come from the mill and are to a large degree transported through the recipient in the surficial water, while the recipient's sediments were a large source of DCHAA and tri- and tetrachlorinated substances. The degree of adsorption can be described as the lipophilicity, which determines the ability to adsorb on a solid state, i.e., on SPM (Xie et al. 1986). Several of the CCS are weak acids and their adsorption potentials are thus strongly affected by pH (Xie et al. 1986). The lipophilicity and the pH of the ambient water will thus have a strong impact on the partitioning between the dissolved and particulate phase of the substances, and thereby influence the turnover in the recipient.

During stratified conditions, the concentrations of 6-CVN and dichlorinated substances below the pycnocline were low compared with periods with mixed conditions, when the concentration below the pycnocline increased. However, the concentrations above the pycnocline did not show any correlation with resuspension periods. The direct outflow (arrow 2, Fig. 3) was 15 times higher than the outflow of resuspended adsorbed outflow (arrow 5, Fig. 3) during stratified conditions. At mixed conditions the outflow was 5 times higher. Therefore, the sediments contributed to a minor part of the export of these substances from the recipient.

On the other hand, tri- and tetrachlorinated substances and DCHAA were largely dependent on the sediment dynamics. The concentration of particulate tri- and tetrachlorinated substances and DCHAA, below the cline, increased during stratified conditions, due to resuspension of surface sediment. When the stratification was weakened or broken, the resuspended substances were distributed in a larger water volume and the concentration decreased below the pycnocline, while the concentration increased in the upper part of the water mass. A considerable export of resuspended material was the result. The sediment was a large source for these substances. This behavior was also demonstrated in a study of BKME contaminated sediments by Remberger *et al.* (1993), who showed that there could be a substantial release of catechols from BKME-contaminated sediments and that over 80% of the released catechols were bound to particulate or colloidal matter.

The net accumulation of the different substances was different due to their different distribution forms and patterns of turnover. It was negative for tri- and tetrachlorinated substances and DCHAA, due to high export of resuspended sediment (Fig. 9). The net accumulation rate of 6-CVN and dichlorinated substances was positive (Fig. 10). These differences probably reflect the fact that the content of tri- and tetrachlorinated substances and DCHAA substances in recent emissions was considerably lower than in the surface sediment, which was contaminated by particles with a high content from the period prior to the mill reconstruction. For 6-CVN and dichlorinated substances the situation was the opposite.

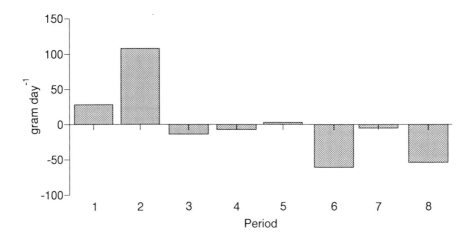

Figure 9. The calculated retention in grams per day of particulate TeCC for different periods in Gårdsfjärden (Table 1).

Except for their different emission history and different tendencies to occur in particulate form, the retention and export of the two investigated groups of substances were determined by the hydrodynamics in the recipient. A weak or absent pycnocline, together with a disturbance of the surface sediment, resulted in a resuspension and distribution of the particles in the whole water column. The horizontal movement of suspended particles to the sea was governed by the water exchange between the recipient and the sea. Both these processes must interact if a major export is to occur.

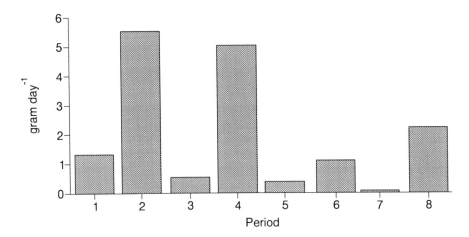

Figure 10. The calculated retention in grams per day of particulate 6-CVN for different periods in Gårdsfjärden (Table 1).

The export of tri- and tetrachlorinated substances and DCHAA was increased by almost 50% in periods with a mixed water column, compared with stratified periods. There were two major circulation periods during 1992 and 1993 that coincided with high water flows, the spring and fall turnover. During other periods of the year, turnovers were rare. In the summer, strong thermal and salinity stratification prevented resuspension and the water exchange was low. The seiche in August 1992, when the water level rose about 50 cm during 30 min, was a rare phenomenon. In the winter, the ice cover prevented wind-induced mixing, but barotrophically forced resuspension might have occurred, since the thermal stratification was weak. Clearly, the turnover of tri- and tetrachlorinated substances and DCHAA was controlled by weather-related forces.

The export to the sea of 6-CVN and dichlorinated substances (Table 3) also depended on the stratification. When the stratification was broken, the sedimentation increased and the export to the sea decreased, which was the opposite situation compared with tri- and tetrachlorinated substances and DCHAA. The increased sedimentation in periods with mixed conditions might have been related to an enhanced adsorption, due to an increased concentration of suspended particles in the water column.

Table 3. Export of 6-CVN and dichlorinated substances, and tri- and tetrachlorinated substances from the recipient to the sea at different hydrological situations.

	% dissolved export of CCS	
Period	6-CVN and dichlorinated substances	Tri- and tetrachlorinated substances
1	80	4
2*	60	7
3	90	7
4*	84	12
5	99	51
6*	98	28
7	99	48
8*	97	26
Average stratified	92	27
Average mixed	85	19

* indicates a mixed situation.

At present, the sediments in Gårdsfjärden are functioning as a sink for 6-CVN and dichlorinated substances. Even though more than 90% of the input passes through the recipient above the pycnocline, the net sedimentation is larger than the export from the resuspended sediment, resulting in an increase of 6-CVN and dichlorinated substances in the surface sediment relative to tri- and tetrachlorinated substances and DCHAA. This is probably a steady situation if emissions and concentrations of BKME are kept at the present level. A reduction of 6-CVN and dichlorinated substances of the same scale as the reduction of the tri- and tetrachlorinated substances and DCHAA will probably change Gårdsfjärden from a sink to a source for the former substances for some time, even though the total amount of exported substance to the Bothnian Sea will be reduced. For tri- and tetrachlorinated substances and DCHAA the sediments in Gårdsfjärden are a source. This situation will continue as long as the content of these substances is higher in resuspended particles than in primary settling particles. This process will take several years due to resuspension and focusing of the sediment (Nilsson in prep.).

Several authors have stressed the importance of sediments and SPM for the cycling of pollutants (see Jaffé 1991 and Olsen et al. 1982) for a review), and especially for BKME (Xie et al. 1986; Remberger et al. 1993) in the aquatic environment. Field studies concerning the connection between the sediment and the cycling of BKME are few. However, there are several studies of the turnover of particulate matter in various environments (Rosa 1985; Brydsten and Jansson 1989; Robbins and Eadie 1991; Lund-Hansen and Skyum 1992; James and Barko 1993) that support the results concerning the cycling of particle-associated substances in this study. Even if CCS are only about 1% of EOCl, which in turn are a few percent of AOX (Södergren 1992), it is likely that variations in the turnover are also representative for other substances in the BKME.

Two important factors determining the retention capacity of a recipient were the morphology of the basin and weather-related forces, controlling the water flow and the possibilities for stratification. They controlled the pycnocline stability and the water exchange rate between Gårdsfjärden and the open sea. The water exchange was determined by the fresh water inflow and barotrophically forced seawater intrusions. The density gradient was governed by salinity and temperature gradients. In the Bothnian Sea, where several pulp and paper mills are situated, the salinity gradient is low and the temperature gradient is pronounced.

In summary, this investigation demonstrates that the turnover of tri- and tetrachlorinated substances and DCHAA is largely controlled by resuspension, induced by the temperature-influenced circulation in the fall and spring, occurring simultaneously with high water flow situations. Resuspended matter becomes an important part of the export to the Bothnian Sea at present, but the importance will diminish at the same rate as the surface concentration of these substances decreases due to lower emissions. 6-CVN and dichlorinated substances largely flow straight through the recipient in dissolved form, on top of the pycnocline, without interacting with the bottom water and the sediment. During mixed situations the sedimentation of these substances increases. Since the factors that control stratification and resuspension depend on forces driven by the weather, the sedimentation and resuspension and thus also the net retention will show considerable variation within a year and between years.

ACKNOWLEDGMENTS

The study was financed by the Swedish Forest Industries Water and Air Pollution Research Foundation (SSVL-Miljö 93).

REFERENCES

Allard, A.-S., M. Remberger, T. Viktor and A.H. Neilson. 1988. Environmental fate of chloroguaiacols and chlorocatechols. Wat. Sci. Tech. 20(2):131-141.

Bengtsson, B.E., Å. Bengtsson and U. Tjärnlund. 1988. Effects of pulp mill effluents on vertebrae of fourhorn sculpin (*Myoxocephalus quadricornis*), bleak (*Alburnus alburnus*) and perch (*Perca fluviatilis*). Arch. Environ. Contam. Toxicol. 17:789-797.

Brydsten, L. and M. Jansson. 1989. Studies of estuarine sediment dynamics using ^{137}Cs from the Tjernobyl accident as a tracer. Estuarine, Coastal and Shelf Science 28:249-259.

Carbol, P. 1993. Speciation and transport of radionuclides from the Chernobyl accident within the Gideå site. Ph.D. Thesis. CTH, Göteborg.

Dahlman, O., R. Mörck, L. Johansson and F. de Sousa. 1991. Environmental fate and effects of bleached pulp mill effluents. SEPA Conference: Environmental fate and effects of bleached pulp mill effluents. Grand Hotel Saltsjöbaden, Stockholm.

Förlin, L., M. Åhlman, B.E. Bengtsson and O. Svanberg. 1985. Studies on the metabolism of 4,5,6-trichloro-(14C-CH3)-guaiacol in fish. Nordic Symposium on Bleaching Effluents. March 26-27, 1985, Åbo Akademi, Åbo, Finland.

Håkansson, L., P. Jonsson, B. Jonsson and K. Martinsen. 1988. Distribution of chlorinated organic substances from pulp mills. Wat. Sci. Tech. 20(2):25-36.

Häggblom, M.M., J.H.A. Apajalahti and M.S. Salkinoja-Salinen. 1988. Hydroxylation and dechlorination of chlorinated guaiacols and syringols by *Rhodococcus chlorophenolicus*. Appl. Env. Microbiol. 54(3):683-687.

Härdig, J., T. Andersson, B.E. Bengtsson, L. Förlin and Å. Larsson. 1988. Long-term effects of bleached kraft mill effluents on red and white blood cell status, ion balance, and vertebral structure in fish. Ecotox. Env. Safety 15:96-106.

Jaffé, R. 1991. Fate of hydrophobic organic pollutants in the aquatic environment: A review. Environ. Pollut. 69:237-257.

James, W.F. and J.W. Barko. 1993. Sediment resuspension, redeposition, and focusing in a small dimictic reservoir. Can. J. Fish. Aquat. Sci. 50:1023-1028.

Landner, L., K. Lindström, M. Karlsson, J. Nordin and L. Sörensen. 1977. Bioaccumulation in fish of chlorinated phenols from kraft pulp mill bleachery effluents. Bull. Env. Cont. Toxic. 18(6):663-673.

Lund-Hansen, L.-C. and P. Skyum. 1992. Changes in hydrography and suspended particulate matter during a barotrophic forced inflow. Oceanologica Acta 15(4):339-346.

Neilson, A.H., A.-S. Allard, P.-Å. Hynning, M. Remberger and L. Landner. 1983. Bacterial methylation of chlorinated phenols and guaiacols: formation of veratroles from guaiacols and high-molecular-weight chlorinated lignins. Appl. Env. Microbiol. 45(3):774-783.

Neilson, A.H., A.-S. Allard, C. Lindgren and M. Remberger. 1987. Transformation of chloroguaiacols, chloroveratrols, and chlorocatechols by stable consortia of anaerobic bacteria. Appl. Env. Microbiol. 53(10):2511-2519.

Neilson, A.H., H. Blanck, L. Förlin, L. Landner, P. Pärt, A. Rosemarin and M. Söderström. 1989. Advanced hazard assessment of 4,5,6-trichloroguaiacols in the Swedish environment. p. 329-374. *In* L. Landner [ed]. Chemicals in the aquatic environment. Springer Verlag, New York.

Nilsson, P. in prep.

Oikari, A.O.J. 1986. Metabolites of xenobiotics in the bile of fish in waterways polluted by pulpmill effluents. Bull. Env. Contam. Toxicol. 36:429-436.

Oikari, A., H.B.E. Ånäs, M. Miilunpalo, G. Kruzynski and M. Castrén. 1985. Ecotoxicological aspects of pulp and paper mill effluents discharged to an inland water system: physiological effects in caged fish (*Salmo gairdneri*). Aquat. Toxicol. 6:219-239.

Olsen, C.R., N.H. Cutshall and I.L. Larsen. 1982. Pollutant-particle associations and dynamics in coastal marine environments: A review. Marine Chemistry 11(1):501-533.

Paasivirta, J. 1985. Polychlorinated phenols, guaiacols and catechols in environment. Chemosphere 14(5):469-491.

Paasivirta, J. 1987. Organochlorine compounds in environment. Papers presented at the second IAWPRC meeting. Tampere, Finland.

Paasivirta, J., J. Knuutinen, P. Maatela, R. Paukku, J. Soikkela and J. Särkkä. 1988. Organic chlorine compounds in lake sediment and the role of the chlorobleaching effluents. Chemosphere 17(1):137-146.

Paasivirta, J., H. Hakala, J. Knuutinen, T. Otollinen, J. Särkkä, L. Welling, R. Paukku and R. Lammi. 1990. Organic chlorine compounds in lake sediments. III. Chlorohydrocarbons, free and chemically bound chlorophenols. Chemosphere 21(12):1355-1370.

Remberger, M., A.-S. Allard and A.H. Neilson. 1986. Biotransformation of chloroguaiacols, chlorocatechols, and chloroveratrols in sediments. Appl. Env. Microbiol. 51(3):552-558.

Remberger, M., P.-Å. Hynning and A.H. Neilson. 1993. Release of chlorocatechols from a contaminated sediment. Environ. Sci. Technol. 27:158-164.

Robbins, J.A. and B.J. Eadie. 1991. Seasonal cycling of trace elements ^{137}Cs, ^{7}Be, and $^{239+240}$Pu in Lake Michigan. J. Geophys. Res. 96(C9):17081-17104.

Rosa, F. 1985. Sedimentation and resuspension in Lake Ontario. J. Great Lakes Res. 11(1):13-25.

Rosemarin, A., M. Notini, M. Söderström, S. Jensen and L. Landner. 1990. Fate and effects of pulpmill chlorophenolic 4,5,6-trichloroguaiacol in a brackish water ecosystem. The Science of the Total Environment 92:69-89.

Södergren, A. 1988. Biologiska effekter av blekeriavlopp. Slutrapport från projektområdet Miljö/Cellulosa I. SNV. 3498.

Södergren, A. 1992. Bleached pulp mill effluents. Composition, fate and effects in the Baltic Sea. SNV. 4047.

Söderström, M., C.A. Wachtmeister and L. Förlin. 1994. Analysis of chlorophenolics from bleach kraft mill effluents (BKME) in bile of perch (*Perca fluviatilis*) from the Baltic Sea and development of an analytical procedure also measuring chlorocatechols. Chemosphere 28:1701-1719.

Van Leeuwen, J.A., B.C. Nicholson and K.P. Hayes. 1993. Distribution of chlorophenolic compounds, from a pulp mill, in Lake Bonney, southern Australia. Aust. J. Mar. Freshwater Res. 44:825-834.

Xie, T.-M., K. Abrahamsson, E. Fogelqvist and B. Josefsson. 1986. Distribution of chlorophenolics in a marine environment. Environ. Sci. Technol. 20(5):457-463.

MODELING THE FATE OF 2,4,6-TRICHLOROPHENOL IN PULP AND PAPER MILL EFFLUENT IN LAKE SAIMAA, FINLAND

Donald Mackay[1], Jeanette M. Southwood[2], Jussi Kukkonen[3], Wan Ying Shiu[1], Debbie D. Tam[1], Dana Varhaníčkova[1] and Rebecca Lun[1]

[1]Pulp and Paper Centre, Department of Chemical Engineering and Applied Chemistry, University of Toronto, Toronto, Ontario, M5S 1A4 Canada
[2]Angus Environmental Limited, 1127 Leslie Street, Don Mills, Ontario, M3C 2J6 Canada
[3]Department of Biology, University of Joensuu, P.O. Box 111, Joensuu, Finland FIN-80101

A water quality model is used to evaluate the fate of 2,4,6-trichlorophenol discharged from a pulp and paper mill into Lake Saimaa in Finland. The lake is characterized as a multi-segment receiving environment consisting of a series of three water columns exposed to the atmosphere, each with a sediment compartment, and five classes of organisms exposed to chemicals present in water and sediment. The model is successful in estimating concentrations in the water column and forage fish with all predictions lying in the range of observations. Although the sediment data are not as well predicted, all estimations are within a factor of two to four of the mean of the measurements. Results from this preliminary modeling exercise are encouraging and suggest that this approach is capable of providing a quantitative link between known loadings of an organochlorine from a specific mill and concentrations in water, sediments, and biota, and thus can contribute to the assessment of the impact of chemical discharges on the receiving aquatic ecosystem.

INTRODUCTION

There is continuing concern that chemicals discharged to the aquatic environment from industrial, municipal, and domestic sources are causing adverse effects on aquatic organisms and may be bioaccumulating to unacceptable levels in fish and wildlife. Much of this attention is focused on the organochlorine chemicals (OCs) discharged by the pulp and paper industry as by-products of bleaching operations. As has been reviewed by Suntio *et al.* (1988), a wide variety of OCs are produced including phenols, guaiacols, veratroles, dibenzo-*p*-dioxins, dibenzofurans, and resin acids, but the structure, properties, and toxicities of much of the OC material remain unknown. Often used to quantify this material is the surrogate parameter, AOX (adsorbable organic halide), but it is recognized that it is not possible to determine the adverse effects of AOX with any reliability.

There are two options for reducing or eliminating these OCs. The first is a ban on chlorine and chlorine components as industrial feedstocks as advocated by the International Joint Commission (1994). The second is a quantitative chemical-by-chemical assessment of sources, fate, and effects which leads to process and product changes and improved treatment, reducing concentrations to acceptable levels with negligible risk of adverse effects.

The latter course of action is generally preferred by industry and most environmental scientists and engineers, but to be successful there must be a capability of linking loadings (e.g., kg y^{-1}) to concentrations (e.g., μg L^{-1} or μg g^{-1}) in water, sediments, and biota of the receiving ecosystem. The only method by which this can be accomplished is by a water quality model in which the ecosystem is segmented into a series of "well-mixed" compartments of air, water, sediments, and organisms. Mass balance equations are written for each compartment expressing the rates of chemical input and output by transport and transformation. The set of equations is then solved to give a comprehensive statement of chemical amounts, concentrations, and process rates in the system. It is usual to compare calculated concentrations with observed values to "validate" the model.

However, despite the obvious incentive to compile and validate models of the fate of OCs and AOX, there have been few published examples of successful models of this type. Most reports have been merely of observations of concentrations and effects which lack the formal rigor inherent in the mass balance approach. The lack of this modeling capability and the implied inability to predict and understand the fate of individual chemicals and groups of chemicals quantitatively is a serious omission and can serve as a stimulus to adopting the first option of banning chlorine as a feedstock. In essence, the absence of a proven quantitative modeling capability renders the pulp and paper industry vulnerable to ill-advised, uneconomic, and disruptive regulations which are, nevertheless, well intentioned in that they seek to protect aquatic ecosystems from contamination. The lack of effort in this area by the industry in Canada is both puzzling and disappointing.

In this paper, the preliminary results are presented of the application of a comprehensive and flexible water quality model which describes the fate of one OC, a chlorinated phenol, when discharged into the receiving environment of the Lake Saimaa system in Finland. The model provides a method by which a known loading of an OC from a specific mill can be used, along with other input information, to deduce amounts, concentrations, process rates in water, sediments, and a range of biota. The more complex task of addressing effects is not attempted here but it seems prudent, as a first step, to understand and manage or control the concentrations and exposures which are believed to be a primary determinant of these effects. Some suggestions are made for future work in this area.

MODEL FORMULATION

The model and the fugacity concept upon which it is based have previously been described in detail (Mackay and Southwood 1992; Mackay 1991) and only a brief overview is presented here.

Fugacity is a thermodynamic property which is related to chemical equilibrium. It is logarithmically related to chemical potential and may be equated to the partial pressure for an ideal gas. Fugacity thus has units of pressure and has been described as an "escaping pressure or tendency".

An often-asked is why use fugacity in environmental models. The answer is that the fugacity concept simplifies the equations used in environmental models. The fugacity of a chemical in a specific compartment describes the tendency of the chemical to escape from that compartment. A relatively high fugacity indicates that a chemical in a particular compartment has a tendency to escape to a compartment where it has a lower fugacity, i.e., a chemical will try to move out of a compartment where it has a high fugacity into a compartment where it has a low fugacity. This difference in fugacity is the driving force behind the diffusive movement of the chemical into and out of different compartments in the environment.

For each environmental compartment such as water or sediment, the concentration, C (mol m^{-3}), is related to fugacity, f (Pa), using a fugacity capacity term, Z (mol m^{-3} · Pa), which is specific to the chemical, the compartment, and the prevailing temperature. The relationship is that C = Zf. Values for Z are calculated from the partitioning and physical-chemical properties of the chemical.

The model is a three-segment version of the Quantitative Water-Air-Sediment-Interaction (QWASI) fugacity model developed by Mackay *et al.* (1983) and applied by Mackay (1989, 1991), Holysh *et al.* (1986), Southwood *et al.* (1989), and others to aquatic systems. The basic unit of the model is a single segment consisting of a well-mixed water column with the atmosphere above, and overlying a well-mixed bottom sediment layer.

Figure 1 is an illustration of processes considered in the QWASI model. Processes which occur in the water column are inflow of water and suspended sediment; inflow of effluent from the paper mill; water-to-air and air-to-water diffusion; wet and dry deposition from the air compartment; water-to-sediment and sediment-to-water diffusion; sediment deposition and resuspension; transformation of chemical by processes such as photolysis, hydrolysis, and biodegradation; and water and suspended sediment outflow. Processes which occur in the sediment compartment are sediment-to-water and water-to-sediment diffusion, sediment resuspension, deposition, burial, and transformation of chemical. These processes can be grouped to represent aggregate rates of chemical transfer into or out of various compartments.

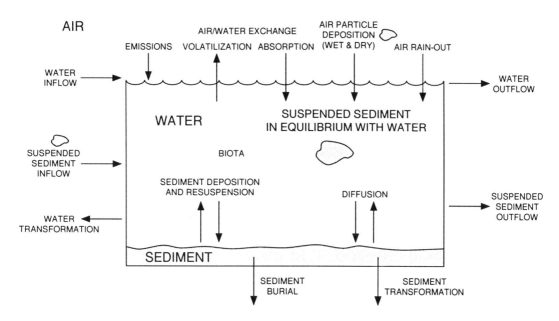

Figure 1. QWASI segment processes.

In the QWASI model, a differential mass balance equation is written in terms of fugacity for each of the water and sediment compartments. One equation describes the rate of change of water fugacity in terms of the rate of chemical entering the water column from the effluent, from the air, and from the sediment less the rate of chemical leaving the water column by reaction, outflow from the water column, and by other removal processes described previously. In the second equation, the rate of change of sediment fugacity is expressed as a function of the rate of chemical entering the sediment from the water column minus the rate of chemical leaving the sediment compartment through removal processes described earlier. In practice, a steady state version of the QWASI model in which inputs are equated to outputs is adequate for assessing the long term behavior of chemicals. The resulting equations are algebraic rather than differential and are readily solved.

To characterize the diluting capacity of the receiving waters, the model which is applied to the Lake Saimaa system is three water-sediment compartment pairs as depicted in Fig. 2. It is believed that a three-segment system can adequately represent the fate of chemicals in many receiving waters. In principle, fewer or more segments can be used as circumstances dictate.

Because one of the most convenient methods of assessing the level of contamination in the vicinity of a pulp and paper mill is to measure concentrations in biota, a simple food chain has been added to the model to calculate the concentrations of the chemicals in six organism classes. At the base of the food chain are plankton and benthos. Benthivores, forage fish, small piscivores, and large piscivores feed on them, each with its own food preferences. Benthivores are assumed to consume only benthos. For the four fish classes, a mass balance equation includes uptake from water and food and loss by gill water, egestion, metabolism, and growth.

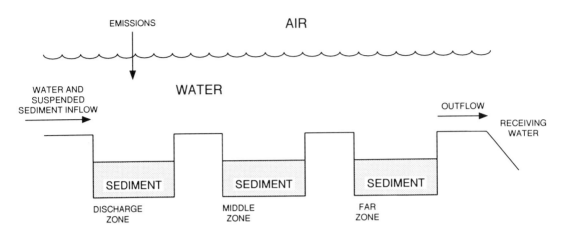

Figure 2. Three QWASI segments in series.

It is assumed that each organism achieves steady state with its environment and that the chemical is not appreciably metabolized. The fugacity of pelagic organisms is assumed to equal the fugacity of the water column. The fugacity of benthic organisms is assumed to equal the fugacity of sediment, i.e., in these cases equilibrium applies.

Figure 3 is a map of Lake Saimaa showing the mill and the sampling stations. Stations 1 and 2 are upstream of the mill and are used to collect data on background concentrations. Stations 3, 4, 5, 5b, 6, 7, 8, and 9 are downstream of the mill.

For modeling purposes, the lake is divided into three segments: stations 3, 4, and 5 are representative of the first segment; stations 5b, 6, and 7 form the second segment; and stations 8 and 9 are in the third segment. The effluent from the paper mill flows into a segment that is termed the discharge zone while the other segments are termed the middle zone and the far zone, respectively (see Fig. 2). Justification for the selection of three segments is largely based on monitored concentrations from stations within each segment being similar. In principle, it is preferable to reduce the number of segments to the minimum consistent with obtaining a reasonable simulation of concentration differences.

For illustrative purposes in this paper, one chemical is modeled: 2,4,6-trichlorophenol (TCP). From work undertaken by Mackay *et al.* (1994), Ma *et al.* (1993), and Suntio *et al.* (1988), values for various physico-chemical properties were obtained (Table 1).

Table 1. Physico-chemical properties of TCP.

Molecular mass (g mol^{-1})	197.45[a]
Solubility (g m^{-3})	708[b]
Vapor pressure (Pa)	1.48[c]
Log K_{OW}	3.75[c]
Half-life - water (h)	500[d]
Half-life - sediment (h)	40,000[d]
Half-life - fish metabolism (h)	700[e]

a - Mackay *et al.* (1994)
b - Ma *et al.* (1993)
c - Suntio *et al.* (1988)
d - estimated from Howard *et al.* (1991)
e - estimated

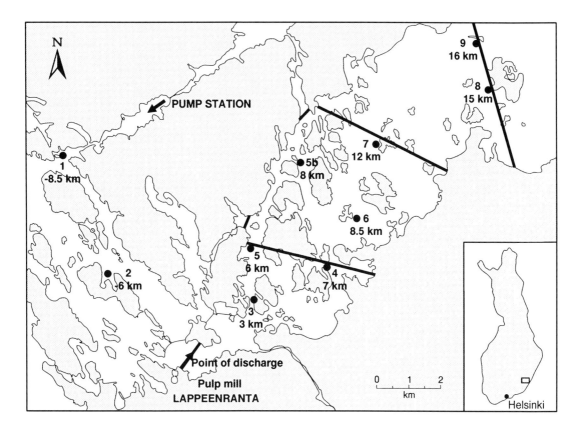

Figure 3. The Lake Saimaa system, segmented for modeling.

The Lake Saimaa system comprises a point source loading of chemical flowing into a lake. Three lake segments are characterized as shown in Table 2. In Table 2 are listed some of the properties of the mill effluent. Aside from the values described as "estimated", data in Table 2 were obtained in 1990 and 1991. Full details are presented elsewhere (Solmasuo *et al.* 1994; Holmbom *et al.* 1992).

RESULTS

Figure 4 is a mass balance for TCP based on preliminary results from the model. On a segment-by-segment basis, important processes are sediment deposition, sediment resuspension, and outflow. Examination of the overall mass balance for this chemical (Fig. 5) reveals that outflow removes 27% of the chemical from the overall system but reaction in water removes 67% of the chemical from the system. It is thus apparent that a simple model describing the dilution of a conservative substance could not reproduce the observed concentrations. Reaction is a major pathway. In the interests of clarity only, net flows of chemicals are shown in Fig. 4, but the model output contains estimates of all processes depicted in Fig. 1.

Figures 6, 7, and 8 illustrate comparisons of measured and modeled concentrations of TCP in water, sediment, and forage fish, respectively. Error bars for measured results are equal to ±2 standard deviations.

Table 2. Properties of the Lake Saimaa system and of mill effluent.

Lake Saimaa System	Segment 1	Segment 2	Segment 3
Water volume (m^3)	40,000,000	75,000,000	65,000,000
Water depth (m)	5	5	5
Sediment depth - active (m, estimated)	0.005	0.005	0.005
f_{oc}[a] in water column particles	0.2	0.2	0.2
f_{oc}[a] in sediment particles (estimated)	0.6	0.6	0.6
River water flow (m^3 h^{-1})	144,000	—	—
TCP concentration in water[b] (µg L^{-1})	0.2 ± 0.04[c]	0.1 ± 0.03[d]	0.07 ± 0.02[d]
TCP concentration in sediment[b] (µg g^{-1}, dry weight)	0.8 ± 0.2	0.6 ± 0.1	0.6 ± 0.1
TCP concentration in fish[b] (µg g^{-1} lipid)	3 ± 2[e]	1 ± 0.6[e]	0.5 ± 0.4[e]
Mill effluent			
Concentration of TCP (g m^{-3})	0.0071		
Flow rate (m^3 h^{-1})	5,400		

a - f_{oc} refers to the fraction of organic carbon in the particles
b - arithmetic mean ±1 standard deviation
c - 12 samples
d - 15 samples
e - 3 samples

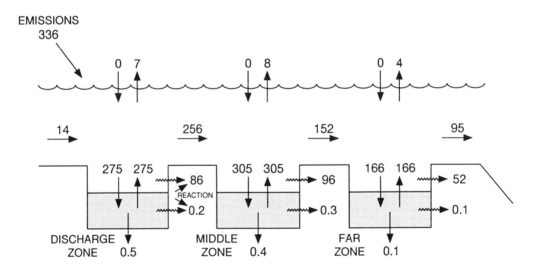

Figure 4. Mass balance for TCP (kg y^{-1}).

DISCUSSION

Given chemical-specific and site-specific data, this model can be used to translate loadings into ecosystem concentrations. The model is very successful in predicting the water column and forage fish concentrations, with all predictions lying in the range of observations. The sediment data are not as well predicted but all predictions are within a factor of two to four of the mean of the measurements. It is possible that the sediment half-lives are in error.

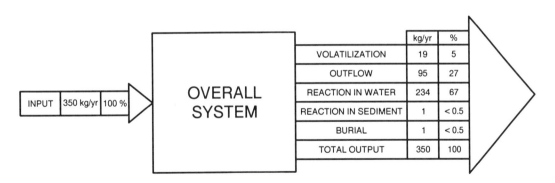

Figure 5. Overall mass balance for TCP.

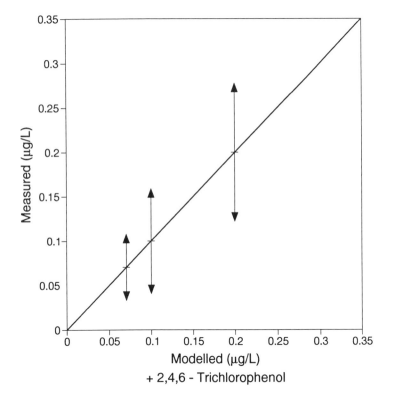

Figure 6. Measured and modeled concentrations of TCP in water.

There is considerable uncertainty about the rates of sediment-water exchange and the rate of degradation in the sediment. It is probable that sorption to sediment is underestimated, explaining the low estimates of sediment concentration. Other complications are the possibility that TCP is formed by reactions from other chlorinated phenolic compounds, that sediment-water exchange can be episodic in nature as a result of storms or high water flow which increases resuspension, and finally that actual conditions are not steady state but represent conditions during a response to changes in mill process. The model is incapable, in its present form, of describing the fate of degradation products.

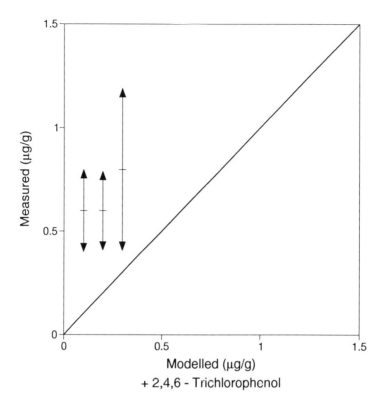

Figure 7. Measured and modeled concentrations of TCP in sediment.

In a "real life" situation, model results would enable the user to focus analysis away from compartments in which concentrations are likely to be very small and below detection levels and, therefore, optimize monitoring programs by avoiding excessive "non-detects". It should be noted that data on the reaction rates in both sediment and water are sparse. To model environmental fate more accurately, further experimentation is needed to characterize chemicals that are present in pulp mill effluents.

The results from this preliminary modeling exercise are encouraging and suggest that the approach illustrated here is capable of providing the quantitative link of loadings to concentration discussed earlier. The model is especially valuable for determining environmental degradation rates in water and sediment. Indeed, it is possible that a model such as this is the only method by which environmentally relevant half-lives can be ascertained. Once successfully applied to a few chemicals of different properties and fate, it can be applied to other chemicals in a predictive mode. Efforts are currently under way in this regard. We hope that further data will become available, especially in Canada, in order that there can be greater assurance that contaminants produced by the pulp and paper industry are adequately assimilated by aquatic ecosystems with no adverse effects.

ACKNOWLEDGMENTS

The authors would like to thank the Natural Sciences and Engineering Research Council of Canada, the Ontario Ministry of Universities and Colleges' University Research Incentive Fund, and the Pulp and Paper Centre at the University of Toronto for funding this work. Special thanks goes to Dr. D.W. Reeve and the industrial members of the Centre's Research Consortium on "Characterization, Treatment, and Fate

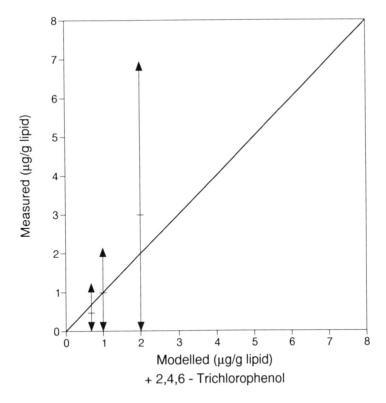

Figure 8. Measured and modeled concentrations of TCP in forage fish.

of Bleach Plant Effluents": Aracruz Cellulose S.A., Boise Cascade Corporation, Champion International Corporation, Georgia-Pacific Corporation, International Paper, James River Corporation, Kymmene Corporation, Nippon Paper Industries Co. Ltd., Potlatch Corporation, Sterling Pulp Chemicals Ltd., and Weyerhaeuser Company. The authors are grateful to Professor Bjarne Holmbom of the University of Åbo Akademi and Professor Aimo Oikari of the Technical University of Helsinki for their generosity in sharing their data and to Brett Ibbotson, Grant Piraine, and Sean Rayman of Angus Environmental Limited for their assistance.

REFERENCES

Holmbom, B., J. Hemming and P. Mäki-Arvela. 1992. Environmental fate of effluent components from the Kaukas pulp and paper mill in South Saimaa lake system, p. 39-52. In M. Viljanen and S. Ollikainen [ed.] Saimaa-seminaari 1992 Tutkimus Saimaalla. Publications of Karelian Institute, No. 103. University of Joensuu, Joensuu, Finland.

Holysh, M., S. Paterson, D. Mackay and M.M. Bandurraga. 1986. Assessment of the environmental fate of linear alkylbenzenesulphonates. Chemosphere 15(1):3-20.

Howard, P.H., R.S. Boethling, W.F. Jarvis, W.M. Meylan and E.M. Michalenko. 1991. Handbook of environmental degradation rates. Lewis Publishers, Chelsea, Michigan. 725 p.

International Joint Commission. 1994. Seventh biennial report on Great Lakes water quality. Windsor, Ontario.

Ma, K.C., W.Y. Shiu and D. Mackay. 1993. Aqueous solubilities of chlorinated phenols at 25°C. Journal of Chemical & Engineering Data 38(3):364-366.

Mackay, D. 1989. Modeling the long term behaviour of an organic contaminant in a large lake: Application to PCBs in Lake Ontario. J. Great Lakes Res. 15:283-297.

Mackay, D. 1991. Multimedia environmental models: the fugacity approach. Lewis Publishers, Chelsea, Michigan. 257 p.

Mackay, D. and J.M. Southwood. 1992. Modeling the fate of organochlorine chemicals in pulp mill effluents. Water Poll. Res. J. Canada 27(3):509-537.

Mackay, D., M. Joy and S. Paterson. 1983. A quantitative water, air, sediment interaction (QWASI) fugacity model for describing the fate of chemicals in lakes. Chemosphere 12(7/7):981-997.

Mackay, D., W.Y. Shiu and K.C. Ma. 1994. Illustrated handbook of physical-chemical properties and environmental fate for organic chemicals. Volume IV: Oxygen, nitrogen, and sulfur-containing compounds. Lewis Publishers, Chelsea, Michigan.

Solmasuo, R., I. Jokinen, J. Kukkonen, T. Petänen, T. Ristola and A. Oikari. 1995. Biomarker responses along a pollution gradient: effects of pulp and paper mill effluents on caged whitefish. Aquatic Toxicology 31:329-346.

Southwood, J.M., R.C. Harris and D. Mackay. 1989. Modeling the fate of chemicals in an aquatic environment: the use of computer spreadsheet and graphics software. Environmental Toxicology and Chemistry 8(11):987-996.

Suntio, L.R., W.Y. Shiu and D. Mackay. 1988. A review of the nature and properties of chemicals present in pulp mill effluents. Chemosphere 17(7):1249-1290.

LOAD FROM THE SWEDISH PULP AND PAPER INDUSTRY (NUTRIENTS, METALS AND AOX): QUANTITIES AND SHARES OF THE TOTAL LOAD ON THE BALTIC SEA

Magnus Enell

Swedish Environmental Research Institute (IVL), P.O. Box 21060, S-100 31 Stockholm, Sweden

The Baltic Sea is one of the most polluted sea areas in the world, with overall eutrophication and oxygen depletion symptoms and local ecological effects of metals and persistent organic compounds. The drainage basin is shared by 14 countries with a total population of 82 million. A variety of anthropogenic activities have resulted in a nutrient and metal load that must be significantly decreased in order to reach an acceptable level for a sustainable Baltic Sea. International agreements, stating a 50% reduction of the nutrient load during the period 1987 to 1995, have been signed. For some metals, a 70% reduction has been agreed. To achieve these goals a comprehensive action program must be conducted. This means that the load from different sources must be quantified and evaluated to arrive at ecologically sound, reliable and cost-effective measures for reducing pollution from different countries. The Swedish pulp and paper industry annually discharges 400 and 3,100 tonnes of total phosphorus (TP) and total nitrogen (TN), respectively. These quantities represent 10 and 2%, respectively, of the total Swedish nutrient load to the Baltic Sea. The total load from all countries and atmospheric deposition on the sea area is 64,000 tonnes TP y^{-1} and 1,528,000 tonnes TN y^{-1} (also including nitrogen fixation). The critical load of phosphorus and nitrogen to the Baltic Sea is estimated to be 30,000 and 600,000 tonnes y^{-1}, respectively. The metal load from the Swedish pulp and paper industry is: zinc 95, cadmium 0.6, copper 7, and lead 4 tonnes y^{-1}. No critical load for metals has been defined. One of the major metal sources for the Baltic Sea is the atmospheric deposition on the sea surface. This means that to control nitrogen and metal loading, comprehensive measures must be taken to decrease emissions to the atmosphere, not only in the drainage area but also worldwide. The load of halogenated organic compounds (analyzed as AOX) from Sweden is significantly lower today (1993) than in previous years. The AOX load from the Swedish pulp and paper industry was ~1,700 tonnes in 1993, compared to ~14,000 tonnes in 1988.

INTRODUCTION

Environmental conditions in waters around the coasts of the Baltic Sea are presently the focus of environmental research, monitoring, and national and international measures. Major remediation efforts are expected in the drainage area to fulfill international agreements. The Baltic Sea Environmental Declaration was signed in September 1990 by the heads of the governments and high political representatives of the Baltic Sea states, the Czech and Slovak Federal Republics, the Kingdom of Norway and the representatives of the European Community. The declaration expresses the firm determination of the parties to:
  "assure the ecological restoration of the Baltic Sea, ensuring the possibility of self-restoration of the marine environment and preservation of the ecological balance".
Further, the Declaration calls for the endorsing parties to:
  "urgently prepare a joint comprehensive programme for decisive reduction of emissions in order to restore the Baltic Sea to a sound balance. The programme shall be based on concrete national plans provided by the countries concerned".

To achieve its objectives, the declaration calls for the establishment of a high level ad hoc task force under the auspices of the Helsinki Commission (HELCOM). A series of pre-feasibility studies of priority areas have been initiated to provide information concerning specific areas and to form a basis for later investment programs. The pre-feasibility studies are also prepared in order to give an overall view of the technical and economical actions that are required within the Baltic Sea drainage area to reduce the environmental pollution to the waters and the air of the Baltic Sea. A joint comprehensive environmental action program was presented and unanimously adopted by the task force.

The objective of this paper is to compile 1993 emission data for the Swedish pulp and paper industry, compare these quantities with emissions from other Swedish sources and to put the information into an international perspective (the Baltic Sea). The production of the Swedish pulp and paper industry in 1993 was 20,000,000 tonnes.

The Baltic Sea

The Baltic Sea is a semi-enclosed, shallow sea area. The area consists of the Gulf of Bothnia, the Gulf of Finland, the Gulf of Riga, the Baltic proper, the Sound and the Belt Sea (the Danish Straits), and the Kattegat (Fig. 1). The drainage area of the Baltic Sea is about 2,150,000 km^2, of which about 415,000 km^2 (19%) consists of sea surface, about 540,000 km^2 (25%) of agricultural land and about 710,000 km^2 (33%) of forest areas.

Nine countries share the Baltic Sea coastline: Sweden, Finland, Russia, Estonia, Latvia, Lithuania, Poland, Germany and Denmark. The drainage area also includes parts of Belarus, Norway, Ukraine, and the Czech and Slovak Federal Republics. About 16 million people live along the coast and around 82 million in the entire drainage area of the Baltic Sea (Helsinki Commission 1992).

The mean depth of the Baltic Sea is only 55 m. In the Baltic proper some deep basins are found (maximum 459 m), each separated by shallow thresholds. As a consequence of limited salt water inflow and a large drainage area, the Baltic Sea water is dominated by fresh water input, which causes layering of the water and a well-defined halocline. The salinity of the surface waters increases from the Gulf of Bothnia southward and out towards Kattegat. The difference in density between surface and bottom waters restricts the exchange between the two layers. This results in semi-stagnant conditions over long periods when little or no oxygenated, more saline water flows in from the North Sea. The intermittent salt water inflow from the North Sea is primarily determined by anemobaric conditions over northern Europe and the North Sea. During this century, major inflows have occurred approximately every 11 years, but have recently been less frequent.

The brackish water of the Baltic Sea plays an important role in its ecology. The water circulation in the Baltic Sea is restricted. Surface water movements are most affected by winds, a significant factor in water mixing and distribution of pollutants. In winter, the Baltic Sea is partly ice-covered, which renders it even more vulnerable to pollution.

The Baltic Sea has always been characterized by the interaction of fresh and salt water sources. The Baltic Sea is a very young sea, which developed after the last deglaciation, about 10,500 years ago. At that time, the Baltic Sea was a large fresh water lake (the Baltic Ice Lake). About 500 years later, sea water from the present North Sea changed the ecosystem to marine conditions (the Yoldia Sea). The land uplift resulted in an isolation of the Yoldia Sea from the marine area in the west - a fresh water lake again developed about 8,000 years ago (the Ancylus Lake). Since the water level rise during this period was faster than the rise of the land, a new marine ecosystem was created (the Littorina Lake), caused by salt water inflow from the North Sea. The greatest extent of this period occurred about 4,500 years ago. The Baltic Sea of today belongs to this latest development stage of the ecosystem.

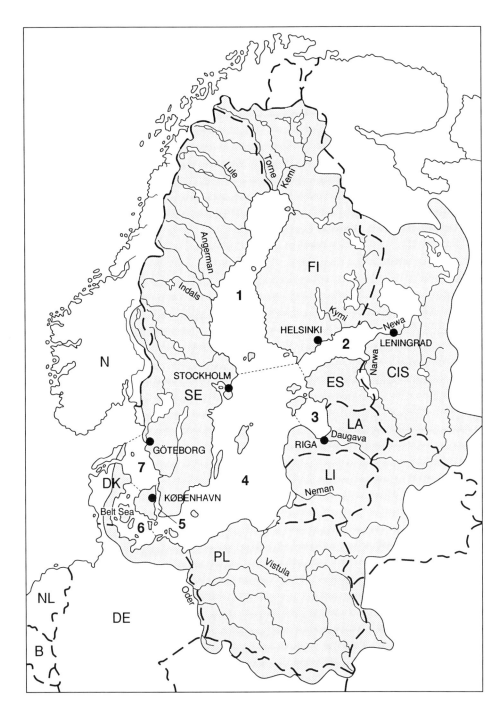

Figure 1. The drainage area of the Baltic Sea. *1) Gulf of Bothnia, 2) Gulf of Finland, 3) Gulf of Riga, 4) Baltic proper, 5) The Sound, 6) Belt Sea and 7) Kattegat.*

RESULTS AND DISCUSSION

Results presented for the Swedish load to the surrounding sea include both direct discharges to the sea and discharges to inland waters (lakes and rivers). For the discharges to inland waters, retentions coefficients (% of the load) have been used: 25% for phosphorus and 50% for nitrogen. For zinc, cadmium, copper, lead and AOX there are no retention coefficients available, and therefore 0% retention has been used.

Nutrients: Phosphorus and Nitrogen

Large loads of nutrients, year after year, have resulted in increased concentrations of phosphorus and nitrogen in all basins of the Baltic Sea. Since most of the nutrients are exported from land to the sea, concentrations are higher in coastal waters than in the open sea. Eutrophication can be regarded as the major environmental problem of the Baltic Sea, followed by local environmental problems caused by metals and persistent organic substances.

In 1993 the load of total phosphorus (TP) and total nitrogen (TN) from Sweden to the sea surrounding Sweden was 4,200 and 132,600 tonnes (Fig. 2), respectively (SCB 1994). These quantities include natural sources (background levels). Despite advanced treatment (P removal) of municipal sewage

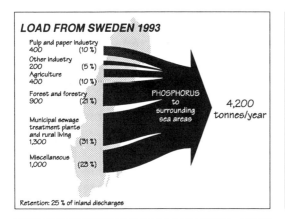

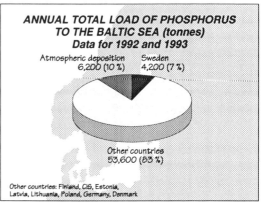

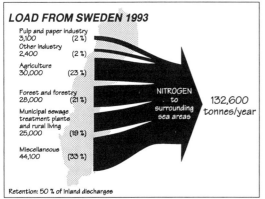

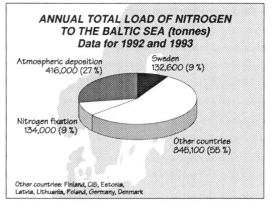

Figure 2. Load (tonnes y^{-1}) of phosphorus (TP) and nitrogen (TN) from Sweden in 1993 to the sea surrounding Sweden (the Baltic Sea), divided into different sources, and the total load (tonnes y^{-1}) to the Baltic Sea, divided into Sweden (1993) other countries (1992) and atmospheric deposition (1992). Source: Naturvårdsverket (1994) and SCB (1994).

water, the TP load from municipal sewage treatment plants and rural households was the dominant source (31% of the total load). The TN load from municipal and rural households was 19% of the total load. This source for nitrogen will decrease since a comprehensive program for nitrogen removal (by denitrification) is now being implemented at a number of sewage treatment plants along the Swedish coast.

Nutrient leaching from agriculture is another important source of nutrients to the Baltic Sea. The quantities of TP and TN, from Swedish agriculture are estimated to be 400 and 30,000 tonnes, respectively. Forest and forestry are responsible for about 900 tonnes of TP and 28,000 tonnes of TN. In 1993 the Swedish pulp and paper industry discharged 400 tonnes of TP and 3,100 tonnes of TN (Naturvårdsverket 1994), representing 10 and 2%, respectively, of the total Swedish load. At present the annual load of TP and TN to the Baltic Sea (Enell 1992b) is ~64,000 and ~1,528,000 tonnes (Fig. 2), respectively. The Swedish share of the load is 7 and 9% for TP and TN, respectively. Atmospheric deposition of phosphorus and nitrogen is significant: 6,200 tonnes of TP and 416,000 tonnes of TN.

With regard to the entire Baltic Sea and its drainage basin, measures to reduce the phosphorus load must focus on sewage water purification for municipalities and rural households. Measures to reduce nitrogen must focus on agriculture. Atmospheric nitrogen deposition must also be reduced, by reducing emissions of NO_x from traffic and combustion, and reduced NH_4 emission from agriculture.

Critical load of nutrients on the Baltic Sea has been estimated at 30,000 tonnes of TP and 600,000 tonnes of TN per year (Miljödepartementet 1990). These loads are assumed to be on the same order as those during the 1950s. To accomplish this, the present loads of TP and TN must be reduced by 55 and 60%, respectively.

Metals: Zinc, Cadmium, Copper and Lead

Critical concentrations and loads for metals have not been defined. The anthropogenic load of metals to the Baltic Sea is derived from many sources, such as discharges from industries and municipal sewage treatment plants, atmospheric deposition and input from inland sources via rivers. The data on metal loads may be unreliable in some cases, but the quantities can be regarded as reasonable estimates rather than an accurate reflection of the situation. Analytical difficulties may be the main reason for difficulties in obtaining accurate data.

The major metal source for the Baltic Sea is atmospheric deposition directly to the sea surface. This means that comprehensive measures must be taken to decrease emissions to the atmosphere, not only in the drainage area but also worldwide.

The total load of metals from Sweden (SEPA 1993; SCB 1994) is zinc 1,350, cadmium 2.9, copper 265 and lead 47 tonnes y^{-1} (Fig. 3). The share of the total Swedish load from the pulp and paper industry is 7% for zinc, 21% for cadmium, 3% for copper and 9% for lead (Enell *et al.* 1992). The major source of metals from industrial activities is leaching from mine wastes (SEPA 1993). Another important source is metals borne by rivers and it must be kept in mind that a large proportion of pollutants transported via rivers originate from municipal sewage treatment plants and industries located inland. The proportion of metals naturally transported via water courses to the sea must be estimated to obtain accurate figures for the anthropogenic quantities.

The Swedish load of cadmium and lead to surrounding sea areas is relatively small (Fig. 3) compared to loads from other countries and atmospheric deposition. Concerning the load of zinc and copper (Fig. 3) the Swedish shares are comparatively high, but small compared to the atmospheric deposition quantities.

Adsorbable Organic Halogens

The quantities of various organic substances that end up in the Baltic Sea are unknown. Toxic substances are often persistent and many are halogenated. Many of these substances are soluble in fat and bioaccumulate. Since they are not broken down, they also undergo biomagnification, i.e., concentrations build up to higher levels in higher animals (top predators). The Baltic Sea is particularly vulnerable to toxic organic pollutants released by anthropogenic activities, since the Baltic Sea is the final destination of discharges and land runoff from a cluster of highly industrialized countries. Moreover, it is a cold sea, with a short reproductive season.

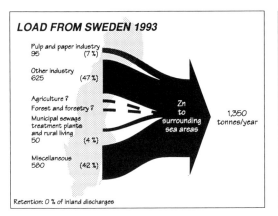
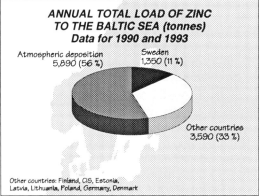
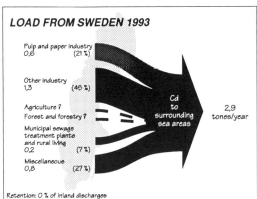
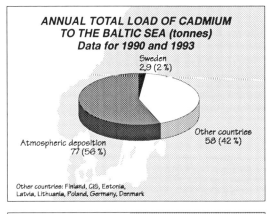
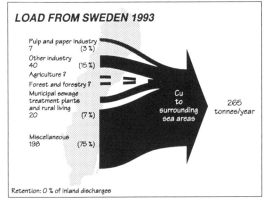
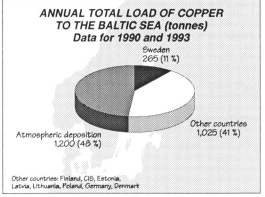

Figure 3. Load (tonnes y^{-1}) of zinc, cadmium, copper and lead from Sweden in 1993 to the sea surrounding Sweden (the Baltic Sea), divided into different sources, and the total load (tonnes y^{-1}) to the Baltic Sea, divided into Sweden (1993), other countries (1990) and atmospheric deposition (1990). Source: Naturvårdsverket SEPA (1993) and SCB (1994).

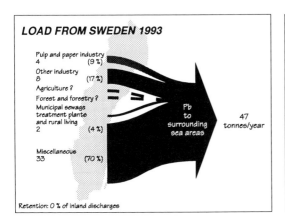
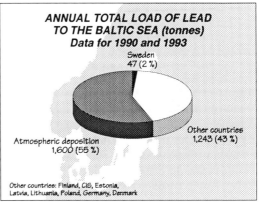

Figure 3. Continued - Load (tonnes y^{-1}) of zinc, cadmium, copper and lead from Sweden in 1993 to the sea surrounding Sweden (the Baltic Sea), divided into different sources, and the total load (tonnes y^{-1}) to the Baltic Sea, divided into Sweden (1993), other countries (1990) and atmospheric deposition (1990). Source: Naturvårdsverket SEPA (1993) and SCB (1994).

AOX has been regarded as a suitable environmental parameter for describing the load of halogenated organic compounds resulting from the bleaching processes in the pulp and paper industry. However, this is not accurate since AOX in nature can come from natural and anthropogenic sources. The AOX loads presented cannot be used for describing and evaluating ecological effects (Enell 1992a), but may be used as a basis for the follow-up of international agreements concerning reduction of external loads of AOX.

The total Swedish load in 1993 of AOX to the sea surrounding Sweden was ~7,300 tonnes (Fig. 4). Of this, 1,700 tonnes came from the pulp and paper industry. About 2,500 tonnes are of natural origin (Enell 1992a) and 500 tonnes from atmospheric deposition on inland waters. The Swedish pulp and paper industry has significantly decreased the load of AOX from ~14,000 tonnes in 1988 (Naturvårdsverket 1994) to 1,700 tonnes in 1993.

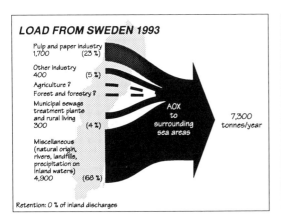
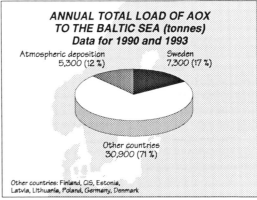

Figure 4. Load (tonnes y^{-1}) of AOX from Sweden in 1993 to the sea surrounding Sweden (the Baltic Sea), divided into different sources, and the total load (tonnes y^{-1}) to the Baltic Sea, divided into Sweden (1993), other countries (1990) and atmospheric deposition (1990). Source: Naturvårdsverket SEPA (1994) and Enell (1992a).

The total AOX load to the Baltic Sea is ~38,600-48,300 tonnes (Enell 1992a), of which about 5,300 tonnes are from atmospheric deposition directly on the sea (Fig. 4). The quantity of AOX in the entire Baltic Sea water volume is ~240,000 tonnes (Enell 1992a).

CONCLUSIONS

In western Europe, awareness of environmental issues has increased and measures to reduce loads from point and non-point sources have been undertaken for more than 30 years. In Sweden, easily identified industrial and municipal pollution sources have been markedly reduced by advanced treatment technology and the water environment close to pollution sources has consequently improved. However, in eastern Europe the environmental situation has in most cases deteriorated and a program for introducing pollution reduction measures is dependent on financing from Western countries.

Measures to reduce pollution must focus on point and non-point sources and consider economy, ecology and technology both for the sources in the country and in other countries. It is important to discuss the environmental situation and the impact of various pollution reduction measures on two levels: coastal systems and open sea areas (Enell and Haglind 1994).

The Swedish pulp and paper industry has conducted comprehensive measures to decrease emissions of pollutants to values acceptable both for the environment and for authorities. The load of nutrients, metals and AOX from the Swedish pulp and paper industry can be considerable locally, but on a national and international perspective, the loads are small. Before further comprehensive emission reduction measures for nutrients and metals are imposed on the pulp and paper industry, accurate information concerning all discharge sources must be compiled and evaluated.

Environmental effects of increased nutrient levels are well known and documented. However, environmental and biological effects of increased metal concentrations are most often lacking, which means that it is not yet possible to define appropriate ecologically based emission reductions. An integrated national and international strategy, with clearly defined priorities and responsibilities, is urgently needed. Such an approach is possible only if an understanding of ecological and economical processes is used to implement a policy that will lead to a sustainable society and development.

ACKNOWLEDGMENTS

The compilation of up-to-date loads for Sweden and countries sharing the Baltic Sea drainage basin was financed by the Swedish Environmental Research Institute (IVL) and the Swedish Forest Industries Water and Air Pollution Research Foundation (SSVL Miljö 93). The author is also grateful to Jonas Fejes at IVL for his excellent work with preparation of illustrative figures.

REFERENCES

Enell, M. 1992a. AOX loadings on sea areas surrounding Sweden - quantities and origins of loadings, p. 57-67. *In* A. Södergren (ed.). Environmental fate and effects of bleached pulp mill effluents. Proceedings of a SEPA conference held at Grand Hotel Saltsjöbaden, Sweden, November 19-21, 1991. SEPA Report 4031.

Enell, M. 1992b. Pollution problems in the coastal areas. Eutrophication and toxic substances. Example: the Baltic Sea. Sustainable development after Rio. Environmental week in the Swedish Pavillion at EXPO '92. Sevilla, Spain, August 24-31, 1992.

Enell, M. and I. Haglind. 1994. Nitrogen, phosphorus and metal loads from Swedish pulp and paper industry on the Gulf of Bothnia - quantities and shares of total loads. Wat. Sci. Tech. 29(5-6):49-59.

Enell, M., J. Henriksson and L. Wennberg. 1992. Massa- och pappersindustrins utsläpp av närsalter, metaller, AOX, SO_2 och NO_x - enskilt och i relation till andra utsläpp. SSVL, Miljö 93, Report no. 13.

Helsinki Commission. 1992. The Baltic Sea Joint Comprehensive Environmental Action Programme (Preliminary Version). Diplomatic Conference on the Protection of the Marine Environment of the Baltic Sea Area. Helsinki, Finland. April 9, 1992. Conference Document No. 5/3, Agenda Item 5.

Miljödepartementet. 1990. Vem förorenar Sverige? Underlagsrapport till utredningen om Sveriges internationella miljösamarbete. Allmänna Förlaget.
Naturvårdsverket (SEPA). 1994. Skogsindustrins utsläpp till vatten och luft. Report 4348.
SCB. 1994. Belastningen på havet 1992 av fosfor, kväve, metaller och organiskt material. No. 29 SM 9401.
Swedish Environmental Protection Agency (SEPA). 1993. Metals and the environment. The environment in Sweden - status and trends. Report 4245.

DEGRADATION OF THE HIGH MOLECULAR WEIGHT FRACTION OF BLEACHED KRAFT MILL EFFLUENT BY BIOLOGICAL AND PHOTOCHEMICAL PROCESSES

Kelly R. Millar[1], John H. Carey[1], B. Kent Burnison[1],
Hung Lee[2] and Jack T. Trevors[2]

[1]National Water Research Institute, P.O. Box 5050, Burlington, Ontario, L7R 4A6 Canada
[2]Department of Environmental Biology, University of Guelph, Guelph, Ontario, N1G 2W1 Canada

The degradation of high molecular weight (HMW) chlorinated organic material isolated from bleached kraft mill final effluent was studied using fungal strains and sunlight. Naturally occurring fungi, isolated from river biofilms, wood, and sediment samples, and two white-rot fungi, *Trametes versicolor* and *Phanerochaete chrysosporium*, were capable of degrading the HMW fraction of bleached kraft mill effluents, resulting in reductions in effluent color, adsorbable organic halogen (AOX), and molecular weight. Size distribution analysis indicated that both high and low molecular weight materials were degraded equally. Photolysis experiments revealed that filter-sterilized HMW material could be photodegraded under natural sunlight conditions, resulting in substantial losses in AOX, color, and molecular weight. Half-lives for AOX mineralization and decolorization were estimated at 10 and 21 d, respectively, during fall months. Irradiation also resulted in the release of chlorinated guaiacols and vanillins. Pretreatment of HMW chlorinated organic material with sunlight did not enhance subsequent fungal degradation.

INTRODUCTION

Prior to 1992, production of bleached pulp in Canada resulted in the annual discharge of over one million tonnes of chlorinated organic waste into the environment (CEPA 1991). Although modern bleaching processes have substantially reduced organochlorine emissions, pulp mills are still significant sources of organically bound chlorine to the environment. The chemical composition of bleach plant effluents is complex and varies between pulp mills that use different wood furnish and bleaching sequences. In mills using older bleaching sequences, over 200 low molecular weight (LMW) chlorinated organic compounds have been identified in bleachery effluents (Gergov et al. 1988; Suntio et al. 1988). More than 80% of the organically bound chlorine, however, is present in high molecular weight (HMW) materials (>1,000 Da) (Kringstad and Lindström 1984).

Although most Canadian bleached pulp mills employ secondary biological treatment to reduce effluent toxicity and the biological oxygen demand associated with dissolved organic materials in the effluent, secondary effluent treatment fails to remove HMW chlorinated organic material and, as a result, large quantities are discharged into the receiving waters.

Increasing concern over possible environmental impacts of chlorinated lignin derivatives in bleach plant effluents on aquatic communities has led to more stringent regulation of mill emissions and research into methods of reducing the organochlorine content of effluents by improved in-plant processes and effluent treatment practices. In the meantime, the fate of HMW chlorinated organic compounds in the receiving environment remains unknown. Factors that may contribute to the removal of these compounds from aquatic systems include sorption, hydrolysis, photolysis, and biological processes. This research focuses on the latter two mechanisms by examining the degradative potential of HMW chlorinated organic material by naturally occurring fungi and sunlight.

MATERIALS AND METHODS

Effluent Collection and Fractionation

Studies were conducted using effluents from a Canadian kraft mill that produced bleached pulp from primarily softwood fibers, using a bleaching sequence of C_DE_ODED, with 50% ClO_2 substitution. Effluent treatment consisted of primary clarification only. Whole mill final effluent was collected at mill discharge using a submersible pump connected to a continuous flow centrifuge (Westfalia, Firing Industries, St. Catherines, ON), to reduce suspended solids. Particulates were further reduced by pressure filtration through 1-µm precombusted glass fiber filters. Filtered effluent was concentrated on site using a maple sap concentrator (Les Equipements Lapierre, St. George de Beauce, PQ), modified to contain a nanofiltration membrane with a MW cutoff of 400 Da (type NF40-40, FilmTec Corp. Minneapolis, MN) (Burnison et al. 1991). Under optimal conditions, approximately 1,000 L of effluent was concentrated per day to a final volume of 20 L.

In the laboratory, field samples were further purified by repeated dilution (6×) and re-concentration by nanofiltration. Samples were diluted 5× using distilled H_2O. The final concentrate was further fractionated by tangential flow filtration using an ultrafiltration membrane having a nominal MW cutoff of 1,000 Da (Millipore Corp., Bedford, MA). The >1,000-Da fraction was diluted and re-concentrated 3× with 10 L organic-free H_2O. Concentrates were stored at $-20°C$ in 1-L polyethylene bottles until needed. As an additional purification measure prior to use, concentrates were dialyzed against 20 L organic-free H_2O for 24 h using dialysis tubing with a MW cutoff of 1,000 Da (Spectrum Medical, Los Angeles, CA).

Sunlight Exposure

Dialyzed HMW (>1,000 Da) material was diluted using organic-free H_2O to original final effluent concentration based on DOC analyses (approx. 200 mg C L^{-1}). The pH of the diluted material was 6.6 and was not adjusted. Diluted material was filter sterilized by successive passage through cellulose acetate filters of 0.8-, 0.45-, and 0.2-µm pore size, then dispensed into 50-mL borosilicate glass tubes with Teflon-lined screw caps. Half of the tubes were covered with foil to protect the contents from solar radiation and serve as experimental controls. Exposure took place over a 21-d period during the months of September and October in an open, shade-free area. The tubes were placed in racks, such that each tube could receive exposure from all angles. The material remained outdoors for 24 h d^{-1} for the duration of the experiment except for two evenings when frost was predicted. To prevent freezing, samples were stored inside at 4°C, in the dark, between the hours of 17:00 and 07:30.

Incoming short-wave solar radiation was measured using an Eppley precision pyranometer (Model 2, Eppley Laboratory Inc., Newport, RI) placed next to the samples. Measurements were recorded at 5-s intervals with totals calculated every 10 min.

Photodegradation was measured by changes in AOX, color, MW distribution, and production of LMW chlorophenolic compounds. Analyses were performed in triplicate. On day 21, sunlight exposure was terminated and all remaining tubes were recovered and stored at 4°C, in the dark, for subsequent use in biodegradation studies.

Microorganisms

River samples consisting of water, wood chips, and biofilm were collected downstream from effluent discharge, as microbial sources. Fungal strains were isolated by enrichment culture techniques using filter-sterilized HMW material as a carbon source, at a final concentration of 50 mg C L^{-1}. Cultures were monitored microscopically on a regular basis with serial transfers to fresh media made at 14-d intervals. Individual strains were isolated by plating 100-µL culture aliquots onto malt extract agar (MEA) and potato dextrose agar (PDA) media (Difco Laboratories, Detroit, MI).

Fungal isolates were screened for wood-degrading enzyme activity (cellulase, laccase, peroxidase) by the methods of Rohrmann and Molitoris (1992) and Smith (1977). Strains with strong activities were then analyzed for the ability to degrade the HMW material.

As positive controls, the white-rot fungi *P. chrysosporium* (ATCC #20696) and *T. versicolor* (kindly supplied by F. Archibald, PAPRICAN, Pointe Claire, PQ) were included in enzyme screening tests and subsequent biodegradation experiments.

Biodegradation Studies

Sunlight-exposed and non-exposed HMW material was tested for biodegradability using the river isolates and two white-rot strains. A defined medium, based on that used by Addleman and Archibald (1993) for pulp bleaching, consisting (L^{-1} medium) of glucose (10 g), asparagine (0.26 g), KH_2PO_4 (0.68 g), $MgSO_4 \cdot 7H_2O$ (0.25 g), $CaSO_4 \cdot 2H_2O$ (15 mg), thiamine $\cdot$ HCl (1 mg), Tween 80 (0.5 g), $FeSO_4 \cdot 7H_2O$ (0.56 mg), $CuSO_4 \cdot 5H_2O$ (0.25 mg), $MnSO_4 \cdot H_2O$ (1.69 mg), $ZnSO_4 \cdot 7H_2O$ (1.44 mg), $CoCl_2 \cdot 6H_2O$ (1.19 mg), $(NH_4)_6Mo_7O_{24} \cdot 4H_2O$ (0.62 mg), was supplemented with either irradiated or non-irradiated HMW material at a concentration of 200 mg C L^{-1}. The pH of the medium was adjusted, maintained at 4.5. Cultures of 55 mL were established in 125 mL flasks fitted with foam stoppers. Through the center of each stopper a 1-mL glass pipet fitted with glass wool was inserted in order to permit culture aeration.

A nutrient medium consisting (L^{-1}) of malt extract (6.0 g), yeast extract (1.2 g), maltose (1.8 g), glucose (6.0 g), and HMW material (100 mg C, non-irradiated) was used for biomass production. Cultures of 100 mL were maintained in 250-mL baffled flasks with stainless steel closures, at 22°C on an orbital shaker at 150 rpm. The primary inocula for biomass production cultures were 5 × 1 cm agar plugs taken from the growing edge of a 5 to 7 d old culture on MEA medium. Cultures were aerated with high-purity, humidified O_2 at 100 mL min^{-1}, for 2 min every other day. Cultures were maintained in the dark at 25°C on an orbital shaker at 150 rpm. Biodegradation was measured by changes in AOX, color, and MW distribution.

Chemical Analyses

Chlorophenolic Compounds

LMW chlorinated phenolic compounds were detected and identified as their acetylated derivatives. Samples of 40 mL were acidified to pH <2 with concentrated H_2SO_4, and extracted for 1 h on a magnetic stirrer with 10 mL of hexane. Extracted chlorophenolics were converted to their acetate esters using 0.1M potassium carbonate and acetic anhydride. Extracts were concentrated to <1 mL under a gentle stream of N_2 and analyzed after solvent exchange into isooctane. Analyses were made using a Hewlett-Packard 5890 Series II gas chromatograph with a split/splitless injector, with injection split between a 30-m, 0.25-mm i.d. DB-5 column and a 30-m, 0.25-mm i.d. DB-17 column (J&W Scientific, Folsom, CA), each with its own electron capture detector. The temperature program was 90°C for 2 min, 3°C min^{-1} to 220°C, 10°C min^{-1} to 280°C, and 280°C for 10 min. Hydrogen was the carrier and argon/methane (95:5), the detector make-up gas. Injector and detector temperatures were 260 and 300°C, respectively.

Organic Carbon

Dissolved organic carbon was measured using a Beckman 915B total organic carbon analyzer (Beckman Industrial, Palo Alto, CA). Standards were prepared using potassium hydrogen phthalate. Prior to analysis, samples were acidified to pH <2 with concentrated HCl and sparged with a stream of high purity N_2 to remove inorganic carbon.

Adsorbable Organic Halogen

Organically bound chlorine was determined using a Mitsubishi TOX-10 organic halogen analyzer (Mitsubishi Kasei Corp., Tokyo, Japan) according to the method of Odendahl et al. (1990).

Color

Effluent color was measured spectrophotometrically at 465 nm using a Shimadzu UV-260 spectrophotometer (Shimadzu Corp., Kyoto, Japan). For biodegradation studies, fungal biomass was removed by centrifugation and filtration onto 0.2-μm cellulose acetate filters prior to analysis.

Molecular Size Distribution

Size exclusion chromatography was conducted using a Waters chromatography system (Millipore Corp., Waters Chromatography Division, Milford, MA) consisting of a model 600E pump, a Satellite WISP automatic injector (model 700), a UV-VIS detector at 280 nm (model 441), a differential refractometer (model 410), and a Waters Baseline chromatography workstation with GPC option software. The system was equipped with two Ultrahydrogel® columns (Waters, models 250 and 120, 6-μm particle size) connected in series, and eluted at 0.4 mL min^{-1} with a mobile phase of 1% $NaNO_2$ and 10% methanol in organic-free H_2O, pH 7. Polystyrene sulfonate and polyethylene glycol standards were used for calibration. Injection volume was 50 μL.

RESULTS AND DISCUSSION

Photodegradation Experiments

Photochemical transformation processes can play a major role in the removal of some organic contaminants from streams and rivers (Hwang et al. 1987; Carey et al. 1988; Brownlee et al. 1992). Photodegradation experiments were conducted on the HMW fraction of bleached kraft mill effluent to determine whether, under sterile conditions, this material could also undergo photochemical transformation. A 21-d exposure of this material to natural sunlight resulted in losses of 44% of the original AOX content (Fig. 1) and 31% of the effluent color (Fig. 2). Prolonged exposure would likely have resulted in further dechlorination and decolorization, since a preliminary experiment which ran over 42 d, from late June to early August, resulted in losses of 63% AOX and 43% color (Millar 1993). In both studies, no changes in AOX or color were observed in the foil-wrapped material that was protected from irradiation.

A simple correlation analysis of AOX mineralization and decolorization suggests that the two processes are related, but that the rates of reaction differ (Fig. 3). Both reactions, however, proceeded rapidly, with almost half of the total losses occurring within the first three days of exposure. To determine the reaction kinetics of photodechlorination and photodecolorization, the natural logarithms of each were plotted against solar radiation (Figs. 4 and 5). A dotted line in each figure represents a first-order plot. Neither photodechlorination nor photodecolorization appears to follow first-order kinetics as the reaction rates decreased over time. Reductions in the rates of AOX mineralization and decolorization may, in part, be explained by inner filtering effects, whereby some of the exciting light that is responsible for such photochemical reactions as AOX and color removal is absorbed by other unreactive light-absorbing materials. In addition, as the primary products of photodegradation may undergo further photochemical reactions, competition for the absorption of incoming irradiation arises with reaction products.

Based on the initial rates of photolysis, the half-lives for HMW AOX and color were estimated at 2,687 and 5,682 Langleys, respectively. With an average daily solar radiation of 265 Langleys, the half-lives in surface waters would be 10 and 21 d for dechlorination and decolorization, respectively. Caron and Reeve (1992) irradiated HMW material isolated from combined C+E-stage bleachery effluents with a xenon-arc lamp and also found significant reductions in AOX with a photolysis half-life equivalent to 35 h for a solar intensity equal to June mid-day sunlight. Caron and Reeve's (1992) calculations were based on initial dechlorination rates, as they had also observed that AOX removal did not follow first-order kinetics and that photolysis rates dropped significantly after one half-life. Decreased rates were attributed to a "recalcitrant" AOX fraction, estimated as 47% of the initial HMW, that was not photolyzable by sunlight. Related studies by Roy-Arcand and Archibald (1993) on sunlight exposed softwood and hardwood bleachery effluent revealed mineralization of 34 and 59% of the HMW AOX and 59 and 47% of the effluent color, respectively. Hardwood effluent was found to be more highly chlorinated and thus more susceptible to photodechlorination. Unlike our studies, Roy-Arcand and Archibald (1993) observed losses of AOX and color in their non-exposed hardwood controls, as much as 38 and 13%, respectively. This difference may be due to the fact that we used final effluent while Roy-Arcand and Archibald used bleachery effluent. Frostell (1988) has shown that the increase in pH that results from combining bleachery effluents results in a reduction in AOX that is likely due to hydrolysis reactions of reactive substituents. Differences in wood furnish, HMW isolation methods, dilution, pH, and incubation time may also contribute to differences observed with respect to the dark controls. In two separate studies, however, we found AOX and color to be extremely stable in the dark controls.

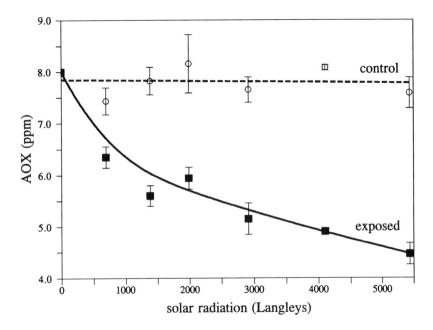

Figure 1. AOX reduction following exposure of sterile HMW material to sunlight.

Molecular size distribution analysis of HMW samples prior to sunlight exposure indicated that the HMW material contained materials within the 1,000 to 10,000 Da range, with the main peak positioned at 3,800 Da, and a shoulder peak at approximately 7,000 Da (Fig. 6A). Exposure of the material to natural sunlight led to a successive removal of the greater than 2,000 Da material. Decreases in overall peak height were accompanied by shifts in molecular size, such that the position of the main peak was at 2,800 Da after 21 d exposure (Fig. 6B). In addition, a new shoulder peak of LMW material (~1,400 Da) had appeared early in the exposure period and had successively increased in magnitude until the end of the experiment. Chromatograms representing dark controls were consistently identical throughout the experiment (Fig. 6A).

Sunlight exposure also resulted in a release of LMW chlorophenolic compounds, primarily chlorinated guaiacols and vanillins, from sterile HMW material (Fig. 7). At the end of 21 d, concentrations of 4,5,6-trichloroguaiacol (4,5,6-TCG), 3,4-dichloroguaiacol (3,4-DCG), 3,4,5-trichloroguaiacol (3,4,5-TCG), and 4,5-dichloroguaiacol (4,5-DCG) were 3.5-6.4 times higher in sunlight-exposed samples than in the dark controls. Concentrations of 6-chlorovanillin (6-CV) increased until day 7, after which they began to fall, possibly as a result of further photodegradation.

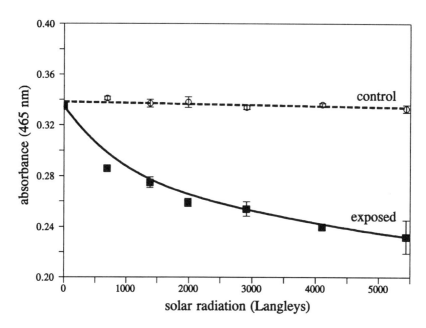

Figure 2. Decolorization of sterile HMW material exposed to sunlight.

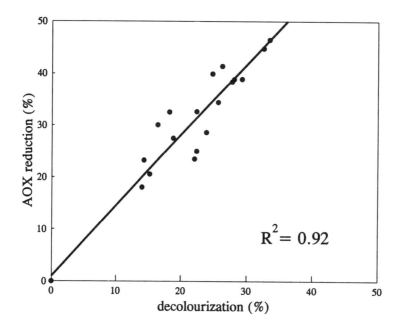

Figure 3. Correlation of AOX reduction and decolorization of HMW material exposed to sunlight.

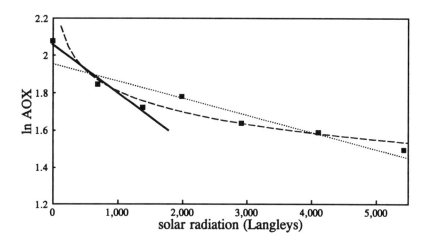

Figure 4. Plot of ln AOX versus solar radiation. Dotted line represents a first-order disappearance over the entire data set; solid line represents over the first three points.

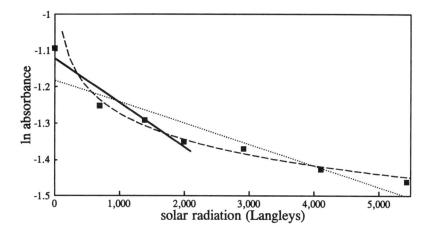

Figure 5. Plot of ln absorbance versus solar radiation. Dotted line represents a first-order disappearance over the entire data set; solid line represents over the first four points.

A very slow release of chlorinated compounds from non-exposed HMW material was also observed. Final concentrations of chloroguaiacols and 6-CV in our dark controls are comparable to concentrations found by Martin (1993) in studies of the release of chlorophenolics from HMW chlorinated organic material upon storage at 22°C, pH 6.5. O'Connor and Voss (1992) also observed a slow release of chlorophenolics from stored HMW bleachery effluents. It is proposed that increased concentrations of LMW chlorophenolics over time are a result of desorption of previously bound or associated compounds rather than a result of chemical instability or biological degradation of the HMW material (O'Connor and Voss 1992; Martin 1993).

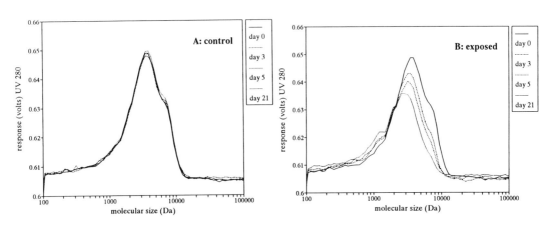

Figure 6. Changes in the molecular size distribution of sterile HMW chlorinated organic material exposed to natural sunlight over a 21 d period.

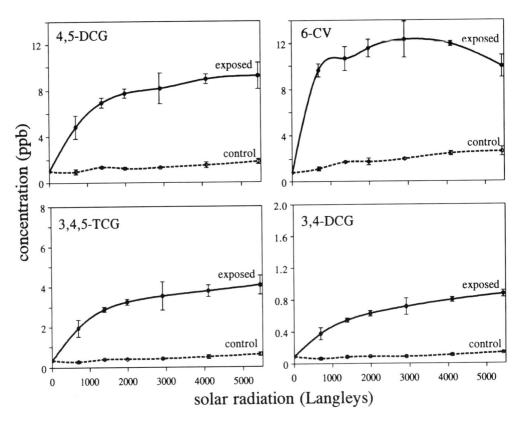

Figure 7. The release of chlorophenolics from sterile HMW chlorinated organic material following exposure to sunlight.

Biodegradation Studies

Biofilms, the consortia of microorganisms that collectively form a slime layer on submerged aquatic surfaces, have been shown in a number of studies to be important for the removal of organic contaminants from aquatic systems (Srinanthakumar and Amirtharaja 1983; Carey et al. 1984; McCarty 1984; Lewis and Gattie 1988). Although biological films comprise a number of different microorganisms, such as bacteria, fungi, algae, and protozoa, members of the fungal component were selectively isolated for use in our biodegradation studies since these microorganisms have been shown to be more efficient degraders of large molecules such as lignin or lignin derivatives (Kirk and Farrell 1987; Vicuña 1988; Zabel and Morrell 1992). Because large compounds are unable to cross biological membranes, biodegradation is limited to those microorganisms capable of producing extracellular enzymes. With the exception of actinomycetes, most bacteria do not produce extracellular enzymes and thus are poor degraders of large, complex materials. Extracellular enzymes are commonly excreted from the growing hyphal tips of filamentous fungi.

Enrichment cultures resulted in a number of filamentous fungal strains from such genera as *Trichosporon*, *Geotrichum*, *Fusarium*, *Pseudeurotium*, *Acremonium*, *Phoma*, *Phialophora*, and *Trichoderma*. Wood-degrading enzymes were produced by each of the fungal strains screened, with cellulase activity common to all but one strain (Millar 1993). Laccases and peroxidases, known to be associated with lignin biodegradation (Buswell and Odier 1987; Bourbonnais and Paice 1990; Roy-Arcand and Archibald 1991; Kirk and Farrell 1987), were common to a number of the river isolates.

Biodegradation experiments were conducted using both the photodegraded HMW material and non-exposed material, to determine whether exposure to sunlight would enhance subsequent microbial degradative processes. Studies by Amador et al. (1989, 1991) demonstrated that compounds bound to humic acids were often more easily mineralized by a soil microbial consortium when the humic acid complexes were previously photodegraded under natural sunlight or artificial light. Photolysis of [^{14}C]glycine- and [^{14}C]phenol-humic acid complexes resulted in a shift in the molecular weight distribution of the radioactivity to a lower molecular weight fraction, thereby enhancing degradation of the labeled substrates by the soil consortium. Other studies have shown that simultaneous treatment of environmental pollutants, with microorganisms and light, can result in an increased degradation over each of the processes alone (Kong and Sayler 1983; Toole and Crosby 1989; Katayama and Matsumura 1991).

Although photolysis of the HMW material resulted in a cleavage of organically bound chlorine, destruction of effluent chromophores, and reduction in the average molecular weight, the photodegraded product did not appear to be more easily degraded than the non-exposed material by the eleven strains tested, including the white-rot strains. Degradation was most extensive by the white-rot strains, with AOX losses of up to 70% (Fig. 8) and color reductions as high as 95% (Fig. 9) regardless of the initial concentrations of AOX or HMW chromophores. White-rot fungi have been shown in a number of studies to be capable of rapid and extensive decolorization of pulp mill bleachery effluent. Archibald et al. (1990) observed that for 5-6 d treatment, with *T. versicolor* could reduce effluent color by approximately 80%. Effluent treatment studies using *P. chrysosporium*, immobilized on a rotating biological contactor, have resulted in color reductions of 65-75% in the first two days (Campbell et al. 1982; Pellinen et al. 1988). Sequential treatment of extraction stage effluent with ozone and *T. versicolor* resulted in a loss of HMW AOX following ozonation; however, this pretreatment did not make the HMW AOX more susceptible to fungal attack (Roy-Arcand et al. 1991). AOX removal in non-ozonized effluent reached approximately 40% in five days.

River isolates were generally found to reduce AOX by 20-35% and color, in a few cases, by 45-70%. Decolorization could not be assessed for certain cultures which were found to release pigments into the medium, thus interfering with the analysis for effluent color (Millar 1993). A study by Prasad and Joyce (1991) may be the only published report of effluent decolorization by a non-white-rotting fungus. Effluent treatment using a species of *Trichoderma*, an imperfect fungus, resulted in a color loss of 57% in three days.

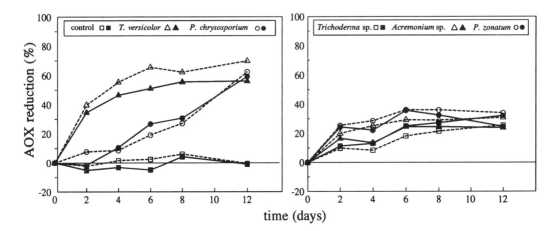

Figure 8. Reduction of AOX in cultures containing HMW material that was previously exposed to sunlight (E, solid lines) or non-exposed (C, dotted lines).

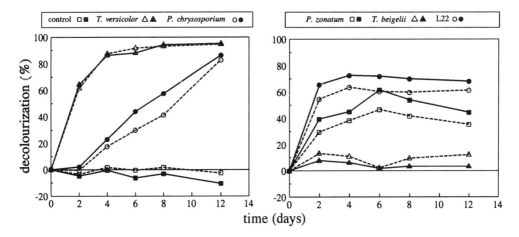

Figure 9. Fungal decolorization of culture medium containing HMW material that was previously exposed to sunlight (E, solid lines) or non-exposed (C, dotted lines).

Molecular size distribution analyses of fungal cultures over the course of incubation indicated a successive removal of HMW material upon treatment with certain strains (Fig. 10). The greatest losses were observed in cultures of *T. versicolor* and *P. chrysosporium*; however, many of the river isolate cultures also sustained significant losses, particularly *Trichoderma* sp. (strain K1), *P. zonatum* (K2), and *Phoma* sp. (K4). LMW peaks appeared in two cultures, possibly as degradation products. Analysis by LC-MS failed to provide enough information for chemical identifications, although it was determined that these compounds were not chlorinated. It is possible that these peaks do not represent HMW degradation products but rather fungal metabolites or cell components.

The molecular size distribution chromatograms of the HMW material before and after fungal treatment indicate that degradation occurred relatively evenly throughout the distribution with no preference for lower molecular weight components. Other studies have confirmed this degradative pattern (Sundman *et al.* 1981; Pellinen *et al.* 1988).

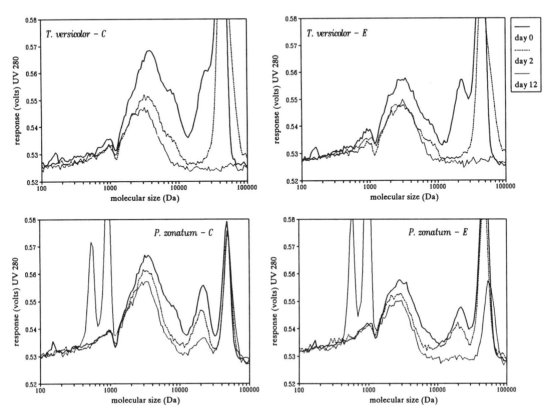

Figure 10. Molecular size distribution of sunlight exposed (E) and non-exposed (C) HMW material during incubation with T. versicolor and P. zonatum.

Most of the river isolates were found to cause equal reductions in the chromatograms of exposed and non-exposed HMW material. Although irradiation resulted in significant degradation of the HMW material, it did not make the remaining material more easily degradable by the fungal strains. It would appear, therefore, that sunlight had attacked substructures within the HMW chlorinated organic material that were fairly resistant to fungal attack and that a combination of both sunlight and fungal treatments results in increased degradation over each of the individual processes.

A number of abiotic processes, such as sorption and photolysis, and biological processes such as biotransformation and biodegradation will determine the fate of HMW chlorinated organic material in the receiving environment. Receiving waters downstream from pulp mills are usually brown in color, dependent upon dilution, and photolysis will be limited to surface waters penetrable by sunlight. In darker and deeper regions, removal mechanisms will consist primarily of sorption to sediments and biodegradation. LMW chlorophenolic compounds, released from the HMW material, will likely be easily transformed or degraded by both bacterial and fungal microorganisms. Biodegradation of HMW material will be limited to fungal species and perhaps certain actinomycetes. Biological activity will probably be the highest in the river biofilms. Further research is required to determine biological rates of disappearance under conditions found in the receiving waters.

ACKNOWLEDGMENTS

Special thanks are extended to B. Brownlee for helpful suggestions on the photodegradation studies, D. Nuttley for technical support, and E. Smith for providing the irradiation data.

REFERENCES

Addleman, K., and F. Archibald. 1993. Kraft pulp bleaching and delignification by dikaryons and monokaryons of *Trametes versicolor*. Appl. Environ. Microbiol. 59:266-273.

Amador, J.A., M. Alexander and R.G. Zika. 1989. Sequential photochemical and microbial degradation of organic molecules bound to humic acid. Appl. Environ. Microbiol. 55:2843-2849.

Amador, J.A., M. Alexander and R.G. Zika. 1991. Degradation of aromatic compounds bound to humic acid by the combined action of sunlight and microorganisms. Environ. Toxicol. Chem. 10:475-482.

Archibald, F., M.G. Paice and L. Jurasek. 1990. Decolorization of kraft bleachery effluent chromophores by *Coriolus (Trametes) versicolor*. Enzyme Microb. Technol. 12:846-853.

Bourbonnais, R., and M.G. Paice. 1990. Oxidation of non-phenolic substrates, an expanded role for laccase in lignin biodegradation. FEBS Lett. 267:99-102.

Brownlee, B.G., J.H. Carey, G.A. MacInnis and I.T. Pellizzari. 1992. Aquatic environmental chemistry of 2-(thiocyanomethylthio)benzothiazole and related benzothiazoles. Environ. Toxicol. Chem. 11:1153-1168.

Burnison, K., V. Martin, A. Rosner, M. Hewitt, K. Millar and T. Williams. 1991. Isolation of high molecular weight organics from kraft mill effluent by nanofiltration. Presented at the Twenty-Sixth Canadian Symposium on Water Pollution Research, February 13-14, Burlington, ON.

Buswell, J.A. and E. Odier. 1987. Lignin biodegradation. CRC Crit. Rev. Biotechnol. 6:1-60.

Campbell, A.G., E.D. Gerrard, T.W. Joyce, H.-M. Chang and T.K. Kirk. 1982. The MyCoR process for color removal from bleach plant effluents: bench scale studies. Proceedings of the Technical Association of the Pulp and Paper Industry (TAPPI) Research and Development Division Conference, August 29-September 1, Asheville, NC.

Carey, J.H., M.E. Fox, B.G. Brownlee, J.L. Metcalfe and R.F. Platford. 1984. Disappearance kinetics of 2,4- and 3,4-dichlorophenol in a fluvial system. Can. J. Physiol. Pharmacol. 62:971-975.

Carey, J.H., M.E. Fox and L.P. Schleen. 1988. Photodegradation of the lampricide 3-trifluoromethyl-4-nitrophenol (TFM). 2. Field confirmation of direct photolysis and persistence of formulation impurities in a stream during treatment. J. Great Lakes Res. 14:338-346.

Caron, R.J. and D.W. Reeve. 1992. Environmental photolysis of chlorinated organic matter discharged in kraft pulp bleaching effluents. Pulp Pap. Can. 93:T209-213.

CEPA (Canadian Environmental Protection Act). 1991. Priority substances list assessment, Report No. 2, Effluent from pulp mills using bleaching. En 40-215/2E, Environment Canada, Ottawa, ON. 60 p.

Frostell, B. 1988. External treatment of bleach plant effluents. Paper presented at the Swedish-Soviet Environmental Symposium, November 14-15, Leningrad, USSR.

Gergov, M., M. Priha, E. Talka, O. Valttila, A. Kangas and K. Kukkonen. 1988. Chlorinated organic compounds in effluent treatment at kraft mills. TAPPI J. 71:175-184.

Hwang, H.-M., R.E. Hodson and R.F. Lee. 1987. Photolysis of phenol and chlorophenols in estuarine water. p. 27-43. *In*: R. G. Zika and W. J. Cooper [ed.] Photochemistry of environmental aquatic systems. American Chemical Society, Washington, DC.

Katayama, A. and F. Matsumura. 1991. Photochemically enhanced microbial degradation of environmental pollutants. Environ. Sci. Technol. 25:1329-1333.

Kirk, T.K., and R.L. Farrell. 1987. Enzymatic "combustion": the microbial degradation of lignin. Ann. Rev. Microbiol. 41:465-505.

Kong, H.-L., and G.S. Sayler. 1983. Degradation and total mineralization of monohalogenated biphenyls in natural sediment and mixed bacterial culture. Appl. Environ. Microbiol. 46:666-672.

Kringstad, K.P., and K. Lindström. 1984. Spent liquors from pulp bleaching. Environ. Sci. Technol. 18:236A-248A.

Lewis, D.L., and D.K. Gattie. 1988. Prediction of substrate removal rates of attached microorganisms and of relative contributions of attached and suspended communities at field sites. Appl. Environ. Microbiol. 54:434-440.

Martin, V. 1993. The formation of chlorophenolics from high molecular weight chlorinated organics (>400 daltons) isolated at a bleached kraft mill. M.Sc. thesis, Univ. Guelph, Guelph, ON. 114 p.

McCarty, P.L. 1984. Biofilm transformations of trace organic compounds in groundwater. *In*: Biofilm processes in ground water research, proceedings of a symposium held in Stockholm, November 17-19, 1983. Ecological Research Committee of NFR, Stockholm.

Millar, K.R. 1993. Degradation of the high molecular weight fraction of bleached kraft mill effluent by biological and photochemical processes. M.Sc. thesis, Univ. Guelph, Guelph, ON. 127 p.

O'Connor, B.I., and R.H. Voss. 1992. A new perspective (sorption/desorption) on the question of chlorolignin degradation to chlorinated phenolics. Environ. Sci. Technol. 26:556-560.

Odendahl, S.M., K.M. Weishar and D.W. Reeve. 1990. Chlorinated organic matter in bleached chemical pulp production. Part II. A review of measurement techniques for effluents. Pulp Pap. Can. 91:T136-140.

Pellinen, J., T.W. Joyce and H.-M. Chang. 1988. Dechlorination of high-molecular-weight chlorolignin by the white-rot fungus *P. chrysosporium*. TAPPI J. 71:191-194.

Prasad, D.Y., and T.W. Joyce. 1991. Color removal from kraft bleach-plant effluents by *Trichoderma* sp. TAPPI J. 74:165-169.

Rohrmann S., and H.P. Molitoris. 1992. Screening for wood-degrading enzymes in marine fungi. Can. J. Bot. 70:2116-2123.

Roy-Arcand, L., and F.S. Archibald. 1991. Direct dechlorination of chlorophenolic compounds by laccases from *Trametes (Coriolus) versicolor*. Enzyme Microb. Technol. 13:194-203.

Roy-Arcand, L., and F.S. Archibald. 1993. Effect of time, daylight, and settling pond microorganisms on the high molecular weight fraction of kraft bleachery effluents. Wat. Res. 27:873-881.

Roy-Arcand, L., F.S. Archibald and F. Briere. 1991. Comparison and combination of ozone and fungal treatments of a kraft bleachery effluent. TAPPI J. 74:211-218.

Smith, R.E. 1977. Rapid tube test for detecting fungal cellulase production. Appl. Environ. Microbiol. 33:980-981.

Srinanthakumar, S. and A. Amirtharajah. 1983. Organic carbon decay in streams with biofilm kinetics. J. Environ. Eng. Div. Am. Soc. Civ. Eng. 109:102-119.

Sundman, G., T.K. Kirk and H.-M. Chang. 1981. Fungal decolorization of kraft bleach plant effluent, fate of the chromophoric material. TAPPI J. 64:145-148.

Suntio, L.R., W.Y. Shiu and D. Mackay. 1988. A review of the nature and properties of chemicals present in pulp mill effluent. Chemosphere 17:1249-1290.

Toole, A.P., and D.G. Crosby. 1989. Environmental persistence and fate of fenoxaprop-ethyl. Environ. Toxicol. Chem. 8:1171-1176.

Vicuña, R. 1988. Bacterial degradation of lignin. Enzyme Microb. Technol. 10:646-655.

Zabel, R.A. and, J.J. Morrell. 1992. Wood microbiology: decay and its prevention. Academic Press, NY. 476 p.

CONCENTRATIONS OF CHLORINATED ORGANIC COMPOUNDS IN VARIOUS RECEIVING ENVIRONMENT COMPARTMENTS FOLLOWING IMPLEMENTATION OF TECHNOLOGICAL CHANGES AT A PULP MILL

S.M. Swanson[1], K. Holley[1], G.R. Bourree[2], D.C. Pryke[3] and J.W. Owens[4]

[1]Golder Associates, 1011 6th Avenue SW, Calgary, Alberta, T2P 0W1 Canada
[2]Weyerhaeuser Canada Ltd., P.O. Bag 1020, Grande Prairie, Alberta, T8V 3A9 Canada
[3]Consultant, RR#1, Erin, Ontario, N0B 1T0 Canada
[4]The Procter & Gamble Company, 5299 Spring Grove, Cincinnati, OH, 45217 U.S.A.

Concentrations of chlorinated organics in the Wapiti/Smoky River system downstream of the Grande Prairie, Alberta kraft pulp mill have declined in all measured compartments. Since 1992, the mill has converted the bleaching process to chlorine dioxide delignification (100% ClO_2 substitution), implemented condensate stripping, upgraded the aerated stabilization basins, and improved effluent monitoring and management. Since these changes, monthly average final mill effluent parameters have been 0.5 kg ADMT^{-1} AOX and <0.1 µg L^{-1} of tri-, tetra- and pentachlorinated phenol compounds. Concentrations of 2,3,7,8-tetrachlorodibenzo-*p*-dioxin (TCDD) and 2,3,7,8-tetrachlorodibenzofuran (TCDF) are now below detection limits of 1.0-5.0 pg L^{-1}. Environmental concentrations have shown a concomitant decrease, which has been rapid in water and sediments. TCDD/TCDF is not detectable in water and is at or near the detection limit in suspended sediments (at both near and far-field sites). Concentrations of TCDD/TCDF have also declined in fish, although at a slower pace which indicates a biological half-life of over 6 months for mountain whitefish (*Prosopium williamsoni*) fillets and >1 yr for burbot (*Lota lota*) livers. Chlorinated phenolic concentrations have paralleled the trends seen for dioxins and furans, with a rapid decrease to non-detectable concentrations in water and concentrations in the very low µg L^{-1} range (for up to 4 compounds only) in suspended sediments. These compounds were rarely detectable in fish fillets even before the mill changes. Extractable organic chlorine (EOCl) was not detectable in water and was generally in the <0.5-2 mg L^{-1} range on suspended sediments. EOCl in fish fillets has declined from 13 to 43 mg L^{-1} before the mill changes to 1-2 mg L^{-1} in September 1993. EOCl in fish bile was not detectable in the current survey.

INTRODUCTION

The Grande Prairie mill was built in 1972 and is located approximately 500 km northwest of Edmonton, Alberta on the Wapiti River. The mill produces 300,000 air dry metric tonnes (ADMT) annually of fully bleached kraft softwood pulp. The fiber make-up of the pulp is lodgepole pine, white spruce, and a small amount of balsam fir. This high grade market pulp is used as a carrier fiber for tissue and towel making and is also used for the manufacture of fine writing papers and other specialty products. Important pulp characteristics are the high and uniform refined and unrefined strength as well as a uniform 90% ISO brightness.

In December 1992, the mill was granted a new 5-yr operating license by Alberta Environment. The license stipulated new monthly average limits for adsorbable organic halogen (AOX), total suspended solids (TSS), biochemical oxygen demand (BOD$_5$), and color (Table 1).

In 1990, an environmental improvement plan was initiated by the mill in expectation of more stringent operating regulations. After extensive evaluation, technologies were selected for implementation by the mill as part of an environmental plan. The technologies selected were chlorine dioxide in the first stage of bleaching, condensate stripping, aerated stabilization basin upgrade, and effluent monitoring and management. The intended result of the technological upgrades was to improve final effluent quality, therefore improving the receiving environment conditions.

Table 1. Operating license limits.

Parameter	Prior Limit kg ADMT^{-1}	New Limit kg ADMT^{-1}	Compliance Date
AOX	3.0	1.5	January 1, 1993
TSS	9.0	5.0	January 1, 1993
BOD$_5$	7.5	5.0	January 1, 1993
		3.0	January 1, 1995
Color	240	160	January 1, 1993
		140	January 1, 1994
		90	July 1, 1997

FINAL EFFLUENT IMPROVEMENTS

The environmental improvements to the mill effluent resulting from the technologies implemented as part of the environmental plan are outlined in Table 2.

Table 2. Environmental impacts resulting from technological upgrades.

Technology Implemented	Date	Environmental Impact
Conversion first to 70% chlorine dioxide substitution. Conversion to 100% chlorine dioxide substitution	Fall 1990 July 1992	• AOX and color reduced in final mill effluent • Reduced chlorinated phenolic and dioxin/furan formation
Foul condensate steam stripper and non-condensable gas system installed	September 1993	• Reduced effluent BOD, sulfides and odor
Aerated stabilization basin upgrade	1991, 1992, 1993	• Increased BOD removal • Improved hydraulic flow characteristics
Effluent monitoring and management		• Dedicated spill recovery system in place • Black liquor spills now contained and reprocessed • Make-up to the liquor cycle has been reduced from 16 to 5.5 kg ADMT^{-1} Na (50 to 17 kg ADMT^{-1} as NA_2SO_4) • Process parameter are optimized for environmental performance • Strategies developed to respond to potential environmental problems

As a result of the improvements to the aerated stabilization basin (ASB) and implementation of steam stripping, BOD_5 removal efficiency has increased from 85% to 94%. At the same time, influent BOD_5 has been decreased by the steam stripping system. Final effluent BOD_5 and TSS have averaged 1.4 and 1.2 kg $ADMT^{-1}$, respectively, since the improvements. These values are below the currently proposed U.S. EPA effluent guidelines of 2.19 and 3.89 kg $ADMT^{-1}$ of product for bleached paper grade kraft (Pryke et al. 1995). Final effluent COD has averaged 40 kg $ADMT^{-1}$ since the start-up of the condensate stripping system.

In 1993 and 1994, AOX levels in final treated effluent have averaged 0.5 kg $ADMT^{-1}$. AOX removal across the ASB is normally 18% (Pryke et al. 1994). The concentrations of 2,3,7,8-tetrachlorodibenzo-p-dioxin (TCDD) and 2,3,7,8-tetrachlorobenzofuran (TCDF) have been consistently less than detection limits in 1994 (detection limits of 1.0-5.0 pg L^{-1}). Concentrations of chlorinated phenolics have been less than 0.1 μg L^{-1} of tri-, tetra- and pentachlorinated compounds.

Improvement in final effluent quality has also had a beneficial impact on the acute and chronic toxicity tests of final mill effluent. Acute toxicity tests for rainbow trout (96-hr LC50) and *Daphnia magna* (48-hr LC 50) have resulted in zero mortality of test species in 100% effluent concentration. Chronic toxicity testing of the final mill effluent has found that the no observable effect concentration (NOEC) is typically $\geq$25% for *Ceriodaphnia* and 100% for fathead minnows.

RECEIVING ENVIRONMENT IMPROVEMENTS

The Wapiti/Smoky River system receives the mill effluent. Monitoring of the river began in 1990, before implementation of the environmental plan. From 1990-1992, a comprehensive ecosystem study was performed that measured the river during the period of conversion to 70% chlorine dioxide substitution and then to 100% chlorine dioxide substitution (Swanson et al. 1993). The Wapiti/Smoky Ecosystem Study assessed the transport of contaminants from the treated effluent in the receiving environment and investigated whether any biological effects had occurred in the river ecosystem. Since the completion of the ecosystem study in 1992, semi-annual follow-up monitoring of water, sediments, and aquatic biota has been continued by the mill. Dioxin/furan congeners have been the primary contaminants monitored during this time. Data on chlorinated phenols, resin and fatty acids, and EOCl are more limited.

Dioxin/Furan Congeners

Results from the ecosystem study (1990-1991) indicated that 2,3,7,8-TCDD and 2,3,7,8-TCDF concentrations in water declined to below detection limits after 70% chlorine dioxide substitution. Concentrations have remained non-detectable since 100% chlorine dioxide substitution, as shown by 1993 and 1994 monitoring in the near field (<5 km from discharge).

In suspended sediments, concentrations of dioxin/furan congeners also declined during the ecosystem study. After the implementation of 100% chlorine dioxide substitution, monitoring of 2,3,7,8-TCDD and 2,3,7,8-TCDF has shown a steady decline of concentrations to at or near the analytical detection limit (Fig. 1).

Declining 2,3,7,8-TCDD and 2,3,7,8-TCDF concentrations were also found in mountain whitefish (*Prosopium williamsoni*) fillets and burbot (*Lota lota*) livers. Levels in mountain whitefish have decreased from above Health and Welfare Canada's limit for human consumption (20 ng kg^{-1} toxic equivalents [TEQ]) to concentrations consistently well below the limit (Fig. 2). Burbot livers have shown a slower decline in 2,3,7,8-TCDD and 2,3,7,8-TCDF concentrations, probably due to their very high lipid content. However, recent data from 1993 and 1994 (after 100% chlorine dioxide substitution) show significant decreases in 2,3,7,8-TCDD and 2,3,7,8-TCDF concentrations in burbot livers (Fig. 3). Other congeners are still detectable in both mountain whitefish fillets and burbot livers including 1,2,3,6,7,8-hexachlorodibenzodioxin, 1,2,3,4,6,7,8-heptachlorodibenzodioxin, 1,2,3,4,6,7,8-hexachlorodibenzofuran, octachlorodibenzodioxin, 1,2,3,7,8-pentachlorodibenzofuran, and 2,3,4,7,8-pentachlorodibenzofuran.

Concentrations of other congeners have declined concomitantly with 2,3,7,8-TCDD and 2,3,7,8-TCDF; concentrations range from 0.2 ng kg^{-1} to 15.0 pg g^{-1} (1994 monitoring data). The results indicate a biological half-life for 2,3,7,8-TCDD and 2,3,7,8-TCDF of over 6 months for mountain whitefish fillets and greater than a year for burbot livers.

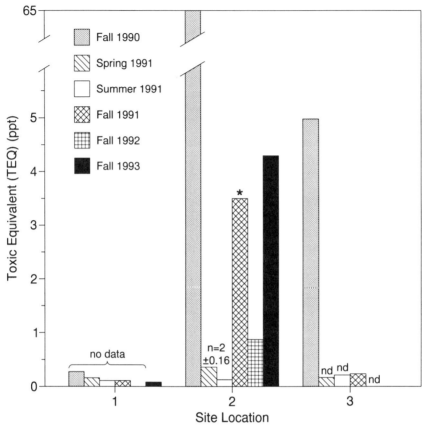

*indicates that TEQ for Fall 1992 = 2.07 ppt
Note: In Spring 1992, 2 suspended samples were collected from Site 2

Figure 1. 2,3,7,8-TCDD/TEQs (ng kg^{-1}) in suspended sediments (n=1 per sampling site) in the Wapiti/Smoky River system, 1990-1993.

Chlorinated Phenol Compounds

Since the implementation of 70% and then 100% chlorine dioxide substitution, concentrations of chlorinated phenolic compounds in river water have declined dramatically. The environmental data parallel the results seen in the effluent.

In 1990, 26 chlorinated phenolic compounds were detected in water at the near-field site, with several concentrations greater than 1 µg L^{-1}. In the spring of 1991, only three chlorinated phenolic compounds were found at the near-field site; concentrations were below 0.1 µg L^{-1}. In the fall of 1991, only one compound, 6-chlorovanillin, was detectable in water. In 1993 and 1994, after 100% chlorine dioxide substitution, data show that concentrations were below detection limits for all compounds except 6-chlorovanillin, which was present at a concentration of 0.10 µg L^{-1} in the spring of 1994.

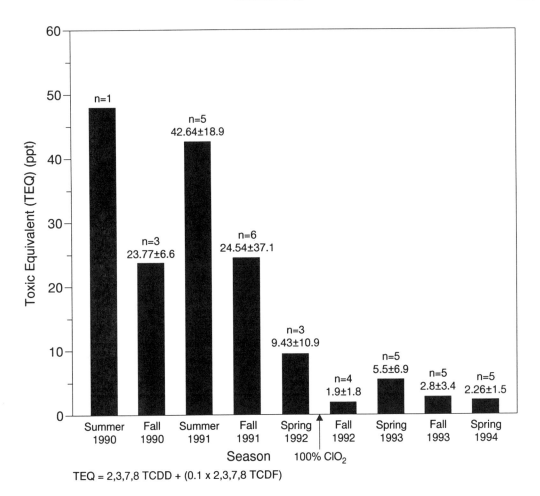

Figure 2. Mean 2,3,7,8-TCDD/TEQs ($\bar{x} \pm SD$) in mountain whitefish fillets in the Wapiti River downstream of the mill, 1990-1994.

During the period just before and after 70% chlorine dioxide substitution (1990-1991), suspended sediments carried a variety of chlorinated phenolic compounds, including the more highly substituted compounds. In 1991, concentrations were generally at or just above detection limits of 0.5 µg L^{-1}, with the exception of the low flow periods when 15 compounds were detected at concentrations ranging from 0.8 to 230 µg L^{-1}. However, after 100% chlorine substitution, fewer compounds were detected. In the fall of 1993, only 4 compounds were above detection limits in suspended sediments; concentrations ranged from 1.1 to 13 µg L^{-1}. The compounds detected were pentachlorophenol, 3,4,5-trichlorocatechol, tetrachlorocatechol, and 6-chlorovanillin.

Extractable Organic Chlorine Compounds

In 1991, EOCl was not detected in water. However, an improvement was made in EOCl detection limits in 1992. Consequently, a concentration of 6.2 µg L^{-1} was reported just downstream of the mill. Upstream of the mill, EOCl was 4.5 µg L^{-1}, indicating other sources of extractable organic chlorine compounds in the watershed. EOCl was not detectable in river water by 1993.

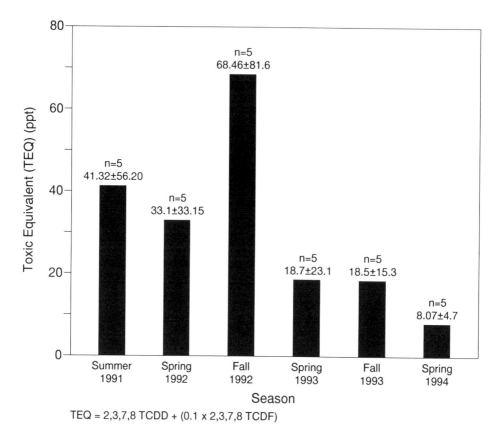

Figure 3. Mean 2,3,7,8-TCDD/TEQs ($\bar{x} \pm SD$) in burbot livers in the Wapiti River downstream of the mill, 1990-1994.

In 1991, the highest EOCl concentrations in suspended sediments were detected upstream of the mill at 1.6 mg L^{-1}; concentrations downstream of the mill ranged from 0.5 mg L^{-1} (detection limit) to 0.89 mg L^{-1}. In 1992, EOCl was non-detectable. A concentration of 12 mg L^{-1} was reported at the near-field site in the fall of 1993. The elevated concentrations in fall 1993 may have resulted from decant from a holding pond containing dredged sludge from the effluent ponds (dredged in June 1993).

Since 1990, mountain whitefish fillets have shown a downward trend in EOCl concentrations. At that time, concentrations ranged from 13 to 43 mg L^{-1}. In 1991, concentrations declined an order of magnitude to 3-7 mg L^{-1}. A further decline of EOCl in fillets was observed in the fall of 1993, when concentrations ranged from 1 to 2 mg L^{-1}.

The decline of EOCl in mountain whitefish bile parallelled the decline of EOCl in mountain whitefish fillets. EOCl concentrations in 1991 ranged from 24 to 192 mg L^{-1} (just downstream of the mill), whereas in 1993 EOCl was not detectable. The absence of EOCl in fish bile indicates either very low or no recent exposure to mill effluent or low exposure with metabolism and excretion exceeding any intake.

Resin and Fatty Acids

In 1990-1991, resin and fatty acid compounds were not detectable upstream and downstream of the mill in water (Swanson *et al.* 1993). Certain compounds were present at higher concentrations downstream of the mill in 1992. The compounds included myristic, palmitic, linoleic, stearic, and arachicidic acids;

concentrations ranged from 0.01 to 49 µg L^{-1}. Two chlordehydroabietic acid compounds (mono, di) were detected in samples from only the downstream sites, at concentrations ranging from 0.025 to 0.038 µg L^{-1}. In the fall of 1993, all resin and fatty acid compounds were non-detectable. In the spring of 1994, palmitic acid was detected at a concentration of 0.02 µg L^{-1}.

In 1991, resin and fatty acids were detectable in both longnose sucker and mountain whitefish bile. Thus, these compounds were being taken up by fish and then metabolized and excreted via the bile. The source of exposure may have been the bottom sediments, where resin and fatty acids were present in detectable quantities in 1991 (Swanson et al. 1993). After 100% chlorine dioxide substitution in the fall of 1993, resin and fatty acids were no longer detectable in fish bile, indicating that exposure to these compounds has been reduced.

DISCUSSION AND CONCLUSIONS

Numerous improvements to the quality and stability of the final mill effluent have resulted from the environmental improvement plan initiated in 1992. Final effluent improvements included non-measurable TCDD and TCDF, in compliance with the Canadian Environmental Protection Act, <0.1 µg L^{-1} concentrations of tri-, tetra-, and pentachlorinated phenols, TSS average <1.5 kg ADMT^{-1}, BOD$_5$ average <1.5 kg ADMT^{-1}, color average <90 kg ADMT^{-1}, AOX average 0.5 kg ADMT^{-1}, and COD average 40 kg ADMT^{-1}.

Improvements to the final effluent improvements have been accompanied by improvements to the receiving environment. Dioxins and furans have declined to levels that are consistently at or below detection limits in water. Some dioxin furan congeners are still detectable in suspended sediments reflecting the high degree of partitioning to solids characteristic of these compounds. The total organic carbon content of the Wapiti River suspended sediments is 3.7 to 4.8%. Estimated suspended sediment partition coefficients from recent Wapiti River data (corrected for organic carbon) are 1.3×10^5 for both 2,3,7,8-TCDD and 2,3,7,8-TCDF. The decline in water and suspended sediment concentrations has been reflected in fish. Both mountain whitefish fillet and burbot liver concentrations have significantly declined. Depuration is slower in burbot liver than in mountain whitefish fillets.

Concentrations of chlorinated phenols in water have also declined to non-detectable levels as a result of mill improvements. Suspended sediments continue to show the presence of some compounds in the very low µg L^{-1} range (4 compounds only). Thus, as for dioxin/furan congeners, suspended sediments appear to be the compartment where chlorinated phenolics will still be found because of their high affinity for solids. Therefore, suspended sediments and fish appear to be the most appropriate compartments for continued monitoring of organochlorine trends.

EOCl is not detectable in water and fish bile and is generally in the <0.5-2 mg L^{-1} range in suspended sediments. In fish fillets, EOCl declined from 13 to 43 mg L^{-1} before the mill changes to 1-2 mg L^{-1} in September 1993. The continuing presence of low levels of EOCl in suspended sediments parallels the dioxin/furan and chlorinated phenolics data.

In conclusion, the mill improvements have had a direct beneficial impact on both final effluent quality and the receiving environment.

REFERENCES

Pryke, D.C., G.R. Bourree, P. Winter and C. Mickowski. 1995. The impact of chlorine dioxide delignification on pulp manufacturing and effluent characteristics at Grande Prairie, Alberta. Pulp & Paper Canada: In press.

Pryke, D.C., G.R. Bouree, S.M. Swanson, J.W. Owens and P.J. Kloepper-Sams. 1994. The impact of chlorine dioxide delignification on pulp manufacturing and effluent characteristics at Grande Prairie: Effluent quality improvements and ecosystem response. Presented at the Non-Chlorine Bleaching Conference. Vancouver, B.C., Nov. 1994.

Swanson, S.M., R. Schryer, B. Shelast, K. Holley, I. Berbekar, P. Kloepper-Sams, J.W. Owens, L. Steeves, D. Birkholz and T. Marchant. 1993. Wapiti/Smoky River Ecosystem Study. Prepared for Weyerhaeuser Canada Ltd.

MESOCOSM SIMULATION ON SEDIMENT FORMATION INDUCED BY BIOLOGICALLY TREATED BLEACHED KRAFT PULP MILL WASTEWATER IN FRESHWATER RECIPIENTS

E.K. Saski[1], A. Vähätalo[2], K. Salonen[2] and M.S. Salkinoja-Salonen[1]

[1]Department of Applied Chemistry & Microbiology, P.O. Box 27, University of Helsinki
FIN-00014, Helsinki, Finland
[2]Lammi Biological Station, University of Helsinki, FIN-16900 Lammi, Finland

Mesocosms (ca. 2 m^3) were filled either with clear or humic lake water and biologically treated bleached kraft pulp mill effluent (BKME). One series of the mesocosms was run for one year from June to March and the other from November to December. The formed sediment was collected from the bottom of the enclosures after 275 or 290 d of incubation. Sediments formed in the mesocosms containing both BKME and lake water were different from those formed in the presence of lake water alone. The molar ratio of carbon to chlorine was lower (300) in the BKME-driven sediment compared to reference sediment (2000 to 4000). The halogen content of the sediment formed *de novo* was dependent on the timing of the start of the experiment. On the average, more sediment-bound halogen per cubic meter of BKME (560 mg Cl) resulted from experiments initiated in the winter than from the summer (140 mg Cl). Sediments in the BKME mesocosms contained tetrahydrofuran-soluble halogens of larger molecular size (1400 g mol^{-1}) than water phase (360 g mol^{-1}) of the same mesocosm after about one year of incubation starting from November. The data indicate that binding of halogen to *de novo* formed sediment was driven by metabolic activity of the biomass rather than by physical precipitation.

INTRODUCTION

Pulp bleaching in western Europe is mainly carried out today without elemental chlorine (Axegård *et al.* 1993), and the process technology is developing towards closing the water circuits. In the future, with zero discharge, the recipient waters will continue to interact with past discharges through the historical sediments.

It is known that organic halogen compounds accumulate to some extent in the sediment downstream from pulp mills (Salkinoja-Salonen *et al.* 1981, 1984; Bryant *et al.* 1988; Grimvall *et al.* 1991a; Parker *et al.* 1991; Martinsen *et al.* 1994). Single compounds and sum parameters have been analyzed from both water and sediment, but the mass balance of organic halogen in recipient waters is only fragmentarily known. Brownlee *et al.* (1977) estimated a half-life of 6 weeks for non-chlorinated dehydroabietic acid in the water column of a lake, but 21 years in the sediment.

All bleached sulfate pulp producing mills in Finland have built biological treatment plants for their waste waters. The aim of this work was to assess the quality and quantity of sediment produced as a response to pulp mill waste water discharged after secondary treatment. Mesocosms have proven to be a useful tool to quantitate degradation of organically bound halogens in natural temperature and light (Saski *et al.* 1991). Mesocosms can be studied around the year which enables monitoring of degradation over four seasons. Two series of mesocosm experiments were conducted *in situ* in a humic and clear lake, one started in the winter and the other in the summer, both lasting one year.

MATERIALS AND METHODS

Two lakes in southern Finland were used for the study: Lake Valkea Mustajärvi with clear water and Lake Mekkojärvi with humic water (61°13″N, 25°08″E). The enclosures (cylindrical in the summer-to-summer run and cylindrical with a conical bottom in the winter-to-winter run) were made of polyethylene (194 g m^{-2}), translucent for ambient light and black for dark controls. The dark enclosures were covered with sheets of black polyethylene (with a few slashes to allow for rain water inflow and evaporation). The light enclosures were not covered.

The enclosures were supported by wooden frames, each divided into eight subframes. Two sets were placed in the clear and two in the humic lake. The frame was backed with styrofoam flotation aids and anchored to trees ashore. Biologically treated bleached kraft pulp mill effluent (BKME) was collected at the outlet of the activated sludge plant (Mill A) and the aerated lagoon (Mill C). Bleaching sequences and sampling dates are presented in Table 1. The enclosures were filled with BKME and lake water in a ratio (v/v) of ca. 1:10 or 1:50. An aliquot of each effluent was stored frozen (-20°C) for future analysis.

Table 1. Characteristics of the BKMEs and the lake water used in the mesocosm experiments.

	Lake Water[a]		BKME			
	Clear	Humic	Mill A May 24, 1991	Mill A Oct 20, 1992	Mill C May 23, 1991	Mill C Oct 16, 1992
Bleaching sequence (pine and spruce)			D-EO-D-E-D	D-EO-D-E-D	O-C(85)D(15)- EO-D-E-D	O-D-EO-D-E-D
Nitrogen (mg N m^{-3})						
NH_4^+ - N	30	32	na	170[b]	na	2000[b]
$NO_2^- + NO_3^-$-N	35	30	118	4	98	35
Total N	350	560	3620	2000	5050	4500
Phosphorus (mg P m^{-3})						
PO_4^{3-} P	1	2	620	140	620	440
Total P	8	12	902	340	1040	880
AOX (g m^{-3})	0.009	0.03	26	17	21	7
EOX (g m^{-3})	na	na	na	16	na	4
TOC (g m^{-3})	6	21	380	300	280	190
Color (g Pt m^{-3})	17	280	3930	2460[b]	3230	1490[b]
pH	6.5	6.0	7.6	7.6	7.6	7.6
Conductivity (mS m^{-1})	3	4	380	460	300	280

na = Not analyzed.
a. An average of several measurements from two parallel enclosures without BKME.
b. Analyzed from frozen samples.

Sediment was pumped as a slurry from the mesocosms using a centrifuge pump, measured for volume, and homogenized by mixing. Two liters were taken for analysis and the remainder returned to the enclosure.

To determine halogen of the water phase, an aliquot of the sediment slurry was filtered (Whatman GF/A glass fiber). AOX was measured microcoulometrically (halogen analyzer Euroglass, Delft, NL) from the filtrate according to the ISO standard 9562 (International Standardization Organization 1989). To release inorganic chlorine from whole cells, 50 mL of the sediment slurry was subjected to alkaline hydrolysis (pH 10 ± 0.2 1 M KOH, +80°C 1 h). The slurry was then cooled to room temperature; 1 mL of this slurry was diluted with 99 mL of distilled water, acidified (pH ~2, concentrated HNO_3), and shaken overnight with 50 mg of activated carbon (Euroglass). Activated carbon and suspended solids were collected on a filter (Nuclepore PC 0.4 μm) and combusted at 1000°C for the analysis of the halogen contents using the halogen analyzer. Halogen of the suspended solids (sediment of the mesocosms), which will be referred to later as bound halogen (BX), was calculated by subtracting halogen of the water phase from the sediment slurry contained halogens.

For assay of carbon and nitrogen, 100 mL of the sediment slurry was centrifuged, washed twice with distilled water, freeze-dried, and analyzed using Leco CHN-analyzer.

For molecular size assays, 10 mL of undiluted BKME or 300 mL samples from the mesocosm water column were freeze-dried (Christ Gamma 2-20). The residues were extracted with 2 to 4 mL of tetrahydrofuran adjusted to pH of ca. 2 (concentrated HNO_3), for 0.5 h on a shaker and 0.5 h in a bath sonicator. The extracts were centrifuged and the supernatants filtered (0.45 μm Nylon Acrodisc 13). Molar size analysis with high performance size exclusion chromatography (HPSEC) was performed as described elsewhere (Jokela and Salkinoja-Salonen 1992). The calibration range was from 58 to 498,000 g mol^{-1}. A total of 25 fractions (1 mL each) were collected within the range from 35 to 300,000 g mol^{-1}.

The suspended solids were harvested from sediment slurry (50 mL) on a glass fiber filter (Whatman GF/D). The filter was freeze-dried, weighed (dry weight), and extracted with 30 mL of tetrahydrofuran spiked with 30 μL of concentrated HNO_3. Extraction was carried out as described above, and a rotary evaporator was used to concentrate the sediment extracts to 3 mL. An aliquot of 0.5 mL of the filtered extract was evaporated to dryness in a stream of N_2. The residue was analyzed for halogen content (EOX) in the Euroglass analyzer after direct combustion (1000°C).

The sum of nitrite and nitrate was determined from water by the cadmium reduction method (U.S. EPA number 353.2). Orthophosphate was determined according to the U.S. EPA standard no. 365.1.

RESULTS

Table 1 summarizes the features of the waste waters and lake waters used. At the time of the 1992 sampling, Mill A and C discharged secondary treated waste water with 0.5 kg of AOX per tonne of pulp.

Table 1 shows that the waste waters introduced into the mesocosms contained 4 to 14 times more nitrogen and 30 to 130 times more phosphorus than the lake waters. A considerable portion of nitrogen (≥55%) and phosphorus (≥30%) was in organic form. The excess sediment generation observed in the mesocosms holding 2 to 13% of waste water (Fig. 1) may indicate that the waste water nutrients were available for biomass growth and sedimentation. When the waste waters were introduced to the light mesocosms in June, virtually all inorganic phosphate was taken up by the biomass within 2 months, but in dark mesocosms there was no net uptake or an exclusive mineralization was observed (see Saski et al. 1991). This means that the phosphorus uptake was light dependent, i.e. connected to the growth of autotrophic biomass. Our preliminary results with the mesocosms showed that in the dark mesocosms of the summer-to-summer run, the waste water nitrogen was also mineralized during the summer (Saski et al. 1991). These observations indicate that the geochemical cycle, mineralization of nutrients, photosynthesis, and nitrification were not inhibited in the mesocosms holding waste waters in concentrations up to 11% (v/v), and hence, the mesocosms may simulate a natural ecosystem of a recipient lake.

The molar ratio of carbon to nitrogen was 12 to 13 in the sediments of summer-to-summer mesocosms and 12 to 15 in the winter-to-winter mesocosms with lake water alone. The ratio of C:N in the sediments generated in the mesocosms holding lake water plus BKME varied from 10 to 17 (mean 12, n =10) in the summer-to-summer and from 12 to 18 (mean 16, n = 11) in the winter-to-winter experiment. In summary, the sediment generated in response to waste water blended with lake water, contained carbon to nitrogen in a ratio similar to sediment formed in the reference mesocosms without waste water.

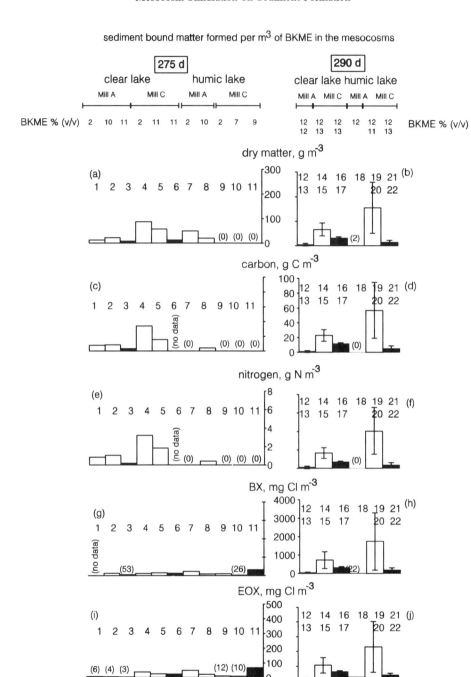

Figure 1. Sediment generation in response to introduction of BKME into the mesocosms (running numbers in each panel). The yield of corresponding component in the sediment of the mesocosms with 0% BKME has been subtracted. The results obtained from the parallel enclosures are joined by vertical lines. Values are given in parentheses above the invisible bars. The relative volume of BKME is given as % (v/v) above the uppermost panel. Total volume was 1.8 m^3 in summer-to-summer and 2.5 m^3 in winter-to-winter mesocosms. BX = bound halogen, EOX = tetrahydrofuran-soluble halogen. Covered mesocosms are black.

During one year, starting either from June or November, from 1 to 5 g of sediment bound dry matter, 0.3 to 2 g of carbon, and 0.03 to 0.2 g of nitrogen were found in the *de novo* formed sediment per mesocosm without BKME. Dry mass and the contents of carbon and nitrogen in the sediment were independent of the start time of the incubation (June or November). As compared to the mesocosms without BKME, usually up to 80 g of sediment matter (mean 9, n = 22) was formed in the mesocosms where the waste waters were introduced. Figure 1 shows the excess sediment generated per cubic meter of BKME. It is noteworthy that more sediment-bound BX and EOX were always formed in the BKME containing mesocosms than in the reference mesocosms, indicated by positive bars in Figure 1 (g-j). The carbon and nitrogen content of the sediment generated as a response to BKMEs in one year was independent of season. In general, however, 1 m^3 of BKME drove more halogen to the sediments when the incubation was started in November as compared to that started in June (Fig. 1). However, in most cases sedimentation of organic halogen explained a minor portion (0.2 to 14%) of AOX removal, which was 20 to 60% of input AOX (see Table 1).

More halogen was found in the sediment of BKME holding mesocosms than in lake water holding reference mesocosms. The natural background accumulation of bound halogen in lake water mesocosms was ≤3 mg Cl calculated as bound halogen (BX) and ≤1.6 mg Cl as EOX. The sediments of the BKME holding mesocosms contained BX from 5 to 42 (summer to summer, 275 d) or 6 to 990 mg Cl per mesocosm (winter to winter, 290 d), and EOX from 0.4 to 11 or 3 to 120 mg Cl per mesocosm, respectively. The molar ratio of carbon to chlorine in the sediment from the BKME holding mesocosms was about 300, which is lower than that of the sediment of the reference mesocosms (2000 to 4000). The BX content of *de novo* sediment in the mesocosms with about 10% waste water was 1 to 14 mg Cl g^{-1} DW (mean 6.2, n = 18), and EOX content was 0.2 to 4 mg Cl g^{-1} DW (mean 1.4, n = 18). The sediments of the reference mesocosms contained 0.2 to 0.6 mg Cl g^{-1} DW (mean 0.4, n = 8) of BX and 0.015 to 0.6 mg Cl g^{-1} DW (mean 0.3, n = 8) of EOX.

The conclusions from the above observations (Fig. 1) are that the halogen content of the sediment increased as a consequence of introduction of BKME to the mesocosms, and more sediment-bound halogen was formed as a response to 1 m^3 of BKME when incubations were started in winter compared to summer.

Figure 2 shows that not only quantity but also quality of sediment-bound halogen was different in the summer-to-summer and winter-to-winter experiments. Figure 2 summarizes the solvent solubility of the sediment-bound halogens in the mesocosms. It shows that when the waste waters were introduced to lake water mesocosms in June, the resulting sediment bound-halogenated compounds were less soluble in tetrahydrofuran than those in the sediment formed after introduction of the waste water in November. Similar differences in tetrahydrofuran solubility of sediment-bound halogens was observed when no waste water was introduced.

Figure 3 shows the molar size distribution of tetrahydrofuran-extractable halogenated and UV-absorbing compounds of the BKME from Mills A and C. On the average, half of the EOX of different sizes of halogenated molecules was lost if the waste water was pre-filtered. The removal of the filterable solids did not selectively remove halogen or UV-absorbing matter from any size class. We have measured the molar size distribution of tetrahydrofuran-soluble halogens from the water column (n = 7) and the sediment (n = 6) after incubation of waste water holding mesocosms over four seasons. The molar size distribution of halogens was closely similar in all water columns (n = 7, 393 d incubation) whether having received waste water from Mill A or C, or whether incubated in clear or humic lakes, and similar to the waste water (Figs. 3 and 4). The sediment-bound tetrahydrofuran-soluble halogens (n = 6, 290 d of incubation) showed molar size distributions very different from those formed in the water columns. The peak of the halogenated molecules shifted from approximately 360 g mol^{-1} in the water phase to ca. 1400 g mol^{-1} in the sediment phase. Representative examples of the size distributions are shown in Fig. 4. The molar sizes of sediment-bound EOX showed two maxima: between 100 and 300 and between 700 and 4000 g mol^{-1}. Therefore, the waste water containing halogens had become polymerized, conjugated or bound to macromolecular cell components, e.g. integral lipids in the sediment.

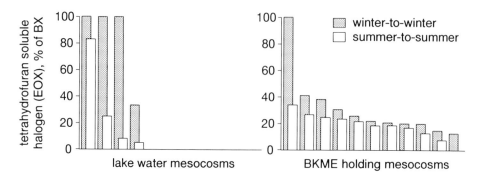

Figure 2. Tetrahydrofuran-soluble halogen (EOX) as percent of the bound halogen (BX) in the mesocosm sediment formed de novo. Each bar stands for one mesocosm. Mesocosm are sorted by descending values.

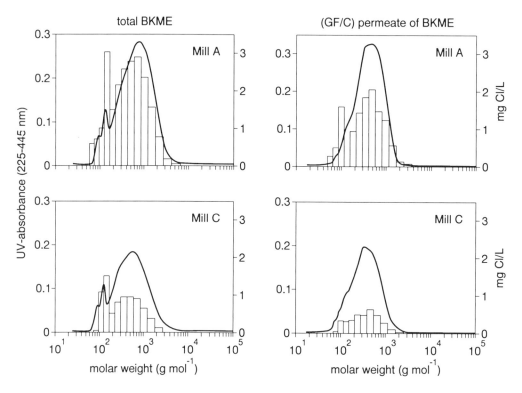

Figure 3. The molar size distribution of halogenated (bars) and UV-absorbing (solid line) compounds in the tetrahydrofuran extracts of BKME (October 1992) from Mills A and C: Crude BKME (left panels) and GF/C permeate (right panels) were extracted into tetrahydrofuran.

DISCUSSION

The aim of this work was to assess the sediment accumulation of BKME containing organic halogen after being discharged to a lake recipient from a biological treatment plant. To study this, we used ca. 2 m^3 mesocosm enclosures incubated *in situ*. The experiments, which lasted one year, were started either in the summer or winter. The quantity and quality of wastewater-generated sediment formation was assessed and compared to mesocosms with no waste water.

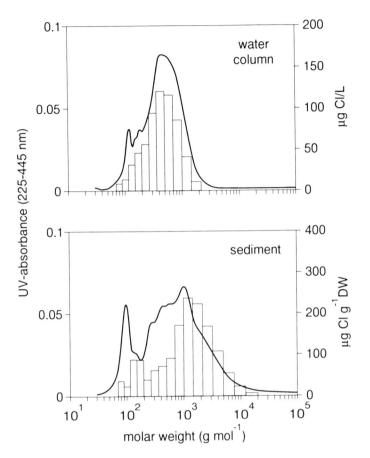

Figure 4. An example of the molar size distribution of tetrahydrofuran-extracted halogens from the water column (upper panel) and the mesocosm sediment (lower panel) formed de novo during the winter-to-winter experiment. In this case, the mesocosm was holding 12% (v/v) BKME from Mill A diluted with clear lake water, and incubated open to light, starting from November for 393 d (water column) and 290 d (sediment phase).

The BKMEs were obtained from the same mills, but at different times. Mill C had substituted elemental chlorine in the bleaching with chlorine dioxide between the samplings. This change in BKME composition may influence the results, but the general trend of the results seems clear. In addition to bleaching, the kraft cooking has been suggested to contribute to the environmental impact of the waste water (Martel et al. 1994).

Biological treatment removes 30 to 50% of AOX and TOC from the bleached kraft mill waste water (Laine et al. 1991; Jokela et al. 1993). It has been shown that AOX removal is mainly dependent on biodegradation, and only a minor fraction (0 to 8%) was recovered from the pressed total sludge (Gergov et al. 1988). We earlier showed that 30 to 60% of AOX of the post-treatment waste water became mineralized when incubated with lake water in the mesocosms for 5 months (Saski et al. 1991). The present study confirmed our earlier results and showed that AOX removal was likely due to degradation because usually less than 10% was recovered from the sediments after one year. Degradation of BKME originating AOX in a large receiving lake was also reported by Grimvall et al. (1991b).

Several earlier studies have shown that organic halogen accumulates in the sediments downstream from pulp mills from 1 to 15 mg g^{-1} DW (Salkinoja-Salonen *et al.* 1981; Paasivirta *et al.* 1988; Maatela *et al.* 1990; Martinsen *et al.* 1994). This is the same order of magnitude to BX (mean 6.2 mg g^{-1} DW) in the sediment formed *de novo* in our mesocosms holding BKME of either 2 or 10% (v/v). Martinsen *et al.* (1994) used extraction methods with cyclohexane and isopropanol, and they were able to extract up to 30% of total organic chlorine. In the BKME holding mesocosms, tetrahydrofuran extracted on the average 26% (n = 21) of BX formed (Fig. 2). In this study, mesocosms were used to obtain quantitative data on the build-up of sediment in response to BKME discharge. Because the carbon to halogen ratio was lower in the *de novo* sediment in the presence of BKME than in the sediment formed without BKME, the halogen may serve as a marker of the presence of wastewater constituents in a recipient sediment. Furthermore, the halogenated compounds of the recipient sediment may show either bioaccumulation or harmful effects on fish even after closing the whole mill (Carlberg *et al.* 1987) or after the mill starts operating closed-loop.

Jokela *et al.* (1993) showed that the halogen of BKME became less tetrahydrofuran extractable in a biological treatment process. In the present investigation, the events subsequent to biological treatment in the mesocosm recipients have been studied. Better solubility of the winter-to-winter compared to summer-to-summer sediment-bound halogen (BX) in tetrahydrofuran demonstrates the different halogen content of the sediment formed in one year depending on the date of the discharge. In the winter-to-winter experiment, all sediment-bound variables, including carbon, nitrogen, and halogen per cubic meter of BKME, correlated positively with the dry weight of the formed sediment per cubic meter of BKME (Fig. 5). In the summer-to-summer experiment, only carbon and nitrogen showed clear correlation; halogen did not. This observation supports the assumption that the halogen quality of sediment resulting from the same sedimentation period (1 year) is different depending on the season at which the waste water is introduced to a lake.

The large halogenated molecules were not more removable than small molecules of the BKMEs by filtration (Fig. 3). This indicates that the accumulation of the large molecules to the sediment in one year in the winter-to-winter mesocosms can not be explained solely by physical precipitation. However, abiotic sorption may play a role in binding since Pellinen (1994) found that larger molecules absorbed to sediment more than smaller molecules. Laine *et al.* (1995) studied the molar weight distribution of halogens accumulated by earthworms that were exposed to sawmill soil. They found larger tetrahydrofuran-extractable halogenated molecules from earthworms than from the soil. Incorporation of the halogenated compounds into tissue lipids may explain both accumulation to earthworms and binding to decaying biomass in the mesocosms.

Mesocosms are attractive tools for mass balance studies to assess the environmental impact of a waste water. Another advantage of the mesocosm as a tool is that mesocosms respond with processes such as phosphate uptake by algae and nitrification which are characteristic of natural lake systems. Mesocosms as used in the present work have disadvantages including the lack of water movement and variation between replicate mesocosms. By enclosing water from the surrounding lake, a new ecosystem is created which may be different from a neighboring enclosure with the same lake water. The results of the sediment producing capacity of BKME obtained with the present mesocosm test should be subjected to verification in an actual receiving lake.

ACKNOWLEDGMENTS

This work was financially supported by the Academy of Finland, Maj and Tor Nessling Foundation, and Ministry of the Environment. The authors wish to thank Ms. Maarit Herranen and Ms. Ann-Christine Eriksson from Metsä-Botnia Oy for their cooperation.

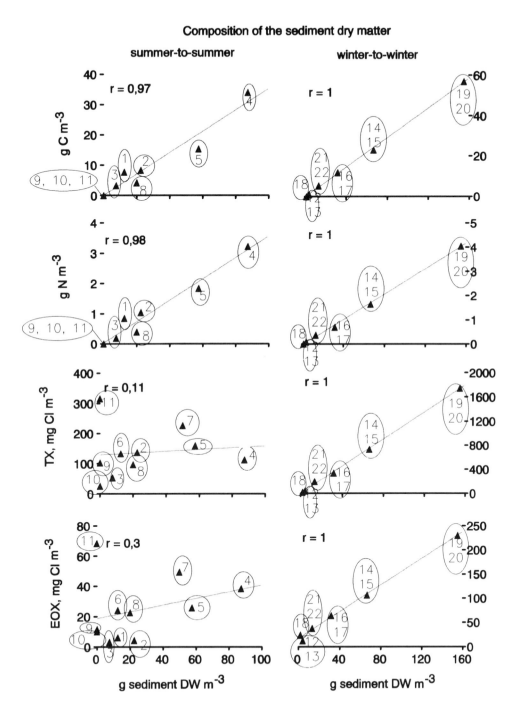

Figure 5. Correlation between de novo sediment dry matter production vs. sediment-bound carbon, nitrogen, and halogen production in the waste water holding mesocosms. The average is plotted when two parallel enclosures were run. Running numbers beside each data point are for mesocosm identification (see Fig. 1).

REFERENCES

Axegård, P., O. Dahlman, I. Haglind, B. Jacobson, R. Mörck and L. Strömberg. 1993. Pulp bleaching and the environment - the situation 1993. Nordic Pulp and Paper Res. J. 4:365-378.

Brownlee, B., M.E. Fox, W.M.J. Strachan and S.R. Joshi. 1977. Distribution of dehydroabietic acid in sediments adjacent to a kraft pulp and paper mill. J. Fish. Res. Board Can. 34:838-843.

Bryant, C.W., G.L. Amy, R. Neill and S. Ahmad. 1988. Partitioning of organic chlorine between bulk water and benthal interstitial water through a kraft mill aerated lagoon. Water Sci. Technol. 20(1):73-79.

Carlberg, G.E., A. Kringstad, K. Martinsen and O. Nashaug. 1987. Environmental impact of organochlorine compounds discharged from the pulp and paper industry. Paperi ja Puu - Papper och Trä 4/1987:337-341.

Gergov, M., M. Priha, E. Talka, O. Valttila, A. Kangas and K. Kukkonen. 1988. Chlorinated organic compounds in effluent treatment at kraft mills. TAPPI J. 71(12):175-184.

Grimvall, A., S. Jonsson, S. Karlsson and R. Sävenhed. 1991a. Organohalogens in unpolluted waters and large bleach-plant recipients, p. 147-154. *In* TAPPI Proc. Environ. Conf. Book 1, 1188 p.

Grimvall, A., H. Borén, S. Jonsson, U. Lundström and R. Sävenhed. 1991b. Long-term accumulation and degradation of bleach-plant effluents in receiving waters. p. 74-84. *In* A. Södergren [ed.] Environmental fate and effects of bleached pulp mill effluents. Swedish Environmental Protection Agency, Stockholm, Report 4031, 394 p.

International Standardization Organization. 1989. Water quality - determination of adsorbable organic halogens (AOX). ISO 9562, Genève. 8 p.

Jokela, J.K. and M. Salkinoja-Salonen. 1992. Molecular weight distributions of organic halogens in bleached kraft pulp mill effluents. Environ. Sci. Technol. 26(6):1190-1197.

Jokela, J.K., M. Laine, M. Ek and M. Salkinoja-Salonen. 1993. Effect of biological treatment on halogenated organics in bleached kraft pulp mill effluents studied by molecular weight distribution analysis. Environ. Sci. Technol. 27(3):547-557.

Laine, M., J. Jokela and M. Salkinoja-Salonen. 1991. Removal of different molecular sizes of organic halogen compounds during biological treatment of bleached kraft pulp mill effluents, p. 113-120. *In* A. Södergren [ed.] Environmental fate and effects of bleached pulp mill effluents. Swedish Environmental Protection Agency, Stockholm, Report 4031, 394 p.

Laine, M., J. Jokela and M. Salkinoja-Salonen. 1995. Biomobility of organic halogen compounds from contaminated soil - earthworms as a tool. *In* M. Munawar [ed.] Chemicals in the arctic-boreal environment. Ecovision, World Monograph Series, Ontario, Canada: In press.

Maatela, P., J. Paasivirta, M.A. Grachev and E.B. Karabanov. 1990. Organic chlorine compounds in lake sediments. V. Bottom of Baikal near a pulp mill. Chemosphere 21(12):1381-1384.

Martel, P.H., T.G. Kovacs, B.I. O'Connor and R.H. Voss. 1994. A survey of pulp and paper mill effluents for their potential to induce mixed function oxidase enzyme activity in fish. Water Res. 28:1835-1844.

Martinsen, K., A. Lund Kvernheim and G.E. Carlberg. 1994. Distribution of organohalogen in sediments outside pulp mills using sum parameters. Sci. Tot. Environ. 144:47-57.

Paasivirta, J., J. Knuutinen, P. Maatela, R. Paukku, J. Soikkeli and J. Särkkä. 1988. Organic chlorine compounds in lake sediments and the role of the chlorobleaching effluents. Chemosphere 17(1):137-146.

Parker, W.J., E.R. Hall and G.J. Farquhar. 1991. Dechlorination of segregated kraft mill bleach plant effluents in high rate anaerobic reactors, p. 787-795. *In* TAPPI Proc. Environ. Conf. Book 2. 1188 p.

Pellinen, J. 1994. Sorption of high molecular weight chlorolignin on sediment. Chemosphere 28(10):1773-1789.

Salkinoja-Salonen, M., M.-L. Saxelin and J. Pere. 1981. Analysis of toxicity and biodegradability of organochlorine compounds released into the environment in bleaching effluents of kraft pulping, p. 1131-1164. *In* L.H. Keith [ed.] Advances in the identification & analysis of organic pollutants in water. Volume 2. Ann Arbor Science Publishers, Michigan.

Salkinoja-Salonen, M.S., R. Valo, J. Apajalahti, R. Hakulinen, L. Silakoski and T. Jaakkola. 1984. Biodegradation of chlorophenolic compounds in wastes from wood-processing industry, p. 668-676. *In* M.J. Klug, and C.A. Reddy [eds.] Current perspectives in microbial ecology. Am. Soc. Microbiol., Washington, D.C.

Saski, E.K., M.S. Salkinoja-Salonen, A. Vähätalo and K. Salonen. 1991. Degradation and environmental fate of bleached kraft pulp mill effluents studied in freshwater mesocosms, p. 183-189. *In* A. Södergren [ed.] Environmental fate and effects of bleached pulp mill effluents. Swedish Environmental Protection Agency, Stockholm, Report 4031, 394 p.

RECENT ADVANCES IN ENVIRONMENTAL FATE OF CHEMICALS FROM PULP MILLS

John S. Gifford

New Zealand Forest Research Institute, Private Bag 3020, Rotorua, New Zealand

A review of recent research on the environmental fate of chemicals from pulp mills is presented. Key issues that affect the fate of chemicals from pulp mills include the chemistry of effluents, the nature of sediments, interactions with dissolved organic matter, biodegradation/biotransformation of compounds, and the effect of modern pulping technologies. Areas for future research are also considered and these may include determining the physical and chemical properties of compounds produced in modern pulping and bleaching sequences, integrated studies of effluents from modern pulping and bleaching processes in recipients, and the turnover of sediment contaminants in aquatic environments resulting from the historic discharge of effluents from pulp mills that used older technology.

INTRODUCTION

The fate of organic compounds from bleached kraft pulp and paper mill effluents (BKME) in receiving environments has been well researched (Grimvall *et al.* 1991; Lindström-Seppä and Oikari 1990; Neilson *et al.* 1991; Owens *et al.* 1994; Robinson *et al.* 1994; Södergren *et al.* 1991). Such studies have investigated the chemical nature of constituents, partitioning between the sediment and water phase, bioaccumulation and biomagnification, persistence, chemical and biodegradation and how chemical fate affects exposure to organism groups and ecotoxicity.

Understanding the fate of organic constituents derived from pulp and paper mills is fundamental for the chronic hazard assessment of these effluents and for developing appropriate on-site effluent treatment and discharge control strategies to minimize environmental impacts in aquatic ecosystems and effects on human health.

The objectives of this review are to:
- provide a brief overview of key issues affecting the environmental fate of pulp mill derived compounds in recipients;
- summarize research presented at the 2nd International Conference on Environmental Fate and Effects of Bleached Pulp Mill Effluents and highlight key advances since the Environmental Fate and Effects conference held in Saltsjöbaden, Stockholm, 1991;
- consider future research issues concerning the fate of pulp mill derived organic compounds.

FATE OF ORGANIC COMPOUNDS FROM PULP AND PAPER MILL EFFLUENTS

The environmental fate of pulp mill effluent constituents is influenced by many factors. The following section briefly considers the current state of knowledge, experimental methods and gaps in the current knowledge base.

Chemistry of Pulp Mill Effluents

Many organic components of environmental concern from pulp and paper mills are hydrophobic (low water solubility) and have a tendency to accumulate with organic substances such as sediments and biological tissues or to volatilize into the atmosphere. The phase that these compounds will partition into depends upon various physico-chemical and biological properties of the compound such as vapor pressure, solubility, lipophilic tendency, chemical stability, and susceptibility to degradation. In addition, the particular characteristics of the aquatic environment, including temperature, pH, oxygen content, dissolved organic matter, organic carbon content, particle size distribution of suspended sediments, hydrodynamics and biological composition, will influence the fate of chemical constituents.

Pulp mill effluents consist of a complex mixture of organic constituents and the composition can be markedly influenced by the type of pulping and bleaching technology employed, wood furnish and the extent of primary and secondary effluent treatment. Typically the effluent will contain wood-derived carbohydrates, lignin derivatives and extractive components. The extractives (compounds readily soluble in organic solvents) include resin and fatty acids, triglycerides, waxes, sterols and phenolic compounds (Kringstad and Lindström 1984). This group, along with lignin derivatives (high molecular weight compounds) and chlorinated organic substances from pulp bleaching, are among the main chemical components of environmental concern (Kringstad and Lindström 1984; Kukkonen 1992; McLeay et al. 1987).

Much research on the fate of pulp mill effluent derived compounds has focused on organochlorine compounds (chlorinated phenols, catechols, guaiacols, dioxins, adsorbable organic halide (AOX), extractable organic halide (EOX) and high molecular weight chlorinated lignin derivatives) and resin acids (de Sousa et al. 1988; Kringstad and Lindström 1984; McLeay et al. 1987; Neilson et al. 1991; Owens et al. 1994; Robinson et al. 1994). However, recently concern has also arisen regarding the potential role of neutral extractives and other compounds such as phytosterols, polyaromatic hydrocarbons (PAHs) (e.g., chlorophenanthrenes) and chlorinated fatty acids (Lehtinen et al. 1991, 1993; Koistinen et al. 1994; Wesén et al. 1991).

Many of the parent compounds present in pulp mill effluents can undergo physico-chemical and biological degradation to either more readily degradable or more persistent components depending on the chemical and biological conditions of the environment. Neilson et al. 1991 reported that O-methylation of chlorinated phenolic compounds may produce chlorinated anisoles and veratroles that are more lipophilic and have bioconcentration factors 10 to 100 times greater than the initial precursors.

Role of Sediments

Sediments markedly affect the transport and fate of compounds in pulp mill effluents. This has already been considered in the review paper by Carlberg and Stuthridge (1996). Sediments may act as sinks and reduce the uptake of BKME compounds by pelagic organisms. However, organisms inhabiting sediments may accumulate hydrophobic substances that can then be transported to the ecosystem by food chain linkages.

Although sediments have a critical role in determining the fate of compounds, a number of issues have been identified that require careful consideration, namely:
- What are the relationships between the concentration of components present in sediments and their ecotoxicological effects?
- How significant are abiotic and biotic degradation processes in sediments?
- How do binding mechanisms in sediments (such as interactions with inorganic and organic constituents) influence adsorption, desorption and bioavailability of parent and degradation products?
- How available are contaminants in deeper sediments (i.e., greater than 5-10 cm) to aquatic ecosystems?
- What are the roles of suspended sediments and their deposition and resuspension on the residence time and bioavailability of attached hydrophobic compounds?

Two of these issues are considered below.

The concentrations of EOX, TOX (total organic halide) and resin acids found in sediments downstream of BKME discharges have been reviewed by Carlberg and Stuthridge (1996). EOX in sediments at New Zealand and Swedish sites ranged between 6-5270 µg g^{-1} dry weight (DW) for sites ≤5 km from secondary treated BKME discharges. At distances greater than 30 km EOX varied between 0.4-250 µg g^{-1}. Although EOX has been readily found in recipient sediments, Lehtinen et al. (1991) reported that extractable organic chlorine (EOCl) is tightly bound and not readily available to organisms such as Tubifex worms nor presumably to fish grazing on invertebrates living in contaminated sediments.

To date little information exists on the availability of pulp and paper effluent derived contaminants adsorbed and integrated with deeper sediments (>5-10 cm). However, it has been found that a tellinid bivalve (*Macomona liliana*), which inhabits estuarine and coastal sediments of New Zealand to approximately 10 cm, can accumulate chlordane from depths >2 cm below the sediment-water interface (Wilcock et al. 1994). The bivalve typically feeds by means of an inhalant siphon that extends to the surface and is used to suck up fine particles. It has been suggested that bioaccumulation of chlordane from deeper sediments occurs by ingestion of contaminated material over a wide range of depths. Furthermore, chlordane losses from sediments were greater when shellfish were present, which indicated that chlordane was being transported by the mollusc to the sediment surface and subsequent solubilizsation into the overlying water.

Interactions with Dissolved Organic Matter

The interaction of hydrophobic contaminants with organic carbon in aquatic environments is an important consideration (Jaffé 1991). Hydrophobic compounds tend to partition into the organic phase of sedimentary materials. However, observations from a number of studies indicate that binding to dissolved organic matter (DOM) enhances the apparent solubility of contaminants and may stabilize these compounds in the aqueous phase. Both the total concentration and nature of DOM influence the adsorption of hydrophobic contaminants (Kukkonen et al. 1990).

Using a series of model compounds (benzo(a)pyrene, 3,3′,4,4′ tetrachlorobiphenyl and 2,3,7,8-tetra chlorodibenzo-*p*-dioxin, Kukkonen (1992) has shown that chlorolignins in pulp mill effluents have a high capacity to bind lipophilic compounds. BaP had the highest partition coefficient to the lignin-derived DOM although its octanol/water partition coefficient is lower than that of the other two model species. In addition, it was shown that kraft lignin had a higher partition coefficient than chlorolignin from BKME. Tests with *Daphnia magna* indicated that adsorption of organic contaminants to kraft lignin and chlorolignins effectively reduced their bioavailability.

Biodegradation/Biotransformation

The biodegradation and biotransformation of chlorinated organic compounds in sediments and aqueous media have been previously investigated (Neilson 1989; Neilson et al. 1991; Seppälä and Kansanen 1988; Häggbolm et al. 1988; Häggblom and Salkinoja-Salonen 1991). Degradation and biotransformation reactions that may occur in aquatic environments are variable and can be influenced by many factors including:
- the presence of co-substrates and concurrent metabolism
- linkages between growth substrates and biotransformation
- substrate concentration
- bacterial cell densities
- oxidative conditions.

The effects of these factors on biotransformation and biodegradation have been previously reviewed by Neilson (1989) and Neilson et al. (1991) and will not be considered further here.

Chlorolignin has been found to be resistant to microbial activity. Eriksson and Kolar (1985) showed that bacteria from an aerated effluent treatment lagoon of a pulp mill were unable to degrade chlorolignin (<4%) during a 3-month period. However a white-rot fungus could remove 35-45% chlorolignin in 2 months. Incubation of chlorolignin, exhaustively extracted with organic solvents, with dense cell suspensions of bacteria has resulted in the synthesis of chloroveratroles (Neilson *et al.* 1983). However, these compounds were not formed in the controls free of bacteria. These observations have led to the conclusion that low molecular weight chlorinated compounds may be formed from chlorolignins as products of microbial metabolism. In contrast, Eriksson *et al.* (1985) observed that breakdown or dissociation products have also been observed under sterile conditions. Other investigators have suggested that low molecular weight compounds are not real degradation products of chlorolignin but were adsorbed to the chlorolignin preparations and could subsequently be dissociated in aqueous solutions (O'Connor and Voss 1992;, Paasivirta *et al.* 1990).

The degradation of AOX has been shown to occur due to physical and/or (bio)chemical reactions between the organic halogen components and humic compounds. In experiments where BKME was incubated in mesocosms, it was observed that about one-fifth of the total AOX was rapidly removed in humic lake water independent of light (Saski *et al.* 1991). Light was also found to promote AOX removal. These results are consistent with the findings of studies by Grimvall *et al.* (1991) on Lake Vättern in Sweden. A substantial proportion (approximately 85%) of AOX input from a bleached kraft mill was degraded in the water column and/or sediments. This conclusion was based on mass balance calculations over a 36 year period.

Modern Pulping Technologies

Modern pulping technologies are markedly influencing the fate of pulp mill derived compounds. In particular, the substitution of chlorine dioxide (ClO_2) for elemental chlorine in the first stage of bleaching results in approximately a 5- to 10-fold decrease in the formation and discharge of organochlorine compounds (Solomon *et al.* 1994). A concurrent decrease in the degree of chlorine substitution reduces the persistence and toxicity of the resultant compounds. The formation of highly hydrophobic compounds, such as dioxins, is virtually eliminated by increasing chlorine dioxide substitution to 100% (Solomon *et al.* 1994).

The reduction in concentration of chlorinated compounds in BKME following chlorine dioxide substitution has been highlighted from the results of studies on the Wapiti/Smoky River system, Alberta, Canada and the Grande Prairie pulp mill (Joshi and Hillaby 1991; Owens *et al.* 1994). Bleach plant process changes to 70% chlorine dioxide decreased AOX in waste streams by 15-20%. In addition, tri-, tetra- and pentachlorinated organic compounds were decreased by 60-80% and the carbon to chlorine ratio increased by 10-20%.

Use of modern bleaching technology in itself will not lead to a marked decrease of non-chlorinated organic compounds such as resin and fatty acids and phytosterols. However, this may be achieved by applying modern pulp washing and screening methods and water recycling. Secondary treatment also substantially reduces resin acid levels (Robinson *et al.* 1994).

Despite these process changes, several issues appear unresolved, namely:
- What is the fate of lower chlorine substituted compounds in aqueous media?
- What new compounds are formed in modern bleaching sequences?
- Will the focus change from the effects of chlorinated organic compounds on aquatic organisms to other extractive components or constituents of pulp mill effluents?
- What are the fate and effects of historic chlorinated contaminant loads on aquatic ecosystems?

RECENT RESEARCH ON THE FATE OF PULP MILL DERIVED COMPOUNDS

In this section, research highlights presented at the 2nd International Conference on Environmental Fate and Effects of Bleached Pulp Mill Effluents are discussed.

Sediments and Biosolids

The role of sediments in determining the fate and ecotoxicological effects of pulp mill derived compounds was considered by Eriksson et al. (1996). In studies where a marine amphipod was exposed to extracts prepared from sediments taken from different depths of the benthic profile downstream of a pulp mill, marked differences were observed in the toxicity between the surface and depth samples for a polyaromatic fraction. The polyaromatic fraction from the 10-20 cm depth had the highest toxicity.

An assessment of tri- and tetrachlorinated substances being transported from a semi-enclosed recipient to the open sea has indicated that the predominant source of these compounds was historic sediments contaminated by a pulp mill discharge (Nilsson et al. 1996). In contrast to these compounds, 6-chlorovanillin and dichlorinated substances, from the modern bleaching sequences, were exported directly to the sea (i.e., were not readily adsorbed to the sediments). The role of biosolids was investigated by Gomm and Lawrence (1994). These investigators suggested that jet/river mixing and effluent interactions with the receiving water may influence the flocculation of biosolids and hence the transport of hydrophobic contaminants attached to these materials.

New Bleaching Technologies

The implementation of new bleaching technologies has significantly reduced the concentration and mass loading of chlorinated organic substances discharged to the environment. Charlet and Claudio-da-Silva (1994) showed that following mill developments, both AOX and EOX in the final discharge were reduced. Concurrent with these changes were marked decreases in effluent COD, color and toxicity. The effect of modern pulping and bleaching technologies was further illustrated by an approximately 70% reduction in the total load of AOX from the Swedish pulping industry during the period 1989 to 1992 (Enell 1996). For a Canadian mill (Grande Prairie, Alberta), changes to the bleaching process (100% ClO_2 substitution), implementation of condensate stripping and upgraded aeration stabilization basins have contributed to a low level of AOX discharge (0.5 kg air dried metric tonnes [ADMt]), dioxin concentrations below the limit of detection (0.8-4 ppq) and a concomitant decrease in recipient water and suspended sediment concentrations of organochlorine compounds (Swanson et al. 1996).

Degradation Processes

The effect of degradation processes on the fate of pulp mill derived compounds has been investigated using both laboratory and field studies. For Lake Vättern in Sweden, which has a hydraulic residence time of approximately 60 years, Grimvall et al. (1994) reported that tri- and tetrachlorophenols were rapidly removed from the water column following the change from elemental chlorine to chlorine dioxide in the bleaching process for a mill discharging into the lake. The effect of seasons on the degradation and sedimentation of chlorolignin was considered by Saski et al. (1996). From studies using 2 m^3 mesocosms placed into lake environments, it was shown that more organohalogen was adsorbed onto *de novo* sediments depending on the seasons during which experiments were undertaken. In addition, these authors suggested that the binding of halogen to *de novo* sediments was possibly due to a combination of physical precipitation and interaction with biomass. Frisk et al. (1994) also used mesocosms to show that light and temperature were important components affecting the degradation of AOX in Finnish lakes.

Millar et al. (1996) considered the degradation of high molecular weight (HMW) material using fungal strains and sunlight. Both naturally occurring fungi and white-rot fungi were found to be capable of degrading HMW fractions of BKME, which in turn would reduce effluent color, AOX and molecular mass. Furthermore, it was observed that irradiation could result in the release of chlorinated guaiacols and vanillins. However, the processes contributing to this were not elucidated. Banerjee et al. (1994) investigated the fate of AOX in an effluent treatment system following the bleaching of pulp with $^{36}ClO_2$.

Differences in the rate of AOX degradation were observed depending on the depth within the treatment lagoon and distance from surface aerators.

Physical Factors

One of the major factors affecting the fate and distribution of pulp mill compounds is the hydrological and physical conditions of the recipient. Nilsson *et al.* (1996) reported that the morphology of an enclosed recipient and weather variations were two important factors affecting the cycling and distribution of chlorinated organic substances. Tri- and tetrachlorinated substances were more susceptible to resuspension during spring and fall circulation of water within the enclosed recipient. On the other hand, lower chlorinated substances were more likely to be removed by sedimentation during this time. Chambers *et al.* (1994) reported that the rate of dissolved oxygen decline in ice-covered rivers due to pulp mill discharges, and other sources, can be predicted based on the dilution of the effluent in the river.

Dioxins from Pulp Mills

Ongoing studies related to the sources and fate of dioxins from pulp mills indicated that environmental concentrations were markedly reduced following changes from elemental chlorine to ClO_2 in bleaching processes. Gobas *et al.* (1994), using a time-dependent model (ECO fate), found that the actual reduction in dioxin levels in a recipient was faster than predicted by the model. Swanson *et al.* (1996) reported that the concentrations of 2,3,7,8-tetrachlorodibenzo-*p*-dioxin and 2,3,7,8-tetrachloro-dibenzofuran in the effluent from the Grande Prairie, Alberta kraft mill declined after conversion of the bleaching process to 100% ClO_2 substitution and other mill upgrades. The decrease in effluent concentration was also reflected in the environment, with reductions in dioxin concentrations in the river water, suspended sediments and biota. Rappe *et al.* (1994) showed that principal component analysis could elucidate possible sources of dioxins. Based on this approach, it was found that for a pulp mill in southern Mississippi, dioxins in soils and sediments upstream and downstream of the mill were coming from sources other than the pulp mill.

Effluent Treatment

Apart from reducing the effluent concentrations of conventional pollutants such as BOD, COD, TOC and suspended solids, effluent treatment systems also affect the nature and concentration of chlorinated organic substances. Charlet and Claudio-da-Silva (1994) found that chloroguaiacols, chlorosyringols and chlorovanillins could be reduced in the treated effluent after mill modernisation but levels of chlorocatechols were increased. These authors also proposed a model of the degradation profiles in different compartments of the effluent treatment system.

FUTURE RESEARCH

This section provides a brief overview of the discussion related to future research issues on the environmental fate of pulp mill derived compounds.

The substitution of elemental chlorine with ClO_2 for pulp bleaching has led to substantial reductions in the discharge of chlorinated organic compounds. This observation, along with the potential role of non-chlorinated compounds to cause sublethal chronic biological responses (e.g., Parrott *et al.* 1994 and Maclatchy and Van Der Kraak 1994), is likely to encourage more research on the fate of natural wood extractive components or analogues produced during pulping. Research into the physical and chemical properties of the parent compounds as well as possible biodegradation or biotransformation products is required along with fundamental work on their adsorption to solids and bioavailability.

Investigations reported by Werker and Hall (1994) on the surfactant nature of resin and fatty acids and their potential to accumulate in surface foams prompted a suggestion that investigations on the fate of pulp mill derived compounds in water surface layers may be warranted. However, it was noted that no evidence currently existed which indicates that this is a significant environmental transport vector.

The influence of site-specific factors on the fate of pulp mill effluent derived compounds in a particular recipient was identified as a limiting factor for developing general models to assess the fate of compounds. Site-specific factors that need to be considered include the hydraulic and mixing characteristics, concentration and composition of organic substances, and food chain relationships. It was suggested that further integrated field studies considering these aspects are required for effluents from modernized bleached kraft mills and other pulping operations.

Based on current information, historical contaminant sinks in recipient sediments may or may not persist for lengthy periods depending on the site-specific conditions. However, it was identified that further research is required on the long- term fate of contaminants in both freshwater and marine situations and in particular the potential availability of contaminants in buried sediments. Such studies should consider the turnover of sediments and the effect of benthic organisms in returning buried contaminants to the surface of sediments.

Several papers indicated that statistical and modeling techniques will enhance our ability to understand possible sources of chemical contamination and processes affecting the fate of chemical contaminants (Rappe *et al.* 1994; Gobas *et al.* 1994; Mackay *et al.* 1996). Future studies should focus on obtaining fundamental chemical data on chemicals present in pulp mill effluents, especially those from modern pulping technologies. The wider application of multivariate analyses would also help to explain environmental observations.

The research issues discussed above may provide a framework for environmental fate research over the next three years. The results of this work will hopefully be presented at the next environmental fate and effects conference in 1997.

REFERENCES

Banerjee, S., C.L. Williams and S.J. Severtson. 1994. Distribution and fate of AOX and model BKME chloro-organics in secondary treatment systems. Presented at the Second International Conference on Environmental Fate and Effects of Bleached Pulp Mill Effluents, November 6-10, 1994, Vancouver, B.C., Canada.

Carlberg, G.E., and T.R. Stuthridge. 1996. Environmental fate and distribution of substances. *In* Environmental Fate and Effects of Pulp and Paper Mill Effluents, M.R. Servos, K.R. Munkittrick, J.H. Carey and G. Van Der Kraak (ed.), St. Lucie Press, Delray Beach, FL.

Chambers, P.A., A. Pietroniro and G.J. Scrimgeour. 1994. Impact of pulp mill effluent on oxygen concentrations in ice-covered rivers. Presented at the Second International Conference on Environmental Fate and Effects of Bleached Pulp Mill Effluents, November 6-10, 1994, Vancouver, B.C., Canada.

Charlet, P., and E. Claudio-da-Silva, Jr. 1994. Study of the effectiveness of aerated lagoons for removal of conventional pollutants, chlorophenolic compounds and acute toxicity. Presented at the Second International Conference on Environmental Fate and Effects of Bleached Pulp Mill Effluents, November 6-10, 1994, Vancouver, B.C., Canada.

de Sousa, F., L.M. Strömberg and K.P. Kringstad. 1988. The fate of spent bleach liquor material in receiving waters: Characterisation of chloroorganics in sediments. Wat. Sci. Tech. 20(2):153-160.

Enell, M. 1996. Load from the Swedish pulp and paper industry - Nutrients, metals and AOX - Quantities and shares of the total load on the Baltic Sea. *In* Environmental Fate and Effects of Pulp and Paper Mill Effluents, M.R. Servos, K.R. Munkittrick, J.H. Carey and G. Van Der Kraak (ed.), St. Lucie Press, Delray Beach, FL.

Eriksson, A.-K., B. Sundelin, D. Broman and C. Naf. 1996. Reproduction effects of HPLC-fractionated extracts of bottom sediments from a pulp mill recipient on *Monoporeia affinis*. *In* Environmental Fate and Effects of Pulp and Paper Mill Effluents, M.R. Servos, K.R. Munkittrick, J.H. Carey and G. Van Der Kraak (ed.) St. Lucie Press, Delray Beach, FL.

Eriksson, K.-E., and M.-C. Kolar. 1985. Microbial degradation of chlorolignins. Environ. Sci. Tech. 19:1086-1089.

Eriksson, K.-E., M.-C. Kolar, P.O. Ljungquist and K.P. Kringstad. 1985. Studies on microbial and chemical conversions of chlorolignins. Environ. Sci. Tech. 19:1219-1224.

Frisk, T., Ä. Bilaletdin and H. Kaipainen. 1994. Modelling decomposition of organic substances in bleached pulp mill effluents. Presented at the Second International Conference on Environmental Fate and Effects of Bleached Pulp Mill Effluents, November 6-10, 1994, Vancouver, B.C., Canada.

Gobas, A.P.C., K. Lien and J. Pasternak. 1994. Development and verification of an environmental fate and food-chain bioaccumulation model of organochlorine emissions in the Fraser-Thompson River basin. Presented at the Second International Conference on Environmental Fate and Effects of Bleached Pulp Mill Effluents, November 6-10, 1994, Vancouver, B.C., Canada.

Gomm, L., and G. Lawrence. 1994. Mixing and transport of pulp mill biosolids in the Fraser River. Presented at the Second International Conference on Environmental Fate and Effects of Bleached Pulp Mill Effluents, November 6-10, 1994, Vancouver, B.C., Canada.

Grimvall, A., H. Borén, S. Jonsson, U. Lundström and R. Sävenhed. 1991. Long term accumulation and degradation of bleach-plant effluents in receiving waters. In Environmental Fate and Effects of Bleached Pulp Mill Effluents. Swedish Environmental Protection Agency Report 4031. Proceedings of a SEPA Conference held at Saltsjöbaden, Stockholm, Sweden, November 19-21, 1991, pp. 74-84.

Grimvall, A., H. Borén, S. Jonsson, I. Pavasars and C. Johansson. 1994. Traces of modern bleached-kraft mill effluents in receiving waters. Presented at the Second International Conference on Environmental Fate and Effects of Bleached Pulp Mill Effluents, November 6-10, 1994, Vancouver, B.C., Canada.

Häggblom, M.M., and M.S. Salkinoja-Salonen. 1991. Biodegradability of chlorinated organic compounds in pulp bleaching effluents. Wat. Sci. Tech. 24(3/4):161-170.

Häggblom, M.M., J.H.A. Apajalahti and M.S. Salkinoja-Salonen. 1988. Degradation of chlorophenolic compounds occurring in pulp mill effluents. Wat. Sci. Tech. 20(2):205-208.

Jaffé, R. 1991. Fate of hydrophobic organic pollutants in the aquatic environment: A Review. Environ. Poll., 69:237-257.

Joshi, B.K., and B. Hillaby. 1991. Effects of process improvements on pulp mill effluent characteristics. In Environmental Fate and Effects of Bleached Pulp Mill Effluents. Swedish Environmental Protection Agency Report 4031. Proceedings of a SEPA Conference held at Saltsjöbaden, Stockholm, Sweden, November 19-21, 1991, pp. 101-109.

Koistinen, J., J. Paasivirta, T. Nevalainen and M. Lahtiperä. 1994. Chlorophenanthrenes, alkylchlorophenanthrenes and alkylchloronaphthalenes in kraft pulp mill products and discharges. Chemosphere 28(7):1261-1277.

Kringstad, K.P., and K. Lindström. 1984. Spent liquors from pulp bleaching. Environ. Sci. Tech. 8:236A-248A.

Kukkonen, J. 1992. Effects of lignin and chlorolignin in pulp mill effluents on the binding and bioavailability of hydrophobic pollutants. Water Research 26(11):1523-1532.

Kukkonen, J., J.F. McCarthy and A. Oikari. 1990. Effects of XAD-8 fractions of dissolved organic carbon on the sorption and bioavailability of organic micropollutants. Arch. Environ. Contam. and Toxicol. 19:551-557.

Lehtinen, K-J., A. Oikari, J. Hemming, K. Mattsson and J. Tana. 1991. Effects on rainbow trout (*Oncorhyncus mykiss*) of food contaminated with solid fraction of bleached kraft mill effluents (BKME). In Environmental Fate and Effects of Bleached Pulp Mill Effluents. Swedish Environmental Protection Agency Report 4031. Proceedings of a SEPA Conference held at Saltsjöbaden, Stockholm, Sweden, November 19-21, 1991, pp. 293-309.

Lehtinen, K.-J., J. Tana, K. Mattsson, J. Härdig, P. Karlsson, C. Grotell, S. Hemming, C. Engström and J. Hemming. 1993. Ecological impact of pulp mill effluents. Part 1: Physiological responses and effects on survival, growth and parasite frequency in fish exposed in mesocosms to treated total mill effluents from production of bleached kraft pulp (BKME), thermomechanical pulp and phytosterols. Published by the National Board of Waters and the Environment, Finland, Report No. A133.

Lindström-Seppä, P., and A. Oikari. 1990. Biotransformation and other toxicological and physiological responses in rainbow trout (*Salmo gairdneri* Richardson) caged in a lake receiving effluents of pulp and paper industry. Aquat. Toxicol. 16:187-204.

Mackay, D., J.M. Southwood, J. Kukkonen, W. Ying Shiu, D.D. Tam, D. Varhaníckova and R. Lun. 1996. Modelling the fate of 2,4,6-trichlorophenol in pulp and paper mill effluent in Lake Saimaa, Finland. In Environmental Fate and Effects of Pulp and Paper Mill Effluents, M.R. Servos, K.R. Munkittrick, J.H. Carey and G. Van Der Kraak (ed.), St. Lucie Press, Delray Beach, FL.

Maclatchy, D.L., and G. Van Der Kraak. 1994. The plant sterol β-sitosterol decreases reproductive fitness in goldfish. Presented at the Second International Conference on Environmental Fate and Effects of Bleached Pulp Mill Effluents, November 6-10, 1994, Vancouver, B.C., Canada.

McLeay and Associates. 1987. Aquatic toxicity of pulp and paper mill effluent: A review. Environment Canada EPS 4/PF/1, 191 pp.

Millar, K.R., J.H. Carey, B.K. Burnison, H. Lee and J.T. Trevors. 1996. Degradation of the high molecular weight fraction of bleached kraft mill effluent by biological and photochemical processes. *In* Environmental Fate and Effects of Pulp and Paper Mill Effluents, M.R. Servos, K.R. Munkittrick, J.H. Carey and G. Van Der Kraak (ed.), St. Lucie Press, Delray Beach, FL.

Neilson, A. 1989. Factors determining the fate of organic chemicals in the environment: The role of bacterial transformations and binding to sediments. *In* Chemicals in the Aquatic Environment: Advanced Hazard Assessment, Lars Landner (ed.), Springer-Verlag.

Neilson, A.H., A.-S. Allard, P-Å. Hynning, M. Remberger and L. Landner. 1983. Bacterial methylation of chlorinated phenols and guaiacols: Formation of veratroles from guaiacols and high molecular weight chlorinated lignin. Appl. Environ. Microbiol. 45:774-783.

Neilson, A.H., A.-S. Allard, P-Å. Hynning and M. Remberger. 1991. Distribution, fate and persistence of organochlorine compounds formed during production of bleached pulp. Toxicol. Environ. Chem. 30:3-41.

Nilsson, P., L. Brydsten, M. Enell and M. Jansson. 1996. Turnover of chloroorganic substances in a Bothnian Sea recipient receiving bleached kraft pulp mill effluents (BKME). *In* Environmental Fate and Effects of Pulp and Paper Mill Effluents, M.R. Servos, K.R. Munkittrick, J.H. Carey and G. Van Der Kraak (ed.), St. Lucie Press, Delray Beach, FL.

O'Connor B.I., and R.H. Voss. 1992. A new perspective (sorption/desorption) on the question of chlorolignin degradation to chlorinated phenolics. Environ. Sci. Tech. 26(3):556-560.

Owens, J.W., S.M. Swanson and D.A. Birkholz. 1994. Environmental monitoring of bleached kraft pulp mill chlorophenolic compounds in a northern Canadian river system. Chemosphere 29:89-109.

Paasivirta, J., H. Hakala, J. Knuutinen, T. Otollinen, J. Särkkä, L. Welling, R. Paukku and R. Lammi. 1990. Organic chlorine compounds in lake sediments. III. Chlorohydrocarbons, free and chemically bound chlorophenols. Chemosphere 21:1355-1370.

Parrott, J.L., B.K. Burison, P.V. Hodson, M.E. Comba and M.E. Fox. 1994. Retene-type compounds - inducers of hepatic mixed function oxygenase (MFO) in rainbow trout (*Oncorhynchus mykiss*)? Presented at the Second International Conference on Environmental Fate and Effects of Bleached Pulp Mill Effluents, November 6-10, 1994, Vancouver, B.C., Canada.

Rappe, C., L.-O. Kjeller, C. Lau and H. Fiedler. 1994. Patterns and sources of polychlorinated dioxins and dibenzofurans found on soils and sediment samples in South Mississippi. Presented at the Second International Conference on Environmental Fate and Effects of Bleached Pulp Mill Effluents, November 6-10, 1994, Vancouver, B.C., Canada.

Robinson, R.D., J.H. Carey, K.R. Solomon, I.R. Smith, M.R. Servos and K.R. Munkittrick. 1994. Survey of receiving-water environmental impacts associated with discharges from pulp mills. 1. Mill characteristics, receiving-water chemical profiles and lab toxicity tests. Environ. Toxicol. Chem. 13(7):1075-1088.

Saski, E.K., M.S. Salkinoja-Salonen, A. Vähätalo and K. Salonen. 1991. Degradation and environmental fate of bleached kraft mill effluents studied in freshwater mesocosms. *In* Environmental Fate and Effects of Bleached Pulp Mill Effluents. Swedish Environmental Protection Agency Report 4031. Proceedings of a SEPA Conference held at Saltsjöbaden, Stockholm, Sweden, November 19-21, 1991, pp. 183-189.

Saski, E., A. Vähätalo, K. Salonen and M. Salkinoja-Salonen. 1996. Mesocosm simulation on sediment formation induced by biologically treated bleached kraft pulp mill waste water in freshwater recipient. *In* Environmental Fate and Effects of Pulp and Paper Mill Effluents, M.R. Servos, K.R. Munkittrick, J.H. Carey and G. Van Der Kraak (ed.), St. Lucie Press, Delray Beach, FL.

Seppälä, J.J., and P.H. Kansanen. 1988. Fate of discharges of total organic chlorine and chlorophenol compounds in Lake Etelä-Saimaa, Finland. Wat. Sci. Tech. 20(2):199.

Södergren, A., M. Adolfsson-Erici, B-E. Bengtsson, P. Jonsson, S. Lagergren, L. Rahm and F. Wulff. 1991. Environmental effects of bleached pulp mill effluents discharged into the Baltic Sea. *In* Environmental Fate and Effects of Bleached Pulp Mill Effluents. Swedish Environmental Protection Agency Report 4031. Proceedings of a SEPA Conference held at Saltsjöbaden, Stockholm, Sweden, November 19-21, 1991, pp. 199-202.

Solomon, K., H. Bergman, R. Huggett, D.B. Mackay and B. Mckague 1994. A review and assessment of the ecological risks associated with the use of chlorine dioxide for the bleaching of pulp. *In* Conference Proceedings of the International Pulp Bleaching Conference, Vancouver, B.C., Canada, June 13-16, 1994, pp. 145-161.

Swanson, S.M, D.C. Pryke, G.R. Bouree and J.W. Owens. 1996. Concentrations of chlorinated organic compounds in various receiving environment compartments following implementation of technological changes at an Alberta pulp mill. *In* Environmental Fate and Effects of Pulp and Paper Mill Effluents, M.R. Servos, K.R. Munkittrick, J.H. Carey and G. Van Der Kraak (ed.), St. Lucie Press, Delray Beach, FL.

Werker, A., and E. Hall. 1994. Surfactancy governing the toxicant fate in pulp mill effluent. Presented at the Second International Conference on Environmental Fate and Effects of Bleached Pulp Mill Effluents, November 6-10, 1994, Vancouver, B.C., Canada.

Wesén, C., K. Martinsen, G. Carlberg and H. Mu. 1991. Chlorinated carboxylic acids are major chloroorganic compounds in fish exposed to pulp bleach liquors. *In* Environmental Fate and Effects of Bleached Pulp Mill Effluents. Swedish Environmental Protection Agency Report 4031. Proceedings of a SEPA Conference held at Saltsjöbaden, Stockholm, Sweden, November 19-21, 1991, pp. 207-218.

Wilcock, R.J., R.D. Pridmore, G.L. Northcott, J.E. Hewitt, S.F. Thrush and V.J. Cummings. 1994. Uptake of chlordane by a deposit feeding bivalve: Does the depth of sediment contamination make a difference? Environ. Toxicol. Chem. 13(9):1535-1541.

SECTION III

BIOACCUMULATION OF SUBSTANCES FROM PULP AND PAPER MILLS TO FISH AND WILDLIFE

Numerous factors influence the distribution and ultimately the bioaccumulation of chemicals from the environment. The chemical speciation, including the interaction with organic matter, will strongly influence the movement of chemicals across biological membranes. The life history, ecological preferences and chemical characteristics (lipid content, metabolism, etc.) of the organisms will also influence the exposure and bioavailability of chemicals. A better understanding of the bioaccumulation process will enhance our ability to predict the concentration of specific chemicals in biota.

BIOACCUMULATION OF BLEACHED KRAFT PULP MILL RELATED ORGANIC CHEMICALS BY FISH

Derek C.G. Muir[1] and Mark R. Servos[2]

[1]Freshwater Institute, Department of Fisheries and Oceans, 501 University Crescent, Winnipeg, Manitoba R3T 2N6 Canada
[2]Great Lakes Laboratory for Fisheries and Aquatic Sciences, Department of Fisheries and Oceans, 867 Lakeshore Road, Burlington, Ontario L7R 4A6 Canada

Bioaccumulation of organic chemicals by fishes is a function of the properties of the chemical, the physiological characteristics of the animal and the physical and biological characteristics of the receiving environment. Most low molecular weight organics in bleached kraft mill effluents (BKMEs), such as chlorophenolics and aliphatics, have log K_{ow} values in the 1 to 5 range and are accumulated mainly via uptake over the gills. Bioaccumulation of these compounds can be predicted with bioconcentration factors (BCF) if dissolved concentrations in water are known, although BCFs may be overestimated because of biotransformation. Very hydrophobic BKME components such as polychlorinated dioxins/furans (PCDD/Fs) are not only dissolved but are associated with particles and dissolved organic carbon in effluent and in receiving waters. PCDD/Fs are accumulated by fish mainly via the diet. Bioaccumulation of PCDD/Fs can be predicted from biota-sediment accumulation factors or with food chain models that include both water-borne and dietary pathways of accumulation. Sediment-water disequilibrium and food chain effects need to be taken into account when using BSAFs and food chain modeling to predict bioaccumulation.

INTRODUCTION

A vast array of organic substances are discharged in bleached kraft pulp and paper mill effluents (BKMEs) and yet only a small fraction of these chemicals have actually been identified in fishes in receiving waters (Servizi *et al.* 1994). This is not surprising because bioaccumulation of organic chemicals by aquatic biota is a function of the properties of the chemical, the physiological characteristics of the animal and the physical and biological characteristics of the receiving environment (Connell 1988). Direct exposure of aquatic animals to BKME-related organics may occur, but if these compounds are not accumulated, or do not reach the site of action, they will not elicit a physiological effect. Knowledge of bioaccumulation potential is therefore very important in assessing both direct exposure of aquatic organisms and exposure via the food chains. Many organics will be rapidly broken down in the environment by physical and microbiological processes in treatment ponds. Other compounds may be persistent and potentially toxic but will not exert a toxic effect because of low environmental concentrations. Others will be accumulated but biotransformed to less toxic products. The actual appearance of individual BKME-related compounds at a given site also depends upon the environmental emissions, i.e., the types and relative abundance of chemicals being released by a specific mill, which are related to the wood furnish, the process, the bleaching sequence and subsequent treatment (Voss *et al.* 1988; Berry *et al.* 1991; Luthe *et al.* 1992).

Our objective is to review the physical/chemical and biological factors which influence the bioavailability and accumulation of BKME-related organics in fish. A fundamental understanding of the bioaccumulation of contaminants in the environment will aid in predicting the tissue residues and potential effects (on biota or on fish consumption) as well as in directing effective remediation efforts and design of field monitoring programs.

BIOACCUMULATION

Bioaccumulation can be simply viewed as the process of transfer of a chemical from water and/or diet into the organism. There are three major routes of exposure: gills, diet and dermal. For large fish, the dermal route of exposure may represent less than a few percent of the uptake relative to gill ventilation, while in small (or larval) fish, which have a much greater surface to volume ratio, the dermal route of exposure can be significant (Saarikoski *et al.* 1986; Rombough and Moroz 1990). Although uptake by reparatory organs (gills) exposed to water is an important route into aquatic animals, the absorption from the diet is particularly important for hydrophobic contaminants.

Bioconcentration is the accumulation of the dissolved phase of the compound in water by the organism via respiratory surfaces or skin (Connell 1988; U.S. EPA 1993). The compound enters the organism by diffusion across the membrane into the blood. This process can be viewed as driven by the thermodynamic equilibrium between the organism and water such that the organic chemical diffuses from phases of high to low fugacities until fugacities in both phases are equal (Mackay 1982). The bioconcentration factor (BCF) for aqueous exposure to biota is defined as:

$$BCF = C_F / C_{WD} \quad [1]$$

where C_F is the concentration in fish and C_{WD} is the (dissolved) concentration in the water. A BCF can be calculated at any time during the process of bioconcentration. The BCF is distinguished from the bioaccumulation factor (BAF), which is the ratio C_F/C_{WD} where the uptake pathway is from both water and food, as would be the case for wild fish. A steady state BCF will be reached at which the BCF approaches a constant value. The processes of uptake and elimination can be modeled with a single compartment model assuming first-order kinetics:

$$dC_F/dt = k_1 \cdot C_{WD} - k_2 \cdot C_F \quad [2]$$

where k_1 is the uptake rate constant (mL g^{-1} d^{-1}), and k_2 is the depuration rate constant (d^{-1}). The uptake rate constant consists of gill and skin uptake rates, while k_2 consists of individual rates for metabolism, fecal and gill elimination as well as a growth rate (U.S. EPA 1993; Gobas 1993). If the water concentration is assumed to be constant, then:

$$C_F(t) = (k_1/k_2) \cdot C_{WD} \cdot (1 - e^{-k_2 \cdot t}) \quad [3]$$

The steady state BCF (C_F/C_{WD}) can then be estimated from the ratio of k_1/k_2 when the term $(1 - e^{-k_2 \cdot t})$ approaches 1 (where $t \to \infty$). If the water concentration is zero, k_2 can be estimated from the slope of the first-order decay curve as concentrations of the chemical depurate over time. The value of the uptake rate constant k_1 is determined by diffusion and flow (e.g., gill ventilation rates). The depuration rate, k_2, is simply k_1/BCF and the time required to achieve equilibrium is therefore BCF/k_1. The rate at which the chemical will approach equilibrium is therefore dependent on the BCF. Very hydrophobic compounds will not achieve equilibrium in short term laboratory exposures but equilibrium BCFs can be estimated from the ratio of k_1/k_2.

For nonpolar compounds the BCF is correlated to the octanol-water partition coefficient (K_{OW}). Octanol is assumed to be a surrogate for lipid and the tissue concentrations are a result of the partitioning of the chemical into the lipid within the organism. Veith *et al.* (1979) developed an empirical relationship of log BCF and log K_{OW} with data from more than 50 compounds. Using the data from Veith *et al.*, Mackay (1982) derived the relationship:

$$\log BCF = \log K_{OW} - 1.32 \pm 0.25 \ (r^2 = 0.95) \text{ or } BCF = 0.048 K_{OW} \quad [4]$$

Connell (1988) concluded that the relationship derived by Mackay (1982) was satisfactory for a range of nonpolar organics with log K_{ow} >2 and <6. The BCF data used by Mackay (1982) were based on wet weight concentrations. Assuming 5% lipid content, the lipid weight BCF approximately equals K_{ow}, i.e., $BCF_L = (0.048/0.05) \cdot K_{ow}$.

For most persistent organochlorine compounds the distribution within the fish is related to lipid fraction of the tissue (Parkerton et al. 1993). Lipid normalized concentrations (C_{FL}) are useful for comparisons among tissues and between fish, and thus lipid-based BAFs are considered less site and species specific:

$$BAF_L = C_{FL}/C_{WD} \qquad [5]$$

Dietary accumulation can be treated with a similar first-order single compartment model (Bruggeman et al. 1981). The change in concentration in fish with time is given by:

$$dC_F/dt = \alpha \cdot F \cdot C_{FOOD} + k_1 \cdot C_w - k_2 \cdot C_F \qquad [6]$$

where α = assimilation (or absorption) efficiency of the chemical from food, F = feeding rate and C_{FOOD} is concentration of pollutant in food. In practice the $\alpha \cdot F \cdot C_{FOOD}$ term is determined in a separate experiment where concentration in water is zero, i.e.:

$$C_F = \alpha \cdot F \cdot C_{FOOD} (1 - \exp(-k_2 \cdot t)/k_2 \qquad [7]$$

and the biomagnification factor (BMF) is obtained as $t \to \infty$:

$$BMF = \alpha \cdot F/k_2 = C_F/C_{FOOD} \qquad [8]$$

Assimilation efficiencies from food can be related to log K_{ow} (Thomann 1989; Gobas et al. 1988) although the relationship is biphasic, tending to increase for log K_{ow} >1 to <6 and decline for very hydrophobic organochlorines with log K_{ow} from 6 to 8. Biomagnification can also be viewed thermodynamically as the situation in which the fugacity of the chemical in the organism is greater than the fugacity in the food (Connolly and Pederson 1988; Gobas et al. 1993). This elevation in fugacity has been shown to occur as a result of food digestion and absorption which increases the fugacity of nonmetabolizable, hydrophobic chemicals in the digestive tract (Gobas et al. 1994b). Thus assimilation involves diffusive transfer across the gut membrane from phases of high fugacity to an internal compartment at lower fugacity.

Food chain bioaccumulation models describe the extent of bioaccumulation as a result of trophic interactions in the food chain as well as uptake from water. The models of Thomann et al. (1992) and Gobas (1993) are kinetics-based food chain models which consist of a series of differential mass balance equations describing chemical uptake from water, uptake from food consumption and elimination at each trophic level. The Gobas model consists of 8 compartments, water, sediment, phytoplankton, benthic invertebrates and 4 fish compartments representing different fish weights. Each fish compartment can feed on various fractions of each compartment (i) i.e., phytoplankton, benthos or other fish. At steady state the concentrations in fish are given by:

$$C_F = (k_1 \cdot C_{WD} + k_D \cdot \Sigma P_i \cdot C_{D,i})/(k_2 + k_E + k_G) \qquad [9]$$

where $k_D = \alpha \cdot F$ = rate constant for chemical uptake from food (kg food kg^{-1} fish d^{-1}), P_i = fraction of the diet of the fish consisting of prey i, $C_{D,i}$ = the chemical concentration in prey i (µg kg^{-1}), k_2 = elimination via the gills to water, k_E = elimination by fecal egestion and k_G = the rate constant of growth dilution. In the Gobas model phytoplankton and zooplankton are assumed to accumulate chemical only from water (i.e., equation 3) and equilibrium partitioning is assumed between sediment organic carbon and benthic organisms.

The Thomann model includes uptake by benthic organisms from sediment particles, pore water or overlying water and does not assume equilibrium partitioning. Forage fish are linked to benthic organisms for a portion of their diet as well as to zooplankton. Chemical concentrations in fishes and zooplankton are calculated with equations similar to equation 9, although the Thomann model does not specifically consider fecal egestion as a route of elimination. Both models derive feeding rates from bioenergetics relationships. Uptake rates from water (k_1) are derived internally by both models from gill ventilation rates and gill uptake efficiency (E_W). But the Thomann model links uptake to respiration rate whereas the Gobas model derives k_1 and E_W from relationships with K_{OW} observed for hydrophobic compounds (Gobas and Mackay 1987). Trophic transfer and magnification are then calculated at steady state from feeding preferences which are also entered into the model. Both models successfully predicted concentrations in the Lake Ontario food chain. The models also demonstrate that chemicals with a log K_{OW} of <4 are primarily taken up directly from water, while chemicals with log K_{OW} >4 have an increasing proportion attributed to uptake via assimilation from food.

FACTORS AFFECTING BIOACCUMULATION OF BKME-RELATED ORGANICS

Pharmacological Availability

The physical/chemical properties of the chemical and the organism's ability to excrete the compound are major factors determining the fate and disposition of organic chemicals within the animal. The pharmacological bioavailability is the dose that reaches the systemic circulation for distribution throughout the body. The toxic response is dependent on the concentration that reaches the site of toxic action and the fraction that interacts with cellular constituents to cause a specific response leading to a cellular change (McCarty 1986; McKim 1994). The dose that reaches the site of action is dependent on many internal processes including absorption, transport, biotransformation and excretion.

As illustrated in Fig. 1, diffusive transfer across gill, gut or internal membranes is influenced by the hydrophobicity and polarity of the molecule as well as by molecular size and shape and biodegradability. Hydrophobic organics which are not metabolized and have a molecular weight less than about 600 daltons are highly bioaccumulative. This group includes polychlorinated biphenyls, organochlorine pesticides and a few BKME-related compounds such as 2,3,7,8-substituted polychlorinated dibenzo-p-dioxin and dibenzofurans (PCDD/F). A large number of non-2,3,7,8-substituted PCDD/Fs are very hydrophobic (log K_{OW} >6) but do not bioaccumulate to the expected extent because of the ability of fishes to biotransform them, via hydroxylation and glucuronide conjugation, to more polar compounds (Opperhuizen and Sijm 1990; Muir et al. 1992a).

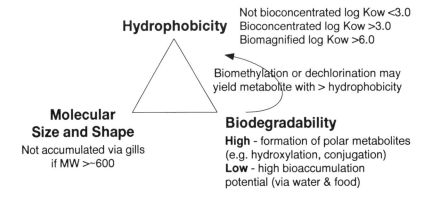

Figure 1. *Physical-chemical factors influencing bioaccumulation of organic chemicals by fish.*

The relationship of BCF to K_{ow} for BKME-related organics is illustrated in Fig. 2 with the data compiled by Healey et al. (1994) for 63 chemicals. Excluding the PCDD/Fs, Healey et al. found 109 laboratory-derived BCFs for fish along with log K_{ow} values for 46 BKME-related phenolic and neutral organics (Fig. 2). A weak but statistically significant relationship exists between log BCF (wet wt) and log K_{ow} for these 109 values: log BCF = -0.03 ± 0.39 + 0.63 (± 0.09) log K_{ow} (r^2 = 0.30; N = 109). The weak relationship with a slope <<1 (slope = 1 is expected if K_{ow} is equal to BCF) is the result of using a data set consisting mainly of chlorophenolics and hydrocarbons with limited Cl substitution which are readily metabolized by fish. The BCF results in Fig. 2 are based on wet weight concentrations; lipid normalization would bring the results closer to a slope of 1. Some of the laboratory-derived BCFs also may not be equilibrium values especially for compounds with log K_{ow} >4.

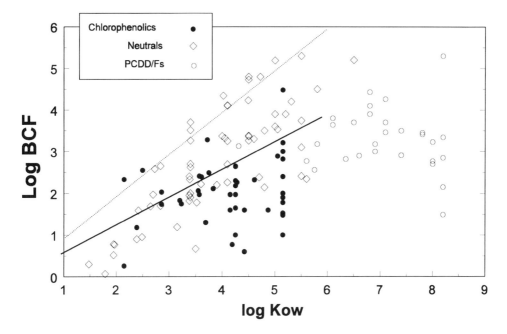

Figure 2. The relationship of BCF to K_{ow} for BKME-related organics with the data compiled by Healey et al. (1994) for 63 chemicals. BCFs are based on wet weight concentrations in fish exposed in static or flow-through laboratory exposures.

Other BCF/K_{ow} correlations, such as the data set used by Mackay (1982) and Connell and Hawker (1988), consisted mainly of BCFs for recalcitrant PCBs and organochlorine pesticides. Chlorophenolics, which are primarily anions in natural waters, are taken up across the gills but are then conjugated and excreted in the bile mainly as glurcuronide and sulfate esters (Oikari and Holmbom 1986; Wachtmeister et al. 1991). The neutral form of most organic chemicals is the species that most efficiently penetrates the cell membranes by diffusion and is well modeled by a simple correlation with K_{ow}. Above the pK_a of a weak acid the compound becomes increasingly ionized and therefore less readily available for transport across the membranes. A decrease of pH therefore causes an increase in the uptake and toxicity of organic acids while it results in a decrease for organic bases (Stehyl and Hayton 1990; Holcombe et al. 1980; Spehar et al. 1985; Doe et al. 1988).

The decline in BCFs of the PCDD/Fs with increasing K_{ow} (Fig. 1) illustrates the effects of molecular size as well as metabolizability on BCF. Octachlorodioxin (OCDD), with log K_{ow} = 8.2, is of sufficiently large molecular size and volume that uptake across the aqueous boundary layers at the gills and diffusion through membranes may be limited (Opperhuizen et al. 1985; Connell and Hawker 1988). It has been argued that these BCF data for very hydrophobic organics are flawed by poor estimates of the dissolved concentrations which are difficult to determine (Geyer and Muir 1993). But assimilation efficiencies for PCDD/Fs across the gut wall during dietary exposure also decline with increasing K_{OW} (Muir et al. 1992a) which suggests that size may be a factor in the low observed BCFs, along with hydrophobicity.

Connell and Hawker (1988) were able to fit a series of log BCF vs. log K_{OW} data with a parabolic distribution similar to that in Fig. 2 using a quadratic relationship. They attributed part of the low bioaccumulation of the high log K_{OW} compounds to decreased lipid solubility. The same authors concluded that k_1 was lipid-phase controlled for compounds with low K_{OW} and aqueous-phase controlled for the very hydrophobic chemicals. The k_2 values were influenced by the same factors as k_1 but were also affected by the decreasing lipid solubility of the very hydrophobic compounds such as OCDD.

Knowledge of BCFs and biotransformation of BKME-related organics is confined to a relatively small number of chlorophenolic compounds, neutrals and PCDD/Fs (Healey et al. 1994). The bioaccumulation and metabolization of the majority of BKME-related organics in fish has not been studied. Nevertheless the bioaccumulation potential of major classes of these components can be assessed on a relative basis from the structure and hydrophobicity (Table 1). We have attempted this for the classes of compounds in Table 1 by considering several rules of thumb for metabolism by fish (Sijm and Opperhuizen 1989). (1) Chlorinated aromatics and aliphatics will be degraded more slowly than non-chlorinated compounds because substitution of oxygen by the P450 mixed function oxidase enzyme system will be hindered by the presence of chlorine on the ring. Of course, the number and positioning of the chlorine, such as the availability of vicinal unsubstituted positions on the ring, will be important. (2) Chlorophenolics (including syringols, vanillins and guaiacols) are likely to be biotransformed more rapidly than neutral aromatics with alkyl substitutents (e.g., cymenes, thiophenes) because the hydroxyl group can be conjugated as a glucoronide or sulfate esters. (3) For fused ring compounds such as PCDD/Fs and alkylchloronaphthalenes and dibenzofurans, positioning of the chlorine and alkyl substituents is likely to be important in determining which components are more resistant to oxidation by the P450 system. But in general these latter groups will have greater bioaccumulation potential because of higher hydrophobicity.

There is evidence from field studies of BKME-related chemicals in fish to support the above assertions. For example, Rantio (1992) found chlorocymenes to have higher bioaccumulation potential in fish than alkylchloronaphthalenes, alkylchlorobibenzyls, alkylchloronaphthalenes or polychlorocymenenes. Alkylchlorodibenzofurans have been reported in sediments but were not observed in fish, suggesting they have lower bioaccumulation potential than the unsubstituted congeners (Kuehl et al. 1987; Buser et al. 1989).

Partitioning and Fate: Physical/Chemical Factors Affecting Bioavailability

The fate and bioavailability of a chemical to fishes will depend on the properties of the chemical and on the physical, chemical and biological environment into which it is released. The extent of disequilibrium between water and sediment-associated compound has also been recognized as an important factor when predicting bioaccumulation (Thomann et al. 1992; U.S. EPA 1993). Bioavailability refers to the extent to which pollutants associated with sediments or suspended or dissolved organic carbon in the water column are available for uptake by biota (Dickson et al. 1994). For most organic compounds it is the unionized truly dissolved fraction that is transported across biological membranes by passive diffusion (Hunn and Allen 1974). Organic ions such as phenolics or carboxylic acids have only a limited ability to cross biological membranes (Saarikoski et al. 1986). Many factors such as dissolved or particulate organic carbon, pH and kinetic limitations may reduce the truly dissolved concentration (which is directly proportional to the fugacity) of a contaminant and therefore reduce the environmental bioavailability (Fig. 3).

Table 1. Physical/chemical characteristics and potential biodegradability by fish of major organic chemical groups in BKME.

Chemical Group	Polarity	log K_{ow}	Biotransformation[1]
Aromatic acids and phenolics	++	1-2	+++
Resin and fatty acids	++	1-2.5	+++
Cymenes and cymenenes	-	?	+++
Chlorophenols and guaiacols	+	2-5	++
Chloroveratroles and anisoles	-	4-6	+/-
Chlorothiophenes	-	3-5	+
Chlorocymenes	-	4-6	+/-
Chlorovanillins	+	1.5-2.5	+
Chlorosyringols	+	3-4	++
Chloroaliphatic acids	+'	?	++
Chlorodimethylsulfones	+	?	++
Chloroaldehydes and ketones	+	?	++
Chloro-(C1-C3)-naphthalenes and dibenzofurans	-	>5	+/-
PCDD/Fs	-	3-8	+/-
Chloroterpenes (Cl1-Cl2)	-	3-4	++
Chloroalkanes and -enes (C1-C6)	-	2-<4	++
Chlorolignins	++	Low	++

[1] Relative rate of biotransformation by fishes. +++ = rapidly transformed, +/- = slow or negligible.

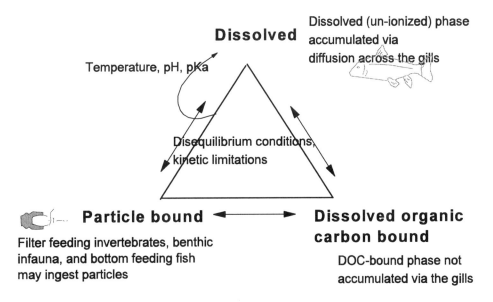

Figure 3. Physical-chemical factors influencing bioavailability of organic chemicals to fish and invertebrates.

The environment into which an effluent is released will vary both spatially and temporally. The effluent itself may affect the physical/chemical environment (e.g., pH, organic carbon, etc.) in the receiving water. The initial dilution and rate of advection will be critical in determining the exposure of biota to the critical components of the effluent, particularly for those compounds which are accumulated directly from the aqueous phase. The movement, characteristics and abundance of suspended organic carbon (POC) and colloidal or dissolved organic carbon (DOC) will strongly influence the fate and bioavailability of the more hydrophobic contaminants in the effluent. Processes which are critical to predicting exposure concentrations in the receiving waters include photolysis, biodegradation, volatilization, sedimentation and resuspension (Mackay and Southwood 1992).

The dissolved concentration of chemical in water is a key parameter for estimating BAFs and the extent of disequilibrium between water, sediments and biota. The total concentration in the water (C_{WT}) is the sum of dissolved, POC (C_{POC}) and DOC-bound (C_{DOC}) fractions:

$$C_{WT} = C_{WD} + POC \cdot C_{POC} + DOC \cdot C_{DOC} \qquad [10]$$

where POC and DOC are the mass fractions of these phases in water. The fraction dissolved can be calculated from the equation:

$$f_D = 1/(1 + POC \cdot K_{POC} + DOC \cdot K_{DOC}) \qquad [11]$$

where K_{POC} and K_{DOC} are the particle organic carbon and dissolved organic carbon partition coefficients, respectively. Several approaches can be used to calculate K_{POC} and K_{DOC}, which are difficult to measure because a dissolved concentration in water is required. The U.S. EPA (1993) recommended that K_{POC} be assumed to be equal to K_{DOC} and K_{OW} be used as a surrogate for the partition coefficient. The f_D for many low molecular weight organics in BKME (Table 1) which have log K_{OW} <5 will be high (i.e., ≈1) unless levels of POC or DOC are elevated. The composition of the organic matter in the environment may differ dramatically in terms of aromaticity, molecular size distribution and hydrophobic acid content, resulting in different affinities of contaminants for organic matter in different environments or individual phases (Servos and Muir 1989; Kukkonen and Oikari 1991; Kukkonen *et al.* 1990).

An illustration of the disequilibrium between suspended solids and water is shown in Fig. 4 for di- to tetrachlorodibenzofurans (PCDFs) emitted to the Athabasca River (Alberta) in a BKME (Pastershank and Muir 1995). These compounds were detected in both suspended solids and centrifugate (from continuous centrifugation) samples. To calculate f_D for PCDFs in river water, Pastershank and Muir (1995) assumed that the K_{POC} calculated between particulate and centrifugate was equal to the K_{DOC}. The f_D declined with distance from the mill following initial dilution of the effluent in the river. The disequilibrium immediately downstream of the mill was evident from the K_{OC} values, which were <K_{OW}. After 116 km K_{OC} values (concentration on POC/C_{WD}) approached K_{OW} values for the PCDFs, indicating that near equilibrium conditions were being achieved.

Reliable measurements of C_{WD} are difficult to obtain for hydrophobic organics, making the use of a BAF problematic. The alternative is to reference the concentrations in biota relative to sediment or suspended sediment. The biota-sediment accumulation factor (BSAF) (Ankley *et al.* 1992; U.S. EPA 1993), also called a bioavailability index, biota-sediment factor (Parkerton *et al.* 1993; DiToro *et al.* 1991) or accumulation factor (Lake *et al.* 1990), can be calculated with C_{FL} and an organic carbon normalized sediment concentration, C_{SOC}.

$$BSAF = C_{FL}/C_{SOC} \qquad [12]$$

A similar relationship, the BSSAF, can be derived with organic carbon normalized concentrations in suspended sediments (U.S. EPA 1993). BSAFs can be viewed as a measure of disequilibrium between biota-sediment and overlying water (U.S. EPA 1993; Thomann *et al.* 1992). Rearranging equation 12 by replacing C_{FL} with $C_{WD} \cdot K_{OW}$ and C_{SOC} with $C_{SD} \cdot K_{OC}$:

$$BSAF = R_{AW} \cdot R_{WS} \cdot (C_{WD} \cdot K_{OW}/C_{SD} \cdot K_{OC}) \qquad [13]$$

where R_{AW} and R_{WS} are the disequilibrium factors between fish and water, and water and sediment, respectively. R_{AW} can be estimated from the ratio of BAF_L/K_{OW} and R_{WS} from dissolved concentrations in overlying water and pore water ($C_{WD}/C_{porewater}$). At equilibrium $R_{AW} \cdot R_{WS} = 1$ and BSAF ≈ 1 assuming $K_{OW} = K_{OC}$. In Lake Ontario, lakewide mean BSAFs for 2,3,7,8-tetrachlorodebenzo-*p*-dioxin (TCDD) in fishes range from 0.03 to 0.20, indicating that fish are generally not in equilibrium with sediments. This has been attributed to disequilibrium between water and sediment because of decreasing TCDD inputs to sediment (R_{WS} <1) and to growth dilution and metabolism (R_{AW} <1).

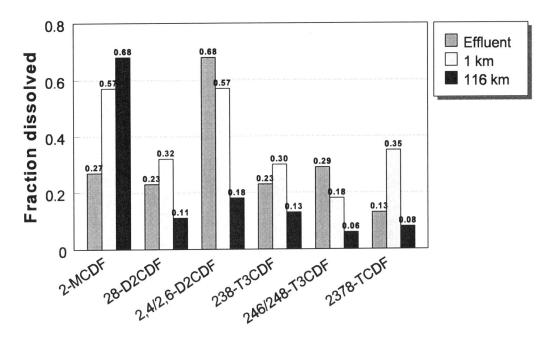

Figure 4. Variation of the fraction of mono- to tetrachloro-PCDFs in the "dissolved" phase with distance from a BKME on the Athabasca River, Alberta, Canada (from data of Pastershank and Muir 1995).

For TCDD and TCDF in fishes from riverine environments near pulp mills, much higher BSAFs and BSSAFs are obtained (Muir *et al.* 1992b; Pastershank and Muir 1995). The range of reported BSAF/BSSAFs is shown in Fig. 5. The high BSAFs/BSSAFs in riverine environments may be related to disequilibrium between particle and dissolved TCDD/F, as mill effluents equilibrate with river waters, and to food chain effects. In the Athabasca River (Alberta), Pastershank and Muir (1995) found that BAF_L values for TCDD/F in mountain whitefish were in the 10^6 to 10^7 range. Values of R_{AW} for TCDD and TCDF in mountain whitefish averaged 1.3 and 2.5, respectively within 176 km downstream of a BKME. The high BSSAFs can be explained by the combination of R_{AW} and R_{WS} >1, in other words, the combined effect of suspended sediment-water disequilibrium and food chain effects making TCDD/F more available in the receiving waters downstream of the mill than predicted on the basis of log K_{OW}. A previous study using fish and sediment data from sites near 18 bleached kraft pulp mills located on rivers in Canada also showed BSAFs >1 (Fig. 5; Muir et al. 1992b). In that study, however, sediments were not collected with the objective of calculating sediment-reference bioaccumulation and may not have been representative of depositional sediments.

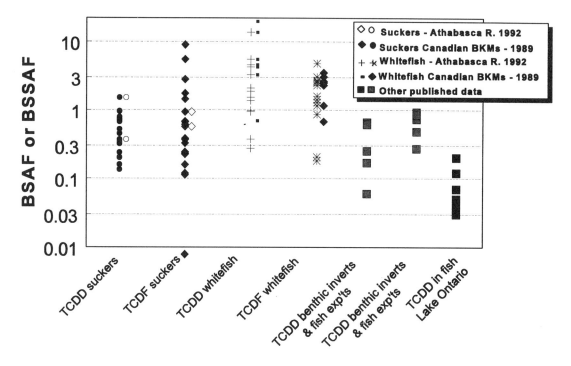

Figure 5. Comparison of biota-sediment (BSAF) or biota-suspended sediment (BSSAF) accumulation factors for 2,3,7,8-TCDD/F in mountain whitefish and longnosed suckers downstream of a BKME on the Athabasca River (Alberta) with results for the same or related fish species at other riverine locations in Canada (Muir et al. 1992b) and with TCDD in fish from Lake Ontario (U.S. EPA 1993).

In the Athabasca River, BSSAFs proved more consistent than BSAFs in characterizing bioaccumulation of TCCD/Fs downstream of a BKME (Pastershank and Muir 1995). Depositional sediments in rivers are difficult to sample and the contaminant burden in them may not be representative of the sediment exposure of fishes or invertebrates. Among the three simple bioaccumulation parameters discussed here, BAF_L, BSAF and BSSAFs, the BSSAF may offer the best approach for predicting the accumulation of hydrophobic BKME-related organic riverine environments provided that disequilibria can be accounted for.

THE EFFECT OF FOOD WEB AND OTHER BIOLOGICAL FACTORS ON BIOACCUMULATION

The characteristics of the organism, including lipid content, habitat preference and trophic status, also restrict or enhance the environmental bioavailability. The characteristics of individual biota are important in controlling or limiting the exposure or bioavailability via each of these routes. The behavior of the species including habitat selection may be such that exposure directly to BKME effluent is limited or transitory. On the other hand, the diet selection or trophic position may be such that individual organisms are exposed to chemicals which are biomagnified through the food chain, resulting in a higher exposure. These food web issues were well illustrated by Owens et al. (1994) in a study of bioaccumulation of

2,3,7,8-TCDD/F in fishes near a pulp and paper mill in Alberta. Mountain whitefish (*Prosopium williamson*) were found to accumulate higher concentrations of TCDD/F due to their diet of filter-feeding insects while bottom-feeding suckers (*Catostomus catostomus*) had much lower concentrations. Biomagnification (i.e., BMF >1) was not observed in omnivores and piscivores, walleye (*Stizostedium vitreum*) and burbot (*Lota lota*), respectively, downstream of the mill. A specific dietary food niche, involving ingestion of filter-feeding *Trichoptera* and *Plecoptera*, explained the elevated levels in mountain whitefish relative to the piscivores. Mah *et al.* (1989) and Muir *et al.* (1992b) observed similar higher levels of TCDD/F in mountain whitefish at other locations near BKME in Alberta and British Columbia.

Another source of uncertainty in predicting bioaccumulation of BKME-related organics is the migration of fish. Owens *et al.* (1994) found longnose suckers were very mobile in their Alberta streams, especially during the spring spawning period. Mountain whitefish undergo a similar annual fall migration. Thus individual fish may spend a relatively short time in the immediate receiving waters of the mill. Knowledge of assimilation and depuration rates may be critical to the modeling of the bioaccumulation of TCDD/F by fish with transitory exposure to the BKME.

Food chain models have been applied to the accumulation of TCDD/F in fish near bleached kraft pulp and paper mills. Gobas *et al.* (1994a) applied a chemical fate and food chain model to predict PCDD/Fs in fish in the Fraser and Thompson River downstream of BKMEs. By segmenting the river into reaches of similar flow and food chains, the authors were able to link emissions and water/sediment concentrations with concentrations of PCDD/Fs in the food web. The model also successfully predicted declining concentrations in biota observed in this river system following major process changes at the mills to reduce PCDD/F emissions.

CONCLUSIONS

The bioaccumulation of the majority of low molecular weight BKME-related organics by fishes can be predicted with well established relationships between log BCF and log K_{ow}. This applies to compounds with log K_{ow} <5 which are primarily in the dissolved phase in receiving waters. In practice the accumulation of many BKME-related phenolics and neutrals will be overestimated with the K_{ow} relationship because of metabolism. An alternative for readily conjugated compounds may be to use a BCF based on concentrations in bile, especially for chlorophenolics and resin acids. The rates and pathways as well as physiological factors influencing metabolism of most BKME-related organics by fish are not well documented. At the present time it is not possible to include a metabolism rate in pharmacokinetic models of BKME-related organics in fish. Further work is also needed to understand accumulation of chlorolignins and other high molecular weight substances by fishes. Much of the extractable organic chlorine (EOCl) in fishes is high molecular weight (>500 daltons) but chlorophenolics are known to contribute very little to the EOCl (Hemming and Lehtinen 1988; Wesen *et al.* 1991). The uptake pathways, depuration and transformation of the higher molecular weight EOCl material need further investigation. For hydrophobic organics such as 2,3,7,8-substituted TCDD/F, BSSAFs are more consistent spatially, and easier to measure, than BSAFs or BAF_Ls for PCDD/Fs in riverine biota.

Future challenges for predicting bioaccumulation of BKME-related organics include the need to develop information on polar, nonchlorinated organics emitted by pulp mills. Many of these compounds (e.g., resin acids, pinenes, cymenes, syringols, sterols) have relatively low K_{ow}s and high rates of biotransformation in fish, making prediction of tissue levels difficult. Yet the ability to predict levels could be useful for assessing exposure and toxicity as well as for addressing issues such as tainting.

REFERENCES

Ankley, G.T., P.M. Cook, A.R. Carlson, D.J. Call, J.A. Swenson, H.F. Corcoran and R. Hoke. 1992. Bioaccumulation of PCBs from sediment by oligochaetes and fishes: comparison of laboratory and field studies. Can. J. Fish. Aquat. Sci. 49:2080-2085.

Berry, R.M., C.E. Luthe, R.H. Voss, P.E. Wrist, P. Axegard, G. Gellerstedt, P-O. Lindbald and I. Popke. 1991. The effects of recent changes in bleached softwood kraft mill technology on organochlorine emissions: an international perspective. Pulp and Paper Canada 92:43-53.

Bruggeman, W.A., L.B.J.M. Martron, D. Kooiman and O. Hutzinger. 1981. Accumulation and elimination kinetics of di-, tri- and tetrachlorobiphenyls by goldfish after dietary and aqueous exposure. Chemosphere 10:811-832.

Buser, H-R., L-O. Kjeller, S.E. Swanson and C. Rappe. 1989. Methyl-, polymethyl-, and alkylpolychlorodibenzofurans identified in pulp mill sludge and sediments. Environ. Sci. Technol. 223:1130-1137.

Connell, D.W. 1988. Bioaccumulation behavior of persistent organic chemicals with aquatic organisms. Rev. Environ. Contam. Toxicol. 101:117-154.

Connell, D.W. and D.W. Hawker. 1988. Use of polynomial expressions to describe the bioconcentration of hydrophobic chemicals by fish. Ecotox. Environ. Safety 16:242-257.

Connolly, J.P. and C.J. Pedersen. 1988. A thermodynamic-based evaluation of organic chemical accumulation in aquatic organisms. Environ. Sci. Technol. 22:99-103.

Dickson, K.L., J.P. Giesy, R. Parrish and L. Wolfe. 1994. Closing remarks: summary and conclusions. p. 221-230. In Bioavailability: Physical, Chemical and Biological Interaction. J.L. Hamelink, P.F. Landrum, H.L. Bergman and W.H. Benson [Eds]. Lewis Publishers, Boca Raton, FL.

DiToro, D.M., C.S. Zarba, D.J. Hansen, W.J. Berry, R.C. Swartz, C.E. Cowan, S.P. Pavlou, H.E. Allen, N.A. Thomas and P.R. Paquin. 1991. Technical basis for establishing sediment water quality criteria for nonionic organic chemicals using equilibrium partitioning. Environ. Toxicol. Chem. 10:1541-1583.

Doe, K.G., W.R. Parker, G.R.L. Julien and P.A. Hennigar. 1988. Influence of pH on the acute lethality of fenitrothion, 2,4-D, and aminocarb and some pH-altered sublethal effects of aminocarb on rainbow trout (*Salmo gairdneri*). Can J. Fish. Aquat. Sci. 45:287-293.

Geyer, H.J., and D.C.G. Muir. 1993. New results and consideration on the bioconcentration of the superlipophilic persistent chemicals octachlorodibenzo-*p*-dioxin (OCDD) and mirex in aquatic organisms. p. 185-197. In Fate and Prediction of Environmental Chemicals in Soils, Plants and Aquatic Systems. M. Mansour [Ed.]. Lewis Publ., Ann Arbor, MI.

Gobas, F.A.P.C. 1993. A model for predicting the bioaccumulation of hydrophobic organic chemicals in aquatic food-webs: application to Lake Ontario. Ecol. Modelling 69:1-17.

Gobas, F.A.P.C., and D. Mackay. 1987. Dynamics of hydrophobic organic chemical bioconcentration in fish. Environ. Toxicol. Chem. 6:495-504.

Gobas, F.A.P.C., D.C.G. Muir and D. Mackay. 1988. Dynamics of dietary bioaccumulation of hydrophobic organic chemicals in fish. Chemosphere 17:943-962.

Gobas, F.A.P.C., J.R. McCorquodale and G.D. Haffner. 1993. Intestinal absorption and biomagnification of organochlorines. Environ. Toxicol. Chem. 12:567-576.

Gobas, F.A.P.C., K. Lien and J. Pasternak. 1994a. Development and verification of an environmental fate and food-chain bioaccumulation model of organochlorin emissions in the Fraser-Thompson River Basin. Report to Environment Canada, Simon Fraser University, Burnaby, B.C.

Gobas, F.A.P.C., X. Zhang and R. Wells. 1994b. Gastrointestinal magnification: the mechanisms of biomagnification and food chain accumulation of organic chemicals. Environ. Sci. Technol. 27:2855-2863.

Healey, J., M.R. Servos and K.R. Munkittrick. 1994. Tracers of exposure of fish to pulp and paper mill effluents - a review of the published literature. Can. Tech. Rep. Fish. Aquat. Sci. 1929, 96 p.

Hemming, J., and K. Lehtinen. 1988. Extractable organic chlorine (EOCl) in fish exposed to combined mill effluents from bleached kraft pulp production. Nordic Pulp & Paper Res. J. 4:185-190.

Holcombe, G.W, J.T. Fiandt and G.L. Phipps. 1980. Effects of pH increases and sodium chloride additions on the acute toxicity of 2,4-dichlorophenol to fathead minnow. Water Research 14:1073-1077.

Hunn, J.B., and J.L. Allen. 1974. Movement of drugs across the gills of fishes. Annu. Rev. Pharmacol. 14:1-27.

Kuehl, D.W., B.C. Butterworth, W.M. DeVita and C.P. Sauer. 1987. Biomed. Mass. Spectrom. 14:443.

Kukkonen, J., and A. Oikari. 1991. Bioavailability of organic pollutants in boreal waters with varying levels of dissolved organic material. Wat. Res. 25:455-463.

Kukkonen, J., J.F. McCarthy and A. Oikari. 1990. Effects of XAD-8 fractions of dissolved organic carbon on the sorption and bioavailability of organic micropollutants. Arch. Environ. Contam. Toxicol. 19:551-557.

Lake, J.L., N.I. Rubenstein, H. Lee II, C.A. Lake, J. Heltshe and S. Pavignano. 1990. Equilibrium partitioning and bioaccumulation of sediment-associated contaminants by infaunal organisms. Environ. Toxicol. Chem. 9:1095-1106.

Luthe, C.E., P.E. Wrist and R.M. Berry. 1992. An evaluation of the effectiveness of dioxin control strategies on organochlorine effluent discharges from the Canadian bleached chemical pulp industry. Pulp and Paper Canada 93:40-49.

Mackay, D. 1982. Correlation of bioconcentration factors. Environ. Sci. Technol. 16:274-278.

Mackay, D., and J.M. Southwood. 1992. Modelling the fate of organochlorine chemicals in pulp mill effluents. Water Poll. Res. J. Canada 27:509-537.

Mah, F.T.S., D.D. Macdonald, S.W. Sheehan, T.M. Tourminen and D. Valiela. 1989. Dioxins and furans in sediment and fish from the vicinity of ten inland pulp mills in British Columbia. Water Quality Branch, Environment Canada, Vancouver, B.C., 77 p.

McCarty, L.S. 1986. The relationship between aquatic toxicity QSARs and bioconcentration for some organic chemicals. Environ. Toxicol. Chem. 5:1070-1080.

McKim, J.M. 1994. Physiological and biochemical mechanisms that regulate the accumulation and toxicity of environmental chemicals in fish. p. 179-201. In Bioavailability: Physical, Chemical and Biological Interaction. J.L. Hamelink, P.F. Landrum, H.L. Bergman and W.H. Benson [Eds]. Lewis Publishers, Boca Raton, FL.

Muir, D.C.G., W.L. Fairchild, A.L. Yarechewski and D.M. Whittle. 1992a. Derivation of bioaccumulation parameters and application of food chain models for chlorinated dioxins and furans. p. 185-208. In F.A.P.C. Gobas and F. MacCorquadale, eds., Chemical Dynamics in Freshwater Ecosystems, Lewis Publishers, Ann Arbor, MI.

Muir, D.C.G., W.L. Fairchild and D. M. Whittle. 1992b. Predicting bioaccumulation of chlorinated dioxins and furans in fish near Canadian bleached kraft mills. Water Poll. Res. J. Can. 27:103-123.

Oikari, A., and B. Holmbom. 1986. Assessment of water contamination by chlorophenolics and resin acids with the aid of fish bile metabolites. p. 252-267. In Aquatic Toxicology and Environmental Fate: Ninth Volume. Poston and Purdy [Eds.]. ASTM STP921, Am. Soc. Test. Mater., Philadelphia.

Opperhuizen, A., and D.T.H.M. Sijm. 1990. Bioaccumulation and biotransformation of polychlorinated dibenzo-p-dioxins and dibenzofurans in fish. Environ. Toxicol. Chem. 9:175-186.

Opperhuizen, A., E.W. Vander Velde, F.A.P.C. Gobas, D.A.K. Liem and J.M.D. Van der Steen. 1985. Relationship between bioconcentration in fish and steric factors of hydrophobic chemicals. Chemosphere 14:1871-1896.

Owens, J.W., S.M. Swanson and D.A. Birkholz. 1994. Bioaccumulation of 2,3,7,8-tetrachlorodibenzo-p-dioxin, 2,3,7,8-tetrachlorodibenzofuran and extractable organic chlorine at a bleached-kraft mill site in a northern Canadian river system. Environ. Toxicol. Chem. 13:343-354.

Parkerton, T.F., J.P. Connolly, R.V. Thomann and C.G. Uchrin. 1993. Do aquatic effects of human health end points govern the development of sediment-quality criteria for nonionic organic chemicals? Environ. Toxicol. Chem. 12:507-523.

Pastershank, G., and D.C.G. Muir. 1995. Polychlorinated dibenzo-p-dioxins and polychlorinated dibenzofurans in fish and other environmental samples collected downstream from kraft pulp and paper mills. Northern River Basins Study Report No. 44, Northern River Basins Study, Edmonton, AB., 84 pp.

Rantio, T. 1992. Chlorocymenes, cyemenes and other chlorohydrocarbons in pulp mill effluents, sludges and exposed biota. Chemosphere 26:505-516.

Rombough, P.J., and B.M. Moroz. 1990. The scaling and potential importance of cutaneous and branchial surfaces in respiratory gas exchange in young chinook salmon (*Ocnorchynchus tshawytscha*). J. Exp. Biol. 145:1-12.

Saarikoski, J., M. Lindstrom, M. Tyynila and M. Viuksela. 1986. Factors affecting the absorption of phenolics and carboxilic acods in the guppy (*Poecilia reticulata*). Ecotoxicol. Environ. Saf. 11:158-173.

Servizi, J.A., R.W. Gordon, D.W. Martens, W.L. Lockhart, D.A. Metner, I.H. Rogers, J.R. McBride and R.J. Norstrom. 1994. Effects of biotreated bleached kraft mill effluent on fingerling chinook salmon (*Oncorhynchus tshawytscha*). Can. J. Fish. Aquat. Sci. 50: In press.

Servos, M.R., and D.C.G. Muir. 1989. Effect of dissolved organic matter from Canadian Shield lakes on the bioavailability of 1,3,6,8-tetrachlorodibenzo-p-dioxin to the amphipod *Crangonyx laurentianus*. Environ. Toxicol. Chem. 8:141-150.

Sijm, D.T.H.M., and A. Opperhuizen. 1989. Biotransformation of organic chemicals by fish: review of enzyme activities and reactions. pp. 163-225. In Handbook of Environmental Chemistry, Vol. 2, Part E. Reactions and Processes. O. Hutzinger (Ed.). Springer-Verlag, Heidleberg.

Spehar, R.L., H.P. Swanson and J.W. Renoos. 1985. Pentachlorophenol toxicity to amphipods and fathead minnow at different test pH values. Environ. Toxicol. Chem. 4:389-397.

Stehyl, G.R., and W.L. Hayton. 1990. Effect of pH on the accumulation of pentachlorophenol in goldfish. Arch. Environ. Contam. Toxicol. 19:464-470.

Thomann, R.V. 1989. Bioaccumulation model of organic chemical distribution in aquatic food chains. Environ. Sci. Technol. 23:699-707.

Thomann, R.V., J.P. Connolly and T.F. Parkerton. 1992. An equilibrium model of organic chemical accumulation in aquatic food webs with sediment interaction. Environ. Toxicol. Chem. 11:615-629.

U.S. Environmental Protection Agency. 1993. Interim report on data and methods for assessment of 2,3,7,8-tetrachlorodibenzo-*p*-dioxin risks to aquatic life and associated wildlife. EPA/600/R-93/055, Environ. Res. Lab., Duluth, MN.

Veith, G.D., D.L. DeFoe and B.V. Bergstedt. 1979. Measuring and estimating the bioconcentration factor of chemicals in fish. J. Fish. Res. Board Can. 36:1040-1048.

Voss, R.H., C.E. Luthe, B.I. Fleming, R.M. Berry and L.H. Allen. 1988. Some new insights into the origins of dioxins formed during chemical pulp bleaching. Pulp and Paper Canada 89:151-162.

Wachtmeister, C.A., L. Förlin, K.C. Arnoldsson and J. Larsson. 1991. Fish bile as a tool for monitoring aquatic pollutants: studies with radioactively labelled 4,5,6-trichloroguaiacol. Chemosphere 22:39-46.

Wesen, C., K. Martinsen, G. Carlberg and H. Mu. 1991. Chlorinated carboxylic acids are major chloroorganic compounds in fish exposed to pulp bleach liquor. pp. 207-219. *In* Environmental Fate and Effects of Bleached Pulp Mill Effluents. Proceedings 1991 SEPA Conference, November 19-21, Report 4031.

PERSISTENT PULP MILL POLLUTANTS IN WILDLIFE

J.E. Elliott[1], P.E. Whitehead[1], P.A. Martin[2], G.D. Bellward[3] and R.J. Norstrom[4]

[1] Canadian Wildlife Service, 5421 Robertson Road, RR1 Delta, British Columbia, V4K 3N2 Canada
[2] RR#8-40-9 Lethbridge, Alberta, T1J 4P4 Canada
[3] Faculty of Pharmaceutical Sciences, University of British Columbia, 2146 East Mall, Vancouver, British Columbia, V6T 1Z3 Canada
[4] National Wildlife Research Centre, Canadian Wildlife Service, 100 Gamelin Blvd, Bldg. #9, Hull, Quebec, K1A 0H3 Canada

Published and unpublished data on levels and effects of pulp mill contaminants in wildlife are reviewed. Polychlorinated dioxins and furans were detected at high concentrations in eggs of various fish-eating bird species from the Strait of Georgia on the west coast of British Columbia. Highest concentrations were near pulp mills: for example TCDD, PnCDD and HxCDD in cormorant (*Phalacrocorax* spp.) eggs were as high as 100, 275 and 950 ng kg^{-1} wet weight, respectively, near a pulp mill in 1986. In contrast, levels of these contaminants in cormorants collected from colonies near pulp mills on the Canadian Atlantic coast were typically <15 ng kg^{-1}. Polychlorinated dioxin and furan levels were also elevated in tissues of fish-eating waterfowl wintering in the Strait of Georgia, but lower in non-piscivorous waterfowl. Whales and porpoises sampled from the Strait of Georgia had lower levels than piscivorous waterbirds. Episodes of poor breeding success during the 1980s in a colony of great blue herons (*Ardea herodias*) near a bleached kraft pulp mill in the Strait of Georgia were associated with sublethal effects on embryos, including edema, reduced body weight and EROD induction. Sublethal responses including CYP1A induction and porphyria were linked to pulp mill contaminant exposure in eagles and cormorants in the Strait of Georgia and to herring gulls breeding near a mill in Quebec.

INTRODUCTION

Polychlorinated dibenzo-*p*-dioxins (PCDDs) and dibenzofurans (PCDFs) occur in the environment as a result of a variety of industrial processes and combustion. Pulp and paper mills that use chlorine in their bleaching process (bleached kraft pulp mills) or utilize chlorophenol-treated wood chips for feedstock are sources of these contaminants (Luthe *et al.* 1990). Tetrachlorodibenzo-*p*-dioxins (TCDD) and tetrachlorodibenzofurans (TCDF) have frequently been reported in fish downstream of pulp mills (Kuehl *et al.* 1987; Mah *et al.* 1989; Servos *et al.* 1994). Effects on fish health, including abnormal sexual development, metabolic dysfunction and deformities, have been reported, although not directly linked to dioxin and furan toxicity (Lindejoo and Thulin 1987; Andersson *et al.* 1988; Munkittrick *et al.* 1991, 1992). Fish- and invertebrate-consuming wildlife are exposed to PCDDs and PCDFs in their diet. PCDDs and PCDFs have been detected, sometimes at very high levels, in aquatic birds in the Strait of Georgia, British Columbia (Elliott *et al.* 1989; Whitehead *et al.* 1992a; Vermeer *et al.* 1993) and throughout the Great Lakes (Stalling *et al.* 1985; Kubiak *et al.* 1989; Hebert *et al.* 1994).

The persistence and toxicity of some pulp mill contaminants raise concerns about the health of fish-eating wildlife near mills. Findings of high PCDD and PCDF have led to fishery closures (Harding and Pomeroy 1990) and wildlife consumption advisories (Whitehead *et al.* 1990) near west coast pulp mills. This paper presents a broad overview of persistent pulp mill pollutants in wildlife, with comparisons to less contaminated reference sites and sites receiving other sources of contamination. Effects on health of exposed wildlife are also briefly summarized. Information was obtained primarily from the published

LEVELS OF PERSISTENT PULP MILL CONTAMINANTS IN WILDLIFE

Contaminants in Pulp Mill Effluents

Effluents from bleached kraft pulp mills contain a wide variety of compounds, ranging from water-soluble and rapidly degraded chlorinated lignins to the persistent, highly bioaccumulative dioxins and furans. Tissues are typically analyzed for lipid-soluble neutral organics, primarily PCDDs and PCDFs. Recently, a wider range of compounds have been measured in fish and wildlife samples, including chlorophenols, catechols and guaiacols. Pentachlorophenol was detected at levels of <5 ng g^{-1} wet weight in 21 of 28 waterfowl samples collected near pulp mills along the British Columbia coast. Of 34 other chlorophenol-related compounds measured, only three appeared with any regularity. These were 3,4,5,6-tetrachloroguaiacol, 5-chloroguaiacol and 4,5-dichloroguaiacol, which were quantifiable in 5, 1 and 2 samples, respectively. Thus, although it is recognized that many contaminants are present in pulp mill effluent, PCDDs and PCDFs are given most attention due to their frequency of occurrence and high levels of toxicity.

Marine and Estuarine Birds

Most information on pulp mill contaminant levels in marine birds comes from surveys in the Strait of Georgia, British Columbia. Initial concerns in the Strait of Georgia resulted from a survey of PCDD and PCDF levels in piscivorous birds in 1982. Norstrom and Simon (1983) showed that PCDD levels in the eggs of great blue herons (*Ardea herodius*) from a colony at the University of British Columbia (UBC) were unusually high (TCDD, pentachlorodibenzo-*p*-dioxins (PnCDD) and 1,2,3,6,7,8-hexachlorodibenzo-*p*-dioxin (HxCDD) levels were 76, 490 and 740 mg kg^{-1}, respectively). The Fraser River delta is industrialized and there are several pulp mills upstream, explaining elevated PCDD concentrations at UBC (Harfenist *et al.* 1993). Contaminant monitoring of great blue heron eggs continued throughout the 1980s, at sites receiving pulp mill effluent (UBC and Crofton) and at some less contaminated reference sites (Sidney and Nicomekl) (Table 1). Overall PCDD levels in eggs from the Crofton and UBC sites were higher than those of the reference sites in 1986 ($p < 0.05$). In 1987 they were highest overall from the Crofton site ($p < 0.001$; Elliott *et al.* 1989). The relationship of egg concentrations with the adjacent bleached kraft pulp mill was evidenced by their substantial decline by 1991 (Fig. 1a) following process and product changes implemented to reduce PCDDs and PCDFs in the effluent ($p < 0.05$; Whitehead *et al.* 1992b). Eggs of double-crested cormorants collected at Crofton in 1987 exhibited PCDD levels approximately 50% those of great blue herons collected at that site the same year (Table 1). Cormorant eggs from Howe Sound, where there are two bleached kraft pulp mills, contained contaminant levels that were similar to those of the Crofton sample from 1987. Levels in eggs from both colonies dropped substantially by 1990 and 1992 (Fig. 1b), probably due to improved effluent quality.

Congener patterns were similar between these two piscivorous waterbird species, with particularly high levels of 1,2,3,6,7,8-HxCDD, indicating pulp mill use of chlorophenol-tainted wood chips (Luthe *et al.* 1990). Up to 1989, penta- and tetrachlorophenol fungicides had long been used in the British Columbian forest industry to protect wood products from decay. They were also a cause of contamination of receiving waters near pulp, lumber and saw mills (Krahn *et al.* 1987). This characteristic congener pattern was evident in the eggs of great blue herons even at sites distant from mills within the Strait of Georgia (e.g., Sidney Island), indicating widespread PCDD contamination from chlorophenol-tainted wood chips in this body of water (Fig. 1a; Elliott *et al.* 1989). Great blue heron eggs collected in 1993 near a

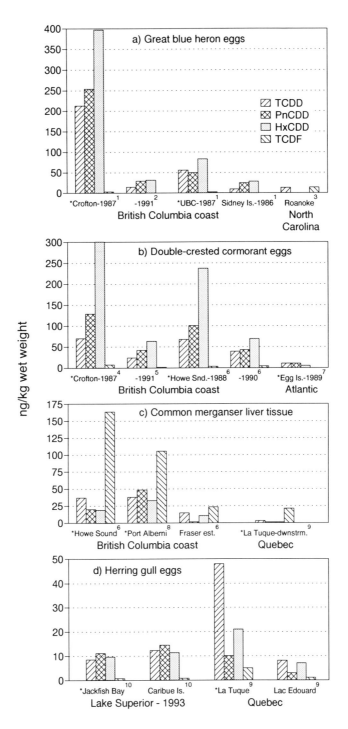

Figure 1. Comparison of PCDD and PCDF levels in tissues of marine and aquatic birds in North America. * indicates associated with pulp mills. Superscript letters indicate source as follows: [1]Elliott et al. 1989; [2]Whitehead et al. 1992b; [3]Beeman et al. 1993; [4]Norstrom et al. 1988; [5]Sanderson et al. 1994a; [6]Whitehead et al. 1992a; [7]Norstrom and Pearce, unpublished data; [8]Whitehead et al. 1990; [9]Champoux 1993; [10]Shutt, unpublished data.

Table 1. Levels of PCDDs and PCDFs (ng/kg, wet weight) in tissues of marine and estuarine birds in North America from pulp mill sources, as well as reference sites. Values are either from single pooled samples or are geometric means (range) of several samples.

Location	Yr[a]	Species	Tissue[b]	No.[c]	% lipid	PCDDs 2378-TCDD	12378-PnCDD	123678-HxCDD	OCDD	PCDFs 2378-TCDF	Total PnCDF	Ref.[d]
BRITISH COLUMBIA												
Strait of Georgia												
Howe Sound	88	Double-crested cormorant	egg	7*	4.5	68	101	237	<10	4	51	1
	89	Double-crested cormorant	egg	10*	5.0	30	23	36	14	<2	12	1
	90	Double-crested cormorant	egg	7	3.7	38.7 (30-65)	42.0 (30-57)	69.1 (40-114)	<10	3.6 (<2-10)	12.4 (8-19)	1
	89	Western grebe	liv	10*	4.0	46	29	77	<11	109	43.5	1
	89	Harlequin duck	liv	3*	4.7	<2	<1	<2.5	<8	36	<1	1
	89	Oldsquaw	liv	3*	3.1	9.2	1.7	14	3.2	9.6	6.8	1
	90	Surf scoter	BM	11*	2.8	2.2	2.4	<3.8	<14	31	<1.5	1
			liv	11*	3.1	8.7	<1.7	4.3	<12	34	<1.6	1
	90	Common goldeneye	BM	12*	3.4	<1.4	<0.6	<1.7	<5.5	30	<0.7	1
			liv	12*	3.6	7.1	<2.4	<4.1	<16	66	4.5	1
	90	Common merganser	BM	9*	3.4	21	7.3	7.8	1.4	116.5	10.8	1
			liv	9*	5.1	37	20	19	<4.1	163	19.1	1
Burrard Inlet	90	Lesser scaup	liv	10*	4.3	14	5.5	19	16	11	6.4	2
	90	Barrow's goldeneye	liv	2*	4.3	14	14	40	25	1.7	13	2
Fraser estuary	90	Dunlin	liv	18*	7.3	6.3	<0.7	8.4	15	13	2	2
	90	Common goldeneye	BM	3*	4.1	<2.1	<1.5	<3.8	<17	9.4	<1.5	2
			liv	3*	5.3	<12	<12	<19	<19	17	<12	2
	90	Bufflehead	BM	1*	3.4	18	<0.8	17	<6.8	9.5	<0.8	2
		Common merganser	BM	4*	4.7	13	<2.3	5.5	<19	18	4	2
			liv	4*	4.8	14	<2.3	9.8	<17	23	4.5	2
	89	Surf scoter	BM	10*	4.1	<1.6	<1.9	<3.1	<11	13	<1.5	2
			liv	10*	4.3	<2.3	<2.7	<4.9	<18	17	<2.1	2
Lower Fraser Valley	90	Bald eagle	egg	6	5.6	53 (42-58)	48 (37-55)	70 (42-112)		75 (23-112)	13 (12-14)	3
University of B.C.	86	Great blue heron	egg	7	6.0	25 (7-63)	74 (28-255)	104 (51-211)	6 (2-47)	3 (1-10)	16 (7-35)	6
	87	Great blue heron	egg	10	6.1	55 (9-204)	49 (23-109)	83 (39-158)	6 (2-19)	17 (5-50)	16 (9-25)	6
Crofton	87	Double-crested cormorant	egg	5	4.7	71.8 (56-100)	132 (68-275)	329 (133-962)	4 (ND-5)	8 (5-15)	23 (11-36)	5
	86	Great blue heron	egg	10	7.0	66 (8-218)	252 (11-1018)	337 (13-1298)	3 (ND-21)	2 (ND-14)	10 (ND-68)	6
	87	Great blue heron	egg	10	6.0	209 (59-444)	257 (73-728)	435 (110-1052)	<10	3.4 (1-17)	38 (10-104)	6
	90	Great blue heron	egg	7		102	229					7
	91	Great blue heron	egg	7		16	30	34		37		7
	90	Greater scaup	BM	9*	3.0	7.2	<0.9	<2.3	7.2	34	<0.9	4
			liv	9*	4.5	<1.5	<1.5	<1.8	<5.5	40	<0.7	4
	90	Surf scoter	BM	10*	2.8	5.1	<0.5	3.3	3	57	1	4
			liv	10*	4.2	6.9	<0.9	6.5	<10	28	<0.9	4
	90	Bufflehead	BM	2*	2.1	9.4	<1.8	29	<11	13	3.5	4
			liv	2*	4.3	<2.3	<3.3	6	<8.3	21	<2.6	4
	90	Bald eagle	egg	3	5.4	107	180	342			30	3

Table 1. Continued.

Location	Yr[a]	Species	Tissue[b]	No.[c]	% lipid	PCDDs 2378-TCDD	PCDDs 12378-PnCDD	PCDDs 123678-HxCDD	PCDDs OCDD	PCDFs 2378-TCDF	PCDFs Total PnCDF	Ref.[d]
Sidney Is.	86	Great blue heron	egg	4	6.6	9 (3-18)	24 (9-59)	30 (6-174)	ND	1 (ND-1)	2 (ND-6)	6
Nicomekl	86	Great blue heron	egg	5	4.9	10 (7-12)	30 (14-44)	52 (21-110)	1 (ND-3)	ND	5 (2-10)	6
N. Vancouver Is.												
Alert Bay	92	Western grebe	BM	2*	3.1	<1.6	<3.2	<3.5	<18	<0.9	<2.4	4
	92	Surf scoter	BM	7*	1.6	<2	<4	<4.7	<15	2.6	<2	4
	92	White-winged scoter	BM	4*	1.7	<2	<6.5	<6	<22	3.2	<2.8	4
	92	Bufflehead	BM	11*	3.7	<1.1	<2.6	<3.4	<9.8	1.4	<1.2	4
W. Vancouver Is.												
Port Alberni	89	Western grebe	liv	5*	7.3	117	385	249	<15	217	38	2
	92	Western grebe	BM	8*	3.8	25	66	64	<22	69	<2.4	4
	89	Greater scaup	liv	3*	3.8	3	7	13	<15	32	<3	2
	89	Surf scoter	liv	5	3.1	24	22	30	11	123	13	2
	89	Barrow's goldeneye	liv	2*	4.7	6	10	6	<15	115	3	2
	89	Common merganser	liv	6*	3.8	38	48	33	<15	105	11	2
	92	Common merganser	BM	6*	2.7	<4.3	<9.5	<6	<11	26	<4.6	4
WASHINGTON STATE												
Columbia R. estuary	80-87	Bald eagle	egg	17		32 (5.1-61)						8
ATLANTIC COAST												
Nova Scotia												
Pictou Is.	88	Double-crested cormorant	egg	5*	4.5	17	12	9	5	<2	6	9
New Brunswick												
Manawagonish	88	Double-crested cormorant	egg	5*	4.7	6	11	10	4	<2	7	9
Egg Is.	89	Double-crested cormorant	egg	5*	4.1	12	12	7	<2	<2	5	9
	89	Herring gull	egg	5*	8.5	5	2	3	<2	<2	<2	9
	89	Common tern	egg	5*	10.2	5	2	2	<2	10	3	9
Bon Ami Rock	89	Double-crested cormorant	egg	5*	4.3	14	15	9	5	<2	9	9
Pinnock Is.	89	Herring gull	egg	5*	9.3	<2	<2	3	<2	<2	<2.4	9
Bathurst Harbour	89	Red-breasted merganser	egg	5*	16.8	34	10	13	10	55	19	9
	89	Common tern	egg	10*	9.2	2	1	3	<2	3	1	9
Quebec												
Pointe St. Pancrace	93	Great blue heron	egg	5*	-	5	9	14	<1	<0.1	10	10

Note: Blank spaces mean the compound was not reported in the reference. ND means not detected.

[a] Yr = year(s) of collection.
[b] liv = liver; BM = breast muscle.
[c] Number of tissues in a pooled sample (*) or number of individual samples.
[d] References:
1 = Whitehead et al. 1992a.
2 = Whitehead et al. 1990.
3 = Elliott, in prep.
4 = Elliott and Whitehead, unpublished data
5 = Norstrom et al. 1988
6 = Elliott et al. 1989
7 = Whitehead et al. 1992b
8 = Anthony et al. 1993
9 = R. Norstrom, pers. comm.
10 = Desgranges, unpublished data

pulp mill in the estuary of the St. Lawrence River in Quebec had a pattern of HxCDD > PnCDD > TCDD at low levels, suggesting a pulp mill and possibly chlorophenol sources (Table 2). In contrast, heron eggs from a site several kilometers from a pulp mill on the Roanoke River in North Carolina contained low TCDD and TCDF levels (Fig. 1a; Beeman et al. 1993); the very low HxCDD levels indicate absence of significant chlorophenol-related contamination.

The eggs of bald eagles from the Strait of Georgia in 1990 exhibited elevated levels of the TCDD, PnCDD and HxCDD congeners, as well as levels of TCDF higher than heron and cormorant samples (Table 1). As with the other two species, eagle samples collected near the pulp mill at Crofton had the highest levels of the chlorophenol-related congeners. Liver samples of western grebes (*Aechmorphorus occidentalis*), taken near pulp mill locations in Howe Sound and Port Alberni on the west coast of Vancouver Island, showed a similar distribution of congeners. Overall concentrations were higher at the Port Alberni site. The Howe Sound sample contained PCDD levels similar to the cormorant sample taken the same year, but had a greatly elevated TCDF concentration (Table 1). In contrast, residues were not detected in a western grebe sample from a reference site (Alert Bay) on north Vancouver Island (Table 1). Whereas the high HxCDD levels in tissues of eagles and grebes likely indicate a chlorophenol source of contamination, the elevated TCDD and TCDF levels suggested additional, independent contamination from bleach kraft pulp mills.

Tissues of waterfowl wintering along coastal British Columbia exhibited a different pattern of PCDD and PCDF congeners than those of the other species examined (see Fig. 1). Typically, only TCDD and TCDF were elevated (Table 1). Of all the species, sampled the primarily piscivorous common mergansers (*Mergus merganser*) had the highest concentrations, with levels of TCDF exceeding 100 ng kg^{-1} in liver tissue of birds collected near pulp mills (Port Alberni and Howe Sound). Other waterfowl species sampled, including Barrow's and common goldeneye (*Bucephala islandica* and *B. clangula*), bufflehead (*B. albeola*), oldsquaw (*Clangula hyemalis*), greater scaup (*Aythya marila*), surf scoter (*Melanitta perspicullata*) and harlequin duck (*Histrionicus histrionicus*), are primarily benthivores and contained low levels of all congeners except TCDF, suggesting bleached kraft pulp mill as the main source. Whitehead et al. (1990) suggested that the detection of elevated TCDF levels in waterfowl and western grebes compared to cormorants and herons could be due to different tissues sampled (liver and breast muscle versus eggs). It has been shown in captive mallard ducks and herring gulls that TCDF is excreted much more rapidly than more highly chlorinated PCDFs and 2,3,7,8-substituted PCDDs; it is therefore seldom detected in avian eggs (Norstrom et al. 1976; Braune and Norstrom 1989). Elliott et al. (1989) suggested that the increase in TCDF levels in the eggs of great blue herons between 1986 and 1987 at the Crofton colony (see Table 1) may have been a result of increases in immediate exposure of females to TCDF in their food source at the time of yolk formation. Elevated TCDF levels in eagle eggs may also be a result of direct deposition to yolk lipids from recent dietary intake. Wintering waterfowl were collected within one kilometer of each pulp mill, therefore, during high contaminant exposure. The waterfowl species, with the exception of the common merganser, consume a large proportion of crustaceans and molluscs in contrast to the fish-eating herons, cormorants and eagles (Ehrlich et al. 1988). Vermeer et al. (1993) reported elevated PCDD and PCDF levels in *Corophium* amphipod prey of the waterfowl collected near the bleached-kraft pulp mill at Port Alberni, B.C. (Table 1). TCDF levels in the hepatopancreas of dungeness crabs were greater than 100 ng kg^{-1} throughout the Strait of Georgia and between 10,000 and 22,000 ng kg^{-1} adjacent to pulp mill effluent outlets (Norstrom et al. 1988; Harding and Pomeroy 1990). Levels in bivalves were much lower; nevertheless, near Crofton, oyster levels ranged from 300 to 600 ng kg^{-1}, prior to the changes to mill processes. In contrast, two sculpin species contained less than 30 ng kg^{-1} TCDF at all sites except Crofton, where levels reached 180 ng kg^{-1}. Harfenist et al. (1993) reported that, although TCDF levels were consistently greater than those of TCDD in small fish that make up the diet of great blue herons in the Fraser estuary collected in 1991, they remained below 25 ng kg^{-1}.

Less information on pulp mill contaminants from other coastal locations in North America is available. Double-crested cormorant, larid and red-breasted merganser (*Mergus serrator*) egg samples collected on the Atlantic coast from sites near pulp mills in the late 1980s were generally low in comparison to Strait of Georgia locations (Table 1). However, like waterfowl samples collected on the west coast, red-breasted merganser eggs contained substantial quantities of TCDF (Table 1). In particular, the PnCDD and HxCDD congeners were under 20 ng g^{-1} in all the Atlantic samples compared to values >100 in many Strait of Georgia samples. For example, comparison of double-crested cormorant eggs from Mandarte Island, a site in the Strait of Georgia remote from direct industrial exposure, to those from Manawagonish Island on the east coast of Canada showed comparatively low TCDD levels at both sites, but much higher PnCDD and HxCDD levels on the west coast, indicating the contamination of the Strait of Georgia food chains from the use of chlorophenol-tainted wood chips. However, a sample of common merganser eggs from an aquatic site downstream of a pulp mill in Quebec showed a pattern, 24:28:40 ng kg^{-1} TCDD:PnCDD:HxCDD (Champoux 1993), similar to that observed in Strait of Georgia waterbird samples in the early 1990s, some years after most mills had severely curtailed use of chlorophenol-tainted wood chips. This pattern is also similar to that in common murre (*Uria aalge*) eggs from the Baltic Sea (Cederberg *et al.* 1991), and may indicate a pulp mill signature in some waterbird food chains.

Marine Mammals

Blubber samples from several cetacean species were collected during the late 1980s in the Strait of Georgia for chlorinated hydrocarbon analysis (Jarman *et al.* in prep.). Harbor porpoise samples collected from the Strait of Georgia exhibited the highest levels of most dioxins (Table 3), while porpoises from the California coast contained no detectable residues of any of the congeners measured. Beluga whale and narwhal samples from the east coast and Baffin Island were, like the California sample, completely lacking in PCDDs, and the St. Lawrence samples contained only TCDF at very low levels (Norstrom *et al.* 1992). The presence of relatively elevated 1,2,3,6,7,8-HxCDD levels as well as the widespread occurrence of PnCDD congeners in many of the British Columbian cetacean samples, and the pentachlorodibenzofuran (PnCDF) and heptachlorodibenzofuran (HpCDF) congeners found in Dall's and harbor porpoise samples suggest contamination from a chlorophenol source, typical of the Strait of Georgia. The almost ubiquitous presence of TCDF in Strait of Georgia samples also indicates an independent source of bleached kraft pulp mill contamination. The detection of non-2,3,7,8-substituted PCDF congeners in Dall's and harbor porpoise samples is unusual, and Jarman *et al.* (in prep.) suggest that some cetaceans may be unable to metabolize these compounds. Their absence in other whale samples may be a function of differences in diet rather than in metabolic capability. Fish are able to efficiently metabolize non-2,3,7,8-substituted PCDF congeners, while invertebrates appear to lack these capabilities (Sijm *et al.* 1993; Rappe and Buser 1989). Thus, the piscivorous orca whales and carnivorous false killer and gray whales are less exposed to these congeners than are the invertebrate-consuming porpoises. Norstrom *et al.* (1992) found very low TCDD and TCDF residues in St. Lawrence River beluga whales, but again, residues of some non-2,3,7,8-substituted PCDFs were detected. The authors concluded that cetaceans, as a group, have superior capabilities to metabolize the more toxic 2,3,7,8-substituted PCDD and PCDFs, but are less able to eliminate the less toxic non-2,3,7,8-substituted congeners.

Aquatic Birds

Whitehead *et al.* (1993) found significant differences ($p < 0.05$) in levels of TCDD and TCDF in the eggs of osprey (*Pandion haliaetus*) sampled up- and downstream of kraft pulp mills on the Thompson and Columbia rivers in central British Columbia, indicating local pulp mill contaminant sources (Table 2). The congener pattern in osprey eggs differed considerably from the other piscivorous raptor species that we studied, the bald eagle, in that PnCDD and HxCDD levels were low and octachlorodibenzo-*p*-dioxin (OCDD) levels were unusually high. Interestingly, hepta- and octa-congeners were very high in

Table 2. Levels of PCDD and PCDF (ng/kg, wet weight) in tissues of aquatic birds in North America from pulp mill sources, as well as reference sites. Values are either from single pooled samples, or are geometric means (range) of several samples.

Location	Yr[a]	Species	Tissue[b]	N[c]	% Lipid	2378-TCDD	PCDDs 12378-PnCDD	123678-HxCDD	OCDD	PCDFs 2378-TCDF	Total PnCDF	Ref[d]
British Columbia												
Riske Creek	82	Barrow's goldeneye	egg			5.9	<1.3	7.3	6.4	42	4.9	1
Kamloops, Thompson R. - upstream	91	Osprey	egg	6		12 (3.4-22)	0.4 (0 - 7.0)	15 (5-38)	374 (22-1200)	0.7 (0-4.4)		1
downstream	91	Osprey	egg	6		47 (23-95)	2.6 (0-13)	22 (6.7-64)	92 (0 - 950)	1.5 (0-15)		1
upstream	92	Osprey	egg	5		3.9 (.-9-13)	3.4 (0-7.1)	15 (10-33)	472 (230-837)	0.9 (0-2.9)		1
downstream	92	Osprey	egg	5		29 (5.5-134)	5.3 (1.1-18)	12 (1.8-25)	94 (6.1-303)	3.9 (1-10)		1
Castlegar, Columbia R. - upstream	91	Osprey	egg	10		0.9 (0-5.7)	1.2 (0-6.3)	8.2 (1.3-36)	41 (5.7-225)	0.3 (0-18)		1
downstream	91	Osprey	egg	9		20 (.-7-53)	5.5 (2.4-16)	6.5 (0-35)	17 (2.4-91)	16 (5-61)		1
downstream	92	Osprey	egg	5		17.4 (4.6-56)	3.0 (1.5-4.9)	6.0 (1.0-17)	37 (3.3-1147)	35 (23-6.8)		1
Ontario												
Jackfish Bay, L. Superior	91	Herring gull	egg			18						2
	93	Herring gull	egg	10*	8.2	8.8	10.7	9.7	13.4	0.2	8.8	8
	93	Herring gull	liv	10*	5.5	12.9	25.1	89.6		0.7	24.5	8
Marathon, L. Superior	93	Herring gull	egg	10*	9.9	13.0	6.2	5.6	5.9	0.2	5.0	8
	93	Herring gull	liv	10*	5.2	5.5	25.1	23.0	99	0.5	10.2	8
Caribou Is., L. Superior	93	Herring gull	egg	10*	10	12.2	14.6	10.8	10.3	0.9	8.1	8
	93	Herring gull	liv	10*	6.5	15.4	33.3	81.6	53	5.0	28	8
Mutton Is., L. Superior	93	Herring gull	egg	10*	9.0	3.1	4.4	5.7	7.7	<0.5	3.1	8
	93	Herring gull	liv	10*	5.1	5.3	13.5	53.1	56	0.3	9.6	8
Granite Is., L. Superior	84-91	Herring gull	egg			15.0	12.6	14.5	3.6	1.5	10.4	3
Saginaw Bay, L. Huron	84-91	Herring gull	egg			85.5	23.2	32.5	18.6	3.6	23	3
Quebec												
Ile Aux Bic	86	Great blue heron	egg	5*	4.0	17	8	10	<3	1	9	7
Ile St-Bernard (pulp mill)	86	Great blue heron	egg	8*	4.3	10	10	2	5	3	8	7
Ile Aux Basques (pulp mill)	86	Great blue heron	egg	6*	3.3	12	28	28	4	1	5	7

Table 2. Continued.

						PCDDs				PCDFs		
Location	Yr[a]	Species	Tissue[b]	N[c]	% Lipid	2378-TCDD	12378-PnCDD	123678-HxCDD	OCDD	2378-TCDF	Total PnCDF	Ref[d]
LaTuque (pulp mill)	89	Herring gull	liv	2*		23	24	25		8	14	4
			egg	4*		48	10	21	11	5	ND	4
	89	Ringed-bill gull	liv	2*		13	11	21		12	33	4
			egg	3*		4	ND	6	10	ND	ND	4
	91	Hooded merganser	egg	8*		26.5	13.0	7.5	ND	204	11	4
	91	Common merganser	egg	2*		24.0	28.0	40.0	14.0	270	81	4
upstream	91	Common merganser	liv	4*		2.5	0.65	0.88	ND	20.8	0.9	4
downstream		Common merganser	liv	4*		0.15	ND	ND	ND	0.2	ND	4
Lac Wayagamac	91	Herring gull	liv	5*		1.6	ND	2.0	6.4	5.6	1.0	4
			egg	5*		14	4	6	ND	1	3	4
Lac Edouard	91	Herring gull	liv	5		ND	ND	ND	ND	ND	ND	4
			egg	5*		8	3	7	8	1	3	4
California central coast	87/88	Peregrine falcon	egg	7	4.3	5.7	11	11	5.3	6.2	7.3	5
North Carolina Roanoke R. - upstream		Great blue heron	egg			11	ND	ND	11	ND	ND	6

Note: Blank spaces mean the compound was not reported in the reference. ND means not detected.
a) Yr = year(s) of collection.
b) Number of tissues in a pooled sample (*) or number of individual samples.
c) liv = liver; BM = breast muscle.
d) References:
1 = Whitehead et al. 1993
2 = Shutt 1993
3 = Hebert et al. 1994
4 = Champoux 1993
5 = Jarman et al. 1993
6 = Beeman et al. 1993
7 = R. Norstrom, Canadian Wildlife Service, Hull, Quebec, pers. comm.
8 = G. Shutt, Canadian Wildlife Service, Hull, Quebec, pers. comm.

most samples; Whitehead et al. (1993) suggested that, as these congeners were not found in fish typical of osprey prey from that system (Mah et al. 1989), contamination may result from exposure on their southern wintering grounds.

Table 3. Mean levels of PCDD and PCDFs (ng kg^{-1}, wet weight) in the blubber of cetacean species from the west coast of North America, 1986-1989 (west coast samples from Jarman et al. in prep.; eastern samples from Norstrom et al. 1992).

Species	Location	N[a]	PCDDs					PCDFs			
			2378-TCDD	12378-PnCDD	123678-HxCDD	1234678-HpCDD	OCDD	2378-TCDF	Total PnCDF	124689/123689-HxCDF	1234689-HpCDF
Risso's dolphin	B.C.	1	<2	<5	<8	<10	<20	4	<5	<8	<10
Dall's porpoise	B.C.	3	3	6	10	3	13	29	23	13	9
Harbor porpoise	B.C.	6	3	9	101	7	8	13	13	15	12
	California	3	<2	<5	<8	<10	<20	<2	<5	<8	<10
Orca	B.C.	6	2	4	6	2	5	18	<5	<8	<10
False killer whale	B.C.	2	1	3	7	1	<20	15	<5	<8	<10
Gray whale	B.C.	2	<2	<5	<8	<10	<20	3	<5	<8	<10
Beluga whale	St. Lawrence R.	10	<1					2			
Beluga whale	Baffin Is.	6	<2					<2			

a) Number of samples included in geometric mean.

Eggs and liver tissue of juvenile herring gulls (*Larus argentatus*) collected in the vicinity of a bleached kraft pulp mill at LaTuque, Quebec, contained higher levels of the TCDD/F, PnCDD/F and 1,2,3,6,7,8-HxCDD congeners than birds collected at a remote lake (Lac Edouard: Table 2: Champoux 1993). PCDD and PCDF levels were considerably higher in herring gulls than ring-billed gulls (*Larus delawarensis*) at the contaminated site, perhaps reflecting a greater dietary reliance on fish by the former species. The TCDD level in herring gull eggs from a colony near a bleached kraft pulp mill at Jackfish Bay, Lake Superior, was similar to the reference site, Granite Rock on Lake Superior (Table 2), and it was suggested these birds were foraging extensively in garbage dumps (L. Shutt, Canadian Wildlife Service, Hull, Quebec, pers. comm.). Relatively elevated HxCDD levels as well as high TCDD levels in the LaTuque gull sample suggest both chlorophenol and bleached kraft pulp mill sources of contamination at this site (Fig. 1d). Although levels of Pn- and HxCDD were low in Lake Superior herring gull eggs (Fig. 1d), the Hx congener in particular was elevated in gull liver samples from those sites (Table 2). Selective retention of higher chlorinated PCDDs in herring gull livers relative to eggs was reported by Braune and Norstrom (1989). Hebert et al. (1994) reported elevated levels of all major congeners in gull eggs collected from Saginaw Bay, Lake Huron (Table 2). In the industrialized lower Great Lakes, PCDDs and PCDFs typically occur as by-products of the manufacture of chlorophenols and some herbicides, or through the combustion of leaded gasoline or the incineration of industrial wastes (Ramel 1978; Czuczwa and Hites 1984; Rappe 1984; Rappe et al. 1987). The watershed of Saginaw Bay is highly industrialized, and contamination results from a variety of industrial rather than pulp mill sources.

Common and hooded merganser (*Lophodytes cucullatus*) eggs and liver tissue from a pulp mill site in Quebec show similar congener patterns to those observed in waterfowl in marine and estuarine environments of coastal British Columbia (Fig. 1c); although overall concentrations were lower, TCDF levels were unusually high. Even a Barrow's goldeneye egg sample, collected from an uncontaminated lake in central British Columbia (Table 2), contained unusually high levels of TCDF, probably accumulated in lipid stores while wintering on the British Columbia coast.

Aquatic Mammals and Reptiles

Mink livers from uncontaminated sites had very low or undetectable levels of PCDDs and PCDFs (Table 4). Even sites associated with pulp mills, including samples from LaTuque, Quebec (Champoux 1993) and Cale Creek, British Columbia (Wilson et al. in prep), had unremarkable levels of TCDD and TCDF. However, the more highly chlorinated congeners, including 1,2,3,6,7,8-HxCDD, 1,2,3,4,6,7,8-heptachlorodizeno-p-dioxins and OCDD were present, suggesting combustion or chlorophenol sources of contamination, as well as perhaps an inability to effectively metabolize these congeners. In contrast, a river otter sample from the Columbia River had slightly elevated levels of TCDD and PnCDF, but low levels of other congeners, suggesting exposure to both bleached kraft pulp mill effluent and polychlorinated biphenyls (PCBs).

Two snapping turtle egg samples were taken downstream from a pulp mill on Thurso River in Quebec. One had relatively unremarkable contaminant levels, whereas a turtle egg sample taken from highly industrialized Hamilton Harbour contained high TCDF levels, and some penta-, hexa- and hepta-CDDs were detectable (Struger et al. 1993). In Hamilton Harbour PCDD sources were undoubtedly many and varied.

BIOLOGICAL EFFECTS OF PULP MILL CONTAMINANTS ON WILDLIFE

Laboratory studies have shown that dioxins cause a wide variety of teratogenic, immunologic, reproductive, carcinogenic and developmental toxic effects in birds and mammals (Peterson et al. 1993). 2,3,7,8-TCDD is considered to be the most toxic congener and is used as the standard to which the toxicity of all other halogenated aromatic hydrocarbons are compared using a system of toxic equivalency factors (TEFs; Safe 1990) which combine to give an overall toxic equivalency (TEQ). Avian embryos appear to be more sensitive to exposure than are adult birds, although considerable interspecies variability in sensitivity is evident in both forms. Acute oral LD50 doses range from 15 to >810 µg kg^{-1} body weight for the northern bobwhite (*Colinus virginianus*) and the ringed turtle-dove (*Streptopelia risoria*), respectively (Hudson et al. 1984); those for avian embryos range from 250 ng kg^{-1} injected into the air cell for the domestic chicken embryo, 1,100 ng kg^{-1} for ring-necked pheasant (*Phasianus colchicus*) embryos and between 1,000 and 10,000 ng kg^{-1} for eastern bluebird (*Sialia sialis*) embryos when injected into the albumin (Allred and Strange 1977; Martin et al. 1989; Nosek et al. 1992). Adult hen ring-necked pheasants injected at 1.0 µg kg^{-1} wk^{-1} over a 10-wk period suffered severe weight loss, resulting in 57% mortality and almost complete reproductive failure. However, dietary levels of 0.3 and 3.0 ng kg^{-1} TCDD fed for 18 wk caused no impacts on the productivity or survivability of embryos in bobwhites (Kenaga and Norris 1983).

Field studies in highly polluted areas of the Great Lakes and elsewhere have reported symptoms in wild birds that resemble those caused by TCDD in laboratory studies. For example, Kubiak et al. (1989) investigated reproduction and contaminant levels in a Forster's tern (*Sterna forsteri*) colony on Green Bay, Lake Michigan. They found that breeding success was greatly reduced at the highly contaminated site relative to a clean reference site, and through experimentation attributed reduced hatchability to both embryotoxicity and decreased parental nest attentiveness. Although levels of PCDD and PCDFs were elevated in tern eggs at the Green Bay site (medians of 37, 37 and 18.5 ng kg^{-1} for TCDD, HxCDDs and total PCDFs, respectively), the high levels of non-*ortho*-PCBs (median of 5,500 mg kg^{-1}) suggest that reproductive problems were caused mainly by PCBs (Table 5). White and Seginak (1994) monitored nest success of wood ducks (*Aix sponsa*) breeding downstream from an EPA-designated industrial waste site known to be contaminated with dioxins. They concluded that TCDD accounted for 70% of the toxicity, although only a limited range of chemical parameters were reported. Hatching success dropped from 90

Table 4. Levels of PCDDs and PCDFs in the livers of aquatic mammals and eggs of reptiles in North America, 1989-1991 from pulp mill sources as well as reference sites.

Species	Location	N[b]	% lipid	PCDDs					PCDFs					Ref[c]
				2378-TCDD	12378-PnCDD	123678-HxCDD	1234678-HpCDD	OCDD	2378-TCDF	Total PnCDF	HxCDF[a]	1234689-HpCDF	OCDF	
Mink	**Quebec**													
	LaTuque - above mill	3*		ND	ND	ND	1.5	ND	ND	4.6	ND	ND	ND	1
	LaTuque - near mill	6*		0.4	0.2	1.3	6.5	50.5	0.2	0.3	1.4	ND	0.6	1
	LaTuque - below mill	2*		0.9	1.8	20.6	35.5	29.4	1.1	0.7	49.6	1.3	ND	1
	B.C. - Fraser R.													
	Upper Fraser R.	6*		ND	ND	3	8	26		1	ND	ND	1	2
	Redrock L.	1		ND	ND	ND	ND	2		ND	ND	ND	ND	2
	Cale Cr. (1)	1		ND	2	22	120	190		3	6	10	3	2
	Cale Cr. (2)	1		ND	ND	2	18	16		ND	ND	ND	ND	2
	Cale Cr. (3)	1		ND	ND	ND	1	4		ND	ND	ND	ND	2
	Stone Cr. canyon	1		ND	ND	ND	ND	1		ND	ND	ND	ND	2
River otter	**B.C. - Columbia R.**													
	Hanna Cr.	1		11	ND	6	tr.	tr.		19	2	ND	ND	2
	B.C. - Fraser R.													
	Shirley Cr.	1		ND	ND	11	16	20		ND	ND	ND	ND	2
Snapping turtles	Quebec - Ottawa Thurso R.	1	6.1	2	2	3	3	2	ND	2	ND	ND		3
	Ontario - Hamilton Harbour	1	7.7	67	6	4	2	ND	ND	14	ND	ND		4

Note: Blank space means value not reported in the reference; ND means not detected; tr. means a definite but unquantifiable peak.
a = In mammals, this value is the sum of 3 congeners: 1,2,3,4,7,8-, 1,2,3,6,7,8-, and 2,3,4,6,7,8-HxCDF; in turtles it is 1,2,3,6,7,8-HxCDF as the other two were not reported.
b = Number of tissues in a pooled sample (*) or number of individual samples.
c = References:
1 = Champoux 1993
2 = Wilson et al. in prep.
3 = CWS 1990
4 = Struger et al. 1993

Table 5. *Examples of North American bird populations reported to have experienced reproductive or sublethal effects associated with pulp mills or other sources of TCDD-like compounds.*

Species	Location	Years	Effect	Cause[a]	Ref.[b]
Double-crested cormorant	Strait of Georgia, B.C.	91	EROD induction, reduced yolk weight	PCBs	1
Herring gull	Jackfish Bay, Lake Superior	91	Reproductive failure EROD induction	Food shortages Unknown	2
Herring gull	LaTuque, Quebec	91	EROD induction Uroporphyrin increases	PCBs TCDD liver levels correlated	3
Great blue heron	Strait of Georgia, B.C.	87-88	Reproductive failure; EROD induction; edema; reduced embryonic weight	Human and predator disturbance TCDD correlated with these parameters	4, 5, 6
		91	EROD induction; reduced body, stomach, intestine weight	Positive regression with TCDD Negative regression with TCDD	7
Bald eagle	Strait of Georgia, B.C.	92	CYP1A induction; uroporphyrin increases	Positive correlation with TCDD	8
Wood duck	Bayou Meto, Arkansas	88-90	Hatching, nest and duckling success reduced; bill deformities, edema	TCDD in combination with PCBs from industrial sources	9
Forster's tern	Green Bay, Lake Michigan	83	75% reduction in nest success; reduced embryo weight	PCBs and TCDD from industrial sources	10

a = Suspected cause of effect.
b = References:
1 = Sanderson et al. 1994a
2 = Shutt 1993
3 = Champoux 1993
4 = Elliott et al. 1989
5 = Bellward et al. 1990
6 = Hart et al. 1991
7 = Sanderson et al. 1994b
8 = Elliott et al. submitted
9 = White and Seginak 1994
10 = Kubiak et al. 1989

to 53% at egg TEQ levels of 20 to 50 and >50 ng kg^{-1}, and duckling success (number leaving the nest) was reduced from 9.3 to 5.6 ducklings per nest. Subcutaneous edema or malformed lower mandibles were found in a few unhatched embryos (Table 5). Wood ducks appear, therefore, to be extremely sensitive to TCDD. By way of comparison, in great blue herons chicks, there were no observable adverse effects (including ethoxyresorufin-O-deethylase (EROD) induction), the most sensitive endpoint determined for that species) at 2,3,7,8-TCDD < 100 ng kg^{-1} (Sanderson et al. 1994b).

There is relatively little information on the impacts of pulp mill contaminants on avian populations. Elliott et al. (1989) reported complete reproductive failure in a great blue heron colony near the bleached kraft pulp mill at Crofton in the Strait of Georgia during 1987, following an almost threefold increase in TCDD (66 to 210 ng kg^{-1}), as well as a smaller increase in HxCDD levels in eggs. Other colonies at less contaminated sites had >50% nest success during both 1986 and 1987. They concluded, however, that disturbance by humans and eagles, rather than TCDD-induced embryo toxicity, could have been the proximal cause of failure of the Crofton colony, as the range in contaminant levels overlapped with those from 1986, when nest success was around 60% (Table 5). The possible mechanism of adult behavioral aberrations in those nesting herons, resulting from dioxin-mediated endocrine disruption, has also been acknowledged (Moul 1990). Bellward et al. (1990) and later Sanderson et al. (1994b) found that EROD activity in heron eggs was highly positively correlated with TCDD concentration ($r = 0.70$, $p < 0.001$, $n = 54$: combined data). Embryo weight and the weight of several body organs were also negatively correlated with TCDD concentrations (r values > 0.48, p values < 0.01; Hart et al. 1991; Sanderson et al. 1994b). In the 1988 study of Crofton herons, prior to the clean-up of mill effluent, Hart et al. (1991) found that one third of heron hatchlings exhibited subcutaneous edema, typical of the syndrome frequently observed in the embryos of piscivorous waterbirds in the Great Lakes, which has been associated with high concentrations of halogenated aromatic hydrocarbons (Gilbertson et al. 1991). Henshel et al. (1993) also reported gross brain asymmetries in the same heron chicks. Sanderson et al. (1994a) assessed morphological and biochemical parameters of double-crested cormorant chicks hatching from eggs collected from three colonies in the Strait of Georgia. Although EROD activities were greater in Strait of Georgia sites than at a reference site in Saskatchewan, they did not correlate significantly with TCDD levels. EROD did, however, correlate with TEQs (EROD: $r = 0.70$, $p < 0.001$, $n = 20$), but the authors calculated that mono-$ortho$ and non-$ortho$ PCB congeners contributed more to TEQ totals than did PCDD or PCDF congeners (Table 5). Induction of CYP1A enzymes has proven to be a useful biomarker for monitoring the exposure to TCDD-like contaminants from both pulp mill and other sources in a variety of other studies (Bosveld et al. 1994; Elliott et al. 1995; Rattner et al. 1994; Van den Berg et al. 1994).

Herring gulls nesting near a pulp mill in Jackfish Bay, Lake Superior, experienced complete reproductive failure in 1991 and 1992 (Shutt 1993). However, artificial incubation indicated that hatchability and viability of chicks were no different at uncontaminated colonies. Levels of PCDDs and PCDFs were low, as were other organochlorine contaminants. EROD activity was less than that of birds at a reference colony, indicating that birds were likely not exposed to the mill effluent. The author concluded that reproductive failure was probably a result of food shortages due to a depletion in fish stocks in the area (Table 5). Champoux (1993) reported significant induction of EROD and increases of uroporphyrin levels in juvenile herring gulls at a colony from which the adults foraged at the pulp mill near LaTuque, Quebec, compared to a more distant, uncontaminated site. However, PCDD levels in the livers of these gulls were very low (1.6 and 5.6 ng kg^{-1} TCDD and TCDF, respectively, see Table 2). There were no differences in overall nest success between those two colonies ($p < 0.05$, $n = 270$; Champoux 1993).

In ranch mink, Hochstein et al. (1988) determined a 28-d LD50 for TCDD of 4.2 µg kg^{-1} body weight. Mortality generally followed a prolonged period of reduction in food consumption and extreme weight loss, characteristic of the "wasting syndrome" observed following TCDD exposure in rats (Hsia and Kreamer 1985). However, mink fed bleached kraft pulp mill effluent in their diets and drinking water for 8 months did not exhibit any impact on reproductive performance (Smits and Schiefer 1993). Nevertheless, mink appear to be almost an order of magnitude more sensitive to TCDD than laboratory rats (LD50s of 4.2 versus 22 to 45 µg kg^{-1} body weight, respectively). In rats, significant reproductive impairment occurred at a dietary concentration of 120 ng kg^{-1} TCDD and a no-effect level of 12 to 30 ng kg^{-1} was determined (Eisler 1986). Given the interspecific difference in sensitivity, reproduction in mink may be affected at a dietary TCDD concentration of 12 ng kg^{-1}, with a no-effect level of 1.2 to 3.0 ng kg^{-1}. Levels much higher than this mink no-effect level, up to 60 ng kg^{-1} 2,3,7,8-TCDD and 704 ng kg^{-1} of 2,3,7,8-TCDF, were found in fillets of fish collected in 1988 downstream of a pulp mill at Kamloops on the Fraser River of British Columbia (Mah *et al.* 1989). Difficulties in trapping mink along the mainstems of the Fraser and Columbia rivers downstream of the pulp mills (Elliott, Whitehead and Henny, unpubl. data) may be partly the result of PCDDs and PCDFs in prey items which are too high for mink survival and/or reproduction.

CONCLUSIONS

From the initial findings of PCDDs and PCDFs in the marine and estuarine birds in the Strait of Georgia and the realization of their relationship with pulp mills in the mid-1980s, efforts have been made to reduce their release into receiving environments. Monitoring should continue post effluent clean-up and include measurement of both contaminant concentrations in tissues and biomarkers such as CYP1A activity. More work on productivity and wildlife population health near pulp mills is recommended. From the limited data on effects of pulp mill contaminants on reproduction, as with laboratory animals, there are large differences among wild species in sensitivity to TCDD-like compounds. Wood ducks were reported to suffer significant reproductive impairment at TCDD egg levels well below the apparent no-effect level for great blue herons of about 100 ng kg^{-1}. If waterfowl are generally more susceptible to dioxin toxicity, then levels measured in the livers of diving ducks in the Strait of Georgia may have caused significant health effects. Further information on interspecific variability is needed for extrapolation in risk assessments. Finally, the determination of levels and effects of pulp mill contaminants other than PCDDs and PCDFs in wildlife is limited and should be broadened.

ACKNOWLEDGMENTS

J.-L. Desgranges, P. Laporte, P. Pearce, and L. Shutt are thanked for use of unpublished data.

REFERENCES

Allred, P.M., and J.R. Strange. 1977. The effects of 2,4,5-trichlorophenoxyacetic acid and 2,3,7,8-tetrachlorodibenzo-p-dioxin on developing chicken embryos. Arch. Environ. Contam. Toxicol. 5:483-489.

Andersson, T., L. Förlin, J. Hardig and A. Larsson. 1988. Physiological disturbances in fish living in coastal water polluted with bleached kraft pulp mill effluents. Can. J. Fish. Aquat. Sci. 45:1525-1536.

Anthony, R.G., M.G. Garrett and C.A. Schuler. 1993. Environmental contaminants in bald eagles in the Columbia River estuary. J. Wildl. Manage. 57:10-19.

Beeman, D.K., T.A. Augsperger and W.J. Fleming. 1993. Productivity and PCDDs/PCDFs in eggs from two great blue heron rookeries near a pulp and paper mill on the Roanoke River, North Carolina. Abstracts, 14th Annual Meeting of Society of Environmental Toxicology and Chemistry, Houston, TX, November, 1993.

Bellward, G.D., R.J. Norstrom, P.E. Whitehead, J.E. Elliott, S.M. Bandiera, C. Dworschak, T. Chang, S. Forbes, B. Cadario, L.E. Hart and K.M. Cheng. 1990. Comparison of polychlorinated dibenzodioxin levels with mixed function oxidase induction in great blue herons. J. Toxicol. Environ. Health 30:33-52.

Bosveld, A.T.C., J. Gradener, A.J. Murk, A. Brouwer, M. van Kampen, E.H. Evers and M. Van den Berg. 1994. Effects of PCDDs, PCDFs and PCBs in common tern (*Sterna hirundo*) breeding in estuarine and coastal colonies in the Netherlands and Belgium. Environ. Toxicol. Chem. 14:99-116.

Braune, B.M., and R.J. Norstrom. 1989. Dynamics of organochlorine compounds in herring gulls. III. Tissue distribution and bioaccumulation in Lake Ontario gulls. Environ. Toxicol. Chem. 8:957-968.

Cederberg, T., E. Storr-hansen, M. Cleeman and J. Dyck. 1991. Organochlorine pollutants in guillemot eggs from the Baltic sea and northern Atlantic - polychlorinated dibenzo-p-dioxins, dibenzofurans, biphenyls and pesticides. Dioxin '91, 23-27 Sept., 1991, Chapel Hill, NC.

Champoux, L. 1993. Contamination et ecotoxicologie de la faune dans la region d'une usine de pate blanchie au chlore a La Tuque (Quebec). Tech. Rep. Ser. No. 187. Canadian Wildl. Serv. St. Foy, PQ, 63 pp.

Czuczwa, J.M., and R.M. Hites. 1984. Environmental fate of combustion-generated polychlorinated dioxins and furans. Environ. Sci. Technol. 18:444-450.

Ehrlich, P.R., D.S. Dobkin and D. Wheye. 1988. The birder's handbook: a field guide to the natural history of North American birds. Simon and Schuster, New York.

Eisler, R. 1986. Hazards of dioxin to fish and wildlife. U.S. Fish Wildl. Serv. Tech. Rep. No. 8.

Elliott, J.E., R.J. Norstrom, L. Lorenzen, L.E. Hart, H. Philibert, S.W. Kennedy, J.J. Stegeman, G.D. Bellward and K.M. Cheng. 1995. Biological effects of polychlorinated dibenzo-p-dioxins dibenzofurans and biphenyls in bald eagle (*Haliaeetus leucocephalus*) chicks. Environ. Toxicol. Chem. (Submitted).

Elliott, J.E., R.W. Butler, R.J. Norstrom and P.E. Whitehead. 1989. Environmental contaminants and reproductive success of great blue herons *Ardea herodias* in British Columbia, 1986-87. Environ. Pollut. 59:91-114.

Gilbertson, M., T. Kubiak, J. Ludwig and G. Fox. 1991. Great lakes embryo mortality, edema, and deformities syndrome (GLEMEDS) in colonial fish-eating birds - similarity to chick edema disease. J. Toxicol. Environ. Health 33:455-520.

Harding, L.E., and W.M. Pomeroy. 1990. Dioxin and furan levels in sediments, fish and invertebrates from fishery closure areas of coastal British Columbia. Environment Canada, North Vancouver, BC, Rep. No. 90-09, 77 pp.

Harfenist, A., P.E. Whitehead, W.J. Cretney and J.E. Elliott. 1993. Food chain sources of polychlorinated dioxins and furans to great blue herons (*Ardea herodias*) foraging in the Fraser River estuary, British Columbia. Tech. Rep. Ser. No. 169. Canadian Wildl. Serv., Delta, BC, 26 pp.

Hart, L.E., K.M. Cheng, P.E. Whitehead, R.M. Shah, R.J. Lewis, S.R. Ruschkowski, R.W. Blair, D.C. Bennett, S.M. Bandiera, R.J. Norstrom and G.D. Bellward. 1991. Dioxin contamination and growth and development in great blue heron embryos. J. Toxicol Environ. Health 32:331-344.

Hebert, C.E., R.J. Norstrom, M. Simon, B.M. Braune, D.V. Weseloh and C.R. Macdonald. 1994. Temporal trends and sources of PCDDs and PCDFs in the Great Lakes: herring gull monitoring, 1981-1991. Environ. Sci. Technol. 28:1268-1277.

Henshel, D.S., K.M. Cheng, R.J. Norstrom, P.E. Whitehead and J.D. Steeves. 1993. Mophometric and histological changes in brains of great blue heron hatchlings exposed to PCDDs: preliminary analyses. ASTM STP #1179, pp. 288-303.

Hochstein, J.R., R.J. Aulerich and S.J. Bursian. 1988. Acute toxicity of 2,3,7,8-tetrachlorodibenzo-p-dioxin to mink. Arch. Environ. Contam. Toxicol. 17:33-37.

Hsia, M.T.S., and B.L. Kreamer. 1985. Delayed wasting syndrome and alterations of liver gluconeogenic enzymes in rats exposed to the TCDD congener 3,3',4,4'-tetrachloroazoxy-benzene. Toxicol. Lett. 25:247-258.

Hudson, R.H., R.K. Tucker and M.A. Haigele. 1984. Handbook of toxicity of pesticides to wildlife. U.S. Fish Wildl. Serv. Resour. Publ. 153. 90 pp.

Jarman, W.M., R.J. Norstrom, D.C.G. Muir, M. Simon and R.W. Baird. Levels of organochlorine compounds, including PCDDs and PCDFs, in cetaceans from the west coast of North American. Mar. Pollut. Bull. (Submitted).

Jarman, W.M., S.A. Burns, R.R. Chang, R.D. Stephens, R.J. Norstrom, M. Simon and J. Linthicum. 1993. Determination of PCDDs, PCDFs, and PCBs in California peregrine falcons (*Falco peregrinus*) and their eggs. Environ. Toxicol. Chem. 12:105-114.

Kenaga, E.E., and L.A. Norris. 1983. Environmental toxicity of TCDD, p. 277-300 *In* R.E. Tucker, A.L. Young, and A.P. Grey [ed.] Human and environmental risks of chlorinated dioxins and related compounds. Plenum, New York.

Krahn, P.K., J.A. Shrimpton and R.D. Glue. 1987. Assessment of storm water related chlorophenol releases from wood protection facilities in British Columbia. Environment Canada, North Vancouver, BC, Rep. No. 87-14.

Kubiak, T.J., H.J. Harris, L.M. Smith, T.R. Schwartz, D.L. Stalling, J.A. Trick, L. Sileo, D.E. Docherty and T.C. Erdman. 1989. Microcontaminants and reproductive impairment of the Forster's tern on Green Bay, Lake Michigan - 1983. Arch. Environ. Contam. Toxicol. 18:706-727.

Kuehl, D.W., B.C. Butterworth, W. DeVita and C.P. Sauer. 1987. Environmental contamination by polychlorinated dibenzo-p-dioxins and dibenzofurans associated with pulp and paper mill discharge. Biomed. Environ. Mass Spectrosc. 14:443-447.

Lindejoo, E., and J. Thulin. 1987. Fin erosion of perch (*Perca fluviatilis*) in a pulp mill effluent. Bull. Eur. Assoc. Fish Pathol. 7:717-749.

Luthe, C.E., R.M. Berry, R.H. Voss and B.I. Fleming. 1990. Dioxins and furans in pulp mill questions - research answers. PAPRICAN, Pointe Claire, PQ, Misc. Rep. MR 175.

Mah, F.T.S., D.D. MacDonald, S.W. Sheehan, T.M. Tuominen and D. Valiela. 1989. Dioxins and furans in sediment and fish from the vicinity of ten inland pulp mills in British Columbia. Environment Canada, North Vancouver, BC, 77 pp.

Martin, S., J. Duncan, D. Thiel, R.E. Peterson and M. Lemke. 1989. Evaluation of the effects of dioxin-contaminated sludges on eastern bluebirds and tree swallows. Nekoosa Papers Inc., Port Edwards, WI.

Moul, I.E. 1990. Environmental contaminants, disturbance and breeding failure at a great blue heron colony on Vancouver Island. Unpubl. M.Sc. thesis, Univ. Brit. Col., Vancouver.

Munkittrick, K.R., C.B. Portt, G.J. Van Der Kraak, I.R. Smith and D.A. Rokosh. 1991. Impact of bleached kraft mill effluent on population characteristics, liver MFO activity, and serum steroid levels of a Lake Superior white sucker (*Catostomus commersoni*) population. Can. J. Fish. Aquat. Sci. 48:1371-1380.

Munkittrick, K.R., G.J. van Der Kraak, M.E. McMaster and C.B. Portt. 1992. Response of hepatic MFO activity and plasma sex steroids to secondary treatment of bleached kraft pulp mill effluent and mill shutdown. Environ. Toxicol. Chem. 11:1427-1439.

Norstrom, R. 1995. Canadian Wildlife Service, Hull, Quebec, pers. comm.

Norstrom, R.J., and M. Simon. 1983. Preliminary appraisal of tetra- to octachlorodibenzodioxin contamination in eggs of various species of wildlife in Canada. p. 165-170. *In* J. Miyamoto [ed.] IUPAC pesticide chemistry: human welfare and the environment. Pergamon Press, Oxford.

Norstrom, R.J., R.W. Riseborough and D.J. Cartwright. 1976. Elimination of chlorinated dibenzofurans associated with polychlorinated biphenyls fed to mallards (*Anas platyrhynchos*). Toxicol. Appl. Pharmacol. 37:217-228.

Norstrom, R.J., D.C.G. Muir, C.A. Ford, M. Simon, C.R. MacDonald and P. Beland. 1992. Indications of P450 monooxygenase activities in beluga (*Delphinapterus leucas*) and narwhal (*Monodon monoceros*) from patterns of PCB, PCDD and PCDF accumulation. Mar. Environ. Res. 34:267-272.

Norstrom, R.J., M. Simon, P.E. Whitehead, R Kussat, C. Garrett and F. Mah. 1988. Levels of polychlorinated dibenzo-p-dioxins (PCDDs) and polychlorinated dibenzofurans (PCDFs) in biota and sediments near potential sources of contamination in British Columbia, 1987. Canadian Wildlife Service Analytical Report CRD-88-5.

Nosek, J.A., S.R. Craven, J.R. Sullivan, S.S. Hurley and R.E. Peterson. 1992. Toxicity and reproductive effects of 2,3,7,8-tetrachlorodibenzo-*p*-dioxin in ring-necked pheasant hens. J. Toxicol. Environ. Health 35:187-198.

Peterson, R.E., H.M. Theobald and G.L. Kimmel. 1993. Developmental and reproductive toxicity of dioxins and related compounds: cross species comparisons. CRC Crit. Rev. Toxicol. 23:283-335.

Ramel, C. [ed.]. 1978. Chlorinated phenoxy acids and their dioxins. Ecol. Bull. 27, Stockholm. 302 pp.

Rappe, C. 1984. Analysis of polychlorinated dioxins and furans. Environ. Sci. Technol. 18:78A-90A.

Rappe, C., and H.R. Buser. 1989. Chemical and physical properties, analytical methods, sources and environmental levels of halogenated dibenzodioxins and dibenzofurans. p. 71-102. *In* R.D. Kimbrough, and A.A. Jensen [ed.] Halogenated biphenyls, terphenyls, naphthalenes, dibenzodioxins and related products, 2nd ed. Elsevi, Amsterdam.

Rappe, C., R. Andersson, P.A. Bergqvist, C. Brohede, M. Hansson, L.O. Kjeller, G. Lindstrom, S. Marklund, M. Nygren, S.E. Swanson, M. Tysklind and K. Wiberg. 1987. Overview on the environmental fate of chlorinated dioxins and dibenzofurans. Sources, levels and isomeric pattern in various matrices. Chemosphere 16:1603-1618.

Rattner, B.A., J.S. Hatfield, M.J. Melancon, T.W. Custer and D.E. Tillitt. 1994. Relation among cytochrome P450, Ah-active PCB congeners and dioxin equivalents in pipping black-crowned night-heron embryos. Environ. Toxicol. Chem. 13:1805-1812.

Safe, S.H. 1990. Polychlorinated biphenyls (PCBs), dibenzo-p-dioxins (PCDDs), dibenzofurans (PCDFs), and related compounds: environmental and mechanistic considerations which support the development of toxic equivalency factors (TEFs). CRC Crit Rev. Toxicol. 21:51-88.

Sanderson, J.T., R.J. Norstrom, J.E. Elliott, L.E. Hart, K.M. Chang and G.D. Bellward. 1994a. Biological effects of polychlorinated dibenzo-*p*-dioxins, dibenzofurans, and biphenyls in double-crested cormorant chicks (*Phalacrocorax auritus*). J. Toxicol. Environ. Health 41:247-265.

Sanderson J.T., J.E. Elliott, R.J. Norstrom, P.E. Whitehead, L.E. Hart, K.M. Cheng and G.D. Bellward, 1994b. Monitoring biological effects of polychlorinated dibenzo-*p*-dioxins, dibenzofurans and biphenyls in great blue heron chicks (*Ardea herodias*) in British Columbia. J. Toxicol. Environ. Health 41:435-450.

Servos, M.R., S.Y. Huestis, D.M. Whittle, G.J. Van der Kraak and K.R. Munkittrick. 1994. Survey of receiving-water environmental impacts associated with discharges from pulp mills. 3. Polychlorinated dioxins and furans in muscle and liver of white sucker (*Catostomus commersoni*). Environ. Toxicol. Chem. 13:1103-1115.

Shutt, J.L. 1993. Reproductive success of herring gulls nesting near a bleached kraft pulp mill. Proceedings, Thirteenth International Symposium on Chlorinated Dioxins and Related Compounds. Vienna, Austria, September, 1993.

Shutt, L. 1995. Canadian Wildlife Service, Hull, Quebec, pers. comm.

Sijm, D.T.H.M., H. Wever and A. Opperhuizen. 1993. Congener-specific biotransformation and bioaccumulation of PCDDs and PCDFs from fly ash in fish. Environ. Toxicol. Chem. 12:1895-1908.

Smits, J.E.G., and B.H. Schiefer. 1993. The biological impact of bleached kraft pulp mill effluent on mink (*Mustela vison*). Abstracts, 14th Annual Meeting of Society of Environmental Toxicology and Chemistry, Houston, TX, November, 1993.

Stalling, D.L., R.J. Norstrom, L.M. Smith and M. Simon. 1985. Patterns of PCDD, PCDF and PCB contamination in Great Lakes fish and birds and their characterization by principal components analysis. Chemosphere 14:627-643.

Struger, J., J.E. Elliott, C.A. Bishop, M.E. Obbard, R.J. Norstrom, D.V. Weseloh, M. Simon and P. Ng. 1993. Environmental contaminants in eggs of the common snapping turtle (*Chelydra serpentina serpentina*) from the Great Lakes-St. Lawrence River basin of Ontario, Canada (1981, 1984). J. Great Lakes Res. 19:681-694.

Van den Berg, M., B.L. Craane, T. Sinnige, S. Van Mourik, S. Dirksen, T. Boudewijn, M. Van der Gaag, I.J. Lutke-Schipholt, B. Spenkelink and A. Brouwer. 1994. Biochemical and toxic effects of polychlorinated biphenyls (PCBs), dibenzo-*p*-dioxins (PCDDs) and dibenzofurans (PCDFs) in the cormorant (*Phalacrocorax carbo*) after in ovo exposure. Environ. Toxicol. Chem. 13:803-816.

Vermeer, K., W.J. Cretney, J.E. Elliott, R.J. Norstrom and P.E. Whitehead. 1993. Elevated polychlorinated dibenzodioxin and dibenzofuran concentrations in grebes, ducks and their prey near Port Alberni, British Columbia, Canada. Mar. Pollut. Bull. 26:431-435.

White, D.H., and J.T. Seginak. 1994. Dioxin and furans linked to reproductive impairment in wood ducks. J. Wildl. Manage. 58:100-106.

Whitehead, P.E., J.E. Elliott, R.J. Norstrom and K. Vermeer. 1990. PCDD and PCDF contamination of waterfowl in the Strait of Georgia British Columbia, Canada: 1989-1990. Proceedings, Dioxin '90, Bayreuth, FRG, Eco-informa Press, pp. 459-462.

Whitehead, P.E., A. Harfenist, J.E. Elliott and R.J. Norstrom. 1992a. Levels of polychlorinated dibenzo-*p*-dioxins and polychlorinated dibenzofurans in waterbirds of Howe Sound, British Columbia. p. 229-238. *In* C.D. Levings, R.B. Turner, and B. Ricketts [ed.] Proceedings of the Howe Sound Environmental Science Workshop. Can. Tech. Rep. Fish. Aquat. Sci. No. 1879.

Whitehead, P.E., Norstrom, R.J. and J.E. Elliott. 1992b. Dioxin levels in eggs of great blue herons (*Ardea herodius*) decline rapidly in response to process changes in a nearby kraft pulp mill. Proceedings, Dioxin '92, Tampere, Finland, Organohalogen Compounds 9:325-328.

Whitehead, P.E., J.E. Elliott, R.J. Norstrom, C. Steeper, J. Van Oostdam and G.E.J. Smith. 1993. Chlorinated hydrocarbon levels in eggs of osprey nesting near pulp mills in British Columbia, Canada. Proceedings, Dioxin '93, Vienna, Austria, Organohalogen Compounds 12:231-2234.

Wilson, L.K., J.E. Elliott and P.E. Whitehead. 1995. Chlorinated compounds in wildlife from the Fraser River basin. Fraser River Action Plan, Technical Report, Environment Canada, Vancouver (In prep).

DIOXINS AND FURANS IN CRAB HEPATOPANCREAS: USE OF PRINCIPAL COMPONENTS ANALYSIS TO CLASSIFY CONGENER PATTERNS AND DETERMINE LINKAGES TO CONTAMINATION SOURCES

Mark B. Yunker[1] and Walter J. Cretney[2]

[1] 7137 Wallace Dr., Brentwood Bay, British Columbia, V8M 1G9 Canada
[2] Ocean Chemistry Division, Institute of Ocean Sciences, Box 6000, Sidney, British Columbia, V8L 4B2 Canada

Principal components analysis (PCA) has been applied to a large data set of the concentrations of chlorinated dibenzo-p-dioxins and dibenzofurans in samples of crab hepatopancreas from British Columbia, Canada (B.C.). These samples have been collected since 1987 from locations in the vicinity of B.C. coastal mills and the principal B.C. harbors. PCA distinguishes harbor samples from mill samples, reveals differences between individual mill sites and reliably classifies the dioxin and furan congener patterns according to chlorine bleaching, digested polychlorinated phenol and pentachlorophenol wood preservative sources. PCA may also be used to assign sources to individual crab samples. Since 1987 both the proportion of toxic 2,3,7,8-chlorinated congeners and the overall dioxin and furan concentrations have generally decreased near B.C. mill sites. The mill-related tetrafurans have been removed faster than the hexadioxins, with the result that composition profiles and PCA projections have become more similar over time at the mill sites. Crab samples from B.C.'s harbors have lower proportions of the toxic 2,3,7,8-chlorinated congeners, but have shown less change over time.

INTRODUCTION

Chlorinated contaminants produced and discharged by pulp mills were documented in British Columbia, Canada (B.C.) coastal waters more than a decade ago (e.g., Voss and Yunker 1983). More recently the presence of chlorinated dioxins and furans in ecologically significant concentrations in sediments, shellfish and groundfish was described by Norstrom *et al.* (1988), both in the vicinity of B.C. mills and in the principal B.C. harbors. As one result, harvesting of prawn, shrimp and crab was restricted near pulp mills in late 1988, and more extensive fisheries closures have followed as the extent of contamination has been more fully delineated.

The major processes that produced dioxins and furans in pulp mill effluent in the past were the chlorination of dioxin precursors during bleaching and the condensation of polychlorinated phenols during pulp digestion (Luthe and Prahacs 1993; Luthe *et al.* 1993). The elimination of dioxin precursors from defoamer products, exclusion of chlorophenol-contaminated wood chips and introduction of chlorine dioxide bleaching have dramatically reduced mill-related contamination (Whitehead *et al.* 1992; Luthe *et al.* 1992). The introduction of dioxins and furans from combustion (both from the mill and ambient deposition) into mill effluent is a more minor process, but has proved to be more difficult to eliminate (Hites 1990; Luthe and Prahacs 1993).

Monitoring programs were initiated in 1989 to document the sources and concentrations of chlorinated dibenzo-p-dioxins and dibenzofurans in B.C. harbors and pulp mill sites. These programs were designed to assess ecosystem impact, address human health concerns due to contamination of the fishery and assist in the formulation of regulations for fisheries management. Aspects such as the geographical extent of contamination, persistence of dioxins and furans over time and trends in the congener composition and proportion of toxic congeners were also important considerations. However, translation of these aspects into monitoring programs has produced a very large data set. For crab hepatopancreas alone, research and monitoring activities since the late 1980s in the vicinity of 10 coastal mills and the principal harbors (Fig. 1; Table 1) had produced dioxin and furan concentrations for nearly 500 samples. A data set this size can only be effectively interpreted with multivariate analysis techniques.

In this work we employ principal components analysis (PCA) to explore the interrelationships and trends in both the dioxin and furan pattern and the toxic congener composition among B.C. mill sites and between harbors and mill sites. The integration of toxic equivalency into the interpretation provides an important window into both ecosystem effects and potential health risks to humans.

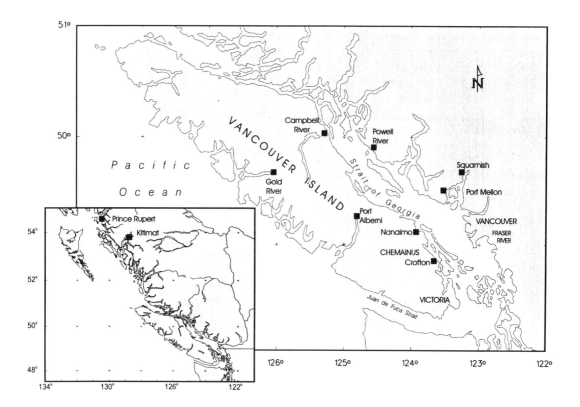

Figure 1. Map of the B.C. coast showing the pulp mills and the principal harbors.

METHODS

Data for the concentrations of chlorinated dibenzo-*p*-dioxins and dibenzofurans in 469 samples of crab hepatopancreas were obtained primarily in electronic form from the Canadian Departments of Fisheries and Oceans and Environment. All data were then reverified against original laboratory reports. Data were primarily for Dungeness crab (*Cancer magister*) hepatopancreas, although 13 hepatopancreas samples from Red Rock crabs (*Cancer productus*), six from Box crabs (*Lopholithodes spp.*) and one sample from a Kelp crab (*Pugettia producta*) were also included.

The data set included 11 baseline samples from 1987 (i.e., prior to any mill process changes) which were obtained from a Canadian Wildlife Service unpublished report (Norstrom *et al.* 1988). Post-1987 monitoring data for the mill sites included 378 samples analyzed by Axys Analytical Ltd. of Sidney, B.C. (formerly Seakem Analytical Ltd.), 15 samples analyzed by Envirotest Ltd. in Edmonton, Alberta and 8 samples analyzed by Wellington Laboratories Ltd. in Guelph, Ontario. The 1989 and 1993 samples from Kitimat and Nanaimo, respectively, were analyzed by Envirotest and the 1993 samples from Gold River were analyzed by Wellington Laboratories. Data for samples from remote locations have principally been produced by the Department of Fisheries and Oceans, with 30 samples from the Burlington, Ontario laboratory and 25 samples from the Sidney, B.C. laboratory.

Table 1. Coastal B.C. sampling sites, their physical environment and the pulping process used by the different mills.

Site	Geographical Location	Physical Environment	Mill Process
Pulp mill sites			
Port Alberni	Fjord	Restricted circulation	Bleached kraft
Crofton	Channel	Well flushed	Bleached kraft
Campbell River (Elk Falls Mill)	Channel	Well flushed	Bleached kraft
Gold River	Fjord	Restricted circulation	Bleached kraft
Howe Sound (Port Mellon and Woodfibre Mills)	Fjord	Restricted circulation	Bleached kraft
Kitimat	Fjord	Restricted circulation	Unbleached kraft
Nanaimo (Harmac Mill)	Channel	Well flushed	Bleached kraft
Powell River	Strait	Well flushed	Bleached kraft
Prince Rupert	Narrow channel	Restricted circulation	Bleached kraft
Harbors and industrialized areas			
Vancouver Harbor (Burrard Inlet)	Fjord	Restricted circulation	
Nanaimo Harbor	Bay	Restricted circulation	
Victoria/Esquimalt Harbors	Bay	Restricted circulation	
Fraser River mouth	Estuary	Well flushed	

Consistent collection, dissection, extraction and analysis procedures have been maintained for the crab samples analyzed by Axys Analytical (more than 80% of the data set). All samples were collected and dissected by Hatfield Consultants and all were subjected to the same extraction protocol by Axys Analytical (e.g., Dwernychuk 1993). All samples were spiked with ^{13}C-labeled surrogate standards for the 2,3,7,8-tetra- to octachlorodioxins plus 2,3,7,8-tetrachlorofuran; as a minimum ^{13}C-labeled 1,2,3,4-tetrachlorodioxin was added as a recovery standard. For samples analyzed by high resolution GC/MS (i.e., 1991 onward), ^{13}C-labeled 2,3,7,8-penta- to heptachlorofuran surrogates and 1,2,3,7,8,9-hexachlorodioxin and 1,2,3,4,6,7,8-heptachlorodioxin recovery standards were also added. Minor differences in method occurred with the other laboratories, but all analyses employed ^{13}C-labeled standards.

Unsupervised PCA was performed using the nonlinear iterative partial least squares (NIPALS) algorithm in a program which allowed a detailed examination of both crab sample and variable (congener) projections (respectively scores and loadings; software was provided by the Chemometrics Clinic, Seattle, WA). Congeners and congener totals which were undetected (ND) were replaced by the limit of detection. Values with questionable ion ratios (NDR) were used only if the concentration was reasonable in the context of similar samples from the site.

For most PCA variables the limit of detection was used to estimate the concentration for less than 15% of the samples (Table 2; Fig. 2). The more chlorinated congeners, particularly octachlorodioxin and the hepta- and octachlorofurans, were undetectable in higher percentages of samples and were used only when it was verified that their presence did not skew the model. The congeners 1,2,3,4,7,8-H6 hexachlorodibenzo-*p*-dioxin and 1,2,3,4,7,8,9-H7 heptachlorodibenzo-*p*-dioxin were undetectable in most samples and were removed. Because the 2,3,7,8-substituted hexafurans were only detectable in a few samples, an average of the detection limits or concentrations for the individual congeners was used. For each congener series a non-2,3,7,8 congener total was obtained by subtracting the 2,3,7,8-substituted congener(s) from the homologue total; if a value of zero resulted for these "other" dioxins, the detection limit for the homologue total was substituted.

Table 2. PCA model abbreviations, percent of samples for which the congeners used in the PCA model were undetectable and 2,3,7,8-toxic equivalent factors for the PCA congeners.

Congener	PCA Abbrev.	Percent Undetectable	TEF
2,3,7,8-T4CDD	4D	8.1	1.0
Non-2,3,7,8 T4CDD	4DO	5.8	0
1,2,3,7,8-P5CDD	5D	11.1	0.5
Non-2,3,7,8 P5CDD	5DO	9.2	0
1,2,3,6,7,8-H6CDD	6D2	1.1	0.1
1,2,3,7,8,9-H6CDD	6D3	12.2	0.1
Non-2,3,7,8 H6CDD	6DO	0.4	0
1,2,3,4,6,7,8-H7CDD	7D	15.4	0.01
Non-2,3,7,8 H7CDD	7DO	12.8	0
O8CDD	8D	39.2	0.001
2,3,7,8-T4CDF	4F	0.6	0.1
Non-2,3,7,8 T4CDF	4FO	0.6	0
1,2,3,7,8-P5CDF	5F1	29.9	0.05
2,3,4,7,8-P5CDF	5F2	14.1	0.5
Non-2,3,7,8 P5CDF	5FO	7.2	0
2,3,7,8-H6CDF mean	6F	66.3	0.1
Non-2,3,7,8 H6CDF	6FO	23.0	0
1,2,3,4,6,7,8-H7CDF	7F1	41.2	0.01
Non-2,3,7,8 H7CDF	7FO	39.4	0
O8CDF	8F	87.2	0.001

Normalization of each sample was employed before PCA, because: 1) concentrations differed by orders of magnitude, particularly between years and 2) a primary goal was to examine trends in the dioxin and furan congeners with a "mill" pattern. Log transformation was also evaluated, but the resulting PCA model was more sensitive to detection limit artifacts. A normalization factor derived from a subset of variables with mid-range standard deviations and means was used to remove negative correlations due to closure without inducing spurious effects in the normalizing variables (Johansson *et al.* 1984; Yunker *et al.* 1995). Data were then autoscaled (mean centered and divided by the variance on a variable-by-variable basis) before PCA.

The use of the limit of detection as an estimation of the actual concentration proved to be a problem for 18 low concentration samples that either were from remote areas or were earlier (1989-91) samples with high detection limits. These samples had only 2 to 6 congeners detectable out of 20 (average 4), and the limit of detection (and hence apparent concentration) was usually noticeably higher for the higher chlorine-numbered congeners. The resulting dominance of hepta- and octachlorinated congeners produced a spurious separation of these samples from all other samples in the first principal component (PC). When these samples were excluded the hepta- and octachlorinated congeners became much less important to the PCA model (see following). To simplify source interpretation, a Varimax axis rotation was then applied to the first three PCs.

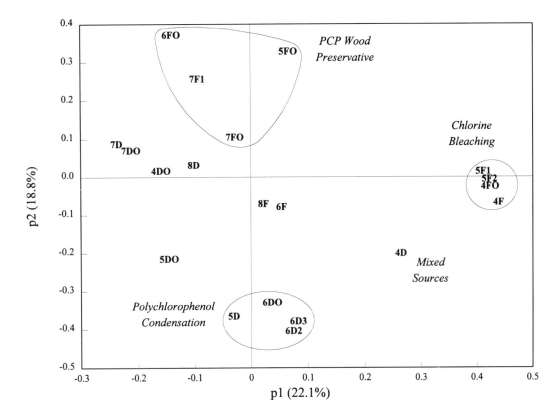

Figure 2. Varimax rotated variable (loadings) plot for dioxin and furan congener projections in the first and second principal components of the PCA model.

RESULTS AND DISCUSSION

The utility of this data set for the interpretation of trends in congener composition both between sites and over time is greatly enhanced by having more than 80% of the samples analyzed by one laboratory. Furthermore, the normalization of samples before PCA removed potential concentration biases and allowed samples from different laboratories to be added to the model. Concentration profiles, and PCA projections for the non-Axys Analytical samples were entirely consistent with the Axys samples from similar sites or the same site in different years.

Overview of the PCA Model

PCA is invaluable for classifying dioxin and furan congener variables according to their primary source (e.g., chlorine bleaching, pentachlorophenol) and for identifying congeners that covary. PCA separates a data matrix into the product of two matrices (one each for samples and variables) with the PCs comprising individual column or row vectors in their respective matrix. Each PC provides interrelated information concerning trends in the samples and the variables. The PCA model is calculated with no *a priori* assumptions about the data structure and completely unbiased projections result. A very readable explanation of the application of PCA to chemical data has been published by Meglen (1992).

The crab hepatopancreas PCA model includes 451 samples and employs 20 dioxin and furan variables. Before Varimax axis rotation the variance accounted for by the model in, respectively, the first, second and third PCs was 27.4, 17.4 and 14.6%, for a total of 59.4%. After axis rotation the variance in the first three PCs was redistributed to 22.1, 18.8 and 18.5%.

In Fig. 2 the chlorinated dibenzo-p-dioxin and dibenzofuran variables project together according to source. Congeners that are characteristic of pulp mill chlorine bleaching project to the far right and include 2,3,7,8-T4CDF, the non-2,3,7,8 tetrafurans and the two 2,3,7,8-P5CDF isomers (e.g., Luthe and Prahacs 1993). Dioxins that are produced by the condensation of polychlorinated phenols during pulp digestion cluster in the bottom of Fig. 2, close to the y-axis, and include 1,2,3,7,8-P5CDD, 1,2,3,6,7,8-H6CDD, 1,2,3,7,8,9-H6CDD and the non-2,3,7,8 H6CDD isomers (Luthe et al. 1993). 2,3,7,8-T4CDD projects midway between these two groups, which suggests mixed sources. Congeners that are present in pentachlorophenol wood preservative but are removed during the pulping process project at the top of Fig. 2 and include 1,2,3,4,6,7,8-H7CDF, the non-2,3,7,8 tetrafurans and hexafurans, and to a lesser extent the non-2,3,7,8 heptafurans (Hagenmaier and Brunner 1987; Luthe et al. 1993). With the exception of 1,2,3,4,6,7,8-H7CDF, all of the above congeners have significant loadings only in the first two PCs.

Congeners at or near the limit of detection (Table 2), such as 2,3,7,8-H6CDF and O8CDF, and the hepta- and octadioxins project close to the origin in the first two PCs but make a strong contribution to the third PC. Thus the contributions of detection limit artifacts and/or combustion have been largely shifted to the third PC. The remaining variable, non-2,3,7,8 P5CDD, makes a contribution to all three PCs and is apparently present in all sample types.

The projection of the toxic 2,3,7,8-tetra- and pentachlorinated dioxins and furans on the right side of Fig. 2 produces a correlation between samples and their toxicity in Fig. 3. The log of the 2,3,7,8-TCDD toxic equivalents (TEQ) is highly correlated ($r^2 = 0.42$; $p < 0.001$) with the projection of each sample in the first PC (t1). The TEQ correlation with the second PC (t2) is lower but is still significant ($r^2 = 0.17$; $p < 0.001$). The correlation of the log TEQ with t1 and t2 is only slightly improved over the correlation with t1 ($r^2 = 0.44$; $p < 0.001$).

There is a direct source correspondence between the positions of the variables in Figure 2 and the positions of the samples in Fig. 3. In Fig. 3 samples with similar distributions of the 20 dioxin and furan congeners and congener totals cluster or project together. Samples that are far apart have a different "fingerprint" of the 20 congeners, and the crabs collected for these samples have been exposed to a different suite of dioxins and furans.

Crab samples projecting to the far right in Fig. 3 have the highest proportion of congeners characteristic of pulp mill chlorine bleaching, while samples in the lower center have been most strongly influenced by polychlorinated phenol condensation. Samples in between these two areas have been exposed to varying amounts of these two inputs.

Crab samples projecting in the upper left of Fig. 3 have been exposed to pentachlorophenol wood preservatives and are from the major B.C. harbors. These samples have low proportions of the toxic congeners, but have high proportions of the non-2,3,7,8 congeners—principally the penta- and hexafurans and the hexadioxins. Samples collected in different years project together, indicating little or no change in composition over time.

Crabs do not appear to take up hepta- and octadioxins and furans to any great extent: these congeners were generally only detectable when concentrations were very high. One important consequence is that the contribution of atmospheric combustion (the typical pattern is predominant octadioxin with lesser amounts of heptadioxin and furan; Hites 1990) can not be evaluated based solely on crab data.

Application of PCA to B.C. Coastal Sites

Crabs from the vicinity of nine B.C. coastal kraft mills at eight sites have been exposed to chlorine bleach plant effluents (Table 1). The tenth mill at Kitimat does not use chlorine bleaching. Crabs from the principal harbors have all been exposed to chlorophenols from wood treatment facilities as well as to dioxins and furans from combustion. Crabs from reference locations, which include Sidney Island (north of Victoria), the Queen Charlotte Islands and the B.C. central coast, were expected to have received minimal exposure to dioxins and furans.

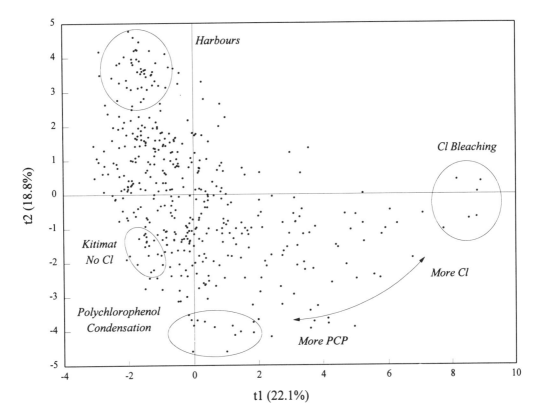

Figure 3. Varimax rotated sample (scores) plot for crab hepatopancreas samples in the first and second principal components of the PCA model.

Averaged congener profiles for crab samples with similar PCA projections that are from locations geographically close to each mill site provide a direct connection between congener composition and the PCA results. Congener profiles are presented for the earliest and most recent sampling times (Figs. 4 and 5). In each case the tip of the arrow indicates the mean projection of the samples that have been used to define each profile. In all cases where 1987 and 1989/90 samples have been averaged there was little or no change in either composition profile or PCA projection over the time interval shown. For Campbell River (Elk Falls) and Powell River the crab sample collected closest to the mill site sometimes had a PCA projection similar to that of the harbors. Because this likely indicates the presence of pentachlorophenol contaminated wood chips, these samples were not included in the profiles.

Examination of Figs. 4 and 5 reveals striking differences between the dioxin and furan profiles of different mill sites. At Gold River, for example, the 1990 profile is made up almost entirely of tetrafurans. In contrast, the 1989 profile from Kitimat shows only inputs from polychlorinated phenol condensation, without any inputs from chlorine bleaching. This condensation was likely a result of heating contaminated wood chips during the kraft pulping process (Luthe *et al.* 1993). In 1987/90 locations such as Nanaimo (Harmac Mill), Port Alberni and Campbell River contained equal proportions of the tetrafurans and the non-2,3,7,8 hexadioxins. The mill at Crofton had a lower proportion of the hexadioxins and the mills at

Powell River, Howe Sound and Prince Rupert were similar to the Gold River mill in having only small proportions of the hexadioxins. The composition of Gold River samples indicates either that very few pentachlorophenol-contaminated chips were used in the bleach plant or, less likely, that digester conditions did not favor hexadioxin formation.

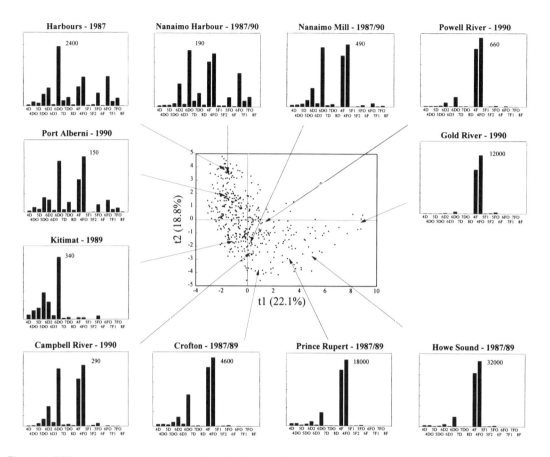

Figure 4. PCA congener concentration maxima (pg/g), profiles and corresponding PCA projections for harbor and mill site crab samples from the earliest sampling times available. The number of samples for each bar plot is the total for the 1987 and 1989/90 samples in Table 3.

Over the four or five years since the mills introduced changes to control the discharge of dioxins and furans, the tetrafurans have been removed faster than the hexadioxins at the mill sites, while the harbor profiles have shown little change over time (Fig. 5). At each mill site an incremental increase in the proportion of the hexadioxins has been accompanied by a corresponding decrease in the tetrafurans. While the mill sites have maintained much of their individual character, the overall effect has been for the composition profiles and PCA projections to become more similar over time.

Examination of profile and PCA results for the Nanaimo area illustrates that different dioxin and furan fingerprints can be obtained for locations which are only ca. 5 km apart (Figs. 4 and 5). In general, crab samples from the Harmac mill site project with similar mill samples (close to the y-axis), while samples from Nanaimo Harbor project close to the Victoria and Vancouver Harbor samples. Nanaimo Harbor samples have a slightly higher proportion of the tetrafurans than the larger harbors, and some contribution of chlorine bleaching effluents from the mill is likely. At a few locations from the boundary area just outside Nanaimo Harbor, crab samples had a mill pattern one year and a harbor pattern in another year; this highlights the utility of PCA in assigning a contaminant source to individual crab samples.

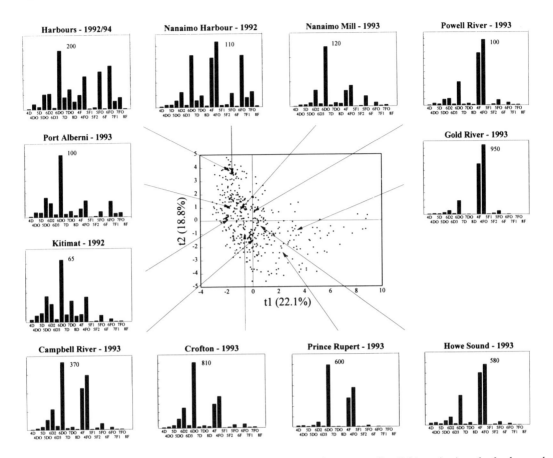

Figure 5. PCA congener concentration maxima (pg/g), profiles and corresponding PCA projections for harbor and mill site crab samples from the most recent sampling times. The number of samples for each bar plot is shown for the 1992/93 samples in Table 3.

In general, 1987 samples had the highest concentrations of toxic congeners (Table 3). In subsequent years both lower proportions of the toxic 2,3,7,8-chlorinated congeners and a decrease in overall dioxin and furan concentrations have been observed. The result has been a decrease in the TEQ at B.C. mill sites.

Mills located on a fjord (where the bottom water exchanges very slowly) or in an area with restricted circulation had the highest proportion of toxic congeners in the earliest years of sampling, but have shown the largest TEQ reduction over time (Tables 1 and 3). The Howe Sound and Gold River mills provide good examples. In quiescent areas burial plays an important role in the removal of the particulate-associated dioxins and furans (cf. Macdonald *et al.* 1992)

In contrast, samples from locations which are well flushed by tidal currents had lower initial proportions of the toxic congeners than samples from inlets but have exhibited a smaller decrease in TEQ over time. The Crofton and Campbell River locations illustrate this trend. In this case burial rates are either too low or resuspension processes are too active for effective removal of contaminants. The principal harbors have been less intensively studied, but they would appear to present a different situation.

These locations are adjacent to metropolitan areas, with the potential for chronic inputs. Furthermore, sedimentation and burial rates are likely to be lower in harbors than they are in fjords that are fed by a major river (e.g., Howe Sound).

Table 3. Dungeness crab hepatopancreas congener total and TEQ averages obtained using samples with similar PCA projections.

Sampling Location	1987			1989/90			1992/94		
	Congener Total pg g^{-1}	Calculated TEQ pg g^{-1}	n^a	Congener Total pg g^{-1}	Calculated TEQ pg g^{-1}	n^a	Congener Total pg g^{-1}	Calculated TEQ pg g^{-1}	n^a
Principal harbors	9100	320	3	-	-	-	990	23	17
Nanaimo Harbor	1900	75	1	450	15	2	520	18	4
Port Alberni	-	-	-	620	23	6	290	11	4
Crofton	37,000	2100	2	2700	130	5	2200	94	3
Campbell River	-	-	-	1000	47	5	1200	53	3
Gold River	-	-	-	23,000	1100	5	2000	99	4
Howe Sound	220,000	12,000	2	17,000	850	6	1700	94	8
Kitimat	-	-	-	790	68	8	140	7.2	41
Nanaimo (Harmac) mill	5600	280	1	1100	55	6	310	11	4
Powell River	-	-	-	1400	70	6	260	12	2
Prince Rupert	150,000	9200	1	4700	260	3	1400	53	4

^{a} n is the number of samples.

It has proved to be difficult to define a background pattern for crabs on the B.C. coast. Most samples from remote areas were analyzed by low resolution mass spectrometry, with too few congeners detectable to provide a reliable PCA projection. Subsequent samples have generally been collected from locations just outside the mill monitoring areas and have been impacted by mill discharges. These samples will be discussed in more detail in an accompanying technical report (Yunker and Cretney, in review). However, even samples from areas more removed from mills and harbors have PCA projections and composition profiles that differ from the usual atmospheric pattern (cf. Hites 1990). This topic requires further study.

CONCLUSIONS

PCA is a valuable method for exploring trends and defining relationships in the large sample set of chlorinated dibenzo-*p*-dioxin and dibenzofuran data for samples of crab hepatopancreas that have been collected from B.C. coastal sites. PCA has proved particularly useful for classifying samples according to principal source.

- Variations in the proportions of tetrafurans from chlorine bleaching and hexadioxins derived from pentachlorophenol have produced striking differences between the dioxin and furan profiles of crabs from different mill sites.
- Since 1987 both the proportion of the toxic 2,3,7,8-chlorinated congeners and the overall dioxin and furan concentrations have generally decreased in crabs collected from the mill sites.
- Mill-related tetrafurans have been removed faster than the hexadioxins: one effect is that crab composition profiles and PCA projections have become more similar over time at the mill sites.
- Crab samples from B.C.'s harbors have lower proportions of the toxic 2,3,7,8-chlorinated congeners, but have shown less change over time.

ACKNOWLEDGMENTS

Collection and data analysis for most of the samples in this study were principally carried out by Hatfield Consultants, Beak Consultants, G3 Consultants and Norecol, Dames and Moore, with chemical analyses done by Axys Analytical, Envirotest and Wellington Laboratories. This work was funded directly by the operators of the B.C. mills. The assistance of W. Knapp and M. Sullivan of the Department of Fisheries and Oceans and M. Hagen of the Department of the Environment in assembling the data set is gratefully acknowledged. H. Rogers, M. Ikonomou, M. Whittle and J. Thompson provided unpublished data from the harbors and remote areas. R. Norstrom is thanked for providing access to the unpublished 1987 data. Funding for this multivariate interpretation was provided by the Toxic Chemicals Green Plan.

REFERENCES

Dwernychuk, L.W. 1993. Dioxin/furan trend monitoring program. Howe Sound 1992. Report prepared for Howe Sound Pulp and Paper, Port Mellon B.C. and Western Pulp Ltd., Squamish B.C. by Hatfield Consultants Ltd., West Vancouver, B.C.

Hagenmaier, H. and H. Brunner. 1987. Isomer specific analysis of pentachlorophenol and sodium pentachlorophenate for 2,3,7,8-substituted PCDD and PCDF at sub-ppb levels. Chemosphere 16:1759-1764.

Hites, R.A. 1990. Environmental behavior of chlorinated dioxins and furans. Acc. Chem. Res. 23:194-201.

Johansson, E., S. Wold and K. Sjödin. 1984. Minimizing effects of closure on analytical data. Anal. Chem. 56:1685-1688.

Luthe, C., and S. Prahacs. 1993. Dioxins from pulp mill combustion processes: implications and control. Pulp Paper Can. 94(8):37-46.

Luthe, C.E., P.E. Wrist and R.M. Berry. 1992. An evaluation of the effectiveness of dioxins control strategies on organochlorine effluent discharges from the Canadian bleached chemical pulp industry. Pulp Paper Can. 93(9):40-49.

Luthe, C.E., R.M. Berry and R.H. Voss. 1993. Formation of chlorinated dioxins during production of bleached kraft pulp from sawmill chips contaminated with polychlorinated phenols. Tappi J. 76(3):63-69.

Macdonald, R.W., W.J. Cretney, N. Crewe and D. Paton. 1992. A history of octachlorodibenzo-*p*-dioxin, 2,3,7,8-tetrachlorodibenzofuran, and 3,3′,4,4′-tetrachlorobiphenyl contamination in Howe Sound, British Columbia. Environ. Sci. Technol. 26:1544-1550.

Meglen, R.R. 1992. Examining large databases: a chemometric approach using principal components analysis. Mar. Chem. 39:217-237.

Norstrom, R.J., M. Simon, P.E. Whitehead, R. Kussat and C. Garrett. 1988. Levels of polychlorinated dibenzo-*p*-dioxins (PCDDs) and dibenzofurans (PCDFs) in biota and sediments near potential sources of contamination in British Columbia, 1987. Canadian Wildlife Service Analytical Report CRD-88-5, 26 pp.

Voss, R.H. and M.B. Yunker. 1983. A study of the potential persistence of chlorinated phenolics discharged into kraft mill receiving waters. Report prepared for the B.C. Council of Forest Industries Technical Advisory Committee, Vancouver, B.C., 131 pp.

Whitehead, P.E., R.J. Norstrom and J.E. Elliott. 1992. Dioxin levels in eggs of great blue herons (*Ardea herodias*) decline rapidly in response to process changes in a nearby kraft pulp mill. *In* Dioxin 92. 12th International Symposium on Dioxins and Related Compounds. Tampere, Finland, pp. 325-328.

Yunker, M.B., R.W. Macdonald, D.J. Veltkamp and W.J. Cretney. 1995. Terrestrial and marine biomarkers in a seasonally ice-covered Arctic estuary—integration of multivariate and biomarker approaches. Mar. Chem. 49:1-50.

STUDIES ON EULACHON TAINTING PROBLEM: ANALYSES OF TAINTING AND TOXIC AROMATIC POLLUTANTS

P. Mikkelson[1], J. Paasivirta[1], I.H. Rogers[2] and M. Ikonomou[2]

[1]Department of Chemistry, University of Jyväskylä, P.O. Box 35, 40351 Jyväskylä, Finland
[2]Institute of Ocean Sciences (IOS), Fisheries & Oceans, 9860 West Saanich Road, Sidney, B.C., Canada V8L 4B2

Eulachon were taken from the Kitimat River downstream of the discharge of a nonbleaching kraft mill, and reference fish from the Kemano River were exposed in tanks to concentrations of mill effluent. Eulachon grease was rendered from exposed and unexposed fish by a traditional method. Gas chromatographic/mass spectrometric analyses were performed on homogenates of whole fish and on eulachon grease to measure the levels of chloroanisoles (PCAs) as well as coplanar and mono-*ortho*-PCBs, dioxins (PCDDs), furans (PCDFs), polyaromatic hydrocarbons (PAHs) and some sulfur aromatics. The concentrations of PCBs, PCDDs, PCDFs and PAHs (excluding dibenzothiophene and its methyl derivatives) were at low background levels without statistically significant differences between exposed and unexposed fish samples. The levels of the sulfur aromatic dibenzothiophene and its mono- and dimethyl derivatives (DBT+C1DBT+C2DBT) were 181 ng g^{-1} fw in exposed and 18 ng g^{-1} fw in unexposed fish grease samples. These substances are persistent and lipophilic components of crude oil. They also occur in oil-based defoamers used in pulp mills and, consequently, in pulp mill biosludge. Their significance as pollutants in exposed eulachons needs to be evaluated. 3,5-Dichloroanisole and 2,4,5-trichloroanisole were detectable (4-70 ng g^{-1} lw) in all tainted and nondetectable (<2 ng g^{-1}) in all nontainted oil samples. Levels of these PCAs in tainted eulachon grease were similar to levels of PCAs and PCVs (polychloroveratroles) in bad-tasting fish from a watercourse receiving pulp mill wastes. The source of the PCAs in the eulachons is unknown but cannot derive from bleaching.

INTRODUCTION

Effluents from pulp mills employing bleaching are among the pollution sources that have been associated with tainting of food and water. Pulp mill-related polychlorinated anisoles (PCAs) and veratroles (PCVs) have unpleasant taste and odor at low concentrations and have been detected in fish from water bodies receiving pulp mill effluent discharges (Paasivirta *et al.* 1987). However, bleach plant effluents are not the only source of compounds with tainting potential in pulp mills. It is important to evaluate contributions from these other sources in order to assess the relative importance of contributions from the bleach plant to tainting problems.

The eulachon (*Thaleichthys pacificus*) is an anadromous fish of the smelt family which spawns annually in rivers of the west coast area of British Columbia, Canada. The eulachon is harvested by aboriginal people, and its flesh and a derived grease form a valued part of their food and cultural traditions. More than a decade ago the grease, prepared from fish in Kitimat River, became nonedible because it was tainted by an intense off-flavor. Studies to find the source of tainting confirmed that pulp mill effluent can taint the characteristic flavor of eulachon (D.A. Kelly, Eurocan Co., P.O. Box 1400, Kitimat, B.C., Canada V8C 2H1, personal communication). First results of sensory and chemical analyses suggested that the terpenes α-pinene, cumene and thujane were associated with flavor impairment (Craig and Stasiak 1993). Another study by integrated sensory and instrumental analysis (Veijanen *et al.* 1994) also showed elevated amounts of bad-tasting terpenes (α-pinene, β-pinene, menthene-2, α-terpinene, and three carenes) in tainted eulachon grease.

In this work chloroanisoles (PCAs), which are extremely potent tainters, were studied in Jyväskylä by a special clean-up procedure and GC/MS analysis. Also, to complete knowledge on possible pollution at Kitimat, toxic aromatic xenobiotics were analyzed by GC/MS in Sidney after liquid chromatographic clean-up. In each series, parallel exposed (Kitimat River) and unexposed (Kemano) fish and grease samples were studied.

MATERIALS AND METHODS

The following samples obtained from Kitimat were studied:
- CF1: Kemano River eulachons exposed to the Kitimat River above the effluent discharge point.
- EF1: Kemano River eulachons exposed to 20% Kitimat pulp mill effluent for 6 d.
- EF2: Kitimat River eulachons.
- CG1: Untainted control grease prepared from Kemano River eulachons.
- CG2: Untainted control grease prepared from Kemano River eulachons.
- CGS: Control grease stripped at 160°C for 4 h.
- EG1: Tainted grease from Kemano River eulachons exposed to 20% Kitimat pulp mill effluent for 6 d.
- EG2: Tainted grease from Kemano River eulachons exposed to 20% Kitimat pulp mill effluent for 4 d.
- EGS: Grease EG2 stripped at 160°C for 4 h.
- CHT: Hexane trap of volatiles purged from the control grease CG2.
- EHT: Hexane trap of volatiles purged from the tainted grease EG2.

Determination of the Chloroanisoles

A weighed amount (ca. 10 g) of fresh fish sample and 40 g of Na_2SO_4 was homogenized. The mixture was transferred to an extraction thimble (33 x 100 mm) and allowed to stand for 48 h at room temperature (+22°C). After addition of internal standards [tetrabromoveratrole (TeBV) and 2,4,6-tribromoanisole (246TBA)] the mixture was extracted (Soxhlet extraction) with 230 mL of a solvent mixture of light petroleum (b.p. 40-60°C), acetone, n-hexane and diethyl ether (9:5.5:2.5:1) for 6 h. The solvent was evaporated first with a Rotavapor to about 10 mL and transferred to a weighed Kimax tube, where the solvent was removed by a nitrogen gas stream. The residue was weighed to give the fat content of the fish sample.

Grease samples (about 120 mg) were weighed and internal standards (TeBV and 246TBA) were added. Then, from these samples or from fat extracts of the fish samples (see above), PCAs and PCVs were separated by column chromatography. The column (Pasteur pipette, 230 mm x 5 mm) was packed with 2 g of neutral Al_2O_3 (activated at 800°C for 8 h and then deactivated with 5% water). The fat residue was transferred into a column and the chloroanisoles were eluted with 3 mL of the solvent mixture of dichloromethane:n-hexane (1:1). The solvent mixture was concentrated to 0.5 mL, and the clean-up step was repeated twice. Before gas-chromatographic (GC) analyses the solvent mixture was concentrated to about 0.3 mL. Internal standards were also added to the hexane samples CHT and EHT, and the solvent was concentrated to about 0.3 mL before GC analysis.

Gas chromatographic/mass fragmentographic (GC/SIM) analyses were carried out using an HP 5970 mass selective detector and an HP-5 quartz capillary column (25 m x 0.2 mm, i.d.). The carrier gas was helium flowing at a rate of about 1 mL min^{-1}. The temperature program was 80°C + 3° min^{-1} to 170°C then + 5° min^{-1} to 220°C and then + 8° min^{-1} to 250°C and hold there 10 min.

The chloroanisoles and -veratroles were identified and quantified with authentic reference compounds. Determination limits for PCAs and PCVs were (in ng g^{-1} lipid weight) 1 for dichloroanisoles (DCAs), 2 for dichloroveratroles (DCVs) and trichloroanisoles (TCAs), 4 for trichloroveratroles (TCVs) and tetrachloroanisoles (TeCAs) and 5 for tetrachloroveratrole (TeCV) and pentachloroanisole (PeCA).

Determination of the Chlorophenolic Compounds

About 300 mg of grease was weighed and the internal standard [2,3,6-trichlorophenol (236TCP)] was added. The internal standard was added to the hexane samples CHT and EHT from PCA and PCV analysis, and then they were prepared as fat samples. Fat was removed with 2 mL of hexane to 50 mL of 0.1 M potassium carbonate solution and shaken for 5 min. The water layer was washed twice with 20 mL of hexane; 1 mL of acetic acid anhydride was added to the water layer and shaken for 5 min. Then 10 mL of hexane was added and shaken for 5 min. The hexane layer was evaporated to 0.3 mL before GC analysis.

GC/SIM analyses were carried out using an HP 5970 mass selective detector and an HP-5 quartz capillary column (25 m x 0.2 mm, i.d.). The carrier gas was helium flowing at a rate of about 1 mL min^{-1}. The temperature program was 80°C + 3° min^{-1} to 170°C then + 5° min^{-1} to 220°C and then + 8° min^{-1} to 250°C and hold there 10 min.

Also, gas chromatographic analyses with electron capture detectors (GC/ECDs) were carried out for chlorophenolics using Orion Analytica Micromat HRGC 412 with dual column operation and with Ni-63 ECD. The columns were of fused silica coated with SE 54 and OV 1701. The carrier gas was helium, 1 mL min^{-1}. The temperature program was 100°C + 4° min^{-1} to 250°C and hold there 10 min.

The chlorophenolics studied were polychlorophenols (PCPs), guaiacols (PCGs) and catechols (PCCs). They were identified and quantified as acetates compared to authentic reference compounds. Determination limits were 10 for dichloro congeners and 1-3 ng g^{-1} lipid weight for other more chlorinated congeners.

Determination of Chloroaromatics

Non-*ortho* PCBs, mono-*ortho* PCBs, polychlorodibenzo-*p*-dioxins (PCDDs) and polychlorodibenzofurans (PCDFs) were determined by modification of the generally applied GC/MS methods. The samples were extracted with dichloromethane. The clean-up steps were gel permeation (Stalling *et al.* 1972), LC in silica column, LC in neutral alumina column and carbon column fractionation. Internal standards for identification, quantitation and recovery were ^{13}C-labeled PCB, PCDD and PCDF authentic reference compounds. High resolution gas chromatography with high resolution mass spectrometric detection was used in the final determination. Details of the analytical procedure will be published in another paper.

Determination of Polyaromatic Hydrocarbons

Polyaromatic hydrocarbon (PAH) analyses were done as a contract study in Axys Environmental Systems Ltd., Sidney, B.C., Canada V8l 3S8. The following procedure was applied. A wet sample was accurately weighed into a 500 mL round bottom flask and spiked with an aliquot of surrogate standard solution containing deuterated homologues of acenaphthene, chrysene, naphthalene, perylene, phenanthrene, pyrene, 2-methyl-naphthalene, dibenz(*a,h*)anthracene, benzo(*g,h,i*)perylene and benzo(*a*)pyrene. Methanol and a potassium hydroxide solution were then added and the mixture was heated under reflux for 1 h, cooled and extracted. Water was then added through the condenser. Refluxing was resumed for an additional hour.

The digest was then transferred to a separatory funnel with methanol rinses and extracted with pentane (3 x 100 mL). The pentane layers were combined, washed with extracted water three times, and dried over anhydrous sodium sulfate. The extract was then concentrated in a Kuderna-Danish flask prior to column clean-up.

The sample was loaded onto a silica gel column and eluted with pentane followed by dichloromethane. The dichloromethane fraction (contains PAHs) was concentrated to a small volume, transferred to a microvial and an aliquot of recovery standard containing deuterated benzo(*b*)fluoranthene,

fluoranthene, and acenaphthylene was added prior to analysis of PAHs by high resolution mass chromatography with low resolution (quadrupole) mass spectrometric detection.

Analysis of the extract was carried out using a Varian 3400 gas chromatograph (GC) with a Finnigan Incos 50 mass spectrometer, a CTC autosampler and a DG 10 Data system. A 30 m DB-5 (0.25 mm i.d. x 0.25 μm film thickness) chromatography column, used for GC separation, was coupled to the MS source. The GC was operated under the following temperature program: injection at 70°C, hold 2 min, ramp at 30°C min^{-1} to 100°C, hold 3 min, ramp at 10°C min^{-1} to 300°C, hold for 10 min.

The mass spectrometer was operated in the EI mode (70 eV) using multiple ion detection to enhance sensitivity, acquiring at least two characteristic ions for each target analyte and surrogate standard. A split/splitless injection sequence was used.

RESULTS AND DISCUSSION

The results of the chloroanisole/veratrole and chlorophenol analyses are listed in Table 1. Only those compounds which occurred above the limit of determination in at least one sample are reported. Chloroanisoles (PCAs) and chloroveratroles (PCVs) were nondetectable except for 3,5-dichloroanisole (35DCA) and 2,4,5-trichloroanisole (245TCA) in the tainted greases (Table 1). Purging of the tainted greases caused these pollutants to evaporate from the grease; they were recovered from the purge gas in hexane.

Table 1. Concentrations of 3,5-dichloroanisole (35DCA), 2,4,5-trichloroanisole (245TCA) and pentachlorophenol (PeCP) in eulachon grease and fish samples (ng g^{-1} lw).

Sample	35DCA	245TCA	PeCP
CF1	<1	<2	NA
EF1	12	4	NA
EF2	15	9	NA
CG1	<1	<2	NA
CG2	<1	<2	5
EG1	27	7	NA
EG2	70	20	15
CGS	<1	<2	5
EGS	<1	<2	15
CHT	<1 ng mL^{-1} ¤	<2 ng mL^{-1} ¤	1 ng mL^{-1} ¤
EHT	24 ng mL^{-1} ¤	2 ng mL^{-1} ¤	3 ng mL^{-1} ¤

¤ Amount in 1 mL of purge hexane.
NA Not analyzed.

35DCA in the GC/SNIFF test gives a "sweet muddy" odor and 245TCA gives an "unpleasant plant" odor. The total PCA/PCV level of about 20 ng g^{-1} in fresh fish has resulted in a significant tainting score in blind panel tests (Paasivirta et al. 1987). Considering that flavors, as a rule, are intensified in lipid, it is possible that the PCAs determined could be significant tainters in the present case. Threshold odor (TOC) and taste (TTC) values of these substances are not properly measured thus far. Reported TOC values for 246TCA and 236TCA in water are 0.05 and 0.0005 ng L^{-1}, respectively (Curtis et al. 1972;

Paasivirta et al. 1983). Accordingly, greases from the exposed fish contained levels of 35DCA and 245TCA which could cause off-taste sensation. They were nondetectable in greases from unexposed fish.

The two anisoles associated with bad taste cannot be derived from chlorophenols from bleaching, because the Kitimat mill does not use this process. Also, 35DCA and 245TCA identified in the present tainted eulachon samples are not typical of microbially methylated chlorobleaching or water chlorination wastes where PCVs are more common (Paasivirta et al. 1987, 1992). Also, they seem not to originate from wood preservative chlorophenols or from combustion, where 2346TeCA, 246TCA and PeCA should be the major components (Curtis et al. 1972). The 245TCA could originate from heavy use of the phenoxyacid herbicide 2,4,5-T in the area. Another possible source could be contamination of the raw material by the use of chlorophenolic fungicides or hexachlorocyclohexanes (HCHs), lindane or technical lindane as insecticide. In case of fungicide uses, 2,4,5-tetrachlorophenol is the main initial product of anaerobic dechlorination of 2,3,4,6-tetrachlorophenol and 3,5-dichlorophenol the stable product of anaerobic dehalogenation of pentachlorophenol (Mikesell and Boyd 1985, 1988). HCHs are known to metabolize to trichloro- and dichlorophenols (Klein and Korte 1970; Tanaka et al. 1977; Salah and Farghaly 1993). The chlorophenol metabolites could then biomethylate in the environment to 35DCA and 245TCA.

Pentachlorophenol (PeCP) was the only chlorophenolic compound that was found (Table 1) at low levels in control samples (CG2 and CGS) about 5 ng g^{-1} lw and in EG2 and EGS samples about 15 ng g^{-1} lw and 3 ng mL^{-1} in EHT. Traces of PeCP were observed in sample CHT.

The PCBs, PCDDs, PCDFs (Table 2), and PAHs except dibenzothiophenes derivatives (Table 3) showed similar patterns and occurred at equally low (background) levels in both exposed and nonexposed samples. This was supported by statistical comparison (one-way ANOVA with Scheffé post hoc multiple comparison and paired samples t test).

Table 2. Lipid percent and toxic chloroaromatics (pg g^{-1} fresh weight).

Variable	CG1	EG1	EF1*	EF1*	EF1*	CF1*	CF1*	EF2*	EF2*
Lipid %	87.57	90.75	13.28	13.73	13.19	13.84	13.24	10.07	12.93
CDiCB	.001	.001	38.7	27.1	21.2	3.6	6.7	5.6	4.1
CTriCB	.001	.001	15.2	14.6	14.3	5.9	8.9	8.8	5.6
CTeCB	.001	.001	14.0	2.5	14.8	13.6	13.6	12.9	13.6
CPeCB	.001	.001	4.1	8.0	2.8	3.1	3.6	3.5	6.7
CHxCB	.001	.001	nd	nd	nd	2.7	nd	nd	nd
MPeCB	6725	5368	948.6	692.3	560.2	549.8	579.6	720.8	760.9
MHxCB	4449	5221	613.4	665.7	620.3	554.2	602.4	549.4	600.0
MHpCB	nd	9.8	1.1	1.2	1.3	0.9	1.0	1.5	1.2
TeCDD	nd	1.10	nd	nd	nd	0.11	nd	nd	nd
PeCDD	nd	0.87	nd	nd	0.12	nd	0.15	0.16	nd
HxCDD	2.14	1.80	0.21	nd	0.63	0.39	0.39	0.65	0.56
HpCDD	1.76	3.62	0.40	0.28	0.22	0.15	0.25	nd	nd
OCDD	3.92	12.79	0.98	nd	0.72	0.77	0.56	0.53	0.33
TeCDF	10.47	11.50	1.19	1.20	1.31	1.07	1.16	1.10	1.34
PeCDF	nd	3.68	0.40	0.64	0.23	0.67	0.24	0.51	0.14
HxCDF	nd	0.85	0.18	0.08	0.11	nd	nd	0.40	nd
HpCDF	0.98	0.41	0.19	nd	0.12	0.15	nd	0.46	nd
OCDF	1.92	0.97	0.49	0.16	0.21	0.22	0.16	0.30	nd

* Duplicate determinations
nd = Not detected
.001 = No analysis result
Substance key: C..PCB = coplanar PCBs; M..PCB = mono-ortho-PCBs; Te = tetra; Pe = penta; Hx = hexa; Hp = hepta; O = octa

Table 3. Lipid percent and PAHs (ng g^{-1} fresh weight).

Variable	CG1	EG1	EF1	CF1	EF2
Lipid %	87.57	90.75	13.73	13.24	12.93
Naphthalene N	1100	200	16	7.8	5.5
Methylnaphthalenes C1N	720	280	26	13	10
Dimethylnaphthalenes C2N	480	410	25	20	16
Trimethylnaphthalenes C3N	460	nd	nd	nd	nd
Acenaphthylene	170	46	1	1	0.8
Acenaphthene	49	250	28	23	17
Fluorene	130	340	28	21	17
Phenanthrene	.001	670	95	.001	85
Monomethylphenanthrenes	nd	330	24	.001	85
Anthracene	.001	59	5.6	.001	4.8
Fluoranthene Fl	.001	180	18	.001	45
Pyrene Py	.001	76	8.6	.001	12
C1Fl/Py	48	16	nd	.001	5.1
C2Fl/Py	57	nd	nd	.001	nd
Dibenzanthracene	21	9.3	nd	nd	0.2
Dibenzothiophene DBT	18	77	8.4	.001	5.3
Monomethyldibenzothiophenes C1DBT	nd	66	5.2	.001	1.7
Dimethyldibenzothiophenes C2DBT	nd	38	2.2	.001	1.6

Substance key: C1 = monomethyl; C2 = dimethyl; C3 = trimethyl

From the aromatic xenobiotics only dibenzothiophene (DBT), methyldibenzothiophenes (C1DBTs) and dimethyldibenzothiophenes (C2DBTs) occurred at significantly higher levels in exposed fish than in Kemano fish. Their source could be an oil-based defoamer used in the Kitimat mill. Koistinen *et al.* (1992) have identified alkylated DBTs in pulp mill effluents, biosludge and in defoamer used in the mill. TOC and TTC concentrations of the alkylated DBTs are unknown. Their measured total level of 181 ng g^{-1} fresh weight in the tainted grease is, in our opinion, probably below the taste threshold. However, this should be studied by sensory testing.

CONCLUDING REMARKS

The analyses reported in this study show that pulp manufacture without bleaching is not any significant source of bioaccumulating toxic chloroaromatic substances in its recipient biota. Also the polycyclic aromatic hydrocarbons were not originated from pulping. However, oil-based defoamers used in the mill could be a source of persistent bioaccumulating alkylated dibenzothiophenes. Tainting of grease manufactured from the recipient fish is a major problem caused by the mill effluent. This tainting was caused, at least partly, from chlorophenolic metabolites of unknown origin. Finding the source of these chlorophenolics could solve the long-lasting tainting problem.

REFERENCES

Craig, G.R., and M. Stasiak. 1993. Fish Tainting Chemicals - Separation, Isolation and Identification of Compounds from Pulp and Paper Effluent. SETAC, 14th Annual Meeting, Houston, Texas, Abstract Book P128.

Curtis, R.F., D.G. Land, N.M. Griffiths, M. Gee, D. Robinson, J.L. Peel, C. Dennis and J.M. Gee. 1972. 2,3,4,6-Tetrachloroanisole association with musty taint in chickens and microbial formation. Nature 235:223-224.

Klein, W. and F. Korte. 1970. Metabolismus von Chlorkohlenwasserstoffen. *In* Chemie der Pflanzenschutz und Schädlingsbekämpfungsmitteln. R. Wegler (Ed.). Band 1. Springer, Berlin.

Koistinen, J., T. Nevalainen and J. Tarhanen. 1992. Identification and level estimation of aromatic coeluates of polychlorinated dibenzo-*p*-dioxins and dibenzofurans in pulp mill products and wastes. Environ. Sci. Technol. 26:2499-2507.

Mikesell, M.D. and S.A. Boyd. 1985. Reductive dechlorination of the pesticides 2,4-D, 2,4,5-T and pentachlorophenol in anaerobic sludges. J. Environ. Qual. 14:337-340.

Mikesell, M.D. and S.A. Boyd. 1988. Enhancement of pentachlorophenol degradation in soil through induced anaerobiosis and bioaccumulation with anaerobic sewage sludge. Environ. Sci. Technol. 22:1411-1414.

Paasivirta, J., J. Knuutinen, J. Tarhanen, T. Kuokkanen, K. Surma-Aho, R. Paukku, H. Kääriäinen, M. Lahtiperä and A. Veijanen. 1983. Potential off-flavour compounds from chlorobleaching of pulp and chlorodisinfection of water. Wat. Sci. Tech. 15:97-104.

Paasivirta, J., P. Klein, M. Knuutila, J. Knuutinen, M. Lahtiperä, R. Paukku, A. Veijanen, L. Welling, M. Vuorinen and P.J. Vuorinen. 1987. Chlorinated anisoles and veratroles in fish. Model compounds. Instrumental and sensory determinations. Chemosphere 16:1231-1241.

Paasivirta, J., A-L. Rantalainen, L. Welling, S. Herve and P. Heinonen. 1992. Organochlorines as environmental tainting substances: taste panel study and chemical analyses of incubated mussels. Wat. Sci. Tech. 25:105-113.

Salah, S.M.A.D. and M. Farghaly. 1993. Behavior of gamma-hexachlorocyclohexane in stored *Vicia faba* beans. Nippon Noyaku Gakkaishi 12:101-103.

Stalling, D.L., R.C. Tindle and J.L. Johnson. 1972. Cleanup of pesticide and PCB residues in fish extracts by gel permeation chromatography. J. Assoc. Off. Anal. Chem. 55:32-38.

Tanaka, K., N. Kurihara and M. Nakajima. 1977. Pathways of chlorophenol formation in oxidative biodegradation of BHC. Agric. Biol. Chem. 41:723-725.

Veijanen, A., K. Villberg and J. Paasivirta. 1994. Studies on eulachon tainting problem. II. Integrated sensory and instrumental analysis. Presented at the Second International Conference on Environmental Fate and Effects of Bleached Pulp Mill Effluents, November 6-10, 1994, Vancouver, B.C., Canada.

MONITORING TRENDS IN CHLOROPHENOLICS IN FINNISH PULP MILL RECIPIENT WATERCOURSES BY BIOACCUMULATION IN INCUBATED MUSSELS

Sirpa Herve[1], Pertti Heinonen[2] and Jaakko Paasivirta[3]

[1] Water and Environment District of Central Finland, P.O. Box 110, 40101 Jyväskylä, Finland
[2] National Board of Waters and the Environment, P.O. Box 250, 00101 Helsinki, Finland
[3] Department of Chemistry, University of Jyväskylä, P.O. Box 35, 40351 Jyväskylä, Finland

Trends in concentrations of chlorophenolic compounds in incubated mussels from two Finnish pulp mill recipient watercourses during the last ten years are discussed. The results are compared with the loading variables (theoretical concentrations of BOD_7, COD_{Cr}, suspended solids and AOX) estimated for the receiving watercourses of the pulp mills concerned. The prevailing discharges in watercourses and the dilution of the waste water loading have also been taken into consideration. The trends of the chlorophenolic compounds originating from the pulping industry as well as the trends of tetrachloroguaiacols in incubated mussels have been decreasing. The reasons for these trends are: 1) the improvement of waste water treatment and 2) the decreased use of chlorine in pulp bleaching. The decreasing trend of the chlorophenolics which originate mainly from wood preservation and combustion is also evident.

INTRODUCTION

In Finland the most remarkable point loading source of pollution has been the chemical wood processing industry. Important loading variables historically included BOD_7, COD_{Cr} and suspended solids (SS). Nowadays in Finland nearly all pulp mills have biological waste water treatment systems and the situation has totally changed. The most interesting monitoring variables from the environmental point of view are now the nutrients phosphorus and nitrogen and especially the large number of organochlorine compounds originating from the pulp bleaching processes. Many of these compounds are persistent and can be found very far from the discharge site.

A practical problem in analyzing these organic chlorocompounds in watercourses, which in many cases in Finland are very shallow lakes with relatively short residence time, is the very low concentrations, near or below the limits of determination, at which they often occur. Concentrations vary widely with time in waste waters and therefore also in the receiving watercourses. In such cases monitoring is too expensive on the basis of chemical analyses from water alone, because the number of analyses needed for statistical processing is too numerous.

Incubated mussels have been used in Finland annually since 1984 in monitoring organochlorine compounds in watercourses, mainly in the recipients of pulp and paper industry discharges (Herve *et al.* 1988; Herve 1991). The common lake mussel (*Anodonta piscinalis*) has proved to be a useful test animal for this type of monitoring because of its ability to survive even under adverse conditions and its high uptake rates of lipophilic persistent pollutants. Mussel incubation for four weeks each August gives more comparable results than averaging the chemical analyses of water samples collected during the month, and thus enables the economical determination of long-term time trends in organochlorine pollution. Altogether 20 freshwater sites downstream of the pulp and paper industry are included as part of the National Monitoring Program of harmful substances. Combined monitoring results from two different areas which received pulp and paper mill effluents, one lake recipient in Central Finland (1984-1992) and another river recipient in southern Finland (1986-1992), are presented.

METHODS

The first area is situated in central Finland (Fig. 1, locality A); the pulp mill is discharging its waste water to the streamlike lake, where the MQ is about 80 m^3 s^{-1} (in sampling station Kuusaankoski). In the very first year of the monitoring program there was only mechanical treatment of waste waters. The activated sludge treatment plant started in March 1985 and the waste water loading, especially the BOD load, decreased drastically. At the same time considerable process alterations were made at the pulp mills.

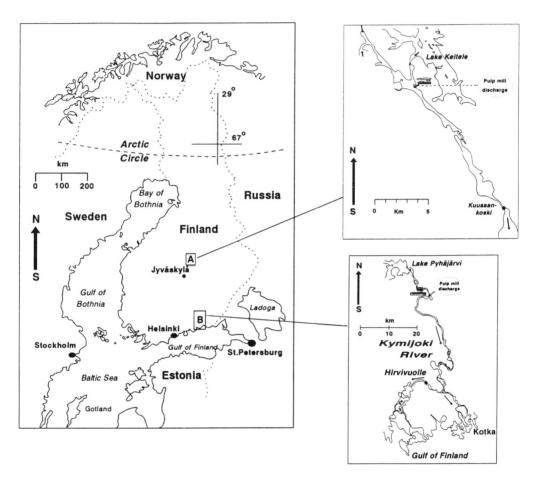

Figure 1. Maps of the study areas. The mussel incubation stations were Kuusaankoski in area A and Hirvivuolle in area B.

The second area (Fig. 1, locality B), the River Kymijoki in southern Finland, receives effluent from one pulp and two paper mills in Kuusankoski. The MQ of this river is some 380 m^3 s^{-1}. Up to 1988 all the waste waters were treated only by mechanical means. A new activated sludge waste water treatment plant started up in the spring of 1989. In this area the AOX loading has been monitored since 1989.

Common lake mussels (*Anodonta piscinalis*) obtained from unpolluted watercourses were preincubated first in aquaria for two weeks to guarantee clean mussels and to eliminate the weak individuals. After the preincubation, the mussels were transferred to the monitoring sites and housed in plastic cages anchored in the epilimnion at a depth of one m. The incubation period at the monitoring areas has been exactly the same every year, four weeks from the beginning of August to the beginning of September.

After incubation the mussels were taken from the cages, transferred in cold boxes to the laboratory and frozen. In the laboratory, composite samples of the soft part homogenates of five mussels were prepared. Before extraction, internal standards were added. The samples were extracted in a Soxhlet apparatus, solvents evaporated and the lipid content determined. Detailed method for analysis has been published previously (Herve et al. 1988; Herve 1991). The results are expressed on a lipid weight (lw) basis.

In data processing, the chlorophenolic compounds (Fig. 2) were divided into two groups as follows (Paasivirta et al. 1980):

A. Chlorophenols originating from wood preservation, combustion, chlorination and pesticide use (S1PCP) = 246TCP + TeCP + PeCP.

B. Chlorophenolics originating mainly from bleaching processes (S2PCP) = 24DCP + 26DCP + 245TCP + 34DCC + 345TCC + TeCC + 45DCG + 345TCG + 456TCG + TeCG + DMP (trichloro-2,6-dimethoxyphenol).

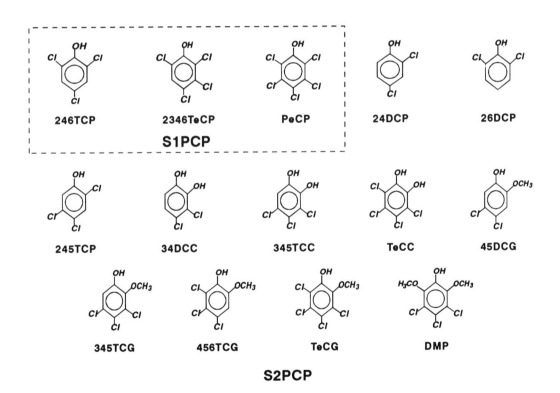

Figure 2. Formula and name abbreviations of the compounds studied.

TeCG has also been presented separately, because it is one of the dominant bioaccumulating components originating especially from the chlorobleaching processes of pulp (Suntio et al. 1988).

RESULTS AND DISCUSSION

In the beginning of the monitoring (1984) the concentrations of S2PCP in incubated mussels in the vicinity of waste water discharges were usually relatively high, in some cases even more than 10 µg g^{-1} (lw). At that time the pulp and paper industry in Finland was not using biological treatment plants. The use of chlorine in bleaching of pulp was also not restricted. In the latter part of the 1980s new biological treatment plants were constructed in both of the pulp and paper mills concerned. The effect of the biological treatment on the concentrations of chlorophenolic compounds was dramatic. The concentrations of S2PCP in incubated mussels decreased significantly, to a level of 1-3 µg g^{-1} (lw). During the early 1990s the S2PCP concentrations further decreased because of substitution of chlorine with chlorine dioxide in bleaching. The concentrations of S2PCP in incubated mussels were usually below 1 µg g^{-1} (lw). The organic loads (expressed as concentrations of BOD, COD and SS) are presented in Figs. 3 and 4. The loads were calculated from discharges of the mills for dilution at the sampling stations. There appears to be a fair correlation between the concentration of S2PCP in incubated mussels and the organic load values.

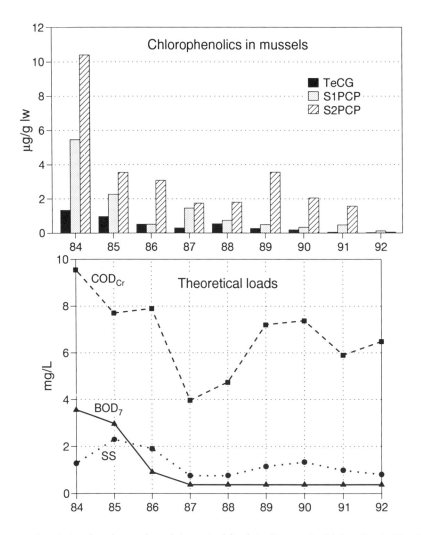

Fig. 3. Concentrations in incubated mussels and theoretical loads in Kuusaankoski (locality A, Fig. 1).

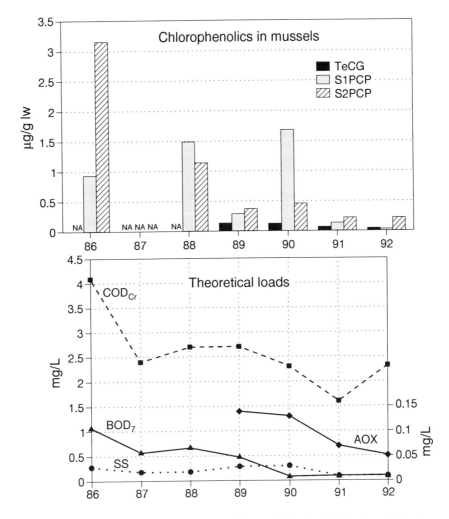

Fig. 4. Concentrations in incubated mussels and theoretical loads in Hirvivuolle (locality B, Fig. 1).

At both areas, trends of the levels of pulp mill originated chlorophenols decreased significantly over time. This is consistent with the changes in bleaching and other pulp manufacturing processes and with construction of new secondary treatment plants. On the contrary, at the same time other organochlorine compounds showed different trends. For example the leakage of PCBs in locality A was at about the same level during the whole period from 1984 to 1992 (Herve *et al.* 1988; Herve and Heinonen unpublished data). In the River Kymijoki, residues of former chlorophenol formulation Ky-5 production (stopped in early 1980s) might have resulted in elevated concentrations of S1PCP (Fig. 4).

In the Kymijoki area the results of the last three years show that the decrease of AOX is not as effective as the decrease of S2PCP during the same period. AOX can occur in waste waters and in watercourses from natural sources (Asplund 1992). That is why AOX would not appear to be a suitable variable for watercourse monitoring of pulp mill recipients. In monitoring it is important to obtain data concerning the occurrence and biological reactions of single compounds (Herve 1991).

REFERENCES

Asplund, G. 1992. On the origin of organohalogens found in the environment. Linköping Studies in Art and Science. 77, 50 p.

Herve, S. 1991. Mussel incubation method for monitoring organochlorine compounds in freshwater recipients of pulp and paper industry. Department of Chemistry, University of Jyväskylä, Research Report 36, 145 p.

Herve, S., Heinonen, P., Paukku, R., Knuutila, M., Koistinen, J. and J. Paasivirta. 1988. Mussel incubation method for monitoring organochlorine pollutants in watercourses. Four-year application in Finland. Chemosphere 17:1945-1961.

Paasivirta, J., Särkkä, K., Leskijärvi, T. and A. Roos. 1980. Transportation and enrichment of chlorinated phenolic compounds in different aquatic food chains. Chemosphere 9:441-456.

Suntio, L.R., Shiu, W.Y. and D. Mackay. 1988. A review of the nature and properties of chemicals present in pulp mill effluents. Chemosphere 17:1249-1290.

BIOACCUMULATION OF PULP CHLOROBLEACHING-ORIGINATED AROMATIC CHLOROHYDROCARBONS IN RECIPIENT WATERCOURSES

Tiina Rantio, Jaana Koistinen and Jaakko Paasivirta

Department of Chemistry, University of Jyväskylä, P.O. Box 35, 40351 Jyväskylä, Finland

Accumulation factor (AF) of persistent aromatic chlorohydrocarbons from pulp mill wastewater to biosludge was measured. The rank order from highest to lowest AF was alkylpolychloronaphthalenes > alkylpolychlorophenanthrenes > polychlorocymenes > alkylpolychlorobibenzyls > polychlorocymenenes. The actual concentrations in mussels (*Anodonta piscinalis*) and fish (*Esox lucius*) showed marked differences most likely due to different metabolic processes. The bioconcentration factors (BCFs) estimated from dilution of the wastewater in recipient and from the concentrations in biota showed fair agreement with the observed AF (alkylpolychloronaphthalenes could not be included in this comparison due to the missing analyses in biota). The results suggest that BCF in the recipient water environment can be roughly estimated from the ratio of concentrations in biosludge and effluent.

INTRODUCTION

Spent liquors of pulp chlorobleaching contain persistent (stable in sulfuric acid treatment) chlorohydrocarbons. The most abundant of these persistent organochlorines have alkylaromatic structures (Kuokkanen 1989; Koistinen *et al.* 1992, 1994). Only minor peaks of nonaromatic chlorohydrocarbons have been detected in spent bleaching liquor, and they were nonidentical with major components of the insecticide toxaphene (chlorination product of monoterpenes) which, in contrast, were found in both recipient and background area fish in Finland (Rantio *et al.* 1993; Paasivirta *et al.* 1993).

The best characterized of these chlorinated alkylaromatics are polychlorocymenes (PCYMS) and polychlorocymenenes (PCYMD) that have been confirmed by synthesis of individual reference substances (Kuokkanen 1989). Recently, alkylpolychloronaphthalenes (R-PCN), alkylpolychlorobibenzyls (R-PCBB) and alkylpolychlorophenanthrenes (R-PCPH) were also identified in waste bleaching liquors by GC/MS (Koistinen *et al.* 1992, 1994). In order to assess the ecotoxicological significance of these chemicals, estimation of their bioaccumulation in the recipient watercourse is important. In this study, levels of these substances in fish and incubated mussels in two watercourses are compared to their estimated exposure concentrations in water and to the calculated distribution factor for these compounds between pulp mill effluent and spent mill biosludge.

MATERIALS AND METHODS

The effluent and sludge samples were from two kraft mills located in Äänekoski, central Finland, and in Kuusankoski on the Kymijoki River, southeast Finland. During the sampling periods in 1990-91, the mill in Äänekoski bleached softwood pulp (sequence D/C-E-D-E-D). The hardwood pulp bleaching data from the Äänekoski mill are not included in this paper. The mill in Kuusankoski bleached both softwood pulp (D/C-E-D-EP-D) and hardwood pulp (D-EO-D-EP-D) in two separate lines. The wastewaters from both lines were combined before the biological treatment. Both mills had activated sludge treatment and there were no significant changes in processes or production volumes during this period.

The sampling place in Äänekoski was the recipient Lake Kuhnamo (1 km), Lake Vatia (12-15 km) and Torronselkä (37 km downstream from the mill discharge point). Fish (pike: *Esox lucius*) were caught in 1990 and lake mussels (*Anodonta piscinalis*) were incubated in 1987-89 as part of the regular monitoring there (Herve *et al.* 1988). In Kymijoki River, fish and incubated mussels included in this 1986-91 study were from Keltti (1 km), Susikoski (35 km) and Hirvivuolle (45 km downstream from the mill).

Structures of the compound groups which were used in this comparison are presented in Fig. 1. Analysis result data for each group used for calculations are sums of the concentrations of individual congeners. The quantitation of the compounds was based on the ratio of their peak areas in reconstructed ion chromatograms in HRGC/MS related to the peak area of the added internal standard. Responses of authentic single standards were used in the case of PCYMS and PCYMD (LRMS), response of 9-chlororetene in the case of C4-PCPH (HRMS) and assumed 1:1 response against internal standard compound in case of the other congener groups (HRMS). The estimated concentration values obtained in the latter cases do not represent absolute amounts, but when they are used in comparisons to determine relative concentrations between compartments, the ratios between levels in samples are exact. Sample data, analytical procedures and measurements have been published previously (Koistinen 1992; Koistinen *et al.* 1993, 1994; Rantio 1992).

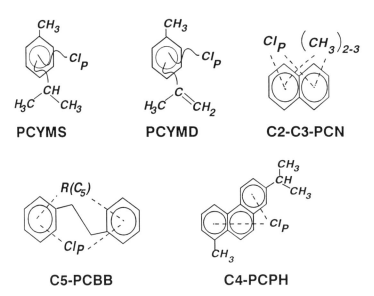

Figure 1. Structures and name abbreviations of the aromatic chlorohydrocarbon groups analyzed. PCYMS = polychlorocymenes, PCYMD = polychlorocymenenes, C2-C3-PCN = alkyl(2-3 carbons)polychloronaphthalenes, C5-PCBB = alkyl(5 carbons)polychlorobibenzyls, C4-PCPH = alkyl(4 carbons)-polychlorophenanthrenes or polychlororetenes.

RESULTS

The measured average concentrations in discharges, sludges, fish and mussels are presented in Tables 1 and 2. Concentrations in recipient water were estimated by dividing the level in discharge by the dilution factor. This estimation method has been validated by multimedia modeling and comparison with observations in water, fish and sediments for some measured organochlorine emissions from the mills to the watercourses in 1986-87 (Paasivirta 1994; Trapp *et al.* 1994). The concentration factors from discharge to biosludge (AF) and bioconcentration factors (BCFs) from recipient water to the lipids of fish and mussels were calculated and are listed in Tables 1 and 2. Comparison of the average BCF and AF values is presented in Fig. 2.

Table 1. Data for Äänekoski kraft mill (dilution factor 133): Concentrations in discharge (C_D), sludge (C_S), water (C_W), fish (CB_F) and mussel (CB_M), accumulation factor (AF) and bioconcentration factors (BCF).

Compound Group	C_D µg L^{-1}	C_S µg kg^{-1} dw	AF sludge	C_W ng L^{-1}	CB_F ng kg^{-1} lw	BCF fish	CB_M ng kg^{-1} lw	BCF mussel
Lipid % fw					0.32		0.6	
PCYMS	7.0	1,426	204	52.6	1,610,000	31,000	316,000	6,000
PCYMD	4.4	450	102	33.1	nd	nd	58,000	1,800
C5-PCBB	0.01	5.1	510	0.075	nd	nd	20,000	267,000
C4-PCPH	0.056	72.0	1290	0.421	na	na	na	na
C2-C3-PCN	0.064	119	1860	0.481	na	na	na	na

AF = C_S/C_D; BCF = CB/C_W
nd = not detected; na = not analyzed; dw = dry weight; lw = lipid weight

Table 2. Data for Kuusankoski kraft mill (dilution factor 240): Concentrations in discharge (C_D), sludge (C_S), water (C_W), fish (CB_F) and mussel (CB_M) accumulation factor (AF) and bioconcentration factors (BCF).

Compound Group	C_D µg L^{-1}	C_S µg kg^{-1} dw	AF sludge	C_W ng L^{-1}	CB_F ng kg^{-1} lw	BCF fish	CB_M ng kg^{-1} lw	BCF mussel
Lipid % fw					0.27		0.6	
PCYMS	0.413	929	2,250	1.72	nd*	nd	285,000	166,000
PCYMD	2.21	470	213	9.21	96,800*	10,500	450,000	49,000
C5-PCBB	0.112	27.0	240	0.470	1,330	2,830	19,800	42,000
C4-PCPH	0.015	61.0	4,070	0.063	nd	nd	89,000	1,410,000
C2-C3-PCN	0.008	51.0	6,380	0.033	nd	nd	na	na

AF = C_S/C_D; BCF = CB/C_W
nd = not detected; na = not analyzed; dw = dry weight; lw = lipid weight
* unpublished results by Rantio

DISCUSSION

Because the mill samples (discharge and sludge) represent short-term process-related values, their comparison with levels in recipient biota (longer exposure periods) includes a rather high degree of uncertainty. In addition, since only softwood pulp requiring high amounts of chlorine chemicals was produced during the sampling in Äänekoski, but both softwood pulp and hardwood pulp (the latter with lower amounts of chlorine chemicals) were produced in Kuusankoski, the results in the Tables 1 and 2 are not quite comparable. Therefore, concentration factors AF and BCF derived from the levels should perhaps be considered as separate data sets from Table 1 and Table 2. The following rank order from highest to lowest accumulation factor was obtained from Tables 1 and 2 (average AF values in parentheses):

alkylpolychloronaphthalenes (4120) > alkylpolychlorophenanthrenes (2680) > polychlorocymenes (1227) > alkylpolychlorobibenzyls (375) > polychlorocymenenes (158).

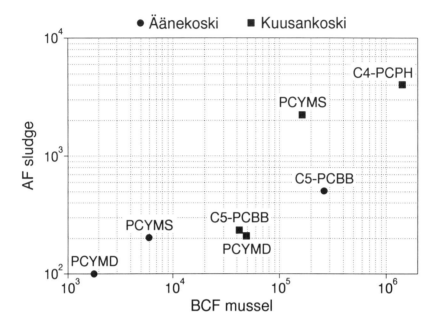

Figure 2. BCF for aromatic chlorohydrocarbons from bleaching to mussels (lw) compared to the ratio AF of their concentrations in sludge (dw) and discharge. Average values from Tables 1 and 2.

Comparison of the mussel BCF values to the AF data (Fig. 2) indicated a fair correspondence. Accordingly, the present results suggest that BCF in recipient water environment could be roughly estimated from the ratio of concentrations in biosludge and effluent. However, due to the missing analyses in biota, the possibly highest BCF values of the alkylpolychloronaphthalenes could not be verified.

All chlorohydrocarbons compared here indicated a rather high bioaccumulation. However, the rank order shown above is based on a limited amount of data and differences between the average BCF values are small. On the other hand, the calculated BCF values for fish and mussel seemed to be significantly different. This has been observed for various aromatic chlorohydrocarbons by Koistinen *et al.* (1993) and noted to indicate different metabolisms of these species.

Given the significance of metabolism, more environmental analyses including levels in recipient waters and biota must be performed to find the out usefulness of determinations in effluent and sludge for estimation of environmental fate of the persistent components there. These estimations can be supported by modeling based on the properties of individual components and recipient watercourse (Paasivirta 1994; Trapp *et al.* 1994).

ACKNOWLEDGMENT

The authors are grateful to the Academy of Finland and to the Maj and Tor Nessling Foundation for financial support.

REFERENCES

Herve, S., P. Heinonen, R. Paukku, M. Knuutila, J. Koistinen and J. Paasivirta. 1988. Mussel incubation method for monitoring organochlorine pollutants in watercourses. Four-year application in Finland. Chemosphere 17:1945-1961.

Koistinen, J. 1992. Alkyl polychlorobibenzyls and planar aromatic chlorocompounds in pulp mill products, effluents, sludges and exposed biota. Chemosphere 24:559-573.

Koistinen, J., T. Nevalainen and J. Tarhanen. 1992. Identification and level estimation of aromatic coeluates of polychlorinated dibenzo-p-dioxins and dibenzofurans in pulp mill products and wastes. Environ. Sci. Technol. 26:2499-2507.

Koistinen, J., J. Paasivirta and M. Lahtiperä. 1993. Bioaccumulation of dioxins, coplanar PCBs, PCDEs, HxCNs, R-PCNs, R-PCPHs and R-PCBBs in fish from a pulp-mill recipient watercourse. Chemosphere 27:149-156.

Koistinen, J., J. Paasivirta, T. Nevalainen and M. Lahtiperä. 1994. Chlorophenanthrenes, alkylchlorophenanthrenes and alkylchloronaphthalenes in kraft pulp mill products and discharges. Chemosphere 28:1261-1277.

Kuokkanen, T. 1989. Chlorocymenes and chlorocymenenes: persistent chlorocompounds in spent bleach liquors of kraft pulp mills. Department of Chemistry, University of Jyväskylä, Research Report 32, 40 p.

Paasivirta, J. 1994. Environmental fate models in toxic risk estimation of a chemical spill. Research Centre of the Defence Forces (Finland) Publications A/4, pp. 11-21.

Paasivirta, J., T. Rantio, J. Koistinen and P.J. Vuorinen. 1993. Studies on toxaphene in the environment. II. PCCs in Baltic and Arctic sea and lake fish. Chemosphere 27:2011-2015.

Rantio, T. 1992. Chlorocymenes, cymenenes and other chlorohydrocarbons in pulp mill effluents, sludges and exposed biota. Chemosphere 25:505-516.

Rantio, T., J. Paasivirta and M. Lahtiperä. 1993. Studies on toxaphene in the environment. I. Experiences on analytical behavior of PCCs. Studies including pulp mill wastes. Chemosphere 27:2003-2010.

Trapp, S., T. Rantio and J. Paasivirta. 1994. Fate of pulp mill effluent compounds in a Finnish Watercourse. Environ. Sci. Pollut. Res. 1:246-252.

SECTION IV

FIELD AND LABORATORY STUDIES OF BIOCHEMICAL RESPONSES ASSOCIATED WITH PULP AND PAPER MILL EFFLUENTS

Numerous recent laboratory and field studies around the world have reported physiological and biochemical responses in biota exposed to pulp and paper mill effluents. The role of different mill processes, bleaching and treatment strategies has been explored. Biochemical responses in biota are often the first indicators of potential impacts of contaminants or effluents on biota. Unfortunately, the usefulness of these measurements is often limited by a lack of a understanding of the processes involved. A mechanistic understanding of the causes and consequences of biochemical responses will lead to an increased ability to detect and interpret meaningful changes.

MIXED FUNCTION OXYGENASE INDUCTION BY PULP MILL EFFLUENTS: ADVANCES SINCE 1991

Peter V. Hodson

National Water Research Institute, Canada Centre for Inland Waters, Environment Canada, Burlington, Ontario, L7R 4A6 Canada

At the first conference on Environmental Impacts of Pulp Mill Effluents, reports of MFO induction in fish exposed to BKME in Scandinavia were confirmed in other countries and extended to mills not using chlorine bleaching. MFO activity was associated in field studies with effects on fish, particularly regulation of steroid hormones. Unknowns arising from the 1991 meeting included: (1) mechanisms linking MFO induction to other toxic effects, (2) sources of inducers within mills, (3) effects of effluent treatment, (4) role of chlorine bleaching, (5) the nature and identity of inducers, and (6) the environmental fate and distribution of inducers in the environment. While known inducers are quite toxic and may have various effects, the links between induction and effects in effluent-exposed fish are not clear. Induction can be caused by effluents or effluent extracts from pulp mills using either chlorine bleaching or no bleaching, and one of the strongest sources within mills is spent cooking liquor. Additional contributions from C- and E-stage effluents imply either further release of inducers from lignin or increased potency of inducers due to chlorination. However, optimized waste treatment processes reduce induction potency of effluents. Inducers isolated from effluents exhibit properties of PAHs rather than of dioxins or furans, but their precise identity remains unknown. Induction downstream of pulp mills can be found at distances of up to 230 km, implying environmental persistence of some inducers, but others appear to be readily metabolized and excreted by fish.

INTRODUCTION

The 1991 conference in Sweden (Södergren 1991) provided an opportunity to share experiences worldwide on effects of pulp mill effluents on fish. One of the most important results was the observation of a common suite of observed effects, including chemical accumulation and contamination of fish, induction of xenobiotic-metabolizing mixed function oxygenase (MFO) enzymes, physiological changes, impaired energy metabolism and reproduction, and gross pathologies. In Canada, changes in steroid hormone regulation in both male and female maturing fish were associated at the population level with increased age to maturity and changes in fecundity-body weight and gonad-body weight relationships.

One of the most consistent effects of bleached kraft mill effluent (BKME) was induction of MFO enzymes. Since MFO induction can be caused by highly toxic co-planar polychlorinated biphenyls (PCBs), polyaromatic hydrocarbons (PAHs), and chlorinated dioxins and furans, a number of unanswered questions were identified:
- Are there mechanisms that link MFO induction to other toxic effects?
- What is the environmental fate and distribution of MFO inducers in aquatic environments?
- What is the role of chlorine bleaching in MFO induction?
- What are the sources within mills of inducing chemicals?
- Does effluent treatment reduce potency for MFO induction?
- What is the nature and identity of MFO inducers?

RELATIONSHIPS OF MFO INDUCTION TO OTHER TOXIC EFFECTS

In fish from waters contaminated by BKME, MFO induction coincides with chemical contamination, pathology, and other physiological and population changes (Adams et al. 1992; McMaster et al. 1992; Munkittrick et al. 1992a, 1993a; Servos et al. 1993). These associations are strong because the severity of all effects decreases simultaneously with effluent dilution (Hodson et al. 1992; Gagnon et al. 1994) and effects co-occur in experimental mesocosms exposed to effluents (Tana et al. 1994).

However, associations or correlations are not the same as a mechanistic link, and MFO induction may not cause other effects as proposed by Lehtinen (1990). MFO induction in fish from receiving waters does not always coincide with effects on serum steroid hormone levels, liver size, and gonad weights (Swanson et al. 1992; Birkholz and Swanson 1993; Munkittrick et al. 1994a), and false positives and false negatives are often observed in comparisons among variables in field studies due to factors of dilution, fish movements, season, and sampling. It is apparent that MFO induction may not be used to predict other responses in all cases. A chronic exposure to BKME of juvenile chinook salmon (*Oncorhynchus tshawytscha*) caused a slight MFO induction but no other definitive effects (Servizi et al. 1993), perhaps because concentrations were always 4% or less. For reproduction, detailed laboratory and field studies of fish steroid hormone regulation indicate a variety of BKME effects unrelated to MFO activity (Munkittrick et al. 1994a). Results of standard toxicity tests of mortality and growth of juvenile fathead minnows (*Pimephales promelas*) and mortality and reproduction of *Ceriodaphnia* also showed no relationship to MFO induction in adult white sucker (*Catostomus commersoni*) from receiving waters (Robinson et al. 1994; Munkittrick et al. 1994a). In regulatory acute toxicity tests, effluent concentrations causing ethoxyresorufin-*o*-deethylase induction were 3-59 times lower than lethal concentrations (Gagné and Blaise 1993).

Fate and Distribution

MFO induction in fish from waters contaminated by BKME occurs at dilutions <1% and at distances of up to 230 km from the source (Hodson et al. 1992; Swanson et al. 1992; Munkittrick et al. 1994b). In the St. Maurice River, a decrease in MFO activity with distance from a mill paralleled a decrease in dioxin equivalents in fish tissue (Hodson et al. 1992), suggesting that inducers are persistent compounds such as tetrachlorodibenzo-*p*-dioxin. However, MFO activity declined rapidly in white sucker downstream of a pulp mill during shutdown, both in free-swimming and in caged fish, and induction was lost rapidly in fish after transfer from BKME to clean water (Munkittrick et al. 1992b, 1994b). The rate of loss accelerated with longer exposures, implying metabolism and excretion of MFO inducers, typical of an inducible excretion mechanism. The inducibility of excretion was confirmed by lab exposures of fish to ß-naphthoflavone, a model MFO inducer, prior to effluent exposure and transfer to fresh water (Munkittrick et al. 1994b). Rapid metabolism and excretion would not be typical of TCDD; it persists in fish tissues and causes persistent MFO induction (Muir et al. 1990; Parrott 1993; Hahn and Stegeman 1994).

The apparent rapid depuration by fish of MFO inducers indicates a need for other ways to monitor their sources and fate. Parrott et al. (1994a) demonstrated that polyethylene dialysis tubes containing one gram of triolein (fish lipid) could accumulate MFO-inducing substances in a laboratory exposure to secondary-treated BKME. The amount of inducers accumulated by these semi-permeable membrane devices (SPMDs) was measured in SPMD extracts by bioassays of MFO induction in fish cell cultures. When the extracts were passed through acid silica columns, all potency for induction was lost, indicating a labile inducer that would not persist in the environment; persistent compounds typically survive passage through these columns.

SPMDs can be deployed in effluent and in surface waters, sampling in conditions that are lethal to fish (e.g., when testing untreated effluent streams to identify sources within mills), or in areas where fish do not reside at the time of the survey. SPMDs accumulated inducers from a stream containing a high percentage of BKME, although the gradient of accumulation was inverse to the dilution gradient (Parrott et al. 1994a). Given the high concentration of effluent, this unusual behavior might be explained by organic complexation of the inducer in concentrated effluent, with subsequent release during dilution.

Laboratory studies of BKME-exposed fish have shown that MFO induction can be caused by whole effluent, effluent filtered through 0.1 µ filters, effluent passed through a 0.2 µm tangential-flow filter, and in methanol extracts of filtered solids (Burnison et al. 1994). The results indicated that MFO inducers were both waterborne and particulate bound, with highest concentrations on solids. As such, they will likely be transferred to sediments where their persistence and availability to food chains are unknown. Munkittrick et al. (1994b) measured induction in white sucker and rainbow trout (*Oncorhynchus mykiss*) caged in a stream receiving high concentrations of BKME. During spring (high flow) and fall (low flow), only fish close to the source showed induction after 4 d exposure, at effluent concentrations between 50 and 100%. Further downstream, at lower concentrations, there was less MFO activity, implying insufficient concentrations of inducers in stream sediments to leach back to the water column. However, fish were not in contact with either the sediment or with the food chain that lives in sediment.

In contrast, Kloepper-Sams et al. (1994) correlated EROD induction in mountain whitefish (*Prosopium williamsoni*) to high concentrations of dioxins in filter-feeding benthic stoneflies (a preferred prey item of whitefish), high dioxin concentrations in whitefish tissues, and high concentrations of the MFO enzyme protein (P4501A1) in cells lining the whitefish intestine during one of their sampling periods. These results point to the possibility of food chain transfer of dioxins and dioxin-induced MFO activity. Servos et al. (1994a) found a decline in MFO activity of fish caged downstream of a pulp mill during a shutdown. Simultaneously, wild fish outside the cages showed sustained activity, suggesting exposure to contaminated sediments or food chains.

ROLE OF CHLORINE BLEACHING

MFO induction in fish exposed to pulp mill effluent was previously associated only with BKME. However, induction in fish is now known to be caused by effluents from sulfite, ground-wood, thermomechanical, chemi-thermomechanical, unbleached and bleached kraft mills, including mills that use a high degree of ClO_2 substitution or that do not use chlorine at all for bleaching (Hänninen et al. 1991; Lindström-Seppä et al. 1992; Gagné and Blaise 1993; Johnsen et al. 1994; Martel et al. 1994, 1996; Munkittrick et al. 1994a). Similar results were found using fish cell cultures (Hänninen et al. 1991; Pesonen and Andersson 1992). However, these results were not entirely consistent because not all effluents caused induction (Gagné and Blaise 1993; Martel et al. 1994), although they support the contention that the potency of effluents for MFO induction cannot be reduced by replacing molecular chlorine with ClO_2 or with peroxide. Martel et al. (1994) proposed that MFO induction was primarily a result of kraft pulping rather than of bleaching.

Typical BKME effects on biota can be replicated in mesocosms simulating brackish water littoral communities, using effluents from mills with chlorine bleaching, different degrees of chlorine dioxide substitution, and no chlorine bleaching at all (Lehtinen et al. 1993; Tana et al. 1994). The greatest impacts on invertebrates were caused by effluents from chlorine bleaching, and the least by effluents from a chlorine-free mill. However, overall ecosystem impacts were unrelated to concentrations of AOX. In fish, MFO activity was induced only by effluents from a mill using chlorine bleaching and a mill using 15% chlorine dioxide substitution. Effluent dilutions were 200- to 1000-fold, so that cause-effect relationships were not easily tested.

There is little association between MFO induction and AOX, reported as either release per tonne of product or concentration in water (Hodson et al. 1992; Martel et al. 1996; Munkittrick et al. 1994a). However, Ahokas et al. (1994) found that EROD induction (but not ECOD [ethoxycoumarin-*o*-deethylase]) in carp (*Cyprinus carpio*) was correlated to total equivalents of tissue dioxin and to concentrations of waterborne AOX in a shallow-water lake in Australia. There were steep gradients of both EROD activity and AOX concentrations over a short distance (5-10 km) in the lake, but it is likely that the gradient was primarily due to dilution. Because AOX persists in the environment, both waterborne AOX and tissue dioxin concentrations may reflect effluent dilution. There is abundant evidence that dioxins could be the inducers, but AOX measurements were likely surrogates for dilution of other inducing agents and not representative of the inducer itself. There were no correlations between MFO induction and concentrations of tissue or sediment EOX.

SOURCES WITHIN MILLS OF MFO INDUCERS

Martel et al. (1994) and Schnell et al. (1993) demonstrated that an important source of MFO inducers among bleached kraft mill waste streams was weak black liquor, reinforcing the conclusion that chlorine is not an essential ingredient for induction. Black liquors generated from softwood pulping appeared more potent than those from hardwood, but high variability in estimates of potency precluded definitive comparisons (Hodson et al. 1994a). Normalizing estimates of potency to dissolved organic carbon content reduced these apparent differences, indicating that potency may be related more to differences in pulping technologies than to differences in wood composition. Differences among mills in potency of final effluents for MFO induction were also reduced by normalizing for dissolved organic carbon (Williams 1993).

Within a kraft mill, the next most potent sources of inducers were effluents from the first bleaching stage with chlorine and chlorine dioxide, followed by the first alkaline extraction stage; the remaining stages of kraft pulping release little, if any (Schnell et al. 1993). The bleaching and extraction effluents were 30-100 times less potent than black liquor, but contributed greater volumes of inducers to final effluent than did black liquor. Inducers in these effluents may represent carryover from pulping, additional chemicals released from lignin by bleaching, or new compounds formed by chlorination (or a combination of all three). Inducers in black liquor are either breakdown products of lignin and cellulose or a wood extractive. MFO induction by spent cooking liquor from the acidic alcohol extraction of lignin would support the notion of wood extractives, assuming no breakdown of lignin or cellulose (Hodson et al. 1994a). Martel et al. (1994) reached similar conclusions because BKME was most potent when wood furnish was hemlock, a wood rich in extractives.

EFFECTS OF EFFLUENT TREATMENT ON POTENCY

Individual waste streams from a bleached kraft mill and simulated whole effluent (volumetric combination of individual waste streams) have been treated by bench-scale activated sludge, facultative stabilization basin, and aerated stabilization basin processes (Schnell et al. 1993). About 89-98% of potency for MFO induction was lost after treatment, as shown by bioassays with trout. An optimized activated sludge pilot plant using actual BKME gave similar success. The cause of reduced potency was not evaluated, but would include microbial metabolism of inducers or absorption to sludge. The demonstrated loss of potency of SPMD extracts on reactive silica columns (Parrott et al. 1994a) supports the hypothesis of labile compounds degraded by microbes.

The value of extended treatment of effluent was shown on Lake Baikal by the lack of MFO induction in fish sampled near a pulp mill. BKME was mixed with municipal waste and treated by anaerobic and aerobic biological, chemical, and mechanical processes, followed by stabilization basins (Hänninen et al. 1991). However, MFO inducers survive even the best secondary treatment in Canada (Hewitt et al. 1996). Fish downstream of fully modernized plants in Alberta and Ontario show significant MFO induction (Servos et al. 1993; Swanson et al. 1993) and the installation of conventional secondary treatment at a mill in Ontario did not remove the MFO induction of fish that was present several years before the changes (Munkittrick et al. 1992b). In a multi-mill study, some of the highest levels of induction were observed in fish downstream of mills with secondary treatment; some mills with primary treatment only had lower levels of induction (Munkittrick et al. 1994a). Some of the differences among mills might be due to differences in chemicals released during different pulping and bleaching processes, in dilution factors, or in concentrations of toxicants that inhibit MFO induction.

NATURE AND IDENTITY OF MFO INDUCERS

Some field studies have demonstrated good correlations between concentrations of dioxins (or total dioxin equivalents) and MFO activity in field-exposed fish (Rogers et al. 1989; Hodson et al. 1992; Birkholz and Swanson 1993; Servizi et al. 1993; Ahokas et al. 1994; Servos et al. 1994b; van den Heuvel et al. 1994). Correlations were best in male fish and could be improved in females if serum estradiol was considered as a co-variate (van den Heuvel et al. 1994), because estradiol modulates MFO induction (Förlin and Haux 1990). The role of dioxins in MFO induction is supported by laboratory bioassays demonstrating strong and persistent induction in fish exposed to either dietary or injected dioxins and furans (Muir et al. 1990; Parrott 1993; Hahn and Stegemann 1994) and by strong correlations between tissue dioxins measured chemically and tissue dioxins measured by induction of MFO enzymes in rat liver cancer cells *in vitro* (van den Heuvel et al. 1994). A reduction in dioxin emissions in BKME was also associated with reductions in MFO activity of whitefish in receiving waters (Birkholz and Swanson 1993). MFO induction in mouse hepatoma cells also demonstrated the presence of high concentrations of dioxins in fly ash from the experimental combustion of sludge from a bleached kraft mill (Kopponen et al. 1993).

Dioxins may not be the only cause of induction. Munkittrick et al. (1994b) observed no decline over a 5-year period in the extent of induction in fish from the receiving waters near a mill that reduced dioxin discharges following ClO_2 substitution. Servos et al. (1994b) showed that MFO activity of fish near a series of pulp mills increased when liver dioxin levels exceeded 16-20 pg g^{-1}, the threshold dose for induction in lab studies (Parrott 1993). However, there was one mill at which activity was elevated when liver concentrations of dioxin equivalents were low, implying the presence of other inducers. MFO induction of fish exposed to BKME also disappeared within a few days of mill shutdown or transfer to clean water (see above). This behavior is more typical of induction by PAHs, compounds rapidly metabolized by MFO enzymes to facilitate excretion. In contrast, mountain whitefish that moved from downstream of a bleached kraft mill to upstream showed the opposite: no reduction in MFO activity (Swanson et al. 1992). These apparently contradictory observations may arise from the interaction between the route of exposure and the chemical characteristics of inducers (Birkholz and Swanson 1993). For suckers exposed to waterborne materials, it is likely that the materials will be relatively hydrophilic and easily metabolized. For dietary exposure, chemicals that are taken up through the food chain are more likely to be hydrophobic and persistent, thereby contributing to persistent induction.

There is a strong correlation between molecular size, molecular shape, and potency for MFO induction. The most potent inducer is TCDD due to its size and the planarity of its ring structure (Safe 1992; Parrott 1993). Increasing or decreasing chlorine substitution and changing patterns of substitution decrease potency. Likewise, some PCBs are more potent inducers than others, depending on the number of chlorine atoms and how their distribution affects co-planarity of the phenyl rings. PAHs that induce MFO activity are multi-ringed, aromatic, and planar, but not chlorinated, so they are less potent than TCDD.

Many chemicals in BKME, particularly the resin acids, appear structurally related to inducers such as PAHs. However, abietic acid, dehydroabietic acid (DHAA), a resin acid mixture, and rosin gum did not induce EROD activity in trout, either by waterborne exposures or by injection (Ferguson et al. 1993). Similar results were reported by Oikari and Lindström-Seppä (1990) for DHAA and tetrachloroguaiacol, but Mather-Mihaich and Di Giulio (1991) found a slight induction with wood rosin at waterborne concentrations about 8 times those used by Ferguson et al. (1993). In contrast, retene, an alkyl-substituted phenanthrene, appears to induce EROD in rainbow trout quite strongly (Parrott et al. 1994b). It is typically found in sediments downstream of pulp mills where it is thought to be formed by the bacterial reduction of DHAA under anaerobic conditions (M.E. Fox, Canada Centre for Inland Waters, Burlington, Ontario, Canada, pers. comm.). It is planar, aromatic, and has structural characteristics typical of other inducers.

Other alkyl-substituted phenanthrenes also induce EROD in fish, but not unsubstituted phenanthrene (Parrott et al. 1994b).

Lehtinen et al. (1993) found a slight but non-significant increase in EROD activity of rainbow trout after exposure to 10 µg L^{-1} of a phytosterol mixture (mostly ß-sitosterol); we observed the same in our lab (Hodson, unpub. data). MFO induction by whole BKME declines progressively through primary and secondary treatment, but ultrafiltration by reverse osmosis to remove high molecular weight compounds (>1000) eliminates potency for MFO induction (Hewitt et al. 1996). The implication is that either high molecular weight compounds are inducers (unlikely since they would not traverse membranes easily) or that inducers were absorbed by either the reverse osmosis membrane or the high molecular weight compounds. Since resin acids, terpenes, chlorophenolics, aliphatic alkanes, plant sterols, and chlorinated dimethyl sulfones survived this process, they could not be considered inducers (Hewitt et al. 1996).

Whole, unfiltered secondary-treated BKME from a mill with a high degree of ClO_2 substitution induced EROD activity in trout. In contrast to reverse osmosis, effluent filtration through 1.0 µ glass-fiber filters followed by a 0.2 µ tangential-flow filter caused no loss in potency for induction. However, methanol extracts of the filtered solids induced greater activity than did the original effluent (Burnison et al. 1994). Filtered BKME lost some potency for induction after flocculation with DEAE cellulose to remove humic acids and color, and residual potency was removed completely by C-18 filtration. Inducers removed by DEAE cellulose and C-18 resins could be recovered by methanol extraction in a quantitative way, as shown by MFO induction in trout (Burnison et al. 1994).

The simultaneous association of BKME inducers with particulates, humates, and the aqueous phase indicates compounds with a low water solubility; above saturation, "excess" inducers become absorbed to DOC and to particles. This absorbed material is not readily soluble in solvents such as pentane, dichloromethane, and acetone (Burnison et al. 1994), indicating that inducers are not highly hydrophobic, as would be typical of chlorinated dioxins and furans. Methanol solubility suggests compounds of intermediate polarity, although solubility occurs only at concentrations above 80%. This property allows the separation of inducers from other compounds by reverse phase HPLC. Using isocratic elution with 80% methanol, followed by a column wash with 100% methanol, Burnison et al. (1994) found that MFO inducers consistently eluted with marker compounds having log octanol-water partition coefficients (K_{ow}) of about 4.5-5.1, somewhat less than would be typical of PCBs, dioxins, and furans. Peaks shown by fluorometric detection included phenanthrene and substituted phenanthrenes.

Based on these properties, waterborne MFO inducers are likely PAH-like compounds of intermediate polarity, consistent with a loss of MFO activity from extracts of SPMDs passed through reactive silica columns (see section on fate). The rapid decline in MFO activity of BKME-exposed white sucker after their transfer to clean water (Munkittrick et al. 1992b) also supports this hypothesis. Maternal transfer to eggs of dioxins, which are persistent and lipophilic, occurs readily and the consequences can be pathology and mortality of offspring in some species (Walker and Peterson 1991). The lack of induction and pathology in offspring of white suckers exposed to effluent just prior to spawning indicates no maternal transfer of inducers (McMaster et al. 1992), as would be typical of a rapidly metabolized compound.

OTHER CONSIDERATIONS

As with most analytical techniques, assays of MFO activity are subject to error and bias. The major error factors include analysts, changes in reagents and stocks of fish, and occasionally unknown factors. There was considerable variation among labs participating in an inter-laboratory round-robin analysis of EROD activity of control or induced fish (Munkittrick et al. 1993b). However, absolute activity is less important than induction potency, as measured by the ratio of activity in treated fish to activity in control fish. In contrast to absolute activity, measurements of induction among labs were remarkably constant (Munkittrick et al. 1993b), and all labs clearly discriminated control from treated fish.

Variations in measurements of MFO activity among treatments or sites are not distributed homogeneously (Hodson et al. 1994b). As mean activity increases under controlled laboratory conditions, so too does the standard deviation, likely because of natural variation in the ability of fish to respond. Scaling factors in fluorometers and spectrophotometers also introduce changes in variance with the size of a response, since relative error remains constant as scale ranges change. Statistical analyses that assume normality of distributions and homogenous variances should not be used without transforming data, e.g., to logarithms. Conclusions about treatment or site effects must be derived from statistical analyses of transformed data that is normally distributed (Hodson *et al.* 1994b). For log transformations, results may be expressed as geometric means (antilog of the mean), with 95% confidence limits (antilog of the calculated confidence limits for $p < 0.05$) as the index of variability.

Bioassays of MFO induction in fish must respect the biomass loading recommendations typical of acute lethality tests. At high ratios of biomass to test effluent volume, MFO induction was reduced compared to lower loading rates, presumably due to exhaustion of the inducing agent by uptake into fish (Hodson *et al.* 1994b). Unfortunately loading requirements and tests with several dilutions require relatively large volumes of effluent, relatively small fish, and costly effluent shipments.

Bioassay and enzyme assay protocols have been refined to permit the routine measurement of MFO induction using 1-3 g trout, based on microplate fluorometers and EROD assays with livers as small as 10-20 mg (Hodson *et al.* 1994b). While microassays allow the measurement of MFO induction in routine lethality tests for effluent regulations, MFO activity of surviving fish can be underestimated due to liver toxicity (Gagné and Blaise 1993). The utility of some bioassay protocols is also limited by exposure conditions. Environmentally "realistic" bioassays restricted to tests of 10% effluent decrease the likelihood of detecting induction in effluents or treatments of low potency. Furthermore, by limiting tests to one dilution, the toxic effects of effluents observed by Gagné and Blaise (1993) may be overlooked, generating "false negatives" due to response inhibition.

A logical alternative to whole fish assays to reduce effluent requirements is the use of fish cells in culture (e.g., Pesonen and Andersson 1992). However, there have been no definitive comparisons of cell cultures with whole fish in terms of sensitivity and the nature of the compounds detected; this comparison is critical before cell lines can replace whole fish.

CONCLUSIONS

1. While MFO induction of fish is often strongly associated with other responses to BKME, there are no defined mechanisms linking induction to specific effects.
2. The environmental fate of MFO inducers, as shown by bioassays, suggests two classes of compounds: those that persist and are distributed widely (dioxins) and those that are labile and disappear quickly (unidentified compounds). The relative persistence in sediments of this latter group and their transfer through food chains is unknown.
3. Chlorine is not an essential component of compounds causing MFO induction, although potency of dioxin and furan congeners varies with the number of chlorine atoms and their distribution. It is likely that characteristics that control the potency of known inducers, such as size, shape, and planarity, would also be shared by inducers in pulp mill effluent.
4. The role of chlorine bleaching in releasing inducers from pulp and in changing their potency through chlorination is unknown. AOX concentrations in effluents and surface waters are unrelated to MFO induction potency.
5. Enhanced effluent treatment can reduce the potency of MFO induction by pulp mill effluents to negligible levels. However, the fate of inducing compounds removed by waste treatment is unknown.
6. Of the two classes of inducers in pulp mill effluent, the persistent, chlorinated class are *likely* no longer a problem in mills that reduce dioxin emissions through process change. However, labile inducers remain in effluents and they share the properties of PAHs. They are found in aqueous, particulate, and dissolved organic carbon phases, are moderately hydrophobic, and appear readily metabolized by fish.

7. Good bioassay methods with whole fish are available to study processes and chemicals causing MFO induction. Cell line assays are now available that will reduce sample volumes, cost, and effort needed for MFO research. However, their use must be validated against whole fish assays.
8. The toxicological and ecological relevance of MFO induction by pulp mill effluents cannot be determined without identifying the inducing compounds. It is imperative that inducers in effluent be isolated, purified, and identified to answer these questions.

REFERENCES

Adams, S.M., W.D. Crumby, M.S. Greeley, L.R. Shugart and C.F. Saylor. 1992. Responses of fish populations and communities to pulp mill effluents: a holistic assessment. Ecotox. Environ. Safety 24:347-360.

Ahokas, J.T., D.A. Holdway, S.E. Brennan, R.W. Goudey and H.B. Bibrowska. 1994. MFO activity in carp (*Cyprinus carpio*) exposed to treated pulp and paper mill effluent in Lake Coleman, Victoria, Australia, in relation to AOX, EOX and muscle PCDD/PCDF. Environ. Toxicol. Chem. 13:41-50.

Birkholz, D.A., and S.M. Swanson. 1993. The relationship between mixed function oxygenase induction and other health parameters for fish exposed to bleached kraft mill effluent. pp. 85-88. Proceedings of the 1993 TAPPI Environmental Conference,

Burnison, B.K., P.V. Hodson, D.J. Nuttley and S. Efler. 1994. Isolation and characterization of a MFO-inducing fraction from BKME. Presented at the Second International Conference on Environmental Fate and Effects of Bleached Pulp Mill Effluents, November 6-10, 1994, Vancouver, B.C., Canada.

Ferguson, M.L., M.R. Servos, K.R. Munkittrick and J. Parrott. 1993. Inability of resin acid exposure to elevate EROD activity in rainbow trout *Oncorhynchus kisutch*. Water Pollut. Res. J. Can. 27:561-574.

Förlin, L., and C. Haux. 1990. Sex differences in hepatic cytochrome P-450 monoxygenase activities in rainbow trout during an annual reproductive cycle. J. Endocrinol. 124:207-213.

Gagné, F., and C. Blaise. 1993. Hepatic metallothionein level and mixed function oxidase activity in fingerling rainbow trout (*Oncorhynchus mykiss*) after acute exposure to pulp and paper mill effluents. Water Res. 27:1669-1682.

Gagnon, M.M., J.J. Dodson, P.V. Hodson, G. Van Der Kraak and J.H. Carey. 1994. Seasonal effects of bleached kraft mill effluent on reproductive parameters of white sucker *Catostomus commersoni* populations of the St. Maurice River, Quebec, Canada. Can. J. Fish. Aquat. Sci. 51:337-347.

Hahn, M.E., and J.J. Stegeman. 1994. Regulation of cytochrome P4501A1 in teleosts: sustained induction of CYP1A1 mRNA, protein, and catalytic activity by 2,3,7,8-tetrachlorodibenzofuran in the marine fish *Stenotomus chrysops*. Toxicol. Appl. Pharmacol. 127:187-198.

Hewitt, L.M., J.H. Carey, D.G. Dixon and K.R. Munkittrick. 1996. Examination of bleached kraft mill effluent fractions for potential inducers of mixed function oxygenase activity in rainbow trout. *In* Environmental Fate and Effects of Pulp and Paper Mill Effluents, M.R. Servos, K.R. Munkittrick, J.H. Carey and G. Van Der Kraak (ed.) St. Lucie Press, Delray Beach, FL.

Hodson, P.V., M. McWhirter, K. Ralph, B. Gray, D. Thivierge, J.H. Carey, G. Van Der Kraak, D.M. Whittle and M.C. Levesque. 1992. Effects of bleached kraft mill effluent on fish in the St. Maurice River, Quebec. Environ. Toxicol. Chem. 11:1635-1651.

Hodson, P.V., M.M. Maj, S. Efler, A. Schnell and J. Carey. 1994a. Kraft black liquor as a source of MFO inducers. Presented at the Second International Conference on Environmental Fate and Effects of Bleached Pulp Mill Effluents, November 6-10, 1994, Vancouver, B.C., Canada.

Hodson, P.V., S. Efler, M.M. Maj and J.Y. Wilson. 1994b. A refined protocol for measuring the potency of effluents, extracts or pure compounds for inducing MFO activity of fish. Presented at the Second International Conference on Environmental Fate and Effects of Bleached Pulp Mill Effluents, November 6-10, 1994, Vancouver, B.C., Canada.

Hänninen, O., P. Lindström-Seppä, M. Pesonen, S. Huuskonen and P. Muona. 1991. Use of biotransformation activity in fish and fish hepatocytes in the monitoring of aquatic pollution caused by pulp industry. pp. 13-20. *In:* Bioindicators and Environmental Management. Academic Press Ltd., New York, NY.

Johnsen, K., K. Mattsson, J. Tana, T.R. Stuthridge, J. Hemming and K-J. Lehtinen. 1994. Uptake and elimination of resin acids and physiological responses in rainbow trout exposed to total mill effluent from an integrated newsprint mill. Environ. Toxicol. Chem. (submitted).

Kloepper-Sams, P., L. Steeves and J. Stegeman. 1994. Environmental fate and effect of bleached pulp mill effluents: follow-up studies on dioxin, EROD induction and the role of dietary pathways. Abstract, p. 151. Proceedings of the Twentieth Annual Aquatic Toxicity Workshop, October 17-21, 1993, Quebec City. Can. Tech. Rept. Fish. Aquat. Sci. 1989, 331 p.

Kopponen, P., J. Tarhanen, J. Ruuskanen, R. Torronen and S. Karenlampi. 1993. Peat induces cytochrome P450IA1 in hepa-1 cell line - comparison with fly ashes from combustion of peat, coal, heavy fuel oil and hazardous waste. Chemosphere 26:1499-1506.

Lehtinen, K-J. 1990. Mixed-function oxygenase enzyme responses and physiological disorders in fish exposed to kraft pulp-mill effluents: a hypothetical model. Ambio 19:259-266.

Lehtinen, K-J., J. Tana, K. Mattson, J. Härdig and J. Hemming. 1993. Ecological impacts of pulp mill effluents. Part 1. Physiological responses and effects on survival, growth, and parasites frequency in fish exposed in mesocosms to treated total mill effluents from production of bleached kraft pulp (BKME), thermomechanical pulp and phytosterols. pp. 3-64. In Study No 10A. National Board of Waters and the Environment, Helsinki, Finland.

Lindström-Seppä, P., S. Huuskonen, M. Pesonen, P. Muona and O. Hanninen. 1992. Unbleached pulp mill effluents affect cytochrome-P450 monooxygenase enzyme activities. Mar. Environ. Res. 34:157-161.

Martel, P.H., T.G. Kovacs., B.I. O'Connor and R.H. Voss. 1994. A survey of pulp and paper mill effluents for their potential to induce mixed function oxidase enzyme activity in fish. Water Res. 28:1835-1844.

Martel, P.H., T.G. Kovacs and R.H. Voss. 1996. Effluents from Canadian pulp and paper mills: a recent investigation of their potential to induce MFO activity in fish. In Environmental Fate and Effects of Pulp and Paper Mill Effluents, M.R. Servos, K.R. Munkittrick, J.H. Carey and G. Van Der Kraak (ed.) St. Lucie Press, Delray Beach, FL.

Mather-Mihaich, E., and R.T. Di Giulio. 1991. Oxidant, mixed-function oxidase and peroxisomal responses in channel catfish exposed to a bleached kraft mill effluent. Arch. Environ. Contam. Toxicol. 20:391-397.

McMaster, M.E., C.B. Portt, K.R. Munkittrick and D.G. Dixon. 1992. Milt characteristics, reproductive performance, and larval survival and development of white sucker exposed to bleached kraft mill effluent. Ecotox. Environ. Safety 23:103-117.

Muir, D.C.G., A.L. Yarechewski, D.S. Metner, W.L. Lockhart, G.R.B. Webster and K.J. Friesen. 1990. Dietary accumulation and sustained hepatic mixed function oxidase enzyme induction by 2,3,7,8-pentachlorobenzofuran in rainbow trout. Environ. Toxicol. Chem. 9:1463-1472.

Munkittrick, K.R., M.E. McMaster, C.B. Portt, G.J. Van Der Kraak, I.R. Smith and D.G. Dixon. 1992a. Changes in maturity, plasma sex steroid levels, hepatic mixed function oxygenase activity, and the presence of external lesions in lake whitefish (*Coregonus clupeaformis*) exposed to bleached kraft mill effluent. Can. J. Fish. Aquat. Sci. 49:1560-1569.

Munkittrick, K.R., G.J. Van Der Kraak, M.E. McMaster and C.B. Portt. 1992b. Response of hepatic MFO activity and plasma sex steroids to secondary treatment of bleached kraft pulp mill effluent and mill shutdown. Environ. Toxicol. Chem. 11:1427-1439.

Munkittrick, K.R., G.J. Van Der Kraak, M.E. McMaster and C.B. Portt. 1993a. Reproductive dysfunction and MFO activity in three species of fish exposed to bleached kraft mill effluent at Jackfish Bay, Lake Superior. Water Pollut. Res. J. Can. 27:439-446.

Munkittrick, K.R., M.R. van den Heuvel, D.A. Metner, W.L. Lockhart and J.J. Stegeman. 1993b. Interlaboratory comparison and optimization of hepatic microsomal ethoxyresorufin o-deethylase activity in white sucker (*Catostomus commersoni*) exposed to bleached kraft pulp mill effluent. Environ. Toxicol. Chem. 12:1273-1282.

Munkittrick, K.E., G.J. Van Der Kraak, M.E. McMaster, C.B. Portt, M.R. van den Heuvel and M.R. Servos. 1994a. Survey of receiving water environmental impacts associated with discharges from pulp mills. II. Gonad size, liver size, hepatic EROD activity and plasma sex steroid levels in white sucker. Environ. Toxicol. Chem. 13:1089-1101.

Munkittrick, K.R., M.R. Servos, K. Gorman, B. Blunt, M.E. McMaster and G.J. Van Der Kraak. 1995. Characteristics of EROD induction associated with exposure to pulp mill effluent. T.W. La Point, F.T. Price and E.E. Little (Eds.). Environmental Toxicology and Risk Assessment: Fourth Volume, ASTM STP 1262. American Society for Testing and Materials, Philadelphia: In press.

Oikari, A., and P. Lindström-Seppä. 1990. Responses of biotransformation enzymes in fish liver experiments with pulp mill effluents and their components. Chemosphere 20:1079-1085

Parrott, J.L. 1993. Relative potency of polychlorinated dibenzo-p-dioxins and dibenzofurans for inducing mixed function oxygenase activity in rainbow trout. Ph.D. Thesis, Dept. of Biology, U. of Waterloo, Waterloo, Ontario.

Parrott, J.L., D.E. Tillitt, P.V. Hodson., D.T. Bennie, J.N. Huckins and J.D. Petty. 1994a. Semi-permeable membrane devices (SPMDs) accumulate inducer(s) of fish mixed function oxygenase (MFO) from pulp mill effluents. Presented at the Second International Conference on Environmental Fate and Effects of Bleached Pulp Mill Effluents, November 6-10, 1994, Vancouver, B.C., Canada.

Parrott, J.L., B.K. Burnison, P.V. Hodson, M.E. Comba and M.E. Fox. 1994b. Retene-type compounds - inducers of hepatic mixed function oxygenase (MFO) in rainbow trout (*Oncorhynchus mykiss*)? Presented at the Second International Conference on Environmental Fate and Effects of Bleached Pulp Mill Effluents, November 6-10, 1994, Vancouver, B.C., Canada.

Pesonen, M., and T. Andersson. 1992. Toxic effects of bleached and unbleached paper mill effluents in primary cultures of rainbow trout hepatocytes. Ecotox. Environ. Safety 24:63-71.

Robinson, R.D., J.H. Carey, K.R. Solomon, I.R. Smith, M.R. Servos and K.R. Munkittrick. 1994. Survey of receiving water environmental impacts associated with discharges from pulp mills. I. Mill characteristics, receiving water chemical profiles and laboratory toxicity tests. Environ. Toxicol. Chem. 13:1075-1088.

Rogers, I.H., C.D. Levings, W.L. Lockhart and R.J. Norstrom. 1989. Observations on overwintering juvenile chinook salmon (*Oncorynchus tshawytscha*) exposed to bleached kraft mill effluent in the upper Fraser River, British Columbia. Chemosphere 19:1853-1868.

Safe, S. 1992. Development, validation and limitations of toxic equivalency factors. Chemosphere 25:61-64.

Schnell, A., P.V. Hodson, P. Steel, H. Melcer and J.H. Carey. 1993. Optimized biological treatment of bleached kraft mill effluents for the enhanced removal of toxic compounds and MFO induction response in fish. Proceedings of the 1993 Environmental Conference of the Canadian Pulp and Paper Association.

Servizi, J.A., R.W. Gordon, D.W. Martens, W.L. Lockhart, D.A. Metner, I.H. Rogers, J.R. McBride and R.J. Norstrom. 1993. Effects of biotreated bleached kraft mill effluent on fingerling chinook salmon (*Oncorhynchus tshawytscha*). Can. J. Fish. Aquat. Sci. 50:846-857.

Servos, M., J. Carey, M. Ferguson, G.J. Van Der Kraak, H. Ferguson, J. Parrott, K. Gorman and R. Cowling. 1993. Impact of a modernized bleached kraft mill on white sucker populations in the Spanish River, Ontario. Water Pollut. Res. J. Can. 27:423-438.

Servos, M.R., S.Y. Huestis, D.M. Whittle, G.J. Van Der Kraak and K.R. Munkittrick. 1994a. Survey of receiving water environmental impacts associated with discharges from pulp mills. III. Polychlorinated dioxins and furans in muscle and liver of white sucker (*Catostomus commersoni*). Environ. Toxicol. Chem. 13:1103-1116.

Servos, M.R., K.R. Munkittrick, E.A. Chisholm, M. Ferguson and G.J. Van Der Kraak. 1994b. Physiological responses in white sucker exposed to treated bleached kraft mill effluent in the Spanish River, Ontario, Canada. Presented at the Second International Conference on Environmental Fate and Effects of Bleached Pulp Mill Effluents, November 6-10, 1994, Vancouver, B.C., Canada.

Swanson, S., R. Shelast, R. Schryer, P. Kloepper-Sams, T. Marchant, K. Kroeker, J. Bernstein and J.W. Owens. 1992. Fish populations and biomarker responses at a Canadian bleached kraft mill site. TAPPI J. 139-149.

Swanson, S.M., R. Schryer, B. Shelast, K. Holley, I. Berbekar, P. Kloepper-Sams, J.W. Owens, L. Steeves, D. Birkholz and T. Marchant. 1993. Wapiti Smoky River ecosystem study. Weyerhaeuser, Canada, Grande Prairie, Alberta, 175 p.

Södergren, A. (Ed.). 1991. Environmental Fate and Effects of Bleached Kraft Mill Effluents. Proceedings of a SEPA Conference., Stockholm, Sweden, November 1991. Swedish Environmental Protection Agency, Stockhom, Report No. 4031, 394 p.

Tana, J., A. Rosemarin, K-J. Lehtinen, J. Härdig, O. Grahn and L. Landner. 1994. Assessing impacts on Baltic coastal ecosystems with mesocosm and fish biomarker tests: a comparison of new and old wood bleaching technologies. Sci. Total Env. 145:213-234.

van den Heuvel, M.R., K.R. Munkittrick, G.J. Van Der Kraak, M.E. McMaster, C.B. Portt, M.R. Servos and D.G. Dixon. 1994. Survey of receiving water environmental impacts associated with discharges from pulp mills. 4. Bioassay derived 2,3,7,8-tetrachlorodibenzo-p-dioxin toxic equivalent concentration in white sucker *(Catostomus commersoni)* in relation to biochemical indicators of impact. Environ. Toxicol. Chem. 13:1117-1126.

Walker, M.K., and R.E. Peterson. 1991. Potencies of polychlorinated dibenzo-para-dioxin, dibenzofuran, and biphenyl congeners, relative to 2,3,7,8-tetrachlorodi-benzo-para-dioxin, for producing early life stage mortality in rainbow trout (*Oncorhynchus mykiss*). Aquat. Toxicol. 21:219-238.

Williams, T.G. 1993. A comparative laboratory based assessment of EROD enzyme induction in fish exposed to pulp mill effluents. M.Sc. Thesis, Dept. of Biology, U. of Waterloo, Waterloo, Ontario. 85 p.

BIOCHEMICAL RESPONSES IN ORGANISMS EXPOSED TO EFFLUENTS FROM PULP PRODUCTION: ARE THEY RELATED TO BLEACHING?

Karl-Johan Lehtinen

European Environmental Research Group (MFG), Finnish Branch, Teknikvägen 12, SF-02150 Esbo, Finland

A great deal of research work worldwide has been devoted to the assessment of the environmental impact of the production of pulp. A large proportion of the research has been performed using different kinds of biomarkers and biological/ecological endpoints. Bleaching of the pulp using chlorine and/or chlorine dioxide has been suspected to be the cause behind observed responses in organisms. Laboratory and mesocosm investigations on biochemical responses in fish do not give evidence that bleaching *per se* is the reason behind hepatic phase I and phase II detoxification enzymes in fish. Neither is it possible to relate other biochemical effects in fish to bleaching. Effluents from unbleached, chlorine bleached and non-chlorine bleached pulp production as well as resin acids and phytosterols caused responses in fish when exposed to these effluents under controlled laboratory and mesocosm conditions.

INTRODUCTION

The aim of this paper is to discuss laboratory and mesocosm results on biochemical responses in fish and if/how they relate to bleaching of pulp with conventional and new techniques.

The environmental research on the aquatic impacts of pulp mill effluents has been concentrated in Scandinavia and North America. In the 1970s, the regulatory framework embraced conventional parameters such as biochemical oxygen demand, suspended solids and nutrients. From the 1980s until today, concern has been directed towards chemical toxicity. Chlorine bleaching of sulfate and sulfite pulp results in the extensive formation of organic chlorinated compounds with varying molecular size. It has been claimed that this organic material would be the specific cause of a number of biological effects. These suspicions have prompted intensive research activity to study relationships between exposure and effects as well as profound process changes within the pulp mill industry. Consequently, the use of chlorine gas as a bleaching medium is practically extinct in Scandinavia. The bleaching of pulp is currently conducted using high levels of chlorine dioxide (elemental chlorine free = ECF), as well as by bleaching technologies not requiring chlorine-containing chemicals at all (totally chlorine free = TCF). It has been anticipated that elimination of the use of chlorine would result in an elimination of biological responses observed in receiving waters.

The process trends in the pulp and paper industry are significant: reduction in the kappa number to the bleachery, improved management leading to fewer upsets and spills, improved washing, better handling of condensates, increased external treatment and an elimination of polychlorinated materials even before the introduction of TCF pulp production.

Chemical toxicity control has been performed at various levels of biological organization: community, population, individual, organ, tissue, cell, subcellular and molecular levels. Many studies performed on pulp mill effluents have focused upon within-organism effects such as physiology and biochemistry (Owens 1991). Despite the profound process changes performed in the industry, there is still evidence of several physiological changes in organisms; however the ecological significance of these changes is unknown. Physiological and biochemical measurements have been referred to as biomarkers and, in the majority of the studies, fish have been used as test organisms (McLeay 1987).

RESPONSES ON LIVER PHASE I AND PHASE II DETOXIFICATION ENZYMES

In the detoxification process, cytochrome P-450 dependent mixed function oxygenase (MFO) enzymes (phase I) act by carrying out a series of oxidation reactions whereby organic compounds are converted into water-soluble metabolites which may be further conjugated (phase II) and excreted in urine or bile. One enzyme of the MFO system is the enzyme 7-ethoxyresorufin-*o*-deethylase (EROD). The activity of this enzyme is most commonly used by researchers in connection with studies using pulp mill effluents as pollutant (Owens 1991). Other MFO enzymes have also sometimes been used as biomarkers for exposure to pulp mill effluents, however (Ahokas *et al.* 1976; Lindström-Seppä 1990; Huuskonen 1994). Finnish scientists have rather frequently used the conjugation enzyme uranosylglucoronsyltransferase (UDP-GT), which eliminates substances from the body (e.g., Oikari *et al.* 1984, 1985; Tana 1988; Lindström-Seppä 1990; Lehtinen *et al.* 1992, 1993).

The use of the EROD enzyme as a biomarker became widespread after the Swedish "Environment-Cellulose" project in the mid-1980s when higher EROD and UDP-GT activities were detected in perch (*Perca fluviatilis*), outside the bleached kraft pulp mill at Norrsundet. It may also be noted that UDP-GT activity was analyzed and found to be increased in the same study (Andersson *et al.* 1989). The mill conditions and receiving water circumstances have been discussed by Owens (1991). After the Norrsundet study it was strongly suspected that chlorinated organic substances were the reason behind the EROD induction.

LABORATORY STUDIES WITH EFFLUENTS AND EFFLUENT FRACTIONS

Short-term (4 d) laboratory studies on EROD activation by Canadian scientists found that secondary treated thermomechanical (TMP) and chemi-thermomechanical (CTMP) effluents did not increase MFO activity (Martel *et al.* 1994) (Table 1). In contrast, MFO activity was significantly induced after exposure to bleached and unbleached kraft mill effluents, as well as after exposure to black liquor. The Canadian data also suggested that replacement of chlorine in the bleaching of kraft pulp by chlorine dioxide or non-chlorine-containing compounds did not significantly alter an effluent's ability to induce MFO activity. New pulping technologies in combination with secondary treatment did not eliminate the MFO response in fish.

Huuskonen (1994) reported inhibitive responses on the EROD activity in rainbow trout exposed to 0.25% v/v diluted biologically treated effluent from a mill producing unbleached kraft pulp, whereas effluent diluted to 0.5% v/v of full strength slightly stimulated the EROD enzyme after an 18 d exposure (Table 1). After 30 d, no significant differences occurred in either dose, although mean EROD activity was still lower in fish exposed to the higher dilution. Moreover, the activity of UDP-GT was significantly stimulated after 3 d and inhibited in 0.5% v/v of effluent after 18 d exposure. In general the activity of UDP-GT was lower than in control fish in both doses after 18 and 30 d exposure.

Simulated unbleached kraft mill effluents prepared from a sulfate soap solution caused a significant decrease in UDP-GT activity in rainbow trout exposed for 11 d to 0.15% of the 96 h LC50 value of the resin acids present (Oikari and Nakari 1982) (Table 1). Other biochemical parameters, such as lower liver bilirubin levels, gave evidence of impaired liver function. Impaired liver function had previously been observed after exposure of rainbow trout to resin acids (Castrén and Oikari 1987).

Effluents from other types of pulping processes, such as sulfite, have not been studied to any extent since sulfite pulp production is now scarce in Scandinavia and North America. However, Ahokas *et al.* (1976) found strongly inhibited MFO activities in pike (*Esox lucius*) caught from an area polluted by sulfate and sulfite pulp mill industrial effluents. The authors concluded that the fish, due to the enzyme inhibition observed, suffered from hepatotoxic effects.

Some chlorinated phenolic compounds directly or indirectly related to pulp bleaching have been studied. Intraperitoneal injections of 4,5-dichloroveratrole, 4,5,6-trichloroguaiacol, 3,4,5-trichloroveratrole,

Table 1. Hepatic phase I EROD and phase II UDP-GT enzyme responses in fish liver and liver cell cultures exposed to various whole pulp mill effluents as well as fractions of different effluents.

Process/Fraction	Treatment	EROD	UDP-GT	Species	Duration of Exp.	Reference
O(C85+D15)	none	Induction	Induction	Perch	Norrsundet, field	Andersson et al. 1989
CDEoDED	yes	Induction	N.A.	Rainbow trout	4 d	Martel et al. 1994
DEoDED	yes	Induction	N.A.	Rainbow trout	4 d	"
O(DC)EoDD	yes	Induction	N.A.	Rainbow trout	4 d	"
Unbl. KME	yes	Induction	N.A.	Rainbow trout	4 d	"
Black liquor	none	Induction	N.A.	Rainbow trout	4 d	"
TMP	yes	No effect	N.A.	Rainbow trout	4 d	"
CTMP	yes	No effect	N.A.	Rainbow trout	4 d	"
TCF	yes	Induction	N.A.	Rainbow trout	4 d	"
Unbl. KME	yes	Inhibition	N.A.	Rainbow trout	18 d	Huuskonen 1994
TMP	yes	No effect	Inhibition	Rainbow trout	8 wk	Lehtinen et al. 1993
Sulfite mill extract	none	Inhibition	N.A.	Primary cell cult.	h	Pesonen 1992
KME mill extract	none	Inhibition	N.A.	"	h	Pesonen 1992
BKME/KME "	yes	Inhibition	N.A.	"	h	Hänninen et al. 1991
Simulated KME	none	N.A.	Inhibition	Rainbow trout	11 d	Oikari and Nakari 1982
QeZ(Eop)ZEP	none	Inhibition	No effect	Rainbow trout	8 wk	Sangfors et al. 1994
"	yes	Inhibition	No effect	Rainbow trout	8 wk	"
D(Eop)DED	none	No effect	No effect	Rainbow trout	8 wk	"
"	yes	No effect	No effect	Rainbow trout	8 wk	"

C = chlorine gas; D = chlorine dioxide; O = oxygen stage; Z = ozone; Eop = alkaline stage + oxygen and peroxide; Qe = enzyme and chelating agent; TMP = thermomechanical pulp; CTMP = chemi-thermomechanical pulp; TCF = totally chlorine free; KME = kraft mill effluent.

Table 2. Hepatic phase I EROD and phase II UDP-GT enzyme responses in fish liver or liver cell cultures exposed to various compounds in pulp mill effluents.

Compound	EROD	UDP-GT	Species	Duration of Exp.	Reference
DHAA[1)]	N.A.	Inhibition	Rainbow trout	80 d	Tana 1988
Trichlorophenol	N.A.	Induction	Rainbow trout	80 d	"
DHAA/TCP[2)]	N.A.	Inhib./induction	Rainbow trout	80 d	"
DHAA	Inhibition	N.A	Primary cell cult.	h	Pesonen 1992
Dichloroveratrole	No effect	No effect	Rainbow trout	Intraperitoneal	Förlin et al. 1989
Trichloroguaiacol	No effect	No effect	Rainbow trout	"	"
Trichloroveratrole	No effect	No effect	Rainbow trout	"	"
Tetrachloroveratrole	No effect	No effect	Rainbow trout	"	"
Tetrachloroguaiacol	No effect	No effect	Rainbow trout	"	"
Phytosterols	No effect	Inhibition	Rainbow trout	8 wk via water	Lehtinen et al. 1993
Phytosterols	No effect	Inhibition	Rainbow trout	8 wk via food	"

1) DHAA = dehydroabietic acid.
2) TCP = trichlorophenol.
N.A. = not analyzed.

tetrachloroveratrole and tetrachloroguaiacol did not induce phase I enzyme activities except for weak induction at high doses (>100 mg) of tetrachloroveratrole (Förlin et al. 1989) (Table 2). No effects were observed on phase II enzyme activities except for high doses of 4,5-dichloroveratrole which inhibited UDP-GT and glutathione transferase in the same study.

Tana (1988) reported inhibition of the transformation enzyme UDP-GT activity in rainbow trout exposed to 5 µg L^{-1} dehydroabietic acid (DHAA) for 60 d. On the other hand, trichlorophenol at the same concentration stimulated the UDP-GT activity. When added together the two compounds caused an oscillating stimulatory and inhibitory reaction of the UDP-GT enzyme (Table 2).

IN VIVO EXPERIMENTS IN THE LABORATORY

Pesonen and Andersson (1992) studied primary cell cultures of rainbow trout hepatocytes and different kinds of pulp mill effluent extracts. It was observed that unbleached sulfite pulp mill effluent more strongly inhibited the EROD enzyme activity than a bleached kraft pulp effluent. It was observed that inhibition occurred after a stimulatory phase. The unbleached sulfite effluent extract caused markedly increased lactate dehydrogenase (LDH) and disruption of the cell monolayer at 1-2 µL mL^{-1} concentrations, indicating cellular damage as the reason behind EROD inhibition. No chemical characterization of the effluent extracts was reported.

Hänninen et al. (1991) found that effluent extract from unbleached kraft pulp production abolished the EROD activity at concentrations higher than 1 µL L^{-1}, whereas extract from bleached pulp production partially inhibited the EROD activity at the same concentration or at higher concentrations tested.

Pesonen and Andersson (1992) tested DHAA on primary cell cultures of rainbow trout hepatocytes. They found that this resin acid decreased the EROD activity at concentrations between 0.1-40 µg mL^{-1}. Leakage of LDH or reduction of cellular glutathione occurred only at 20-40 µg mL^{-1}, however.

In a study by Råbergh et al. (1992) the cytotoxic action on isolated rainbow trout hepatocytes by resin acids was studied. Exposure to dehydroabietic and isopimaric acid inhibited bile acid uptake, confirming that resin acids cause impaired liver function in fish and can be major contributors to pulp mill effluent toxicity. The resin acid concentrations used by Råbergh et al. (1992) were much higher, 30-97 mg L^{-1}, than those used by Oikari and Nakari (1982) in a 3-11 d experiment with rainbow trout and simulated unbleached kraft mill effluent (70-150 µg L^{-1}). However, the isolated cell culture test serves as a tool to study principal mechanisms of toxic compounds present in pulp mill effluents.

LONG-TERM EXPOSURES IN MESOCOSMS

Untreated and pilot activated sludge treated ECF and TCF whole mill effluents from production of hardwood pulp were studied using freshwater mesocosms (Sangfors et al. 1994). Significant reduction of the hepatic EROD activity occurred in rainbow trout exposed to untreated and treated effluent (0.05 and 0.25% v/v) from TCF pulp production (Table 1). However, the EROD activity in rainbow trout exposed to untreated and treated effluent (0.05 and 0.25% v/v) from production of ECF bleached pulp from the same mill did not differ from control fish. No correlation between conjugated compounds in bile and the EROD inhibition was noted.

Lehtinen et al. (1993) obtained significantly inhibited UDP-GT activity in fish exposed to treated effluent from production of TMP and newsprint after an 8 wk exposure period to 0.05% v/v effluent in a mesocosm experiment (Table 1). However, 0.25% v/v effluent did not cause significant inhibition. Such negative dose-response relationships have previously been observed by Lehtinen (1990). Time-dependent stimulation and inhibition of the UDP-GT activity in rainbow trout simultaneously exposed to DHAA and trichlorophenol was also observed by Tana (1988) (Table 2). Such responses may be caused by compounds with different mode of action at the subcellular level so that some compounds such as trichlorophenol stimulate conjugation processes, whereas DHAA damages membranes and partly inhibits enzyme activities (Oikari and Nakari 1982).

Biomarker effects have been observed with several wood-related compounds when these compounds were extracted and tested independently. Phytosterols, mainly ß-sitosterol, inhibited the UDP-GT activity when administered via both food and water (Lehtinen et al. 1993) (Table 2). Phytosterols are compounds present in hardwood. Sitosterol is very much like cholesterol, which, among other things, serves as a substrate for steroid hormone production. A hormonal effect of ß-sitosterol was demonstrated by Denton et al. (1985) and Krotzer (1990) who reported masculinization of female mosquitofish (*Gambusia affinis*) exposed to phytosterols together with a bacterium, which metabolizes plant sterols to steroids. The effects and consequences of these compounds on fish are presently not well known.

Biochemical Responses Other than Enzymatic Detoxification in Fish in Mesocosm and Long-Term Laboratory Studies

In a series of mesocosm tests during 1989-1993, fish studies were conducted with a number of different effluents (Table 3) (Lehtinen et al. 1992, 1993; Tana et al. 1994; Sangfors et al. 1994). Exposure to untreated bleach plant effluents from conventional chlorine bleaching (CEHDED), effluents with higher substitution of elementary chlorine combined with oxygen reinforced peroxide stage [(C60+D40)(Eo)D] and effluents from the chlorine gas-free sequence [OPD (O = oxygen, P = peroxide, D = chlorine dioxide)] caused increased liver glycogen values in fish. The increase in liver glycogen was accompanied by increased or unaltered hepatic EROD activity (Tana et al. 1994). Higher liver glycogen values in exposed fish were also noted after exposure to untreated and treated effluents from production of hardwood [bleaching sequences (D80+C20)(Eop)DED and O(D27,C68+D5)(Eop)D(Ep)D] or pure phytosterols (Lehtinen et al. 1992). On the other hand, Lehtinen et al. (1993) reported decreased glycogen values in fish exposed to phytosterols via food and water. The decreased liver glycogen was accompanied by significantly decreased hepatic UDP-GT activity and decreased liver-somatic indices (LSI) after an 8 wk exposure. Exposure to phytosterols via water also increased the oxygen consumption in the fish by 90% (measured as rate of oxygen consumption under static conditions) as compared with control fish. Tana (1988) reported decreased liver glycogen values in fish exposed to DHAA for 80 d (Table 3).

Table 3. Liver glycogen levels and hepatic phase I (EROD) and phase II (UDP-GT) enzyme responses in fish exposed to different pulp mill effluents, phytosterols and DHAA.

Process/Subst.	Treatment	Liver Glycogen	EROD	UDP-GT	Duration of Exp.	Reference
CEHDED	none	Increased	Induction	No effect	8 wk	Tana et al. 1994
O(C85D15)(Eo)DED	none	No effect	No effect	No effect	8 wk	"
O(C60+D40(Eo)D +(C60+D40)(Eo)DED	none	Increased	No effect	No effect	8 wk	"
OPD	none	Increased	No effect	No effect	8 wk	"
(D80+C20)(Eop)DED	none	Increased	No effect	No effect	8 wk	Lehtinen et al. 1992
"	yes	Increased	No effect	No effect	8 wk	"
O(D23,C68+D5)(Eop)D(Ep)D	none	Increased	No effect	No effect	8 wk	"
"	yes	Increased	No effect	No effect	8 wk	"
Phytosterols	none	Decreased	No effect	Inhibition	8 wk/water	Lehtinen et al. 1993
Phytosterols	none	Decreased	No effect	Inhibition	8 wk/food	"
DHAA	none	Decreased	N.A.	Inhibition	80 d	Tana 1988
(C84+D16)(Eo)DED	yes	No effect	No effect	No effect	8 wk	Lehtinen et al. 1993
QZ(Eop)ZEP	none	Decreased	Inhibition	No effect	8 wk	Sangfors et al. 1994
"	yes	Decreased	Inhibition	No effect	8 wk	"
D(Eop)DED	none	Decreased	No effect	No effect	8 wk	"
"	yes	No effect	No effect	No effect	8 wk	"

Exposure to a treated effluent from production of bleached softwood [(C84+D16)(EO)(DED)] did not cause significant liver glycogen decreases after the same period, nor did the exposure cause inhibition of UDP-GT. Furthermore, Sangfors et al. (1994) observed decreased liver glycogen contents in fish exposed to untreated and treated whole mill effluent from production of hardwood TCF bleached pulp. In contrast, only untreated effluent from production of hardwood ECF bleached pulp caused a decrease in glycogen of fish exposed for 8 wk in freshwater mesocosms. The low liver glycogen in the TCF effluent-exposed fish was accompanied by lower LSI values, whereas fish exposed to the ECF effluents lacked correlation between liver glycogen levels and LSI. Moreover, the low liver glycogen values in TCF effluent-exposed fish were accompanied by lower EROD values.

From the data above, the tentative interpretation is that low liver glycogen values were often accompanied by enzyme inhibition. Increased or non-altered values to control fish seemed to be accompanied by unaltered or induced enzyme activities. There is no apparent correlation to bleaching with chlorine-containing chemicals and both phytosterols and DHAA alone are capable of producing similar responses as whole mill effluents.

Bengtsson et al. (1989) exposed juvenile fourhorn sculpin (*Myoxocephalus quadricornis*) and bleak (*Alburnus alburnus*) to untreated whole mill effluents from production of A) unbleached hardwood, B) bleached hardwood with 100% chlorine dioxide, C) bleached softwood with 48% chlorine dioxide in the CD stage and finally D) externally treated whole mill effluent from softwood pulp production with a conventional bleaching sequence with chlorine gas as bleaching medium [C84+D16]. The authors stressed a possible link between competition for ascorbic acid for detoxification processes and ascorbic acid dependent cartilage synthesis.

The fourhorn sculpin exposure to (A) caused changes in 8 out of 13 parameters related to vertebral cartilage strength and biochemical composition, bleaching with 100% chlorine dioxide (B) caused changes in 6 parameters, softwood pulp bleached with 48% chlorine dioxide (C) caused changes in 9 and treated conventionally bleached softwood pulp (D) caused changes in 2 parameters (Table 4). These data suggest that digestion products from hardwood pulp production produce substances acting on vertebral structure and biochemical composition and further that bleaching with chlorine dioxide may partly destroy such compounds or that they do not enter the effluent from the bleach plant. It should be kept in mind however, that the effluents originated from two separate mills and therefore cannot be fully compared. The results from exposure to softwood pulp cannot be compared directly with the results from the experiment with hardwood since no unbleached alternative was tested. However, external waste treatment in connection with the process but with high elementary chlorine in the bleach plant largely eliminated compounds producing vertebral anomalies.

Table 4. Number of statistically significantly deviating vertebral characteristics in fourhorn sculpin, M. quadricornis, and bleak, A. alburnus, exposed for 4.5 months to different pulp mill effluents (Bengtsson et al. 1989). Total number of parameters studied = 13. Percentage given in parentheses.

	Unbleached Hardwood Untreated	Bleached Hardwood D100 Untreated	Bleached Softwood D48 Untreated	Bleached Softwood D16 External treatment
Fourhorn sculpin	8 (62)	6 (46)	9 (69)	2 (15)
Bleak	4 (31)	3 (23)	7 (54)	1 (8)

The effects observed in bleak were fewer but in the same direction: (A) produced effects in a higher number of parameters than (B), (C) produced the highest number and (D) the fewest.

DISCUSSION

The crucial question with biomarkers is whether a "response" is a negative "effect" on vital processes within the individual. If so, will the number of individuals responding be high enough to affect the population? At their best, biomarkers such as EROD and UDP-GT are considered to be sensitive warning signals of stressors and should provide quantifiable measures of an organism's response to either natural or anthropogenic stressors. Preferably a biomarker response should also relate to exposure of substances present in an effluent. The question arising in this connection is whether induction of biomarkers such as EROD and UDP-GT and effects on other metabolic parameters in the "historical" perspective were related to bleaching with chlorine. Undoubtedly effects occurred and still do occur in bleached kraft mill receiving waters, and effects are also observed in laboratory experiments (Robinson et al. 1994; Munkittrick et al. 1994; Servos et al. 1994). The effects on enzyme biomarkers occur, however, both when the pollutant originates from chemical (sulfate, sulfite) or thermomechanical and (treated/untreated) bleached or unbleached pulp (Lindström-Seppä et al. 1992; Pesonen and Andersson 1992; Gagne and Blaise 1993; Servizi et al. 1993; Huuskonen 1994). Another fact making conclusions on the severity of enzyme responses difficult is the varying direction of the response (i.e., induction or inhibition) and the interpretation of the responses (Lehtinen 1990). From the work by Råbergh et al. (1992), Pesonen and Andersson (1992), Tana (1988), Oikari et al. (1983) and Oikari and Nakari (1982), strong evidence is given that extractives such as resin acids are cytotoxic and enzyme inhibitors and that such responses would be expected in fish exposed to effluents containing dominating levels of extractives. When it comes to specific chlorinated compounds such as chlorophenolics it seems that trichlorophenol would be stimulatory on UDP-GT (Tana 1988). The oscillatory response of UDP-GT when trichlorophenol and DHAA were added together by Tana (1988) clearly indicates an antagonistic mode of behavior of these substances. On the other hand, tri- and tetrachloroguaiacol and some chlorinated veratroles did not stimulate EROD or UDP-GT (Förlin et al. 1989). The work by Pesonen and Andersson (1992) shows that low levels of a complex pollutant may induce hepatic MFO enzymes. A successive inhibition is reached with increasing pollutant concentrations. The problem arising here is that a level where the enzyme activity is similar to the control is sometimes reached and may cause a false negative result (Jimenez et al. 1990). Jimenez et al. (1990) tested this hypothesis on fish by using a known hepatotoxin and a subsequent challenge of the fish with benzo(a)pyrene. The administration of the hepatotoxin significantly reduced the ability of the liver to induce EROD activity as a result of B(a)P exposure.

The observations by Jimenez et al. (1990) are highly important regarding studies on pulp mill effluents, both in the laboratory and in the receiving waters. Low EROD or any other enzyme activity in exposed fish may indicate two things: absence of inducing substances or hepatotoxic levels of contaminants. Consequently, comparison of several biomarker responses is needed to provide information needed to correctly interpret MFO responses. Such biomarkers would include general metabolic ones reflecting the nutritional status (liver glycogen, liver lipids, etc.), histopathology and serum enzyme levels (Lehtinen et al. 1992, 1993; Johnsen et al. 1994) and biomarkers related to reproduction (Munkittrick et al. 1992). Finally, biochemical tests must have a defined linkage to the level of exposure in order to be useful and need to be associated with particular chemicals or toxicity syndromes. Otherwise, the significance of any observation will be unknown and useless for directing technical changes in mill operations.

In conclusion, the material reviewed here gives no strong evidence that chlorine bleaching or bleaching *per se* would have been, or is, the main reason behind effects observed. Both Canadian initial comparisons of ECF and TCF and the work by Sangfors et al. (1994) are also indications that bleaching procedures as such are not necessarily the main effect-causing process of pulp manufacture. The production of pulp and the composition of the effluents involves a multitude of factors which together act on the effect picture starting from a) the wood species and storage in the forest, b) type of pulping process, c) degree of closure and handling of condensates, d) black liquor carry-over to the bleach plant, e) type of effluent treatment, etc. It may be expected that mills with well managed processes, adequate spill control and external treatment would cause low toxic environmental impact regardless of whether or not chlorine-containing chemicals are used in the bleach plant.

REFERENCES

Ahokas, J.T., N.T. Kärki, A. Oikari and A. Soivio. 1976. Mixed function monooxygenase of fish as an indicator of pollution of aquatic environment by industrial effluent. Bull. Environ. Contam. Toxicol. 3:270-274.

Andersson, T., L. Förlin, J. Härdig and Å. Larsson. 1989. Physiological disturbances in fish living in coastal water polluted with bleached kraft pulp mill effluents. Can. J. Fish. Aquat. Sci. 45:1525-1536.

Bengtsson, B-E., Å. Bengtsson and U. Tjärnlund. 1989. Effects of pulp mill effluents on vertebrae of fourhorn sculpin, *Myoxocephalus quadricornis*, bleak, *Alburnus alburnus* and perch, *Perca fluviatilis*. Arch. Environ. Contam. Toxicol. 15:62-71.

Castrén, M., and O. Oikari. 1987. Changes of the liver UDP-glucoronosyltransferase activity in trout (*Salmo gairdneri* Rich.) acutely exposed to selected aquatic toxicants. Comp. Biochem. Physiol. 86C:357-360.

Denton, T.E., W.M. Howell, J.J. Allison, J. McCollum and B. Marks. 1985. Masculinization of female mosquitofish by exposure to plant sterols and *Mycobacterium smegmatis*. Bull. Environ. Contam. Toxicol. 35:627-632.

Förlin, L., T. Andersson and C.A. Wachtmeister. 1989. Hepatic microsomal 4,5,6-trichloroguaiacol glucoronidation in five species of fish. Comp. Biochem. Physiol. 3:653-656.

Gagne, F., and C. Blaise. 1993. Hepatic metallothionein level and mixed function oxidase activity in fingerling rainbow trout (*Oncorhynchus mykiss*) after acute exposure to pulp and paper mill effluents. Water Res. 27:1669-1682.

Huuskonen, S. 1994. Effects of effluents from unbleached pulp production on xenobiotic metabolism in the perch and rainbow trout. MSc thesis. Univ. of Kuopio, Kuopio, Finland. 73 p. (in Finnish)

Hänninen, O., P. Lindström-Seppä, M. Pesonen, S. Huuskonen and P. Muona. 1991. Use of biotransformation activity in fish and fish hepatocytes in the monitoring of aquatic pollution caused by pulp industry. *In:* Bioindicators and Environmental Management. Academic Press. pp. 13-20.

Jimenez, B.D., A. Oikari, S.M. Adams, D.E. Hinton and J.F. McCarthy. 1990. Hepatic enzymes as biomarkers: Interpreting the effects of environmental physiological and toxicological variables. Lewis 123-142.

Johnsen, K., K. Mattsson, J. Tana, T.R. Stuthridge, J. Hemming and K-J. Lehtinen. 1994. Uptake and elimination of resin acids and physiological responses in rainbow trout exposed to total mill effluent from an integrated newsprint mill. Environ. Toxicol. Chem.: In press.

Krotzer, J. 1990. The effects of induced masculinization on reproductive and aggressive behaviours of the female mosquitofish, *Gambusia affinis affinis*. Env. Biol. Fish. 29:127-134.

Lehtinen, K-J. 1990. Mixed function oxygenase enzyme responses and physiological disorders in fish exposed to kraft pulp-mill effluents: A hypothetical model. Ambio 5:259-265.

Lehtinen, K-J., J. Tana, J. Härdig, K. Mattsson, J. Hemming and P. Lindström-Seppä. 1992. Effects on survival, growth, parasites and physiological status in fish exposed in mesocosms to effluents from bleached hardwood kraft pulp production. Publ. Finnish Bd. Water Env. Ser. A 105:54 p.

Lehtinen, K-J., J. Tana, K. Mattsson, J. Härdig and J. Hemming. 1993. Physiological responses and effects on survival, growth and parasite frequency in fish exposed in mesocosms to treated total mill effluents from production of kraft pulp (BKME), thermomechanical pulp and phytosterols. Publ. Finnish Bd. Water Env. Ser. A 133:64 p.

Lindström-Seppä, P. 1990. Biotransformation in fish: Monitoring inland water pollution caused by pulp and paper mill effluents. PhD thesis. Orig. Rep. Univ. of Kuopio 8:69 p. Kuopio, Finland.

Lindström-Seppä, P., S. Huuskonen, M. Pesonen, P. Muona and O. Hänninen. 1992. Unbleached pulp mill effluents affect cytochrome P450 monooxygenase enzyme activities. Mar. Env. Res. 34:157-161.

Martel, P.H., T.G. Kovacs, B.I. O'Connor and R.H. Voss. 1994. A survey of pulp and paper mill effluents for their potential to induce mixed function oxidase enzyme activity in fish. Water Res. 28:1833-1844.

McLeay, D. 1987. Aquatic toxicity of pulp and paper mill effluent: A review. Env. Prot. Ser. Rep. EPS 4/PF/1. 191 p.

Munkittrick, K.R., M.E., McMaster, C.B. Portt, G.J. Van Der Kraak, I.R. Smith and D.G. Dixon. 1992. Changes in maturity, plasma sex steroids levels, hepatic mixed-function oxygenase activity, and the presence of external lesions in lake whitefish (*Coregonus clupeaformis*) exposed to bleached kraft mill effluent. Can. J. Fish. Aquat. Sci. 49:1560-1569.

Munkittrick, K.R., G.J. Van der Kraak, M.E. McMaster, C.B. Portt, M.R. Van den Heuvel and M.R. Servos. 1994. Survey of receiving-water environmental impacts associated with discharges from pulp mills. 2. Gonad size, liver size, hepatic EROD activity and plasma sex steroid levels in white suckers. Environ. Toxicol. Chem. 7:1089-1101.

Oikari, A., and T. Nakari. 1982. Kraft pulp mill effluent cause liver dysfunction in trout. Bull. Environ. Contam. Toxicol. 28:266-270.

Oikari, A., B-E. Lönn, M. Castrén, T. Nakari, B. Snickars-Nikinmaa, H. Bister and E. Virtanen. 1983. Toxicological effects of dehydroabietic acid (DHAA) on the trout, *Salmo gairdneri* Richardson, in fresh water. Water Res. 17:81-89.

Oikari, A., T. Nakari and B. Holmbom. 1984. Sublethal actions of simulated kraft pulp mill effluents (KME) in *Salmo gairdneri*: Residues of toxicants, and effects on blood and liver. Ann. Zool. Fennici 21:45-53.

Oikari, A., M. Nikinmaa, S. Lindgren and B-E. Lönn. 1985. Sublethal effects of simulated pulp mill effluents on the respiration and energy metabolism of rainbow *Salmo gairdneri*. Ecotoxicol. Environ. Safety 9:378-384.

Owens, J.W. 1991. The hazard assessment of pulp and paper effluents in the aquatic environment: A review. Environ. Toxicol. Chem. 10:1511-1540.

Pesonen, M. 1992. Xenobiotic metabolizing enzymes in rainbow trout (*Oncorhynchus mykiss*) kidney and liver characterization and regulation by xenobiotics. PhD thesis. Univ. Gothenburg, Sweden. 39 p.

Pesonen, M., and T. Andersson. 1992. Toxic effects of bleached and unbleached paper mill effluents in primary cultures of rainbow trout hepatocytes. Ecotoxicol. Environ. Safety 24:63-71.

Robinson, R.D., J.H. Carey, K.R. Solomon, I.R. Smith, M.R. Servos and K.R. Munkittrick. 1994. Survey of receiving-water environmental impacts associated with discharges from pulp mills. 1. Mill characteristics, receiving water chemical profiles and lab toxicity tests. Environ. Toxicol. Chem. 7:1075-1088.

Råbergh, C.M.I., B. Isomaa and J.E. Eriksson. 1992. The resin acids dehydroabietic acid and isopimaric acid inhibit bile acid uptake and perturb potassium transport in isolated hepatocytes from rainbow trout (*Oncorhynchus mykiss*). Aquat. Toxicol. 23:169-180.

Sangfors, O., J. Tana, J. Härdig and C. Grotell. 1994. Effects of pilot treated bleach plant effluents in mesocosms. Publ. Finnish Bd. Water Env. Ser. A (in Finnish, English summary: In Press.

Servizi, J.A., R.W. Gordon, D.W. Martens, W.L. Lockhart, D.A. Metner, I.H. Rogers, J.R. McBride and R.J. Norstrom. 1993. Effects of biotreated bleached kraft mill effluent on fingerling chinook salmon (*Oncorhynchus tsawytscha*). Can. J. Fish. Aquat. Sci. 50:846-857.

Servos, M.R., S.Y. Huestis, D.M. Whittle, G.J. Van der Kraak and K.R. Munkittrick. 1994. Survey of receiving-water environmental impacts associated with discharges from pulp mills. 3. Polychlorinated dioxins and furans in muscle and liver of white sucker (*Catostomus commersoni*). Environ. Toxicol. Chem. 7:1103-1115.

Tana, J. 1988. Sublethal effects of chlorinated phenols and resin acids on rainbow trout (*Salmo gairdneri*). Wat. Sci. 2:77-85.

Tana, J., A. Rosemarin, K-J. Lehtinen, J. Härdig, O. Grahn and L. Landner. 1994. Assessing impacts on Baltic coastal ecosystems with mesocosm and fish biomarker tests: A comparison of new and old wood pulp bleaching technologies. Sci. Tot. Environ. 145:213-234.

EVALUATION OF BIOMARKERS OF EXPOSURE TO 2-CHLOROSYRINGALDEHYDE AND SIMULATED ECF BLEACHED EUCALYPT PULP EFFLUENT

C.M. Brumley, V.S. Haritos, J.T. Ahokas and D.A. Holdway

Key Centre for Applied and Nutritional Toxicology, RMIT-University, GPO Box 2476V,
Melbourne, 3001 Australia

Substitution of chlorine dioxide for molecular chlorine as a bleaching agent in hardwood pulp mills has lowered the organochlorine content of effluents, but produces several chlorinated phenolics with unknown effects on the environment. This study examines the effects of 2-chlorosyringaldehyde (2-CSA), quantitatively the major chlorinated phenol found in elemental chlorine free (ECF) bleached kraft eucalypt pulp effluent, on a suite of biomarkers in sand flathead. Fish received an intraperitoneal injection of 2-CSA at doses of 0.15, 1.5, 15 or 75 mg kg^{-1}. After 4 d, no changes with dose were found in the activities of the detoxification enzymes ethoxycoumarin-o-deethylase (ECOD), ethoxyresorufin-o-deethylase (EROD) or UDP-glucuronosyltransferase (UDPGT). Cytochrome P450 content and serum sorbitol dehydrogenase (SSDH) activity were also unchanged with dose. A subsequent experiment applied the same suite of biomarkers to sand flathead exposed to simulated ECF bleached eucalypt pulp effluent, at concentrations of 0.5, 2 and 8% (v/v). Similarly, there were no significant differences between effluent-exposed and control fish for ECOD, EROD, P450, UDPGT or SSDH. Liver somatic index was also not significantly different between exposed and control fish. Results indicate that liver detoxification enzyme activities, which are widely used biomarkers of pulp mill effluent exposure, may not be useful biomarkers for effluents from mills using ECF bleaching.

INTRODUCTION

In the late 1980s, the Commonwealth Government of Australia recognized that the development of bleached eucalypt kraft pulp mills would add significant value to eucalypt pulp wood, but that such development could have serious environmental implications. To address these issues, environmental guidelines for new bleached eucalypt kraft pulp mills were released in 1989 (Commonwealth of Australia 1989). The main objective of the guidelines was to ensure environmental protection from emissions generated by bleached eucalypt kraft pulp mills. To achieve this end, a suite of biochemical and physiological indicators of exposure ("biomarkers") were included in the guidelines to be measured in both baseline and ongoing environmental monitoring programs. One of these biomarkers was the measurement of the hepatic mixed function oxidase (MFO) system in an Australian marine fish. MFOs facilitate the excretion of lipophilic xenobiotics by incorporating an atom of molecular oxygen into the lipophilic substrate, thus producing a more water-soluble product. The activities of MFOs in teleosts have been used extensively to monitor pulp mill effluent exposure in North America and Scandinavia (Lindström-Seppä and Oikari 1988; Munkittrick et al. 1992).

Before the environmental impacts of bleached kraft eucalypt pulp effluent could be effectively monitored, two issues needed to be addressed. First, no native Australian marine fish had been validated for use as a bioindicator organism of pollution exposure, and second, the constituents of bleached mature eucalypt pulp effluent were not known, limiting the expansion of an appropriate biomarker suite.

The sand flathead (*Platycephalus bassensis*) represents a possible fish species for monitoring elemental chlorine free (ECF) bleached eucalypt pulp effluent exposure, as it is demersal, relatively non-migratory and common to south-eastern Australian waters (Dix and Martin 1975). In addition, the MFO system in sand flathead has already been shown to be a sensitive biomarker of exposure to industrial and urban pollution (Holdway *et al.* 1994). A study by Brumley *et al.* (1995) measured a suite of biomarkers in sand flathead that had been exposed by intraperitoneal (i.p.) injection to the model inducing agent, Aroclor 1254. MFO activity, as measured by ethoxyresorufin-*o*-deethylase (EROD), and the activity of the conjugation enzyme UDP-glucuronosyltransferase (UDPGT) were both found to significantly increase with dose of Aroclor 1254. In addition, conjugated metabolites of polychlorinated biphenyls were found in the bile. As this study indicated that MFO and UDPGT activities in sand flathead could be increased in a dose-related manner with xenobiotic exposure, it was concluded that future use of the sand flathead for monitoring ECF bleached eucalypt pulp effluent was appropriate.

Research into the types and levels of chlorinated phenolics found in ECF bleached eucalypt pulp effluent indicates that 2-chlorosyringaldehyde (2-CSA, Fig. 1) is quantitatively the major chlorinated phenol produced by the ECF bleaching treatment of eucalypt pulp (Smith *et al.* 1993). Chlorinated syringaldehydes and syringols are found only in the effluents of bleached hardwood pulp, with the degree of chlorine substitution dependent on the amount of chlorine bleach used. Thus, 2,6-dichlorosyringaldehyde dominates when there is a high proportion of molecular chlorine used, and 2-CSA dominates when chlorine dioxide is used as the bleaching agent. Little is known of the environmental fate and effects of chlorinated syringaldehydes in aquatic systems; however, Stauber *et al.* (1994) have found that 2-CSA is far less acutely toxic to *Nitzschia closterium*, an Australian marine diatom, than more highly substituted phenols and guaiacols.

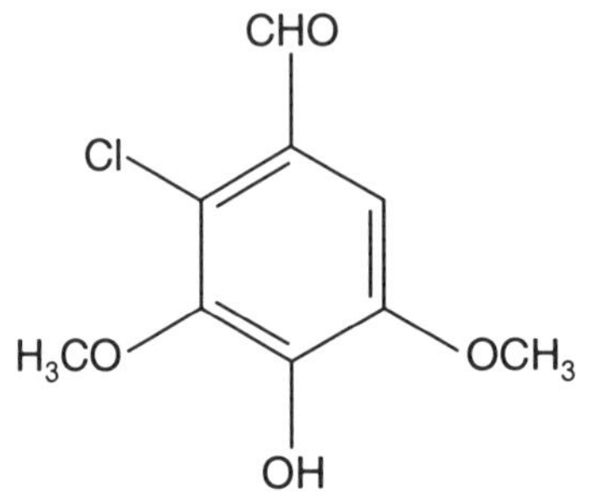

Figure 1. Structure of 2-chlorosyringaldehyde.

The aim of this study was to evaluate the responses of a suite of biomarkers in sand flathead exposed to simulated ECF bleached eucalypt pulp effluent. Biomarkers of exposure that had already been successfully used in sand flathead, MFOs, UDPGT and the blood enzyme serum sorbitol dehydrogenase (SSDH), were measured for change with dose after the fish were exposed by i.p. injection to 2-CSA. In a subsequent experiment, sand flathead were exposed to untreated, simulated ECF bleached eucalypt pulp effluent, and the effects on the above biomarkers were measured for change with effluent exposure concentration.

MATERIALS AND METHODS

Fish

Sand flathead were caught by hook and line from the northern end of the Loelia Channel in Port Phillip Bay, Australia in May and September 1993. All fish caught were of reproductive age. The mean ($\pm$ SD) length and weight of the fish were 303 ± 24 mm and 166 ± 48 g, respectively, in May ($n = 50$), and 297 ± 19 mm and 149 ± 33 g ($n = 72$) in September. In May, 33% of the fish were male; in

September, 19% of the fish were male. In both experiments, fish were acclimated for at least 8 d in the laboratory in flow-through seawater tanks prior to the exposure.

Effluent

Simulated bleachery effluent was supplied by the CSIRO Division of Forest Products Laboratories in Clayton, immediately prior to the start of the experiment. Mature *Eucalyptus* sp. woodchips were pulped using a kraft process and the resulting pulp oxygen delignified to kappa number 10.4.

The eucalypt kraft-oxygen pulp was bleached using a four- stage sequence involving 100% chlorine dioxide bleaching (Do), alkaline extraction with sodium hydroxide (EO), followed by two 100% chlorine dioxide bleach treatments (DD) [Do(EO)DD]. Effluents from each stage were obtained by washing and hand-squeezing the bleached pulp. On completion of the final bleaching stage, all effluents were combined to give a total volume of 5.0 L per 100 g oven-dried pulp. Stocks of effluent were stored in polyethylene drums in the dark at 4°C throughout the experiment.

Treatment

2-CSA Exposure

Fish were anesthetized in a solution of MS-222 (70 mg L^{-1}) prior to being given an i.p. injection of 2-CSA, which had been dissolved in DMSO (4% total volume) and then diluted in corn oil. A logarithmic dosing range of 0.15, 1.5, 15 and 75 mg kg^{-1} (at 1 mL kg^{-1}) was chosen, as the effects of this compound were not known. 2-CSA would not stay in solution at the proposed highest dose (150 mg kg^{-1}); thus half that value was chosen (75 mg kg^{-1}). Control fish received carrier alone. Four days after treatment, fish were anesthetized in MS222 to enable sampling to be carried out. Previous studies (Brumley *et al.* 1995; Brumley, unpublished data) have shown that MS222 anesthesia prior to sampling has no effect on sand flathead MFO activity, as measured by EROD and ethoxycoumarin-*o*-deethylase (ECOD) activities. Under anesthesia, a blood sample was removed by cardiac puncture. Fish were then killed by a blow to the head and the livers were removed and stored in liquid nitrogen until processed. The blood sample was centrifuged, and the resulting serum was stored in liquid nitrogen until processed.

Simulated ECF Bleached Eucalypt Pulp Effluent Exposure

Fish were held in 80 L aquaria (6 fish in each) with flow-through sea water at a rate of 61 ± 1.2 mL min^{-1}, giving a 50% molecular turnover in 15 h (Sprague 1969). Untreated simulated ECF bleached eucalypt pulp effluent was added to the aquaria via a Watson-Marlow peristaltic pump to give concentrations of 0.5, 2 and 8% (v/v). Fish that received sea water alone were used as controls. Each effluent/seawater concentration was made up daily in 340 L polyethylene tubs, which then fed via the pump to three aquaria of that concentration. Temperature, pH, dissolved oxygen, conductivity and ammonia concentration were measured twice a day throughout the exposure period; CO_2 concentration was measured twice a day for the last 2 d.

Four days after the exposure started, fish were immersed in an anesthetic solution of MS222 and a blood sample was removed by cardiac puncture. Fish were then killed by a blow to the head and the livers were removed to liquid nitrogen until processed.

Tissue Analyses

Livers were removed from liquid nitrogen, weighed while still frozen and processed to a microsomal suspension. Samples was assayed for ECOD activity, based on a modification of the method of Ullrich and Weber (1972), as described in detail in Brumley *et al.* (1995). EROD activity was assayed by a modification of the method of Burke and Mayer (1974), as described in detail in Brumley *et al.* (1995). Both ECOD and EROD activities were measured on a Hitachi F-4500 fluorescence spectrophotometer. UDPGT activity was measured by a spectrophotometric method on a Hitachi U2000 UV spectrophotometer using *p*-nitrophenol as the substrate (Castrén and Oikari 1983). Assay conditions of time, temperature, pH and substrate concentrations were previously optimized for sand flathead (Brumley *et al.* 1995). Cytochrome P450 content was analyzed by the dithionite difference spectrum of carbon monoxide bubbled samples according to the method of Greim *et al.* (1970), with modifications from Rutten *et al.* (1987) on a Shimadzu UV-3000 dual wavelength/double beam spectrophotometer. An extinction coefficient of 104 mM^{-1} cm^{-1} was used (Matsubara *et al.* 1976). The method is described in detail in Brumley *et al.* (1995). Protein content of the microsomes was determined by the method of Lowry *et al.* (1951) using bovine serum albumin as standard protein. SSDH activity, a marker of acute liver injury, was measured at 340 nm on a Hitachi U-2000 spectrophotometer using the Sigma Diagnostics kit 50-UV (Sigma, U.S.A.). Liver somatic index (LSI) was calculated according to liver weight (100)/fish weight.

Statistical Analyses

Data were assessed for normality and homogeneity of variance using partial residual plots. Non-normal data were log transformed. LSI values were arcsine transformed. Data were analyzed using one and two-way analyses of variance (ANOVA) using the software package SuperANOVA. A two-way ANOVA involving sex and dose with each interaction term was initially performed. If no differences existed between sexes, the data were pooled and a one-way ANOVA was run. When significant differences were found, Fisher's protected LSD test was used to distinguish the differences between the means.

RESULTS

2-CSA Exposure

As no differences between the sexes were detected for any of the biomarkers measured, sexes were pooled for subsequent analyses. Injection of 2-CSA into sand flathead had no significant effect ($p > 0.05$), compared with mean control levels, on the detoxification enzymes ECOD, EROD or UDPGT (Table 1). P450 content of the liver was also not significantly different between control and exposed fish, with mean control levels at 0.128 nmol mg^{-1} protein. SSDH activity of exposed fish was not significantly different from controls; however variability across all samples was high, making differences hard to detect statistically (Table 1).

Simulated ECF Bleached Eucalypt Pulp Effluent Exposure

As there were no differences between the sexes for any of the biomarkers measured, sexes were pooled for subsequent analyses. There were no significant differences between control and exposed fish for ECOD or UDPGT activity or P450 content (Table 2). EROD activity in exposed fish was not significantly different from controls; however both 0.5 and 8% effluent-exposed fish had significantly lower activity than the 2% effluent-exposed fish ($p = 0.024$) (Fig. 2).

LSI was not significantly different between control and exposed fish (Table 2). SSDH activity was not elevated in exposed over reference fish, with a mean control value of 17.3 mIU.

Table 1. Liver detoxification enzyme activities in sand flathead exposed by i.p. injection to 2-CSA in corn oil. Control fish received carrier alone. Values are expressed as mean (n, SE). There were no significant differences between values within columns, based on Fisher's protected LSD test ($p \leq 0.05$).

Dose 2-CSA mg kg^{-1}	ECOD nmol min^{-1} mg protein^{-1}	EROD nmol min^{-1} mg protein^{-1}	UDPGT U mg protein^{-1}	SSDH mIU
Control	0.087 (10, 0.012)	0.024 (10, 0.005)	0.546 (6, 0.133)	5.349 (9, 0.807)
0.15	0.079 (10, 0.009)	0.017 (10, 0.004)	0.860 (6, 0.142)	3.577 (10, 0.811)
1.5	0.113 (10, 0.017)	0.024 (10, 0.003)	0.862 (6, 0.035)	6.157 (10, 0.751)
15	0.095 (10, 0.011)	0.014 (10, 0.003)	0.787 (5, 0.035)	3.255 (9, 0.640)
75	0.098 (9, 0.010)	0.019 (9, 0.003)	0.788 (6, 0.162)	5.601 (9, 0.880)

ECOD = ethoxycoumarin-o-deethylase; EROD = ethoxyresorufin-o-deethylase; UDPGT = UDP-glucuronosyltransferase, SSDH = serum sorbitol dehydrogenase.

Table 2. ECOD and UDPGT activities, P450 content and LSI values of sand flathead exposed to simulated ECF bleached eucalypt pulp effluent. Control fish were exposed to sea water alone. Values are expressed as mean (n, SE). There were no significant differences between values within columns, based on Fisher's protected LSD test ($p \leq 0.05$).

Effluent concentration (v/v)	ECOD nmol min^{-1} mg protein^{-1}	UDPGT U mg protein^{-1}	P450 nmol mg protein^{-1}	LSI
Control	0.145 (17, 0.011)	1.519 (9, 0.112)	0.118 (17, 0.008)	0.869 (17, 0.043)
0.5%	0.130 (18, 0.015)	1.215 (9, 0.096)	0.102 (18, 0.008)	0.916 (18, 0.037)
2%	0.140 (18, 0.011)	1.227 (9, 0.136)	0.109 (18, 0.005)	0.953 (18, 0.063)
8%	0.125 (18, 0.017)	1.264 (8, 0.135)	0.101 (18, 0.009)	0.953 (18, 0.047)

ECOD = ethoxycoumarin-o-deethylase; UDPGT = UDP-glucuronosyltransferase; P450 = cytochrome P450; LSI = liver somatic index.

DISCUSSION

The i.p. injection of 2-CSA in sand flathead allowed the examination of effects of the major chlorinated phenolic of ECF bleached eucalypt pulp effluent, independent of the potential influences of other compounds in the effluent. The absence of effects of 2-CSA on either of the MFO enzymes measured in this study, or on total P450 content, is consistent with results from a study on 2-CSA metabolism by Haritos et al. (1995). In that study, bile samples were removed and analyzed from sand flathead exposed by i.p. injection to 2-CSA. Haritos et al. (1995) found that the primary fate of 2-CSA was reductive metabolism to the corresponding alcohol (2-chloro-4-hydroxy-3,5-dimethoxy-benzylalcohol; 2-CB-OH), followed by conjugation with glucuronic acid or sulfate. Some 2-CSA was conjugated without reduction, and a very small amount of 2-CSA was oxidized to 2-chloro-4-hydroxy-3,5-dimethoxybenzoic acid. These results indicated little role for the P450 system in the metabolism of 2-CSA. The formation of 2-CB-OH may be catalyzed by the enzyme family of aldehyde reductases; cytoplasmic enzymes concentrated in the liver and kidney of vertebrates of which aromatic aldehydes are model substrates (von Wartburg and Wermuth 1980).

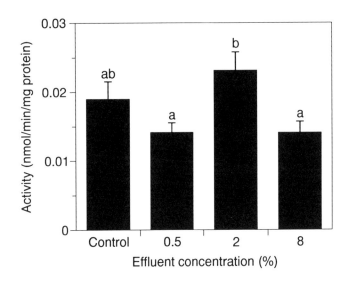

Figure 2. EROD activity in sand flathead exposed to simulated ECF bleached eucalypt pulp effluent at concentrations of 0.5, 2 and 8% (v/v). Control fish received sea water alone. Values are expressed as mean (± SE); bars without a letter in common are significantly different based on Fisher's protected LSD test ($p \leq 0.05$).

Although about 99% of metabolized 2-CSA excreted into the bile of sand flathead has been found to be conjugated with glucuronic acid or sulfate (Haritos et al. 1995), the present study found no increases in UDPGT activity, even at doses of 2-CSA as high as 75 mg kg^{-1}. Such high levels of 2-CSA exposure are unlikely to occur in the environment, suggesting that UDPGT activity cannot be used to monitor 2-CSA levels in ECF bleached eucalypt pulp effluent.

The measurement of increased levels of SSDH activity has been used to indicate hepatic injury in fish (Dixon et al. 1987; Ozretic and Krajnovic-Ozretic 1993). As SSDH activity in exposed fish in this study was not significantly elevated over control values, 2-CSA does not appear to be acutely hepatotoxic to sand flathead. The SSDH data also confirm that the absence of increases in MFO and UDPGT activity were not secondary effects due to liver injury. Holdway et al. (1994) have suggested that MFO activity in fish livers experiencing direct toxicity may not be increased to the same extent as in healthy livers. Thus a marker of liver toxicity such as SSDH should be used in conjunction when using MFOs as biomarkers of exposure.

The absence of any significant differences in MFO activity between male and female flathead in this study is not unexpected. Previous work involving the MFO activity measurement of large numbers of sand flathead (n > 760) consistently failed to find any sex differences in ECOD, while EROD was found to be significantly different during only one sampling period (Holdway et al. 1994).

The measurement of MFO activity in fish is perhaps the most widely used and sensitive biomarker of exposure of a wide range of pulp mill effluents, arising from different wood stocks and treatment processes (Lehtinen et al. 1990; Ahokas et al. 1994; Munkittrick et al. 1994). Therefore the absence of any effect of simulated ECF bleached eucalypt pulp effluent on sand flathead MFOs, even at concentrations of effluent as high as 8%, is somewhat unusual. The MFO system in sand flathead has previously been shown to be a sensitive biomarker of xenobiotic exposure in both field and laboratory studies (Brumley et al. 1995; Holdway et al. 1994); thus it appears that simulated ECF bleached eucalypt pulp effluent does not contain MFO-inducing compounds. Two factors may contribute to this property of simulated ECF bleached eucalypt pulp effluent. First, bleached hardwood pulps, such as eucalypts, produce effluents with a total organochlorine load several-fold lower than a softwood equivalent (Voss et al. 1980). Secondly, the increased substitution of chlorine dioxide for molecular chlorine as the bleaching agent of wood pulp reduces the amount of highly substituted organochlorines found in effluent (Kachi et al. 1980). Potent MFO-inducing compounds such as 2,3,7,8-tetrachloro-p-dibenzodioxin (TCDD) are found only in effluents bleached with a high percentage of molecular chlorine (Chung and Halliburton 1990). Lehtinen et al. (1990), in a study on six effluents with different amounts of chlorine bleach, found no MFO induction in rainbow trout exposed to the effluent with the highest amount of chlorine dioxide substitution.

The use of chlorine as a bleaching agent and the induction of fish MFOs is not, however, a simple relationship. Both Canadian (Martel et al. 1994) and Scandinavian groups (Lindström-Seppä et al. 1992; Pesonen and Andersson 1992) have found significant MFO induction in fish and fish liver preparations exposed to unbleached effluents of the kraft and sulfite processes. In recent studies Canadian researchers compared the effects of 12 mill effluents on a range of physiological parameters and found no relationship between the amount of chlorine used in bleaching and the degree of MFO induction (Munkittrick et al. 1994). Clear interpretations were difficult to make for the mill employing ECF bleaching, as neither the degree of MFO induction nor the levels of TCDD toxic equivalents in fish liver were reported (Munkittrick et al. 1994). Significant MFO induction was present at the mill employing 70% chlorine dioxide substitution, but there were also substantial levels of TCDD toxic equivalents present in the fish livers (Servos et al. 1994; van den Heuvel et al. 1994). As such high levels of TCDD toxic equivalents would significantly increase hepatic MFO activity, these results cannot be compared with ECF bleached eucalypt filtrate where the levels of 2,3,7,8-TCDD and 2,3,7,8-tetrachlorodibenzofuran were below detection limits (Nelson 1994). In order for the relationships between ECF bleaching and MFO induction to be fully understood, laboratory experiments using simulated effluent with increasing amounts of chlorine dioxide bleach will need to be carried out.

Simulated ECF bleached eucalypt pulp effluent exposure appeared to affect EROD activity very slightly in sand flathead, but as the data did not exhibit a dose-response relationship, it was difficult to interpret. There was no evidence of the several-fold increase in activity that would be considered induction and that has been shown in sand flathead in controlled laboratory experiments using a model inducing agent (Brumley et al. 1995).

UDPGT in fish has been used as a biomarker of pulp mill effluent exposure, but the enzyme does not appear to be as sensitive to increases in activity as MFOs (Förlin et al. 1985; Lindström-Seppä and Oikari 1990). As there was no induction of MFO activity in this study, increases in UDPGT activity would not be expected. Inhibition of UDPGT activity can occur without changes in MFOs and has been found in fish exposed in the field to pulp mill effluents (Oikari et al. 1985). Laboratory experiments have shown that resin acids are potent inhibitors of UDPGT activity (Oikari et al. 1984); however the effluent used in this study was produced from a eucalypt pulp which contains no resin acids. Inhibition of UDPGT activity would therefore not be expected.

The SSDH values for sand flathead exposed to simulated ECF bleached eucalypt pulp effluent indicated that no acute liver damage had occurred, and therefore the absence of MFO and UDPGT induction were not secondary effects due to liver injury. Hodson et al. (1991) found no changes in SSDH activity in white sucker (Catostomus commersoni) exposed in the field to primary treated effluent.

Although the flow rate in this experiment was not high and resulted in decreased water quality, fish health does not appear to have been compromised. SSDH activity, a marker of liver integrity, was not altered, and cytochrome P450 content was similarly unaffected. LSI, which can be used to indicate gross alterations in liver structure or osmoregulatory disturbances such as water accumulation (Matsoff and

Oikari 1987), was not significantly different between exposed and control fish, and there was no inhibition of detoxification enzyme activities, measured as EROD and ECOD.

2-CSA and its source, ECF bleached eucalypt pulp effluent, represent the results of improvements in bleaching technologies, which are designed to reduce the organochlorine load discharged from bleaching plants. None of the biomarkers measured in these studies responded to exposure to 2-CSA or simulated ECF bleached eucalypt pulp effluent in a dose-related manner. Even the hepatic MFO system, which has been found to be induced in fish found as far as 40 km from bleached pulp mill discharges (Balk *et al.* 1993), was unaffected. Only the measurement of the biliary metabolite 2-CB-OH has clearly indicated 2-CSA exposure in a dose-related manner (Haritos *et al.* 1995). Bile samples were removed from fish exposed to simulated ECF bleached eucalypt pulp effluent, and these are currently under analysis. Preliminary results confirm the presence of 2-CB-OH in the bile of sand flathead exposed to the highest dilution of effluent.

As cleaner pulping and bleaching technologies become standard for mills around the world, the types of biomarkers used to monitor bleached effluent exposure may need to change.

ACKNOWLEDGMENTS

This work was supported by the National Pulp Mills Research Program (Phase 2) and the Australian Research Council. The authors would like to thank Rod Watson and Sue Brennan for their technical assistance. C.M.B. is the recipient of an Australian Postgraduate Award.

REFERENCES

Ahokas, J.T., D.A. Holdway, S.E. Brennan, R.W. Goudey and H.B. Bibrowska. 1994. MFO activity in carp (*Cyprinus carpio*) exposed to treated pulp and paper mill effluent in Lake Coleman, Victoria, Australia, in relation to AOX, EOX, and muscle PCDD/PCDF. Environ. Toxicol. Chem. 13:41-50.

Balk, L., L. Förlin, M. Söderstrom and Å. Larsson. 1993. Indications of regional and large-scale biological effects caused by bleached pulp mill effluents. Chemosphere 27:631-650.

Brumley, C.M., V.S. Haritos, J.T. Ahokas and D.A. Holdway. 1995. Validation of sand flathead as a bioindicator species of marine pollution exposure. Aquat. Toxicol. 31:249-262.

Burke, M.D. and R.T. Mayer. 1974. Ethoxyresorufin: Direct fluorometric assay of a microsomal O-dealkylation which is preferentially inducible by 3-methylcholanthrene. Drug Metab. Dispos. 2:583-588.

Castrén, M. and A. Oikari. 1983. Optimal assay conditions for liver UDP-glucuronosyltransferase from the rainbow trout, *Salmo gairdneri*. Comp. Biochem. Physiol. 76C:365-369.

Chung, A. and D. Halliburton. 1990. Measures to control chlorinated dioxin and furan formation and releases at Canadian bleached chemical pulp mills. Chemosphere 20:1739-1746.

Commonwealth of Australia. 1989. Pulp and Paper Industry Package, December 1989. Prepared by the Departments of Industry, Technology and Commerce; Art, Sports, the Environment, Tourism and Territories; and Primary Industries and Energy; with technical assistance from CSIRO.

Dix, T.G. and A. Martin. 1975. Sand flathead (*Platycephalus bassensis*), an indicator species for mercury pollution in Tasmanian waters. Mar. Pollut. Bull. 6:142-143.

Dixon, D.G., P.V. Hodson and K.L.E. Kaiser. 1987. Serum sorbitol dehydrogenase activity as an indicator of chemically induced liver damage in rainbow trout. Environ. Toxicol. Chem. 6:685-696.

Förlin, L., T. Andersson, B.-E. Bengtsson, J. Härdig and Å. Larsson. 1985. Effects of pulp bleach plant effluents on hepatic xenobiotic biotransformation enzymes in fish: Laboratory and field studies. Mar. Environ. Res. 17:109-112.

Greim, H., J.B. Schenkman, M. Klotzbücher and H. Remmer. 1970. The influence of phenobarbital on the turnover of hepatic microsomal cytochrome b5 and cytochrome P-450 hemes in the rat. Biochim. Biophys. Acta 201:20-25.

Haritos, V.S., C.M. Brumley, D.A. Holdway and J.T. Ahokas. 1995. A unique metabolite of 2-chlorosyrigaldehyde in fish bile, a potential biomarker of exposure to bleached hardwood pulp effluent. Xenobiotica: In press.

Hodson, P.V., D. Bussieres, M.M. Gagnon, J.J. Dodson, C.M. Couillard and J.C. Carey. 1991. Review of biochemical, physiological, pathological and population responses of white sucker *(Catostomus commersoni)* to BKME in the St. Maurice River, Quebec. Proceedings of the Environmental Fate and Effects of Bleached Pulp Mill Effluents Conference, Nov. 19-21, 1991, Stockholm, Sweden. pp. 269.

Holdway, D.A., S.E. Brennan and J.T. Ahokas. 1994. Use of hepatic MFO and blood enzyme biomarkers in sand flathead *(Platycephalus bassensis)* as indicators of pollution in Port Phillip Bay, Australia. Mar. Pollut. Bull. 28:683-695.

Kachi, S., N. Yonese and Y. Yoneda. 1980. Identifying toxicity from bleached hardwood mills. Pulp Paper Can. 81:T287-T291.

Lehtinen, K-J., A. Kierkegaard, E. Jakobsson and A. Wändell. 1990. Physiological effects in fish exposed to effluents from mills with six different bleaching processes. Ecotox. Environ. Saf. 19:33-46.

Lindström-Seppä, P. and A. Oikari. 1988. Hepatic xenobiotic biotransformation in fishes exposed to pulp mill effluents. Water Sci. Technol. 20:167-170.

Lindström-Seppä, P. and A. Oikari. 1990. Biotransformation activities of feral fish in waters receiving bleached pulp mill effluents. Environ. Toxicol. Chem. 9:1416-1424.

Lindström-Seppä, P., S. Huuskonen, M. Pesonen, P. Muona and O. Hänninen. 1992. Unbleached pulp mill effluents affect cytochrome P450 monooxygenase enzyme activities. Mar. Environ. Res. 34:157-161.

Lowry, O.H., N.J. Rosebrough, A.L. Farr and R.J. Randall. 1951. Protein measurement with the Folin phenol reagent. J. Biol. Chem. 193:143-164.

Martel, P.H., T.G. Kovacs, B.I. O'Connor and R.H. Voss. 1994. A survey of pulp and paper mill effluents for their potential to induce mixed function oxidase enzyme activity in fish. Water Res. 28:1835-1844.

Matsoff, L. and A. Oikari. 1987. Acute hyperbilirubinaemia in rainbow trout *(Salmo gairdneri)* caused by resin acids. Comp. Biochem. Physiol. 88C:263-268.

Matsubara, T., M. Koike, A. Touchi, Y. Tochino and K. Sugeno. 1976. Quantitative determination of cytochrome P-450 in rat liver homogenate. Anal. Biochem. 75:596-603.

Munkittrick, K.R., G.J. Van Der Kraak, M.E. McMaster and C.B. Portt. 1992. Response of hepatic MFO activity and plasma sex steroids to secondary treatment of bleached kraft pulp mill effluent and mill shutdown. Environ. Toxicol. Chem. 11:1427-1439.

Munkittrick, K.R., G.J. Van Der Kraak, M.E. McMaster, C.B. Portt, M.R. van den Heuvel and M. Servos. 1994. Survey of receiving water environmental impacts associated with discharges from pulp mills. 2. Gonad size, liver size, hepatic MFO activity and plasma sex steroid levels in white sucker. Environ. Toxicol. Chem. 13:1089-1101.

Nelson, P.J. 1994. Dioxin measurements in relation to the Australian pulp and paper industry. National Pulp Mills Research Program Technical Report Series No. 6, August 1994.

Oikari, A., E. Ånäs, G. Kruzynski and B. Holmbom. 1984. Free and conjugated resin acids in the bile of rainbow trout, *Salmo gairdneri*. Bull. Environ. Contam. Toxicol. 33:233-240.

Oikari, A., B. Holmbom, E. Ånäs, M. Miilunpalo, G. Kruzynski and M. Castrén. 1985. Ecotoxicological aspects of pulp and paper mill effluents discharged to an inland water system: Distribution in water, and toxicant residues and physiological effects in caged fish *(Salmo gairdneri)*. Aquat. Toxicol. 6:219-239.

Ozretic, B. and M. Krajnovic-Ozretic. 1993. Plasma sorbitol dehydrogenase, glutamate dehydrogenase, and alkaline phosphatase as potential indicators of liver intoxication in grey mullet *(Mugil auratus* Risso). Bull. Environ. Contam. Toxicol. 50:586-592.

Pesonen, M. and T. Andersson. 1992. Toxic effects of bleached and unbleached paper mill effluents in primary cultures of rainbow trout hepatocytes. Ecotoxicol. Environ. Saf. 24:63-71.

Rutten, A.A.J.J.L., H.E. Falke, J.F. Catsburg, R. Topp, B.J. Blaauboer, I. van Holsteijn, L. Doorn and F.X.R. van Leeuwen. 1987. Interlaboratory comparison of total cytochrome P-450 and protein determinations in rat liver microsomes. Arch. Toxicol. 61:27-33.

Servos, M.R., S.Y. Huestis, D.M. Whittle, G.J. Van Der Kraak and K.R. Munkittrick. 1994. Survey of receiving-water environmental impacts associated with discharges from pulp mills. 3. Polychlorinated dioxins and furans in muscle and liver of white sucker *(Catostomus commersoni)*. Environ. Toxicol. Chem. 13:1103-1115.

Smith, T.J., R.H. Wearne and A.F.A. Wallis. 1993. Effect of wood sample and bleaching sequence on the levels of chlorinated phenols formed during bleaching of oxygen-delignified eucalypt kraft pulps. Proceedings of the 47th APPITA Annual General Conference, 1993, New Zealand. pp. 491-497.

Sprague, J.B. 1969. Measurement of pollutant toxicity to fish. I. Bioassay methods for acute toxicity. Water Res. 3:793-821.

Stauber, J., L. Gunthorpe, J.G. Deavin, B.L. Munday and M. Ahsanullah. 1994. Application of new marine bioassays for assessing toxicity of bleached eucalypt kraft mill effluents. Appita 47:472-476.

Ullrich, V. and P. Weber. 1972. The O-dealkylation of 7-ethoxycoumarin by liver microsomes. Hoppe-Seyler's Physiol. Chem. 353:1171-1177.

van den Heuvel, M.R., K.R. Munkittrick, G.J. Van Der Kraak, M.E. McMaster, C. Portt, M.S. Servos and D.G. Dixon. 1994. Survey of receiving-water environmental impacts associated with discharges from pulp mills. 4. Bioassay-derived 2,3,7,8-tetrachlorodibenzo-p-dioxin toxic equivalent concentration in white sucker *(Catostomus commersoni)* in relation to biochemical indicators of impact. Environ. Toxicol. Chem. 13:1117-1126.

von Wartburg, J.-P. and B. Wermuth. 1980. Aldehyde reductase, pp. 249-260. *In* W.B. Jakoby [ed.] Enzymatic basis of detoxification. Academic Press, New York.

Voss, R.H., J.T. Wearing, R.D. Mortimer, T. Kovacs and A. Wong. 1980. Chlorinated organics in kraft bleachery effluents. Paperi ja Puu 12:809-816.

RAINBOW TROUT (*ONCORHYNCHUS MYKISS*) MIXED FUNCTION OXYGENASE RESPONSES CAUSED BY UNBLEACHED AND BLEACHED PULP MILL EFFLUENTS: A LABORATORY-BASED STUDY

Todd G. Williams[1,2,3], John H. Carey[1], B. Kent Burnison[1],
D. George Dixon[2] and Hing-Biu Lee[1]

[1]National Water Research Institute, Environment Canada, 867 Lakeshore Road, Box 5050,
Burlington, Ontario, L7R 4A6 Canada
[2]Department of Biology, University of Waterloo, Waterloo, Ontario, N2L 3G1 Canada
[3]Present Address: T.G. Williams & Associates Ltd., 3333 New Street, Unit #37,
Burlington, Ontario, L7N 1N1 Canada

Exposures of rainbow trout (*Oncorhynchus mykiss*) to effluents from kraft pulp mills resulted in measurable increases in hepatic mixed function oxygenase (MFO) activity (induction) in all cases, regardless of bleaching or effluent treatment technology. Effluent concentration thresholds for MFO induction ranged from 0.33 to 9.1%. Comparison of chemical analyses and effluent concentration thresholds showed poor correlations between MFO induction and effluent concentrations of adsorbable organic halogen (AOX), dissolved organic carbon (DOC), chlorophenolics or resin and fatty acid concentrations. Mill comparisons of MFO induction thresholds normalized for flow and pulp production suggested that the bleaching process is a major contributor to MFO induction at older mills, but that other process sources can contribute to MFO induction at modern mills or mills with no bleaching. Induction at a mill not employing bleaching and the lack of a correlation between MFO induction and concentration of organochlorines, including AOX, suggest that regulatory control of AOX is not a satisfactory strategy to eliminate the MFO-inducing potential of pulp mill effluents.

INTRODUCTION

One of the most sensitive sublethal physiological responses used to assess pollution of surface waters by organic chemicals is the induction of the hepatic MFO enzyme system in fish (Payne *et al.* 1987). Studies in both Scandinavia and Canada have shown MFOs to be sensitive indicators of exposure to chemicals associated with effluent from pulp mills (Södergren *et al.* 1988; Rogers *et al.* 1989; Södergren 1989; Servizi *et al.* 1990; Munkittrick 1991; Munkittrick *et al.* 1991, 1992a,b, 1994; Hodson *et al.* 1992; Servos *et al.* 1992). In fish populations exposed to pulp mill effluents, MFO induction has consistently occurred in conjunction with other physiological, pathological and population effects. The rapid increase in MFO activity (within 24-48 h) after exposure to pulp mill effluents provides an indication of possible effects long before other symptoms of toxicity develop. As a result, MFO induction is seen as a useful tool in field and laboratory experiments to evaluate the relative potency of pulp mill effluents and their components (Carey *et al.* 1993). One of the more sensitive MFO measurements is ethoxyresorufin-*o*-deethylase (EROD) activity (Stegeman 1981; Kleinow *et al.* 1987; Payne *et al.* 1987).

Although many studies have demonstrated the presence of sublethal biochemical responses, including increases in hepatic EROD activity in fish communities from pulp mill effluent receiving waters, the responsible processes or chemicals have not been identified (Munkittrick *et al.* 1994). In the field, differences in effluent volume, receiving water dilution ratios and processes at mills (pulping vs. pulping and paper) make it difficult to make comparisons among different sites since exposure conditions are not controlled. Such comparisons can only be made under conditions of controlled exposure in the laboratory.

We have applied a laboratory exposure protocol (Williams 1993) to determine the relative potency of five kraft mill effluents with respect to their ability to induce hepatic EROD activity in rainbow trout. Dose-response experiments were used to investigate whether MFO activity could be related to known chemical constituents of effluents such as AOX, DOC, chlorinated organics and resin and fatty acids.

MATERIALS AND METHODS

Fish Acclimation

Rainbow trout (weight 5-10 g, $\bar{x}$ = 8 g) were purchased from Rainbow Springs Fish Hatchery (Thamesford, ON) and acclimated for a minimum of 14 d prior to use. Holding tank temperatures ranged between 12 and 15°C. Fish were fed daily until 2 d prior to exposures and were not fed during effluent exposures.

After collection at the pulp mills (for mill descriptions, see Table 1) and transport to the laboratory, effluent samples were stored in 20 or 100 L polyethylene carboys at 4°C. Exposures of fish to these effluents were started within 24 h after their arrival at the lab and in all cases were within a week of sample collection (all experiments were conducted and completed between January and August 1992).

Table 1. Description of effluent treatment and production characteristics at the five kraft mills studied.

Mill	Treatment	Production (yearly averages)	Flow ($m^3\ d^{-1}$)	Total Production (during sampling)	Wood Furnish	Bleaching Process
Mill A (Old Technology Mill)	primary	400 t d^{-1} bleached pulp 900 t d^{-1} kraft paperboard and bleachboard	199,419	965 t d^{-1}	softwood	50% ClO_2
Mill B (Modernized Mill)	secondary (6-8 d lagoon)	1,000 t d^{-1} bleached kraft pulp 200 t d^{-1} fine and specialty papers	141,825	1,059 t d^{-1}	hardwood/ softwood (60:40)	ClO_2 (60% hardwood; 30% softwood) O_2 delig.
Mill C (Modernized Mill)	secondary	1,200 t d^{-1}	108,096	1,234 t d^{-1}	softwood/ hardwood	50% ClO_2
Mill D (Modern Mill)	secondary	1,000 t d^{-1} bleached kraft pulp 500 t d^{-1} newsprint	70,800	1,208 t d^{-1}	softwood	O_2 delig. 100% ClO_2
Mill E (Unbleached Mill)	secondary (7 d lagoon)	400 t d^{-1} unbleached kraft paper	41,694	423 t d^{-1}	softwood	none

Experimental Design and Exposure Conditions

All experiments were performed in an environmental chamber maintained at 13-14°C and a photoperiod of 16 h light:8 h dark. Laboratory controls were performed with every exposure tested using the same laboratory reference water that was used to dilute effluents. Ten fish were placed in each aquarium (13 L aquaria with 10 L volumes of exposure solutions). Fish were exposed to a series of concentrations of effluent depending on the mill tested and the amount of effluent received. Effluents (grab samples collected and characterized by the parameters identified in Table 1) were exchanged on a daily static renewal basis over a 6 d period to correspond with field caging studies (Williams et al. 1991). Bioassay tanks were continuously aerated and exposure conditions were monitored. Exposure temperatures ranged between 13 and 15°C and dissolved oxygen concentrations ranged between 7.5 to 9.8 mg L^{-1}.

Conductivity of holding tank water and pulp mill effluent ranged between 0.25 to 2.2 mS cm^{-1} and pH ranged between 7.1 to 8.2 depending on which pulp mill was studied. No significant differences between the various effluents were noted for each of the parameters measured other than differences in conductivity at higher effluent concentrations. Different conductivity values can be attributed to differing amounts of inorganic salts resulting from the pulping and bleaching processes. All fish were sacrificed after 6 d of exposure and hepatic MFO activity was determined as activity of EROD using ethoxyresorufin as the substrate (Hodson et al. 1991; Williams 1993). For each exposure experiment, one aquarium contained fish injected with β-naphthoflavone (βNF) as a positive control to confirm that each batch of fish was capable of induction. For these positive controls, rainbow trout were given an intraperitoneal injection with 0.5 mg kg^{-1} βNF and sacrificed after 4 d. Fish were analyzed individually and an EROD value (mean ± standard error of the mean) was calculated for each treatment.

To further characterize each effluent, production and flow data were tabulated to determine the flow per tonne of production (Table 1).

Water Chemistry Parameters

On arrival at the lab, grab samples of effluent were taken for AOX, DOC, chlorophenolics and resin and fatty acid analysis. Samples were stored at 4°C in the dark until analyzed. AOX was analyzed on a Mitsubishi Tox 10 analyzer according to the method of Odendahl et al. (1990). Dissolved organic carbon was measured in samples acidified to pH <2 with HCl on a Beckman Industrial Model 915-B organic carbon analyzer according to the method of Beckman (1987). The samples for chlorophenolic analysis were analyzed as their acetates using gas chromatography with a 30 m DB-5 (J & W) capillary column and an electron capture detector (ECD) for quantitation and mass spectrometry for confirmation as described in Lee et al. (1989). Resin and fatty acid analysis was performed according to Lee et al. (1990). Acids were extracted from effluent samples at pH 8 by methyl tert-butyl ether and converted into their respective pentafluorobenzyl ester derivatives. Samples underwent silica gel column cleanup and extracts were analyzed by gas chromatography with an ECD using a 30 m DB-17 (J & W) column. For AOX and DOC analysis, the mean of the 3 to 5 subsamples was reported. Effluents were mixed thoroughly in the 100 L containers prior to obtaining grab samples for the individual analyses.

Statistical Analysis

Linear regressions were performed on the log transformed data of EROD activity versus effluent concentration using the Systat statistical package (Systat Inc. 1990). For each effluent, threshold effluent concentrations for EROD enzyme induction were determined from each regression as the effluent concentration at which the EROD value was 0.9 pmol mg^{-1} protein min^{-1} (i.e., the upper 95% confidence limit for control water). To compare the regressions from the five mills, an analysis of covariance (ANCOVA) was performed. Prior to ANCOVA analysis, the assumption of equality of slopes was verified. Differences among the elevations were calculated by a multiple comparison test (Tukey test).

Based on the threshold effluent concentrations and chemical analysis of whole effluents, concentrations of individual chemical parameters (AOX, DOC, chlorophenolics and resin and fatty acids) at threshold effluent concentrations were calculated for each effluent. Coefficients of variation for the computed concentrations were then calculated to determine variability about the mean. Smaller coefficients of variability indicate increased correlation between that chemical parameter and EROD activity.

RESULTS

Five kraft mills representing different bleaching and effluent treatment technologies were utilized for this study (Table 1). The mills are characterized as follows: Mill A was considered an old technology mill which utilized only primary effluent treatment although 50% ClO$_2$ substitution was employed in its bleaching process. Mill B was considered a modernized mill which utilized oxygen prebleaching, 60% ClO_2 substitution and extensive primary and secondary effluent treatment (6-8 d aerated lagoon). Mill C was also a modernized mill with a secondary effluent treatment system that contained an aerated stabilization basin and utilized a low chlorine dioxide substitution ratio (50% ClO_2). Mill D utilized modern technologies. This mill had an extensive secondary treatment system, high chlorine dioxide substitution (100% ClO_2) and oxygen prebleaching. The final mill studied was mill E which had a secondary effluent treatment system but used no chlorine or chlorine dioxide since it produced only unbleached pulp.

Production at the mills was mainly pulp although kraft paperboard and bleachboard were produced at mill A and mills B, D and E produced paper in addition to pulp. At mill D, the newsprint furnish is primarily thermomechanical pulp, with a small amount of bleached kraft added for strength, whereas the total pulp and paper production at mill E was unbleached. Yearly averages typical for the five mills are presented in Table 1. The total daily production at the time of sampling (1992) was approximately 1,000 t d^{-1} for the four bleached kraft mills and approximately 400 t d^{-1} for the unbleached kraft mill. All mills used a softwood wood furnish, with mills B and C utilizing hardwood as well.

Effluent Chemistry

Each of the kraft mill effluents was characterized according to AOX, DOC, chlorophenolics and resin and fatty acids. Values for each of these parameters are reported for 100% effluent on the left side of Table 2. Chemistry parameters which were measured as non-detectable were assigned a value of one half the detection limit. Detection limits for chlorophenolic analysis were determined to be 0.5 µg L^{-1} for all except the monochlorinated phenolics which had a detection limit of 1 µg L^{-1} (based on a 50 mL sample) (Lee et al. 1989). For resin and fatty acid analysis the detection limit was 1 µg L^{-1}, based on a 25 mL sample (Lee et al. 1990).

AOX values ranged from a high of 17 mg L^{-1} at mill D to 0.07 mg L^{-1} at the unbleached mill (mill E). DOC ranged from 49 mg L^{-1} at mill B to 228 mg L^{-1} at mill A for the bleached kraft mills. The unbleached kraft mill had a value of 14 mg L^{-1}. Values of the total chlorophenolics ranged from a low of approximately 18.0 µg L^{-1} at the modernized mills to a high of 126 µg L^{-1} at the modern mill. The unbleached mill showed background levels for total chlorophenolics at 6.24 µg L^{-1}. Total resin acids ranged from a low of 4.40 µg L^{-1} at mill D to a high of 12,100 µg L^{-1} at mill A, while total fatty acids ranged from 68.5 µg L^{-1} at mill B to 3,520 µg L^{-1} at mill A. The high values at mill A may be attributed to the lack of secondary effluent treatment at the mill.

Fish Exposures

Exposures of rainbow trout to all five kraft mill effluents resulted in EROD induction (Fig. 1). Fish exposed to mill A effluent (old technology mill) showed acute toxicity (i.e., fish died) within 2 d in the 50 and 100% effluent exposures. In the remaining four effluent concentrations, ranging between 1 and 20%, the majority (≥80%) of the fish survived for the full 6 d exposure period.

Fish exposed to the effluents from three of the remaining four mills survived in 100% effluent for the duration of the exposure period. For mill C, 80% of the fish survived in 100% effluent and there was 100% survival in the remaining effluent concentrations (i.e., 1, 2.5, 5, 7.5, 10, 20, 30, 50 and 75% effluent concentrations). No acute toxicity was observed in fish exposed to the other mill effluents up to and including the 100% effluent concentration.

MFO induction thresholds were calculated and ranged between 0.33 and 9.1% effluent concentrations for the five mills tested (Fig. 1 and Table 3). Every effluent examined was tested at concentrations ranging between 1 and 100% effluent with the exception of mill D which only ranged between 1 and 20%. An insufficient volume of effluent was available to test 50 and 100% effluent exposure conditions for this mill. However, as seen in Fig. 1, threshold effluent concentrations for induction were calculated at the

EROD value which represents the upper 95% confidence limit for the control water (0.9 pmol mg^{-1} protein min^{-1}). An upper range of 20% effluent concentration was still sufficient to determine a threshold value for this mill.

Table 2. Concentrations of individual chemicals ($\mu g\ L^{-1}$ except as noted) measured in 100% effluent.

	WHOLE EFFLUENT					THRESHOLD EFFLUENT CONCENTRATIONS	
	Mill A	Mill B	Mill C	Mill D	Mill E	Mean ± STD[1]	C.V. (%)[2]
dissolved organic carbon (mg L^{-1})	228	49	95	170	14	1.1 ± 0.59	56
tetrachlorocatechol	1.44	1.51	1.51	1.63	(0.01)[3]	0.0093 ± 0.0061	66
5,6-dichlorovanillin	12.6	1.51	3.75	5.86	1.23	0.055 ± 0.038	69
4,5,6-trichloroguaiacol	3.53	1.58	5.27	4.30	(0.01)	0.020 ± 0.015	76
3,4,5-trichlorocatechol	9.03	1.65	3.70	2.37	(0.01)	0.021 ± 0.017	80
total chlorinated phenolics	81.3	18.0	34.9	126	6.24	0.55 ± 0.47	85
6-monochlorovanillin	21	1.47	9.59	7.73	(0.03)	0.051 ± 0.045	89
adsorbable organic halogen (mg L^{-1})	8.0	5.0	14	17	0.07	0.067 ± 0.064	95
3,4,5-trichlorosyringol	0.26	0.93	0.19	0.32	(0.01)	0.0031 ± 0.0030	95
arachidic acid	56.6	1.60	4.00	3.90	3.20	0.14 ± 0.14	102
3,4,5-trichloroguaiacol	4.51	0.64	4.83	19.9	1.65	0.084 ± 0.087	103
tetrachloroguaiacol	0.66	1.99	0.24	0.61	(0.01)	0.0062 ± 0.0065	105
total fatty acids	3,520	68.5	81.6	83.2	146	7.0 ± 8.2	116
palmitic acid	406	29.4	46.0	44.1	67.8	1.9 ± 2.3	122
stearic acid	138	25.7	21.3	21.0	39.1	0.98 ± 1.3	134
4,5-dichloroguaiacol	24.9	0.65	1.05	3.69	(0.01)	0.039 ± 0.054	138
sandaracopimaric acid	163	1.70	(0.01)	0.60	3.50	0.25 ± 0.36	141
4,6-dichloroguaiacol	0.91	5.32	1.94	2.28	3.36	0.079 ± 0.12	146
myristic acid	18.9	5.30	5.20	4.00	12.7	0.28 ± 0.44	161
pimaric acid	440	1.40	(0.01)	0.90	3.20	0.56 ± 0.98	173
lauric acid	7.80	5.00	2.50	2.90	14.9	0.30 ± 0.53	178
isopimaric acid	753	0.90	(0.01)	(0.01)	4.40	0.94 ± 1.7	179
oleic acid	1,100	0.70	2.70	4.00	4.20	1.3 ± 2.5	184
4,5-dichlorocatechol	2.44	0.80	2.79	77.6	(0.01)	0.18 ± 0.34	191
total resin acids	12,100	23.5	8.30	4.40	21.7	14 ± 27	192
linoleic acid	1,800	0.80	(0.01)	3.30	2.40	2.1 ± 4.1	194
abietic acid	3,950	17.4	2.80	1.60	4.10	4.6 ± 9.0	194
dehydroabietic acid	4,340	2.20	(0.01)	1.30	6.40	5.1 ± 9.8	194
neoabietic acid	842	(0.01)	5.50	(0.01)	(0.01)	0.96 ± 1.9	199
palustric acid	1,560	(0.01)	(0.01)	(0.01)	(0.01)	1.8 ± 3.6	200

[1] Normalized averages and standard deviation values were calculated using the threshold effluent concentration for each mill.
[2] Coefficient of variation (C.V. %).
[3] For parameters below the limit of detection, a value of one-half the detection limit (shown in parentheses) was used to calculate the average and standard deviations.

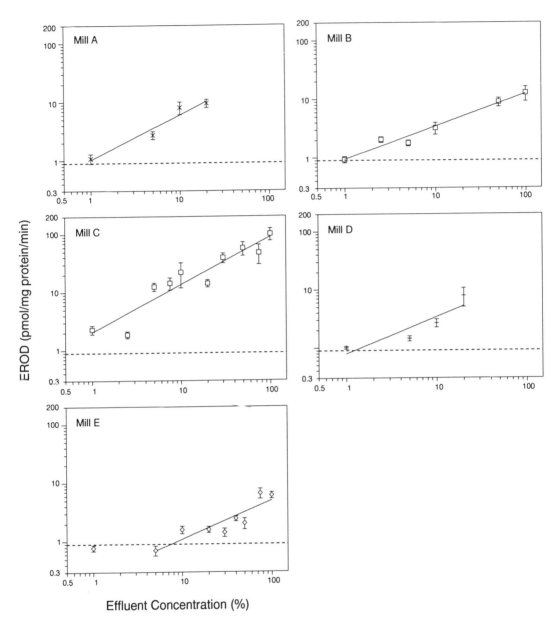

Figure 1. Dose-response regressions showing the relationship between EROD and effluent concentration at five different kraft mills: mills A, B, C, D and E (0.9 pmol mg^{-1} protein min^{-1} represents the threshold for significant EROD induction above control lab water).

DISCUSSION

This study focused on assessing and characterizing five kraft mill effluents employing chlorine and non-chlorine-based bleaching technologies using the sublethal response of EROD enzyme induction as a bioindicator in laboratory studies. By studying effluents in the laboratory as opposed to the field, a greater amount of experimental control on the exposures was gained. Using the upper 95% confidence limit for significant induction in laboratory tests of effluent, an estimate of threshold concentrations was determined.

Table 3. Summary data for regression of EROD activity relative to effluent concentration. The threshold concentration (%) is determined from the intercept of the regression line and the 95% upper confidence limit of the EROD activity in fish exposed to control water.

Mill	Regression Equation	Standard Error of the Estimate	r^2	p	Threshold Effluent Concentration (%)	Flow t^{-1} Production (m^3 ADt^{-1})	Production Normalized Threshold (% m^{-3} ADt^{-1})
Mill A	log Y = 0.098 + 0.597 (log X)	0.124	0.91	0.044	0.57	207	0.0028
Mill B	log Y = -0.031 + 0.471 (log X)	0.100	0.94	0.001	0.93	134	0.0069
Mill C	log Y = 0.298 + 0.708 (log X)	0.205	0.85	<0.001	0.33	88	0.0038
Mill D	log Y = -0.067 + 0.434 (log X)	0.111	0.88	0.064	1.12	59	0.0190
Mill E	log Y = -0.678 + 0.658 (log X)	0.174	0.76	0.005	9.1	98	0.0928

Dilution series experiments resulted in thresholds for MFO induction ranging between 0.33 to 9.1% effluent concentrations at the five kraft mills. At the old technology mill, acute toxicity was observed at 50 and 100% effluent concentrations, yet a threshold for the sublethal response was observed at 0.57%. In contrast, a threshold occurred at a concentration of 0.93% effluent for mill B. This suggests that fish are capable of being induced at low effluent concentrations even at a modernized pulp mill with secondary treatment. Comparing mill A to mill C, secondary treatment appears to have eliminated acute toxicity problems, but an elevated EROD response indicates that EROD-inducing chemicals were still present, even after secondary treatment.

An ANCOVA analysis, performed on the data from the dilution series to determine whether the dose-responses for the effluents were similar, indicated that the slopes of the dose-responses were not different from each other (p = 0.38) but that the overall responses for the five mills were different (p < 0.001). In multiple comparisons using a Tukey test, the responses for mills A, B and D were not significantly different from one another but were different from the responses for either mill C or mill E. The responses for mill C and E were also independent from each other. Similar slopes for the different mills could indicate that the same inducer(s) was/were present at all five mills and that the difference in responses between mills was due to different absolute concentrations of inducer(s) in the effluents. In such a case, if the response data were normalized to a common level of response such as the threshold effluent concentration, the inducer should be present at comparable concentrations.

Effluents were characterized according to AOX, DOC, chlorophenolics and resin and fatty acids. AOX is a bulk parameter that measures the amount of halogen in a sample that is bound to organic matter; it is a parameter currently being used for environmental regulation around the world (Chung *et al.* 1991). In pulp mill waste, essentially all of the halogen is chlorine (Sprague *et al.* 1992).

To relate EROD to the water chemistry, the chemistry parameters were corrected to threshold effluent concentrations identified by regression analysis. By using the threshold effluent concentration from the regression analysis and chemical analysis of whole effluent, chemical parameters were *normalized to threshold levels* as a means of comparing chemistry data. Similar chemistry data, once corrected for threshold EROD induction, would indicate a possible relationship between the sublethal response of EROD induction and the concentration of the chemical present. Means, standard deviations and coefficients of variation were calculated for each of the individual chemistry parameters for the five mills studied. The data were then ranked in ascending order according to the coefficients of variation. Coefficients of variation showed large sample variability relative to the mean (Table 2). The smallest value of 56% was seen for the DOC parameter and a high of 200% was observed for palustric acid. AOX and DOC were ranked eighth and first, respectively, according to rank order. Variations of 56% relative to the mean are very large. This suggests that all of the parameters measured do not correlate well to EROD.

AOX and DOC parameters differed substantially among the five mills. AOX concentrations were minimal at mill E because this mill did not produce bleached pulp. Mill D had high AOX values at threshold, yet induction occurred at a higher effluent concentration relative to the three other bleached kraft mills. DOC was found to be highest in effluent from the old technology mill (mill A) and lowest in effluent from the unbleached mill (mill E).

Resin acids and extractives present in cooking liquors have been known in the past to be major contributors to toxicity (Rogers and Mahood 1974; McLeay et al. 1979a,b; Kovacs 1986; Carey et al. 1993). Resin and fatty acid levels are normally reduced to sublethal concentrations with the implementation of secondary treatment (Servizi et al. 1986; McLeay 1987). This likely explains the acute toxicity observed only at the old technology mill (mill A). This mill has the highest levels of resin and fatty acids. The total resin acids at mills employing secondary treatment were approximately 500 (23.5 µg L^{-1}, mill E) to 2,700 (4.40 µg L^{-1}, mill C) times lower than the levels at mill A (12,100 µg L^{-1}). Fatty acids were reduced 24-fold (146 µg L^{-1}, mill E) to 51-fold (68.5 µg L^{-1}) at mill B relative to mill A (3,520 µg L^{-1}). The main resin acids contributing to toxicity in pulp mill effluents are dehydroabietic, abietic, isopimaric and pimaric acid (Chung et al. 1979; McLeay 1987; Sprague 1992). Acute toxicity levels for these resin acids are between 0.4 and 1.8 mg L^{-1} (Chung et al. 1979). Fish exposed to mill A effluent at the 50 and 100% levels in the laboratory exposures were exposed to resin acid concentrations well above that range.

Total chlorinated phenolics ranged from a low of 6.24 µg L^{-1} at mill E to a high of 126 µg L^{-1} at the modern mill (mill D). Although mill D utilizes modern technology (oxygen prebleaching and 100% ClO$_2$ substitution), chlorinated phenolics were still present at higher levels.

The parameters measured for this study were assumed to be a good starting point to identify or refute which chemicals may be responsible for MFO induction. Unfortunately, none of the chemical groups studied seem to be strongly related to EROD activity. These parameters (AOX, DOC, chlorophenolics and resin and fatty acids) can likely be eliminated from the list of candidate chemicals responsible for MFO induction. Other studies have also shown that resin acids and chlorinated phenolics are not major contributors to MFO induction (Oikari et al. 1988; Mather-Mihaich and Di Giulio 1991).

A second approach for comparing the responses between mills involves correcting the EROD threshold response for flow and production rates at each mill when the samples were taken. To do this, we divided the threshold effluent concentration for each mill by the whole mill discharge and production rates to give a parameter we call the production-normalized threshold (PNT) that has units of percent m$^{3,-1}$ADt^{-1}. A lower PNT indicates a proportionately greater rate of production of the MFO inducers per tonne of pulp produced. The resulting PNT values are listed in Table 3. The PNT values for the three mills whose total production is bleached were roughly comparable, ranging from 0.0028 to 0.0069% m$^{3,-1}$ADt^{-1}. These were much lower than that for mill E (0.0928% m$^{3,-1}$ ADt^{-1}) which produces only unbleached pulp. Mill D, which produces a mix of bleached and unbleached pulp, had a PNT of 0.0190% m$^{3,-1}$ADt^{-1}, intermediate between the bleached and unbleached pulp mills. These results strongly suggest that although the inducer(s) may be produced in unbleached mills, the bleaching process is also a major contributor.

At first glance, our identification of the bleaching process as a major contributor of MFO inducers may seem contradictory to our observations that AOX and other chlorinated parameters were not related to induction. However, the bleaching process is also a major contributor of unchlorinated organics to final effluents. The source of these organic compounds is residual lignin in the unbleached pulp that is liberated by oxidation in the bleaching process. This is reflected in our data. The range of DOC in the effluent samples from the bleached pulp mills was 49-228 mg L^{-1} while that of the unbleached mill was 14 mg

L^{-1}. Thus, the production of organics was higher in the mills with bleaching than in the mill producing unbleached pulp. The coefficient of variation of DOC was the lowest of all parameters we measured. To further support this contention, laboratory studies by Schnell *et al.* (1993) characterized various in-plant process waters and indicated that the bleach plant was a major contributor to the MFO induction potential of treated and untreated bleached kraft mill effluents.

Our results led us to speculate that the MFO inducers in pulp mill effluents are generated primarily from lignin that is liberated from the cellulose in either the bleaching process or in the pulping process and that the discharge of pulping liquors can be a source of inducers in mills with no bleaching. A limited amount of literature data supports this speculation. Field data from ten different pulp mills including non-chlorine bleaching technologies and non-kraft processes without bleaching showed elevated EROD enzyme activity in white sucker in all cases (Munkittrick *et al.* 1994). Laboratory studies (Martel *et al.* 1992, 1993) have shown that the use of chlorine for kraft pulp bleaching did not appear to be *the sole determinant factor* in the ability of final mill effluents to cause MFO induction in fish. These authors also observed that secondary treated effluent from an unbleached mill induced EROD activity.

If this speculation proves correct, it implies that changes in bleaching processes in response to regulatory control of chlorinated parameters such as AOX will not eliminate MFO inducers from pulp mill effluent, particularly if the changes still result in substantial amounts of oxidized lignin being discharged, as would be the case in switching to totally chlorine free bleaching. A strategy based on directing as much of the lignin-derived organics as possible from the effluent to the recovery process would likely be more effective.

In summary, all effluents tested in the lab were capable of inducing EROD activity in a dose-response manner. MFO induction threshold levels indicated that differences existed among mills in the amount of EROD induction observed. MFO induction was apparent even at an unbleached mill, suggesting that *some* MFO-inducing substances in pulp mill effluents are not exclusively associated with the chlorine bleaching process. When effluents were characterized according to AOX, DOC, chlorophenolics and resin and fatty acids, large coefficients of variation occurred when concentrations at the threshold for EROD enzyme induction were compared. This suggested that these parameters were not major contributors to EROD induction. However, by correcting EROD activity to flow and production data from the mills, production-normalized thresholds indicated that the bleaching process is a major contributor to EROD induction. Further studies are required to help identify what compounds may be responsible for the elevated EROD activity in fish when they are exposed to pulp mill effluents.

ACKNOWLEDGMENTS

The support of Environment Canada, the Department of Fisheries and Oceans and the cooperation and assistance of the five mill managers who provided us with effluent samples was greatly appreciated. A special thank you to Don A. Metner, for training TGW the EROD methodology, Sara J. Ramer for providing excellent technical support and to Tom E. Peart for assistance in the analysis of chlorophenolics and resin and fatty acids. Additional financial support was provided to TGW with the IAGLR/Mott fellowship.

REFERENCES

Beckman. 1987. Operation of the Beckman 915-B Total Organic Carbon Analyzer. Beckman Industrial Corp., La Habra, California.

Carey, J.H., P.V. Hodson, K.R. Munkittrick and M.R. Servos. 1993. Recent Canadian studies on the physiological effects of pulp mill effluent on fish. 20 pp.

Chung, L.T.K., H.P. Meier and J.M. Leach. 1979. Can pulp mill effluent toxicity be estimated from chemical analyses? TAPPI 62:71-74.

Chung, A., H. Cook and D. Halliburton. 1991. An overview of pulp and paper mill effluent control in North America and Scandinavia. *In:* Environmental Fate and Effects of Bleached Pulp Mill Effluents, Swedish EPA Report 4031 (A. Södergren, Ed.), Proceedings of a Swedish EPA conference in Stockholm, Sweden, November 19-21, 1991, pp. 7-9.

Hodson, P.V., P.J. Kloepper-Sams, K.R. Munkittrick, W.L. Lockhart, D.A. Metner, L. Luxon, I.R. Smith, M.M. Gagnon, P. Martel, M. Servos and J.F. Payne. 1991. Protocols for measuring mixed function oxygenases of fish liver. Can. Tech. Rep. Fish. Aquat. Sci. 1829:49 pp.

Hodson, P.V., M. McWhirter, K. Ralph, B. Gray, D. Thivierge, J. Carey, G. Van Der Kraak, D.M. Whittle and M.-C. Levesque. 1992. Effects of bleached kraft mill effluent on fish in the St. Maurice River, Quebec. Environ. Toxicol. Chem. 11:1635-1651.

Kleinow, K.M., M.J. Melancon and J.J. Lech. 1987. Biotransformation and induction: implications for toxicity, bioaccumulation and monitoring of environmental xenobiotics in fish. Environ. Health Perspect. 71:105-119.

Kovacs, T.G. 1986. Effects of bleached kraft mill effluents on freshwater fish: a Canadian perspective. Wat. Pollut. Res. J. Can. 21:91-118.

Lee, H.-B., R.L. Hong-You and P.J.A. Fowlie. 1989. Chemical derivatization analysis of phenols. Part VI. Determination of chlorinated phenolics in pulp and paper effluents. J. Assoc. Offic. Anal. Chem. 72(6):979-984.

Lee, H.-B., T.E. Peart and J.M. Carron. 1990. Gas chromatographic and mass spectrometric determination of some resin and fatty acids in pulp mill effluents as their pentafluorobenzyl ester derivatives. J. Chromatogr. 498:367-379.

Martel, P.H., T.G. Williams, T.G. Kovacs, B.I. O'Connor, R.H. Voss, J.H. Carey, D.G. Dixon and K.R. Solomon. 1992. Inter-laboratory comparison of a laboratory approach for screening of pulp and paper mill effluents with respect to their ability to cause elevated mixed function oxygenase (MFO) activity in fish. Presented at the 19th Annual Aquatic Toxicity Workshop, Edmonton, Alberta.

Martel, P.H., T.G. Kovacs, B.I. O'Connor and R.H. Voss. 1993. A survey of pulp and paper mill effluents for their potential to induce mixed function oxidase enzyme activity in fish. Proc. of the 79th Annual Meeting, Tech. Section, Can. Pulp Pap. Assoc., Montreal, Quebec, pp. A165-A177.

Mather-Mihaich, E. and R.T. Di Giulio. 1991. Oxidant, mixed-function oxidase and peroxisomal responses in channel catfish exposed to a bleached kraft mill effluent. Arch. Environ. Contam. Toxicol. 20:391-397.

McLeay, D.J., C.C. Walden and J.R. Munro. 1979a. Influence of dilution water on the toxicity of kraft pulp and paper mill effluent, including mechanisms of effect. Water Res. 13:151-158.

McLeay, D.J., C.C. Walden and J.R. Munro. 1979b. Effect of pH on toxicity of kraft pulp and paper mill effluent to salmonid fish in fresh and seawater. Water Res. 13:249-254.

McLeay, D.J. (D. McLeay and Associates Ltd.). 1987. Aquatic toxicity of pulp and paper mill effluent: a review. Env. Can. Rep. EPS 4/PF/1, Ottawa, 191 pp.

Munkittrick, K.R. 1991. Recent North American studies of bleached kraft mill impacts on wild fish. *In:* Environmental Fate and Effects of Bleached Pulp Mill Effluents, Swedish EPA Report 4031 (A. Södergren, Ed.), Proceedings of a Swedish EPA conference in Stockholm, Sweden, November 19-21, 1991, pp. 347-356.

Munkittrick, K.R., C. Portt, G.J. Van Der Kraak, I. Smith and D. Rokosh. 1991. Impact of bleached kraft mill effluent on liver MFO activity, serum steroid levels and population characteristics of a Lake Superior white sucker population. Can. J. Fish. Aquat. Sci. 48:1371-1380.

Munkittrick, K.R., G.J. Van Der Kraak, M.E. McMaster and C.B. Portt. 1992a. Relative benefit of secondary treatment and mill shutdown on mitigating impacts of bleached kraft mill effluent (BKME) on MFO activity and serum steroids in fish. Environ. Toxicol. Chem. 11:1427-1439.

Munkittrick, K.R., M.E. McMaster, C.B. Portt, G.J. Van Der Kraak, I.R. Smith and D.G. Dixon. 1992b. Changes in maturity, plasma sex steroid levels, hepatic MFO activity and the presence of external lesions in lake whitefish exposed to bleached kraft mill effluent. Can. J. Fish. Aquat. Sci. 49:1560-1569.

Munkittrick, K.R., M.R. Servos, M.E. McMaster, G.J. Van Der Kraak, C. Portt and M.R. van den Heuvel. 1994. Survey of receiving water environmental impacts associated with discharges from pulp mills. II. Gonad size, liver size, hepatic MFO activity and plasma sex steroid levels in white sucker. Environ. Toxicol. Chem. 13 (7):1089-1101.

Odendahl, S.M., K.M. Weishar and D.W. Reeve. 1990. Chlorinated organic matter in bleached chemical pulp production. Part II: A review of measurement techniques for effluents. Pulp Pap. Can. 91(4):60-66.

Oikari, A., P. Lindström-Seppa and J. Kukkonen. 1988. Subchronic metabolic effects and toxicity of a simulated pulp mill effluent on juvenile lake trout, *Salmo trutta m.lacustris*. Ecotoxicol. Environ. Saf. 16:202-218.

Payne, J.F., L.L. Fancey, A.D. Rahimtula and E.L. Porter. 1987. Review and perspective on the use of mixed-function oxygenase enzymes in biological monitoring. Comp. Pharm. Physiol. 86C(2):233-245.

Rogers, I.H. and H.W. Mahood. 1974. Removal of fish-toxic organic solutes from whole kraft effluent by biological oxidation and tne role of wood extractives. Fish. Res. Board Can. Tech. Rep. 434, 43 pp.

Rogers, I.H., C.D. Levings, W.L. Lockhart and R.J. Norstrom. 1989. Observations on overwintering juvenile chinook salmon (*Oncorhynchus tshawytscha*) exposed to bleached kraft mill effluent in the Upper Fraser River, British Columbia. Chemosphere 19:1853-1868.

Schnell, A., P.V. Hodson, P. Steel, H. Melcer and J.H. Carey. 1993. Optimized biological treatment of bleached kraft mill effluents for the enhanced removal of toxic compounds and MFO induction response in fish. 1993 Conf. of the CPPA, pp. 97-111.

Servizi, J.A., D.W. Martens, R.W. Gordon, J.P. Kutney, M. Singh, E. Dimitriadis, G.M. Hewitt, P.J. Salisbury and L.S.L. Choi. 1986. Microbiological detoxification of resin acids. Wat. Pollut. Res. J. Can. 21:119-129.

Servizi, J., R. Gordon, D. Martens, L. Lockhart, D. Metner, I. Rogers, J. McBride and R. Norstrom. 1990. Effects of biotreated bleached kraft mill effluent on fingerling Chinook salmon. Presented at the 25th Canadian Symposium on Water Pollution Research, CCIW Burlington, Ontario, Feb. 15, 1990.

Servos, M.R., J.H. Carey, M.L. Ferguson, G. Van Der Kraak, H. Ferguson, J. Parrott, K. Gorman and R. Cowling. 1992. Impact of a modern bleached kraft mill on white sucker populations in the Spanish River, Ontario. Wat. Pollut. Res. J. Can. 27(3):423-437.

Sprague, J.B. (J.B. Sprague Associates Ltd.). 1992. Concepts for developing Canadian water quality guidelines for effluents from bleached pulp mills. 129 pp.

Stegeman, J.J. 1981. Polynuclear aromatic hydrocarbons and their metabolism in the marine environment. *In:* Polycyclic Hydrocarbons and Cancer, Volume 3, (H. Gelboin and P.O.P. Ts'o, Eds.), Academic Press, New York, pp. 1-60.

Systat Inc. (Wilkinson, Leland). 1990. SYSTAT: The System for Statistics. Evanston, Illinois.

Södergren, A., B.-E. Bengtsson, P. Jonsson, S. Lagergren, Å. Larson, M. Olsson and L. Renberg. 1988. Summary of results from the Swedish project "Environment/Cellulose". Water Sci. Technol. 20(1):49-60.

Södergren, A. (Ed.). 1989. Biological effects of bleached pulp mill effluents. National Swedish EP Board Rep. 3558, Final Report, 139 pp.

Williams, T.G. 1993. A comparative laboratory based assessment of EROD enzyme induction in fish exposed to pulp mill effluents. M.Sc. Thesis, University of Waterloo, Waterloo, Ontario, 85 pp.

Williams, T.G., M.R. Servos, J.H. Carey and D.G. Dixon. 1991. Biological response monitoring of pulp mill effluents. *In:* Environmental Fate and Effects of Bleached Pulp Mill Effluents, Swedish EPA Report 4031 (A. Södergren, Ed.), Proceedings of a Swedish EPA conference in Stockholm, Sweden, November 19-21, 1991, pp. 391-394.

LETHAL AND SUBLETHAL EFFECTS OF CHLORINATED RESIN ACIDS AND CHLOROGUAIACOLS IN RAINBOW TROUT

C.J. Kennedy[1], R.M. Sweeting[2], J.A. Johansen[2], A.P. Farrell[1] and B.A. McKeown[1]

[1]Department of Biological Sciences, Simon Fraser University, Burnaby, B.C., V5A 1S6 Canada
[2]BioWest Environmental Research Consultants, Burnaby, B.C., V5G 1M7 Canada

Resin acids and guaiacols are common constituents in pulp mill effluent and are considered to contribute the greatest effect to the acute toxic actions of mill discharge to fish. The objectives of this study were to examine the sublethal toxicity of 14-monochlorodehydroabietic acid (MCDHAA), 12,14-dichlorodehydroabietic acid (DCDHAA) and 3,4,5,6-tetrachloroguaiacol (TeCG) to rainbow trout, *Oncorhynchus mykiss*. A biological indicator approach was used involving a suite of indicators which spanned several levels of biological complexity, as well as having ecological relevance. These included biochemical responses and effects on swimming performance and disease resistance. The experimentally determined 96-h LC50 values for juvenile rainbow trout were 0.37, 1.1 and 0.9 mg L^{-1} for TeCG, MCDHAA and DCDHAA, respectively. Using these values as a guideline, sublethal exposure to each chemical for 24 h resulted in a classical stress response and included significant primary (secretion of corticosteroids), secondary (hyperlacticemia, hyperglycemia) and tertiary (reduced swimming performance and lowered disease resistance) effects. Chronic exposure of fish to TeCG for 25 d resulted in most parameters being at control levels with the exception of leucocrit which was elevated, cortisol which was depressed, and disease resistance which remained impaired. This study indicates that the selected parameters may be useful indicators of acute exposure to pulp mill effluent; however, under chronic exposure conditions only disease resistance was a reliable indicator of persistent toxicant exposure.

INTRODUCTION

Bleached kraft pulp mill effluent (BKME) is a complex mixture of inorganic and organic compounds, which is both acutely and chronically toxic to a variety of aquatic organisms including fish. Concern over the effect of pulp mill effluent on fish populations has resulted in numerous studies dealing with these effects; however, the results indicate that the toxicity of BKME varies greatly from site to site and even daily or hourly within the same mill (Oikari *et al.* 1984a). This may be due to changes in the overall chemical composition of the effluent which depends upon several factors including the particular pulping methods, the age and type of wood being utilized, the consistency of waste material processing and the integration of treatment methods. In addition, alterations in water quality parameters such as pH, temperature and salinity in the receiving environment will modify an organism's response to these effluents. Due to the complexity and temporal variation of BKME, very few toxicant-specific actions of BKME components on fish have been described (Kennedy *et al.* 1993). In this regard it has been suggested that as much as 90% of the total toxicity of untreated BKME results from resin acids and guaiacols (Leach and Thakore 1975; Pearson 1980; McKague 1981).

Resin acids (RAs) are a group of diterpene acids extracted from the cellulose of wood fiber during processing. Of the 10 major resin acids, dehydroabietic acid (DHAA) typically comprises 5-10% of the total RAs in raw pulp (Oikari *et al.* 1984b; Mattsoff and Oikari 1987) and is considered one of the most toxic and persistent (Brownlee and Strachan 1977; Fox 1977; Oikari *et al.* 1983). Under certain conditions, RAs may be chlorinated at the bleaching stage (Leach and Thakore 1975; Kutney and Dimitriadis 1982), and as a result, DHAA and its chlorinated derivatives have been predominant in bleached pulp effluents

(Leach and Thakore 1975). The chloroguaiacols (CGs) are chlorinated phenolic compounds formed from lignin residues during chlorine bleaching of wood pulps (Rosemarin *et al.* 1990).

Due to the predominance of RAs and CGs in BKME, their potential environmental impacts and the limited information available on the sublethal toxicity of these compounds, the objectives of this research were to examine the effects of two chlorinated RAs (14-monochlorodehydroabietic acid [MCDHAA] and 12,14-dichlorodehydroabietic acid [DCDHAA]) and one CG (3,4,5,6-tetrachloroguaiacol [TeCG]) in fish. In this study, a suite of biological indicators which span several levels of biological complexity and ecological relevance were used. These indicators included various biochemical responses, swimming performance and disease resistance.

MATERIALS AND METHODS

Fish

Juvenile rainbow trout (*Oncorhynchus mykiss*), weighing 8-12 g, were obtained from West Creek Trout Farms, Aldergrove, British Columbia. Fish were kept in dechlorinated municipal water at ambient temperature and pH 6.7-6.9, O_2 saturation >95% and hardness 5.2-6.0 mg L^{-1} $CaCO_3$. Fish were fed ad libitum until one day prior to an experiment.

Chemicals and Exposure

MCDHAA (99% purity), DCDHAA (99% purity) and TeCG (99% purity) were purchased from Helix Biotech Corp. (Richmond, British Columbia). Trout were exposed to MCDHAA, DCDHAA or TeCG using a flow-through exposure system as described by Johansen and Geen (1990). Using computer-controlled pumps and solenoid valves, the appropriate amount of toxicant stock solution and fresh water was delivered to each aquarium every 30 min, replacing 175 L of water daily while maintaining the desired toxicant concentration. All trials were conducted according to bioassay guidelines outlined by the American Public Health Association (1976) and Sprague (1973). In this study there were three categories of toxicity tests: (1) LC50 determinations, (2) acute sublethal tests and (3) chronic sublethal tests. Before any experiment, fish were placed in each vessel or aquarium and allowed to acclimate under continuous flow conditions for 72 h before toxicant exposure began. The LC50 value for each toxicant was determined separately. For this, 10 fish were stocked in each 8-L glass aquarium receiving nominal toxicant concentrations. In each acute study (for biochemical, swimming and disease resistance effects), 10-15 fish were exposed to a toxicant for 24 h in each 65-L glass aquarium at nominal sublethal concentrations. Using the determined 96-h LC50 value as a guideline, there were three concentrations used for the resin acids (0.0, 0.21 and 0.82 mg L^{-1} for MCDHAA; 0.0, 0.18 and 0.73 mg L^{-1} for DCDHAA) and four concentrations used for TeCG (0.0, 0.1, 0.2, 0.3 and 0.4 mg L^{-1}). All of these concentrations were sublethal in a 24-h exposure as determined from the 96-h LC50 trials. For chronic exposure trials, 50 fish were exposed for 25 d to sublethal nominal TeCG concentrations in each 144-L glass aquarium using the same dosing apparatus. The concentrations used in this experiment were 0.0, 0.05, 0.1 and 0.2 mg L^{-1}. These doses did not cause any mortality in the 25-d exposure period. At the end of all exposure trials, fish were either sacrificed for biochemical analysis, or transferred for either swimming performance or disease resistance studies. Two control aquaria were typically used: one was supplied with fresh water only and the second contained the toxicant carrier (0.4% ethanol and NaOH) with no chemical.

Biochemical Analysis

To minimize handling stress, fish were rapidly (<1 min) anesthetized with 2-phenoxyethanol (Janz *et al.* 1991). Fish length and weight were measured and blood sampled via the caudal vasculature using heparinized capillary tubes. An aliquot of whole blood was immediately assayed for hemoglobin (Sigma, St. Louis, MO) and the remaining blood was centrifuged to determine hematocrit and the plasma stored at -20°C pending subsequent assays. The liver and spleen were removed and weighed. Individual livers were immediately frozen on dry ice and stored at -20°C until further analysis. Plasma cortisol levels were determined using a commercial radioimmunoassay kit (IncStar Corp., Stillwater, MN). Plasma was analyzed spectrophotometrically for lactate, glucose (except acute TeCG; Sigma) and plasma protein (except acute TeCG; BioRad Lab. Ltd., Mississauga, Ont). Liver protein, liver glycogen, muscle protein and muscle glycogen (TeCG only) levels were determined as in Sweeting (1989).

Swimming Performance

Fish raceways as described by Farrell et al. (1991) were used to determine maximal prolonged swimming performance (critical swimming speed; Ucrit). Briefly, the apparatus consisted of a 1,500-L ovoid, fiberglass raceway tank equipped with two variable-output propulsion motors. The velocity profile within the enclosed cylindrical testing chambers was smoothed by a series of straightening vanes, screens and contraction cones placed upstream of the chambers. Water velocity was controlled by regulating voltage output to the propulsion motors. A portable current meter was used to determine water velocity within the chambers at various voltages before the initiation of swim trials. Control and exposed fish were placed together in the swimming apparatus immediately following removal from the exposure chambers and allowed to acclimate in still water for 3 h before the beginning of the trial. The different groups of fish were identified by an adipose fin clip. The initial velocity was 0.15 m s^{-1} (1.9 body lengths s^{-1}, bl s^{-1}), and the speed was increased in increments of 0.08 m s^{-1} (1.0 bl s^{-1}) at 30-min intervals until all fish had fatigued. Fatigued fish were individually removed from the test chamber and time to exhaustion, fish fork length and weight were recorded. Ucrit was calculated using the method of Farrell et al. (1990). A fish was considered exhausted when it rested against the rear grid and did not respond to mechanical stimulation.

Disease Challenge

Primary isolates of virulent *Aeromonas salmonicida*, the bacterium responsible for the highly infectious fish disease furunculosis, were cultured from coho salmon (*Oncorhynchus kisutch*) known to have died from the disease. Bacterial suspensions in sterile peptone-saline (0.85% NaCl, 0.10% peptone) were prepared from 4-d-old *A. salmonicida* cells grown on TSA agar plates at 10°C. Prior to the disease challenge, control and exposed fish were fin-clipped for identification and combined. Fish were challenged by immersion into an aerated peptone-saline suspension of *A. salmonicida* containing 5×10^5 cells mL^{-1} for 15 min. Fish were then placed into a single tank to a fish density of approximately 2 g L^{-1} that was supplied with fresh-flowing water and were fed ad libitum three times a week. Mortalities were monitored daily and dead fish were frozen at -20°C for necropsy at a later date. The criteria for death by furunculosis were according to Bullock et al. (1983). Preceding the disease challenge experiments, unexposed fish from a stock tank were examined for the presence of the bacterium.

Calculations and Statistics

The LC50 and 95% confidence limits were calculated according to Finney (1971). All biochemical and swimming performance values are reported as mean ± S.E. These data were compared using either the Student's t-test or ANOVA (Zar 1974) to test for significant differences ($p < 0.05$) between control and exposed groups. All percent mortality data from the disease challenge experiments were arcsin transformed prior to analysis. Disease challenge mortality curves were compared to controls using the Mantel (1966) and Breslow (1970) test statistics ($p < 0.05$).

RESULTS

LC50 Determination

The 96-h LC50 values and their 95% confidence limits determined by the probit method for MCDHAA, DCDHAA and TeCG were 1.03 (0.72, 1.48), 0.91 (0.70, 1.21) and 0.37 (0.30, 0.44) mg L^{-1}. For the sublethal acute exposure trials, fish were exposed to 20 and 80% of the determined LC50 values for MCDHAA and DCDHAA. Using these values, the low sublethal doses used in these acute exposures were 0.21 and 0.18 mg L^{-1} and the high sublethal doses were 0.82, 0.73 and 0.30 mg L^{-1} for MCDHAA and DCDHAA. Acute sublethal doses of TeCG were 0.1, 0.2, 0.3 and 0.4 mg L^{-1}. The sublethal doses of TeCG used in the 25-d chronic exposure study were 0.0, 0.02, 0.1 and 0.2 mg L^{-1}.

Biochemical Analysis

The results of the biochemical analyses of the acute sublethal exposure studies are presented in Tables 1, 2 and 3. Significant differences between the high dose of MCDHAA and the control were seen in the following variables: plasma lactate, leucocrit, hematocrit, plasma cortisol, plasma glucose and liver protein and glycogen (Table 1). Significant differences were also seen with spleen somatic index (SSI), plasma cortisol and plasma glucose at the low dose of MCDHAA. Significant differences between the high dose of DCDHAA and the controls were seen in the following variables: plasma lactate, plasma cortisol, plasma protein, plasma glucose and liver glycogen (Table 2). Acute exposure to TeCG at the highest dose significantly affected SSI, lactate, hemoglobin, cortisol, liver glycogen and liver protein (Table 3). Following a 25-d chronic exposure to the lowest TeCG concentration (0.02 mg L^{-1}), the following parameters were significantly affected: leucocrit and cortisol (Table 4). Plasma protein and SSI were significantly affected at the highest TeCG concentration. It is noted that a typical dose-response relationship was seen only in the cortisol values for the acute studies with MCDHAA, DCDHAA and TeCG.

Table 1. *Effects of a 24-h acute exposure on several parameters in rainbow trout at several sublethal MCDHAA concentrations.*

MCDHAA Conc. mg L^{-1}	HSI	SSI	Lactate mg dL^{-1}	Hb g dL^{-1}	Hematocrit %	Leucocrit %
0.0	0.86 ± 0.02	0.10 ± 0.01	22.12 ± 3.22	9.16 ± 0.37	42.5 ± 0.89	0.91 ± 0.09
0.21	0.75 ± 0.07	0.07 ± 0.01*	20.08 ± 2.94	9.13 ± 0.25	44.9 ± 1.04	1.00 ± 0.14
0.82	0.85 ± 0.07	0.10 ± 0.02	35.44 ± 3.43*	8.71 ± 0.40	46.6 ± 1.21*	0.65 ± 0.09*
	Cortisol µg dL^{-1}	Plasma Glucose mg %	Plasma Protein g %	Plasma Cholesterol mg dL^{-1}	Liver Glycogen mg g^{-1}	Liver Protein mg g^{-1}
0.0	0.96 ± 0.19	137.43 ± 13.38	2.69 ± 0.70	137.3 ± 4.4	55.4 ± 7.3	167.1 ± 9.8
0.21	3.56 ± 1.03*	90.43 ± 11.53*	3.54 ± 0.52	137.8 ± 3.9	49.4 ± 9.4	186.2 ± 13.0
0.82	13.83 ± 4.29*	175.91 ± 13.57*	3.88 ± 0.62	145.2 ± 2.6	35.5 ± 7.0*	237.2 ± 9.1*

HSI = hepatosomatic index, SSI = spleen somatic index, Hb = hemoglobin. Values are means ± standard errors, n = 10.
* denotes a significant difference from controls at p < 0.05 level.

Swimming Performance

The Ucrit of control fish in the acute MCDHAA, DCDHAA and TeCG trials were 6.48 ± 1.30, 6.08 ± 0.20 and 4.85 ± 0.58 bl s^{-1}, reflecting a sustained velocity of 0.85, 0.79 and 0.83 m s^{-1}, respectively.

Critical swimming speed values of fish exposed to MCDHAA, DCDHAA or TeCG for 24 h are given in Table 5. There were no significant differences between critical swimming speeds of control and RA-exposed fish. Mean critical swimming speeds were significantly reduced by 32 and 30% compared with control fish in the 0.1 and 0.2 mg L^{-1} concentration TeCG groups, respectively, following 24-h exposure (Table 5). However, following chronic exposure to TeCG, mean critical swimming speeds in treated fish were not significantly different from controls (Table 5).

Table 2. *Effects of a 24-h acute exposure on several parameters in rainbow trout at several sublethal DCDHAA concentrations.*

DCDHAA Conc. mg L^{-1}	HSI	SSI	Lactate mg dL^{-1}	Hb g dL^{-1}	Hematocrit %	Leucocrit %
0.0	1.25 ± 0.08	0.09 ± 0.01	15.58 ± 2.43	8.40 ± 0.41	43.81 ± 1.88	1.33 ± 0.13
0.18	1.10 ± 0.08	0.07 ± 0.02	19.30 ± 1.90	10.04 ± 0.25	46.21 ± 1.18	1.05 ± 0.07
0.73	1.20 ± 0.09	0.09 ± 0.02	32.42 ± 2.67*	9.41 ± 0.51	43.83 ± 1.88	1.04 ± 0.09
	Cortisol μg dL^{-1}	Plasma Glucose mg %	Plasma Protein g %	Plasma Cholesterol mg dL^{-1}	Liver Glycogen mg g^{-1}	Liver Protein mg g^{-1}
0.0	1.46 ± 0.31	121.02 ± 15.92	5.08 ± 0.76	261.88 ± 25.88	31.3 ± 4.1	213.0 ± 8.5
0.18	2.23 ± 0.52	114.23 ± 9.94	3.37 ± 0.58	270.22 ± 14.20	23.4 ± 3.7	215.8 ± 10.0
0.73	3.74 ± 0.86*	141.44 ± 6.99*	14.61 ± 1.82*	333.05 ± 32.88	18.3 ± 1.9*	198.3 ± 3.8

HSI-hepatosomatic index, SSI-spleen somatic index, Hb-hemoglobin. Values are means ± standard errors, n=10. * denotes a significant difference from controls at $p < 0.05$ level.

Table 3. *Effects of an acute 24-h exposure on several parameters in rainbow trout at several sublethal TeCG concentrations.*

TeCG Conc. mg L^{-1}	HSI	SSI	Lactate mg dL^{-1}	Hb g dL^{-1}	Hematocrit %	Leucocrit %
0.0	0.87 ± 0.10	0.12 ± 0.02	25.74 ± 5.04	8.84 ± 0.54	44.6 ± 1.87	1.32 ± 0.18
0.1	0.98 ± 0.06	0.12 ± 0.02	15.95 ± 3.51	9.06 ± 0.38	46.0 ± 1.45	0.72 ± 0.05*
0.2	1.17 ± 0.16	0.11 ± 0.02	20.90 ± 2.53	7.45 ± 0.28	42.9 ± 1.54	1.07 ± 0.18
0.3	0.84 ± 0.05	0.10 ± 0.03	24.48 ± 5.44	7.16 ± 0.40*	43.6 ± 1.52	1.04 ± 0.14
0.4	0.74 ± 0.08	0.07 ± 0.02*	52.93 ± 11.53*	5.89 ± 0.30*	39.6 ± 2.02	1.10 ± 0.15
	Cortisol μg dL^{-1}	Muscle Glycogen mg g^{-1}	Muscle Protein mg g^{-1}	Liver Glycogen mg g^{-1}	Liver Protein mg g^{-1}	
0.0	1.93 ± 0.43	3.00 ± 1.05	11.07 ± 0.64	95.06 ± 27.88	272.8 ± 75.9	
0.1	1.29 ± 0.21	1.78 ± 0.81	12.02 ± 0.70	73.44 ± 14.16	184.2 ± 14.9	
0.2	1.06 ± 0.16	1.44 ± 0.81	11.61 ± 0.33	71.33 ± 28.67	157.3 ± 20.6*	
0.3	1.73 ± 0.45	1.25 ± 1.34	12.38 ± 0.47	11.94 ± 8.12*	214.8 ± 12.6	
0.4	9.76 ± 3.38*	1.00 ± 1.10	12.38 ± 0.60	0.89 ± 0.69*	155.8 ± 11.2*	

LC50 = median lethal concentration, HSI = hepatosomatic index, SSI = spleen somatic index, Hb = hemoglobin. Values are means ± standard errors, n = 9.
* denotes a significant difference from controls at $p < 0.05$.

Disease Resistance

In acute MCDHAA-exposed fish, significant disease-related mortalities continued to occur until day 46, when maximum cumulative moralities reached 20% in the 0.21 mg L^{-1} group of fish (Fig. 1). In acute DCDHAA-exposed fish, cumulative mortalities were 26.7% by day 30 in the 0.18 mg L^{-1} group of fish. In each of these experiments, control fish experienced fewer mortalities than exposed fish by the end of the experiment. Acute or chronic exposure to TeCG at 0.2 mg L^{-1} resulted in significantly more mortalities than in control fish. At higher concentrations of TeCG (0.3 and 0.4 mg L^{-1}) in the acute trial, fewer mortalities occurred than in the 0.2 mg L^{-1} group. There are several explanations for this; for example, high concentrations of TeCG may invoke increased mucous production at the gills which led to a decrease in uptake of the pathogen. Although not within the scope of this study, direct injections of pathogens may overcome such problems.

Table 4. *Effects of a 25-d chronic exposure on several parameters in rainbow trout at several sublethal TeCG concentrations.*

TeCG Conc. mg L^{-1}	HSI	SSI	Lactate mg dL^{-1}	Hb g dL^{-1}	Hematocrit %	Leucocrit %	Cortisol µg dL^{-1}
0.0	0.91 ± 0.07	0.10 ± 0.01	24.09 ± 2.07	7.66 ± 0.36	38.70 ± 1.32	0.50 ± 0.05	1.36 ± 0.45
0.02	0.80 ± 0.08	0.14 ± 0.03	22.78 ± 3.41	7.75 ± 0.28	38.42 ± 1.04	0.87 ± 0.07*	0.48 ± 0.15*
0.10	0.80 ± 0.06	0.13 ± 0.02	29.40 ± 3.14	7.63 ± 0.31	38.76 ± 1.13	0.83 ± 0.07*	0.50 ± 0.18*
0.20	0.81 ± 0.06	0.15 ± 0.02*	28.77 ± 2.59	7.50 ± 0.36	40.23 ± 1.66	1.03 ± 0.08*	0.60 ± 0.11*

	Muscle Glycogen mg g^{-1}	Muscle Protein mg g^{-1}	Liver Glycogen mg g^{-1}	Liver Protein mg g^{-1}	Plasma Glucose mg%	Plasma Protein g%	
0.0	4.10 ± 1.76	12.21 ± 0.64	63.10 ± 8.55	187.73 ± 8.75	89.14 ± 11.03	2.83 ± 0.24	
0.02	6.60 ± 3.02	14.64 ± 1.37	41.55 ± 16.86	201.00 ± 8.38	81.12 ± 6.07	2.69 ± 0.26	
0.10	6.60 ± 1.81	13.24 ± 0.41	44.85 ± 16.60	178.07 ± 22.0	101.64 ± 8.89	3.24 ± 0.26	
0.20	2.50 ± 1.50	13.04 ± 0.64	50.95 ± 13.22	173.49 ± 7.81	114.85 ± 12.85	3.80 ± 0.24*	

HSI = hepato somatic index, SSI = spleen somatic index, HB = hemoglobin.
Values are means ± standard errors, n = 9.
*denotes a significant difference from controls at $p < 0.05$.

Table 5. *Critical swimming speeds of juvenile rainbow trout following acute sublethal 24-h exposures to MCDHAA, DCDHAA or TeCG or a chronic 25-d exposure to TeCG.*

Acute 24-h Exposures						Chronic 25-d Exposure	
MCDHAA Conc.	Ucrit bl s^{-1}	DCDHAA Conc.	Ucrit bls	TeCG Conc.	Ucrit bl s^{-1}	TeCG Conc.	Ucrit bl s^{-1}
0.0	6.48 ± 1.31	0.0	6.08 ± 0.20	0.0	6.31 ± 0.44	0.0	4.85 ± 0.58
0.21	5.97 ± 1.26	0.18	6.34 ± 0.34	0.1	4.22 ± 0.44*	0.02	4.69 ± 0.51
0.82	6.55 ± 1.02	0.73	5.83 ± 0.29	0.2	4.43 ± 0.29*	0.10	5.12 ± 0.29
				0.3	5.42 ± 0.50	0.20	5.05 ± 0.30
				0.4	5.05 ± 0.79		

Values are means ± SE of 10 fish.
* denotes a significant difference from controls at $p < 0.05$.

The majority of dead fish displayed typical signs of external pathology associated with A. salmonicida infection, which included areas of focal necrosis in the muscle as a swelling under the skin which in some cases developed into an abscess opening to the surface, appearing as a crateriform lesion. In some fish, necrotic lesions at the base of pelvic and pectoral fins were evident. In the laboratory diagnosis, *A. salmonicida* was isolated from kidney tissue and identification of the bacterium was made according to the criteria outlined previously.

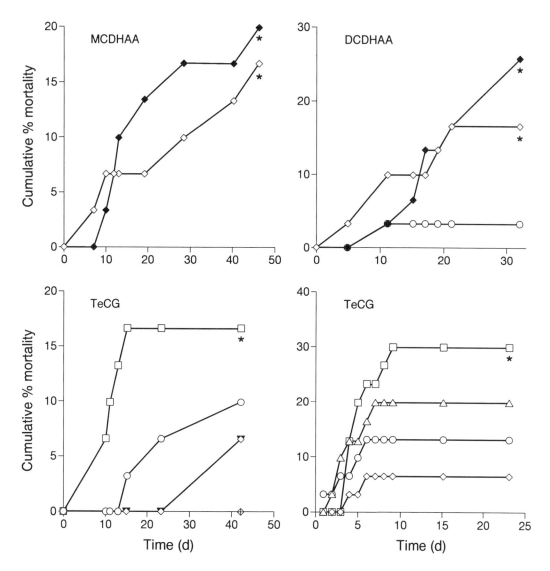

*Figure 1. Cumulative percent mortality of 30 rainbow trout following a challenge to the bacterium Aeromonas salmonicida. Fish had previously been exposed to an acute sublethal 24-h exposure to (a) 0.0 (no mortalities: no symbol), 0.21 (♦) and 0.82 (◊) mg L^{-1} MCDHAA; (b) 0.0 (○), 0.18 (♦) and 0.73 (◊) mg L^{-1} DCDHAA; (c) 0.0 (○), 0.1 (◊), 0.2 (□), 0.3 (▼) and 0.4 (⊕) mg L^{-1} TeCG; or a chronic sublethal 25-d exposure to (d) 0.0 (○), 0.02 (△), 0.1 (◊) and 0.2 (□) TeCG. Each point represents the mean of duplicate experiments. * denotes a significant difference in the mortality curve between exposed and control fish at a significance level of $p < 0.05$.*

DISCUSSION

Exposure of fish to sublethal levels of contaminants may impose considerable stress on their physiological systems, resulting in a number of manifestations such as reduced growth, impaired reproduction, predisposition to disease, reduced locomotory performance or reduced capacity to tolerate subsequent stress (Adams 1990). This study employed a suite of selected stress responses spanning several levels of biological organization as indicators of acute or chronic sublethal exposure of juvenile rainbow trout to MCDHAA, DCDHAA and TeCG. The indicators used in this study were various biochemical parameters which have been identified as indicators of stress in fish. These measurements were all possible on blood and tissue samples taken from a single juvenile salmonid. Assessment of swimming performance and disease resistance required additional experiments but these measures are considered to have a broader ecological relevance.

The results of this study are similar to others which show that exposure to BKME or constituents may cause a rapid disruption of organism homeostasis, i.e., stress, followed by a period of compensation. Of the various blood and tissue parameters used to assess biochemical fitness, plasma cortisol, plasma lactate and liver glycogen were the most responsive and consistent parameters in assessing acute stress. In other studies, increased plasma cortisol levels have been used to measure the primary endocrine responses to noxious stimuli and various pollutants (Schreck 1981; Donaldson et al. 1984; Thomas *et al.* 1987). Once the primary stress response, e.g., corticosteroid hormone release, is triggered, a variety of biochemical and physiological responses called secondary stress responses can follow. Typical secondary stress responses elicited by increases in plasma cortisol levels include hyperglycemia, depletion of tissue glycogen reserves, catabolism of muscle protein and altered blood levels of protein and cholesterol. Sublethal exposure to MCDHAA, DCDHAA or TeCG clearly evoked secondary stress responses. These were hyperlacticemia and depletion of liver glycogen. The primary and secondary stress responses observed occurred rapidly (<24 h). Other secondary effects could become evident with somewhat longer sublethal exposures. Clearly, the sublethal exposure duration could have direct implication in the selection of biochemical parameters used as indicators of aquatic contamination and should be the focus of further research.

With acute or chronic exposure to stressors, tertiary stress responses can be evoked. These tertiary responses are reflected in various physiological and behavioral functions and are often triggered by the primary and secondary stress responses. Consequently, tertiary stress responses incorporate several levels of biological organization, are integrative in nature and provide direct links to the survival of the organism (Schreck 1990). In this study, two performance tests, swimming performance (Ucrit) and disease resistance, were used as measures of tertiary stress responses to RA or CG exposure in trout. Since corticosteroids have been shown to decrease the performance capacity of fish (Specker and Schreck 1980; Schreck 1981), the increase in circulating corticosteroids coupled with the increased energy demands brought about during exposure and resistance to stressors suggests that the swimming performance of trout in this study may be adversely affected. However, the measure of the swimming capacity provided equivocal results. Neither of the RA exposures had an effect on Ucrit, whereas two of the lower but not higher doses of TeCG in the acute exposure reduced swimming performance. That chronic exposure to TeCG had no effect on Ucrit suggests a compensation mechanism.

The most consistent and reliable measure of stress, i.e., it occurred in response to both acute and chronic exposure to the RAs or TeCG, was an alteration in disease resistance. In the past, changes in leucocrit and plasma cortisol levels have been used as primary stress responses that might be related to immune dysfunction. An increase in circulating corticosteroid hormones is generally thought to compromise the immune system of fish since a close association between high circulating levels of corticosteroid hormones, immunosuppression and increased susceptibility to disease has been demonstrated in several studies (Ellis 1981; Pickering and Pottinger 1987; Thomas and Lewis 1987). In addition, Ellis (1981) has shown that high corticosteroid levels may decrease the number of circulating white blood cells in fish. In this study, MCDHAA, but not DCDHAA or TeCG, exposure reduced the leucocrit significantly at the highest concentration. Leucocrit was therefore a poor predictor of the more ecologically relevant measure of disease resistance to *A. salmonicida*. Changes in plasma cortisol were somewhat equivocal. For acute exposures, an increase in plasma cortisol was associated with decreased disease resistance for all three test chemicals. However, for the chronic exposure to TeCG, a decrease in plasma cortisol was associated decreased disease resistance. This suggests that direct measures of immune dysfunction, such as disease challenge with *A. salmonicida*, be used more routinely for the assessment of sublethal toxicity to fish. In the case of these three important components of pulp mill effluent tested here, trout would appear to be at an increased risk to mortality resulting from secondary bacterial/fungal infections even though the toxicant concentrations are themselves sublethal. However, it should be noted that in these experiments high concentrations of TeCG resulted in fewer mortalities in the disease challenge experiments than control fish. There may be several explanations for this including an increase in mucous production at the gills due to TeCG exposure, which prevented bacterial entry or possibly antibacterial activity of TeCG in the trout tissues, which prevented an infection.

The indicators of stress examined in this study have shown that rainbow trout exhibit many of the classic primary (secretion of corticosteroids), secondary (hyperlacticemia, hyperglycemia, lowered energy stores) and even tertiary (lowered disease resistance) stress responses after a 24-h exposure to a sublethal dose of MCDHAA, DCDHAA or TeCG. Following long-term exposure to TeCG, however, fish appeared to compensate for the effects of sublethal exposure to this chemical, with most indicators returning to pre-exposure levels. This indicates that knowledge of the duration of contaminant exposure must be taken into account when selecting biological indicators. The similarities between the results of the present study and those obtained in studies examining whole bleached kraft mill effluent support the suggestions that RAs and CGs are major contributors to BKME fish toxicity (Leach and Thakore 1975; Pearson 1980; McKague 1981). In this regard, additional research appears to be warranted to more fully examine the contribution of resin acids and guaiacols in effluents to the deleterious effects seen in fish populations.

ACKNOWLEDGMENT

This work was supported by a grant to CJK, APF and BAM from the Science Council of British Columbia under the Environmental Research Fund.

REFERENCES

Adams, S.M. 1990. Status and use of biological indicators for evaluating the effects of stress on fish. Am. Fish. Soc. Symp. 8:1-8.

American Public Health Association. 1976. Standard methods for the examination of water and waste water. 14th edition. New York.

Breslow, N. 1970. A generalized Kruskal-Wallis test for comparing k samples subject to unequal patterns of censorship. Biometrics 30:89-99.

Brownlee, B. and W.M.J. Strachan. 1977. Distribution of some organic compounds in the receiving waters of a kraft pulp and paper mill. J. Fish. Res. Board Can. 34:830-837.

Bullock, G.L., R.C. Cipriano and S.F. Snieszko. 1983. Furunculosis and other diseases caused by *Aeromonas salmonicida*. Fish Disease Leaflet 66. U.S. Department of Interior, Washington, DC.

Donaldson, E.M., U.H.M. Fagerlund and J.R. McBride. 1984. Aspects of the endocrine stress response to pollutants in salmonids. *In:* V.W. Cairns, P.V. Hodson and J.O. Nriagu, Eds., Contaminant Effects on Fisheries. Wiley, New York, pp. 213-221.

Ellis, A.E. 1981. Stress and the modulation of defense mechanisms in fish. *In:* A.D. Pickering, Ed., Stress and Fish. Academic Press, London, pp. 147-169.

Farrell, A.P., J.A. Johansen, J.F. Steffensen, C.D. Moyes, T.G. West and R.K. Suarez. 1990. Effects of exercise training and coronary ablation on swimming performance, heart size and cardiac enzymes in rainbow trout, *Oncorhynchus mykiss*. Can. J. Zool. 68:1174-1179.

Farrell, A.P., J.A. Johansen and R.K. Suarez. 1991. Effects of exercise-training on cardiac performance and muscle enzymes in rainbow trout, *Oncorhynchus mykiss*. Fish Physiol. Biochem. 9:303-312.

Finney, D.J. 1971. Probit analyses. Cambridge University Press, London.

Fox, M.E. 1977. Persistence of dissolved organic compounds in kraft pulp and paper mill effluent plumes. J. Fish. Res. Board Can. 34:798-804.

Janz, D.M., A.P. Farrell, J.D. Morgan and G.A. Vigers. 1991. Acute physiological stress responses of juvenile coho salmon (*Oncorhynchus kisutch*) to sublethal concentrations of Garlon 4, Garlon 3A and Vision herbicides. Environ. Toxicol. Chem. 10:81-90.

Johansen, J.A. and G.H. Geen. 1990. Sublethal and acute toxicity of the ethylene glycol butyl ether ester formulation of triclopyr to juvenile coho salmon, *Oncorhynchus kisutch*. Arch. Environ. Contam. Toxicol. 19:610-616.

Kennedy, C.J., R.M. Sweeting, J.A. Johansen, A.P. Farrell and B.A. Mckeown. 1993. Stress effects of pulp mill effluent and its components in fish. *In:* A.P. Farrell, Ed., The aquatic resources research project: towards environmental risk assessment and management of the Fraser River basin. 450 pp.

Kutney, J.P. and E. Dimitriadis. 1982. Studies related to biological detoxification of kraft pulp mill effluent. V. The synthesis of 12- and 14-chlorodehydroabietic acids and 12,14-dichlorodehydroabietic acid, fish-toxic diterpines from kraft pulp mill effluent. Helv. Chim. Acta 65:1351-1363.

Leach, J.M. and A.N. Thakore. 1975. Isolation and identification of constituents toxic to juvenile rainbow trout, *Salmo gairdneri*, in caustic extraction effluents from kraft pulp mill bleach plants. J. Fish. Res. Board Can. 32:1249-1257.

Mantel, N. 1966. Evaluation of survival data and two new rank order statistics arising in its consideration. Cancer Chemother. Rep. 50:163-170.

Mattsoff, L. and A. Oikari. 1987. Acute hyperbilirubinaemia in rainbow trout, *Salmo gairdneri*, caused by resin acids. Comp. Biochem. Physiol. C 88:263-268.

McKague, A.B. 1981. Some toxic constituents of chlorination-stage effluents from bleached kraft pulp mills. Can. J. Fish. Aquat. Sci. 38:739-743.

Oikari, A., B-E. Lonn, M. Castren, T. Nakari, B. Snickars-Nikinmaa, H. Bister and E. Virtanen. 1983. Toxicological effects of dehydroabietic acid (DHAA) on the trout, *Salmo gairdneri* Richardson, in fresh water. Water Res. 17:81-89.

Oikari, A., T. Nakari and B. Holmbom. 1984a. Sublethal actions of simulated kraft pulp mill effluents (KME) in *Salmo gairdneri*: residues of toxicants, and effects on blood and liver. Ann. Zool. Fenn. 21:45-53.

Oikari, A., E. Anas, G. Kruzynski and B. Holmbom. 1984b. Free and conjugated resin acids in the bile of rainbow trout, *Salmo gairdneri*. Bull. Environ. Contam. Toxicol. 33:233-240.

Pearson, T.H. 1980. Marine pollution effects of pulp and paper industry wastes. Helgolander Meersunters. 33:340-365.

Pickering, A.D. and T.G. Pottinger. 1987. Poor water quality suppresses the cortisol response of salmonid fish to handling and confinement. J. Fish Biol. 30:363-374.

Rosemarin, A., M. Notini, M. Soderstrom, S. Jensen and L. Landner. 1990. Fate and effects of pulp mill chlorophenolic 4,5,6-trichloroguaiacol in a model brackish water ecosystem. Sci. Tot. Environ. 92:69-89.

Schreck, C.B. 1981. Stress and compensation in teleostean fishes: response to social and physical factors. *In:* A.D. Pickering, Ed., Stress and Fish. Academic Press, New York, pp. 295-321.

Schreck, C.B. 1990. Physiological, behavioral, and performance indicators of stress. Am. Fish. Soc. Symp. 8:29-37.

Specker, J.L. and C.B. Schreck. 1980. Stress response to transportation and fitness for marine survival in coho salmon (*Oncorhynchus kisutch*) smolts. Can. J. Fish. Aquat. Sci. 37:765-769.

Sprague, J.B. 1973. The ABC's of pollutant bioassay using fish. Biological methods for the assessment of water quality. American Society for Testing and Materials, STP 528:6-30.

Sweeting, R.M. 1989. Aspects of growth hormone in the physiology of smoltification and sea water adaptation of coho salmon, *Oncorhynchus kisutch*. Ph.D. Thesis, Simon Fraser University, British Columbia.

Thomas, P. and D.H. Lewis. 1987. Effect of cortisol on immunity in red drum, *Sciaenops ocellatus*. J. Fish. Biol. 31 (Supplement A):123-127.

Thomas, R., R.S. Carr and J.M. Neff. 1987. Biochemical responses and alterations of tissue ascorbic acid and glutathione content. *In:* C.S. Giam and L.E. Ray, Eds., Pollutant Studies in Marine Animals. CRC Press, Boca Raton, FL, pp. 155-180.

Zar, J.H. 1974. Biostatistical analysis. Prentice-Hall, Englewood Cliffs, New Jersey.

EFFLUENTS FROM CANADIAN PULP AND PAPER MILLS: A RECENT INVESTIGATION OF THEIR POTENTIAL TO INDUCE MIXED FUNCTION OXYGENASE ACTIVITY IN FISH

P.H. Martel, T.G. Kovacs and R.H. Voss

Pulp and Paper Research Institute of Canada, 570 St. John's Boulevard, Pointe Claire, Quebec, H9R 3J9 Canada

Technological innovations such as modified and extended kraft cooking, improved chemical recovery, reduced water usage, elementa -chlorine-free bleaching and biological treatment of effluents have been widely implemented by the Canadian pulp and paper industry. This survey was undertaken to assess the mixed function oxygenase (MFO)-inducing potential of effluents discharged by mills at the current state of operating technology. Between December 1993 and June 1994, 46 effluent samples were obtained from 33 different pulp and paper mills. In the laboratory, rainbow trout (*Oncorhynchus mykiss*) were exposed to a 10% (v/v) concentration of each effluent for 96 h. Hepatic ethoxyresorufin-*o*-deethylase (EROD) activity was assayed in effluent-exposed and control trout. Overall, 17 of the 46 samples caused no statistically significant increases in EROD activity, 10 samples caused statistically significant but ≤2-fold EROD inductions, while the remainder of the samples caused up to 15.5-fold induction. The EROD-inducing potential of effluents could not be linked to a particular bleaching or pulping process. For example, increased MFO activity was measured in trout exposed to some effluents from both unbleached and bleached kraft mills as well as thermomechanical and chemi-thermomechanical mills. Presently, the precise combinations of operating conditions that control the EROD-inducing potential of effluents from these mills are unclear.

INTRODUCTION

Presently, the major concerns regarding the environmental impact of pulp and paper mill effluents are based on reports of increased liver mixed function oxygenase (MFO) activity and symptoms of altered reproductive capacity in fish (Carey *et al.* 1993; Södergren *et al.* 1989). A recent Canadian field survey suggested that these effects occur at virtually every site studied (Munkittrick *et al.* 1994).

Our previous survey of final effluents from Canadian mills, for their ability to cause MFO induction, was performed in 1991-1992 (Martel *et al.* 1994). The survey focused on the effects of bleaching process changes from conventional (i.e., elemental chlorine-based bleaching) to elemental chlorine-free (ECF) and totally chlorine-free (TCF) bleaching. Secondary-treated effluents from mills using conventional, ECF and TCF bleaching were found to cause significant MFO induction in trout, irrespective of the bleaching technology used. It was also reported that treated unbleached kraft (UK) mill effluents and some thermomechanical pulp (TMP) mill effluents exhibited potential for MFO induction, although in a less pronounced and consistent fashion (Martel *et al.* 1994).

While a total of 31 effluents had been tested in the 1991-1992 survey, they originated from only five kraft mills, two TMP mills and one chemi-thermomechanical pulp (CTMP) mill. The eight mills in the survey represented less than 10% of the total number of mills in Canada. Thus, one could legitimately question whether the results of the survey were truly representative of the Canadian industry. In addition, since that time, technological changes have been implemented at many mills. These have included modified and extended kraft cooking, improved chemical and condensate recovery, reduced water usage and ECF bleaching, as well as improved and more widespread usage of secondary treatment facilities.

Consequently, a new survey of the ability of pulp and paper mill effluents to cause MFO induction was undertaken. The objective of the survey was to study effluents from as many mills with different processes as possible and to include at least one from each province where mills are currently in operation. Special attention was given to examining the MFO-inducing potential of non-kraft processes such as groundwood/sulfite (GWS), TMP and CTMP operations. It was felt that such an expanded survey would give a better picture regarding the extent and magnitude of the MFO-inducing potential of pulp and paper mill effluents at the current state of operating technology in Canada.

MATERIALS AND METHODS

Effluents

Grab samples of primary- or secondary-treated combined mill effluents were taken by mill personnel and shipped to the Paprican laboratory in Pointe Claire, Quebec. At least one sample was collected from every province where a pulp mill was currently in operation. The samples arrived at the laboratory within 1 to 7 d and they were stored at 4°C. The tests were initiated within 3 d of arrival of the effluents. In all, 46 samples, collected over a 6-month span, from 17 different bleached kraft (BK) mills, 2 UK mills, 7 TMP mills, 4 CTMP mills, 2 GWS mills and 1 mill concurrently producing groundwood (GW) and BK pulps were tested. Tables 1 and 2 give information for secondary- and primary-treated effluent samples, respectively, regarding mill operating conditions such as wood furnish, bleaching sequence, type of secondary treatment (where applicable) and water usage.

Fish Exposure Procedure

A procedure based upon direct exposure of fish to effluents under controlled laboratory conditions was used (Martel *et al.* 1994, 1995) to assess the ability of effluent samples to induce MFO enzyme activity in fish. Sexually immature rainbow trout weighing 34 to 101 g were purchased from a local fish hatchery (Arthabaska Inc., Chesterville, Quebec). Prior to use, the trout were acclimated for at least 2 wk to laboratory well water at a temperature of $13 \pm 1°C$, a pH of 8.3, hardness of 170 mg L^{-1} (as $CaCO_3$) and alkalinity of 275 mg L^{-1} (as $CaCO_3$). The same well water was used for effluent dilution purposes. The trout were kept in fiberglass tanks under flow-through conditions. Dissolved oxygen concentrations were maintained above 90% saturation with oil-free compressed air. The photoperiod was 16 h light and 8 h dark. The trout were fed a commercial diet at a daily rate of 1% body weight. Feeding was stopped 48 h before the beginning of each exposure.

Exposures lasting 96 h were conducted in glass aquaria of 165-L capacity at a temperature of $13 \pm 1°C$. For each experiment, six to ten fish of approximately the same weight and length were placed in aquaria containing either 10% effluent or well water which served as control. The fish were not fed during the test period. All exposures were conducted under static conditions with 80% daily renewals of test solutions. The test volumes were 100 or 150 L resulting in initial loading densities (i.e., fish-to-volume ratios) of 2.1 to 5.7 g L^{-1}. Parameters such as pH (7.0 to 8.5), dissolved oxygen (7.8 to 11.8 mg L^{-1}) and temperature (12 to 14°C) were monitored in each aquarium daily.

Hepatic EROD Activity Assay

At the end of the 96-h exposure, fish were stunned with a blow to the head and rapidly weighed and measured. Subsequently, all procedures were performed in a cold room at 4°C, and all reagents, vials and tubes were kept on ice. The liver was dissected, carefully removing the gall bladder to avoid puncture. The liver was rinsed with pH 7.4, 50 mM Tris-HCl, 0.15 M KCl buffer, dried with an absorbent paper, weighed and homogenized in buffer for six seconds using a Brinkmann homogenizer equipped with a 10-mm probe. The homogenate volume was adjusted to 25% weight/volume with pH 7.4, 50 mM Tris-HCl,

0.15 M KCl buffer and spun for 20 min at 10 000 g in a refrigerated centrifuge. The post-mitochondrial supernatant (PMS) was carefully removed and stored in 500-µL aliquots at -80°C until assayed for ethoxyresorufin-o-deethylase (EROD) activity.

Table 1. Characteristics of mills with secondary treatment.

Process	Sample	Wood furnish[1]	Bleaching sequence[2] and % ClO$_2$ substitution in the first stage	Type of secondary treatment[3] and retention time (d)	Water usage (m^3 ADt^{-1})
Bleached kraft	A	SW/HW‡	CdWEoDED, 45	ASB,7.5	108
	B	SW	DcEoDED, 60	ASB,5 + QS,15	116
	C-1	SW	XDcWEoDED, 75	ASB,8	115
	C-2	SW	XDcWEoDED, 79	ASB,8	112
	D	SW	DEoDED ODEEoDED D/CEoDED, 100§	AS,0.5	118
	E	SW	DEoDED, 100	ASB,7	62
	F-1	SW	DEopDED, 100	ASB,12	65
	F-2	SW	DEopDED, 100	ASB,12	60
	G-1	SW	DEEpo(DE)D, 100 DEEpDED, 100¶	ASB,6	98
	G-2	SW	DEEpo(DE)D, 100 DEEpDED, 100¶	ASB,6	98
	H	SW	ODEop(DED), 100	ASB,5	96
	I-1	SW	ODEopDED, 50/100*	ASB,7.5	94
	I-2	SW	ODEopDED, 100	ASB,7.5	110
	J-1	SW	ODEpopD$_N$D, 100	AS,1.5	59
	J-2	SW	ODEpopD$_N$D, 100	AS,1.5	71
	J-3	SW	ODEpopD$_N$D, 100	AS,1.5	87
	K-1	SW	ODEopDED, 100	ASB,9	93
	K-2	SW	ODEopDED, 100	ASB,9	83
	L	SW	O,Multi-stage P*	AS,1	65
	M-1	HW	DcEoHDP, 55	ASB,5	44
	M-2	HW	DcEo(HD)p, 55	ASB,5	44
	N	HW	CDEopDED, 50	ASB,15	106
	O-1	HW	OD$_N$DEopD, 100	AS,1	46
	O-2	HW	OD$_N$DEopD, 100	AS,1	46
Unbleached kraft	P	SW	NA	ASB,3	78

Table 1. (Continued).

Process	Sample	Wood furnish[1]	Bleaching sequence[2] and % ClO$_2$ substitution in the first stage	Type of secondary treatment[3] and retention time (d)	Water usage (m^3 ADt^{-1})
Groundwood & bleached kraft	Q†	HW/SW & SW	NA, DcEoHD, 70	ASB,10	29
Groundwood/ sulfite	R	SW	NA	AS,0.05	58
Thermomechanical	S-1	SW	Hydrosulfite	AS,2.5 + ASB,2.5	24
	S-2	SW	Hydrosulfite	AS,2.5 + ASB,2.5	24
	T-1	SW	Hydrosulfite	ASB,9.5	38
	T-2	SW	Hydrosulfite	ASB,9.5	38
	U-1	SW	Hydrosulfite	AS,1.4	45
	U-2	SW	Hydrosulfite	AS,1.4	26
Chemi-thermomechanical	V	SW/HW††	Peroxide	eaAS,10	22
	W	SW	Peroxide	AN + AS,2.25	23
	X	SW	Peroxide	ASB,5	23

NA = Not applicable.
[1] Abbreviation for wood furnish where SW stands for softwoods and HW for hardwoods. Classified as softwood (or hardwood) when composed of >80% of that wood furnish.
[2] Abbreviation for multi-stage bleaching sequence where X stands for xylanase pre-treatment, O for oxygen delignification, C for chlorine, W for Papricycle washing stage, D for chlorine dioxide, Cd and Dc for first bleaching stages where chlorine dioxide represents a minor and a major part, respectively, of the total bleaching chemical, E for extraction with alkali, Eo for extraction with alkali and the addition of elemental oxygen, Eop for extraction with alkali and the addition of elemental oxygen and peroxide, Epop for pressurized oxidative extraction stage, D_N for chlorine dioxide with neutralization, P for peroxide and H for hypochlorite.
[3] Abbreviation for secondary treatment where AS stands for activated sludge, ASB for aerated stabilization basin, QS for quiescent stabilization, eaAS for extended aeration activated sludge and AN for anaerobic treatment.
‡ This mill alternated between 5 d of softwood and 2 d of hardwood pulping.
§ This mill operated three bleaching lines where the ClO$_2$ substitution varied between 50 and 100% at the time of sampling.
¶ This mill concurrently operated two pulping and bleaching lines.
♦ Mill shifted from 50 to 100% ClO$_2$ substitution. Due to the retention time of secondary treatment, the sample collected contained mill effluents from both operations.
* TCF bleaching sequence.
† 75% of the effluent originated from the groundwood line and 25% from the bleached kraft line.
†† This mill concurrently operated softwood and hardwood lines.

The EROD assays were conducted within 2 wk after the fish were sacrificed. EROD analysis was of the fixed-time type and was based primarily on the fluorometric method described by Hodson et al. (1991). Final concentrations of substrate and cofactors in a 2.0-mL reaction volume were 1.0 µM ethoxyresorufin, 0.12 mM NADPH. Dicumarol was added at a final concentration of 10 µM to inhibit possible phase II metabolic enzymes which may be present, since they could interfere with the EROD assay (Lubet et al. 1985). Each PMS was thawed on ice and assayed in triplicate, with one blank. Variation in triplicate EROD assays was less than 5% in all cases. Each blank contained 1 µM

ethoxyresorufin. Incubation time was 5 min at 25°C in an oscillating (160 cycles min^{-1}) water bath. The reaction was stopped by the addition of 1 mL of 5% $ZnSO_4$ and 1 mL of saturated $Ba(OH)_2$ to precipitate protein. The assay tubes were centrifuged at 3400 g for 15 min. 1 mL of the deproteinized supernatant was mixed with 2 mL of a 0.5 M glycine-NaOH buffer, pH 8.5.

Table 2. Characteristics of mills with primary treatment only.

Process	Sample	Wood furnish[1]	Bleaching sequence[2] and % ClO_2 substitution in the first stage	Water usage (m^3 ADT^{-1})
Bleached kraft	AA-1	HW	CdEDED, 37	136
	AA-2	HW	CdEDED, 37	136
	BB	HW	DcEoDED, 70	74
Unbleached kraft	CC	SW/HW	NA	87
Groundwood/Sulfite	DD	SW	NA	75
Chemi-thermomechanical	EE	SW	Hydrosulfite	40
Thermomechanical	FF	SW	Hydrosulfite	21
	GG	SW	Hydrosulfite	55
	HH*	SW	Hydrosulfite	42
	II	SW	Hydrosulfite	73

NA = Not applicable.
[1] Abbreviation for wood furnish where SW stands for softwoods and HW stands for hardwoods.
[2] Abbreviation for multi-stage bleaching sequence where C stands for chlorine, D for chlorine dioxide, Cd and Dc for first bleaching stages where chlorine dioxide represents a minor and a major part, respectively, of the total bleaching chemical, E for extraction with alkali and Eo for extraction with alkali and the addition of elemental oxygen.
* Mill HH also produced a small amount of groundwood/sulfite pulp at the time of sampling, but will be considered as a TMP mill for the purpose of this survey.

Fluorescence was read using a Turner 112 filter fluorometer equipped with a narrow bandpass combination excitation filter effective at 546 nm and an emission sharp cut-off filter effective at 595 nm. The light source was a F4T5 green lamp with maximum energy output between 520 and 565 nm. Standards of known resorufin (Sigma Chemicals Co.) concentration were used to convert fluorescence units to resorufin concentration. EROD activity was expressed as picomoles of resorufin produced per minute per milligram of total protein in the PMS sample.

Total protein concentration of each PMS sample was assayed in duplicate by the method of Bradford (1976) using the Bio-Rad reagent (Bio-Rad Laboratories Ltd., Ontario, Canada). The assay is based on the shift in absorbency maximum from 465 to 595 nm of Coomassie Brilliant Blue G-250 when binding to proteins occurs. All protein analyses were performed on the same day as EROD assays. The protein concentration was determined from a standard bovine serum albumin preparation (Sigma Diagnostics, St. Louis, MO, U.S.A.).

Statistical Analysis

We tested for differences in average control and average effluent-exposed fish EROD activities by two-tailed comparisons of means using the t-test (Zar 1984) at the 5% significance level. The assumption of normally distributed data was verified by the Chi-square method. In some cases however, there were significant deviations from the assumption of equal variance between averages. Using non-parametric statistical methods to test for differences did not change any conclusions regarding any significant differences between means.

RESULTS

Overall, 17 of the 46 samples caused no statistically significant increases in EROD activity, ten samples caused statistically significant but ≤2-fold EROD inductions, while the remainder of the samples caused up to 15.5-fold induction (see Figs. 1 and 2).

Biologically Treated Effluents

Of the 24 biologically treated bleached kraft mill effluents (BKME) tested, 14 caused a statistically significant increase in EROD activity resulting in relative inductions of 1.4- to 8.6-fold when compared to controls (see Figs. 1A, 1B and 1C). One fish died during exposure to BKME sample F-1. Of the samples causing statistically significant EROD induction, eight were obtained from six mills bleaching softwood pulps at 45 to 100% chlorine dioxide substitution (see Fig. 1A). Significant EROD inductions were also measured following exposure to effluents sampled from kraft mills pulping a softwood furnish followed by oxygen delignification and bleached at 100% chlorine dioxide substitution (Fig. 1B). In addition, effluent sampled at Mill L, which bleached an oxygen delignified softwood pulp under TCF conditions, caused statistically significant EROD induction (see Fig. 1B).

The three effluent samples from two BK mills pulping predominantly hardwood furnishes without oxygen delignification and bleached at 50 to 55% ClO_2 substitution did not cause significant EROD inductions. Sample O-1 collected at a hardwood BK mill using oxygen delignification before bleaching at 100% ClO_2 substitution caused a low but statistically significant EROD induction of 1.4-fold, whereas the second sample (O-2) collected at the same mill did not have statistically significant effects (see Fig. 1C). The single samples collected from a UK mill (sample P) and a mill concurrently pulping softwood by the GWS and BK processes (sample Q) caused 2.8- and 2.9-fold EROD inductions, respectively.

Effluent obtained from a GWS mill (sample R) caused three fish to die during exposure to 10% effluent concentration. No significant EROD induction was measured from the remaining fish (see Fig. 1D). The six effluent samples obtained from three TMP mills with secondary treatment all caused less than 2-fold relative EROD inductions. However, the induction was statistically significant for three of the samples. Of the three biotreated CTMP mill effluents sampled, only one (sample W) caused a significant increase in EROD activity.

Effluents were collected more than once at 11 mills. The average EROD activities of effluent-exposed fish exposed to effluents from the same mills were within 2.5 pmoles min^{-1} mg^{-1} PMS of each other except for samples from Mills J and K (see Fig. 1).

Primary-Treated Effluents

Primary-treated effluent samples were obtained from two BK, one UK, one GWS, one CTMP, and four TMP mills. With the exception of one BK mill effluent sample, all others caused statistically significant EROD inductions of 1.7- to 15.5-fold (see Fig. 2). Two effluents (samples EE and FF) had to be tested at concentrations of 5 and 1%, respectively, due to acute lethal toxicity of these samples at the 10% concentration. Two effluents (samples BB and DD) caused four and one of eight fish to die, respectively, at 10% concentration.

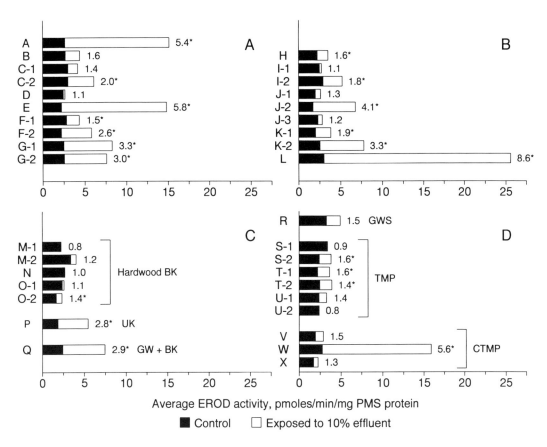

Figure 1. Average EROD activity of control and treatment groups of fish exposed to secondary-treated effluents sampled at (A) Seven BK mills where the wood furnish was predominantly (>80%) made up of softwoods and where no oxygen (O_2) delignification was performed before bleaching; (B) Five BK mills where the wood furnish was predominantly softwood and where O_2 delignification was performed before bleaching; (C) Three bleached kraft mills where the wood furnish was predominantly hardwood, one UK mill (softwood) and one mill where BK and GW lines operated concurrently; and (D) One GWS mill, three TMP and CTMP mills. Numbers beside bars are the ratios of treatment to control values (* = statistically significant).

DISCUSSION

Industry Significance

The most significant finding of this work was that 17 of the 46 effluents tested did not cause any statistically significant increase in the hepatic EROD activity of trout exposed to 10% effluent concentrations. The second most significant finding was that 26 of the 46 effluents elicited <2-fold relative induction. Actually, of the 36 secondary-treated effluents tested, only 33% caused enzyme activity to be ≥2 times the controls. Since by 1995, most mills in Canada will have some form of secondary treatment, we can assume that most effluents will not have the capacity to cause more than 2-fold increases in EROD activity.

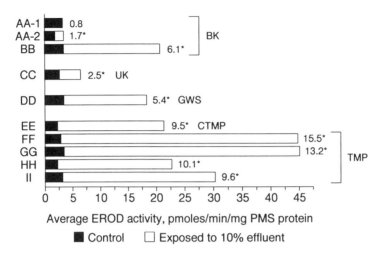

Figure 2. Average EROD activity of control and treatment groups of fish exposed to primary-treated effluents from: two BK mills, one UK mill, one GWS mill, one CTMP mill and four TMP mills., Numbers beside bars are the ratios of treatment to control values (* = statistically significant).

In this report we chose to distinguish between less than and greater than 2-fold inductions. All inductions which were >2-fold were found to be statistically significant. In contrast, when the relative induction was between 1- and 2-fold, statistically significant differences could not always be established. Even more important, when two different effluent samples caused the same level of enzyme activity, on one occasion the result was found to be statistically significant, but on the other occasion it was not, as illustrated by samples C-1 and O-2, causing 1.4-fold relative inductions (see Figs. 1A and 1C). Clearly, in such cases, it is the extent of variation around the means which affected the probability of finding statistical differences. Whether within-group variations are due to test conditions, experimental error or intrinsic differences in individual fish responses is unknown. However, enough is known about some biological factors to assist in the discussion of the biological significance of MFO activities which are <2-times greater than the controls. As we and others reported previously (Martel et al. 1995; Viganò et al. 1993), trout fed commercial food could exhibit >2-fold EROD activities over starved fish. The hepatic EROD activity of the control trout in this study and in our previous two studies (Martel et al. 1994, 1995) could also be different by at least a factor of two. In addition, the scope for EROD induction is very wide in this species. For example, trout exposed under the same conditions to some secondary-treated effluents and weak black liquor or subjected to intraperitoneal injections of β-naphthoflavone have been found to exhibit 30 to 50 times greater hepatic EROD activity than corresponding control fish (Martel et al. 1994, 1995). Collectively, these findings suggest that <2-fold inductions fall within the range of biological variation and thus we conclude that effluents causing <2-fold increases are marginal inducers at best.

The results of this survey can be considered to be the most extensive of the Canadian industry. Despite the extensive nature of the survey, it is only a snapshot representation of operations between December 1993 and May 1994, with only one effluent sample tested from most mills. In some cases, effluents collected from the same mills on different occasions had different EROD-inducing potential which could lead to contradictory interpretations. This variation of EROD-inducing potential was reported earlier (Martel et al. 1994). On the basis of the results of this survey, no clear trends regarding manufacturing processes and effluent treatment were evident, which could explain why the effluent from one mill could cause significant EROD induction while the effluent from another could not. While the operating conditions presented in Tables 1 and 2 do not adequately illustrate the actual complexity of the pulp manufacturing processes, the information in the tables clearly illustrates that no two mills using the same basic technology (e.g., bleached kraft pulp) are the same.

Since the use of chlorine in the bleaching of kraft pulp has been at the center of the controversy regarding the origin of EROD-inducing compounds, special attention was focused on data from such operations. It is evident that over the range of BK mills examined, there were no obvious indications that oxygen delignification, in combination with ECF and TCF bleaching, will discharge effluents either with or without EROD-inducing potential. This observation is consistent with previous reports (Andersson et al. 1987; Carey et al. 1993; Förlin et al. 1985; Gagné and Blaise 1993; Lehtinen et al. 1990; Lindström-Seppä and Oikari 1988; Martel et al. 1994, 1995; Mather-Mihaich and Di Giulio 1991). Nonetheless, one may be tempted to speculate that the type of wood furnish used by BK mills may be an important factor. Of the five effluents from mills using a hardwood furnish and having secondary treatment, none caused EROD activity to increase, by more than 2-fold. In fact, only one effluent was found to cause a statistically significant increase and the induction in this case was only 1.4-fold. However, further investigation is required to determine whether hardwood furnish will always result in effluents with negligible EROD-inducing potential.

The database for UK and GWS mill effluents is much more modest than for BKME. Nevertheless, effluents from mills utilizing these processes can cause significant MFO induction as indicated previously (Carey et al. 1993; Martel et al. 1994). Effluents from such mills, however, could also be non-inducing, as was reported earlier for UK mill effluent (Martel et al. 1994) and as was reported here for secondary-treated effluent (i.e., effluent sample R, see Fig. 1D) from a GWS mill.

In this survey, there were a total of nine secondary-treated effluents tested from TMP and CTMP mills. While four of the effluents caused statistically significant inductions, fish exposed to eight of the effluents caused hepatic EROD activities ≤ 1.6 times the hepatic activities of the control fish. Only one of the eight effluents (i.e., effluent W) had the ability to clearly affect the EROD activity of the fish. These results are in agreement with the low-level inducing capability of such effluents reported previously (Martel et al. 1994).

Although new regulations will require most mills in Canada to operate some type of secondary effluent treatment, at present some mills are operating with only primary effluent treatment. These were included in this survey to give a true picture of the situation as it exists today and to perhaps get a preliminary indication of the effect of secondary treatment. A superficial examination of the data may lead one to speculate that secondary treatment is reducing the EROD induction potential of effluents because these generally caused lower relative inductions compared to the primary-treated effluents. However, firm conclusions in this regard can not be made from our data. Our survey was not designed to investigate the role of secondary treatment on effluent quality and the effluents for the project were not obtained from the outfall of primary and secondary treatment facilities at the same mills.

From this survey, it is also not possible to determine if one type of secondary treatment is more effective in reducing or eliminating EROD induction potential than another. Significant EROD inductions were observed in fish exposed to effluents from mills with aerated lagoons of 5 to 12 d retention time or activated sludge plants with typically short (i.e., <2 d) retention times. Previous publications on the role of secondary treatment in the ability of effluents to cause increased MFO activity are contradictory. A field study (Munkittrick et al. 1992) found that the installation of secondary treatment at a mill site did not reduce EROD induction in wild fish. In contrast, recent work by Schnell et al. (1993) showed that optimized secondary treatment could substantially reduce the EROD-inducing potential of a BKME. More work is required to better understand the role of secondary treatment on effluent quality.

Ecological Significance

The relevance of the laboratory approach used for this survey to monitor the MFO-inducing potential of pulp and paper mill effluents has been discussed extensively elsewhere (Martel et al. 1994, 1995).

Overall, this survey was not designed to compare the EROD-inducing potency of different mills but simply to assess their EROD-inducing potential. For this reason, the use of a single test concentration of 10% (v/v) was both practical and relevant, as discussed elsewhere (Martel et al. 1994). To gain better insight into the potency of different effluents, test conditions would have to be identical and normalization of water usage by the mills would be required.

In this study, test conditions for each effluent were not identical, although every effort was made to keep them as consistent as possible between tests. Thus, while two-thirds of the tests were conducted at initial fish loading densities (fish to volume ratios) between 2.1 and 3.0 g L^{-1}, the remaining were conducted at initial loading densities between 3.1 and 4.3 g L^{-1}, with one exception at 5.7 g L^{-1} (sample T-1). Although a minimum of variation regarding test volume and loading density would have been preferred, this was not always possible because the same size fish were not available throughout the study. Nevertheless, the conditions reported here are within the range used in earlier work (Martel et al. 1994, 1995) which has been shown to be appropriate for detecting EROD-inducing potential.

We examined the possible relationships between water usage in Table 1 and EROD induction reported in Figs. 1 and 2, and it is evident that water usage is not the governing parameter which controls the EROD-inducing potential of an effluent. For example, samples E, F-1 and L were collected from different BK mills with comparable water usage (62 to 65 m^3 ADt^{-1}). The relative inductions for these samples were 5.8-, 1.5- and 8.6-fold, respectively. Obviously in these cases, factors other than water usage were responsible for differences in levels of EROD activity. As reported earlier, the relationship between water usage and EROD induction was also not supported by the results of repeated samplings conducted at the same mill over a period of 12 months (Martel et al. 1994).

The laboratory assay would be most interesting if it was predictive of effects noted in field-captured fish. We compared our results to the findings of a recently published (Munkittrick et al. 1994) study which examined the hepatic MFO status of white suckers captured from waters receiving the discharges of selected pulp mills in Ontario. In our previous survey (Martel et al. 1994) and this survey, we tested effluents from five of the mills included in the field survey. In all cases there was good agreement between the results obtained in this laboratory and the results reported by the field study with respect to the ability of an effluent to cause hepatic MFO induction.

Presently, there is still much debate about the ecological significance of increased MFO activity resulting from exposure to pulp and paper mill effluents. Some investigations have reported that fish with increased MFO activity exhibited other biological effects (Carey et al. 1993; Munkittrick et al. 1994; Södergren et al. 1989), while others reported MFO induction in fish but no other apparent effects (Borton 1993; Kloepper-Sams and Benton 1994). Therefore, for now, this biomarker response can only be used to indicate that an organism has been exposed to an effluent, and decisions regarding biological significance require further research.

CONCLUSIONS

The effluents from some currently operating BK, TMP, CTMP and GWS mills tested at 10% did not cause statistically significant EROD induction in rainbow trout. Furthermore, at least half of the effluent samples tested in this Canada-wide survey elicited <2-fold relative inductions in trout exposed to a 10% effluent concentration for 96 h, which is similar to the variation in hepatic EROD activities found in control fish sampled at different times or fed certain diets. Thus, the results of this survey question the claim that all pulp mill effluents have the potential to cause MFO induction.

No particular manufacturing processes could be linked with the absence of MFO-inducing potential as measured by the EROD assay, although we speculate that pulping a hardwood furnish may result in little or no MFO-inducing potential. Conversely, we still do not know what are the key factors in the manufacturing processes which result in effluents causing an increase in MFO activity. To answer this question, the contribution of specific process plumes from various types of pulp and paper operations should be assessed along with secondary treatment options and the role of various additives used by the industry. Hopefully, this will demonstrate which combinations of existing technologies permit the operation of an open system mill that discharges effluent with no potential to cause MFO induction.

ACKNOWLEDGMENTS

The dedicated technical assistance of Maria V. Ricci and Dany Boudrias is gratefully acknowledged. The technical support of Michel Simard, Brian Walker, Jim Larocque and Laura Sciascia is also acknowledged. We thank the mills and their staff for their participation in this survey. The authors thank Mr. Mike Paice of Paprican for his review of the manuscript. This work was supported by the maintaining members of the Pulp and Paper Research Institute of Canada and funded in part by Industry Canada.

REFERENCES

Andersson, T., B.E. Bengtsson, L. Förlin, J. Härdig and Å. Larsson. 1987. Long-term effects of bleached kraft mill effluents on carbohydrate metabolism and hepatic xenobiotic biotransformation enzymes in fish. Ecotoxicol. Environ. Saf. 13:53-60.

Borton, D.L. 1993. Survival, growth, production and biomarker responses of fish exposed to biologically treated bleached kraft mill effluent in experimental streams. Abstract Book of the 14th Annual Meeting of the Society of Environmental Toxicology and Chemistry, November 14-18, 1993, Houston, Texas.

Bradford, M.M. 1976. A rapid and sensitive method for the quantization of microgram quantities of protein utilizing the principle of protein-dye binding. Anal. Biochem. 72:248-254.

Carey, J.H., P.V. Hodson, K.R. Munkittrick and M.R. Servos. 1993. Recent Canadian studies on the physiological effects of pulp mill effluent on fish. Environment Canada-Fisheries and Oceans Special Report, Burlington, Ontario, Canada.

Förlin, L., T. Andersson, B.-E. Bengtsson, J. Härdig and Å. Larsson. 1985. Effects of pulp bleach plant effluents on hepatic xenobiotic biotransformation enzymes in fish: laboratory and field studies. Mar. Environ. Res. 17:109-112.

Gagné, F. and C. Blaise. 1993. Hepatic metallothionein level and mixed function oxidase activity in fingerling rainbow trout (*Oncorhynchus mykiss*) after acute exposure to pulp and paper mill effluents. Water Res. 27:1669-1682.

Hodson, P.V., P.J. Kloepper-Sams, K.R. Munkittrick, W.L. Lockhart, D.A. Metner, P.L. Luxon, I.R. Smith, M.M. Gagnon, M. Servos and J.F. Payne. 1991. Protocols for measuring mixed function oxygenases of fish liver. Canadian Technical Report of Fisheries and Aquatic Sciences 1829, Fisheries and Oceans Canada, Mont-Joli, Quebec.

Kloepper-Sams, P.J., and E. Benton. 1994. Exposure of fish to biologically treated bleached-kraft effluent. 2. Induction of hepatic cytochrome P4501A in mountain whitefish (*Prosopium williamsoni*) and other species. Environ. Toxicol. Chem. 13:1483-1496.

Lehtinen, K-J., A. Kierkegaard, E. Jackobson and A. Wandell. 1990. Physiological effects in fish exposed to effluents from mills with six different bleaching processes. Ecotoxicol. Environ. Saf. 19:33-46.

Lindström-Seppä, P. and A. Oikari. 1988. Hepatic xenobiotic biotransformation in fishes exposed to pulp mill effluents. Wat. Sci. Technol. 20:167-170.

Lubet, R.A., R.W. Nims, R.T. Mayer, J.W. Cameron and L. Schechtman. 1985. Measurement of cytochrome P-450 dependent dealkylation of alkoxyphenoxazones in hepatic S-9s and hepatocyte homogenates, effects of dicumarol. Mutat. Res. 142:127-132.

Martel, P.H., T.G. Kovacs, B.I. O'Connor and R.H. Voss. 1994. A survey of pulp and paper mill effluents for their potential to induce mixed function oxidase enzyme activity in fish. Water Res. 28:1835-1844.

Martel, P.H., T.G. Kovacs, B.I. O'Connor and R.H. Voss. 1995. A laboratory exposure procedure for screening the potential of pulp and paper mill effluents to cause increased mixed function oxidase activity in fish. Environ. Pollut.: In Press.

Mather-Mihaich, E., and R.T. Di Giulio. 1991. Oxidant, mixed-function oxidase and peroxisomal responses in channel catfish exposed to a bleached kraft mill effluent. Arch. Environ. Contam. Toxicol. 20:391-397.

Munkittrick, K.R., G.J. Van Der Kraak, M.E. McMaster and C.B. Portt. 1992. Response of hepatic MFO activity and plasma sex steroids to secondary treatment of bleached kraft mill effluent and mill shutdown. Environ. Toxicol. Chem. 11:1427-1439.

Munkittrick, K.R., G.J. Van Der Kraak, M.E. McMaster, C.B. Portt, M.R. van den Heuvel and M.R. Servos. 1994. Survey of receiving water environmental impacts associated with discharges from pulp mills. 2. Gonad size, liver size, hepatic EROD activity and plasma sex steroid levels in white sucker. Environ. Toxicol. Chem. 7:1089-1101.

Schnell, A., P.V. Hodson, P. Steel, H. Melcer and J.H. Carey. 1993. Optimized biological treatment of bleached kraft mill effluents for the enhanced removal of toxic compounds and MFO induction response in fish. Preprints of the 1993 Environment Conference of the Canadian Pulp and Paper Association, pp. 97-111.

Södergren, A., P. Jonsson, A. Larsson, L. Renberg, M. Olsson, S. Lagergren and B.-E. Bengtsson. 1989. Summary of results from the Swedish Project Environment/Cellulose. Water Sci. Technol. 20:49-60.

Vigano, L., A. Arillo, M. Bagnasco, C. Bennicelli and F. Melodia. 1993. Xenobiotic metabolizing enzymes in uninduced and induced rainbow trout (*Oncorhynchus mykiss*): Effects of diets and food deprivation. Comp. Biochem. Physiol. 104C:51-55.

Zar, J.H. 1984. Biostatistical Analysis. Second edition. Prentice Hall, Englewood Cliffs, New Jersey.

THE USE OF STABLE ISOTOPIC ANALYSES TO IDENTIFY PULP MILL EFFLUENT SIGNATURES IN RIVERINE FOOD WEBS

Leonard I. Wassenaar and Joseph M. Culp

National Hydrology Research Institute, 11 Innovation Boulevard, Saskatoon, Saskatchewan, S7N 3H5 Canada

Because stable isotopes of carbon and sulfur in pulp mill effluent (PME) can be incorporated through food sources into biological tissues, we hypothesized that stable isotopes may be used to quantify exposure of riverine biota to PME. We tested this hypothesis by assessing the C, N and S isotopic compositions of the waters and food web of the Thompson River and effluent samples from the Weyerhaeuser pulp mill in Kamloops, BC. Although the $\delta^{34}S$ isotopic composition of the PME was distinct, it could not be used reliably to trace the fate of PME due to the mixing pattern of waters of the North and South Thompson River. In contrast, the food web had previously been thought to be supported by C fixed by riverine algae. However, an isotopic mass balance model suggested two C sources support this food web: (1) the algal biofilm which utilizes dissolved inorganic C from the river and (2) a source with an $\delta^{13}C$ isotopic signature similar to terrestrial plants. Similarly, $\delta^{15}N$ isotopic composition of the food web indicated a substantial input from terrestrial plants. This striking and unexpected input of terrestrial C and N is hypothesized to be from PME, suggesting that C and N loadings from PME may play an important role in supporting downstream food webs.

INTRODUCTION

Treated effluents from bleached kraft pulp mills have the potential to modify aquatic food webs by affecting the biota of primary and secondary producer trophic levels. These environmental impacts can be attributed to the effects of organic and nutrient enrichment or to chemical toxicity caused by contaminants contained within effluents. For example, nutrients added by these effluents can produce major changes in the level of algal productivity (Bothwell *et al.* 1992) and species diversity (Amblard *et al.* 1990). In severe cases organic enrichment from pulp mills can decrease species diversity of benthic invertebrates (Poole *et al.* 1978), but under moderate enrichment effluent addition may lead to increased invertebrate productivity (Swanson *et al.* 1992). Impacts on fish include reduced recruitment, inhibited growth of gonads and induction of hepatic mixed function oxygenase (Munkittrick *et al.* 1992; Servos *et al.* 1992).

Although biological responses to pulp mill effluents (PME) can be measured, often it is difficult to quantify exposure of the biota to PME because of significant dilution in the receiving waters. Following the widely used stressor-exposure-response model of Hunsaker and Carpenter (1990), the quantitative establishment of exposure of biota to PME stressors is critical to developing cause-and-effect relationships among PME stressors and biological responses. A promising technique for establishing such exposure is the use of stable isotopic analyses, since stable isotopes of carbon, nitrogen and sulfur incorporated through food sources into biological tissues may serve as markers, or signatures, of the spatial extent of PME exposure (ingestion) in downstream riverine food webs.

Naturally occurring stable isotopes of carbon, nitrogen or sulfur have been used in a variety of terrestrial studies to follow the fate of anthropogenic pollutants in natural systems (e.g., sewage, nitrate contamination, sour gas), but to a much lesser extent in aquatic environments and food chain studies (Gearing *et al.* 1991; Van Dover *et al.* 1992). The basis of the natural abundance stable isotope approach is that pollutants entering an ecosystem may have isotopic signatures of ^{15}N, ^{13}C or ^{34}S that are distinct from the unpolluted environment (Van Dover *et al.* 1992). Organisms that ingest these pollutants will assimilate, along with their other food sources, the isotopic signals of the contaminant into their body parts. In addition, for food chain studies, biological fractionation (or discrimination) against a particular isotope with increasing trophic status must also be accounted for (Fry and Sherr 1984; Minagawa and Wada 1984). Fortunately, significant isotopic discrimination appears to be limited to nitrogen (e.g., about 3‰ isotopic enrichment per trophic level) and is much less for carbon or for sulfur (about 0-1‰; Minagawa and Wada 1984; Fry and Sherr 1984; Rundel *et al.* 1988; Fry 1991). As a result, carbon and sulfur isotopes may be used to track the fate of pollutants into the food chain, and nitrogen isotopes can further be used to establish trophic status. To our knowledge, no studies have attempted to trace the fate of pollutants into the food chain in terrestrial aquatic systems using stable isotope techniques.

Our working hypothesis was that PME would contribute higher than background levels of dissolved or particulate organic carbon to the river, which could subsequently oxidize and contribute isotopically "light" CO_2 ($\delta^{13}C$ of -25 to -30‰; terrestrial organic matter) to the dissolved inorganic carbon (DIC) pool of the Thompson River. Further, sodium sulfate (saltcake) used in the pulp treatment process could add higher than background sulfate levels to the river, possibly with a unique $\delta^{34}S$ signature. The stable isotopes of carbon, nitrogen and sulfur in aquatic organisms could be used to establish food web patterns and trace the flow of PME-derived carbon and sulfur into the aquatic food web of the Thompson River.

The specific objectives were (1) to assess whether stable isotopes of carbon and sulfur could be used to characterize and trace the fate of PME carbon and sulfur into the waters of the Thompson River and (2) to determine the extent to which stable isotopes could be used to trace PME-derived carbon and sulfur into the Thompson riverine food web. These objectives were to be met by (1) chemically and isotopically characterizing PME carbon (DIC) and sulfur (SO_4) discharging into the Thompson River; (2) establishing natural background levels and the isotopic composition of DIC, sulfate and nitrate in the North and South Thompson and Thompson rivers; and (3) measuring the carbon, nitrogen and sulfur isotopic compositions aquatic organisms in the form of biofilm, invertebrates and fish. It was anticipated *a priori* that the results of the study could be compromised by mixing of waters of different chemistries and isotopic composition from the North and South Thompson rivers and further complicated by the presence of sewage discharge from the Kamloops sewage treatment plant just upstream from the pulp mill.

SAMPLING AND ANALYTICAL METHODS

A single field trip for the purpose of collecting river water, epilithic biofilm, invertebrates and fish from the Thompson River was conducted between March 1 to 5, 1993. Sampling stations included sites on the North Thompson, South Thompson and Thompson Rivers, as well as effluent samples from the Kamloops sewage treatment plant and the Weyerhaeuser pulp mill (Fig. 1). Due to the extensive presence of ice cover on the North and South Thompson rivers, only water samples could be collected from all of the stations. A total of 52 water samples were collected, 4 from each of the 13 sampling stations; 21 samples of epilithic biofilm were collected at 5 sites along the Thompson River, and samples of invertebrates and fish were collected at only 2 sites along the Thompson River.

Water Samples

River water samples were collected from 11 sites along the Thompson River during the week of March 1-5, 1993 (Fig. 1). Water samples for oxygen and hydrogen isotope analyses were collected in tightly capped 50 mL plastic bottles. These samples for water isotopes were collected to aid in distinguishing mixing proportions of waters from the North and South Thompson rivers, the sewage treatment plant and the pulp mill. Water samples for selected nutrient ion concentration analyses (SO_4, NO_3) were collected in 50 mL plastic bottles, stored at 5°C and analyzed within 1 wk using standard ion chromatography techniques. Dissolved sulfate for $\delta^{34}S$ analyses were precipitated as $BaSO_4$ from 1 L samples as by the addition of excess barium chloride. River water DIC samples were collected in tightly sealed 1 L plastic bottles and stored at 5°C until further processing.

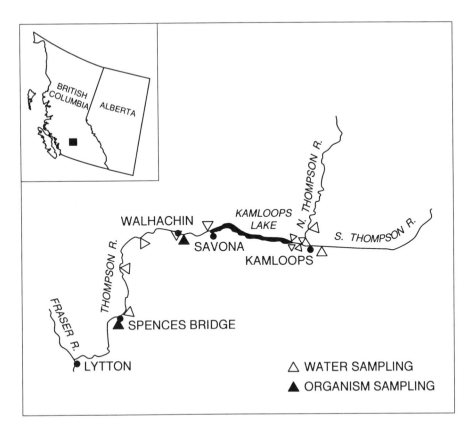

Figure 1. Location of the water and organism sampling sites on the South Thompson, North Thompson, and Thompson rivers, BC.

Stable carbon isotopic analyses of DIC from river water and sewage effluent were performed on 50 mL aliquots, injected via septum into an evacuated 500 mL flask containing 100 mL 85% H_3PO_4. Carbon isotope analyses of DIC could not be done on PME because of its lower viscosity. The CO_2 produced from the DIC extraction was cryogenically purified for $\delta^{13}C$ analyses. Sulfate isotopic analyses ($\delta^{34}S$) were conducted at the Department of Physics, University of Calgary. Water samples (2 L) for hydrogen isotopic analyses were reduced to H_2 gas by reaction with zinc in 6 mm sealed Pyrex tubes at 500°C (Coleman *et al.* 1982). Water samples for oxygen isotope analyses were prepared using a VG Isoprep18™ water-CO_2 equilibrator. Stable isotope measurements for H, C, O and N were performed using a VG Optima™ stable isotope ratio mass spectrometer at the National Hydrology Research Institute (NHRI). All stable isotope data are reported relative to the pertinent international standard (carbon - PDB; hydrogen - SMOW; oxygen - SMOW; nitrogen - air; sulfur - CDT) in the typical delta (‰) notation (Fritz and Fontes 1980). Soluble nutrient ion (NO_3, SO_4) concentrations were conducted on water samples using standard liquid ion chromatography techniques at NHRI.

Aquatic Organisms

Samples of epilithic biofilm were collected from five sites along the Thompson River by using a metal scalpel blade to scrape the epilithic mat into a collecting jar. Each sample represented a composite of material from 2-3 large stones (64-256 cm^2). Stones with a thick biofilm mat were selected for sampling to facilitate collection of large amounts of biofilm.

Invertebrates and fish could be collected only from two of the biofilm sampling sites (Walhachin and Spences Bridge) due to the large investment of time required to collect sufficient biomass for the analyses. Invertebrates were sampled by placing a U-net (Scrimgeour *et al.* 1993) with 250 μm mesh downstream of an area of substratum and vigorously agitating the streambed to a depth of 5 cm. All animals collected in the net were placed in sorting trays on the streambank and the study animals live sorted by taxa. Fish were collected by electrofishing and catching the stunned fish in a minnow net held downstream of the shocking area. In order to clear their guts, each taxonomic group of invertebrates or fish was held separately in aerated aquaria at 10°C for 18-24 h before being placed into vials and frozen. Because organic debris, biofilm and mucus was found to adhere to some animals, considerable time was spent removing this material from individual specimens to reduce the possibility of sample contamination.

In the laboratory, freeze-dried epilithic biofilm, invertebrates and fish were analyzed for their stable carbon and nitrogen isotopic compositions. Fish samples were freeze-dried and ground. For ^{13}C, 2 mg samples of whole biofilm and invertebrates were placed in 20 cm Vycor™ breakseal tubes along with 2 g CuO and 1 g Ag wire (Boutton *et al.* 1983). Samples were evacuated to <10^{-3} torr and the tubes sealed using a flame torch. Samples were combusted at 850°C for 2 h, followed by slow cooling at 0.8°C per h. The CO$_2$ gas produced was cryogenically purified and analyzed for δ^{13}C as described above. For ^{15}N, 10-40 mg samples of biofilm, insects and fish were prepared using the CaO combustion method described by Kendall and Grim (1990). Purified dinitrogen was analyzed for δ^{15}N on a VG Optima™ as described above.

RESULTS AND DISCUSSION

Results of all isotopic analyses of waters, dissolved inorganic carbon and sulfate are reported in Table 1. The measurements taken reflect a single snapshot in time, and it should be noted that the data presented cannot be over-interpreted, as the effect of seasonal variations in water chemistry, isotopes, effluent and water volumes was not within the scope of the project.

Effluent Inputs to the Thompson River

The waters of the North and South Thompson rivers were isotopically distinct in their oxygen isotopic composition, -18.3 and -17.2‰, respectively (Table 1). These isotopic end-members could be used to calculate the downstream mixing proportion of the rivers using a simple two-component isotope mass balance:

$$\delta^{18}O_{downstream} = X (\delta^{18}O_{N.\ Thompson}) + Y (\delta^{18}O_{S.\ Thompson}) \qquad (1)$$

where x + y = 1. Using a downstream (Savona and below) average of -17.6‰ (Table 1), the mixing proportion of the North and South Thompson rivers was about 36 and 64%, respectively. Hydrogen isotope mass balance calculations yielded a similar value.

Both the sewage effluent and pulp mill effluent (Table 1) were slightly isotopically heavier than either the North or South Thompson rivers for δ^{18}O and δ^{2}H. This suggested minor addition of isotopically heavier waters, possibly from reservoir storage or additional water from sources other than the Thompson River.

The δ^{18}O data from the Thompson River below the confluence of the North and South (Fig. 1) and just below the sewage treatment plant and the pulp mill indicated incomplete mixing of the North and South Thompson rivers. The water sample collected from the left bank yielded δ^{18}O values closer to that of the North Thompson (-17.6 to -17.9‰), whereas the water samples collected near the right bank more closely resembled the isotopic signature of the South Thompson River. The Savona site and other downstream occur below Kamloops Lake (Fig. 1) and are presumably well mixed.

Table 1. Isotopic and chemical composition of water, dissolved inorganic carbon (DIC) and dissolved nutrients at sampling sites on the South Thompson, North Thompson, and Thompson Rivers, BC, March 1-5, 1993. TR indicates the site was on the Thompson River.

Sampling Site	$\delta^{18}O$ Water	δ^2H Water	$\delta^{13}C$ (DIC)	$\delta^{34}S$ ($SO_4^=$)	NO_3^- mg L^{-1}	$SO_4^=$ mg L^{-1}
North Thompson	-18.3	-140.0	-7.5	3.2	0.7	12.8
South Thompson	-17.2	-131.9	-8.1	-0.0	0.3	10.0
Sewage Effluent	-16.9	-131.8	-13.9	2.1	<.16	89.5
Right Bank Below Sewage (TR)	-17.3	-133.0	-9.0	3.3	0.4	17.9
Left Bank Below Sewage (TR)	-17.6	-138.3	-9.3	2.2	0.5	15.4
Pulp Effluent	-16.3	-130.7	-	1.3	<0.8	307.0
Left Bank Below Pulp Mill (TR)	-17.9	-138.3	-12.9	2.6	0.6	40.3
Right Bank Below Pulp Mill (TR)	-17.3	-133.5	-8.6	2.3	0.4	13.2
Savona (TR)	-17.6	-135.5	-7.1	3.7	0.5	10.0
Walhachin (TR)	-17.6	-134.8	-7.8	4.2	0.5	9.9
Ashcroft (TR)	-17.8	-134.3	-7.4	3.3	0.5	10.9
Highland Valley (TR)	-16.9	-134.9	-7.6	2.2	0.6	11.6
Spences Bridge (TR)	-17.6	-135.3	-7.5	0.2	0.6	13.0

Nutrients and Isotopes of Carbon and Sulfur

Background concentrations of NO_3 along the Thompson River ranged between 0.3 and 0.6 mg L^{-1}, and were below detection in both the sewage effluent and PME (Table 1). Background sulfate concentrations ranged between 9.9 and 13 mg L^{-1} in the Thompson Rivers; however, sulfate concentrations 8-30 times higher than background in the sewage effluent (89.5 mg L^{-1}) and the PME (307 mg L^{-1}) indicated an anthropogenic source of sulfate in these effluents. The $\delta^{34}S$ isotopic composition of the sewage effluent and the PME (2.1 and 1.3‰, respectively) was isotopically distinct from both the North and South Thompson rivers (3.2 and 0.0‰) (Table 1). Unfortunately, the 36/64 mixing of the North and South Thompson rivers yielded a background $\delta^{34}S$-SO_4 that was virtually identical to both PME and the sewage effluent. As a result, $\delta^{34}S$ could not be reliably used to trace the fate PME or sewage sulfur into the Thompson River food web.

Background $\delta^{13}C$ data for DIC in the Thompson River ranged between -7.1 and -8.1‰ (Table 1). These values are typical for a well-mixed bicarbonate-dominated river in equilibrium with atmospheric CO_2 (LaZerte and Szalados 1982). The sewage effluent, however, was significantly isotopically depleted in ^{13}C relative to the background in the river. Values of -13.9‰ for the sewage effluent indicated the influence of terrestrial-derived organic carbon (-25 to -30‰) to the DIC pool. The negative shift in DIC was also observed in the Thompson River DIC just below the sewage plant diffusers. Although we were unable to measure the δ^{13}C-DIC of the PME, a strong carbon isotopic depletion in the Thompson River DIC just below the PME diffusers indicated that PME contributed a significant terrestrial carbon input to the Thompson riverine DIC pool. The more depleted carbon isotope values resulted from the oxidation of terrestrial organic matter (in sewage and PME) to DIC:

$$CH_2O + O_2 \rightarrow CO_2 + H_2O \rightarrow H_2CO_3 \qquad (2)$$

The DIC produced from the oxidation of terrestrial organic matter would have a similar carbon isotopic signature as its source (-25 to -30‰, Deines 1980). The isotopically depleted DIC produced, however, is diluted with background DIC in the PME, sewage effluent and the Thompson River. For example, if we assumed a PME DIC carbon isotopic value of -20 to -25‰, and an upstream DIC isotopic value of -9.3 (below sewage plant), then using equation 1 and substituting for carbon, we calculate that for the sample near the left bank below the PME diffusers, 23 to 34% of the riverine DIC at this site would be derived from PME. Thus, the δ^{13}C data suggest that carbon isotopes may be a good tracer for following the fate of PME-derived carbon into a riverine food chain. Unfortunately, during the single sampling trip, biofilm or aquatic organisms could not be obtained from sites near the pulp mill or sewage treatment plant (see below).

Isotopes and Trophic Structure of Aquatic Organisms

The food web of the lower Thompson River is thought to be supported by autochthonous (i.e., within the ecosystem) primary production and consist of three trophic levels, namely algae, grazing insects and insectivorous fish (Bothwell *et al.* 1992; Bothwell and Culp 1993). Filter-feeding insects are omnivorous, consuming algal and animal material trapped in their feeding nets. Of the potential autochthonous carbon sources, production by diatoms has been identified as the major component of plant productivity. Conventional stream ecology theory suggests that in an open canopy river such as the Thompson, allochthonous carbon from terrestrial sources, including particulate (POM) and dissolved (DOM) organic matter, should be a relatively unimportant component of energy flow through the ecosystem (Vannote *et al.* 1980).

A dual-isotope tracer study is a useful approach for checking conventional ideas about food web structure (Fry 1991). Carbon composition can be used to indicate the source of plant material for higher trophic levels because metabolism of the biota does not alter the ^{13}C/^{12}C ratio significantly (Rosenfeld and Roff 1992). Nitrogen is fractionated during metabolism such that the ^{15}N/^{14}N ratio is shifted by 2.5-4.0‰ with each successive trophic level (Minagawa and Wada 1984; Fry 1991). In addition, δ^{34}S can be used to trace food sources.

^{13}C of Carbon Sources and Consumers

Algal biofilms in the Thompson River had a carbon composition of -16.6 to -13.7‰ (Table 2; Fig. 2). The dominant carbon source for algae in this river appeared to be DIC. Isotopic fractionation of carbon on the order of 8-12‰ (LaZerte and Szalados 1982) from DIC (-7‰) indicated that the algal carbon source was indeed from riverine DIC. However, we did not attempt to measure other sources of carbon, such as organic detritus (e.g., leaves, grasses, DOM from PME, etc.), that might be available to animal consumers.

Table 2. Carbon and nitrogen isotopic composition of biofilm, invertebrates and fish at Thompson River stations, March 1-5, 1993.

Sample Type	Site	$\delta^{13}C$ (PDB)	$\delta^{15}N$ (Air)
Periphyton	Savona	-14.8	3.8
Periphyton	Walhachin	-15.1	2.7
Periphyton	Ashcroft	-13.7	3.8
Periphyton	Highland Valley	-16.6	0.8
Periphyton	Spences Bridge	-14.3	-
FISH TAXA			
Sculpin	Walhachin	-22.9	9.1
Sculpin	Walhachin	-22.8	8.6
Rainbow Trout	Walhachin	-24.1	9.7
Longnose Dace (large)	Walhachin	-25.8	8.8
Longnose Dace (small)	Walhachin	-21.8	8.5
Longnose Dace (large)	Spences Bridge	-23.7	7.8
Longnose Dace (small)	Spences Bridge	-19.1	6.8
Sculpin	Spences Bridge	-20.1	8.5
INSECT TAXA			
Chironomid	Walhachin	-18.2	2.2
Ephemerella	Walhachin	-22.5	4.2
Ephemerella (small)	Walhachin	-21.9	3.9
Ephemerella (small)	Walhachin	-21.5	3.2
Baetis	Walhachin	-21.7	4.7
Perlodidae	Walhachin	-22.8	4.8
Hydropsyche	Walhachin	-26.5	6.8
Chironomid	Spences Bridge	-18.3	1.8
Ephemerella	Spences Bridge	-22.0	3.5
Ameletus	Spences Bridge	-21.3	4.8
Baetis	Spences Bridge	-22.0	3.7
Heptageniidae	Spences Bridge	-23	4.8
Perlodidae	Spences Bridge	-21.9	4.8

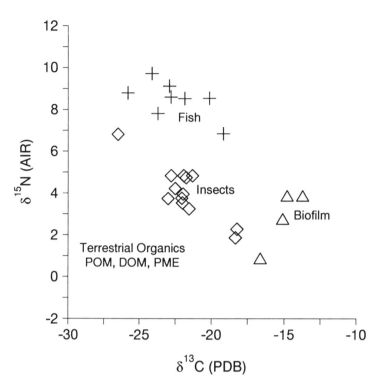

Figure 2. Carbon and nitrogen isotopic composition of epilithic biofilm (△), insects (◊) and fish (+). Terrestrial organics range after Deines (1980).

In the Thompson River all benthic insects were depleted by about 4‰ relative to the algal biofilm (Table 2; Fig. 2), a shift which could not be attributed to fractionation alone. Because gut analyses of insects such as mayflies and chironomids demonstrate that these animals consume large quantities of the biofilm (Culp and Glozier unpubl. data), it is likely that in addition to algae, these insects consume a more isotopically depleted carbon source. This unknown food source is likely terrestrial in origin and may include leaf and wood detritus, POM from upstream headwaters or DOM from PME ($\delta^{13}C$ of -25 to -30‰). Assuming that the unmeasured terrestrial food source had a $\delta^{13}C$ of -25 to -30‰, biofilm an average $\delta^{13}C$ of -14.5‰, the insects an average $\delta^{13}C$ of -22.5‰, no trophic level carbon isotopic fractionation, and substituting into a simple two-component isotope mass balance,

$$\delta^{13}C_{insects} = X(\delta^{13}C_{biofilm}) + Y(\delta^{13}C_{terrestrial\ organic}) \qquad (3)$$

it would appear that most of the insects consumed between 24-48% biofilm carbon and 51-76% terrestrially derived organic carbon. Nitrogen isotopes were used to further support this estimate (see below).

Of the insects examined, chironomids had $\delta^{13}C$ values closest to that of their presumed carbon source, the algal biofilm. The mayflies, *Baetis tricaudatus, Ameletus,* and *Ephemerella*, had more depleted $\delta^{13}C$ values, indicating a greater dependency on the unmeasured terrestrial carbon source. Finally, the $\delta^{13}C$ of the caddisfly filter-feeder, *Hydropsyche* (-26.5‰), was within the -30 to -25‰ range often measured for C_3 terrestrial plants (Fry 1991).

All of the fish sampled are thought to be predominantly insectivorous (Scott and Crossman 1973; Culp 1989; Nelson and Paetz 1992). In fact, the $\delta^{13}C$ values corroborate this conventional view since the carbon composition of fish overlapped completely with that of aquatic insects (Table 2; Fig. 2). Small longnose dace (<50 mm; *Rhinichthys cataractae*) and sculpin (*Cottus* sp.) had less depleted $\delta^{13}C$ values (-22.9 to -19.1‰) than rainbow trout (*Oncorhynchus mykiss*) and large longnose dace (-25.8 to -23.7‰). The wide isotopic variability in dace (>6‰) likely reflected size-dependent variability in feeding preferences.

^{15}N of Nitrogen Sources and Consumers

Riverine food webs are supported by energy flow from allochthonous detritus and autochthonous primary production (Vannote *et al.* 1980), and consumers of this plant tissue should have an isotopic signature approximately 2.5-3.5‰ higher than the primary producer trophic level (Minagawa and Wada 1984; Fry 1991). In the Thompson River, the $\delta^{15}N$ values of the algal biofilm (0.8 to 3.8‰), chironomidae (1.8-2.2‰) and mayflies (3.2-4.8‰) were similar, suggesting that none of the insect consumers feed solely on algal biofilm (Table 2; Fig. 2). At the Walhachin site, the only location where biofilm and insect consumers where sampled simultaneously, mayfly $\delta^{15}N$ values were 0.5-2.0‰ higher than the biofilm. In contrast, chironomid $\delta^{15}N$ values (2.2‰) were depleted relative to algal biofilm (2.7‰). Thus, both mayflies and chironomids appear to consume algal biofilm and an unmeasured, isotopically depleted nitrogen source. This unknown source could be POM ($\delta^{15}N$ of 0-1‰) imported from upstream headwaters, or DOM in the PME. As with carbon, a simple two-component nitrogen isotope mass balance was constructed to estimate relative proportions. The $\delta^{15}N$ values of terrestrial organic matter commonly range between 0 and 1‰. Assuming an average insect $\delta^{15}N$ of +4‰, an average biofilm $\delta^{15}N$ of 3.5‰ and a single trophic level fractionation of 3.5‰,

$$\delta^{15}N_{insects} = [X(\delta^{15}N_{biofilm}) + Y(\delta^{15}N_{terrestrial})] + 3.5_{trophic\ fractionation} \qquad (4)$$

it would appear that the insects consumed about 20-40% biofilm and 60-80% terrestrial food sources. These estimates were supported by the carbon isotope data presented earlier. In the higher trophic levels, nitrogen isotopic composition of the filter-feeding insect, *Hydropysche*, and all of the fish species indicate that aquatic insects are their dominant food source. *Hydropysche* may also consume algal biofilm.

Conclusions on the Effects of PME on the Food Web

In summary, contrary to the single carbon-source model which has been assumed to be appropriate for the food web of the Thompson River (Bothwell *et al.* 1992; Bothwell and Culp 1993), stable isotope analyses suggest the food web of this river is supported by two distinct carbon sources: (1) the algal biofilm which utilizes DIC from the river and (2) an unmeasured source with an isotopic signature similar to that of terrestrial plants. Based upon nitrogen composition, the river has at least three trophic levels including algal biofilm, insect grazers and insectivorous fish. However, this trend in trophic structure was partially obscured by the apparent input of an unmeasured terrestrial source. Future dual-isotope research on this riverine food web must isolate the unmeasured source of carbon and nitrogen, which we hypothesize to be terrestrial plant input from upstream headwaters and PME.

Using stable isotopes to trace the fate of PME carbon and sulfur was not completely successful due to time and sampling limitations, the complicated nature of the incomplete mixing at sites along the two rivers and the presence of a sewage treatment plant near the PME diffusers. Nevertheless, the data clearly indicated that $\delta^{13}C$ of DIC derived from PME was isotopically distinct from riverine background values and that carbon isotopes may be a useful PME tracer, particularly at a less complicated site. Similarly for $\delta^{34}S$, PME contributed significant anthropogenic sulfate inputs to the river, although unfortunately its sulfur isotopic signal was indistinguishable from a mixture of the two Thompson rivers. Despite the limitations encountered, we are convinced that stable isotopes of carbon and sulfur have excellent potential to trace PME-derived carbon and sulfur from effluent source into riverine aquatic food webs. We recommend that a similar study be undertaken at a site that does not suffer from the complications of multiple effluent inputs and the mixing of two rivers. Furthermore, our study suggests that terrestrially derived carbon (perhaps from PME) plays an important role in supporting the food web of the Thompson River downstream of Kamloops Lake.

ACKNOWLEDGMENTS

We wish to thank Randy George, Nancy Glozier, Daryl Halliwell and Garnet Richards for their technical assistance in the field and the laboratory. George Derkson kindly provided us with a sample of pulp mill effluent solids. This work was funded by the National Hydrology Research Institute of Environment Canada and the Fraser River Action Plan.

REFERENCES

Amblard, C., P. Couture and G. Bourdier. 1990. Effects of a pulp and paper mill effluent on the structure and metabolism of periphytic algae in experimental streams. Aquat. Toxicol. 18:137-162.

Bothwell, M.L. and J.M. Culp. 1993. Sensitivity of the Thompson River to phosphorus studies on trophic dynamics. National Hydrology Research Institute Contribution No. 93006, Saskatoon, SK, Canada.

Bothwell, M.L., G. Derksen, R.N. Nordin and J.M. Culp. 1992. Nutrient and grazer control of algal biomass in the Thompson River, British Columbia: A case history of water quality management. In: Aquatic ecosystems in semi-arid regions: Implications for resource management, R.D. Robarts and M.L. Bothwell (Eds.), National Hydrology Research Institute Symposium Series 7, Environment Canada, Saskatoon, Saskatchewan, Canada, pp. 253-266.

Boutton, W., W. Wong, D. Hachey, L. Lee, M. Cabrera and P. Lein. 1983. Comparison of quartz and pyrex tubes for combustion of organic samples for stable carbon isotopic analysis. Anal. Chem. 55:1832-1833.

Coleman, M.L., T.J. Shepherd, J.J. Durham, J.E. Rouse and G.R. Moore. 1982. Reduction of water for hydrogen isotope analysis. Anal. Chem. 54:993-995.

Culp, J.M. 1989. Nocturnally constrained foraging of a lotic minnow (*Rhinichthys cataractae*). Can. J. Zool. 67:2008-2012.

Deines, P. 1980. The isotopic composition of reduced carbon. In: Handbook of environmental isotope geochemistry - The terrestrial environment. P. Fritz, and J.Ch. Fontes (Eds.), Elsevier, pp. 329-405.

Fritz, P. and J.Ch. Fontes. 1980. Handbook of environmental isotope geochemistry. The terrestrial environment. Elsevier, 545 p.

Fry, B. 1991. Stable isotope diagrams of freshwater food webs. Ecology 72:2293-2297.

Fry, B., and E. Sherr. 1984. $\delta^{13}C$ measurements as indicators of carbon flow in marine and freshwater ecosystems. Contrib. Mar. Sci. 27:13-47.

Gearing, P.J., J.N. Gearing, J.T. Maughan and C.A. Oviatt. 1991. Isotopic distribution of carbon from sewage sludge and eutrophication in the sediment and food web of estuarine ecosystems. Environ. Sci. Tech. 25:295-301.

Hunsaker, C.T., and D.E. Carpenter. 1990. Ecological indicators for the environmental monitoring and assessment program. EPA 600/3090/060, U.S. Environmental Protection Agency, Office of Research and Development, Research Triangle Park, North Carolina.

Kendall, C. and E. Grim. 1990. Combustion tube method for measurement of nitrogen isotope ratios using calcium oxide for total removal of carbon dioxide and water. Anal. Chem. 62:526-529.

LaZerte, B. and J. E. Szalados. 1982. Stable carbon isotope ratio of submerged freshwater macrophytes. Limnol. Oceanogr. 27:413-418.

Minagawa, M., and E. Wada. 1984. Stepwise enrichment of ^{15}N along food chains: Further evidence and the relation between $\delta^{15}N$ and animal age. Geochim. Cosmochim. Acta 48:1135-1140.

Munkittrick, K.B., G.J. Van Der Kraak, M.E. McMaster and C.B. Portt. 1992. Reproductive dysfunction and MFO activity in three species of fish exposed to bleached kraft mill effluent at Jackfish Bay, Lake Superior. Water Poll. Res. J. Can. 27:439-446.

Nelson, J.S., and M.J. Paetz. 1992. The fishes of Alberta. 2nd edition. University of Alberta Press, Edmonton, Alberta, Canada.

Poole, N.J., D.J. Wildish and D.D. Kristmanson. 1978. The effects of the pulp and paper industry on the aquatic environment. Crit. Rev. Env. Control 8:153-195.

Rosenfeld, J.S., and J.C. Roff. 1992. Examination of the carbon base in southern Ontario streams using stable isotopes. J. North Am. Benthol. Soc. 11:1-10.

Rundel, P.W., J.R. Ehleringer and K.A. Nagy. 1988. Stable isotopes in ecological research. Springer Verlag, New York.

Scrimgeour, G.J., J.M. Culp and N.E. Glozier. 1993. An improved technique for sampling lotic invertebrates. Hydrobiologia 254:65-71.

Scott, W.B., and E.J. Crossman. 1973. Freshwater fishes of Canada. Fish. Res. Board Can. Bull. 184.

Servos, M.R., J.H. Carey, M.L. Ferguson, G.J. Van Der Krak, H. Ferguson, J. Parrott, K. Gorman and R. Cowling. 1992. Impact of a modern bleached kraft mill on white sucker populations in the Spanish River, Ontario. Water Poll. Res. J. Can. 27:423-438.

Swanson, S., R. Shelast, R. Schryer, P. Kloepper-Sams, T. Marchant, K. Kroeker, J. Bernstein and J.W. Owens. 1992. Fish populations and biomarker responses at a Canadian bleached kraft mill site. TAPPI J. 75:139-149

Van Dover, C.L., J.F. Grassle, B. Fry, R.H. Garritt and V.R. Starczak. 1992. Stable isotope evidence for entry of sewage derived organic material into a deep-sea food web. Nature 360:153-156.

Vannote, R.L., G.W. Minshall, K.W. Cummins, J.R. Sedell and C.E. Cushing. 1980. The river continuum concept. Can. J. Fish. Aquat. Sci. 37:130-137.

DETECTION OF STEROID HORMONE DISRUPTIONS ASSOCIATED WITH PULP MILL EFFLUENT USING ARTIFICIAL EXPOSURES OF GOLDFISH

M.E. McMaster[1], K.R. Munkittrick[2], G.J. Van Der Kraak[1], P.A. Flett[2] and M.R. Servos[2]

[1]Department of Zoology, University of Guelph, Guelph, Ontario, N1G 2W1 Canada
[2]GLLFAS, Department of Fisheries & Oceans, Burlington, Ontario, L7R 4A6 Canada

From 1988 to 1992, we have shown that several species of wild fish collected from Jackfish Bay, Lake Superior have decreased reproductive steroid levels associated with exposure to pulp mill effluent. The cause of the steroid reductions are multi-focal, with at least some of the reduced steroid synthetic capacity explained by substrate limitations and altered enzyme activity within the biosynthetic pathway. We exposed goldfish (*Carassius auratus*) to pulp mill effluent under either field or laboratory conditions, to examine circulating steroid levels and *in vitro* steroid production. Testing during May 1993 demonstrated that reduced steroid production can occur within 4 days of exposure to pulp mill effluent. Further exposures during July and November 1993 did not detect as strong of an effluent response. Goldfish exposure to effluent during May 1994 showed no evidence of depressed steroid production, confirming the apparent recovery found in late 1993. The installation of secondary treatment in 1989 had no detectable effect on the steroid response in wild fish. Apparent recovery of steroid function in artificially exposed goldfish coincides with in-plant process changes in June 1993; the exact nature of these process changes has not been released by mill personnel.

INTRODUCTION

A number of studies have shown that circulating steroid levels are reduced in wild fish exposed to pulp mill effluent (McMaster *et al.* 1991; Munkittrick *et al.* 1991, 1992a,b,c, 1994a; Adams *et al.* 1992; Hodson *et al.* 1992; Servos *et al.* 1992; Gagnon *et al.* 1994a,b). These reductions correlate with other signs of reproductive impairment such as reduced gonadal development, reduced expression of secondary sexual characteristics, delayed maturity, reduced egg size and a reduced fecundity with age relative to a reference population (McMaster *et al.* 1991; Munkittrick *et al.* 1991, 1992a,b,c). These consistent reductions in circulating steroid levels led to studies to determine the underlying mechanisms of bleached kraft mill effluent (BKME) actions on reproductive function; these studies indicated multiple sites of dysfunction within the pituitary-gonadal axis (Van Der Kraak *et al.* 1992a). Impacts included pituitary disruptions (reduced circulating levels of gonadotropin and diminished responsiveness to a gonadotropin-releasing hormone analog), reduced ovarian steroid biosynthetic capacity (lower *in vitro* basal steroid production and diminished responsiveness to human chorionic gonadotropin [hCG]) and altered peripheral steroid metabolism (depressed levels of glucuronidated testosterone) (Van Der Kraak *et al.* 1992a). *In vitro* production of steroids by ovarian follicles from reference and BKME-exposed fish paralleled the differences found in circulation, suggesting that alterations in ovarian steroid production represent a major site of BKME actions at Jackfish Bay (Van Der Kraak *et al.* 1992a; McMaster *et al.* 1994).

Initial studies at Jackfish Bay (1988-89) were conducted when the mill utilized only primary treatment of its effluent (McMaster *et al.* 1991; Munkittrick *et al.* 1991). Detailed studies following the installation of a secondary treatment aeration stabilization basin failed to find any recovery in either the reproductive responses or the ethoxyresorufin-*o*-deethylase (EROD) induction potential in wild fish (Munkittrick *et al.* 1992b,c). However, samples collected two weeks after a planned mill maintenance shutdown showed reduced EROD induction in all three species of fish, but impacts of BKME on circulating steroid levels in female white sucker (*Catostomus commersoni*) and longnose sucker

(*Catostomus catostomus*) persisted (Munkittrick *et al.* 1992a). Expanded studies have shown that white sucker collected downstream of most large Ontario mills exhibited increased EROD activity, decreased steroid levels or smaller gonadal sizes in female fish (Munkittrick *et al.* 1994a). These studies suggest that the reproductive responses may be widespread and are not dependent on a specific pulping, bleaching or treatment technology and that effects on steroids may not be directly related to EROD induction. Robinson (1994) has recently duplicated these reproductive responses using a life cycle exposure to fathead minnows (*Pimephales promelas*). Given the rapid changes that have been taking place in mill process and treatment technologies, it would be beneficial to develop short-term laboratory-based protocols to estimate the potential of effluents to disrupt reproductive function in fish.

Within the present studies, mature goldfish (*Carassius auratus*) were exposed for 4, 8 and 14 d in both a field and laboratory setting to BKME from a mill known to be associated with reproductive changes. The impacts of exposure on circulating steroid levels and the *in vitro* production of steroids by gonadal tissue were measured.

MATERIALS AND METHODS

All studies reported in this paper were conducted with effluent from a mill located in Terrace Bay, Ontario. This site has been used extensively for the study of BKME impacts on fish populations, and the study site and mill characteristics are fully described in McMaster *et al.* (1991) and Munkittrick *et al.* (1991). Selective effluent tracers for the receiving water during some of the study periods are shown in Table 1. During mid-May (11th-19th) 1993, mature male and female goldfish (25-40 g) were transported to the field location and exposed directly to the effluent after discharge into Blackbird Creek ($\approx 50\%$ BKME) for 4 and 8 d. Fish were split into four groups, two groups to be sampled following four days and two groups to be sampled on day 8. Each group consisted of 16 fish which were randomly selected and placed in cages. The cages consisted of two plastic 20 L laundry baskets secured together and anchored in place in Blackbird Creek. Reference fish were suspended in cages in the uncontaminated waters of Sawmill Creek for the same duration. Following exposure, fish were immobilized in a foam block and bled prior to spinal severance. Blood was collected via caudal puncture into 1 mL syringes and quickly placed into 5 mL heparinized vacuum tubes, stored on ice and centrifuged prior to the collection of plasma. Plasma samples were immediately frozen in liquid nitrogen and returned to the laboratory where they were stored at -20°C pending analysis by radioimmunoassay (RIA). Following collection of blood samples, fish were sampled for length, weight, gonad weight and liver for EROD analysis. Following weighing, gonadal tissue was quickly placed into excess Medium 199 buffer (M199, containing Hank's salts without bicarbonate; GIBCO, Burlington, ON) which was supplemented with 25 mM Hepes, 4.0 mM sodium bicarbonate, 0.01% streptomycin sulfate and 0.1% bovine serum albumin (pH 7.4) prior to *in vitro* incubation.

In Vitro Steroid Production

An *in vitro* steroid production bioassay based on the methods of Van Der Kraak *et al.* (1992a) and Wade and Van Der Kraak (1992) and outlined in detail in McMaster *et al.* (1995) was used to determine the steroidogenic capacity of both ovarian follicles and testicular pieces. In brief, follicles were gently teased apart and intact ovarian follicles were added in groups of 20 per well to a polystyrene tissue culture plate (Falcon 3047; Fisher Scientific Co., Toronto, ON). For testicular incubations, testes were cut into small blocks and rinsed with media to remove much of the sperm. Two testis pieces of approximately equal size, weighing a total of 18-25 mg, were added to test tubes with incubation media. Only gonads from fish which were similar in development were used for *in vitro* incubations. Immediately prior to the beginning of the incubation period, the medium was replaced with fresh M199 alone or in combination with the gonadotropin analog, hCG (10 IU mL^{-1}). hCG is a potent analog of GtH I and GtH II in both male and female goldfish (Van Der Kraak *et al.* 1992b). The final incubation volume was 1 mL. Ovarian follicles and testicular pieces were incubated for 18 h at 16-18°C, after which the incubation medium was collected and immediately frozen in liquid nitrogen. Samples were later stored at -20°C prior to steroid production determination by RIA. Incubations were completed on five fish of each sex from each treatment in triplicate for the May field exposures.

Table 1. Selected effluent tracers ($\bar{x}$ ± S.E.) from the receiving water during a number of the study periods.

	MAY 1993			OCT 1993				MAY 1994		
	Sawmill Creek	Blackbird Creek	Effluent	Mountain Bay	Blackbird Creek	Effluent		Sawmill Creek	Blackbird Creek	Effluent
Chlorophenols (ng L^{-1})										
2,4-CP	3	263 ± 68	1443 ± 128	nd	193	368 ± 171		11	312 ± 16	749 ± 47
3,5-CP	nd	130 ± 75	241 ± 26	nd	265	123 ± 14		11	68.7 ± 2.8	190 ± 35
2,4,6-CP	0.5	680 ± 172	2711 ± 244	nd	363	985 ± 356		0.5	778 ± 69	2067 ± 164
2,4,5-CP	nd	nd	5.00 ± 5.00	nd	nd	2.33 ± 2.3		nd	21.0 ± 4.4	45.0 ± 4.0
4,6-CG	nd	274 ± 54	1392 ± 317	nd	209	449 ± 108		16.5	394 ± 11	917 ± 33
3,4,5-CP	3.5	15.5 ± 10.0	39.0 ± 3.0	nd	4.75	30.0 ± 15.0		nd	74.0 ± 37.0	59.7 ± 30.1
4,5-CG	nd	1866 ± 393	5212 ± 464	nd	591	2868 ± 112		nd	3248 ± 204	4451 ± 235
2,3,5,6-CP	nd	39.0 ± 15.0	92.0 ± 49.0	nd	101	73.0 ± 7.0		nd	43.0 ± 1.5	66.0 ± 4
6-MCV	1	5937 ± 1863	2007 ± 688	nd	3301	6601 ± 2031		nd	23565 ± 869	14912 ± 2014
3,4,5-CG	1.5	3052 ± 348	8707 ± 1510	nd	8837	4177 ± 719		4.0	991 ± 19	1620 ± 327
4,5,6-CG	nd	2697 ± 334	7919 ± 253	nd	10005	4166 ± 728		6.5	466 ± 233	1186 ± 593
5,6-CV	nd	1781 ± 532	530 ± 151	nd	835	1500 ± 315		nd	3727 ± 1010	1854 ± 167
3,4,5,6-CG	0.5	135 ± 25	467 ± 37	nd	537	179 ± 32		2.5	239 ± 10	471 ± 21
Sample Size	2	4	3	2	2	3		2	3	3
Major Ions (mg L^{-1})										
Cl	0.76	251 ± 0.4	411 ± 3	1.55	236	363 ± 5				
SiO$_2$	3.78	7.08 ± 0.02	9.05 ± 0	2.36	6.25	6.79 ± 0.01				
SO$_4$	7.9	77.8 ± 5.2	162 ± 9	3.6	106	163 ± 0.3				
K	0.48	6.10 ± 0.39	11.3 ± 0.3	0.43	6.21	11.3 ± 0.3				
Na	0.62	197 ± 8	342 ± 3	1.36	209	359 ± 6				
Sample Size	2	4	3	2	2	3				
Sampling Date	May 14	May 14	May 18	Sep 29	Oct 26	Oct 26		May 26	May 26	May 26

NA = not detectable.

Effluent from Blackbird Creek collected at the time of the field exposures (May 1993) was returned to the laboratory, and goldfish from the same stock were exposed for a similar duration in the laboratory using a static exposure protocol. Goldfish (25-40 g) were acclimated for 7 d in eight 25 L experimental tanks supplied with aerated dechlorinated lab water maintained at 15°C under a constant photoperiod (14 h dark:10 h light); there were 6 fish per tank. At the start of the treatment, water from four of the tanks was replaced with effluent. Fish were not fed during the exposure period and effluent was changed completely every second day. Twenty-four fish were sampled following 4 and 8 d of exposure (12 exposed, 12 control) in the lab. Fish of similar gonadal development were used for the *in vitro* studies.

To determine if goldfish could be used throughout the year to establish effluent effects on reproductive responses, similar laboratory exposures were repeated with effluent collected from the site in July (16th-29th) and early November (2nd-15th), 1993. The exposure protocol in July was identical to the lab exposures in May except for the length of exposure (from 4 and 8 d to 8 and 14 d) and 100% effluent collected directly at the discharge point from the aeration stabilization basin, was used. For the November exposure, only male fish were used, as female fish showed minimal gonadal development; the exposure protocol was similar to that for July.

In May 1994 goldfish were again transported to the field (May 11th-18th) and exposed on-site for 8 d in an effort to duplicate the May 1993 results. Exposures were conducted in large 100 L garbage pails with the effluent being changed daily. Approximately 80 goldfish were split randomly into one of four pails, with two pails containing effluent and the other two control water from Sawmill Creek. Fish were sampled in a similar manner following 8 d of exposure. For *in vitro* studies, 12 males and 12 females were used from each treatment (control and exposed).

Radioimmunoassay

Circulating steroid levels in the plasma were measured by RIA following ether extraction by methods described in McMaster *et al.* (1992b). Descriptions of the antisera used to measure testosterone in both sexes, 17ß-estradiol in females and 11-ketotestosterone in males are reported, respectively, in Van Der Kraak and Chang (1990), Van Der Kraak *et al.* (1990) and Wade and Van Der Kraak (1992). All plasma samples were assayed in duplicate and interassay variability was less than 15% in all three RIA systems. *In vitro* production of steroids was measured in a similar fashion, however media samples did not require ether extraction prior to analysis.

Mixed Function Oxygenase Analysis

Following removal, livers were immediately frozen in liquid nitrogen. Samples were later thawed on ice, weighed, then homogenized in buffer (0.5% KCl) and EROD activity was measured using the substrate ethoxyresorufin as described in McMaster *et al.* (1992a).

Chemical Analysis

A number of chemical parameters were measured in water at each of the sites to define exposure (Table 1). Chlorophenolics were analyzed using a method modified from Carey *et al.* (1988). A 1 L water sample collected in amber glass bottles was acidified to pH 2 with concentrated H_2SO_4 and extracted 3 times with methylene chloride. The organic layer was rotary evaporated and transferred into 50 mL hexane and extracted 3 times with 0.1 M K_2CO_3. The base layer was acetylated using 1.5 mL acetic anhydride and 8 mL hexane and then shaken for 1 h. The organic layer was evaporated under nitrogen and transferred to isooctane. Chlorophenols and chloroguaiacols were quantitated using a Hewlett Packard gas chromatograph with an electron capture detector, a 25 µm, 30 m DB5 column and splitless injection. Major ions were measured according to the procedures of the Environment Canada National Water Quality Laboratory (Burlington, ON).

Statistical Analysis

Blood and EROD data were analyzed by non-parametric Kruskal Wallis analysis of variance by site and sex. *In vitro* incubations were also analyzed by Kruskal Wallis analysis by site and hCG stimulation.

RESULTS AND DISCUSSION

Both male and female goldfish exposed under field and laboratory conditions during 1993 failed to show induction of EROD activity. For this reason, liver samples were not analyzed for EROD activity from goldfish exposed in the field during May 1994. Earlier studies in our laboratory indicate that induction in this species is possible following injection of PCB 126 (0.05 µg g^{-1}) (Kidd *et al.* 1993). Similar exposures of immature white sucker and rainbow trout (*Onchorynchus mykiss*) to this effluent during these sampling periods resulted in strong induction in both species (Munkittrick *et al.* 1994b), indicating that inducing compounds were present in the effluent. Field collections of white sucker in late September and October 1993 indicated that EROD activity was induced to levels similar to that reported at this site over the last 6 years (Munkittrick unpubl. data).

Although BKME exposure has been associated with a number of reproductive responses, short-term exposures to BKME were not expected to alter gonadal sizes or age to maturation in exposed goldfish. However, as these reproductive changes are correlated with reductions in circulating steroids levels (Munkittrick *et al.* 1992c; Robinson 1994), it was hypothesized that these exposures would reduce circulating steroid levels and *in vitro* steroid production by gonadal tissue. Circulating steroid levels were monitored for both male and female goldfish following the laboratory exposures to BKME in May 1993. The major biologically active steroids in male goldfish during this stage of reproduction are testosterone and 11-ketotestosterone, and in female fish testosterone and 17ß-estradiol. Exposure failed to reduce circulating levels of any of these steroids in effluent-exposed fish following 4 or 8 d of exposure (Fig. 1). Although a trend to reduced circulating levels was evident in exposed fish, high variability between fish within treatment may have contributed to the lack of treatment differences. Similar variation within groups during July and November was also found, possibly contributing to the absence of changes in circulating steroid levels following BKME exposure. The large differences even in control exposures may reflect different degrees of sexual development between fish. A number of studies attempting to control this variability are now underway and include: (a) bleeding individual fish prior to the beginning of the experiments and determining circulating steroid differences in fish before and after exposure, (b) removing fish in poor reproductive condition as determined by pre-bleeding, (c) altering and standardizing reproductive state with photoperiod and temperature regimes, (d) stimulating steroid production through injections with Ovaprim (a gonadotropin-releasing hormone analog in combination with a dopamine inhibitor) prior to sampling and (e) increasing the sample sizes used for these studies.

Although no consistent responses of short-term BKME exposure on circulating steroid levels were found in these studies, it was possible that these exposures would reduce production of steroids by gonadal tissue. These reductions would then suggest that other reproductive responses to exposure may occur under longer exposure periods, as Robinson (1994) found a direct correlation between steroid production and reproductive performance in the fathead minnow during laboratory life cycle flow-through studies with BKME. Analysis of *in vitro* incubations of both male and female goldfish gonadal tissue indicated significant reductions in steroid production by both testicular pieces and ovarian follicles in the field exposures during May 1993 (Figs. 2 and 3). Although basal testosterone production by testicular pieces was not reduced following 4 d of exposure, it was reduced following 8 d (Figs. 2A). hCG increased production of testosterone at both sites; however, effluent resulted in reduced stimulated testosterone production following both 4 and 8 d of exposure (Fig. 2A). Stimulated 11-ketotestosterone production was also reduced following 8 d (Fig. 3A), although no other treatment differences were found. In female goldfish, testosterone production by ovarian follicles was consistently reduced under both basal and hCG-stimulated conditions following both 4 and 8 d of exposure in the field setting (Fig. 2B). Again hCG resulted in increased production of testosterone at both sites; however, the magnitude of the stimulation was reduced following effluent exposure.

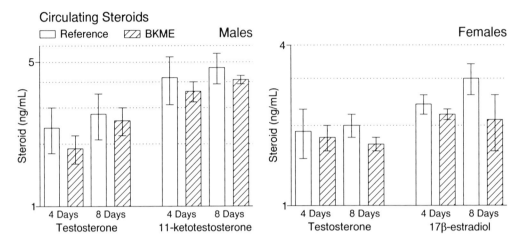

Figure 1. Circulating levels of testosterone and 11-ketotestosterone in male goldfish and testosterone and 17β-estradiol in female goldfish (ng mL^{-1} ± S.E.) exposed to BKME (hatched bars) or to control water (open bars) in the laboratory during May 1993 for 4 and 8 d. There were no statistical differences between treatments for either sex.

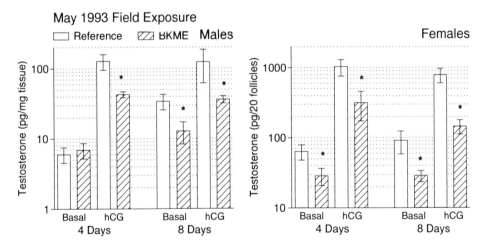

Figure 2. In vitro production of testosterone by male goldfish testicular fragments and female goldfish ovarian follicles in response to hCG (10 IU/mL) ($\bar{x}$ ± S.E., n = 5) exposed in the field to BKME (hatched bars) or reference water (open bars) for 4 or 8 days during May 1993. Asterisks indicate significant differences from the comparable reference incubation (p < 0.05).

Similar impacts of BKME exposure on *in vitro* steroid production have been demonstrated in white sucker throughout the reproductive season at Jackfish Bay (Van Der Kraak *et al.* 1992a; McMaster *et al.* 1994). These reductions in goldfish steroid production are the first indication that short-term exposures of laboratory fish to BKME are capable of identifying forms of reproductive responses similar to those documented in wild fish populations. Although these responses have now been demonstrated in fish caged

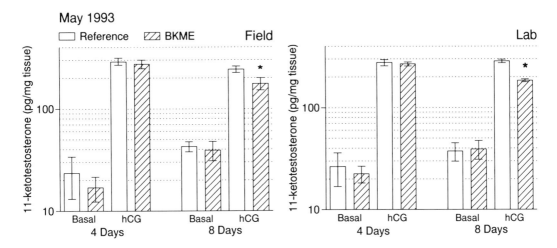

Figure 3. In vitro production of 11-ketotestosterone ($\bar{x}$ ± S.E., n = 6) by male goldfish testicular fragments in response to hCG (10 IU/mL) exposed in the field or in the laboratory to BKME (hatched bars) or reference water (open bars) during May 1993. Asterisks indicate significant differences from the comparable reference incubation (p < 0.05).

in the field setting, it would be beneficial if similar responses could be measured in the laboratory with effluent transported back from the field. This would make the protocol suitable for testing large numbers of effluents for reproductive responses and for examining the efficacy of changes in pulping and treatment technologies for eliminating these responses. Fish exposed in the laboratory to effluent under static conditions also showed reductions in *in vitro* steroid production; however, the differences were not as pronounced or as frequent (Fig. 3 and 4). Similar to the field exposures, the only site difference in 11-ketotestosterone production was found under hCG-stimulated conditions following 8 d of exposure (Fig. 3B). Male testosterone production by testicular pieces was reduced following effluent exposure under basal conditions following both 4 and 8 d of exposure (Fig. 4A). Stimulation with hCG resulted in increased production; however, testosterone production was lower following 8 d of exposure to BKME. These static exposures also reduced ovarian follicular testosterone production under basal and stimulated conditions (Figure 4B). In the field exposures these differences were evident following both 4 and 8 d; however, in the laboratory exposures, treatment differences were present only following 8 d of exposure. Although the responses seen in field exposures were transferable to the laboratory, some of the potency of the effluent on the reduction of *in vitro* steroid production appeared to be lost. Loading densities in the laboratory exposures would have been high relative to the flow-through field exposures and the influence of this factor is currently under investigation.

The short-term exposure of goldfish to effluent in both the field and laboratory setting was capable of reducing steroid production by gonadal tissue during May 1993. One of the problems associated with monitoring steroid depressions in wild fish is related to the seasonal reproductive cycle; white sucker have a postspawning steroid hormone shutdown which lasts from May to mid-August. To examine the ability of goldfish exposures to monitor reproductive responses during the summer, exposures were conducted during July 1993. Since the transportation of large volumes of effluent can represent a logistical problem, the duration of the exposures was extended in an attempt to compensate for the decreased potency seen during the May laboratory exposures. Testosterone production was measured for both male and female goldfish exposed for 8 and 14 d. During this stage of development in the goldfish, basal production of testosterone was undetectable in both sexes in both treatments. hCG stimulation resulted in detectable production levels of testosterone; however, the only significant site differences occurred in males following 8 d of exposure (Fig. 5). It was unclear whether the failure to see large reductions in testosterone production similar to those in May were a result of differences in the reproductive state of the goldfish or due to a change in effluent quality. However, differences in both circulating steroid levels and *in vitro* steroid production in white sucker from Jackfish Bay are relatively consistent regardless of the reproductive state (McMaster *et al.* 1991, 1994; Munkittrick *et al.* 1991; Van Der Kraak *et al.* 1992a). Levels of 11-ketotestosterone produced by male testes were reduced under basal conditions following both 8 and 14 d of exposure and following hCG stimulation following 14 d (Fig. 6). These treatment differences in 11-ketotestosterone relative to the May exposures suggest that the lack of differences in testosterone production during this time of the year may have been related to the reproductive state of the goldfish.

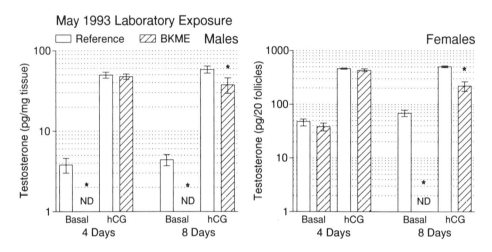

Figure 4. In vitro production of testosterone ($\bar{x} \pm$ S.E., n = 6) by male goldfish testicular fragments and female goldfish ovarian follicles in response to hCG (10 IU/mL) exposed in the laboratory to BKME (hatched bars) or reference water (open bars) during May 1993. Asterisks indicate significant differences from the comparable reference incubation (p < 0.05) and ND indicates that steroid production was non-detectable in the media for those incubations.

Exposures were also conducted in the fall of 1993 to determine whether similar reductions in steroid production occurred. Because gonadal development in the female goldfish was negligible during this time of the year, studies were only conducted on male goldfish. Although steroid production was detectable for all incubations and hCG resulted in increased steroid production under both treatments, BKME failed to reduce steroid production in male fish under any of the treatment regimes (Fig. 7). This apparent recovery in reproductive performance also coincides with in-plant process changes in June 1993; however, the exact process changes have not been released by mill personnel. We know that the installation of secondary treatment showed no improvement of the reproductive alterations in wild fish populations (Munkittrick *et al.* 1992a,c) and that mills which utilize different pulping, bleaching and treatment strategies also show altered reproductive responses (Munkittrick *et al.* 1994a). During the latest process changes at Terrace Bay (June 1993), the mill apparently upgraded it is bleaching sequence to include 70% chlorine dioxide substitution. Depressions in steroid hormones have been seen at other sites where mills have more than 70% substitution (Nickle *et al.* 1994), and at mills which did not use chlorine (Munkittrick *et al.* 1994a), and it is unclear whether the change in bleaching sequence contributed to the apparent recovery in this study.

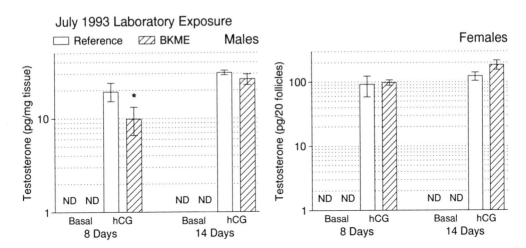

Figure 5. In vitro production of testosterone ($\bar{x} \pm S.E.$, n = 6) by male goldfish testicular fragments and female goldfish ovarian follicles in response to hCG (10 IU/mL) after exposure to BKME (hatched bars) or reference water (open bars) in the laboratory during July 1993. Asterisks indicate significant differences from the comparable reference incubation ($p < 0.05$) and ND indicates that steroid production was non-detectable in the media for those incubations.

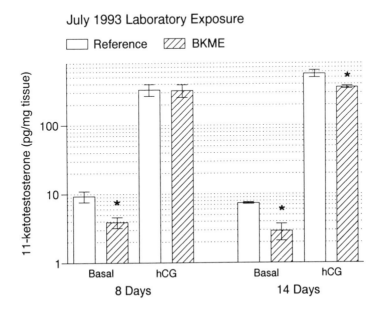

Figure 6. In vitro production of 11-ketotestosterone ($\bar{x} \pm S.E.$, n = 6) by male goldfish testicular fragments in response to hCG (10 IU/mL) after exposure to BKME (hatched bars) or reference water (open bars) in the laboratory during July 1993. Asterisks indicate significant differences from the comparable reference incubation ($p < 0.05$).

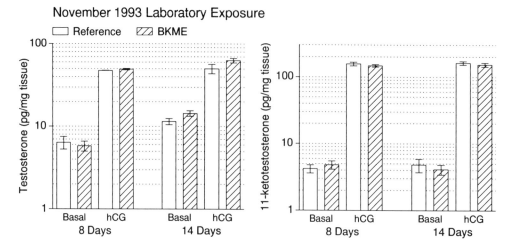

Figure 7. In vitro production of testosterone and 11-ketotestosterone ($\bar{x} \pm S.E.$, n = 6) by male goldfish testicular fragments in response to hCG (10 IU/mL) after exposure to BKME (hatched bars) or reference water (open bars) in the laboratory during November 1993.

It was necessary to confirm the status of steroid hormone production in artificially exposed goldfish during May 1994. Both July and November exposures were only conducted in the laboratory; the optimal exposure protocol for laboratory determination of steroid hormone responses to industrial effluents has yet to be established. Furthermore, the goldfish during July and November may not have been at an optimal reproductive state and only males were used in November exposures. The May 1994 exposures confirmed the apparent recovery of steroid hormone production in goldfish. *In vitro* production of both testosterone and 11-ketotestosterone by mature male testicular tissue under basal or stimulated conditions was not reduced by exposure to effluent (Fig. 8); however, steroid production was increased following hCG stimulation for both treatments. Similar to male fish, *in vitro* production of testosterone and 17ß-estradiol by ovarian follicles was similar between treatments (Fig. 9). In fact, higher production of testosterone was found in exposed fish following hCG stimulation. Both results suggest the loss of the ability of the effluent from this mill to cause steroid reductions under short-term exposures.

Although steroid production by ovarian follicles was similar in 1994 to that in 1993, steroid production levels in male fish were significantly lower under both treatments in 1994. Whether differences in the reproductive stages of the fish existed between years and whether differences of this nature could confound effects of effluent exposure are unknown. Field studies on the white sucker collected during early September 1993 also showed signs of reproductive improvement. Although gonadal size differences were still evident, no site differences in circulating steroid levels (Jardine 1994) or *in vitro* steroid production (McMaster unpubl. data) were found. However, circulating and *in vitro* differences were present in samples collected in late September and late October 1993 at this site (Munkittrick and McMaster unpubl. data). While collecting fish in May 1994 to address this recovery at the BKME site, severe weather patterns resulted in white sucker at the reference location releasing their eggs and milt in the lake prior to their migration up the spawning streams. This resulted in a failure to collect suitable samples from this site during this period and prevented site comparisons.

In May 1993, wild white sucker showed decreased steroid levels (Jardine 1994), as did the goldfish exposures, suggesting that the laboratory assay has potential for predicting effects found in the field setting. Similar field studies have now been conducted on white sucker at other mill locations. Females captured downstream of a pulp mill on the Winnipeg River showed reduced *in vitro* production of both testosterone and 17ß-estradiol (McMaster, Van Der Kraak, Friesen and Munkittrick upubl. data), as well as reduced circulating steroid levels relative to upstream fish (Friesen et al. 1994). White sucker collected during September at mills located in Kapuskasing and Smooth Rock Falls showed no alterations in *in vitro* steroid production; however, reduced circulating steroid levels and reduced gonadal development were found (Nickle, Van Der Kraak and Munkittrick unpubl. data). Robinson (1994) tried to validate the short-term exposures and showed differences at some mills, but found some sites where wild fish showed decreased circulating steroids (Nickle, Van Der Kraak and Munkittrick unpubl. data) but laboratory exposed fish did not (Robinson 1994). This suggests that the presence of steroid effects in the laboratory assay can predict field effects but that more work is needed to thoroughly understand short-term exposure results and to optimize the laboratory protocol.

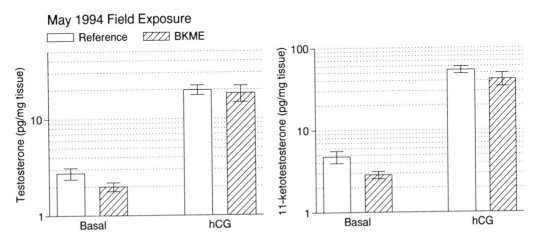

Figure 8. In vitro production of testosterone and 11-ketotestosterone ($\bar{x} \pm S.E.$, n = 12) by male goldfish testicular fragments in response to hCG (10 IU/mL) after exposure to BKME (hatched bars) or reference water (open bars) for 8 days in the field during May 1994.

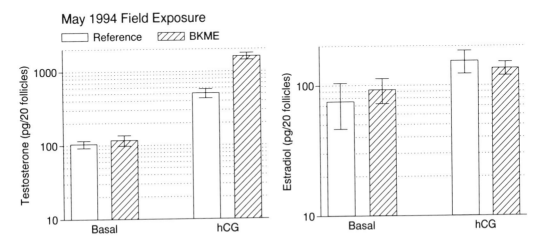

Figure 9. In vitro production of testosterone and 17β-estradiol ($\bar{x} \pm S.E.$, n = 12) by female goldfish ovarian follicles in response to hCG (10 IU/mL) after exposure to BKME (hatched bars) or reference water (open bars) for 8 days in the field during May 1994.

CONCLUSIONS

1) It is possible to produce reductions in *in vitro* steroid production following the short-term exposure of mature goldfish to effluent in field or laboratory exposures. These depressions paralleled differences seen in wild fish and in fathead minnow exposed during life cycle tests.
2) These reductions were seen within 4 d of exposure.
3) Laboratory renewal tests at high loading densities show decreased potency relative to field testing, indicating that an optimal laboratory exposure protocol should be developed.
4) There appears to have been some recovery of steroid function at Jackfish Bay coincident with some uncharacterized June 1993 process changes. The interpretation of normal steroid production during artificial exposures needs to be clarified since wild fish collected at Jackfish Bay late in the fall of 1993 still showed some reproductive responses.
5) It appears that pulp mills can use process changes to reduce the potential of their effluent to alter steroid production. Depressed steroid production has been previously documented with mills which do not use chlorine, at a site with 100% chlorine dioxide substitution and with mills using secondary treatment. Follow-up studies will attempt to refine the laboratory assay and attempt to identify mill processes associated with an increased risk of altering reproductive performance in fish.

REFERENCES

Adams, S.M., W.D. Crumby, M.S. Greeley Jr., L.R. Shugart and C.F. Saylor. 1992. Responses of fish populations and communities to pulp mill effluents: A holistic approach. Ecotoxicol. Environ. Saf. 24:347-360.

Carey, J.H., M.E. Fox and J.H. Hart. 1988. Identity and distribution of chlorophenolics in the north arm of the Fraser River estuary. Wat. Poll. Res. J. Can. 23:31-44.

Friesen, C., W.L. Lockhart and S.B. Brown. 1994. Results from the analysis of Winnipeg River water, sediment and fish. First annual report to Department of Indian Affairs and Northern Development, Winnipeg, pp. 10.

Gagnon, M.M., J.J. Dodson and P.V. Hodson. 1994a. Ability of BKME (bleached kraft mill effluent) exposed white suckers (*Catostomus commersoni*) to synthesize steroid hormones. Comp. Biochem. Physiol. 107C:(2) 265-273.

Gagnon, M.M., J.J. Dodson and P.V. Hodson. 1994b. Seasonal effects of bleached kraft mill effluent on reproductive parameters of white sucker (*Catostomus commersoni*) populations of the St. Maurice River, Quebec, Canada. Can. J. Fish. Aquat. Sci. 51:337-347.

Hodson, P.V., M. McWhirter, K. Ralph, B. Gray, D. Thivierge, J. Carey, G. Van Der Kraak, D.M. Whittle and M.C. Levesque. 1992. Effects of bleached kraft mill effluent on fish in the St. Maurice River, Quebec. Environ. Toxicol. Chem. 11:1635-1651.

Jardine, J.J. 1994. Impacts of physical stressors on physiological fitness of white sucker at a bleached kraft pulp mill site. M.Sc. thesis, University of Guelph, Guelph, Ontario, 92 p.

Kidd, K., G. Van Der Kraak and K. Munkittrick. 1993. The effects of PCB126 on hepatic mixed function oxidase activity and steroidogenic capacity of the goldfish (*Carassius auratus*). Proc. 19th Aquatic Toxicity Workshop, October 4-7, 1992, Edmonton, Alberta, E.G. Baddaloo, S. Ramamoorthy and J.W. Moore (eds.), Can. Tech. Rept. Fish. Aquat. Sci. 19:470-475.

McMaster, M.E., G.J. Van Der Kraak, C.B. Portt, K.R. Munkittrick, P.K. Sibley, I.R. Smith and D.G. Dixon. 1991. Changes in hepatic mixed function oxygenase (MFO) activity, plasma steroid levels and age at maturity of a white sucker (*Catostomus commersoni*) population exposed to bleached kraft pulp mill effluent. Aquat. Toxicol. 21:199-218.

McMaster, M.E., C.B. Portt, K.R. Munkittrick and D.G. Dixon. 1992a. Milt characteristics, reproductive performance and larval survival and development of white sucker exposed to bleached kraft mill effluent. Ecotoxicol. Environ. Saf. 23:103-117.

McMaster, M.E., K.R. Munkittrick and G.J. Van Der Kraak. 1992b. Protocol for measuring circulating levels of gonadal sex steroids in fish. Can. Tech. Rept. Fish Aquat. Sci. 1836, 29 p.

McMaster, M.E., G.J. Van Der Kraak and K.R. Munkittrick. 1994. Evaluation of the steroid biosynthetic capacity of ovarian follicles from white sucker exposed to bleached kraft mill effluent (BKME). Proc. 20th Aquatic Toxicity Workshop, October 17-21, 1993, Quebec City, Quebec, R. van Coillie, Y. Roy, Y. Bios, P.G.C. Campbell, P. Lundahl, L. Martel, M. Michaud, P. Riebel and C. Thellen (eds.), Can. Tech. Rept. Fish. Aquat. Sci. 1989:259.

McMaster, M.E., K.R. Munkittrick, J.J. Jardine, R.D. Robinson and G.J. Van Der Kraak. 1995. Protocol for measuring *in vitro* steroid production by fish gonadal tissue. Can. Tech. Rept. Fish Aquat. Sci. 1961, 78 p.

Munkittrick, K.R., C.B. Portt, G.J. Van Der Kraak, I.R. Smith and D.A. Rokosh. 1991. Impact of bleached kraft mill effluent on population characteristics, liver MFO activity and serum steroid levels of a Lake Superior white sucker (*Catostomus commersoni*) population. Can. J. Fish. Aquat. Sci. 48:1371-1380.

Munkittrick, K.R., G.J. Van Der Kraak, M.E. McMaster and C.B. Portt. 1992a. Response of hepatic mixed function oxygenase (MFO) activity and plasma sex steroids to secondary treatment and mill shutdown. Environ. Toxicol. Chem. 11:1427-1439.

Munkittrick, K.R., M.E. McMaster, C.B. Portt, G.J. Van Der Kraak, I.R. Smith and D.G. Dixon. 1992b. Changes in maturity, plasma sex steroid levels, hepatic MFO activity and the presence of external lesions in lake whitefish exposed to bleached kraft mill effluent. Can. J. Fish. Aquat. Sci. 49:1560-1569.

Munkittrick, K.R., G.J. Van Der Kraak, M.E. McMaster and C. Portt. 1992c. Reproductive dysfunction and MFO activity in three species of fish exposed to bleached kraft mill effluent at Jackfish Bay, Lake Superior. Water Pollut. Res. J. Can. 27(3):439-446.

Munkittrick, K.R., G.J. Van Der Kraak, M.E. McMaster, C.B. Portt, M.R. van den Heuvel and M.R. Servos. 1994a. Survey of receiving water environmental impacts associated with discharges from pulp mills. 2. Gonad size, liver size, hepatic EROD activity and plasma sex steroid levels in white sucker. Environ. Toxicol. Chem. 13:1089-1101.

Munkittrick, K.R., M.R. Servos, K. Gorman, B. Blunt, M.E. McMaster and G.J. Van Der Kraak. 1994b. Characteristics of EROD induction associated with exposure to pulp mill effluent. 4th Symposium on Environmental Toxicology and Risk Assessment, American Society for Testing and Materials, Montreal, Quebec, April 11-13, 1994.

Nickle, J.C., G.J. Van Der Kraak, K.R. Munkittrick and C.B. Portt. 1994. Effects of BKME or CTMP mill effluent on plasma sex steroid and gonadotropin levels, and MFO activity in white sucker. Presented at the Second International Conference on Environmental Fate and Effects of Bleached Pulp Mill Effluents, November 6-10, 1994, Vancouver, B.C., Canada.

Robinson, R.D. 1994. Evaluation and development of laboratory protocols for estimating reproductive impacts of pulp mill effluent on fish. Ph.D. thesis, University of Guelph, Guelph, Ontario, 150 p.

Servos, M.R., J.H. Carey, M.L. Ferguson, G.J. Van Der Kraak, H. Ferguson, J. Parrott, K. Gorman and R. Cowling. 1992. Impact of a modern bleached kraft mill with secondary treatment on white suckers. Water Pollut. Res. J. Can. 27:423-437.

Van Der Kraak, G. and J.P. Chang. 1990. Arachidonic acid stimulates steroidogenesis in goldfish preovulatory ovarian follicles. Gen. Comp. Endocrinol. 77:221-228.

Van Der Kraak, G., P.M. Rosenblum and R.E. Peter. 1990. Growth hormone-dependent potentiation of gonadotropin-stimulated steroid production by ovarian follicles of the goldfish. Gen. Comp. Endocrinol. 79:233-239.

Van Der Kraak, G.J., K.R. Munkittrick, M.E. McMaster, C.B. Portt and J.P. Chang. 1992a. Exposure to bleached kraft pulp mill effluent disrupts the pituitary-gonadal axis of white sucker at multiple sites. Toxicol. Appl. Pharmacol. 115:224-233.

Van Der Kraak, G., K. Suzuki, R.E. Peter, H. Itoh and H. Kawauchi. 1992b. Properties of common carp gonadotropin I and gonadotropin II. Gen. Comp. Endocrinol. 85:217-229.

Wade, M.G., and G. Van Der Kraak. 1992. The control of testicular steroidogenesis in the goldfish: Effects of activators of different intracellular signalling pathways. Gen. Comp. Endocrinol. 83:337-344.

FIELD AND LABORATORY STUDIES OF BIOCHEMICAL RESPONSES ASSOCIATED WITH PULP MILL EFFLUENTS: STATUS IN 1991, 1994 AND BEYOND

P.J. Kloepper-Sams

Procter & Gamble, European Technical Center, Temselaan 100, B-1853 Strombeek-Bever, Belgium

In this contribution, the state of the science in applying biochemical responses to the study of pulp mill impacts will be briefly reviewed. The first international conference on the Environmental Fate and Effects of Pulp Mill Effluents in 1991 illustrated the utility of biochemical and physiological responses in determining exposure to pulp mill effluents. At that time, many questions were raised about the role of organochlorine compounds as dominant (or sole) sources of the responses observed. By the second conference in 1994, additional field sites, new techniques, and further pulping technologies had been assessed. Some progress in fundamental understanding of processes underlying biomarker responses was also made. However, the promise of biomarkers as "early warning indicators" has not yet been fulfilled. Finally, some speculations are made about remaining challenges and potential future advances.

INTRODUCTION

The Second International Conference on the Environmental Fate and Effects of Pulp Mill Effluents focused on responses of aquatic organisms to the release of pulp mill effluents at the subcellular, whole organism, and population/community level. This contribution will focus on subcellular responses, or "biomarkers". Because different researchers use this and related terms in slightly different ways, environmental biomarkers will be defined here broadly as subcellular responses due to exposure to an external stressor.

Biomarkers measured in fish exposed to pulping effluents include biochemical, physiological, and histopathological parameters, as well as blood chemistry, indicators of energy metabolism, and reproductive endpoints such as circulating sex steroid levels and the in vitro response of gonads to hormonal stimulation. Further discussions of the definition and application of biomarkers can be found in Kloepper-Sams and Owens (1993), Peakall and Shugart (1993), and Huggett et al. (1992). A recent review of a specific biomarker, cytochrome P4501A (CYP1A) induction, illustrates the predominance which pulp mill sites have played in the application of biomarkers in field monitoring (Bucheli and Fent 1994).

The great promise of biomarkers is their potential ability to act as inexpensive biological integrators of exposure and/or effects due to stressors in the environment. Basic biological and biochemical research builds our understanding of the fundamental mechanisms of these responses, but the key environmental advantage - to provide "early warning indication" of damage at the population, community, or even ecosystem level remains elusive. Variability in the response of individual organisms due to environmental or biological factors as well as the role which phenotypic adaptation may play in the complex response to stressors tend to make it difficult to determine which and what magnitude of change signifies a meaningful biological response which may precede serious impacts at the whole organism, population, or community level.

In examining the impacts of pulping effluents on aquatic ecosystems, some questions concerning biomarker application include:
1. What biochemical responses have been employed?

2. Which biochemical responses have shown consistent or dramatic responses?
3. What do these results indicate about the types and fate of responsible compounds and about the degree of recovery observed (and possible) as mills modernize?

The challenge is to further develop existing or novel biomarkers to be used in field or controlled laboratory situations which will act as endpoints of pulp mill exposure and effects at the level of the organism and, the most rewarding but difficult task, as predictive endpoints of population-level impacts.

DISCUSSION

The first major application of biomarkers in fish exposed to pulping effluents occurred in Norrsundet, Sweden, in the mid 1980s as part of the Environment Cellulose project. This mill was chosen as an example of "state of the art" pulping technology at that time. It is important to note that fish population-level effects were documented at Norrsundet, as were impacts on key benthic invertebrate and algal communities (Karås et al. 1991;, Södergren 1989). Gross deformities and increased fin erosion were also observed (Lindesjöö and Thulin 1987). A comprehensive suite of fish biomarkers was examined. Numerous biomarker values showed statistically significant changes compared to those of reference site fish; most of these changes decreased in fish caught at sites farther from the area of effluent release (Andersson et al. 1988). Operational upsets may have contributed to some of the most extreme impacts observed at this site (Owens 1991). In any case, these results set the "standard" for all subsequent pulp mill-related field monitoring efforts.

By 1991, several sites in Canada had been investigated using a similar suite of biomarkers. Additional endpoints, particularly measurement of circulating sex steroids and histopathological examination of multiple tissues, were added as investigators sought to better understand the mechanisms of toxicity observed at some sites. At the first International Conference on the Environmental Fate and Effects of Pulp Mill Effluents, about one-half of the oral presentations on biological effects (n = 10 of 19) dealt predominantly with fish biomarkers in response to pulp effluent exposures. Several of the other biological effects papers dealt with results of traditional acute and chronic laboratory bioassays with standard test organisms. In hindsight, similarities and, as importantly, differences in responses across sites in Canada, insights into the role of P450 induction (is it a direct indicator of toxicity?), and an evolving focus on reproductive endpoints and processes were key to future biomarker advances. Information on individual mill processes, feedstock types, or operating conditions was generally available only in a few cases (e.g., Joshi and Hillaby 1991), which hindered the ability to compare across studies or better understand potential generalities or differences derived from these factors. Many of these studies have since been considerably expanded (e.g., Hodson et al. 1991; Grahn et al. 1991; Munkittrick et al. 1991, 1992; Servos et al. 1992; McMaster et al. 1992; Van der Kraak et al. 1992; Kloepper-Sams and Benton 1994; Kloepper-Sams et al. 1994a).

By 1994, the dramatic impacts on the population and whole organism level reported in 1991 were reduced or no longer observed at these and other sites (e.g., Sandström 1996). It is difficult to document the exact reason for these improvements, as at most study sites mills implemented several process modifications simultaneously or within short time periods, such that biological "recoveries" due to specific process changes were difficult to identify. At the same time, some residual loadings due to older processes (e.g., presence of fiber mats, anoxic zones, accumulated persistent compounds in sediments or biota) may have hampered recovery at some sites.

The response of specific biomarkers has not proven predictive of population-level impacts. Whether this is due to differences in fish species, receiving water quality, supportive food webs, dilution factors, or a combination of these and other factors is unclear. For example, some of the strongest biomarker responses (Kloepper-Sams and Benton 1994) are not correlated with discernible population-level effects (Swanson et al. 1994), despite relatively high levels of organochlorine body burdens in these fish (Owens et al. 1994). The status of effluent quality has certainly improved; for example, 50% of Canadian mills had acutely toxic effluents in 1991 (Halliburton et al. 1991). Besides the positive impacts of wastewater treatment, dramatic reductions in chlorinated organic compound emissions have occurred, driven by a desire to reduce the release of chlorinated dioxins and furans. Concurrent reductions in chlorinated phenolics, resin acids, and other toxic compounds were environmental benefits of the programs initiated to reduce dioxin emissions (e.g., Pryke et al. 1995).

Perhaps reflecting reductions in measurable whole organism responses, the recent trend has been to investigate more subtle subcellular responses. Two plenary presentations discussed important aspects of biomarker research. The first described in detail the systematic search for P4501A inducer(s) in pulp effluent by use of a 4-d standardized lab bioassay (Hodson 1996). The second raised some thought-provoking concerns about the interpretability of biochemical responses. The clearest example was the differential response of individual rainbow trout to toxicants due to differing physiological status, not because of external factors but because of the social hierarchy of individuals which is established under laboratory holding conditions, resulting in differing feeding rates by dominants and subordinates (Lehtinen 1996).

Nearly 30 posters in the session on Biochemical Responses presented biomarker research, but this topic permeated other areas such as bioaccumulation of substances and integrated monitoring approaches. Two methods not presented at the 1991 conference are highlighted below.

The first method employs chemical markers. Chemical markers, especially bile metabolites, have been increasingly recognized as important tools for establishing exposure history (Oikari and Holmbom 1986; Kloepper-Sams et al. 1994a). Some controversy remains about the quantitative value of such measurements. However, a very different use of chemical markers can provide significant insights into cycling of nutrients and energy in ecosystems subjected to pulping discharges. Stable isotopes are widely used in biogeochemistry and have been employed in tracking food chains in such diverse locations as Arctic environments and deep sea vents. Here, the contribution of pulp mill effluents to riverine food webs was investigated by assessing carbon, nitrogen, and sulfur isotopic compositions of water, biota, and effluent samples (Wassenaar and Culp 1996). An unexpected terrestrial signature indicated that the effluent may play a significant role in this food web.

The second method employs immunological endpoints to begin to address the immune status of organisms, which may have an impact on their disease resistance and overall fitness. Phagocytic activity, as measured by enhanced whole blood chemiluminescence emission, was examined in rainbow trout exposed to elemental chlorine free (ECF) and totally chlorine free (TCF) pulp mill effluents (Marnila et al. 1994). In the case of mink fed fish exposed to a pulping effluent, no responses were observed at the whole organism (lethality, reproductive performance, kit growth and survival, body and organ weight) level, but minor P4501A induction was observed, and delayed-type hypersensitivity was higher in the exposed mink (Smits et al. 1994). Comparison to similar studies on fish-eating marine mammals may aid in the interpretation of the individual response *and* the potential to bridge this response to population-level effects.

Besides extension of "classical" biomarkers to different mill types (e.g., chemi-thermomechanical pulp [CTMP], TCF) and target organisms (e.g., Australian flathead), newer techniques have improved our mechanistic understanding of responses to pulping effluents. For example, PCR was employed to measure very low levels of P4501A mRNA in liver, gill, and gonads (Campbell and Devlin 1994) of juvenile salmon exposed to prototype inducers. Immunohistochemical techniques were used to examine subcellular localization of P4501A in BKME-exposed fish (Kloepper-Sams et al. 1994b) in a preliminary investigation of the potential role of different uptake routes on expression of P4501A protein. Studies have shown that secondary treatment may reduce, but not abolish, the potential of an effluent to induce P4501A activity. Induction may be related to the presence of organochlorine compounds in some, but not all, cases. In fact, high degrees of induction due to dioxin-like materials may mask a lower level of induction due to non-chlorinated substances, which have been reported for several years (e.g., Lindström-Seppä et al. 1992) but only recently received wider attention (Munkittrick et al. 1994). The potentially "site-specific" nature of inducers, due to the presence of either a few strong or potentially many weak inducers, any of which may vary in concentration and proportion due to feedstock, washing efficiencies, and other mill-internal parameters, as well as the environmental fate and availability of the compounds, indicates that caution should be exercised in attempts to "regulate" P4501A inducibility.

In the study of reproductive impacts, research is proceeding at the molecular as well as the population level. In vitro assays for estrogen binding affinity (Tremblay and Van Der Kraak 1994) and detailed analysis of the steroid biosynthetic pathway (McMaster *et al.* 1994) provided methods for deepening understanding of the types of compounds, and modes of action, that may cause reproductive impairment in fish exposed to pulping effluents. Given the historical findings of recruitment problems at some sites (e.g., Karås *et al.* 1991) and the role reproductive impairment may play at the population level, an increased focus on reproductive endpoints can be expected in the future. However, the potential for wide natural fluctuations in reproductive success and recruitment will add to the challenges in this area.

There is no scientific consensus on the utility of environmental biomarkers in the study of pulp mill impacts. Some participants felt that as effluent quality has improved, we have been searching for ever more sensitive biomarkers while simultaneously moving away from any hopes of linking these responses to the populations we are trying to protect. Others felt that further development of such markers is crucial, as only with suites of markers can we hope to (1) catch impacts before they become full-blown environmental problems and (2) identify new environmental hazards that may not be present today. Thus, the debate on top-down versus bottom-up approaches continues.

In the realm of biomarker interpretation, one participant suggested that understanding is lacking: we need a "diagnosis" rather than just an endpoint or series of endpoints. This is the strength of human biomarkers, because of the knowledge behind, for example, routine blood chemistry analysis. Several participants discussed the use of risk-based approaches, citing a need to strengthen the paradigm of risk assessment and to focus on specific assessment and measurement endpoints in applying risk assessment principles. A need to understand fundamental similarities and differences between human health and environmental approaches to risk assessment was also expressed. We seek to protect populations, rather than individuals, in the environment. An understanding of the individual's continuum of response, from normal homeostasis to compensatory mechanisms to deleterious responses that may lead to sickness, disease, or a reduction in fitness that results in a population-level response, is needed to adequately protect populations and ecosystems. An integrated approach (discussed later in the conference) indicates that integrated *understanding*, not just multiple measurements, is needed. Iterative processes which combine field observations with development of hypotheses and laboratory testing of such hypotheses, followed by further rounds of observation, hypothesis development, and testing, should become a more standard approach. Life cycle analysis was also suggested as a tool in this regard.

A key question for the industry and regulators is: how far do we as a society go in improving effluent quality? It was noted that despite the promise of "effluent-free" processes, there are no waste-free industrial processes, as yields are never 100%. Do we understand the potential risks of alternative approaches? Iif aquatic ecosystems are the best understood, what are the implications for shunting wastes to other environmental compartments? Also, the current pace of technological change is outstripping the ability of environmental scientists to document and interpret ecosystem responses. A question at the core of the conference was put to the pPanel: Is chlorine off the hook or, formulated differently, is the use of chlorine dioxide as a bleaching agent an environmentally benign technology? Most panel members did not want to offer a definitive answer, but noted the improvements seen to date.

The audience did not want to publicly address the issue of whether biomarkers are useful in population- or ecosystem-level effects studies. However, it was clear from informal discussions and from examination of the recent literature and the results presented at the Second Conference that biomarkers have played a pivotal role in these studies.

CONCLUSIONS

An ultimate goal of applied biomarker research is to establish appropriate endpoints which provide information about the potential for population (and community) effects to occur. Important points for researchers to consider as they design new field or laboratory investigations are as follows. If effects can be identified in the lab, are they or similar ones observed under similar conditions in the field? If not, why not? Can we design lab or microcosm/mesocosm exposures during the development of new processes such that potential impacts of new operations can be assessed at least in a preliminary way prior to full-scale conversion (e.g., how to avoid the chlorate problem of the mid-1980s)?

There is an urgent need for continued work to understand how and when biochemical and physiological responses translate to higher levels of biological organization. Increased involvement of population biologists, fisheries experts, endocrinologists, biochemists, and analytical chemists, along with linked lab and field investigations and more rigorous hypothesis testing, are needed. The community of scientists interested primarily in the fate and effects of pulping effluents should broaden their contacts to scientists investigating impacts of other stressors, including other anthropogenic chemicals as well as simultaneous environmental factors such as habitat modification.

It is not prudent to wait for obvious population, community, or ecosystem effects to occur before remedial action is initiated. Thus, biomarkers have found and will maintain a key role in integrated studies of pulp mill effluent fate and effects. However, minimal environmental benefits will accrue if the goal is no biochemical response at all. A rational balance is called for. Environmental biomarkers continue to possess a unique potential. However, this potential will only be realized if carefully designed, executed, and interpreted studies are performed. Hopefully, at the next conference, improved scientific understanding will set the stage for better defining the advantages and the limitations of biomarkers within the realm of larger integrated studies.

REFERENCES

Andersson, T., L. Förlin, J. Härdig and Å. Larsson. 1988. Physiological disturbances in fish living in coastal water polluted with bleached kraft pulp mill effluents. Can. J. Fish. Aquat. Sci. 45:1525-1536.

Bucheli, T.D., and K. Fent. 1994. Biomarkers for assessing exposure and effects of environmental contaminants: a review on field studies. EAWAG, Zürich, Switzerland, 85 pp.

Campbell, P.M., and R.H. Devlin. 1994. Induction of P4501A1 in livers, gills and gonads of juvenile Chinook: measurement by RT-cPCR. Presented at the Second International Conference on Environmental Fate and Effects of Bleached Pulp Mill Effluents, November 6-10, 1994, Vancouver, B.C., Canada.

Grahn, O., J. Tana, C. Monfelt, J. Härdig and K-J. Lehtinen. 1991. Environmental impact of two Swedish bleached kraft pulp mills as determined by field surveys. Swedish Environmental Protection Agency Report 4031:248-256, Solna, Sweden.

Halliburton, D., S.A. Jones, J.H. Carey, D.B. Carlisle, A.G. Colodney, A. Myres, W.L. Lockhart and I.H. Rodgers. 1991. Priority substances list assessment report no. 2: Effluents from pulp mills using bleaching. Environment Canada, Ottawa, Ontario, 59 pp.

Hodson, P.V. 1996. MFO induction by pulp mill effluents - advances since 1991. In Environmental Fate and Effects of Pulp and Paper Mill Effluents, M.R. Servos, K.R. Munkittrick, J.H. Carey and G. Van Der Kraak (ed.), St. Lucie Press, Delray Beach, FL.

Hodson, P.V., M. McWhirter, K. Ralph, B. Gray, D. Thivuerge, J.C. Carey, G. Van Der Kraak, D.M. Whittle and M.-C. Levesque. 1991. Effects of bleached kraft mill effluent on fish in the St. Maurice River, Quebec. Environ. Toxicol. Chem. 11:1635-1651.

Huggett, R.J., R.A. Kimerle, P.M. Mehrle, Jr. and H.L. Bergman. 1992. Biomarkers: biochemical, physiological and histological markers of anthropogenic stress. Lewis Publishers, Chelsea, MI.

Joshi, B.K., and B.D. Hillaby. 1991. Effects of process improvement on pulp mill effluent characteristics. Swedish Environmental Protection Agency Report 4031:101-109, Solna, Sweden.

Karås, P., E. Neuman and O. Sandström. 1991. Effects of a pulp mill effluent on the population dynamics of perch, *Perca fluviatilis*. Can. J. Fish. Aquat. Sci. 4:28-34.

Kloepper-Sams, P.J., and E. Benton. 1994. Exposure of fish to biologically treated bleached kraft effluent. II: Induction of hepatic cytochrome P4501A in mountain whitefish (*Prosopium williamsoni*) and other species. Environ. Toxicol. Chem. 13:1483-1496.

Kloepper-Sams, P., and J.W. Owens. 1993. Environmental biomarkers as indicators of chemical exposure. J. Haz. Mat. 35:283-294.

Kloepper-Sams, P.J., S.M. Swanson, T. Marchant, R. Schryer and J.W. Owens. 1994a. Exposure of fish to biologically treated bleached kraft effluent. I: Biochemical, physiological, and pathological assessment of Rocky Mountain whitefish (*Prosopium williamsoni*) and longnose suckers (*Catostomus catostomus*). Environ. Toxicol. Chem. 13:1469-1482.

Kloepper-Sams, P., E. Benton, T. Marchant, R. Schryer, J.J. Stegeman and S.M. Swanson. 1994b. Biochemical responses of whitefish and longnose suckers at Grande Prairie: P450 and sex steroid findings, 1990-1994. Presented at the Second International Conference on Environmental Fate and Effects of Bleached Pulp Mill Effluents, November 6-10, 1994, Vancouver, B.C., Canada.

Lehtinen, K-J. 1996. Biochemical responses in organisms exposed to effluents from pulp production - are they related to bleaching? *In* Environmental Fate and Effects of Pulp and Paper Mill Effluents, M.R. Servos, K.R. Munkittrick, J.H. Carey and G. Van Der Kraak (ed.), St. Lucie Press, Delray Beach, FL.

Lindesjöö, E., and J. Thulin. 1987. Fin erosion of perch (*Perca fluviatilis*) in a pulp mill effluent. Bull. Eur. Assoc. Fish Pathol. 7:717-749.

Lindström-Seppä, P., S. Huuskoen, M. Pesonen, P. Muona and Ö. Hänninen. 1992. Unbleached pulp effluents affect cytochrome P-450 monooxygenase enzyme activities. Mar. Environ. Res. 34:157-161.

Marnila, P., K. Koivula, K. Mattsson, J. Hardig and E.-M. Lilrus. 1994. A new method for monitoring the effects of bleached pulp mill effluents on immune system of rainbow trout. Presented at the Second International Conference on Environmental Fate and Effects of Bleached Pulp Mill Effluents, November 6-10, 1994, Vancouver, B.C., Canada.

McMaster, M.E., C.B. Portt, K.R. Munkittrick and D.G. Dixon. 1992. Milt characteristics, reproductive performance, and larval survival and development of white sucker exposed to bleached kraft mill effluent. Ecotoxicol. Environ. Saf. 23:103-117.

McMaster, M.E., G.J. Van Der Kraak and K.R. Munkittrick. 1994. Reproductive fitness in white sucker: the impacts of exposure to BKME. Presented at the Second International Conference on Environmental Fate and Effects of Bleached Pulp Mill Effluents, November 6-10, 1994, Vancouver, B.C., Canada.

Munkittrick, K.R., C.B. Portt, G.J. Van Der Kraak, I.R. Smith and D.A. Rokosh. 1991. Impact of bleached kraft mill effluent on population characteristics, liver MFO activity, and serum steroid levels of a Lake Superior white sucker (*Catostomus commersoni*) population. Can. J. Fish. Aquat. Sci. 48:1371-1380.

Munkittrick, K.R., G.J. Van Der Kraak, M.E. McMaster and C.B. Portt. 1992. Response of hepatic MFO activity and plasma sex steroids to secondary treatment of bleached kraft pulp mill effluent and mill shutdown. Environ. Toxicol. Chem. 11:1427-1439.

Munkittrick, K.R., G.J. Van Der Kraak, M.E. McMaster, C.B. Portt, M.R. van den Heuvel and M.R. Servos. 1994. Survey of receiving water environmental impacts associated with discharges from pulp mills. 2. Gonad size, liver size, hepatic EROD activity, and plasma sex steroid levels in white sucker. Environ. Toxicol. Chem. 13:1289-1301.

Oikari, A., and B. Holbom. 1986. Assessment of water contamination by chlorophenolics and resin acids with the aid of fish bile metabolites. *In* Aquatic Toxicology and Environmental Fate. 9th Vol. T.M. Poston and R. Purdy [eds.], ASTM STP 921, pp. 252-267.

Owens, J.W. 1991. The hazard assessment of pulp and paper effluents in the aquatic environment: a review. Environ. Toxicol. Chem. 10:1511-1540.

Owens, J.W., S.M. Swanson and D.A. Birkholz. 1994. Bioaccumulation of 2,3,7,8-tetrachlorodibenzo-p-dioxin and extractable organic chlorine in a northern Canadian river system. Environ. Toxicol. Chem. 13:343-354.

Peakall, D.B., and L.R. Shugart. 1993. Biomarkers: research and application in the assessment of environmental health. NATO ASI Series H: Cell Biology, Vol. 68, 119 pp.

Pryke, D.C., G.R. Bourree, S. Swanson, W. Owens and P.J. Kloepper-Sams. 1995. The impact of chlorine dioxide delignification on pulp manufacturing and effluent characteristics at Grande Prairie: effluent quality improvements and ecosystem response. Pulp and Paper Canada: In press.

Sandström, O. 1996. *In situ* assessments of pulp mill effluent impact on life-history. *In* Environmental Fate and Effects of Pulp and Paper Mill Effluents, M.R. Servos, K.R. Munkittrick, J.H. Carey and G. Van Der Kraak (ed.), St. Lucie Press, Delray Beach, FL.

Servos, M., J. Carey, M. Ferguson, G. Van Der Kraak, H. Ferguson, J. Parrott, K. Gorman and R. Cowling. 1992. Impact of a modernised bleached kraft mill on white sucker populations in the Spanish River, Ontario. Wat. Pollut. Res. J. Can. 27(3):423-437.

Smits, J., G. Wobester and H.B. Schiefer. 1994. Chronic exposure of mink (*Mustela vison*) to bleached pulp mill effluent: physiological, reproductive, immunological and pathological responses. Presented at the Second International Conference on Environmental Fate and Effects of Bleached Pulp Mill Effluents, November 6-10, 1994, Vancouver, B.C., Canada.

Swanson, S.M., R. Schryer, R. Shelast, P.J. Kloepper-Sams and J.W. Owens. 1994. Exposure of fish to biologically treated bleached kraft effluent. III: Fish habitat and population assessment. Environ. Toxicol. Chem. 13:1497-1507.

Södergren, A. (ed.) 1989. Biological effects of bleached kraft mill effluents. Final Report from the Environment/Cellulose Project I. National Swedish Environmental Protection Board, Bratts Tryckeri AB, Jököping, Sweden.

Tremblay, L.A. and G.J. Van Der Kraak. 1994. Development of assays to determine the estrogenic activity of xenobiotic compounds in fish. Presented at the Second International Conference on Environmental Fate and Effects of Bleached Pulp Mill Effluents, November 6-10, 1994, Vancouver, B.C., Canada.

Van Der Kraak, G.J., K.R. Munkittrick, M.E. McMaster, C.B. Portt and J.P. Chang. 1992. Exposure to bleached kraft pulp mill effluent disrupts the pituitary-gonadal axis of white sucker at multiple sites. Toxicol. Appl. Pharmacol. 115:224-233.

Wassenaar, L.I. and J.M. Culp. 1996. The use of stable isotope analyses to identify pulp mill effluent signatures in riverine food webs. *In* Environmental Fate and Effects of Pulp and Paper Mill Effluents, M.R. Servos, K.R. Munkittrick, J.H. Carey and G. Van Der Kraak (ed.), St. Lucie Press, Delray Beach, FL.

SECTION V

FIELD AND LABORATORY STUDIES OF WHOLE ORGANISM RESPONSES ASSOCIATED WITH PULP AND PAPER MILL EFFLUENTS

Recent work has attempted to link biochemical changes with changes at higher levels of organization. A number of laboratory and field studies have recently documented a variety of whole organism effects in biota exposed to pulp and paper mill effluents. Although the number of studies is limited, these impacts have been observed regardless of the mill process, bleaching or treatment strategy. These recent observations have challenged the traditional points of view and generated a number of yet unanswered but critical questions.

IN SITU ASSESSMENTS OF THE IMPACT OF PULP MILL EFFLUENT ON LIFE-HISTORY VARIABLES IN FISH

Olof Sandström

National Board of Fisheries, Institute of Coastal Research, Gamla Slipvägen 19, S-740 71 Öregrund, Sweden

Whole organism field studies on fish exposed to pulp mill effluents are summarized in this review. The major part of the information was produced in Canada and Sweden. Many reports document significant impact on growth and reproduction. Stimulated growth rate and increased body condition were often seen while sexual maturation was delayed and gonad size reduced. Impaired sexual performance was observed in populations suffering from recruitment deficits, verifying the ecological significance of whole organism studies. The reports cover a range of pulp production, bleaching techniques, and external effluent treatment methods. However, response patterns providing guidance for the technical measures to reduce mill effluent toxicity could not be distinguished. The variables studied evidently were sensitive and environmentally relevant indicators of effluent exposure. However, growth and reproduction are plastic in fluctuating environments, also reacting to natural changes. This emphasizes the importance of comprehensive applications of life-history theory in pollution studies. The recent Canadian and Swedish guidelines for environmental effects monitoring at pulp and paper mills provide examples of a systematic use of whole organism characteristics in fish to assess ecosystem health.

INTRODUCTION

Environmental realism and high predictive ability make studies at the organism level useful in field assessments of pollution effects. The routes by which an environmental influence on whole organism variables may act on the population can be summarized in life-history models. The models predict the relations between growth rate, energy storage, age at maturity, fecundity, and mortality and the numerous possible trade-offs involved to maximize fitness. Phenotypic plasticity is a key factor in life-history modeling. In fish, it is well known that growth rate has a strong impact on reproduction processes (Alm 1959; Stearns and Crandall 1984; Thorpe 1986; Roff 1991; Beverton 1992). It is commonly observed that fast juvenile growth will result in early sexual maturation at a small size (Sandström et al. 1995), although other options are also possible (Stearns and Crandall 1984).

Field studies demonstrating impact on growth and reproduction in fish exposed to pulp mill effluents began in the early 1980s at a Baltic coastal site (Sandström et al. 1988). Responses at lower levels, including increased activity of the liver detoxification system, confirmed that the sampled population was reacting to a chemical contamination (Andersson et al. 1988). Due to the paucity of supporting studies in North America, the results were met with considerable scepticism (Sprague and Colodey 1989). As late as 1989, the official opinion was that bleached kraft mill effluent (BKME), after secondary treatment, may well be associated with healthy and productive ecosystems (NCASI 1989). However, a considerable effort was soon devoted to verify the Scandinavian studies, mainly in Canada. Tested models for a holistic approach to pollution impact assessments using fish population characteristics were already developed (Munkittrick and Dixon 1988, 1989) for an immediate introduction to pulp mill research. Results soon appeared supporting the earlier observations of biochemical and physiological impact, reduced investment in reproduction, increased prevalence of external malformations, etc. in wild populations of fish (McMaster et al. 1991; Munkittrick et al. 1991, 1992; Hodson et al. 1992). Today it is more widely accepted that pulp mill effluents may constitute a threat to fish populations in receiving waters.

This paper reviews data reported on whole organism responses related to growth and reproduction in fish exposed *in situ* to effluents from pulp mills. A considerable number of mill site studies in Sweden were not published in journals available for international readers. Some of these results have been added to this review together with a few recent, still unpublished, observations. Finally, the recently updated programs in Sweden and Canada for monitoring environmental effects at pulp mills will be examined briefly.

MATERIALS AND METHODS

General

Reports on whole organism responses in fish were available from the U.S., Canada, and Sweden. Although pulp mill research has a long history in Finland, there are no field studies reported on growth and reproduction in fish. Canadian and U.S. studies were conducted in inland waters, and rivers as well as lakes. All Swedish sites visited are situated on the Baltic. The majority of data were collected at kraft pulp mills; only two sulfite mills were included. Some of the mills had integrated pulp and paper production. Although elemental chlorine was still used to bleach the pulp, many mills, especially in Sweden, were introducing other chemicals at the time of the studies or had already reached 100% chlorine substitution. Unbleached kraft mills were also covered by the investigations reviewed.

Collections of fish were made using gill nets of one or two mesh sizes, multi-mesh nets, hoop nets, and electroshock. The effect of net selectivity was sometimes used to provide a size-specific sample from the population. The sample sizes varied considerably (Table 1) and sometimes relatively few fish were collected for the analyses.

Age determinations were made on scales or operculae. Length at age relations were usually calculated to indicate growth rate. Back-calculations have been made in Swedish studies, as they provide an opportunity to make retrospective growth comparisons for previous calendar years and thereby increase the number of observations considerably. The weight-length relation, indicating the energetic state of the fish, was estimated as the condition factor (Cf).

Changes in age at maturity were calculated from the mean age of mature fish (Swanson *et al.* 1992), from the mean age of the youngest mature fish sampled (Munkittrick *et al.* 1991), or by comparing the frequencies of the youngest spawning age classes (McMaster *et al.* 1991). As it may be argued that recruitment variations could cause errors in these calculations, an additional way to indicate effects on maturation is tested in this paper on recent and previously published Swedish data on perch (*Perca fluviatilis*). Data from the Karlsborg mill in the northern Bothnian Bay are compared with earlier observations at Norrsundet (Sandström *et al.* 1988; Sandström 1994) and Piteå (Bergelin 1987a) to test whether size or age at first maturity was affected or if the observed lower frequency of fish with visible gonad growth was the result of a general depression of gonadal recrudescence in all adult female perch.

Relative gonad size was analyzed by calculating the gonadosomatic index (GSI) or by relating gonad weight to fish length. GSI is sensitive not only to gonad growth but also to changes in somatic energy stores, etc. Thus, either a length-related index or a gonad weight-fish length regression analysis may be preferred if differences in Cf are known (Sandström *et al.* 1988; Sandström 1994). GSI comparisons are relevant only for that part of the population preparing for the next spawning. It is, however, not always clearly stated whether or not immature fish were excluded before the comparisons were made.

GSI and similar indices are only crude indicators of gonad failure. Fecundity is a more precise measure. Only few studies of fecundity are, however, reported, probably due to the high costs involved. Gonad studies are made in order to assess the risk of reduced reproduction. In some studies, estimates were made of egg or sperm viability or embryo quality and survival as a link between the individual's performance and the population.

Table 1. Summary of the reviewed studies.

Mill	Ref.	Process	Year	Species	Sample size	MFO	Growth	Cf	Age at m.	GSI	Fec.	E/s v.
U.S.												
Pigeon River	Adams et al. 1992	bl. kraft	1989	Readbr. sunfish	15-20	+	+	+			0	-
Canada												
Terrace Bay	Munkittrick et al. 1991	bl. kraft	1988	White sucker	33-101	+	+	+	+	-	-	
	McMaster et al. 1991, 1992		1989	White sucker	14-31	+	-	+	+	-	-	-
	Munkittrick et al. 1992		1989/90	Lake whitefish	6-31	+	0	0	+	-	+	-
	Munkittrick et al. 1994		1991	White sucker	11-26	+	0	+	+	-		
Red Rock	Munkittrick et al. 1994	unbl. kraft[1]	1991	White sucker	12	+	+	0		-		
Fort Frances	Munkittrick et al. 1994	bl/unbl. kraft	1991	White sucker	15-19	+	0	+		-		
Smooth Rock F.	Munkittrick et al. 1994	bl. kraft	1991	White sucker	12-15	+	0	+		-		
Espanola	Munkittrick et al. 1994	bl. kraft	1991	White sucker	15-31	+	0	0		+		
St. Maurice R.	Hodson et al. 1992	bl. kraft	1990	White sucker	15	+		0		0		
	Gagnon et al. 1994		1990/91	White sucker	>300	+	+	+	+	0		
Wap./Smoky R.	Swanson et al. 1992	bl. kraft,	1990/91	Longn. sucker	4-22	+	0	+?	-	0	0	
	Swanson et al. 1992	70% ClO$_2$	1990/91	Mount. whitefish	1-24	+	0	-	-	0		
Port Harmon	Khan et al. 1992	bl. kraft	1990	Winter flounder	26-46			0				
Birchy Cove	Khan et al. 1992	bl. kraft	1990	Winter flounder	26-47			0				
Kenora	Munkittrick et al. 1994	bl. sulfite	1991	White sucker	18-20	+	0	0		-		
Kapuskasing	Munkittrick et al. 1994	TPM/sulfite	1991	White sucker	16-17	0	0	0		-		
Sweden												
Norrsundet	Sandström et al. 1988	bl. kraft	1983	Roach	15-66					-		
	Andersson et al. 1988		1983-87	Perch	10-125	+	+	+	+	-	-	-
	Karås et al. 1991											
	Förlin et al. 1991	100% ClO$_2$	1989-92	Perch	20-76	+		+	0	0		
	Sandström 1994											
Husum	Förlin et al. 1985	bl. kraft	1985	Perch	12-79	+	+?	+	+?	-		
Karlsborg	Bergelin 1987b, 1988	bl. kraft	1985-87	Perch	56-127		+?	-	+	0		
	Sandström 1985											
	Perä 1994	100% ClO$_2$	1993	Perch	152-153			0	+	0		
	Förlin et al. 1994		1993	Perch	20-23					0		
Munksund	Bergelin 1987a	unbl. kraft	1985/86	Perch	42-116		0	0	+	-		
Lövholmen	Bergelin 1987a	unbl. kraft	1985/86	Perch	14-116		0	0	+			
Mönsterås	Landner et al. 1994	bl. kraft	1989	Perch	9-40			0		0		

+ = significant increase, - = significant decrease, 0 = no effect, ? = not significant at the 95% level. E/s v. = Egg or sperm viability. Fec. = fecundity. Age at m. = age at maturity. MFO = mixed-function oxygenase.
[1]Red Rock mill had a small bleached pulp production, 57 ADMT compared to 819 ADMT unbleached (Munkittrick et al. 1994).

The reactions of the individuals, of course, depend upon the exposure situation. A commonly used indicator of mill effluent exposure is an increased activity in the liver detoxification system, usually measured as mixed-function oxygenase (MFO) or more specifically ethoxyresorufin-o-deethylase (EROD) activity. Although beyond the scope of this review, supporting EROD/MFO data are added to the list of observations.

Unpublished Reports and New Data

Whole organism studies on perch have been made at several Swedish coastal sites since 1985. Reports based on sufficiently large samples and otherwise of comparable quality were included in the review. Perch were sampled by gill nets and the collections were made during the appropriate period in autumn, well after the start of visible gonad growth. The sites visited at Karlsborg and Piteå are situated at the northern part of the Bothnian Bay, where the water is almost fresh. The Karlsborg mill produced chlorine bleached kraft during the 1980s, but was fully modified to 100% chlorine dioxide bleaching in 1991. At Piteå there are two mills, Munksund and Lövholmen, with unbleached integrated kraft and paper production. Studies have also been made at Husum, a bleached kraft mill on the Bothnian Sea.

SUMMARY OF THE REPORTED OBSERVATIONS

Growth Rate and Condition Factor

Growth rate was reported in 19 studies covering 15 mills and 6 species (Table 1). Most studies could not reveal any difference among samples. The Swedish studies on perch, however, often demonstrated a positive effect of the exposure to BKME. When Cf data are added to the growth observations, the picture of a stimulated somatic energy allocation becomes clearer. There were 25 reported Cf estimations made, covering all 18 sites. Increased Cf was noted in 12 studies (sometimes, however, only in one sex) and there were only two negative observations. Fry growth was studied at one site. As a positive reaction was also seen in the juveniles, the growth and Cf observations could not only be explained as increased somatic allocation made possible by the simultaneously noted lower commitment to reproduction (Sandström et al. 1988).

A reason why many studies report significant effects on Cf values may be that small samples allow accurate estimates of mean values and even small differences can be detected at high levels of significance. Growth rate, on the other hand, is notoriously difficult to estimate accurately. Temperature and feeding conditions influence growth as well as the reproductive state of the individual. Length-at-age correlations are often biased by between-year differences in the basic conditions for fish growth, and a long-time exposure to the effluent is needed before a growth response can be detected. A certain error should always be expected in age determinations and when estimations of back-calculated growth are made; moreover, there is often a considerable natural variation among individuals in growth rate. Large samples are consequently needed for growth analyses. Significant differences were often noted when more serious attempts were made to study growth rate, and inconclusive results were often associated with small data sets.

Sexual Maturation and Gonad Development

Sexual maturation was negatively affected in 8 of the 10 populations studied (Table 1). The maturation of white sucker (*Catostomus commersoni*) and lake whitefish (*Coregonus clupeaformis*) in Lake Superior was reported to be delayed 2-4 years and to occur at a smaller size in white sucker and increased size in lake whitefish (McMaster et al. 1991; Munkittrick et al. 1991, 1992). Swedish data on perch show that effluent-exposed fish had to reach a larger size before maturation occurred (Fig. 1). Although the

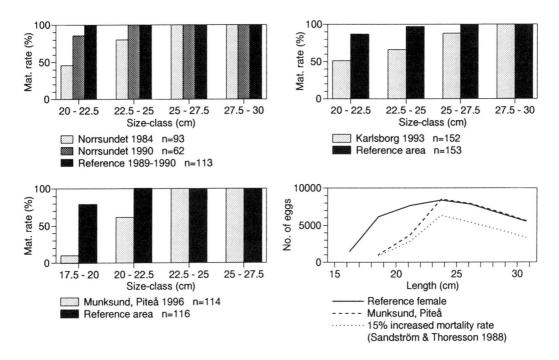

Figure 1. The percentage of mature female perch in different size classes and the calculated average egg production related to fish length.

available age data were too limited for an adequate calculation of differences in age at maturity, a delay of 1-2 years in females was suggested.

At Norrsundet, fast growth and a high Cf were accompanied by a significant delay in maturation in 1984, a combination not suggested by available knowledge on perch life history. Growth was not estimated in the 1990 follow-up study (Sandström 1994), but a higher allocation to somatic growth was indicated by the significant Cf difference (Table 1). However, in 1990 no significant effect on the sexual maturation pattern in perch could be detected.

Although there is an obvious trade-off between early reproduction, minimizing mortality risks, and a later spawning at a larger size, ensuring higher fecundity, fish generally mature as early as possible (Policansky 1983). When presented with good growth conditions, as in a cooling water effluent area (Sandström et al. 1995), female perch can mature during their second summer (as early as physiologically possible) at a size of only about 12 cm. This example indicates that fish may gain from maturing early, and delays caused by pollution may be significantly negative. To illustrate this effect, the expected lifetime fecundity for an average female perch sampled at one of the Piteå mills was compared with a reference female fecundity. By combining maturation data (Fig. 1), natural mortality rate estimates, and fecundity-length relations (Sandström et al. 1995), it is evident that the delay in peak reproductive performance as well as the reduction in lifetime egg production is considerable (Fig. 1). If mortality is increased in exposed adults, as was observed at Norrsundet in 1983-1985 (Sandström and Thoresson 1988), the impact will of course be even greater (Fig. 1).

Growth rate generally has a strong impact on the maturation process in fish (Thorpe 1986). In a comparison of the BKME-contaminated St. Maurice River and a reference river, growth in white sucker was enhanced downstream (Gagnon et al. 1994). An anticipated increased allocation of resources to reproduction, including maturation at a smaller size and at a younger age, was observed in reference fish but failed to appear in the polluted river, and the authors concluded that effluent prevented natural plastic responses from appearing.

Reduced relative gonad size was observed in 4 of the 6 species studied (Table 1). The exceptions, longnose sucker and mountain whitefish, were sampled from the Wapiti/Smoky River system, where effluent dilution and consequently exposure vary greatly according to seasonal river flows. In all, there are 24 reported studies on gonad development. A significantly negative impact was recorded in 14 cases. Reduced gonad performance consequently seems to be a common observation in fish exposed to pulp mill effluent.

Fecundity studies were only done occasionally (Table 1). The results generally support GSI data, with one exception. At the Terrace Bay mill, Lake Superior, lake whitefish experienced reduced gonad size although fecundity was increased (Munkittrick et al. 1992), suggesting a marked reduction in egg size. A reduced egg size was also documented in the same area in white sucker (McMaster et al. 1992) and was suggested by a microscopic analysis of perch ovaries collected at the Norrsundet mill (Sandström et al. 1988).

Studies on egg or sperm viability and embryo quality were done on three species at three sites (Karås et al. 1991; Adams et al. 1992; McMaster et al. 1992; Sandström 1994). Jackfish Bay male white sucker produced spermatozoa with reduced motility. However, this had no impact on the fertilization potential and the production of viable gametes. Larval survival was high, but growth rates were decreased compared to reference larvae. At Norrsundet, perch embryos were significantly smaller than reference fish and the prevalence of malformations was higher, coupled with a very high larval mortality and a significant recruitment deficit (Karås et al. 1991; Sandström 1994).

Most of the reviewed studies were supported by observations of increased EROD or MFO activities, indicating an effluent exposure of the sampled populations (Table 1).

DISCUSSION

The introduction of fish life-history variables in mill site research has contributed considerably to the understanding of effluent toxicity and the risk associated with effluent discharge from pulp production. The development was a significant step towards a higher environmental realism and more efficient integrated systems for monitoring pollution impact. A continued search for relevant biomarkers at low levels of organization coupled with whole organism variables should still be a tempting option. Much is also to be gained from a more precise application of life-history modeling. Compared to other measurements at lower levels, the effort devoted to analyze growth rates and sexual maturation was rather small in the investigations reviewed. Age and size at first maturity and the chances for continued future reproduction are strongly linked to the development of the population. Evidently, according to this review, these variables are sensitive to pulp mill effluent exposure. A proper evaluation of the observed deviations, however, should be made in the context of life-history theory. Although growth rate, or more specifically the energetic state of the animal during a short and defined time period, may be the decisive factor for maturation and gonadal recrudescence (Thorpe 1986), growth is often treated rather superficially using samples that are too small. Temperature, the most potent regulator of growth, is for instance rarely considered in the comparisons.

Some general remarks can be made from the reviewed papers. Growth and condition factor stimulations as well as gonadal disturbances were documented for different kinds of pulp production, including unbleached pulp, in a recent Canadian survey (Munkittrick et al. 1994). Secondary effluent treatment could not eliminate those responses. Studies conducted during 1988 and 1989 at the Terrace Bay mill documented several effects on fish exposed to primary treated effluent. Secondary treatment facilities were installed, but follow-up studies failed to detect any consistent improvements. White sucker still exhibited decreased gonadal size, depressed plasma sex steroid levels, and induced EROD activity.

Swedish observations support most of the conclusions drawn by Munkittrick et al. (1994). Effects on sexual maturation and relative gonad size were seen at mills producing chlorine bleached as well as unbleached pulp, and the lack of responses at some mills, e.g., the Mönsterås mill (Landner et al. 1994), may well be explained by the high dilution level.

Although life-history processes evidently are sensitive to pulp mill effluents, the reported observations consequently can not provide clear guidance for environmental measures at pulp mills. The Canadian survey (Munkittrick et al. 1994) also led to the conclusion that the dominant factor determining the presence or absence of responses appeared to be the dilution level. However, the fish population at one of the mills (Espanola) was judged to be healthy, although effluent dilution was not excessive when compared to other sites.

The observations could be due to a very unspecific response to a full range of chemical influences or to some common and very basic component in pulp mill effluent, such as natural wood sterols. Early observations on masculinization of female mosquitofish (Denton et al. 1985) and later mesocosm experiments including exposure of juvenile three-spined stickleback indicate that these compounds merit future attention. Growth stimulations were also observed in effluents containing low levels of resin acids and chlorophenols, and similar responses were produced in plant sterol exposure experiments (Lehtinen et al. 1992; Tana et al. 1994). However, follow-up studies at mills substituting elemental chlorine with chlorine dioxide or chlorine-free bleaching chemicals have so far only been reported from the Swedish mill at Norrsundet. These long-term studies documented a significant recovery in sexual maturation and gonad growth in perch simultaneous with a marked reduction in organochlorine waste production while other effluent components changed only moderately (Sandström 1994). In this case the investments in new techniques seemed to produce a significant positive environmental feedback.

Recently, Canada and Sweden adopted new guidelines for environmental effects monitoring at pulp mills (Environment Canada 1993; Naturvårdsverket 1994). Monitoring of toxic effects is concentrated on sentinel species of fish. Although the monitoring strategies differ somewhat, both programs are influenced by the observations summarized in this review. The Canadian Adult Fish Survey is focused on identifying alterations in morphological and life-history characteristics by sampling a minimum of two sentinel species during a first monitoring cycle. Length, weight, age, sex, weight of gonads, and fecundity will be measured to provide data for calculations of growth, reproductive investment, age structure, and energy stores. When deviations are detected, the strategy is to make analytical and confirming surveys during subsequent monitoring cycles.

The Swedish guidelines suggest two levels of monitoring. Population and community indicators are monitored annually, together with EROD measurements on one sentinel species to check the exposure of the sampled population. Whole organism-level measurements are integrated with the EROD sampling program. As in Canada, length, weight, growth, sex, and gonad weight are measured to indicate changes in energy stores and reproductive performance. The age structure of the population is used to analyze differences in recruitment patterns, indicating reproduction failure. Besides this basic program, there is the possibility of arranging special investigations to analyze observed deviations, document the geographical extent of the damage, or test whether observations at other sites are also valid at the monitored mill.

ACKNOWLEDGMENT

Unpublished material from the Bothnian Bay was put at my disposal by Ingemar Perä at the National Board of Fisheries, Regional Office, Luleå, Sweden, which is gratefully acknowledged.

REFERENCES

Adams, S.M., W.D. Crumby, M.S. Greeley, Jr., L.R. Shugart and C.F. Saylor. 1992. Responses of fish populations and communities to pulp mill effluents: a holistic approach. Ecotoxicol. Environ. Saf. 24:347-360.

Alm, G. 1959. Connections between maturity, size and age in fishes. Rep. Inst. Fresh Wat. Res., Drottningholm 40:5-145.

Andersson, T., L. Förlin, J. Härdig and Å. Larsson. 1988. Physiological disturbances in fish living in coastal water polluted with bleached kraft pulp mill effluents. Can. J. Fish. Aquat. Sci. 45:1525-1536.

Bergelin, U. 1987a. Undersökning avseende effekter på abborre av utsläpp från cellulosaindustrier i Pite skärgård 1986. Medd. nr 1 - 1987. Fiskeristyrelsen. Utredningskontoret i Luleå. Unpublished report in Swedish. 20 p.

Bergelin, U. 1987b. Effekter på abborre av utsläpp från ASSI-Karlsborgs massaindustri i Kalix skärgård 1986. Medd. nr 2 - 1987. Fiskeristyrelsen. Utredningskontoret i Luleå. Unpublished report in Swedish. 17 p.

Bergelin, U. 1988. Individstudier av abborre i anslutning till ASSI-Karlsborgs cellulosaindustri i Kalix skärgård 1987. Medd. nr. 1 - 1988. Fiskeristyrelsen. Utredningskontoret i Luleå. Unpublished report in Swedish. 17 p.

Beverton, R.J.H. 1992. Patterns of reproductive strategy parameters in some marine teleost fishes. J. Fish. Biol. 41:137-160.

Denton, T.E., M.W. Howell, J.J. Allison, J. McCollum and B. Marks. 1985. Masculinization of female mosquitofish by exposure to plant sterols and *Mycobacterium smegmatis*. Bull. Environ. Contam. Toxicol. 35:627-632.

Environment Canada. 1993. Technical guidance document for aquatic environmental effects monitoring related to federal Fisheries Act requirements. Environment Canada, Department of Fisheries and Oceans, Version 1.0, April 1993.

Förlin, L., T. Andersson, J. Härdig, Å. Larsson and O. Sandström. 1985. Prövning av fiskekologiska och fiskfysiologiska metoder för recipient kontroll vid massafabrik (Husum). Slutrapport 1985-12-20. NBL Rapp 190, Swedish EPA. Unpublished report in Swedish.

Förlin, L., T. Andersson, L. Balk and Å. Larsson. 1991. Biochemical and physiological effects of pulp mill effluents on fish. *In:* A. Södergren (ed.). Environmental Fate and Effects of Bleached Pulp Mill Effluents. Proc. of a SEPA Conference held at Grand Hôtel Saltsjöbaden, Stockholm, Sweden, Nov. 19-21, 1991. Swedish EPA Report 4031.

Förlin, L., S. Blom, M. Adolfsson-Erici and E. Lindesjöö. 1994. Fiskfysiologiska undersökningar i recipienten för ASSI Karlsborg. Dept. of Zoophysiology, Univ. of Gothenburg. Unpublished report in Swedish. 13 p.

Gagnon, M.M., J. Dodson, P.V. Hodson, G. Van Der Kraak and J.H. Carey. 1994. Seasonal effects of bleached kraft mill effluent on reproductive parameters of white sucker (*Catostomus commersoni*) populations of the St. Maurice River, Quebec, Canada. Can. J. Fish. Aquat. Sci. 51:337-347.

Hodson, P.V., M. McWhirter, K. Ralph, B. Gray, D. Thivierge, J.H. Carey, G. Van Der Kraak, M. Whittle and M.-C. Levesque. 1992. Effects of bleached kraft mill effluent on fish in the St. Maurice River, Quebec. Environ. Toxicol. Chem. 11:1635-1651.

Karås, P., E. Neuman and O. Sandström. 1991. Effects of a pulp mill effluent on the population dynamics of perch, *Perca fluviatilis*. Can. J. Aquat. Sci. 48(1):28-34.

Khan, R.A., D. Barker, R. Hooper and E.M. Lee. 1992. Effect of pulp and paper effluent on a marine fish, *Pseudopleuronectes americanus*. Bull. Environ. Contam. Toxicol. 48:449-456.

Landner, L., O. Grahn, J. Härdig, K.-J. Lehtinen, C. Monfelt and J. Tana. 1994. A field study of environmental impacts at a bleached kraft pulp mill site on the Baltic Sea coast. Ecotoxicol. Environ. Saf. 27:128-157.

Lehtinen, K.-J., J. Tana, J. Härdig, K. Mattsson, J. Hemming and P. Lindström-Seppä. 1992. Effects on survival, growth, parasites and physiological status in fish exposed in mesocosms to effluents from bleached hardwood kraft pulp production. Vatten-och Miljöstyrelsens Publ. Serie A, 105. Helsinki, Finland.

McMaster, M.E., G.J. Van Der Kraak, C.B. Portt, K.R. Munkittrick, P.K. Sibley, I.R. Smith and D.G. Dixon. 1991. Changes in hepatic mixed function oxygenase (MFO) activity, plasma steriod levels and age at maturity of a white sucker (*Catostomus commersoni*) population exposed to bleached kraft pulp mill effluent. Aquat. Toxicol. 21:199-218.

McMaster, M.E., C.B. Portt, K.R. Munkittrick and D.G. Dixon. 1992. Milt characteristics, reproductive performance, and larval survival and development of white sucker exposed to bleached kraft mill effluent. Ecotoxicol. Environ. Saf. 23:103-117.

Munkittrick, K.R. and D.G. Dixon. 1988. Growth, fecundity, and energy stores of white sucker (*Catostomus commersoni*) from lakes containing elevated levels of copper and zinc. Can. J. Fish. Aquat. Sci. 45:1355-1365.

Munkittrick, K.R. and D.G. Dixon. 1989. A holistic approach to ecosystem health assessment using fish population characteristics. Hydrobiologia 188/189:123-135.

Munkittrick, K.R., C.B. Portt, G.J. Van Der Kraak, I.R. Smith and D.A. Rokosh. 1991. Impact of bleached kraft mill effluent on population characteristics, liver MFO activity, and serum steriod levels of a Lake Superior white sucker (*Catostomus commersoni*) population. Can. J. Fish. Aquat. Sci. 48:1371-1380.

Munkittrick, K.R., M.E. McMaster, C.B. Portt, G.J. Van Der Kraak, I.R. Smith and D.G. Dixon. 1992. Changes in maturity, plasma sex steriods levels, hepatic mixed-function oxygenase activity, and the presence of external lesions in lake whitefish (*Coregonus clupeaformis*) exposed to bleached kraft mill effluent. Can. J. Fish. Aquat. Sci. 49:1560-1569.

Munkittrick, K.R., G.J. Van Der Kraak, M.E. McMaster, C.B. Portt, M.R. van den Heuvel and M.R. Servos. 1994. Survey of receiving-water environmental impacts associated with discharges from pulp mills. 2. Gonad size, liver size, hepatic EROD activity and plasma sex steriod levels in white sucker. Environ. Toxicol. Chem. 13(7):1089-1101.

Naturvårdsverket. 1994. Vattenrecipientkontroll vid skogsindustrier. Naturvårdsverket, Allmänna råd 94:2.

NCASI. 1989. Pulping effluents in the aquatic environment - part I: A review of the published literature. NCASI Technical Bulletin No. 572.

Perä, I. 1994. Undersökning av fiskbestånden samt abborrens kondition och könsutveckling i ASSI-Karlsborgs recipient. Fiskeriverket, Utredningskontoret i Luleå. Unpublished report in Swedish.

Roff, D.A. 1991. The evolution of life-history variation in fishes, with particular reference to flatfishes. Neth. J. Sea Res. 27:197-207.

Policansky, D. 1983. Size, age and demography of metamorphosis and sexual maturation in fishes. Am. Zool. 23:57-63.

Sandström, O. 1985. Tillväxt, kondition och gonadutveckling hos abborre i Kalixälvens mynningsområde 1985. Swedish EPA, Coastal Water Section. Unpublished report in Swedish.

Sandström, O. 1994. Incomplete recovery in a coastal fish community exposed to effluents from a modernized Swedish bleached kraft mill. Can. J. Fish. Aquat. Sci. 51(10)2195-2202.

Sandström, O. and G. Thoresson. 1988. Mortality in perch populations in a Baltic pulp mill effluent area. Mar. Pol. Bull. 19(11):564-567.

Sandström, O., E. Neuman and P. Karås. 1988. Effects of a bleached pulp mill effluent on growth and gonad function in Baltic coastal fish. Water Sci. Technol. 20:107-118.

Sandström, O., E. Neuman and G. Thoresson. 1995. Effects of temperature on life history variables in perch, *Perca fluviatilis* L. J. Fish. Biol.: In press.

Sprague, J.B. and A.G. Colodey. 1989. Toxicity to aquatic organisms of organochlorine substances in kraft mill effluents. Report for Renewable Resources, Extraction and Processing Division, Industrial Programs Branch, Conservation and Protection, Environment Canada, Ottawa, Ontario, June 1989. 53 p.

Stearns, S.C. and R.E. Crandall. 1984. Plasticity for age and size at sexual maturity: a life-history response to unavoidable stress. *In:* G.W. Potts and R.J. Wootton (ed.). Fish Reproduction: Strategies and Tactics. Academic Press, London.

Swanson, S.M., R. Shelast, R. Schryer, P. Kloepper-Sams, T. Marchant, K. Kroeker, J. Bernstein and J.W. Owens. 1992. Fish populations and biomarker responses at a Canadian bleached kraft mill site. TAPPI J. 76:139-149.

Tana, J., A. Rosmarin, K.-J. Lehtinen, J. Härdig, O. Grahn and L. Landner. 1994. Assessing impacts on Baltic coastal ecosystems with mesocosm and fish biomarker tests: a comparison of new and old wood pulp bleaching technologies. Sci. Tot. Environ. 145:213-234.

Thorpe, J.E. 1986. Age at first maturity in Atlantic salmon, *Salmo salar*: freshwater period influences and conflicts with smolting. *In:* D.J. Meerburg (ed.). Salmonid Age at Maturity. Canadian Special Publication of Fisheries and Aquatic Sciences 89:7-14.

LABORATORY RESPONSES OF WHOLE ORGANISMS EXPOSED TO PULP AND PAPER MILL EFFLUENTS: 1991-1994

Tibor G. Kovacs[1] and Stanley R. Megraw[2]

[1]PAPRICAN, 570 St. John's Blvd., Pointe Claire, Quebec, H9R 3J9 Canada
[2]Domtar Innovation Center, 22025 Trans Canada Highway West, Senneville, Quebec, H9X 3L7 Canada

The results of recent short-term and long-term laboratory tests with a variety of pulp and paper mill effluents were reviewed. In particular, attention was focused on the relationship between new bleaching technologies and the chronic toxicity of final mill effluents. The available data indicated that no generalizations could be made regarding the role of any specific pre-bleaching or bleaching process. Rather, there was limited evidence that the pulping side of mill operations could be a source of residual effluent toxicity. The extent to which laboratory tests have the ability to predict the actual impact of effluents on aquatic ecosystems is still an open question. Nevertheless, such approaches remain a cost-effective means of assessing effluent quality, especially effluents from technologies still in the developmental stage.

INTRODUCTION

Laboratory tests used to assess the toxicity of effluents can be broadly categorized according to the duration of exposure (e.g., acute, chronic), the level of response (e.g., lethal, sublethal) and the type of test organism (Rand and Petrocelli 1985). In addition, Sprague (1971, 1988) has distinguished between whole-organism and within-organism effects. Whole-organism tests include life cycle, partial life cycle and early life stage assays, whereas within-organism tests involve studies of biochemical and physiological effects.

The major purpose of whole-organism aquatic toxicity tests is to establish the concentration of an effluent and the duration of exposure to the effluent that clearly jeopardizes the long-term ability of a species to survive (Sprague 1988). In turn, it is hoped that information from such tests can be used to predict the potential impact of an effluent in the recipient, if the concentration of the effluent in the recipient is known. The environmental significance of within-organism effects is less obvious (Sprague 1971, 1988) due to our poor understanding of normal processes at this level, the considerable degree of natural variability and the difficulty of relating these effects to the well-being of the whole organism. Consequently, the best application of within-organism tests may be as early warning indicators of sublethal stress and, possibly, for elucidating the reasons for whole-organism responses.

During the 1980s, standardized protocols were issued in the United States for life cycle experiments with fathead minnows (Benoit 1982) and for short-term tests (i.e., partial life cycle and early life stage assays) with several different aquatic organisms (Weber *et al.* 1989). These methods were applied to a variety of industrial effluents including pulp and paper mill effluents. For example, at the first International Conference on Environmental Fate and Effects of Bleached Pulp Mill Effluents (Stockholm, Sweden, 1991), O'Connor *et al.* (1993) presented the findings of a study which used short-term tests to evaluate the sublethal toxicity of effluents with different levels of AOX (adsorbable organic halogens). The results of other laboratory tests with mill effluents, as well as the results of model ecosystem experiments and field surveys completed prior to 1991, have been reviewed by Kovacs (1986), McLeay (1987) and Owens (1991).

The objective of this paper is to summarize and integrate the findings of laboratory whole-organism tests with pulp and paper mill effluents conducted since 1991, with particular emphasis on bleached kraft mill effluents (BKMEs). Included is an examination of the effects described by both short-term tests and long-term life cycle experiments, the consequences of advanced bleaching technologies on effluent toxicity and a discussion of the relevance of laboratory tests for predicting environmental impact. The review focuses on the results of chronic tests because pulp and paper mill effluents subjected to proper secondary treatment are typically not acutely lethal (McLeay 1987). A glossary of bleaching terms and abbreviations used in this paper is included in Table 1.

Table 1. Glossary of bleaching terms and abbreviations.

Bleaching Terms	
Oxygen delignification	A pre-bleaching step practiced by many kraft mills to reduce by up to 50% the amount of lignin entering the bleach plant. The filtrate from this stage is usually recycled to the mill's chemical recovery system.
Conventional bleaching	Refers to the bleaching of pulps by the sequential use of: i. chlorine with no more than 10% substitution with chlorine dioxide; ii. sodium hydroxide; iii. chlorine dioxide or sodium hypochlorite or both; iv. sodium hydroxide; and v. chlorine dioxide.
Chlorine dioxide substitution	Refers to the partial (>10%) substitution of chlorine dioxide for elemental chlorine in the first stage of bleaching.
Elemental chlorine-free (ECF) bleaching	Refers to the use of 100% chlorine dioxide substitution of elemental chlorine in the first stage of bleaching.
Totally chlorine-free (TCF) bleaching	Refers to bleaching with agents other than chlorine and chlorine dioxide, such as ozone and hydrogen peroxide.
ECF 0.1	Refers to ECF bleached pulp with an AOX discharge of 0.1 kg t^{-1}.
ECF 0.2	Refers to ECF bleached pulp with an AOX discharge of 0.2 kg t^{-1}.
Lignox ECF	Refers to bleaching with peroxide and chlorine dioxide after pre-treatment with EDTA.
Bleaching Abbreviations	
Pre-bleaching	
O	Oxygen delignification
TCF bleaching	
Z	Ozone
P	Peroxide
Q	Chelation stage with EDTA
Chlorine-based or ECF bleaching	
C	Elemental chlorine
D	Chlorine dioxide
H	Hypochlorite
(Cd) or (Dc)	First bleaching stages where % chlorine (C or c) and % chlorine dioxide (D or d) are represented by subscripts
(C/D) or (D/C)	First bleaching stages where the % chlorine (C) and % chlorine dioxide (D) are not indicated in the reference
E	Extraction with sodium hydroxide
Eo	Extraction with sodium hydroxide and the addition of elemental oxygen
Eop	Extraction with sodium hydroxide and the addition of elemental oxygen and peroxide

ADVANCES IN TOXICOLOGICAL ASSESSMENT OF BKMEs – 1991 to 1994

Approaches

Short-Term Tests

Since 1991, several short-term tests (Table 2) using algae, invertebrates and fish were used in Europe (most notably Sweden), North America and Australia to assess the toxicity of BKMEs. The results of the tests were usually expressed in terms of a toxicity threshold value (TTV) which included the no observable effect concentration (NOEC) and the lowest observable effect concentration (LOEC) or the effluent concentration causing a 25% inhibition (IC25) for a particular endpoint. The algae (e.g., *Selenastrum*) test results were most frequently expressed in terms of EC50, the effective concentration of an effluent causing a 50% reduction in growth. Some investigators calculated toxicity emission factors (TEF), where

$$\text{TEF} = \frac{100}{\text{TTV}} \times \frac{\text{m}^3 \text{ (effluent)}}{\text{air dried tonne of product}}$$

to account for differences in water usage and production.

Table 2. Short-term tests used to assess the toxicity of BKMEs since 1991.

Species	Endpoints	Test Duration
• Algae		
– *Selenastrum capricornutum**	Growth	4 d
– *Nitzschia closterium*†	Growth	3 d
– *Chlorella protothecoides**	Growth	3 d
– *Macrocystis pyrifera* (giant kelp)†	Germination/growth	2 d
• Invertebrates		
– *Ceriodaphnia dubia**	Survival/reproduction	7 d
– *Dendraster excentricus*† (sand dollar)	Egg fertilization	40 min
– *Strongylocentrotus purpuratus* (purple sea urchin)†	Egg fertilization	40-80 min
– *Haliotus rufescens* (red abalone)†	Larval (shell) development	2 d
– *Crassostrea gigas* (Pacific oyster)†	Embryo survival, shell development	2 d
• Fish		
– *Pimephales promelas** (fathead minnow)	Larval survival/growth	7 d
- *Pimephales promelas**	ELST	8 d
– *Brachydanio rerio** (zebra fish)	Egg hatching/larval survival	7 d
– *Cyprinodon variegatus*† (sheepshead minnow)	Larval survival/growth	7 d
– *Menidia beryllina*† (inland silverside)	Larval survival/growth	7 d
– *Oncorhynchus mykiss** (rainbow trout)	Egg survival	7 d

ELST Embryo-larval survival and teratogenecity.
*   Freshwater.
†   Marine.

Long-Term Tests

Since 1991, three long-term tests were completed in Canadian laboratories with effluents from bleached kraft and thermomechanical pulp mills. These involved life cycle experiments with fathead minnows, where the fish were exposed from the egg stage to sexual maturity and reproduction. The major whole-organism endpoints in these tests were survival, growth, reproduction (number of spawns and number of eggs per female) and survival of the F_1 generation (i.e., the offspring resulting from the parental generation used to start the experiment).

Findings

Short-Term Tests

In general, the objective of the studies using the short-term tests was to assess the effect of process changes on effluent toxicity, in particular changes in bleaching technology. Several investigators (Brunsvik et al. 1991; O'Connor et al. 1994; Renberg 1992) observed that the toxicity of bleaching effluents was found to decrease (Tables 3 and 4) with increased chlorine dioxide substitution. Comparisons of ECF and TCF bleaching effluents did not reveal consistent responses (Table 3). Chirat and Lachenal (1993) reported that TCF bleaching effluents (OZP, OPZ) were less toxic to algae than an ECF bleaching effluent (ODEoD) while O'Connor et al. (1994) found that TCF effluents (OQP, OZEP) were more toxic to *Ceriodaphnia* than ECF (DE, ODE) effluents and were less toxic to fathead minnows than all chlorine and chlorine dioxide based bleaching effluents ($C_{90}d_{10}E$, $D_{50}c_{50}E$, DE) except for an ODE effluent.

Table 3. Short-term test results with effluent from laboratory bleaching of softwood and eucalypt* kraft pulp.

Effluent	Biotest		Reference
	Selenastrum, EC50, %		Chirat and Lachenal 1993
OZP	>90		
OPZ	81		
ODEoD	34		
$O(C_{50}d_{50})E$	10		Renberg 1992
ODE	58		
	Ceriodaphnia IC25, %	Fathead minnow, IC25, %	O'Connor et al. 1994
$(C_{90}d_{10})E$	15	10	
$(D_{50}c_{50})E$	13	16	
DE	40	16	
DE	50	24	
ODE	60	>100	
OQP	4	76	
OZEP	0.7	39	
	Chlorella, IC25, %	*Nitzschia*, EC50, %	Stauber 1993
$D_{70}C_{30}$*	19	2.5	
Eo*	3	65	
D_1*	75	20	
D_2*	>100	5	

Table 4. Short-term test results with softwood kraft mill bleach plant and pulp washing effluents.

Effluent	Biotest		Reference
	Selenastrum, EC50, %		
Mill A			Renberg 1992
ECF	39¶		
Mill B			
Lignox ECF	>100		
	Zebra fish, threshold value*, % (TEF)†		
	Egg hatching	Larval survival	
Mill C			Brunsvik et al. 1991
$O(C_{85}d_{15})EoDED$	–	15 (230)	
O(ECF) 0.2	90	45 (80)	
O(ECF) 0.1	>100	90 (40)	
Pulp washing	8	3 (250)	

¶   EC50 estimated from Figure 7 in Renberg (1992).
*   Threshold value derived by calculating the geometric mean of the LOEC and NOEC.
†   Numbers in parentheses are TEF equivalents of the threshold values.
–   No data provided in the reference.

The data of Lövblad and Malmström (1994) showed that the *Selenastrum* and the *Ceriodaphnia* tests could not distinguish between the toxicity of primary-treated combined mill effluents (CMEs) during TCF and ECF bleaching (Table 5). The CME includes wastewaters from both the bleaching and pulping operations of a mill and therefore represents the total discharge from the mill into the receiving waters. The authors claimed that, on the basis of TEFs, the TCF effluents were less toxic because less water was used. However, the use of TEFs was not appropriate for the comparison because the toxicity thresholds of the effluents during ECF bleaching were >100% on all occasions. Calculating TEF values is not justified when threshold concentrations exceed 100%. However, Lövblad and Malmström (1994) did find that effluent from the mills during TCF bleaching was less toxic in zebra fish tests than effluent during ECF bleaching.

The effects of chlorine dioxide substitution (including ECF bleaching), oxygen delignification and TCF bleaching on the toxicity of CMEs were also investigated. Using *Ceriodaphnia* and fathead minnow tests, it was found (Table 6) that CMEs after secondary treatment (biotreatment) were less toxic when oxygen delignification preceded bleaching (Borton et al. 1991; Gibbons et al. 1995), although the exact role of this pre-bleaching step on effluent toxicity could not be clearly delineated. Increasing chlorine dioxide substitution in the bleach plant corresponded to either increased (Mill H) or decreased (Mill I) CME toxicity to algae (Gibbons et al. 1995; Table 6). In contrast to the results with the *Selenastrum* test, increased chlorine dioxide substitution was not associated with any change in the CME toxicity thresholds (Table 6) for *Ceriodaphnia* and fathead minnow assays (Hall et al. 1994; Gibbons et al. 1995). Based on a series of marine and freshwater (Table 6) tests, Hall et al. (1994) concluded that secondary treatment of the CME was more beneficial than chlorine dioxide substitution. Similarly, Hickman et al. (1993) reported that activated sludge treatment was more effective in reducing the chronic toxicity of CME than physical/chemical treatment of CME or individual process sewers (Table 7).

Table 5. Short-term freshwater test results with untreated CMEs from bleached kraft mills.

Mill Conditions			Biotest			Reference
Effluent	Wood Furnish	Bleaching				
			Chlorella, EC50, %			Stauber 1993
Mill D	Not given	Not given	>100			
			Selenastrum, EC50, %	*Ceriodaphnia*, LOEC, %	Zebra fish, LOEC, %*	Lövblad and Malmström 1994
Mill E	Softwood	ECF	>100	>100	55	
Mill E	Softwood	TCF	>100	>100	90	
Mill F	Hardwood	TCF	73	100	89	

* Calculated from TEF and flow (m^3 ADT^{-1}) data given in the reference.

Fathead minnow and *Ceriodaphnia* tests were conducted (Robinson et al. 1994) with receiving water samples from upstream (reference) and downstream of effluent discharges at 11 pulp and paper mill sites in Ontario, Canada. The larval growth of minnows or the survival of *Ceriodaphnia* was not affected by any of the water samples downstream from mills when compared to the reference waters. The survival of minnow larvae was reduced at 4 of the 11 downstream sites and *Ceriodaphnia* reproduction was reduced at 2 downstream sites. These sites all received mill effluents which had been subjected to only primary treatment. In contrast, *Ceriodaphnia* produced significantly more neonates when exposed to 6 of the 11 downstream samples, indicating that the effluents at these sites might have caused organic enrichment rather than toxicity.

Testing of CMEs using the *Selenastrum*, *Ceriodaphnia* and fathead minnow larval survival/growth assays is a requirement for mills discharging into freshwaters as part of regulatory environmental effects monitoring studies in Canada (Environment Canada and Department of Fisheries and Oceans 1992). In provinces where fathead minnows are not indigenous, a seven-day trout egg test must be used (Environment Canada 1992). Consequently, for mill effluents, Gibbons et al. (1995) compared the sensitivity of the minnow and trout tests as well as the sensitivity of the fathead minnow larval survival/growth test and embryo-larval survival/teratogenicity (ELST) test (Table 8). The minnow ELST test was found to be more sensitive than the minnow larval growth test for three of the five effluents tested. The trout embryo survival test was less sensitive than the minnow growth test for the BKME, whereas for the TMP effluent, neither the minnow nor the trout were affected by 100% effluent.

Long-Term Tests

The two long-term life cycle tests with biotreated BKMEs revealed that effluent-related effects on egg hatching, survival and growth of the parent generation and the hatching and larval survival of the F_1 generation were virtually non-existent (Table 9). In both studies, reduced egg production by the minnows was the most sensitive consequence of exposure. Kovacs et al. (1995a) estimated that the IC25 and LOEC of the effluent for reduced egg production were 1.7 and 2.5%, respectively. Robinson (1994) did not report a threshold concentration for this effect, but on the basis of the lowest effluent concentration which caused a ≥25% reduction in fecundity, we estimated the LOEC for reduced egg production to be 12%. Since the water usage per tonne of pulp was approximately equal at both mills (approximately 100 m^3 t^{-1}) these results indicate that the effects of BKMEs on reproduction can vary considerably.

Table 6. Short-term test results with secondary-treated CMEs from bleached and unbleached kraft mills*.

Effluent	Mill Conditions Wood Furnish, % Softwood/% Hardwood	Bleaching	Biotest			Reference
			Selenastrum, EC50, %	Zebra fish, Threshold value, %		Flink et al. 1994
				Egg hatching	Larval survival	
Mill G (1)	†	$O(C_{85}d_{15})EoDED$	26	75	10-20	
Mill G (1)	†	O(ECF)	30	35	35	
			Selenastrum, EC50, %	*Ceriodaphnia*, IC25, %	Fathead minnow, IC25, %	Gibbons et al. 1995
Mill H (1)	100/0	$(C_{90}d_{10})EoDED$	NT	7	20	
Mill H (1)	100/0	$(D_{50}c_{50})EoDED$	4	7	30	
Mill H (1)	100/0	DEoDED	0.5	10	21	
Mill I (3)	100/0	$O(C_{55}d_{45})EoDED$	3 to 14	22 to 59	>100	
Mill I (2)	100/0	ODEoDED	48 and >100	48 and 56	63 and 75	
Mill J (2)	100/0	Unbleached	2.5(1)	12 and 90	>100	
			Ceriodaphnia IC25, %	Fathead minnow IC25, %		Borton et al., 1991¶
Mill K (8)	100/0	O(D/C)EoD	30 to 63	84 to ≥100		
Mill L (3)	30/70	O(C/D)ED	72 to ≥100	≥100		
Mill M (2)	50/50	O(C/D)EoD	43 and 73	82 and ≥100		
			Ceriodaphnia, Mean IC25, %	Fathead minnow, Mean IC25, %		Hall et al. 1994
Mill N (7)	100/0	CEHD, CEHHD	42	83		
Mill N (5)	100/0	$(D_{70}c_{30})EHD$	47	95		
			Sand dollar, mean IC25, %	Inland silverside, mean IC25, %	Sheepshead minnow, mean IC25, %	Hall et al. 1994
Mill N	100/0	CEHD, CEHHD	2.5 (6)	>67 (4)	>70 (6)	
Mill N	100/0	$(D_{70}c_{30})EHD$	0.6 (8)	>70 (1)	70 (3)	

NT Not tested.
* Numbers in parentheses represent number of samples tested.
† Mill uses both softwood and hardwood in unspecified proportions.
¶ Borton et al. (1991) also tested 21 effluents from mills with no oxygen delignification. For these effluents, the average IC25s in *Ceriodaphnia* and fathead minnow tests were 38 and 75%, respectively.

On the basis of secondary sexual characteristics, the two BKMEs also affected the sexual development of the fish although not in a similar manner. Robinson (1994) found that the effluent reduced the secondary sexual characteristics of males, and using the same 25% criteria described above, we estimated the LOEC for this effect to be the 25% effluent concentration. Kovacs et al. (1995a) observed that, at effluent concentrations ≥5%, the proportion of males and females was significantly different from the proportion of males and females in the control. Specifically, increasing effluent concentrations resulted in an increased proportion of males in the minnow population.

Table 7. The effect of various treatment options on the quality of effluents from two bleached kraft mills as determined by short-term marine biotests (Hickman et al. 1993).

Treatment	Effluent	Mill	Average % reduction in toxicity			
			Echinoderm Test	Red Abalone Test	Pacific Oyster Test	Giant Kelp Test
Steam stripping	Foul condensate	Mill O	86	63	NT	84
Polymer coagulation & precipitation	Bleach plant effluent	Mill O	51	23	NT	24
Metabisulfite	Combined mill effluent	Mill O	83	Toxicity increased	NT	Toxicity increased
		Mill P	42	Toxicity increased	NT	Toxicity increased
Activated sludge	Combined mill effluent	Mill O	85	No change	No change	Inconsistent
		Mill P	75	Inconsistent	NT	Reduced

NT Not tested.
Mill O Softwood mill; O(C/D)EopDEpD bleaching.
Mill P Softwood/hardwood mill; O(C/D)EoPDED bleaching.

Table 8. A comparison of the sensitivity of the minnow larval survival/growth (LSG) test to the sensitivity of the minnow embryo-larval survival/teratogenicity (ELST) test and the rainbow trout embryo survival (ES) test (Gibbons et al. 1995). All the tests were conducted with secondary-treated CMEs.

Mill Conditions			Biotest		
Effluent	Process	Bleaching	Fathead Minnow LSG, IC25, %	Fathead Minnow ELST, IC25, %	Rainbow Trout ES, IC25, %
Mill H	Kraft	$(C_{90}d_{10})$EoDED	49	NT	>100†
Mill H	Kraft	$(D_{50}c_{50})$EoDED	30	23	NT
Mill I	Kraft	$O(C_{55}d_{45})$EoDED	>100†	53	NT
Mill I	Kraft	$O(C_{55}d_{45})$EoDED	>100	>100	NT
Mill I	Kraft	ODEoDED	63*	46	NT
Mill J	Kraft	Unbleached	>100†	62	NT
Mill Q	TMP	Hydrosulphite	>100	NT	>100

* IC25 significantly different, as determined by the standard error of differences method at p < 0.05 (Sprague and Fogels 1977).
† Proper statistical testing for significant difference could not be done because one of the IC25s was >100%. Nevertheless, the two IC25s were assumed to be significantly different.

Table 9. The results of fathead minnow life cycle tests with CMEs from two bleached kraft mills and one TMP mill. All three effluents were sampled after secondary treatment.

	BKME-1 (Robinson 1994), LOEC, %*	BKME-2 (Kovacs et al. 1995a), LOEC, %	TMP (Kovacs et al. 1995b), LOEC, %
WOOD FURNISH	55% softwood 45% hardwood	100% softwood	100% softwood
BLEACHING	$O(D_{55}C_{45})EoHD$ $O(C_{80}d_{20})EHD$	$O(D_{45}C_{55})EoDED$	Hydrosulfite
ENDPOINT			
Egg hatching	>50	>20	>20
Survival (larvae to adult)	>50	>20	>20
Growth			
8-d wt	ND	20	>20
30-d wt	>50	>20	>20
60-d wt	>50	ND	ND
90-d wt	†	ND	ND
Final wt males	>50	>20	>20
Final wt females	>50	>20	>20
Reproduction			
Eggs/female	12	2.5	>20
No. of spawns	25	2.5	>20
Spawning delay	†	10	>20
Sexual development¶			
Maturity, males	25	ND	ND
Maturity, females	>50	ND	ND
Male/female ratio	>50	5‡	>20
F_1 generation			
% egg hatch	>50	>10	>20
Larval survival	>50	>10	>20

* The data were tested for the significance of concentration-response relationship by regression analysis. When no significance was reported, the LOEC was judged to be greater than 50% in this review. When a significant relationship was reported, the reviewers, when possible, calculated the LOEC as the concentration that would cause a ≥25% decrease (or increase in the case of spawning delay) from the control.

† The effluent was reported to cause a significant effect on the endpoint. However, the 25% decrease from the control (i.e., LOEC in this review) could not be calculated by the reviewers. For the 90-d wts, even the 50% effluent concentration did not cause a ≥25% decrease.

¶ Based on secondary sexual characteristics.

‡ Greater proportion of males than females.

ND Not determined.

In contrast, a biotreated TMP effluent (Table 9) showed no long-term effects on minnows at concentrations up to 20% (Kovacs et al. 1995b). Because the water usage at the TMP mill was about seven times lower than at the bleached kraft mills, the TMP effluent could be judged to be less harmful than the BKMEs. However, as with between-mill variability in the toxicity of BKMEs, it should not be assumed that the toxicity of all TMP effluents is similar. In fact, Kovacs et al. (1995b) showed that, on the basis of short-term tests, the TTVs of the TMP effluent used in the life cycle test were considerably higher than those found for TMP effluents from other mills.

OVERVIEW AND FUTURE RESEARCH NEEDS

Relationship Between Laboratory Responses and Mill Processes

Advanced Bleaching Technologies and Effluent Quality

Several authors (Berry et al. 1991; Axegård et al. 1993; Shimp and Owens 1993) have noted a significant improvement in the chemical composition and toxicity of effluents as a result of recent technological innovations in the bleached kraft sector of the pulp and paper industry. However, the recent work reviewed in this paper has made it clear that it is difficult to draw general conclusions regarding the effects of specific process modifications, particularly changes in bleaching processes, on the toxicity of CMEs. For instance, although increased chlorine dioxide substitution was shown to have significant benefits regarding the toxicity of *bleaching* effluents (Tables 3 and 4), this benefit was not apparent when the CMEs were tested (Hall et al. 1994; O'Connor et al. 1993). Moreover, studies (Borton et al. 1991; Chirat and Lachenal 1993; Lövblad and Malmström 1994; O'Connor et al. 1994) examining the effects of oxygen delignification or the relative merits of TCF versus ECF bleaching on the CME toxicity were, in our opinion, inconclusive. The life cycle studies by Kovacs et al. (1995a) and Robinson (1994) also indicated that effluents from mills using very similar processes can have significantly different effect thresholds for altering egg production by minnows.

One reason it is difficult to generalize on the potential for newer bleaching technologies to alter the quality of CMEs is that factors in addition to bleaching (e.g., type of wood furnish, increased yield, better chemical recovery, secondary treatment systems) could be affecting toxicity (O'Connor et al. 1994). Another reason for our inability to generalize is that our current knowledge is based on studies which used different experimental strategies. For example, investigators either used different types of biotests (with different sensitivities), tested bleaching or CMEs (from the same mill or from different mills) or in most instances based their conclusions on only one effluent sample. Therefore, as suggested by Axegård et al. (1993) and Flink et al. (1994), each mill should be considered unique and, in our opinion, judgments regarding the effects of process changes should be made on the basis of a consistent approach. This approach should ensure, for example, that a minimum of three composite effluent samples are tested from the same mill under stable operating conditions both before and after the process modification. In addition, each effluent should be tested for toxicity by a battery of biotests using different species and endpoints.

Effluent Quality and Factors External to the Bleach Plant

The majority of research activities related to the toxicity of pulp and paper mill effluents in recent years has focused on the chronic toxicity of BKMEs. The premise has been that bleaching in general, and elemental chlorine in particular, have been responsible for the toxicological properties of the CMEs. However, we now have limited evidence that perhaps there should be more emphasis put on other causes or sources of chronic toxicity in BKMEs. For example, the toxicities of bleached and unbleached CMEs from kraft mills are similar (O'Connor et al. 1993; Gibbons et al. 1995), and as shown in Table 4, a pulping effluent was found to be more toxic than bleaching effluents (Brunsvik et al. 1991). These results suggest that further attention should be given to the toxicity contributed by the pulping operations of a bleached kraft mill. Secondary treatment has also been shown to be an important factor which can influence the toxicity of effluents from kraft mills with bleaching. O'Connor et al. (1993) and Hickman et al. (1993), for example, found that aerated lagoons and activated sludge treatment could reduce the chronic toxicity of effluents to freshwater and marine organisms by 40 to 90%. Such reductions in effluent toxicity have not been achieved by in-plant process modifications or chemical treatment (Hickman et al. 1993; Hall et al. 1994).

Role of Laboratory Tests in Impact Assessment

For a mill which has recently adopted new manufacturing technologies, it is uncertain whether the environmental quality of the recipient as determined in field surveys reflects historical process operations, current process operations or a combination of the two. Environmental scientists working in the pulp and paper sector are also faced with the challenge of assessing the impact of effluents from processes still in the developmental or pilot stage. Therefore, model ecosystem studies as described by Lehtinen (1992) and laboratory biotests as discussed in this paper are probably the only means of evaluating the potential environmental impact of effluents under these conditions.

The major concern with laboratory biotests is that the constant and artificial conditions in the laboratory, which are valuable for establishing clear cause-effect relationships, can be disadvantageous when attempting to relate the laboratory results to a complex recipient situation. Furthermore, there is no assurance that the results of single-species studies can be extended to other species, even when they are phylogenetically similar or that the short-term tests are good predictors of long-term effects such as those noted in life cycle experiments. Consequently, while biotests may furnish critical information regarding the effect of effluents on the survival, growth and reproduction of a fish or an aquatic invertebrate, any extrapolations to higher levels of biological organizations (e.g., at the population or community level) must be made with great caution (Rand and Petrocelli 1985). Based on the work reviewed in this paper, we have attempted to highlight these concerns and thus put the role of laboratory tests for evaluating the impact of pulp and paper mill discharges into better perspective.

Species Sensitivity

The different species used in laboratory experiments were found to respond differently to effluents. For instance, O'Connor et al. (1994) reported that the *Ceriodaphnia* test was more sensitive than the fathead minnow test for TCF and ODE effluents, but the reverse was true for all the other effluents from chlorine and chlorine dioxide based bleaching trials. Gibbons et al. (1995) showed that the fathead minnow larval growth test could be either equally sensitive or significantly more sensitive than a trout embryo survival test (Table 8). Stauber (1993) found that the marine diatom *Nitzschia* was either more sensitive or less sensitive than the freshwater algae *Chlorella protothecoides,* depending upon which effluent in the bleaching sequence was assayed. These results highlight the difficulties associated with drawing conclusions regarding process changes and impact assessment based solely on the results of single-species biotests and illustrate the need for using several different species in toxicity evaluation studies.

Short-Term Versus Long-Term Tests

Kovacs et al. (1995a,b) assessed the ability of short-term tests to predict effects in long-term tests. In both type of tests, BKME and TMP effluents caused no increased mortality or reductions in the weight gained by minnows. The lower neonate production by *Ceriodaphnia* in the short-term tests with BKME corresponded to a decrease in egg production in the long-term study with fathead minnows, albeit at different effluent concentrations. Finally, both short-term tests with *Ceriodaphnia* and long-term tests with minnows showed that a TMP effluent affected neither neonate production nor fecundity of fish at equivalent effluent concentrations (i.e., 20%). These results demonstrate that there may be some merit in using short-term tests as surrogates of long-term effects in the laboratory, although the present data base for this suggestion is very limited. The availability of a meaningful short-term test which could assess the effects of effluents on the ability of fish to reproduce would be very useful and efforts to develop such a test should be a priority.

Laboratory Tests versus Field Studies

In an effort to provide some perspective on this subject, Robinson *et al.* (1994) compared the short-term test results of receiving water samples with historical benthic invertebrate community data from the same areas. They concluded that water collected downstream of several pulp and paper mills caused deleterious effects in the minnow and *Ceriodaphnia* tests and that these effects could be correlated with highly impacted benthic communities. However, the "correlation" between the findings of the laboratory tests and field surveys may have been fortuitous. As acknowledged by the authors, benthic community effects attributed to pulp and paper mill effluents, in particular effluents subjected only to primary treatment, are typically associated with such factors as oxygen depletion, organic enrichment and solids deposition and not direct toxicity as determined in laboratory tests. Therefore, in the case of mill effluents, the short-term toxicity tests may not be appropriate for characterizing the non-toxic impact of the discharges.

Due to a paucity of existing data on fish populations and communities in receiving waters near pulp and paper mill sites, Robinson *et al.* (1994) could not establish if the results of short-term tests were predictive of effects at higher levels of biological organization in the field. Therefore, these investigators attempted to relate whole-organism effects observed in the laboratory to the within-organism (e.g., morphological, physiological) status of wild fish reported by Munkittrick *et al.* (1994). Based on the available data, they concluded that the short-term laboratory tests with receiving water samples did not predict the apparent within-organism differences in fish attributed to the discharge of mill effluents. This lack of correlation is not surprising, since it is within-organism effects which should be used to predict whole-organism effects and not vice versa (Sprague 1971, 1988).

Some insight into the usefulness of life cycle tests for assessing the potential impact of effluents in the receiving waters can be obtained from work done in Ontario, Canada. The life cycle test was done with the effluent from a mill which has an average long-term concentration of 0.9% in the receiving river (Robinson *et al.* 1994). Although the effluent reduced egg production of minnows in the life cycle test (Robinson 1994), this effect occurred at effluent concentrations $\geq 12\%$, well above concentrations which would have been seen in the recipient. Indeed, when white suckers (*Catostomus commersoni*) were sampled above and below the mill's effluent discharge, minimal differences were found in morphological parameters (e.g., gonad size, body size) and plasma sex steroid levels (Munkittrick *et al.* 1994).

In summary, these findings illustrate that the value of laboratory tests for predicting field impacts of pulp and paper mill effluents has not been thoroughly investigated. Therefore, the results from the laboratory should be interpreted with the following points in mind. First, laboratory toxicity test results should not be used to predict the effects of effluent suspended solids, organic enrichment properties and oxygen depletion characteristics. Second, the relevance of responses observed at one level of biological organization (e.g., individual) to successively higher levels of organization (e.g., population, community) should be made with caution. Finally, it is important to realize that effluent concentrations used in laboratory tests must realisticaaly reflect conditions in the receiving environment in order for the results to be meaningful.

CONCLUSIONS

In our opinion, for reasons of cost effectiveness and practicality, laboratory tests will continue to play an important role in directing management decisions regarding the environmental consequences of mill process modifications. A greater effort towards understanding the effects of bleaching and non-bleaching process changes and effluent treatment on effluent quality, as well as a better understanding of the relationship between laboratory and field responses, will unquestionably lead to more defensible conclusions regarding future strategies for effluent management by the pulp and paper industry.

REFERENCES

Axegård, P., O. Dahlman, I. Haglind, B. Jacobson, R. Mörck and L. Strömberg. 1993. Pulp bleaching and the environment-the situation 1993. Nordic Pulp Pap. Res. J. 8(4):365-378.

Benoit, D. 1982. User's guide for conducting life-cycle chronic toxicity tests with fathead minnows (*Pimephales promelas*). U.S. Environmental Protection Agency, Report EPA-600/8-81-011, Environmental Research Laboratory, Duluth, Minnesota.

Berry, R.M., C.E. Luthe, R.H. Voss, P.E. Wrist, P. Axegård, G. Gellerstedt, P.-O. Lindblad and I. Pöpke. 1991. The effects of recent changes in bleached softwood kraft mill technology on organochlorine emissions: an international perspective. Pulp Pap. Can. 92(6):43-55.

Borton, D.L., W.R. Streblow and W.K. Bradley. 1991. Biological characterization studies of oxygen delignification effluents using short-term chronic toxicity tests. pp. 135-145. *In* Proceedings, 1991 TAPPI Environmental Conference, Book 1. TAPPI Press, Atlanta, Georgia.

Brunsvik, J.-J., E. Månsson and L. Landner. 1991. To CD or not to CD, that is the question. pp. 159-170. *In* Proceedings, 1991 TAPPI Pulping Conference. TAPPI Press, Atlanta, Georgia.

Chirat, C. and D. Lachenal. 1993. Ozone bleaching is the key to alternative bleaching technology. pp. 175-184. *In* Proceedings EUCEPA International Environmental Symposium. Centre Technique du Papier, Grenoble, France.

Environment Canada. 1992. Biological test method: toxicity tests using early life stages of salmonid fish (rainbow trout, coho salmon, or atlantic salmon). Report EPS 1/RM/28, Environment Canada, Ottawa, Ontario, Canada.

Environment Canada and Department of Fisheries and Oceans. 1992. Aquatic environmental effects monitoring requirements. Annex 1: Aquatic environmental effects monitoring requirements at pulp and paper mills and off-site treatment facilities regulated under the pulp and paper effluent regulations of the Fisheries Act. Report EPS 1/RM/18, Environment Canada, Ottawa, Ontario, Canada.

Flink, J., R. Grundelius and B. Swan. 1994. Moderne ECF- und TCF-bleiche. Das Papier 48:519-525.

Gibbons, J.S., C. Hayes and T.G. Kovacs. 1995. The chronic toxicity of secondary-treated pulp and paper mill effluents as determined by a battery of short-term tests. In prep.

Hall, T.J., R.K. Haley, T.M. Bousquet and D.L. Borton. 1994. The use of chronic bioassays in characterizing effluent quality changes in a mill conversion to increased chlorine dioxide substitution. pp. 115-121. *In* 1994 International Pulp Bleaching Conference - Papers to be Presented. Canadian Pulp and Paper Association, Montreal, Quebec, Canada.

Hickman, G.T., J.P. Miller and A.R. Amoth. 1993. Toxicity reduction in bleached kraft pulp mill effluent: pilot testing. pp. 261-279. *In* Proceedings 1993 TAPPI Environmental Conference. TAPPI Press, Atlanta, Georgia.

Kovacs, T. 1986. Effects of bleached kraft mill effluent on freshwater fish: a Canadian perspective. Water Poll. Res. J. Canada. 21:91-117.

Kovacs, T.G., J.S. Gibbons, L.A. Tremblay, B.I. O'Connor, P.H. Martel and R.H. Voss. 1995a. The effects of a secondary-treated bleached kraft mill effluent on aquatic organisms as assessed by short-term and long-term laboratory tests. Ecotox. Environ. Saf.: In press.

Kovacs, T.G., J.S. Gibbons, P.H. Martel, B.I. O'Connor and R.H. Voss. 1995b. The effects of a secondary-treated thermomechanical pulp mill effluent on aquatic organisms as assessed by short- and long-term laboratory tests. J. Toxicol. Environ. Health 44:485-502.

Lehtinen, K.-J. 1992. Environmental effects of chlorine bleaching - facts neglected? Pap. Puu 74(9):715-719.

Lövblad, R. and J. Malmström. 1994. Biological effects of kraft pulp mill effluents - a comparison between ECF and TCF pulp production. Paper No.11-2. *In* Proceedings, 1994 International Non-chlorine Bleaching Conference. Miller Freeman Inc., San Francisco, California.

McLeay, D.J. 1987. Aquatic toxicity of pulp and paper mill effluent: a review. Report EPS4/PF/1, Environment Canada, Ottawa, Ontario, Canada.

Munkittrick, K.M., G.J. Van Der Kraak, M.E. McMaster, C.B. Portt, M.R. van den Heuvel and M.R. Servos. 1994. Survey of receiving-water environmental impacts associated with discharges from pulp mills. 2. Gonad size, liver size, hepatic EROD activity and plasma sex steroid levels in white sucker. Environ. Toxicol. Chem. 13:1089-1101.

O'Connor, B.I., T.C. Kovacs, R.H. Voss and P.H. Martel. 1993. A study of the relationship between laboratory bioassay response and AOX content for pulp mill effluents. J. Pulp Pap. Sci. 19:J33-J39. Also see erratum in J. Pulp Pap. Sci. 19:J275.

O'Connor, B.I., T.G. Kovacs, R.H. Voss, P.H. Martel and B. Van Lierop. 1994. A laboratory assessment of the environmental quality of alternative pulp bleaching effluents. Pulp Pap. Can. 95(3):47-56.

Owens, J.W. 1991. The hazard assessment of pulp and paper effluents in the aquatic environment: a review. Environ. Toxicol. Chem. 8:1511-1540.

Rand, G.M. and S.R. Petrocelli. 1985. Introduction. p. 1-28. In G.M. Rand and S.R. Petrocelli (eds.). Fundamentals of aquatic toxicology. Hemisphere Publishing, Washington, D.C.

Renberg, L. 1992. The use of cost-effective chemical and biological tests for the estimation of the environmental impact of bleaching plant effluents. pp. 317-329. In Proceedings, 1992 Environmental Conference. TAPPI Press, Atlanta, Georgia.

Robinson, R.D. 1994. Evaluation and development of laboratory protocols for estimating reproductive impacts of pulp mill effluent on fish. Ph.D. Thesis, Univ. of Guelph, Guelph, Ontario, Canada, 150 p.

Robinson, R.D., J.H. Carey, K.R. Solomon, I.R. Smith, M.R. Servos and K.R. Munkittrick. 1994. Survey of receiving-water environmental impacts associated with discharges from pulp mills. 1. Mill characteristics, receiving-water chemical profiles and lab toxicity tests. Environ. Toxicol. Chem. 13:1075-1088.

Shimp, R.J. and J.W. Owens. 1993. Pulp and paper technologies and improvements in environmental emissions to aquatic environments. Toxicol. Environ. Chem. 40:213-233.

Sprague, J.B. 1971. Measurement of pollutant toxicity to fish - III. Sublethal effects and "safe" concentrations. Water Res. 5:245-266.

Sprague, J.B. and A. Fogels. 1977. Watch the Y in bioassay. pp. 107-118. In Proceedings, 3rd Aquatic Toxicity Workshop. Environmental Protection Service Technical Report No. EPS-5-AR-77-1, Halifax, Nova Scotia, Canada.

Sprague, J.B. 1988. Aquatic toxicology. pp. 491-528. In C.B. Schreck and P.B. Moyle (eds.). Methods for fish biology. American Fisheries Society, Bethesda, Maryland.

Stauber, J. 1993. Development and application of algal bioassays for determining the toxicity of bleached eucalypt kraft mill effluents. pp. 789-796. In Proceedings 47th Appita Annual General Conference, Book 2. Australian Pulp and Paper Industry Technical Association, Parkside, Victoria, Australia.

Weber, C.I., W.H. Peltier, T.J. Norberg-King, W.B. Horning, F.A. Kessler, J.F. Menkedick, T.W. Neiheisel, P.A. Lewis, D.J. Klemm, Q.H. Pickering, E.L. Robinson, J.M. Lazorchak, L.H. Wymer and R.W. Freyberg. 1989. Short-term methods for estimating the chronic toxicity of effluents and receiving waters to freshwater organisms. Report EPA/600/4-89/001, U.S. Environmental Protection Agency, Cincinnati, Ohio.

SURVIVAL, GROWTH, PRODUCTION AND BIOMARKER RESPONSES OF FISH EXPOSED TO HIGH-SUBSTITUTION BLEACHED KRAFT MILL EFFLUENT IN EXPERIMENTAL STREAMS

Dennis L. Borton[1], William R. Streblow[1], W. Kenneth Bradley[1], Terry Bousquet[2], Peter A. Van Veld[3], Richard E. Wolke[4] and Alexander H. Walsh[5]

[1]National Council of the Paper Industry for Air and Stream Improvement, P.O. Box 12868, New Bern, NC, 28560 U.S.A.
[2]National Council of the Paper Industry for Air and Stream Improvement, P.O. Box 458, Corvallis, OR 97339 U.S.A.
[3]The College of William and Mary, School of Marine Science, Virginia Institute of Marine Science, Gloucester Point, VA, 23062 U.S.A.
[4]University of Rhode Island, Kingston, RI, 02812 U.S.A.
[5]Pfizer Central Research, Groton, CT, 06340 U.S.A.

Largemouth bass (*Micropterus salmoides*), channel catfish (*Ictalurus punctatus*), bluegill (*Lepomis macrochirus*) and golden shiners (*Notemigonus crysoleucas*) were exposed to a high-substitution bleached kraft mill effluent (BKME) at 0, 4 and 8% by volume in experimental streams for 263 d. The total production of all fish from the 8% effluent was nearly double the production in other streams, and the production of the four fish species in the stream receiving 4% effluent was very similar to the production in the two control streams. Fish were evaluated using several biomarkers (EROD, P450IA, hematocrit, leucocrit, condition factor, LSI, SSI) and histopathology but elevated liver P450IA content and EROD activity were the only responses that could be concluded to be caused by effluent exposure. The large number of significant differences between biomarkers of fish from the two controls or between an effluent exposure and only one of the two controls limited the interpretation of other biomarkers. Since the survival, growth and production of fish with elevated liver P450IA and EROD activity were not adversely affected by the effluent, these biomarkers did not correspond to the population level parameters measured in this study.

INTRODUCTION

The compatibility of effluents from pulp and paper mills with the continued survival, growth and production of fish populations or with other parameters measuring the responses of aquatic communities has been investigated using a variety of techniques. The results of many of these studies were recently summarized by Owens (1991). Investigators have used biochemical, physiological or histological biomarkers as indicators of exposure to pulp mill effluents as well as possible adverse effects (Lehtinen *et al.* 1990; Munkittrick *et al.* 1991; Kloepper-Sams and Swanson 1982; Swanson *et al.* 1993). Studies comparing these biomarkers with population-level parameters of fish with a known history of exposure to pulp mill effluents are limited. The studies described here provide information comparing the concentrations of a high-substitution bleach kraft pulp mill effluent (HS BKME) to the survival, growth and production of four fish species and allow these population parameters to be related to several biomarkers.

Experimental streams were used to determine the effects of pulp mill effluent on the survival and production of several species of fish (NCASI 1983, 1993; Hall and Haley 1991). The studies reported to date have been with effluents from mills that used primarily elemental chlorine in the first bleaching stage. The recent use of ClO_2 as a substitute for Cl_2 in pulp bleaching reduces the production of chlorinated organic compounds (Berry et al. 1991). Changes in the use of Cl_2 and ClO_2 in the bleaching process of the mill at the Southern Experimental Streams site has allowed comparisons of population-level effects and bioaccumulation of tetrachlorodibenzo-p-dioxin (TCDD) and tetrachlorodibenzofuran (TCDF) to be made before and after conversion to high substitution of ClO_2 (NCASI 1991, 1995; NCASI unpublished). Additional biomarker studies conducted after the conversion allow the comparison of these biomarkers with the survival, growth and production of largemouth bass (*Micropterus salmoides*), channel catfish (*Ictalurus punctatus*), bluegill (*Lepomis macrochirus*) and golden shiners (*Notemigonus crysoleucus*) exposed to 4 and 8% of the HS BKME in the experimental streams.

MATERIALS AND METHODS

Effluent from a bleached kraft pulp mill, producing approximately 700 t d^{-1} of market pulp from a furnish consisting of 25% hardwood and 75% softwood species, was added to experimental streams. The pulp bleaching sequence was D/CE$_{op}$ Wash DE$_o$D (D = chlorine dioxide, C = chlorine, E$_{op}$ = extraction with oxygen and peroxide) during this exposure, reducing elemental chlorine use by 70% compared to earlier studies. The 105,000 m^3 d^{-1} of effluent was treated for 14 d in an aeration stabilization basin. Effluent was pumped continuously from the basin discharge canal to the experimental streams site where it was sampled for analysis. A 24-h composite or grab sample of effluent was taken weekly. The effluent was collected in specially cleaned glass containers and shipped to the NCASI organic analysis laboratory for analysis of resin acids, chlorinated resin acids, chlorinated phenolic compounds and dioxins. Methods for these analyses are described elsewhere (NCASI 1986a,b). The effluent was characterized by personnel of the mill environmental laboratory 4 to 5 times each week for BOD, suspended solids and pH and weekly for total solids and COD.

Four experimental streams were used for this study. These experimental streams were 110 m long with square pools 8 m wide and 1 m deep at the center alternating with riffle areas 9 m long, 2 m wide and 0.1 m deep at the center. Each stream received 17 L s^{-1} of water pumped from the Neuse River, or approximately 10 exchanges d^{-1}. A weir box at the head of each stream allowed the water flow to be monitored. The temperature of the water entering and exiting one of the experimental streams was continuously recorded. Dissolved oxygen was measured at approximately 0900 each weekday using a portable dissolved oxygen meter.

In addition to receiving river water, two of the experimental streams, hereafter termed treatment streams, received effluent. The flow of the effluent to the treatment streams was controlled by valves and measured in headboxes using V-notch weirs. Streams 1 and 3 were never exposed to effluent and are referred to as control streams, while streams 2 and 4 were treatment streams. During these studies stream 4 was designated to receive 4% by volume (%v/v) of effluent and stream 2 was designated to receive 8% v/v. Conductivity measurements of the effluent, water and treatment streams were used to calculate the actual concentration of effluent in each treatment stream.

Fish stocked in the streams were purchased from local suppliers. The streams were stocked with 250 golden shiners (5.5 g average weight) 3 wk before the other fish species to allow bioaccumulation of organic compounds in this potential prey fish. Each stream was stocked with 119 largemouth bass (43 g average weight), 125 channel catfish (66 g average weight) and 132 bluegill (44 g average weight). Periodically fish were seined from the streams, measured and weighed and returned to the streams, except those fish used for bioaccumulation, liver P450IA and EROD samples. On two occasions the streams were pumped down to be certain that all fish were collected. With the exception of the final sampling date when all fish were removed, sampling was limited to when the water temperature was less than 20°C to minimize stressing the fish. Estimates of the survival, biomass, growth rates and production of each species were calculated from the numbers of fish surviving and their weight changes as described in Chapman (1971).

Five bass or catfish used for P450IA measurements were removed from one control stream and one or both of the treatment streams on two to four sampling dates during the study. Twelve bass or catfish were sampled from each stream on the final sampling date. Sampling and handling of fish for the measurement of liver ethoxyresorufin-*o*-deethylase (EROD) activity and P450IA concentrations are described elsewhere (Bankey *et al.* 1994). On the final day of exposure 12 bass and 12 catfish were sampled from each stream for organ somatic indices, hematocrit, leucocrit and histopathological analysis; 12 bluegill were sampled from stream 3 (control) and stream 4 (4%), but only 10 bluegill were sampled from stream 2 (8%) and 2 from stream 1 (control) because of lower survival in those streams. Fish sampled for histopathology were placed in 10% neutral buffered formalin. One or two sections of 20 tissues from each fish were examined for lesions. Gonads were examined for maturity and were rated on a 5-point scale. No indication of gonad differentiation was rated as a 1 and a sexually mature gonad was given the rating of 5. The methods used to examine the fish as well as the statistical methods are described in Hall *et al.* (1992).

Fish condition factors (Cf), liver somatic index (LSI), and spleen somatic index (SSI) were calculated for fish from the final sampling date (Nielsen and Johnson 1983). Blood hematocrits and leucocrits were measured as described by McLeay and Gordon (1977). Condition factors are based on the total fish population collected on the final sampling date; all other indices are calculated from the samples of fish used for histopathological analysis. The percent mortalities of each fish species in each stream were compared for significant differences using Fisher's Exact Test ($p \leq 0.05$). The fish removed for samples from each stream were included in the number of survivors. Males and females were tested for significant differences within each group for each biomarker. Only the gonad maturity was found to be significantly different between sexes; thus this parameter is reported for each sex. All other biomarkers are reported using samples with combined sexes. Since variances between groups were frequently significantly different using Bartlett's test and the sample size did not allow an evaluation of the distribution, the average fish weights, Cf, LSIs, SSIs, hematocrits and leucocrits were compared between streams using Kruskal-Wallis analysis of variance followed by the Mann-Whitney test for significant differences ($p \leq 0.05$) between each stream. The Bonferroni test ($p \leq 0.05$) was used to compare P450IA and EROD responses (Bankey *et al.* 1994). The numbers of each specific lesion in each organ found during the histopathological evaluation were compared for significant differences using Fisher's exact test ($p \leq 0.05$).

RESULTS

The dilution water hardness averaged about 25 mg L^{-1}, alkalinity was 20 mg L^{-1} and pH ranged from 6.4 to 7.5 (USGS 1991). The temperature and dissolved oxygen in the experimental streams followed the seasonal variation common to streams of the southeastern U.S. During October 1990, the earliest portion of this study, the temperature was approximately 7°C, reached a low of 0°C in January and rose to 15°C by mid-April. The temperature was approximately 30°C by the final sampling date in early July 1991. The dissolved oxygen rose to over 12 mg L^{-1} in the winter and declined to approximately 6 mg L^{-1} by July. Streams receiving effluent frequently had dissolved oxygen concentrations that were 0.5 to 1 mg L^{-1} less than the control streams.

The average concentrations (37 samples) of 13 mg L^{-1} for BOD, 343 mg L^{-1} for COD, 10 mg L^{-1} for total suspended solids, 16 μg L^{-1} for total fatty acids (4 compounds) and 80 μg L^{-1} for total resin acids (8 compounds) in the BKME did not differ significantly from earlier studies with this effluent (NCASI unpublished). However, the average concentrations of total chlorocatechols (11 μg L^{-1}, 9 compounds), total chloroguaiacols (μg L^{-1}, 10 compounds), total chlorophenols (0.3 μg L^{-1}, 7 compounds) and total chlorinated resin acids (11 μg L^{-1}, 12 compounds) were all significantly reduced by factors of 4 to 6 times in the HS BKME compared to earlier studies. Total chlorinated phenolic compounds (55 μg L^{-1}, 32 compounds) were not significantly reduced, but this was primarily due to an increase in 6-chlorovanillin in the HS BKME.

The average weights of bass were not significantly different between all control and treatment streams after 263 d of exposure to the effluent (Table 1) and mortality was also not significantly different among streams. The average weight of the catfish from control stream 3 was significantly less than the average weights of fish from the other three streams, including the other control stream. The average weights of catfish from the other three streams were not significantly different from each other, but the mortality of catfish in the 8% effluent was significantly reduced compared to the other 3 streams. The average weight of bluegill was greatest in stream 1. However, only two bluegill remained on the final sampling date; thus stream 1 was not included in other analyses. The average weight of bluegill from the 8% effluent concentration was also significantly less than from the other two streams. The average weights of golden shiners were significantly different between the two control streams (stream 1 = 27 g, stream 3 = 42 g), while the average weight of shiners from the treatment streams were between the 2 controls, but were significantly different from both control groups. The mortality of shiners was significantly less in control stream 1 compared to all other streams (Table 1).

Table 1. Average weight, number and production of four fish species exposed for 263 d to HS BKME. SDs are in parenthesis. Numbers within rows are not significantly different if followed by the same letter(s). Numbers followed by letters but having no letters in common are significantly different.

	Control						Treatment					
	Stream 1			Stream 3			Stream 4 (4%)			Stream 2 (8%)		
Species	Avg. wgt.[a] (g fish^{-1})	N.	Prod. (g m^{-2})	Avg. wgt. (g fish^{-1})	N.	Prod. (g m^{-2})	Avg. wgt. (g fish^{-1})	N.	Prod. (g m^{-2})	Avg. wgt. (g fish^{-1})	N.	Prod. (g m^{-2})
Bass	161.6a (40.5)	18b[b]	6.1	153.7a (48.4)	19b	5.2	160.2a (32.6)	23b	5.6	142.3a (49.4)	38b	8.1b
Channel Catfish	216.2a (82.0)	42c	13.6	175.5b (55.8)	35c	8.6	220.8a (55.7)	42c	12.6	228.0a (75.3)	80d	28
Bluegill	158.6 (58)	2c	2.4	110.3 (36.9)	28c	5.5	113.3a (34.5)	17c	5.3	75.5b (55.2)	10c	2.9
Golden Shiners	27.1a (7.9)	99d	6.4	42.0b (8.8)	18e	3.4	35.5c (8.6)	39e	3.9	33.7c (12.3)	27e	3.9

[a]Abbreviations: Avg. wgt = average weight, N. = number; Prod. = Production.
[b]Letters after numbers in this category indicate groups with similar or significantly different mortality. Sampled fish were included with survivors.

Production of bass was similar in the two control streams and 4% treatment stream, all being near 5 to 6 g m^{-2}. The production of bass in the 8% treatment was about 30% greater than the other streams. The production of catfish was at least 30% less in control stream 3 than in any other stream. Production of catfish in control stream 1 and the 4% treatment was similar. Production of catfish in the 8% treatment stream was nearly double that of any of the other streams. The production of bluegill and golden shiners in both of the treatment streams was within the range of production of the same species in the two control streams.

A total of 88 comparisons between the averages of 6 biomarkers (Cf, LSI, SSI, hematocrit, leucocrit, maturity index) from catfish, bass and bluegill after 263 d in the 4 experimental streams are shown in Table 2. Since only 2 bluegill were available from stream 1 this stream was not included with the bluegill comparisons. Leucocrits did not clearly separate in most samples from bass and bluegill from all streams; therefore these parameters were not reported. Of the 88 comparisons, 28 were statistically significant; 5 of the 28 statistically significant comparisons were between the 2 control streams, 12 of the comparisons were significant between a treatment stream and only 1 of the 2 control streams (or the average of the biomarker from the treatment stream was significantly different from both controls but was between the averages of the 2 control streams) and 2 of the significant comparisons were between the 4% effluent concentrations and both controls but the same parameter was not significantly different when this comparison was made to the 8% effluent concentrations. Thus at least 5 of the 28 significant comparisons were not due to the presence of effluent and another 17 significant comparisons were probably not caused by the effluent. The Cf of bass from 4 and 8% effluent were significantly reduced compared to both controls, and LSIs of bluegills from both treatment streams were also reduced compared to bluegill from control stream 3.

Table 2. Average [± SD (n)] condition factor, liver somatic index, spleen somatic index, hematocrit, leucocrit and maturity index of three fish species exposed to HS BKME for 263 d. Numbers within rows are not significantly different when followed by the same letter(s). Numbers followed by letters but having no letter in common are significantly different.

Parameter	Species		Controls		Treatments	
			Stream 1 Average	Stream 3 Average	Stream 4 (4%) Average	Stream 2 (8%) Average
Cf[a]	LB		1.40a ± 0.05 (18)	1.42a ± 0.09 (19)	1.30b ± 0.06 (23)	1.31b ± 0.10 (38)
	CC		0.94a ± 0.14 (42)	0.88b ± 0.07 (35)	0.95a ± 0.11 (42)	0.90b ± 0.07 (80)
	BG			2.41a ± 0.15 (28)	2.77a ± 0.18 (17)	2.44a ± 0.27 (10)
LSI	LB		1.09ab ± 0.28 (12)	1.23b ± 0.13 (12)	0.92a ± 0.23 (12)	1.15ab ± 0.40 (12)
	CC		2.22a ± 0.27 (12)	1.53b ± 0.37 (12)	1.94c ± 0.30 (12)	1.54b ± 0.24 (12)
	BG			1.67b ± 0.24 (12)	1.29a ± 0.48 (12)	1.34a ± 0.22 (10)
SSI	LB		0.05a ± 0.01 (12)	0.06ab ± 0.01 (12)	0.05a ± 0.02 (12)	0.07b ± 0.02 (12)
	CC		0.05a ± 0.01 (12)	0.06a ± 0.02 (12)	0.06a ± 0.02 (12)	0.06a ± 0.01 (12)
	BG			0.09a ± 0.02 (12)	0.06b ± 0.03 (12)	0.10ab ± 0.08 (10)
%HC	LB		37a ± 5 (12)	30b ± 7 (12)	35ab ± 6 (11)	36a ± 6 (12)
	CC		26a ± 4.5 (12)	33b ± 2.7 (12)	25a ± 6.3 (12)	36b ± 2.8 (12)
	BG			32a ± 8.7 (9)	34a ± 6.8 (12)	31a ± 7.3 (9)
%LC	CC		2.5a ± 0.8 (12)	1.4b ± 0.8 (12)	2.1a ± 0.8 (12)	2.0ab ± 0.5 (12)
MI[a]	LB	m	2.8a ± 0.7 (6)	3.0a ± 0.0 (3)	3.3a ± 0.7 (6)	2.6a ± 0.5 (8)
		f	4.4ab ± 0.5 (5)	4.0a ± 0.9 (7)	4.5ab ± 0.5 (6)	5.0ab ± 0.0 (4)
	CC	f	3.8a ± 0.6 (8)	4.1a ± 0.3 (8)	4.2a ± 0.4 (5)	4.2a ± 0.4 (8)
	BG	m		4.0a ± 0.3 (6)	4.6a ± 0.5 (5)	3.9a ± 0.7 (7)
		f		5.0a ± 0.0 (6)	5.0a ± 0.0 (7)	4.5 ± 0.5 (2)

[a]Abbreviations: C = condition factor, LSI = liver somatic index, SSI = spleen somatic index, %HC = hematocrit, %LC = leucocrit, m = male, f = female, MI = maturity index, LB = largemouth bass, CC = channel catfish, BG = bluegill.
[b]Immature fish: stream 1 = 1 LB, 3 CC; stream 3 = 2 LB, 4 CC, 0 BG; Stream 4 = 0 LB, 4 CC, 0 BG; stream 2 = 0 LB, 2 CC, 0 BG.

The maturity of gonads of the male and female fish of three species were similar, but a few fish had undifferentiated gonads. The maturity of gonads from the two treatments was similar to at least one control for all three species.

The complete results of liver P450IA concentration and EROD activity analyses have been reported by Bankey et al. (1994). Liver P450IA and EROD activity were consistently significantly elevated in fish from the treatment streams. Liver EROD activities were elevated by 2 to 5 times the control levels in bass from treatment streams and 2 to 12 times the control levels in the channel catfish. P450IA concentrations demonstrated a more consistent effluent dose response than the EROD activity levels and ranged from multiples of 12 to 54 times the control levels in bass and 9 to 97 times the control levels in channel catfish. Both parameters were variable over 263 d of exposure but were always significantly elevated.

Lesions found in individual organs of each of the three species were compared between fish from control and treatment streams and no lesion from treatment streams was found to be significantly greater ($p \leq 0.05$) than those from both control streams. Trematodes, cestodes and some other parasites were common lesions in all species of fish. Fatty change in the liver was also common in all three species.

DISCUSSION

During this study with the HS BKME, the survival, growth and production of all four fish species from the treatment stream with 4% effluent were near or between the two control streams, but the survival and production of the bass and catfish in 8% effluent were greater than the other three streams, and this increased the total production of fish in 8% effluent. Some studies have found both greater growth and production in effluent concentrations of BKME or HS BKME from 1 to 5% effluent (Hall and Haley 1991; NCASI 1993), but other studies have found reduced growth and survival at much lower concentrations (Södergren 1989; Lehtinen 1990). Sprague (1991) and Owens (1991) have discussed possible reasons for differences in such results. The results of the studies reported here are somewhat similar to the findings of Hall et al. (1992) because production was greatest at the highest effluent concentration.

The channel catfish were added to the streams as a benthic fish species for bioaccumulation studies. Channel catfish were stocked during only two investigations: this study and the previous study with the BKME (NCASI 1995, unpublished). During both studies the production of channel catfish in 8% effluent was at least 100% greater than in the control or 4% effluent. This may be coincidental since differences between even control streams often occur. However the elevation of catfish production in the stream receiving the effluent was consistent over two studies and the effluent probably contributed to this increase. The addition of color in the shallow streams may provide better habitat for catfish, or nutrients and dissolved solids in the effluent may increase the productivity of the streams. Channel catfish may also be a major competitor for food with both bass and bluegill. The stomach contents of all three species had a high percentage of amphipods, a common benthic organism in the streams (NCASI 1995). Bass survival, growth and production was within that found in earlier studies in these streams, but the bluegill survival and production were not as great in all four streams as in earlier studies and may have been affected by the addition of catfish (NCASI 1983). Also, the average weight of bluegill from the effluent treatment was significantly lower than the other streams. This weight reduction may have been caused by competition for prey organisms from the catfish and bass which had 50% greater biomass in that stream compared to the other three streams. In earlier studies with the BKME bluegill were not adversely affected by concentrations of effluent that were greater than 8% (NCASI 1983).

Many studies have reported differences in condition factors in fish in the vicinity of pulp mill effluents. Most have reported increases in condition factors of effluent-exposed fish (Södergren 1989; Munkittrick et al. 1991; Swanson et al. 1993; NCASI 1993) and some have hypothesized that this may be due to altered metabolism (Munkittrick et al. 1991, 1994; Lehtinen 1990). Hodson (1992) found that condition factors were reduced although this was over 90 km from the entrance point of the effluent. In this study the condition factors of bass in both treatments were significantly reduced from controls. However, overall bass growth and production were equal to or better than controls. Considering that the condition factors of catfish were significantly different between the two control populations and bluegill condition factors were not significantly different between groups, the bass condition factors probably fall within the normal range for these streams. LSIs have also frequently been reported as being greater in fish exposed to pulp mill effluents (Södergren 1989; Lehtinen et al. 1990; Munkittrick et al. 1991), although others did not observe increased LSIs in fish exposed to effluent (Hall et al. 1992; NCASI 1993; Swanson et al. 1993) as was found for bass and catfish in this study. The reduction in LSIs for bluegill from treatment streams is similar to the results of Adams et al. (1992). However the observed difference in the LSIs of catfish between the two control streams indicates that caution in the interpretation of these results is needed since all of the LSIs may be within the overall range for these streams. The hematocrit and leucocrit have been found to be altered in some studies (McLeay and Gordon 1977), but not in others, (Swanson et al. 1993). In this study the only significant differences in hematocrits or leucocrits were between fish from the two controls. Again these are probably within the normal range for these streams and these fish species.

Alteration of the time to sexual maturation has been suggested by some authors (Munkittrick et al. 1991, 1992, 1994). The results of this study agree with others when found that although most of the fish were not yet mature, the fish in effluent concentrations of 4 or 8% (this study) and 1 or 5% HS BKME (NCASI 1993) were progressing toward maturity at a similar rate. Since most of the fish were juveniles this study would not allow an evaluation of reproductive fitness.

Some studies have found that lesions may be present in significantly greater numbers in fish exposed to pulp mill effluent (Andersson 1988; Södergren 1989; Lehtinen et al. 1990). The findings in these studies are similar to those that found some parasites or other lesions in the fish but no differences in lesions due to effluent exposure (Hall and Haley 1991). Many studies have found increased EROD activity in fish exposed to pulp mill effluent (Owens 1991; Munkittrick et al. 1991; Kloepper-Sams and Swanson 1992; Swanson et al. 1993). The results of these studies are similar in that bass and catfish exposed to 4 and 8% effluent had elevated P450IA levels and EROD activity. These were the only biomarkers measured that were consistently altered in a dose-response manner with the concentrations of effluent. Although the EROD activity of bass and catfish exposed to 4% effluent was elevated, the survival, growth and production of these fish was approximately equal to or greater than control streams. The highest elevation of EROD activity and P450IA concentrations was found in catfish and bass from the 8% concentration. These fish also had the highest survival and production compared to fish from all four of the streams. Thus EROD activity was not an indicator of adverse effects on survival, growth, production, other biomarkers or gonadal development of fish from these streams. These results are similar to those reported in another recent study (NCASI 1993).

In summary, the survival, growth and production of all four fish species were not adversely affected at the 4% effluent compared to the two control streams during 263 d of exposure to the HS BKME. At 8% effluent bluegill growth was less than the other streams but production and survival were not adversely affected compared to controls. The survival and production of channel catfish and largemouth bass were greatest in the stream receiving 8% effluent, while the survival and production of bluegill and golden shiners were within the range of those parameters for bluegill and golden shiners from the two control streams. The total production of all fish from the 8% effluent was nearly double the other three streams and the production of the four fish species from the stream receiving 4% effluent was very similar to the production of fish in the two control streams. Although parasitic lesions were frequently noted in the catfish and bass, no effects associated with effluent exposures of 4 or 8% were found. The Cf of largemouth bass and the LSIs of bluegills from both effluent streams were reduced compared to the same parameters in fish from the control streams. However the number of significant differences in biomarkers when fish from the two control streams were compared indicates that the differences between control and effluent exposed fish may all be within the natural range for these fish in these streams. The number of significant differences in biomarker measurements of fish from the two control streams indicates that knowledge of the normal ranges of these parameters is probably more important than statistically significant differences between populations. Significant differences in liver P450IA and EROD activity content between bass and catfish from the control streams and streams receiving 4 or 8% effluent were noted on all sampling dates. Since the survival, growth and production of these fish were not adversely affected where significant differences in P450IA or EROD activity or other biomarkers measured during the study were found, they could not be associated with the population-level parameters measured in this study.

REFERENCES

Adams, S.M., W.D. Crumby, M.S. Greeley, Jr., L.R. Shugart and C.F. Saylor. 1992. Responses of fish populations and communities to pulp mill effluents: A holistic assessment. Ecotoxicol. Environ. Saf. 24:347-360

Andersson, T. 1992. In vivo and in vitro methods to assess the toxicity of pulp and paper effluents to fish. in Proceedings, Seventh Colloquium on Pulp and Paper Effluents. December 15-16, University of Toronto, Toronto, Ontario, 33 p.

Bankey, L.A., P. Van Veld, D. Borton, L. LaFleur and J. Stegeman. 1994. Responses of cytochrome P450IA in freshwater fish exposed to bleached kraft mill effluent in experiment stream channels. Can. J. Fish. Aquat. Sci.: In press.

Berry, R.M., C.E. Luthe, R.H. Voss, P.E. Wrist, P. Axegard, G. Gellerstedt, P.O. Linbland and I. Popke. 1991. The effects of recent changes in bleached softwood kraft mill technology on organochlorine emissions: An international perspective. Pulp Paper Can. 92:43-55.

Chapman, D.W. 1971. Production *In*: Methods for Assessment of Fish Production in Fresh Waters. W.E. Ricker (Ed.). IBP Handbook No.3, Second Edition.

Hall, T.J. and R.K. Haley. 1991. Effects of biologically treated bleached kraft mill effluent on cold water stream productivity in experimental streams. Environ. Toxicol. Chem. 10:1051-1060.

Hall, T.J., R.K. Haley, D.L. Borton, A.H. Walsh and R.E. Wolke. 1992. Histopathology of rainbow trout *(Oncorhyncus mykiss)* after long-term exposure to biologically treated bleached kraft mill effluent in experimental stream channels. Can. J. Fish. Aquat. Sci. 49:939-944.

Hodson, P.H., M. McWhirter, K. Ralph, B. Gray, D. Thivierge, J.H. Carey, G. Van Der Kraak, D.M. Whittle and M. Levesque. 1992. Effects of bleached kraft mill effluent on fish in the St. Maurice River. Environ. Toxicol. Chem. 11:1635-1651.

Kloepper-Sams, P., and S. Swanson. 1992. Bioindicator field monitoring: Use of fish biochemical parameters at a modern bleached kraft pulp mill site. Mar. Environ. Res. 34:163-168.

Lehtinen, K.J. 1990. Mixed-function oxygenase enzyme responses and physiological disorders in fish exposed to kraft pulp mill effluents: A hypothetical model. Ambio 19:259-265.

Lehtinen, K.J., A. Keirkegaard, E. Jakobsson and A. Wandell. 1990. Physiological effects in fish exposed to effluents from mills with six different bleaching processes. Ecotoxicol. Environ. Saf. 19:33-46.

McLeay, D.J. and M.R. Gordon. 1977. Leucocrit: A simple hematological technique for measuring acute stress in salmonid fish, including stressful concentrations of pulp mill effluent. J. Fish. Res. Board Can. 34:2164-2175.

Munkittrick, K.R., C.B. Portt, G.J. Van Der Kraak, I.R. Smith and D.A. Rokosh. 1991. Impact of bleached kraft mill effluent on population characteristics, liver MFO activity, and serum steroid levels of a Lake Superior white sucker *(Catostomus commersoni)* population. Can. J. Fish. Aquat. Sci. 48:1371-1380.

Munkittrick, K.R., M.E. McMaster, C.B. Portt, G.J. Van Der Kraak, I.R. Smith and D.G. Dixon. 1992. Changes in maturity, plasma sex steroid levels, hepatic MFO activity, and the presence of external lesions in lake whitefish exposed to bleached kraft mill effluent. Can. J. Fish. Aquat. Sci. 49:1560-1569.

Munkittrick, K.R., G.J. Van Der Kraak, M.E. McMaster, C. Portt, M.R. van den Heuvel and M.R. Servos. 1994. Survey of receiving-water environmental impact associated with discharges from pulp mills. 2. Gonad size, liver size, hepatic EROD activity, and plasma sex steroid levels in white sucker. Environ. Toxicol. Chem. 13:1089-1101.

NCASI. 1983. Effects of biologically stabilized bleached kraft mill effluent on warm water stream productivity in experimental streams - 3rd progress report. NCASI Tech. Bull. No. 414, National Council of the Paper Industry for Air and Stream Improvement, New York.

NCASI. 1986a. NCASI methods for the analysis of chlorinated phenolics in pulp industry wastewaters. NCASI Tech. Bull. No. 498, National Council of the Paper Industry for Air and Stream Improvement, New York.

NCASI. 1986b. Procedures for the analysis of resin and fatty acids in pulp mill effluent. NCASI Tech Bull. No. 501, National Council of the Paper Industry for Air and Stream Improvement, New York.

NCASI. 1991. Observations on the bioaccumulation of 2,3,7,8-TCDD and 2,3,7,8-TCDF in channel catfish and largemouth bass and their survival or growth during exposure to biologically treated bleached kraft mill effluent in experimental streams. NCASI Tech. Bull. No. 611, National Council of the Paper Industry for Air and Stream Improvement, New York.

NCASI. 1993. Aquatic community effects of biologically treated bleached kraft mill effluent before and after conversion to increased chlorine dioxide substitution: Results from an experimental streams study. NCASI Tech. Bull. No. 653, National Council of the Paper Industry for Air and Stream Improvement, New York.

NCASI. 1995. Bioaccumulation of 2,3,7,8-TCDD and 2,3,7,8-TCDF in channel catfish and largemouth bass exposed to biologically treated kraft mill effluent in experimental streams after high chlorine dioxide substitution. NCASI Tech. No. 692, National Council of the Paper Industry for Air and Stream Improvement, New York.

NCASI. Unpublished. Characterization of a bleach kraft mill effluent with high substitution of chlorine dioxide for chlorine using laboratory bioassays and experimental streams. Parts I and II. NCASI Tech. Bull., National Council of the Paper Industry for Air and Stream Improvement, New York.

Nielsen, L.A. and D.L. Johnson (Eds.). 1983. Fisheries Techniques. American Fisheries Society, Bethesda, MD.

Owens, J.W. 1991. The hazard assessment of pulp and paper effluents in the aquatic environment: A review. Environ. Toxicol. Chem. 10:1511-1540.

Sprague, J.B. 1991. Contrasting findings from Scandinavia and North America on toxicity of BKME. Introductory comments. pp. 652-663 in Proceedings of the Seventeenth Annual Aquatic Toxicity Workshop. November 5-7, 1990, Vancouver, B.C., P. Chapman, F. Bishay, E. Power, K. Hall, L. Harding, D. McLeay, M. Nassichuk and W. Knapp (Eds.). Can. Tech. Rep. Fish. Aquat. Sci. No. 1774.

Swanson, S.M., R. Schryer, B. Shelast, K. Holley, I. Berbekar, P. Kloepper-Sams, J.W. Owens, L. Steeves, D. Birkholz and T. Marchant. 1993. Wapiti/Smokey River ecosystem study. Weyerhaeuser Canada, Grande Prairie, Alberta.

Södergren, A. 1989. Biological effects of bleached pulp mill effluents. Report 3558. Final Report. National Swedish Environmental Protection Board, Solna, Sweden.

USGS (U.S. Geological Survey). 1991. Water Data Report Nos. NC 86-1, NC 87-1, NC 88-1, NC 89-1 (1986 through 1989). Water Resources Data for North Carolina.

EFFECT MONITORING IN PULP MILL AREAS: RESPONSE OF THE MEIOFAUNA COMMUNITY TO ALTERED PROCESS TECHNIQUE

Brita Sundelin and Ann-Kristin Eriksson

Institute of Applied Environmental Research, Laboratory for Aquatic Ecotoxicology, Stockholm University, S-611 82 Nyköping, Sweden

A great number of pulp mills border the coast of the northern Baltic, particularly along the Gulf of Bothnia. Surveys of the benthic macrofauna, which have regularly been carried out in the receiving waters during the last 20 years, have generally failed to detect changes more than 2-3 km from the pulp mill. In contrast we found effects on the meiofauna community several kilometers beyond the area earlier reported as affected when macrofauna density and biomass were used. Density and biomass of nematodes were low at all stations except one when compared with the reference station and a strong negative correlation ($p < 0.01$) occurred between nematodes and sediment concentrations of EOCl. The most abundant Baltic ostracod (*Paracyprideis fennica*) appeared particularly sensitive. Its density and biomass were generally reduced when compared with the reference and also showed a strong negative correlation with EOCl ($p < <0.001$). Kinorhynchs were reduced in density and biomass at all stations except the southernmost but showed no clear correlation with the concentration of EOCl, suggesting other factors were responsible for the reduction. Density of harpacticoids on the innermost station was higher than other sites in the area and there was no correlation between density and EOCl. The pulp mill changed the bleaching process (from chlorine to elemental chlorine free) between 1987 and 1991. This resulted in substantially decreased discharges of COD and chlorinated substances. Meiobenthic responses at the inner station indicated a recovery of most taxa and species except the most sensitive ostracod species, *Paracyprideis fennica*, which was still absent close to the mill.

INTRODUCTION

Soft sediments that dominate the bottoms of the Baltic Sea represent an important component of the Baltic ecosystem. The sediment is involved in the recycling and processing of nutrients and xenobiotics, such as metals and organic compounds. Due to physical adsorption and chemical bonds, pollutants become highly enriched in sediments compared to the overlying water. Once sorbed in the sediment, noxious substances may be slowly released back into the water column. Since the deep reducing layer of many sediments retards oxidative decomposition, sediments often represent a sink for metals and organic pollutants. This can then cause long-lasting or chronic contamination of the benthos (Giere 1993). Sediment-dwelling invertebrates are often exposed to two sources of contamination: the substrate, which is often their food, and the interstitial and overlying water (Moore et al. 1979).

Benthic meiofauna, i.e., Metazoa and Foraminifera small enough to pass a 0.5 or 1 mm sieve and large enough to be retained on a 0.042 or 0.063 mm sieve (Giere 1993), are important in the energy flow of marine benthic ecosystems (Gerlach 1971). In the northern Baltic proper, meiofauna contribute about one third of the summed production of the benthic meio- and macrofauna (Ankar and Elmgren 1978). Further north in the Bothnian Sea, the relative importance of meiofauna increases in comparison with the macrofauna (Elmgren *et al.* 1984). Effective utilization of sparse and finely particulate food by the meiofauna and lessened interference and predation from the macrofauna have been suggested as possible explanations for the greater importance of the meiofauna in the food-limited benthic environment of the northern Baltic (Elmgren 1978).

Meiofauna might respond quickly to pollution and thus have potential for effects biomonitoring (ICES 1978), but there have been few applications of meiofauna biomonitoring programs, mainly due to the high degree of taxonomic expertise required for the identification of sensitive genera or species (Raffaeli and Mason 1981). Warwick *et al.* (1988), Herman and Heip (1988) and Bett (1988) discussed the possibility of using major meiofaunal taxa instead of species in environmental impact assessment. Discrimination to higher taxonomic levels than species led to little loss of information with respect to pollution responses. Theoretical arguments for the use of meiofauna include their benthic sedentary habit, direct benthic development, short generation times (which means faster response time to disturbance), high abundance and diversity, smaller size samples and a presumed lower natural variability from year to year (Bett 1988; Gee *et al.* 1992).

Many pulp mills border the coast along the Gulf of Bothnia and discharges from these industries are mainly responsible for the pollution impact on the Bothnian Sea ecosystem (Thorman 1987). In areas polluted by pulp mill effluents, the most noticeable effect is an increase in organic loading of the receiving waters, which results in a high oxygen demand in both the water column and sediments (Leppäkoski 1968; Bagge 1969; Pearson 1975; Rosenberg 1976; Millner 1980). There are few scientific investigations concerning the toxic effects of pulp mill effluents on Baltic bottom fauna and, in particular, few attempts to distinguish between stimulatory effects caused by the organic enrichment and inhibitory effects caused by toxic substances in the mill effluents. In addition to their toxic effects, bleached kraft pulp mill effluents (BKME) stimulated the growth of benthic macrofauna, such as the amphipod *Monoporeia affinis* and the bivalve *Macoma balthica* (usually regarded as indicators of healthy conditions in pulp mill areas) in the Baltic (Sundelin 1988, 1989). Cederwall and Blomqvist (1986) found a negative correlation between the biomass of *Monoporeia* and the distance from bleaching pulp mills, which indicates organic enrichment effects from the effluents. This means that stimulatory effects might confound inhibitory effects, such as, for example, effects on reproduction (Sundelin 1989, 1992). In contrast, despite increased nematode abundance during exposure to BKME in microcosm experiments with meiofauna from the Baltic proper, the total biomass was unchanged (Sundelin 1989). Widbom and Elmgren (1988) reported little response to organic carbon by the benthic meiofauna community of Narraganset Bay. Nematode abundance increased, while kinorhynchs, ostracods and harpacticoids decreased and the total biomass was unchanged.

The main purpose of this study was to examine the usefulness of the meiofauna community and select particularly sensitive taxa or species as variables for monitoring effects in pulp mill areas. During the last decade, discharges of organic wastes and chlorinated substances from pulp mills have decreased substantially, and as a consequence, there is a demand for effect variables that respond quickly to reduction of discharges. The pulp mill receiving area was examined with respect to the meiofauna community before and after a change in the bleaching process at the mill. The period between the two surveys involved a large investment which changed the bleaching process at the mill from chlorine to mixed oxygen and chlorine dioxide. This resulted in substantially decreased total wastewater and COD discharges.

MATERIALS AND METHODS

Surveys were performed in a pollution gradient of a sulfate pulp mill located on the coast of the Bothnian sea in the same week of August in 1987 and 1991 (see also Sundelin 1992). A preliminary sediment survey conducted to ensure uniform grain size and organic contents of the sediments resulted in different locations for some stations in the 1991 study. The same innermost station in the two surveys was compared between surveys to investigate whether the fauna had recovered after the altered process technique. The reference station east of the area in 1987 was replaced in 1991 by another station near the coast, 40 km north of the mill. This station was within the only uncontaminated area within a distance of 100 km north and south of the pulp mill having sediment characteristics matching those of stations in the impacted area.

Sampling

Meiofauna were sampled with an improved modified Kajak corer (Blomqvist and Abrahamsson 1985) within two pollution gradients, one extending southwards, including 4 stations, and the other extending eastwards, comprising 3 stations (Fig. 1). Five gravity core samples per station were taken. The top 4 cm of sediment was taken for analysis of fauna and organic content, measured as loss on ignition (Dybern et al. 1976). Samples for meiofauna analysis were preserved in 4% formaldehyde buffered with hexamine, with Rose bengal added as stain. The meiofauna in each corer were sorted following the techniques of Elmgren (1973) and Sundelin and Elmgren (1991). Individuals were determined to major taxon (Foraminifera, Turbellaria, Kinorhyncha, Nematoda, Oligochaeta) or to species (Ostracoda, Harpacticoida). Dry weight biomasses of meiofauna groups were estimated using sieve mesh- specific conversion factors from Widbom (1984).

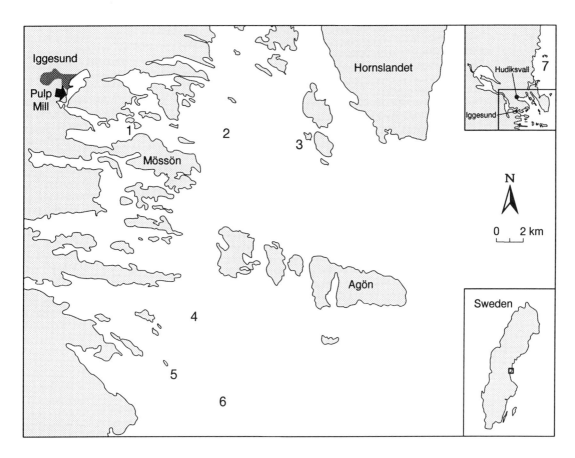

Figure 1. The investigation area. The reference station (7) on the inserted map is situated 40 km north of the pulp mill area.

Statistical evaluations were made by one-way ANOVA followed by multiple comparisons with Dunnet tests (two-tailed) ($\alpha = 0.05$) comparing theoretical mean with other group means (Zar 1984). Dose-response relationships for extractable organic chlorine (EOCl), a parameter for extractable chlorinated organics and density (ind m^{-2}) (4 stations) were tested for significance by linear regression analysis. The relationship between distance, organic content of the sediment, depth and abundance (all stations expect the reference) also was tested by multiple regression followed by Holm's procedure, a simple sequentially rejective multiple test procedure (Holm 1979), which resulted in rejection of p > 0.013 (linear regression) and p > 0.003 (multiple regression). Abundance data were $\log_{10} (x + 1)$ transformed, to normalize variance. EOCl is not directly associated with the pulp mill effluents, but nevertheless, has been reported as strongly negatively correlated with distance from the pulp mill and varies positively with specific markers of bleaching effluents in the area, such as chlorinated guaiacols, polychlorinated dibenzo-*p*-dioxins and dibenzofurans and alkylated polychlorinated dibenzofurans (available only at two stations), indicating the same source and transportation regime (Jonsson *et al.* 1993). Mann-Whitney U-test (two-tailed) was used for the two-year comparisons (Daniel 1978) ($\alpha = 0.05$). No dose-response relationships were analyzed for biomass and concentration of EOCl, since the biomass was estimated using conversion factors and the relationship would be redundant.

RESULTS

Abiotic Factors

In 1991, the organic content of sediments was between 7 and 12% at all stations (Table 1). Southern stations were deeper than those east in the mill, and measured EOCl concentrations varied from 6270 µg g^{-1} IG at the innermost station 1 to 528 µg g^{-1} IG at the more distant station 4 (Jonsson *et al.* 1993; Håkansson *et al.* 1988) (Table 1).

Table 1. Depth, organic content of the sediment (loss on ignition) and concentration of EOCl outside a bleaching pulp mill located along the coast of the Bothnian Sea.

Station	Depth (m)	Organic Content (%)	EOCl (µg g^{-1} IG)
1	15	12.4	6270
2	25	6.9	528
3	29	9.1	589
4	46	10.5	-
5	56	10.0	1170
6	61	9.2	-
7	58	8.7	-

Meiofauna Analyses

Density (individuals m^{-2}) of total meiofauna and nematodes was possibly affected by the discharges from the pulp mill as indicated by significant correlations between meiofaunal density and sediment concentration of EOCl with distance (p < 0.01). Negative correlations were found between organic content and total meiofauna and nematode density (both p < 0.01) (Fig. 2). These relationships illustrate the strong correlation for EOCl with organic content (r = 0.94). When compared with the reference site, 40 km north of the mill, significantly lower abundances of total meiofauna and nematodes were demonstrated at all stations in the impacted area, except station 4 (Fig. 2). The kinorhynchs were less abundant at all near-field stations, except the most distant southern stations (Fig. 3) when compared with the reference station, but no significant correlation was found with distance from the mill (p < 0.05) and no clear relationship was demonstrated when correlated with concentration of EOCl. The total density of ostracods (three species) was not correlated with concentration of EOCl, but ostracod densities were significantly reduced at stations along the north-south transect relative to the reference station (Fig. 4). *Paracyprideis fennica*, the most abundant and pollutant-sensitive ostracod species (Blanck *et al.* 1989; Sundelin 1989; Sundelin and Elmgren 1991), was significantly less abundant at all stations relative to the reference station except at the eastern station 3 and was completely absent at the innermost station 1 (Fig. 4). This species correlated strongly with concentrations of EOCl and negatively with organic content (both $p < <0.001$), reflecting this species' high susceptibility to pollutants. The ostracod *Candona neglecta* was only present on shallower bottoms in the east and seemed to be little affected by the mill effluents. The third species, *Heterocyprideis fennica*, was nearly absent at all stations in 1991. Harpacticoids (two species) showed no clear dose-response relationship to mill effluents when repressed against EOCl, but were marginally negatively correlated with station depth ($p < 0.01$) (Fig. 3) and the shallower station 1 showed moderately high density in comparison with deeper distant stations. No difference in response to measured variables was found between the two present species *Pseudobradya* sp. and *Microarthridion littorale*. Turbellarians were reduced at the inner station 1, whereas they were not different at other stations in the impacted area compared to the reference station. No correlation between turbellarians and the local level of pollutants was observed.

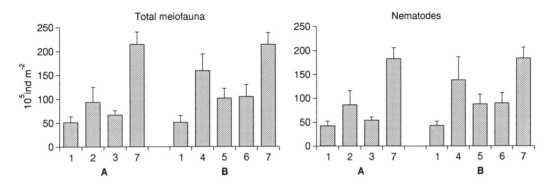

Figure 2. Density (ind m^{-2}) of total meiofauna and nematodes in two pollution gradient of a pulp mill. A = extending eastward and B = extending southward. The reference station (7) is situated 40 km north of the pulp mill area. Data are given as mean ± SE of five replicates.

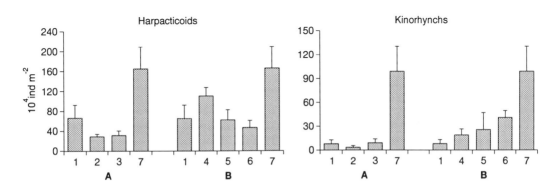

Figure 3. Density (ind m^{-2}) of kinorhynchs and harpacticoids in two pollution gradients of a pulp mill. A = extending eastward and B = extending southward. The reference station (7) is situated 40 km north of the pulp mill area. Data are given as mean ± SE of five replicates.

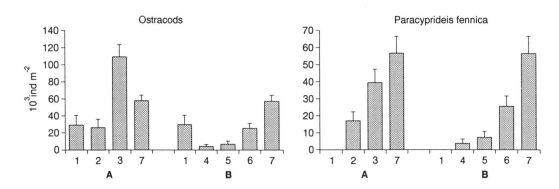

Figure 4. Density (ind m⁻²) of total ostracods and the ostracod Paracyprideis fennica in two pollution gradients of a pulp mill. A = extending eastward and B = extending southward. The reference station (7) is situated 40 km north of the pulp mill area. Data are given as mean ± SE of five replicates.

The trend for biomass was similar to that for density. For total meiofauna, nematodes, kinorhynchs and ostracods, the biomass showed the same pattern as abundance (Figs. 5 to 7). *Paracyprideis fennica* and kinorhynchs showed significantly lower biomass at all stations in comparison with the reference station (Figs. 6 to 7).

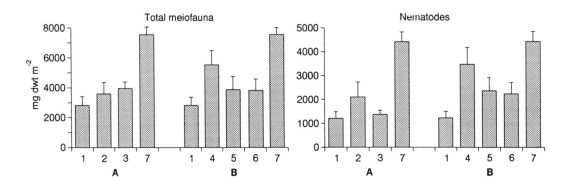

Figure 5. Biomass (mg dwt m⁻²) of total meiofauna and nematodes in two pollution gradients of a pulp mill. A = extending eastward and B = extending southward. The reference station (7) is situated 40 km north of the pulp mill area. Data are given as mean ± SE of five replicates.

Between Years

A comparison between the two years showed a trend of increased abundances for all taxa at station 1 (Figs. 8 and 9) and the abundance increased significantly for ostracods ($p < 0.05$) (Fig. 9). *Candona neglecta* had recovered from a very low abundance in 1987, indicating improved conditions in the sediments, while *Paracyprideis fennica* was absent both years. Biomass of total meiofauna increased ($p < 0.05$) at station 1 (Fig. 10) as did the ostracods ($p < 0.01$) (Fig. 11). There also was a trend towards increased biomass for other taxa, such as nematodes, harpacticoids and kinorhynchs (Figs 10 and 11).

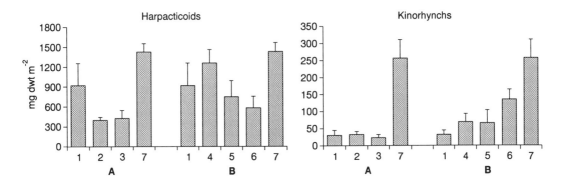

Figure 6. Biomass (mg dwt m^{-2}) of kinorhynchs and harpacticoids in two pollution gradients of a pulp mill. A = extending eastward and B = extending southward. The reference station (7) is situated 40 km north of the pulp mill area. Data are given as mean ± SE of five replicates.

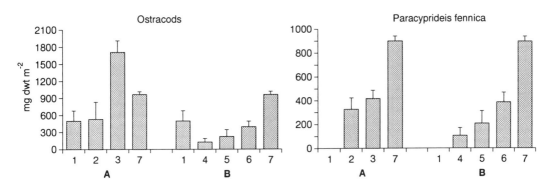

Figure 7. Biomass (mg dwt m^{-2}) of total ostracods and the ostracod *Paracyprideis fennica* in two pollution gradients of a pulp mill. A = extending eastward and B = extending southward. The reference station (7) is situated 40 km north of the pulp mill area. Data are given as mean ± SE of five replicates.

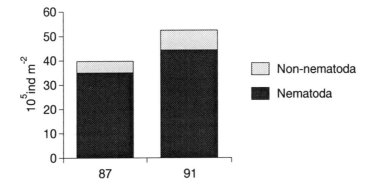

Figure 8. Density (ind m^{-2}) of total meiofauna (nematodes and non-nematodes). A comparison between 1987 (left bars) and 1991 (right bars).

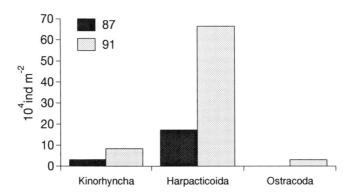

Figure 9. Density (ind m^{-2}) of kinorhynchs, harpacticoids and ostracods. A comparison between 1987 (left bars) and 1991 (right bars).

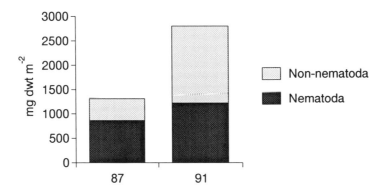

Figure 10. Biomass (mg dwt m^{-2}) of total meiofauna (nematodes and non-nematodes). A comparison between 1987 (left bars) and 1991 (right bars).

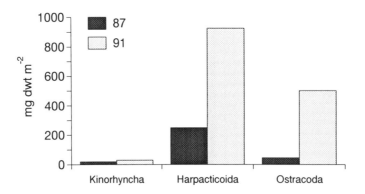

Figure 11. Biomass (mg dwt m^{-2}) of kinorhynchs, harpacticoids and ostracods. A comparison between 1987 (left bars) and 1991 (right bars).

DISCUSSION

Sensitivity of Different Groups

The 1987 survey showed lower abundance of total meiofauna within 3.5 km from the outlet, while the total abundance of sensitive groups, such as ostracods, was significantly lower at all eastern stations within 27 km of the discharge point (Sundelin 1992). The 1991 study indicates that the area is still affected by the pulp mill. Densities of total meiofauna and nematodes were relatively low at most stations in the impacted area and were correlated strongly with sediment concentration of EOCl. The negative correlation between abundance and organic content demonstrated by total meiofauna and nematodes probably reflects the strong concomitant correlation between EOCl and organic content rather than negative effects caused by high organic load. Nematodes have been reported as comparatively resistant to low oxygen conditions (Elmgren 1973) and are favored by moderately enriched organic carbon (Widbom and Elmgren 1988; Sundelin 1989). Furthermore, nematodes have been reported as resistant to pollutants, particularly heavy metals (Vranken et al. 1985), although conflicting results exist (Howell 1984). For sediment-associated fenvalerate, nematodes showed the highest acute sensitivity, followed by two copepod species (Chandler et al. 1994). Nematodes usually dominate all meiofauna samples, both in density and biomass, and are globally the most abundant metazoan taxon (Giere 1993). In the study area 85-95% of individuals and 40-70% of biomass are made up by nematodes (Sundelin 1992). Occurring in every kind of substrate, sediment and climatic zone, they are of considerable ecological importance.

A low abundance and biomass of kinorhynchs occurred at most stations in the impacted area when compared with the reference site. The lack of correlation between kinorhynchs and EOCl was surprising, since kinorhynchs were seriously affected by pulp mill contaminated sediment in microcosm experiments (Sundelin 1989). However, the abundance of this taxon is generally low in the area and showed great dispersion, which may decrease the sensitivity of kinorhynch abundance as an index of effluent impact.

The abundance of harpacticoids was lower at most stations in the impacted area compared with the reference station. In contrast, the innermost station showed moderately high density in comparison with more distant stations. There was no correlations with EOCl concentration, indicating no responses to the mill effluents for this taxon. The negative correlation ($p < 0.01$) between depth and harpacticoid abundance was rejected for significance after application of Holm's procedure, but nevertheless the highest harpacticoid abundance is encountered in shallower bottoms (Giere 1993) and this density-depth relationship likely confounds any effects of the mill effluents. Harpacticoid copepods are considered as sensitive to organic pollutants and the ratio nematodes/copepods was first suggested by Raffaeli and Mason (1981) as a variable for monitoring organic pollution. Results from the two surveys showed lower effects on harpacticoids compared with nematodes. Microcosm studies with the Baltic meiofauna community, however, have demonstrated high sensitivity of the harpacticoid species, *Pseudobradya sp.* and *Microarthridion littorale,* to heavy metals such as cadmium and contaminated sediments from pulp mill areas (Sundelin 1989; Sundelin and Elmgren 1991). Warwick et al. (1988) reported higher response of copepods exposed to copper and hydrocarbon dosed in undisturbed sediment cores, while nematodes were virtually unaffected. Since there was no difference in sediment levels between treatments, while pore water concentrations showed a dosing gradient, the higher response of copepods might indicate that copepods were mainly epibenthic and nematodes mainly burrowing. If the ratio pore water/sediment concentrations determines the sensitivity difference between nematodes and copepods, one could expect higher toxicity responses for copepods when exposed in the laboratory. Equilibrium usually exists between sediment and pore water *in situ*, while artificially contaminated sediments often result in higher pore water concentrations. Another reason for the weak correlation to pollutants for harpacticoids in field studies could be the aggregate distributional pattern which was reported for 5 of 6 harpacticoids (Heip and Engels 1977).

The most dominating ostracod species in the Baltic, *Paracyprideis fennica,* was found to be particularly sensitive and both density and biomass were low relative to the reference station. Both ostracods and kinorhynchs responded to high organic carbon loading with lower density and biomass than did nematodes (Widbom and Elmgren 1988), suggesting reduced problems with confounding stimulatory effects for these taxa.

Comparison with Microcosm Experiment

Observed patterns from the two surveys agree well with effects from microcosm experiments using sediments and effluents from the pulp mill, both in responses of particularly sensitive species and the total meiofauna community (Sundelin 1989). For some taxa, such as harpacticoids, there were however conflicting results. This might partly be explained by differences in the exposure situation. Artificially dosed effluents might result in higher pore water/sediment concentrations, which means relatively high doses for epibenthic species. Another reason for lower effects in field surveys might be the density-depth relationship observed for this taxon.

Before and After Altered Process Technique

The total abundance of meiofauna did not change between years, but total biomass did increase, significantly suggesting improved conditions between 1987 and 1989. The density and biomass of the ostracod fauna were higher in 1991, due to a recovery of *Candona neglecta*. The most abundant ostracod species at the reference station, *Paracyprideis fennica*, was still absent at station 1 in 1991. *Candona neglecta*, the only ostracod species at the innermost station, has a one-year life cycle in the Baltic, which could explain its more rapid recruitment relative to *Paracyprideis fennica*, which has a generation time of two years. Besides the differences in generation time, *Paracyprideis fennica* was the most sensitive ostracod species when exposed in microcosms to cadmium, arsenic and pulp mill contaminated sediments (Sundelin 1989; Blanck *et al.* 1989; Sundelin and Elmgren 1991). Our results suggest that improved conditions in sediments allowed the colonization of less sensitive ostracod species, like *Candona*, while the more sensitive and long-lived species, like *Paracyprideis*, required even better sediment conditions or more time to recover.

Mean number increased for harpacticoids, whereas nematode abundance remained unchanged in 1991. The differences may be explained in terms of their vertical distribution in the sediments and differential immigration/recruitment rates when linked to increased retention rate of contaminants in deeper layers of the sediments (Gee *et al.* 1992). Olafsson and Moore (1992) reported rapid colonization of harpacticoids in azoic sediments. Almost all harpacticoid copepods, kinorhynchs and ostracods were found in the uppermost flocculent bottom layer. These characteristics facilitate rapid population recovery of taxa with high reproductive rates in surface sediments which have reduced contaminant levels (Leaver *et al.* 1987). Nematodes, on the other hand, penetrate much deeper into the sediment where contaminants persist for longer periods and nematodes also migrate principally by movement through the sediment. Thus nematode assemblages may take much longer to recover after the removal of a pollution source. Kinorhynchs are associated with the uppermost sediment surface layer but do not use freshly produced detritus as effectively as harpacticoids (Rudnick 1989). If kinorhynchs depend on older reservoirs of detritus, one might expect slower recovery of this taxon.

CONCLUSION

In comparison with corresponding macrofauna variables, the meiofauna seem to be sensitive to pulp mill effluents. Effects on the total meiofauna community were detected several kilometers outside the area earlier reported to be affected by pulp mill effluents based on macrofaunal variables (Landner *et al.* 1977). The main reason for the differences between dominant macrofauna and meiofauna taxa is the stimulatory response of macrofauna to enriched organic load (Sundelin 1988, 1989, 1992), which may counter inhibitory toxic effects.

Many taxa and some species of meiofauna have been found to be more sensitive to pulp mill effluents than dominant macrofauna species. Nematode abundance was significantly reduced, the ostracod species *Paracyprideis fennica* was particularly sensitive and it is possible that species determination of single taxa (e.g., ostracods) could be a routine variable in environmental monitoring. Kinorhynchs and harpacticoid copepods showed weaker response to the mill effluents, and other factors, such as depth, contributed to the variation in abundance, while turbellarians were comparatively unaffected by discharges from the mill.

Meiofaunal recovery occurred at the innermost station as a possible result of altered process technique to elemental chlorine-free bleaching, both in terms of ostracods and total biomass of meiofauna. Since the reference station was not identical in the two surveys, one can not exclude the generally higher production of the benthos in 1991 as an explanation for the increased abundance at the inner station in 1991.

ACKNOWLEDGMENT

This work was supported by grants from the National Swedish Environmental Protection Agency to Brita Sundelin. Torsten Öhrn sorted meiofauna, Ragnar Elmgren gave valuable advice and criticized the manuscript, and Ulf Larsson and Stig Johan Wiklund gave valuable comments on the statistical evaluation.

REFERENCES

Ankar, S., and R. Elmgren. 1978. The benthic macro- and meiofauna of the Askö-Landsort area (Northern Baltic Proper). A stratified random sampling survey. Contr. Askö Lab., Univ. of Stockholm 11, 115 pp.

Bagge, P. 1969. Effects of pollution on estuarine ecosystems. Merentutkimuslait. Julk./Havsforskningsinst. Skr. 228.

Bett, B. 1988. Monitoring with meiofauna. Mar. Pollut. Bull. 19(6):293-295.

Blanck, H., K. Holmgren, L. Landner, H. Norin, M. Notini, A. Rosemarin and B. Sundelin. 1989. Advanced hazard assessment of arsenic in the Swedish environment, pp. 85-122. In L. Landner [ed.] Chemicals in the aquatic environment. Advanced Hazard Assessment. Springer Ser. Environ. Quality Comm. CM E 58.

Blomqvist, S., and B. Abrahamsson. 1985. An improved kajak-type gravity core sampler for soft bottom sediments. Schweiz. Z. Hydrol. 47(1):81-84

Cedervall, H., and M. Blomqvist. 1986. Relationer mellan mjukbottenfauna och sediment i Bottenhavet, speciellt med avseende på blekeriutsläpp. Naturvårdsverket rapport 3338. Övervakning av mjukbottenfauna i Bottniska vikens kustområden, H. Cedervall, K. Leonardsson. Eng. summary.

Chandler, G.T., B.C. Coull and J.C. Davis. 1994. Sediment- and aqueous-phase fenvalerate effects on meiobenthos: implications for sediment quality criteria development. Mar. Environ. Res. 37:313-327.

Daniel, W.W. 1978. Applied nonparametric statistics. Houghton Mifflin, Boston, 503 pp.

Dybern, B.I., H. Ackefors and R. Elmgren [eds.]. 1976. Recommendations on methods for marine biological studies in the Baltic Sea. The Baltic Marine Biologists Publ. 1:1-98.

Elmgren, R. 1973. Methods of sampling sublittoral soft bottom meiofauna. Oikos Suppl. 15:112-120.

Elmgren, R. 1978. Structure and dynamics of Baltic benthos communities, with special reference to the relationship between macro- and meiofauna. Kieler Meereforsch. Sonderh. 4:1-22.

Elmgren, R., R. Rosenberg, A.B. Andersin, S. Evans, P. Kangas, J. Lassig, E. Leppäkoski and R. Varmo. 1984. Benthic macro-meiofauna in the Gulf of Bothnia (northern Baltic). Finn. Mar. Res. 250:3-18.

Gee, J.M., M. Austen, G. De Smet, T. Ferraro, A. McEvoy, S. Moore, D. Van Gausbeki, M. Vincx and R.M. Warwick. 1992. Soft sediment meiofauna community responses to environmental pollution gradients in the German Bight and at a drilling site off Dutch coast. Mar. Ecol. Prog. Ser. 91:289-302.

Gerlach, S.A. 1971. On the importance of marine meiofauna for benthos communities. Oecologia (Berl.) 6:176-190.

Giere, O.G. 1993. Meio-benthology, the microscopic fauna in aquatic sediments, Springer-Verlag, Berlin, 328 pp.

Heip, C., and P. Engels. 1977. Spatial segregation in copepod species from a brackish water habitant. J. Exp. Mar. Biol. Ecol. 26:77-96.

Herman, P.M.J., and C. Heip. 1988. On the use of meiofauna in ecological monitoring: who needs taxonomy? Mar. Pollut. Bull. 19(12):665-668.

Holm, S. 1979. A simple sequentially rejective multiple test procedure. Scand. J. Statist. 6:65-70.

Howell, R. 1984. Acute toxicity of heavy metals to two species of marine nematodes. Mar. Environ. Res. 11:153-161.
Håkansson, L., B. Jonsson, P. Jonsson and K. Martinsen. 1988. Påverkansområden för klorerat organiskt material från massablekerier. Natuvårdsverket rapport 3522. Eng. summary.
ICES. 1978. On the feasibility of effects monitoring. Cooperative Research Report No. 75.
Jonsson, P., C. Rappe, L.-O. Kjeller, A. Kierkegaard, L. Håkansson and B. Jonsson. 1993. Pulp-mill related polychlorinated organic compounds in Baltic Sea sediments. Ambio 22(1):37-43.
Landner, L., K. Nilsson and R. Rosenberg. 1977. Assessment of industrial pollution by means of benthic macrofauna surveys along the Swedish baltic coast. Vatten. 3:324-379.
Leaver, M.J., D.J. Murison, J.M. Davies and D. Rafaelli. 1987. Experimental studies of the effects of drilling discharges. Phil. Trans. R. Soc. Lond. B316:625-640.
Leppäkoski, E. 1968. Some effects of pollution on the benthic environment of the Gullmarsfjord. Helgoländer Wiss. Meeresunters. 17:291-301.
Millner, S.R. 1980. Pulp and paper mill waste pollution in the Swale, a tidal channel on the east coast of England. Helgoländer Meeresunters. 33:366-376.
Moore, J.W., V.A. Beaubien and D.J. Sutherland. 1979. Comparative effects of sediment and water contamination on benthic invertebrates in four lakes. Bull. Environ. Contam. Toxicol. 23:840-847.
Olafsson, E., and C.G. Moore. 1992. Effects of macroepifauna on developing nematode and harpacticoid assemblages in a subtidal muddy habitant. Mar. Ecol. Prog. Ser. 84:161-171.
Pearson, T.H. 1975. The benthic ecology of Loch Linnhe and Loch Eil, a sea-loch system on the west coast of Scotland. IV. Changes in the fauna attributable to organic enrichment. J. Exp. Mar. Biol. Ecol. 20:1-41.
Raffaelli, D.G., and C.F. Mason. 1981. Pollution monitoring with meiofauna, using the ratio of nematodes to copepods. Mar. Pollut. Bull. 12:158-163.
Rosenberg, R. 1976. Benthic faunal dynamics during succession following pollution abatement in a Swedish estuary. Oikos 27:414-427.
Rudnick, T.D. 1989. Time lags between the deposition and meiobenthic assimilation of phytodetritus. Mar. Ecol. Prog. Ser. 50:231-240.
Sundelin, B. 1988. Effects of sulfate pulp mill effluents on soft bottom organisms - a microcosm study. Wat. Sci. Tech. 20 (2):175-177.
Sundelin, B. 1989. Ecological effect assessment of pollutants using Baltic benthic organisms. Ph.D. thesis, University of Stockholm.
Sundelin, B. 1992. Effect monitoring in pulp mill areas using benthic macro- and meiofauna. In A. Södergren [ed.] Environmental fate and effects of bleached pulp mill effluents. Sweden Env. Prot. Agency report 4031.
Sundelin, B., and R. Elmgren. 1991. Meiofauna of an experimental soft bottom ecosystem - effects of macrofauna and cadmium exposure. Mar. Ecol. Prog. Ser. 70:245-255.
Thorman, S. 1987. Miljökvalitetsbeskrivning av Bottniska viken och dess kustområden. Nat. Sweden Environ. Prot. Bd. Rep. 3363.
Vranken, G., R. Vanderhagen and C. Heip. 1985. Toxicity of cadmium to free-living marine and brackish water nematodes *Monhystera microphthalma, Monhystera disjuncta, Pellioditis marina*. Dis. Aquat. Org. 1:49-58.
Warwick, R.M., M.R. Carr, K.R. Clarke, J.M. Gee and R.H. Green. 1988. A mesocosm experiment on the effects of hydrocarbon and copper pollution on a sublittoral soft-sediment meiobenthic community. Mar. Ecol. Prog. Ser. 46:181-191.
Widbom, B. 1984. Determination of average individual dry weights and ash-free dry weights in different sieve fractions of marine meiofauna. Mar. Biol. 84:101-108.
Widbom, B., and R. Elmgren. 1988. Response of benthic meiofauna to nutrient enrichment of experimental marine ecosystems. Mar. Ecol. Prog. Ser. 42:257-268.
Zar, J.H. 1984. In Biostatistical analysis, second edition. Prentice-Hall, Englewood Cliffs, N.J.

THE WESTERN MOSQUITOFISH AS AN ENVIRONMENTAL SENTINEL: PARASITES AND HISTOLOGICAL LESIONS

Robin M. Overstreet[1], William E. Hawkins[1], and Thomas L. Deardorff[2]

[1]Gulf Coast Research Laboratory, P.O. Box 7000, Ocean Springs, MS, 39566 U.S.A.
[2]International Paper, Erling Riis Research Laboratory, P.O. Box 2787, Mobile, AL, 36652 U.S.A.

Combining parasitological with histopathological data from the western mosquitofish inhabiting sites above, below, and in an integrated pulp and paper mill effluent canal along the Sulphur River in Texas and Arkansas demonstrates that the western mosquitofish serves as a good sentinel. Fish from the canal habitat receiving direct effluent as well as those downstream to the mill effluent canal were healthy in terms of species richness of parasites and of being free from pathogenic parasites known to be associated with stress. Histopathologic examination of mosquitofish specimens supported by additional tissues known to respond to toxicant challenge from large fish specimens collected at the same sites revealed no cancerous or other lesions or abnormalities that could be related to anthropogenic chemical toxicants. In fact, from the 4,324 slides of 816 fish examined, only one lesion was found that resembled a neoplastic lesion. That lesion, a "pre"-neoplastic one and one representing no more than background level, was in the liver of a freshwater drum from a tributary to the river. The most stressed site, based on low species richness of particular parasites, invasion into host tissues by a ciliate, a high prevalence of macrophage aggregates in the spleen, and high prevalence of vacuolated hepatocytes, occurred upstream from the effluent canal.

INTRODUCTION

This report demonstrates that the western mosquitofish, *Gambusia affinis*, serves as a good sentinel or indicator of general environmental health, including the health of fish associated with effluents from pulp and paper mills. Presence of enough fish to sample shows that a habitat is healthy enough to maintain a reproducing population. In many cases, the mosquitofish's parasites, most of which are links in complicated life cycles, indicate the richness in the habitat of a variety of invertebrates and vertebrates. That indication is because specific groups of animals all have to be present over a relatively long period to complete the parasites' specific life cycles. Heavy infections of other parasites, usually species that reproduce on or in the fish host, can indicate a stressed individual fish, often resulting from a stressed environment. In addition to the presence of parasites in the mosquitofish, chronic and acute histopathological lesions in the fish can indicate general or, in some instances, specific environmental stress factors as well as the extent of contaminant exposure (e.g., Overstreet 1993).

The mosquitofish serves as a good sentinel, or early warning system, for assessing the influence of pulp and paper mills on the environment because it meets the criteria for a sentinel species as characterized by Lower and Kendall (1990). It has a wide geographic and ecological distribution; it has a restricted home range of individuals, not migrating far; it is present in important locations; it has available significant biological endpoints; and it is easy to identify. Moreover, it forms relatively large populations from which relatively large subsamples can be assessed, serves as an intermediate or final host for a large number of parasites, can rapidly develop neoplasms when induced with carcinogens, is small enough to allow histological preparations to be produced easily and economically, and can be bred in captivity and used for experimental studies.

This study examined mosquitofish collected in a single sampling period from five sites along the Sulphur River in Texas and Arkansas, including those above, below, and in an effluent canal of a bleached kraft mill, International Paper Texarkana Mill, Domino/Texarkana, Texas. The Texarkana Mill operated two bleach lines. For softwood (pine), it used a two-stage oxygen delignification system followed by high substitution of chlorine with chlorine dioxide. For hardwoods, it used only the high substitution system. About 1,500 tonnes of pulp was produced each day. That resulted in over 130×10^6 L of effluent passing through an aerated stabilization basin each day to a holding pond. An amount of that water (standing about 1 yr) is discharged from the pond into the outflow canal and is equivalent to 2 to 6% (usually 3%) of the flow through Wright Patman Dam. The canal water then flows into the Sulphur River. Wright Patman Lake is a reservoir separated from the effluent canal by Wright Patman Dam and 29 km of river. Mosquitofish from the location sites were sectioned whole (except for the region posterior to the body cavity and in some cases the ovary) so that most of their tissues could be observed histologically for parasites and lesions using a combination of sections along the median and along two parasagittal (off-center) planes.

In addition to investigating mosquitofish, we conducted histological analyses of the health of large fish based on a few select tissues because some of the fish have value as game species (e.g., largemouth bass, several sunfish species) and can be used as environmental monitors to detect neoplastic (cancerous) lesions and other lesions that are typically considered to result from exposure to chemical toxicants. Also, some have different preferred habitats (e.g., catfish dwell near sediments where toxicants sometimes accumulate) and all can provide results to compare with those from mosquitofish. For those fishes, we examined the (1) liver because it is the major gland involved in metabolism of toxicants and nutrients and it reflects the metabolic health of an organism, (2) spleen because it is a principal immunological organ, (3) gills because they comprise the primary respiratory organ directly at the interface between organism and its environment, and (4) lesions, if visible to the naked eye.

MATERIALS AND METHODS

Sites

Samples of mosquitofish and other fishes were collected from the five locations (Fig. 1) described in Table 1. The two additional locations (Wright Patman Lake and Days Creek) are described because large fish or other small fish species that are used for comparisons were collected there. At each of the seven locations, measurements of water and its quality were recorded. These included water depth, air and water temperature, dissolved oxygen concentration measured with a YSI Model 57 oxygen meter to the nearest 0.1 mg L^{-1}, pH measured with a Model 57 digital mini pH meter to the nearest 0.1 unit, conductivity measured with a Model 33 meter, and turbidity with the white side of a Wildco Model 58-AZ5 20-cm diameter Secchi disk to the nearest 1 cm.

Fish Collections

Western mosquitofish were usually associated with floating or submerged vegetation and were caught using hand-held kick nets. An attempt to collect and fix at least 60 specimens per site resulted in only 39 fish from the site 16.4 km downstream. From that site, we also collected 22 small shiners (*Notropis* sp.). No mosquitofish were observed in Days Creek, so we also sampled 41 blackspotted topminnows (*Fundulus olivaceus*). These three small fish species all were injected with fixative to assure good fixation.

"Large fishes," species consisting mostly of those eaten or sold by local recreational and commercial fishermen, were sampled by electroshocking, inspected visually for lesions and abnormalities, and fixed for histological preparations of portions of their liver, spleen, gills, and observable lesions. Fifty specimens of these large fish, consisting of as many different species of fish at each site as were available, were sampled from the general area of each of the five mosquitofish sites as well as from Wright Patman Lake and Days Creek. After fish were shocked, they were maintained live in water until examination. These fishes consisted of *Alosa mediocris* (hickory shad, 1), *Ameiurus nebulosus* (brown bullhead, 5),

Aplodinotus grunniens (freshwater drum, 25), *Carpiodes carpio* (river carpsucker, 6), *Cyprinus carpio* (common carp, 2), *Dorosoma cepedianum* (gizzard shad, 6), *Ictalurus furcatus* (blue catfish, 15), *Ictalurus punctatus* (channel catfish, 72), *Ictiobus bubalus* (smallmouth buffalo, 9), *Lepomis cyanellus* (green sunfish, 1), *Lepomis gulosus* (warmouth, 9), *Lepomis macrochirus* (bluegill, 40), *Lepomis megalotis* (longear sunfish, 15), *Lepomis microlophus* (redear sunfish, 9), *Micropterus salmoides* (largemouth bass, 58), *Morone chrysops* (white bass, 30), *Morone saxatilis* (striped bass, 1), *Pomoxis nigromaculatus* (black crappie, 43), and *Pylodictis olivaris* (flathead catfish, 3).

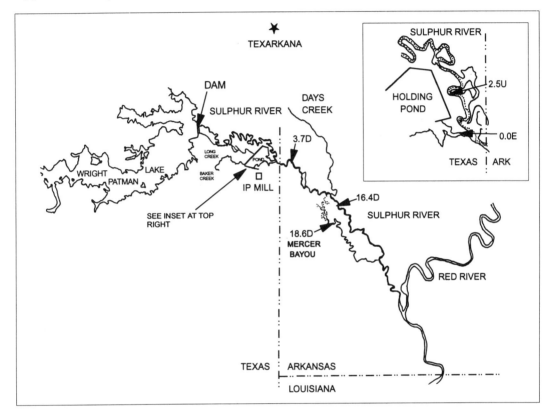

Figure 1. Sites where western mosquitofish were sampled from the Sulphur River near the Texas/Arkansas border in October 1992. Sites are designated by their distances in river kilometers upstream (U) and downstream (D) from the effluent canal (E) of an integrated pulp and paper mill.

Histological Methods

Mosquitofish were injected with Lillie's fixative (i.e., 85% picric acid, 10% formalin, and 5% formic acid) in the field and then immersed in specimen cups with additional Lillie's fixative (at a volume ratio of 1 part tissue to 20 parts fixative). When brought to the laboratory, each whole individual's standard length (distance in millimeters of the snout to the base of the caudal fin) and weight in grams were recorded. External morphological observations, including skeletal abnormalities and any external pathological lesions, were recorded for each fish. The body cavity was opened by making a ventral slit in the carcass from the anus to the opercular region, and gross observations of the internal organs were recorded. If embryos were well developed, the ovaries were removed to improve sectioning of the fish proper. Each fish was placed in a separate plastic histologic embedding cassette. To make room for large

Table 1. Summary of sample sites and environmental conditions during collection of western mosquitofish, topminnows, shiners, and large recreational fishes from the Sulphur River on 21-23 October 1992.

Sample Site	Site Description	Species Sampled	Water Temperature (°C) Surface	Water Temperature (°C) Bottom	Air Temperature (°C)	Dissolved Oxygen (mg L^{-1}) Surface	Dissolved Oxygen (mg L^{-1}) Bottom	Conductivity (mho) Surface	Conductivity (mho) Bottom	pH Surface	pH Bottom	Secchi disk depth (cm)
Wright Patman Lake (Reservoir)	29.1 km above mill outflow canal, upstream of and adjacent to Wright Patman Dam; recreational lake; low turbidity at time of sample; 1.8 m water depth in area of sample	large fish	21.9	21.4	30.2	7.6	7.5	290	950	8.2	7.6	310
2.5 km U Sulphur River	2.5 km upstream from mill outflow canal; more vegetative cover and underwater structures than other sites; water level approximately 1 m below normal at time of sample	western mosquitofish, large fish	22.0	21.3	30.0	8.6	8.6	200	150	7.7	7.8	510
0.0 km E Outflow Canal	In mill outflow canal, 60 m upstream from its confluence with the Sulphur River; volume of discharge in canal is regulated to equal 2 to 6% of flow through Wright Patman Dam; discharge slight at time of sampling; 1.8 m water depth in area of sample	western mosquitofish, large fish	21.5	22.0	30.0	7.6	7.5	290	950	8.2	7.6	310
3.7 km D Sulphur River	3.7 km downstream from its confluence with the outflow canal; deep-water and shallow-water stretches with large detritus and acutely inclined, undercut banks	western mosquitofish, large fish	19.5	19.3	17.5	8.5	9.3	200	200	7.2	7.1	310
Days Creek	0.8 km upstream from its confluence with Sulphur River, 13.9 km downstream from outflow canal; 3- to 6 m wide ditch with steep banks and average water depth of 1 m	blackspotted topminnow, large fish	18.0	18.0	29.0	8.2	8.2	210	200	6.2	6.0	410
16.4 km D Sulphur River	16.4 km downstream of outflow canal; at this location the river was straight and deep with little debris or cover for large fish	western mosquitofish, shiner, large fish	19.5	18.9	29.0	9.3	9.1	200	190	7.1	7.0	350
18.6 km D Mercer Bayou	18.6 km downstream of outflow canal; area does not always receive water from outflow canal; cypress bayou receiving water from Sulphur River during periods of high water; at time of sample, water level was low such that there was no water received from the river	western mosquitofish	19.2	19.8	30.0	8.7	8.3	150	150	8.0	6.5	415

mosquitofish in the embedding cassette, the caudal fin and posterior portion of each fish were removed by an angular cut beginning near the anus and continuing anteriorly to the dorsal side.

Large fish from which tissues were to be removed were identified in the field, and data similar to those obtained from mosquitofish were obtained. The selected tissues (two pieces of gills, one piece of liver, one piece of spleen, and skin if gross lesions were present) were then dissected from the fish in the field and immersed in 10% neutral buffered formalin.

Whole mosquitofish and large fish tissues were processed using the same methodology with the exception that whole fish were decalcified with Shandon TBD-2 (a commercial product) to soften scales and bone for easier tissue sectioning. These specimens and tissues were then stored in 70% ethanol until processed in an automated Shandon Hypercenter II for dehydrating, clearing, and embedding. Whole fish were individually embedded in paraffin blocks lateral side down, and gill tissue of large fish was embedded so that sections would show the length of the filaments. Liver and spleen tissues were oriented with their largest surface area down.

Sectioning was accomplished with an AO 820 rotary microtome, a Zeiss Microm HM 350, or an LKB Historange microtome. Blocks were chilled, and 5-µm-section ribbons were cut for both whole fish and large fish tissues. Paraffin sections of whole fish were cut at three levels: a mid-lateral plane, a sagittal plane, and a mid-lateral plane on the opposite side of the fish. Paraffin sections from each large fish tissue were cut full face at two levels. Slides were stained with hematoxylin and eosin using an automated Shandon Varistain.

Data Analysis

Data analyses involved several indices to describe the parasite community structure. These were species richness, diversity described as Shannon index (Shannon-Wiener diversity index) using log 10, and Shannon evenness (Zar 1984).

Definitions for parasitic infections follow those of Margolis *et al.* (1982): (1) "prevalence" = number of individual fish hosts infected with a particular parasite species divided by the number of fish examined and expressed as percentage, (2) "mean intensity" = mean number of individuals of a particular parasite species per infected host in a sample, (3) "abundance" (relative density) = mean number of individuals of a particular parasite species per host examined, and (4) "mean number of helminth individuals (not species) per host" has no special term.

RESULTS

Parasites

A total of 1,209 slides of sections from 403 mosquitofish collected at the 5 sites included 18 different parasite species. Of these, the protozoans *Apiosoma* sp. from the gills and buccal cavity, *Tetrahymena corlissi*(?) from several invaded anterior tissues, and *Myxidium* sp. from the gall bladder are new host records, previously unreported from the western mosquitofish. Table 2 lists the prevalence and mean intensity for all species at each site.

Species richness was 10 at and above the effluent canal, with a value of 12 for species occurring 3.7 km downstream and a value less than 10 at sites further downstream (Table 2). If the parasite species are restricted to the 11 internal helminths (worms), the highest number (7) occurred in mosquitofish from the canal (Table 2). Values for diversity and evenness for all parasites by site also appear in Table 2.

Numbers of individuals, prevalence of infection, and mean intensity all serve to indicate the general faunal richness and composition of the sites. Figure 2 shows that the mean number of internal helminths per mosquitofish was highest for fish from the effluent canal and lowest upstream. The two most abundant parasite species, the ciliate *Apiosoma* sp. on the gills and the digenean *Ornithodiplostomum ptychocheilus* encysted in the eyes and ovary, demonstrate two kinds of parasitic indicators (Fig. 3). The ciliate was least

Table 2. Prevalence and mean intensity of individual parasite species from histological sections of the western mosquitofish in the Sulphur River on 21–23 October 1992 plus species richness, species diversity, and evenness of those infections by locality upstream (U) and downstream (D) from the effluent canal (E) of an integrated pulp and paper mill.

Sample Site	Site 2.5 km U		Site 0.0 km E		Site 3.7 km D		Site 16.4 km D		Site 18.6 km D Mercer Bayou	
Sample Size	n = 99		n = 99		n = 104		n = 39		n = 62	
Average Standard Length	18.7 mm (14–30)		20.1 mm (15–30)		18.1 mm (12–26)		17.5 mm (13–26)		15.5 mm (10–30)	
Parasite	Prevalence (%)	Mean Intensity	Prevalence (%)	Mean Intensity	Prevalence (%)	Mean Intensity	Prevalence (%)	Mean Intensity	Prevalence (%)	Mean Intensity
Ciliophora										
Apiosoma sp.	20.2	1.5††	2.02	1††	21.15	1††	25.58	1††	35.48	1††
Tetrahymena corlissi?	2.02	4††								
Trichodina sp.					1.92	1††			1.61	2††
Myxozoa										
Myxidium sp.	10.1	4††	16.16	3.5††	6.73	4††	4.65	3.5††	11.29	3.5††
Myxobolus pharyngeus	1.01	4††			1.92	4††				
Bivalvulida	1.01	4††			1.92	2††				
Monogenea										
Dactylogyridae			1.01	1	7.69	1.12			1.61	1
Cestoda										
Bothriocephalus acheilognathi							2.33	1		
Proteocephalus sp.			1.01	1						
Digenea										
Allocreadiidae									32.26	1.3
Clinostomum marginatum	1.01	1	1.01	1	1.92	2.5			1.61	1
Diplostomulum scheuringi	2.02	1.5	52.53	1.9	9.62	1.1			1.61	1
Ornithodiplostomum ptychocheilus	3.03	1	3.03	1	6.73	1			4.84	1
Posthodiplostomum minimum							4.65	1.5	12.9	1.25
Macroderoides spinifera							25.64	2.8		
Rhipidocotyle papillosa			4.04	1	8.65	1.11	10.26	1.25		
Nematoda										
Contracaecum sp.	1.01	1	1.01	1	1.92	1				
Spiroxys sp.	7.07	1	11.11	1.09	7.69	1			13.95	1.33
Species richness	10		10		12		7		9	
Species diversity	0.937		0.395		0.983		0.592		0.640	
Evenness	0.937		0.395		0.911		0.701		0.671	
Species richness of internal helminths†	5		7		6		5		5	

* Immature stage.
† The tapeworms (Cestoda), flukes (Digenea), and roundworms (Nematoda) comprise the internal helminths.
†† For mean intensity of protozoans, 1 = very light, 2 = light, 3 = moderate, 4 = heavy.

abundant on fish in the effluent canal, perhaps indicating the least amount of bacteria and digestible organic material available as food for it in the canal. On the other hand, the encysted metacercaria in the eyes indicate that the highest number of the infected first intermediate snail host probably also occurred in the canal.

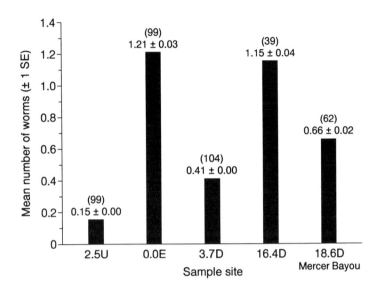

Figure 2. Mean number of internal helminth individuals per sectioned western mosquitofish collected in the Sulphur River near the Texas/Arkansas border. Sites are designated by their distances in river kilometers upstream (U) or downstream (D) from the mouth of the effluent canal (E) of an integrated pulp and paper mill. Numbers above bars are sample size (in parentheses) and mean with standard error.

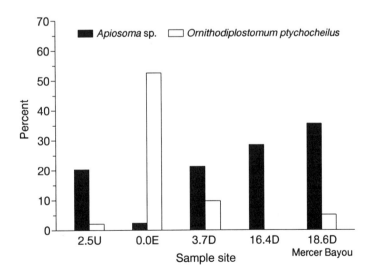

Figure 3. Prevalence of Apiosoma sp. and Ornithodiplostomum ptychocheilus in sectioned western mosquitofish collected in the Sulphur River near the Texas/Arkansas border. Sites are designated by their distances in river kilometers upstream (U) or downstream (D) from the mouth of the effluent canal (E) of an integrated pulp and paper mill.

A third kind of parasitic indicator is represented by the ciliate *Tetrahymena corlissi*(?), which invaded the internal tissues of the head region in one fish and the musculature, brachial chamber, and pericardial sac of another from the site located 2.5 km upstream from the canal and not influenced by its effluent. The presence of the ciliate likely indicates lowered resistance in at least the infected individuals. A fungal infection, not listed in Table 2, also occurred in the peritoneal cavity of a different individual from only that site.

Lesions

Table 3 lists the lesions and other histologic conditions in mosquitofish relative to the five sites. There was no neoplasia or skeletal abnormalities in any mosquitofish from any site. The spleen in fish from the 2.5-km upstream site had a greater abundance of macrophage aggregates than that in fish from other sites. Vacuolated livers in fish from the upstream site were also more common. Other indications of pathological alterations were not common at any site. What is apparently an abnormally high number of mitotic figures was observed in hepatocytes of several mosquitofish, mostly from Mercer Bayou.

From the 2,926 slides of select gill, liver, spleen, and skin tissues of 50 large fish examined from each of the 7 sites plus 123 slides of whole specimens of 41 blackspotted topminnows from Days Creek and 66 slides of 22 shiners from 16.4 km downstream, there were numerous parasites and few chronic lesions. None of these fish exhibited neoplasia, lesions, or abnormalities, including skeletal abnormalities, that could be associated with effluent from the mill or attributed to exposure to any toxicant.

In the liver of a freshwater drum from Days Creek there was a lesion that resembled an altered focus, generally considered a "pre"-neoplastic lesion, or one that might at some point develop into a neoplasm. In a brown bullhead from Mercer Bayou, a proliferative lesion was noted that consisted mainly of fibroblasts and could have resulted from an exuberant reaction to an injury or been a fibroma. At Wright Patman Lake, there were two cases of thickened skin, at least one of which was associated with a secondary fungal infection. In the canal, there was a hyperplastic epithelial lesion in the mouth of one white bass. Eight shiners from 16.4 km downstream demonstrated mitotic activity in hepatocytes, a condition also observed in mosquitofish from other sites. In three blackspotted topminnows from Days Creek, some unusual inclusions occurred in liver cells. At all sites, but especially at the site 2.5 km upstream from the effluent canal, there was considerable variation in the level of storage products (glycogen and fat) in the hepatocytes, especially in fish such as the white and striped bass.

Almost every large fish specimen exhibited some degree of parasitism. These parasites included fluke metacercariae of strigeoids and other digeneans, adult monogeneans, a tapeworm metacestode, a subadult cestode, a juvenile spirurid roundworm, a juvenile ascaridoid nematode, helminth eggs, a ciliate, a microsporan, different myxosporans, other protozoan cysts, an unidentified prokaryotic cyst, copepods, fungi, and unidentifiable degenerating parasites. None of these demonstrated a parasitic or host response that compromised the health of the host.

DISCUSSION

The presence of parasites does not necessarily imply an unhealthy condition for the host or locality. In fact, the presence of many species indicates a healthy environment. As long as infections are not too extensive, the health of the individual host may not be impaired (Overstreet 1993).

Gambusia affinis is known to host a wide range of parasites, including several in Texas (e.g., Hoffman 1967; Meade and Bedinger 1972; Davis and Huffman 1977, 1978; Wright and Boyce 1986). Of the 18 species of parasites identified from the mosquitofish in the Sulphur River, 3, *Apiosoma* sp., *Tetrahymena corlissi*(?), and *Myxidium* sp., had not been previously reported from the host.

Table 3. Selected histologic features of western mosquitofish collected from sites along the Sulfur River relative to the effluent canal of a paper mill. Values indicate numbers and percentage of sample.

	Site 2.5 km U	Site 0.0 km E	Site 3.7 km D	Site 16.4 km D	Site 18.6 kmD
Total no. examined	99	99	104	39	62
No. ♂s examined	39/39%	31/31%	36/35%	13/33%	20/32%
No. ♀s examined	47/47%	62/62%	54/52%	11/28%	30/48%
No. undetermined sex examined	13/13%	6/6%	14/13%	15/38%	12/19%
Macrophage aggregates: spleen	22/22%	8/8%	14/13%	3/8%	4/6%
Macrophage aggregates: other organs	12/12%	8/8%	15/14%	4/10%	2/3%
Vacuolated hepatocytes	22/22%	1/1%	3/3%	1/3%	0/0%
Eosinophilic infiltration in liver	7/7%	5/5%	3/3%	0/0%	0/0%
Mitotic figures in liver	7/7%	9/9%	8/8%	0/0%	19/18%
Apoptotic bodies in liver	0/0%	0/0%	3/3%	0/0%	0/0%
New renal tubule formation	0/0%	4/4%	1/1%	0/0%	0/0%
Intestinal enteritis	1/1%	1/1%	2/2%	1/3%	0/0%
Atrophy of intestinal epithelium	1/1%	1/1%	1/1%	2/5%	0/0%

Heavy infestations of some species of *Apiosoma* may be associated with mortalities (Lom and Dyková 1992) and with high levels of organic matter in the water (Calenius and Bylund 1980). In an article on eight identified and additional unidentified species of *Apiosoma*, none identical to that reported here, Calenius and Bylund (1980) noted heavy infestations in early spring and September. They referenced an article by Hakala (1974), which reported high levels of suspended organic particles to occur also in May and September from the same areas as they encountered the ciliates. These ciliates feed on bacteria and organic particles and not on the host.

Tetrahymena corlissi is a free-living ciliate that has been reported to invade fish in aquaria and hatcheries and cause pathological alterations in and even kill their hosts (Hoffman et al. 1975). Hoffman et al. (1975) could not experimentally infect fish with that ciliate, but they suggested that it could have invaded the hosts following a wound, microbial infection, malnourishment, organic material in the water, or other environmental influences. We questioned our identification because some specimens were larger than the range given by Hoffman et al. (1975). Otherwise, the sections of specimens agreed with those illustrated in those authors' review of facultative ciliates in freshwater fishes. Whether the species is or is not closely related to *Tetrahymena corlissi*, its facultative relationship with the mosquitofish is probably similar to those reported for other hosts.

We plan to investigate the identification and biology of *Myxidium* sp. in the future. The identification of some of the digeneans was partly based on similarities to species identified by Davis and Huffman (1977), who studied live and mounted specimens from mosquitofish near San Marcos, Texas. Not all of the individuals of parasites infecting a given fish can be seen in selected sections. Three planes of two or three serial sections of fish clearly do not incorporate every specimen present on or in the fish. Based on unpublished data from a comparison of observations on "skip-sectioned" material from three planes with those from fresh specimens of cohorts examined under a dissecting microscope, the sections gave an approximate indication of the infections present in the population. Also, since encysted specimens of the digenean *Ornithodiplostomum ptychocheilus* occurred in the ovary and *Macroderoides spinifera* and *Rhipidocotyle papillosa* occurred in tail musculature, both structures that were removed from many or all the specimens, the numbers are low but similarly reduced for material from all locations.

High values for species richness of all parasites and of only internal helminths from the canal site indicated a rich associated fauna at that site. The diversity and evenness, however, are both relatively low in the canal, primarily because of the strong influence on the values by *Ornithodiplostomum ptychocheilus*. Similarly, infections by *Macroderoides spinifera* in mosquitofish from the site 16.4 km downstream, the only site where that parasite occurred, influenced the relatively low diversity and evenness values for that site. That digenean as well as *Rhipidocotyle papillosa*, the second most common helminth at that downstream site, was present as an encysted metacercaria which matures in gars that eat the mosquitofish. No gars were collected in this study because members of that primitive family of fishes were not as sensitive as most of the locally abundant large fishes to electroshocking under the conditions we used.

Richness was also expressed as the mean number of species per fish at the canal site. The sites 16.4 and 18.6 km downstream had smaller sample sizes of fish, but they both contained two species not observed at the other sites. The relatively high number of parasites we report in mosquitofish relative to those reported regionally in many other fishes (e.g., Hoffman 1967) and the abundance of this species exemplify the major influence that mosquitofish have on the environment. This portrayal is especially true for the internal helminths, with their highly evolved specific life cycles.

Both records of species as well as the prevalence and mean intensity of those species indicate what types of animals had inhabited the different sites over a relatively long period. A heavy infection of a species typically indicates an abundance of the other appropriate hosts in its cycle. Because of knowledge of the components of life cycles of the specific parasites or of closely related species, the presence of the component invertebrate and vertebrate hosts in the sites is inferred. The recorded parasitic worms from Sulphur River indicate the presence of copepods, insects, crustaceans, snails, bivalves, and oligochactcs as intermediate hosts. Based on the same helminths infecting those invertebrates, the mosquitofish are eaten by a variety of birds, fishes, and mammals as well as at least one species of turtle. A concurrent evaluation of macrobenthos from the same general sites (Rakocinski *et al.* 1996) corroborates the abundance of oligochaetes, insects, and bivalves in the Sulphur River, especially in the outflow canal. Snails and crustaceans were present in low numbers from some benthic sites where mosquitofish were sampled. The snails and crustaceans were often associated with floating or submerged vegetation, but were not evaluated.

As an example, the high prevalence of *Ornithodiplostomum ptychocheilus* indicates the abundance in the effluent canal of (1) physid snails in which larvae that infect the mosquitofish are produced and (2) herons and egrets that readily feed on the fish and promote maturation of the worms, which ultimately produce eggs that are deposited into the bird's intestine and voided into the water when the bird defecates. The larval stage in the egg then infects the snail and undergoes considerable asexual reproduction, and the released free-living larva infects the mosquitofish and completes the cycle. This situation is in contrast to another in Finland where mills had different processes than in the U.S. In Finland, the prevalence of two related flukes in the eyes of perch (*Perca fluviatilis*) and roach (*Rutilus rutilus*) was less near the mill than in those fish from other localities examined presumably because of a small or absent population of the snail host resulting from the effluents (Valtonen *et al.* 1987).

The large fish tissues contained several parasites different than those which infected mosquitofish. There was not, however, a large enough sample size of any of those fish species, even when including the shiner and topminnow, to assess for use as a sentinel. Also, few tissues were examined for each species of large fish, and most of those fish probably would migrate from the study locations in response to environmental conditions. Consequently, these fish would probably not serve as indicators as well as the mosquitofish. There was no evidence suggesting that health of any of the hosts was compromised by parasites.

When parasites that have multiple hosts in their cycle are absent from some sites in a study, this suggests that one or more of the necessary hosts are missing from the site. The reason for the absence of the other hosts can result from contaminants or some other environmental condition. Some parasites have other means of indicating contaminants or stress in a site. For example, many species of external monogeneans and attached ciliates on the gills and skin are known to indicate a detrimental environment (Overstreet and Howse 1977; Overstreet 1993).

Monopisthcotylid monogeneans typically feed on mucus and epithelial tissue and are abundant in contaminated habitats because the fish hosts produce an abundance of mucus and epithelium in response to presence of the contaminants. Moreover, the immunological response of fish exposed to certain effluents of pulp and paper mills may become weakened (Valtonen and Koskivaara 1989). Infestations on mosquitofish were not common in any sites along the Sulphur River, and no individual had a heavy infestation. Other fishes had infestations of different monogenean species, but no individual had a heavy infestation. In contrast, the frequency of two species of monogeneans on roach from the Gulf of Bothnia (Thulin et al. 1988) and from Central Finland (Valtonen and Koskivaara 1989) was affected in such a way that it increased with distance from pulp mills. This gradation in infestation, however, was nullified when a mill initiated a more efficient method of purifying effluents (Koskivaara et al. 1991).

Mobile peritrich ciliates belonging to *Trichodina* and related genera feed on a diet similar to that of the monogeneans, and heavy infestations often indicate a stressed environment. Lehtinen et al. (1984) conducted an experiment exposing lightly infested flounder with effluents from three Swedish pulp mills that used different processes involving bleached and unbleached pulp. Heaviest infestations resulted on experimental fish exposed to effluent from a mill using chlorine bleaching with a pre-bleaching step with oxygen as well as on wild fish from near the discharge of the same mill (200 times heavier infestations than controls). Infestations of these ciliates from Sulphur River were uncommon and observed on mosquitofish from sites 3.7 and 18.6 km downstream only.

Individual mosquitofish with internal ciliate or fungal infections 2.5 km upstream from the effluent canal in the Sulphur River were not the same as those with abundant splenic macrophage aggregates. This finding of both spreading infections and the inflammatory response in the same location suggests many or all of the individuals at this location were stressed. The occurrence, abundance, and size of macrophage aggregates in the spleen appear to be good indicators of cumulative stresses in contaminated habitats (Blazer et al. 1987, 1993; Wolke et al. 1985).

Some parasites can make fish more susceptible to the effects of toxicants, and some toxicants can make the fish more susceptible to pathologic invasions by parasites (e.g., Overstreet 1993). We observed no histological signs indicating such relationships.

Histological lesions and conditions can serve as useful indicators of the acute and chronic health status of aquatic organisms, and histopathological changes in key organs and systems have been frequently used to identify and evaluate deleterious aquatic environmental conditions (e.g., Hinton et al. 1992). Several authors have associated skeletal abnormalities, fin erosion and deformation, and gill abnormalities with pulp and paper effluents (e.g., Lehtinen and Oikari 1980; Lehtinen et al. 1984; Thulin et al. 1988; Khan et al. 1992). Nevertheless, histological examinations of fish species exposed to pulp and paper effluents in laboratory and field studies have not demonstrated a specific pathological "marker" of exposure to those effluents (see Kloepper-Sams et al. 1991; Owens 1991). Those studies, however, dealt mainly with selected tissues of a limited number of specimens of large fish species. As mentioned above, the use of small fish species such as the mosquitofish has several advantages including the opportunity to examine numerous tissues and organs of many specimens economically on a few histologic slides. The lack of such acute and chronic conditions in our samples suggests the lack of the toxic substances or conditions that produce those effects. Future studies should exploit the statistical advantages of using the small fish along with using a table of lesion definitions similar to the one described by Reimschuessel et al. (1992).

Reported differences in types and presence of lesions in fish from different locations apparently result from different causes. For example, a high percentage of perch near one pulp mill in the Gulf of Bothnia exhibited fin erosion, but those from near a similar mill with a different bleaching process had normal fins (Lindesjöö and Thulin 1987). Moreover, those authors also noted an absence of fin erosion with introduction of an improved purification process in combination with natural high water exchange. In other cases where lesions are reported, effluents from sources other than pulp and paper mills occur nearby and attributing cause should be considered carefully (e.g., Almaça 1986; Khan et al. 1992).

Of paramount concern is whether effluents from pulp and paper mills or other sources induce neoplastic lesions, or cancers, in exposed organisms. Numerous types of neoplastic lesions have been reported to occur in wild fishes; however, only liver neoplasms have been shown convincingly to have a chemical etiology (Harshbarger et al. 1993). In addition to noting a strong epizootic association of liver cancer with environmental pollutants, those authors listed nine other lines of evidence dealing with experimental studies, biochemistry, molecular biology, lack of virus, and risk assessment that support such an etiology. To our knowledge, no studies have associated effluents from pulp and paper mills with histologically confirmed neoplasms in exposure fish although unusual tissue masses were seen in kidneys of 6 of 15 winter flounder exposed to effluent from a pulp and paper mill at Port Harmon, Newfoundland (Khan et al. 1992). The lack of neoplasms in mosquitofish and other fishes suggests that if carcinogens were present in the effluent, they were not present in high enough concentration to induce cancer.

Similar to other organisms, all fish species are not equally susceptible to the effects of carcinogen exposure (Hawkins et al. 1988). Mosquitofish, however, should be considered carcinogen-susceptible because Law et al. (1994) demonstrated that exposing methylazoxymethanol acetate to *Gambusia affinis* induced altered foci, hepatocellular adenomas, hepatocellular carcinomas, and cholangiocellular carcinomas after a relatively short latency period. The mosquitofish is as sensitive or more sensitive to neoplasia induction as any of seven other small species tested (Hawkins et al. 1988) and more sensitive than large species (Couch and Harshbarger 1985).

Of the two proliferative lesions seen in the large fish species, the altered focus in the freshwater drum liver from Days Creek might be significant because the presence of altered hepatic foci is correlated with the eventual development of advanced neoplasms in wild fish species (Myers et al. 1991). From our experience, the proliferative lesion in the brown bullhead from Mercer Bayou was most likely an exuberant fibroblastic reaction to an injury and not a neoplasm. Regardless, the occurrence of one case of any particular lesion in a sample size as large as the one in this study could only be considered as the reflection of a background incidence.

The wide range of nonneoplastic lesions we list from mosquitofish indicates the susceptibility of that species to a response induced by a variety of contaminants or foreign matter. In this study, however, the only site where fish responded significantly to such stimulation, as judged by the occurrence of macrophage aggregates, appeared to be the upstream site.

The nature of the unusual inclusions in liver cells of blackspotted topminnows in Days Creek was not determined, but the inclusions could have been associated with infectious agents, such as bacteria, viruses, or protozoa, or they could be unusual storage products. The significance of mitotic activity in hepatocytes of mosquitofish and shiners is not clear at this time, but, based on the geographic distribution, it does not seem related to effluent from the mill.

The considerable variation in the level of storage products in the hepatocytes was probably a normal condition dependent on species, site, sex, period in the sexual cycle, and diet. The size of the fish downstream 16.4 and 18.6 km averaged 1 to 3 mm shorter than those upstream and were more difficult to collect. The sort of normal-appearing vacuolation we saw, however, was probably unlike that seen by Lehtinen et al. (1984) who noted greater vacuolation with nuclear condensation, suggesting a toxic response, in fish exposed experimentally and naturally to effluents of chlorinated bleaching with a pre-bleaching treatment with oxygen than to control water or to effluent from mills using conventional bleaching or unbleached pine.

Sectioned mosquitofish were good tools because they demonstrated both parasites and lesions. Fresh specimens would provide more accurate information on numbers and presence of parasites, but they would not provide adequate information on lesions (none or few of the lesions we observed would have been evident to the naked eye).

In summary, combining the parasitological with the histopathological data from the mosquitofish produces a good sentinel for assessing effluents from the pulp and paper mill. At least for the period of sampling and several months prior to that, the combination of data shows that the habitat near and downstream to the mill effluent was relatively rich in terms of species richness and the fish were free from lesions known to be associated with some other mills and other contaminated waters. The mosquitofish, probably better than any other available species, served as the best sentinel in the area because it met both the criteria for a sentinel species as characterized by Lower and Kendall (1990) and the additional criteria presented in our introduction.

ACKNOWLEDGMENTS

This study represents the efforts of several people. We thank Jean Jovonovich Alvillar, Pam Monson, Susan Powell, Nate Jordan, John Hanson, Marie Wright, Ronnie Palmer, Helen Gill, Retha Edwards, Vivian Heitzman, Rena Krol, Robert Allen, Rosemary Jacobs, Christa Waller, Wanda Tillman, and Lisa Ortego, all of the Gulf Coast Research Laboratory; Kendall K. Brown of Ecological Research and Management Incorporated; Terry Birdwell, Keith Hogue, and Felix Jenkins of International Paper, Texarkana; Brad DeVore then at Jones, Day, Reavis & Pogue; and the Arkansas Game and Fish Commission and Texas Parks and Wildlife Department. It was funded by International Paper.

REFERENCES

Almaça, C. 1986. Fish and their environment in large European river ecosystems. Sciences de L'eau 7(1):3-19.

Blazer, V.S., R.E. Wolke, J. Brown and C.A. Powell. 1987. Piscine macrophage aggregate parameters as health monitors: effect of age, sex, relative weight, season and site quality in largemouth bass (*Micropterus salmoides*). Aquat. Toxicol. 10:199-215.

Blazer, V.S., J.W. Fournie, L.A. Courtney, J.K. Summers and D.E. Facey. 1993. Comparison of macrophage aggregate parameters and macrophage function as indicators of contaminant stress. *In* J.S. Stolen [ed.] Modulators of fish immune responses: models for environmental toxicology/biomarkers, immunostimulators. SOS Publications, Fair Haven, NJ.

Calenius, G., and G. Bylund. 1980. Parasites of fish in Finland. IV. Ciliates of the genus *Apiosoma*. Acta Acad. Abo. Ser. B 40:1-12.

Couch, J.A., and J.C. Harshbarger. 1985. Effects of carcinogenic agents on aquatic animals: an environmental and experimental overview. Environ. Carcinogenesis Rev. 3:63-105.

Davis, J.R., and D.G. Huffman. 1977. A comparison of the helminth parasites of *Gambusia affinis* and *Gambusia geiseri* (Osteichthyes: Poeciliidae) from the upper San Marcos River. Southwest. Nat. 22:359-366.

Davis, J.R., and D.G. Huffman. 1978. Some factors associated with the distribution of helminths among individual mosquitofish, *Gambusia affinis*. Tex. J. Sci. 30:43-53.

Hakala, I. 1974. Sedimentaatio Pääjärvessä. Luonnon Tutkija 78:108-110.

Harshbarger, J.C., P.M. Spero and N.M. Wolcott. 1993. Neoplasms in wild fish from the marine ecosystem emphasizing environmental interactions, chapter 6, pp. 157-176. *In* J.A. Couch and J.W. Fournie [ed.] Pathobiology of marine and estuarine organisms. CRC Press, Boca Raton, FL.

Hawkins, W.E., R.M. Overstreet and W.W. Walker. 1988. Carcinogenicity tests with small fish species. Aquat. Toxicol. 11:113-128.

Hinton, D.E., P.C. Baumann, G.R. Gardner, W.E. Hawkins, J.D. Hendricks, R.A. Murchelano and M.S. Okihiro. 1992. Histopathologic biomarkers, chapter 4, pp. 155-209. *In* R.J. Huggett, R.A. Kimerle, P.M. Mehrle, Jr. and H.L. Bergman [ed.] Biomarkers: biochemical, physiological, and histological markers of anthropogenic stress. Lewis Publishers, Chelsea, MI.

Hoffman, G.L. 1967. Parasites of North American freshwater fishes. University of California Press, Berkeley, CA. 486 p.

Hoffman, G.L., M. Landolt, J.E. Camper, D.W. Coats, J.L. Stookey and J.D. Burek. 1975. Disease of freshwater fishes caused by *Tetrahymena corlissi* Thompson, 1955, and a key for identification of holotrich ciliates of freshwater fishes. J. Parasitol. 61:217-223.

Khan, R.A., D. Barker, R. Hooper and E.M. Lee. 1992. Effect of pulp and paper effluent on a marine fish, *Pseudopleuronectes americanus*. Bull. Environ. Contam. Toxicol. 48:449-456.

Kloepper-Sams, P., T. Marchant, J. Bernstein and S. Swanson. 1991. Use of fish biomarkers and exposure measures to assess fish health at a Canadian bleached kraft mill site. Swedish Environmental Protection Agency Report 4031:283-292.

Koskivaara, M., E.T. Valtonen and M. Prost. 1991. Dactylogyrids on the gills of roach in central Finland: features of infection and species composition. Int. J. Parasitol. 21:565-572.

Law, J.M., W.E. Hawkins, R.M. Overstreet and W.W. Walker. 1994. Hepatocarcinogenesis in western mosquitofish (*Gambusia affinis*) exposed to methylazoxymethanol acetate. J. Comp. Pathol. 110:117-127.

Lehtinen, K.-J., and A. Oikari. 1980. Sublethal effects of kraft pulp mill waste water on the perch, *Perca fluviatilis*, studied by rotary-flow and histological techniques. Ann. Zool. Fenn. 17:255-259.

Lehtinen, K.-J., M. Notini and L. Landner. 1984. Tissue damage and parasite frequency in flounders, *Platichtys [sic] flesus* (L.), chronically exposed to bleached kraft pulp mill effluents. Ann. Zool. Fenn. 21:23-28.

Lindesjöö, E., and J. Thulin. 1987. Fin erosion of perch (*Perca fluviatilis*) in a pulp mill effluent. Bull. Eur. Ass. Fish Pathol. 7:11-13.

Lom, J., and I. Dyková. 1992. Protozoan parasites of fishes. Developments in aquaculture and fisheries science. Vol. 26, Elsevier Science Publishers BV, Amsterdam, The Netherlands. 315 p.

Lower, W.R., and R.J. Kendall. 1990. Sentinel species and sentinel bioassay, chapter 18, pp. 309-331. *In* J.F. McCarthy and L.R. Shugart [ed.] Biomarkers of environmental contamination. Lewis Publishers, Boca Raton, FL.

Margolis, L., G.W. Esch, J.C. Holmes, A.M. Kuris and G.A. Schad. 1982. The use of ecological terms in parasitology. J. Parasitol. 68:131-133.

Meade, T.G., and C.A. Bedinger, Jr. 1972. Helminth parasitism in some species of fresh water fishes of eastern Texas. Southwest. Nat. 16:281-295.

Myers, M.S., J.T. Landahl, M.M. Krahn and B.B. McCain. 1991. Relationships between hepatic neoplasms and related lesions and exposure to toxic chemicals in marine fish from the U.S. West Coast Environ. Health Perspect. 90:7-15.

Overstreet, R.M. 1993. Parasitic diseases of fishes and their relationship with toxicants and other environmental factors. Chapter 5, pp. 111-156. *In* J.A. Couch and J.W. Fournie [ed.] Pathobiology of marine and estuarine organisms. CRC Press, Boca Raton, FL.

Overstreet, R.M., and H.D. Howse. 1977. Some parasites and diseases of estuarine fishes in polluted habitats of Mississippi. *In* H.F. Kraybill, C.J. Dawe, J.C. Harshbarger and R.G. Tardiff [ed.] Aquatic pollutants and biologic effects with emphasis on neoplasia. Ann. NY Acad. Sci. 298:427-462.

Owens, J.W. 1991. The hazard assessment of pulp and paper effluents in the aquatic environment: a review. Environ. Toxicol. Chem. 10:1511-1540.

Rakocinski, C.F., M.R. Milligan, R.W. Heard and T.L. Deardorff. 1996. Comparative evaluation of macrobenthic assemblages from the Sulphur River Arkansas in relation to pulp mill effluent. *In* M.R. Servos, K.R. Munkittrick, J.H. Carey and G. Van Der Kraak (Eds.). Environmental fate and effects of pulp and paper mill effluents. St. Lucie Press, Delray Beach, FL.

Reimschuessel, R., O. Bennett and M.M. Lipsky. 1992. A classification system for histological lesions. J. Aquat. Anim. Health 4:135-143.

Thulin, J., J.Höglund and E. Lindesjöö. 1988. Diseases and parasites of fish in a bleached kraft mill effluent. Water Sci. Technol. 20:179-180.

Valtonen, E.T., and M. Koskivaara. 1989. Effects of effluent from a paper and pulp mill on parasites of the roach in central Finland, pp. 163-167. *In* O.N. Bauer [ed.] Parasites of freshwater fishes of north-west Europe. Petrozavodsk: Karel'skii filial AN SSSR.

Valtonen E.T., M. Koskivaara and H. Brummer-Korvenkontio. 1987. Parasites of fishes in central Finland in relation to environmental stress. Biol. Res. Rep. Univ. Jyväskylä 10:129-130.

Wolke, R.E., R.A. Murchelano, C.D. Dickstein and C.J. George. 1985. Preliminary evaluation of the use of macrophage aggregates (MA) as fish health monitors. Bull. Environ. Contam. Toxicol. 35:222-227.

Wright, S.A., and K.W. Boyce. 1986. A survey of mosquitofish *Gambusia affinis* parasites in Sacramento County. Proceedings of the Annual Conference of the California Mosquito and Vector Control Association 54:79-85.

Zar, J.H. 1984. Biostatistical analysis. 2nd ed., Prentice-Hall, Englewood Cliffs, NJ. 718 p.

ABNORMALITIES IN WINTER FLOUNDER (*PLEURONECTES AMERICANUS*) LIVING NEAR A PAPER MILL IN THE HUMBER ARM, NEWFOUNDLAND

R.A. Khan, D.E. Barker, K. Ryan, B. Murphy and R.G. Hooper

Department of Biology and Ocean Sciences Center, Memorial University of Newfoundland,
St. John's, Newfoundland, A1C 5S7 Canada

Winter flounder were sampled near a sulfite-bleaching pulp and paper mill (Birchy Cove) located in a marine inlet (Humber Arm) that also receives urban and industrial discharges. Samples were also obtained from other sites (Summerside and Meadows) in the Humber Arm and from Norris Point situated in another inlet located northward. A comparison was made of blood values, body and organ indices, macroscopic and microscopic lesions and parasitic levels. Macroscopic lesions including epidermal ulcers and liver discoloration occurred more often in samples taken at Birchy Cove than at the three other sites. Although condition factor and infestation with metacercariae of the trematode *Cryptocotyle lingua* were similar among two samples from the Humber Arm, liver and ovarian somatic indices were significantly greater in fish from Birchy Cove than Summerside. These differences are probably related to delayed spawning in fish from Birchy Cove. Lower hemoglobin and lymphocyte levels, higher prevalence of liver and gill lesions and elevated levels of hemosiderin in the spleen and kidney in samples from Birchy Cove relative to the other sites suggest that fish health near the mill was impaired. It is likely that the observed lesions might be associated with sulfite-laden sediment in which the affected flounder were submerged.

INTRODUCTION

Several studies have reported that effluent discharged from pulp and paper mills has an adverse effect on fish living in the vicinity of the discharge (Oikari *et al.* 1985; Andersson *et al.* 1987; McLeay *et al.* 1987; Bengtsson *et al.* 1988; Sandström and Thoresson 1988; Myllyvirta and Vuorinen 1989; Lindstrom-Seppä and Oikari 1990; Munkittrick *et al.* 1991). Since most mills use one or more bleaching agents such as chlorine dioxide, sodium hydrosulfite or derivatives of these in paper (newsprint) production, the outfall usually contains a variety of chemicals such as dioxins, furans and resin acids that are known to be toxic and/or carcinogenic to fish (McLeay *et al.* 1987; Waldichuk 1990). Chronic exposure might culminate in the uptake and accumulation of these chemicals into various tissues and may eventually result in a variety of changes in blood chemistry, detoxifying liver enzymes, sex hormone levels and histopathology and have a profound effect on growth and reproduction (Whittle and Flood 1977; Lehtinen *et al.* 1984; Andersson *et al.* 1988; Couillard *et al.* 1988; Härdig *et al.* 1988; Axelsson and Norgren 1991; McMaster *et al.* 1991; Bucher *et al.* 1992). Munkittrick *et al.* (1992a) noted a reduction in liver size in white sucker (*Catostomus commersoni*) following installation of a secondary treatment aeration stabilization basin but other impacts were not changed. There are also reports suggesting that fish population structure including abnormal size distribution and age structure is affected following exposure (Neuman and Karås 1988; Adams *et al.* 1992). However, Hall *et al.* (1991, 1992) observed a greater production of rainbow trout (*Oncorhynchus mykiss*) in streams receiving bleached kraft mill effluent (BKME) than in reference locations. Moreover, examination of fish health values such as blood variables, liver somatic indices, condition factor and histopathology revealed no differences between exposed and control groups (Hall *et al.* 1991, 1992).

Two pulp and paper mills in the province of Newfoundland use sodium hydrosulfite as the bleaching agent in the production of paper. The effluent is discharged into the marine environment which is inhabited by winter flounder (*Pleuronectes americanus*) and a variety of plant and animal life. Previous studies have shown that flounder living in one inlet (Port Harmon) under the influence of the outfall had a lower condition factor and blood values, histopathological changes, delayed spawning and increased parasitism relative to samples collected at a reference site (Khan et al. 1992, 1994). Anomalies were also observed in sculpins (*Myoxocephalus* spp.) inhabiting the same marine inlet (Barker et al. 1994). Less pronounced effects were also noted in winter flounder sampled in the vicinity of a second sulfite pulp and paper mill located in the Humber Arm (Khan et al. 1992). A more detailed study was conducted in 1992 and 1993 on flounder inhabiting this inlet to determine the extent of these abnormalities and our results are reported herein.

SITE DESCRIPTION

The sulfite pulp and paper mill is located in a marine inlet in the Humber Arm in western Newfoundland (49°32'N, 57°52'W) (Fig. 1). The mill has been operating since the 1920s and produces about 329,000 tons of newsprint annually by thermomechanical and sulfite-pulping processes using local softwood, mainly black spruce (*Picea mariana*) and balsam fir (*Abies balsamea*). The untreated effluent which contains suspended solids (76-93 mg L^{-1}), tannin and the resin acids (~27 mg L^{-1}), especially dehydroabietic (600-1000 µg L^{-1}) and abietic acids (600-1200 µg L^{-1}), fans out after discharge and is clearly visible at the surface up to 2 km from the source (Environment Canada unpubl. data). Its spread is influenced by wind and tide. Prior to dry-debarking process, pulp logs were floated and stored next to the mill. As a result of wet debarking, a considerable amount of wood and bark has been deposited in the area for a number of years. During heavy rainfall, bark chips are washed into the inlet and carried by the current in the direction of Birchy Cove and it has been estimated that about 3 to 5 km^2 of the bottom is densely coated with hydrogen-sulfide-laden sludge, fiber and bark deposits in the vicinity of the mill. The area adjacent to the mill is devoid of marine life but an increasing number of organisms are found towards the ocean. Analysis of sediment, water and fish tissues has revealed that polychlorinated biphenyls, petroleum aromatic hydrocarbons (<5 mg L^{-1}) and heavy metals (0.02 µg g^{-1}, with the exception of zinc) were below background levels. At several sites near and up to 10 km from the mill in the Humber Arm, zinc levels in cunner (*Tautogolabrus adspersus*) which reside in the area varied from 20 to 70 µg g^{-1} (Ledrew et al. 1989) and were independent of proximity to the mill.

The Humber Arm receives industrial and urban waste from about 23 untreated sewer outlets, discharges from two oil depots, cement manufacturing plants and a gypsum loading facility (Ledrew et al. 1989). Other sources of contaminants originate from a fish processing facility, land runoff and shipping. Fresh water enters the inlet from the Humber River and flow of the current is clockwise. Low-salinity water, low oxygen saturation, high levels of dissolved carbon and turbidity occur at the surface (Ledrew et al. 1989). A comparison of water characteristics including temperature, salinity, conductivity, oxygen and pH in the Humber Arm at the surface and bottom (8 m) during May and June in 1992 at two sites (Birchy Cove and Summerside) and Norris Point revealed no major differences. Ledrew et al. (1989) also reported that water characteristics were similar between Birchy Cove and Summerside.

MATERIALS AND METHODS

Winter flounder were captured by Scuba divers in late May-early June (1992 and 1993) at depths of 5 to 10 m at Birchy Cove less than 3 km from a pulp and paper mill (Fig. 1). A reference site, Summerside, lying about 4 km obliquely across the inlet also receives untreated sewage from the community. We were not aware that this location was not a suitable reference site, based on the report of Ledrew et al. (1989) until the study was completed. A limited number (12) of samples were also obtained from a second reference site in the Humber Arm at Meadows primarily for histopathology. A third reference site, Norris Point (40°32' N, 57°52' W) is situated in another inlet, Bonne Bay. In contrast

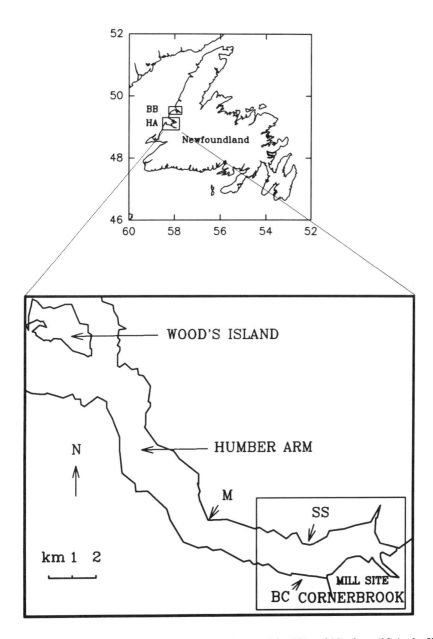

Figure 1. Location of collection sites at Birchy Cove (BC), Summerside (SS) and Meadows (M) in the Humber Arm and Norris Point, Bonne Bay (BB), Newfoundland.

to the dark, anaerobic and highly organic bottom substrate at Birchy Cove, clear sand with cobblestone and traces of bark and fiber predominated at the reference sites at Summerside, Meadows and Norris Point. After collection, the flounder were bled from the dorsal aorta with a needle and heparinized syringe for determination of hemoglobin and lymphocyte levels. Blood smears were fixed with methanol and stained with Giemsa. Hemoglobin was determined by means of a hemoglobinometer (American Optical TM). Total body length and eviscerated weight as well as those of the liver, gut and gonad were recorded for each specimen. Tissues (gill, liver, spleen, kidney and gonad) were fixed in 10% buffered formalin for light microscopy, processed using conventional histological methods, stained with hematoxylin and eosin and examined by light microscopy. Sections of liver and spleen were also stained with oil red O and Perl's Prussian blue, monospecific stains to detect lipid and hemosiderin, respectively. Concentration of the latter in tissue sections was estimated by digital image analysis as outlined by Khan and Nag (1993). The viscera of the flounder were examined macroscopically for cysts of the microsporan *Glugea stephani* (Takvorian and Cali 1984). The stomach of each fish was also examined for food contents. Afterwards, the entire gastrointestinal tract of a subsample of flounder from each site was dissected and examined for parasites. Food and parasites were identified subsequently. Encysted metacercariae on the fins were classified into three categories, after enumeration, as low (0-50), moderate (51-100) and high (>101) (Barker 1993).

The data including condition (K) factor (w/L^3), organ somatic indices (organ weight/eviscerated weight), hemoglobin levels, lymphocyte numbers per 1000 erythrocytes, parasitic abundance and hemosiderin concentration were compared among samples obtained from Birchy Cove, Summerside and Norris Point for significant differences using the one-way ANOVA and Duncan's multiple range test (Sokal and Rohlf 1987). The sample size from Meadows was too small for a valid statistical comparison with the other groups. Prevalence of parasitism between groups was compared by the G test. Differences were considered significant when $p \leq 0.05$.

RESULTS

Macroscopic Observations

External and internal macroscopic lesions occurred more often in flounder captured at Birchy Cove than at Summerside but not at Meadows or Norris Point, the three reference sites. Epidermal ulcers (Fig. 2, Plate 2) were observed in 17% of the fish taken at Birchy Cove whereas the lesions were rare (3%) at Summerside and absent at Norris Point and Meadows. Similarly, fin necrosis (5%) was seen only in samples from Birchy Cove. Two (11%) male flounder from Birchy Cove had a testicular tumor (Fig. 2, Plate 3) but none was observed in samples taken at the reference sites. Similarly, three (13%) female flounder originating from Birchy Cove harbored xenomas of the microsporan *Glugea stephani* in the liver, heart and kidney (Fig. 2, Plate 4) whereas the infection in one fish from Summerside occurred only on the wall of the gastrointestinal tract. Flounder at the two sites in the Humber Arm were infested with metacercariae of a digenetic trematode, *Cryptocotyle lingua*, which appeared as dark spots mainly on the fins (Fig. 2, Plate 5). Prevalence of moderate to high infestations was similar in samples from Summerside and Birchy Cove while none was seen in fish taken at Norris Point or Meadows. Following autopsy, 60% of the liver samples from Birchy Cove showed evidence of discoloration appearing as opaque nodules 2 to 6 mm, in contrast to 10% at Summerside and none at Norris Point or Meadows.

Comparison of Condition Factor, Organ Somatic Indices and Blood Values

Only male fish taken from Birchy Cove were greater in length and weight than flounder from Summerside but k-factor values were not significantly different (Table 1). However, liver somatic indices (LSI) for both males and females were greater in samples from Birchy Cove than Summerside. Additionally, ovarian indices (GSI) were significantly greater in flounder taken at Birchy Cove as 6 (26%) of 23 fish were in a pre-spawning condition compared to the post-spawning stage of fish captured at Summerside and Norris Point. Males from the three sites were spent. Hemoglobin and lymphocyte levels were also significantly lower in samples obtained from Birchy Cove than from either Summerside or Norris Point (Fig. 3).

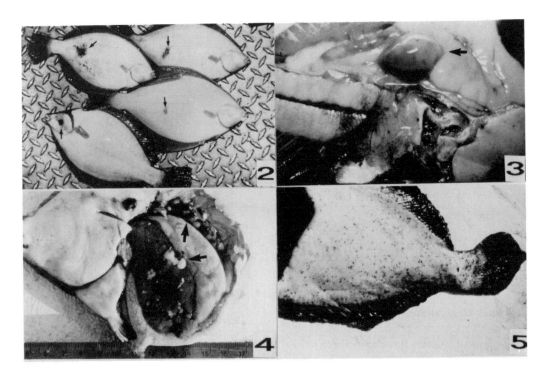

Figure 2. Macroscopic lesions in winter flounder captured at Birchy Cove. Plate 2: Epidermal ulcers (arrows). Plate 3: Tumor in testis (arrow). Plate 4: Foci of xenomas (arrows) of the microsporan Glugea stephani in the liver and kidney. Plate 5: Skin with heavy infection of metacercariae of the digenetic trematode C. lingua.

Table 1. Comparison of variables ($x \pm S.E.$) in winter flounder, P. americanus, collected in the immediate vicinity of a pulp and paper mill at Birchy Cove and Summerside (both in the Humber Arm) and at Norris Point, a reference site located in another oceanic inlet, Bonne Bay, Newfoundland.

Variable	Birchy Cove	Summerside	Norris Point
n ♂	19	23	17
Length (cm) ♂	24.3 ± 1.0	18.9 ± 0.7*	24.6 ± 0.8
Weight (g) ♂	185.4 ± 19.9	89.7 ± 8.4 *	189.2 ± 10.1
K-factor (x10^{-2}) ♂	1.25 ± 0.07	1.29 ± 0.07	1.26 ± 0.06
LSI (x10^{-2}) ♂	2.49 ± 0.23	1.60 ± 0.09*	2.81 ± 0.1
GSI ♂	2.40 ± 0.73	3.87 ± 0.63	2.81 ± 0.2
n ♀	23	17	21
Length (cm) ♀	23.8 ± 1.0	22.0 ± 1.0	25.2 ± 1.2
Weight (g) ♀	191.8 ± 16.9	134.5 ± 13.8*	234.1 ± 12.0
K-factor (x10^{-2}) ♀	1.45 ± 0.09	1.24 ± 0.06	1.42 ± 0.07
LSI (x10^{-2}) ♀	3.33 ± 0.19	2.56 ± 0.17*	3.61 ± 0.2
GSI ♀	4.35 ± 0.38	2.29 ± 0.33*	2.41 ± 0.2*

*$p \leq 0.05$

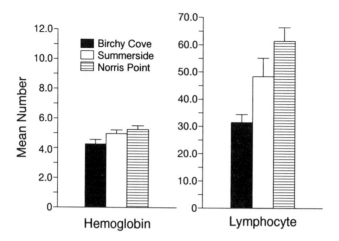

Figure 3. Mean (± S.E.) hemoglobin (g%) and number of lymphocytes (per 1000 erythrocytes) in winter flounder captured at Birchy Cove, Summerside and Norris Point.

Microscopic Observations

The livers of winter flounder at Birchy Cove and Summerside displayed microscopic evidence of liver lesions whereas none was observed in samples from Norris Point or Meadows. Lesions are divided into three categories: non-specific, pre-neoplastic and neoplastic lesions after Myers *et al.* (1987). In contrast to densely stained liver cells in samples taken at the three reference sites, pale-staining hepatocytes with peripherally displaced nuclei and oil red O positive vacuoles occurred in fish (50%) from Birchy Cove. In addition, there was evidence of pericholangial fibrosis in 14% of the samples from Birchy Cove. The most common disorder was the presence of multifocal hemosiderosis characterized by melanomacrophage centers (MMCs) which occurred equally in samples from Birchy Cove (67%) and Summerside (74%). These centers were always located adjacent to blood vessels and variable in shape and size. Although the number of MMCs appeared to be more abundant in samples from Birchy Cove than at Summerside, estimation by image analysis was not possible because of the low concentration. Pre-neoplastic lesions were identified as clear cell foci. These consisted of well-defined areas of enlarged vacuolated cells with peripherally displaced nuclei that spread outward and became confluent. These foci were seen initially as isolated areas, eventually assuming a curvilinear array, and appeared to coalesce. No discrete lumena were observed within the foci nor was there any evidence of circumscribing connective tissue. Cell foci occurred more often in larger (>27 cm in length) than in smaller (<25 cm) winter flounder. Thirty-three percent of the samples from Birchy Cove had clear cell foci in contrast to 13% from Summerside. The livers of three (7%) flounder also taken at the same site had hepatocellular carcinomas. These tumors were composed of clones of basophilic cells with vacuoles within them. Between these aggregates were clear cell foci and some apparently normal hepatocytes.

Lesions were also observed in the spleen and kidney and identified as an accumulation of MMCs. The latter were more abundant in the spleen and estimation by image analysis revealed that the concentration was greater in samples from Birchy Cove than Summerside or Norris Point (Fig. 4). No MMCs were seen in samples from Meadows. Hemosiderin concentration in the kidney collected from samples taken from two localities (Birchy Cove and Summerside in the Humber Arm) were lower than that in the spleen but were more concentrated in samples taken from Birchy Cove than Summerside. Two (5%) female flounder taken at Birchy Cove also had granulomas in the kidney. The core was composed of a mass of vacuolated cells with an infiltrate of mononuclear cells. The latter were also present between the pale-staining core and an external fibrous capsule.

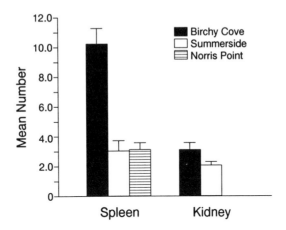

Figure 4. Percent of melanomacrophage centers/mm^2 in sections of spleen and kidney of winter flounder from Birchy Cove, Summerside and Norris Point.

The testes of two flounder harbored tumors. These are characterized by mesenchymal cells interspersed between islets of seminiferous tubules. The latter in both fish contained spermatozoa. In spite of replacement of most of the testicular tissue, both fish appeared macroscopically normal.

Gill filaments of flounder captured in the Humber Arm at Birchy Cove and Summerside showed evidence of hyperplasia but none was observed in samples from Norris Point or Meadows. Moderate to severe hyperplasia, characterized by thickening of the epithelium lining the secondary gill lamellae, was evident more often in samples from Birchy Cove (42%) than Summerside (10%). Hyperplasia was observed more in the distal than in the proximal parts of the secondary lamellae. This change was more pronounced in fish samples exceeding 25 cm in length (67% of the affected fish at Birchy Cove) than in smaller fish. Infestation of the gill lamellae by metacercariae of *C. lingua*, which caused distortion of the primary lamellae, occurred more often in samples taken at Birchy Cove (93%) than at Summerside (60%). The parasite was not seen in fish captured at Norris Point or Meadows.

Histological examination of the ovaries of six winter flounder that had not spawned and had originated from Birchy Cove revealed ova in a variety of developmental stages. There was a greater percentage (64%) of small and immature to larger, mature ova. Following the five stages of classification of mature ova by Harmin and Crim (1992), most of the mature ova were in stages 1 and 2, i.e., the germinal vesicle (nucleus) of each was located either in the center or slightly off the central position.

Flounder captured in the Humber Arm also harbored infections of the microsporan parasite *Glugea stephani*. The latter was recognized by the large xenomas in three infected fish taken at Birchy Cove in the liver, heart, kidney and gonad. Moreover, examination of histological sections revealed spores of the parasite. Only one flounder captured at Summerside was infected and xenomas were observed only on the wall of the gastrointestinal tract. None was seen in fish taken at Norris Point or Meadows.

Food and Enteric Parasites

Variations in the frequency of prey consumed were observed in winter flounder inhabiting sites in the Humber Arm and at Norris Point. Slightly more plant material was noted in flounder sampled at Birchy Cove (55%) than in fish at Summerside (48%). The frequency of occurrence of amphipods (~68%), bivalve molluscs (~35%) and polychetes (~30%) was similar at both sites. Winter flounder taken at Norris Point fed more often on polychetes (72%) and decapod crustaceans (52%) than on amphipods (12%) and bivalve molluscs (6%) or fish (2%). Polychetes (78%) and bivalves (46%) were the predominant food of flounder captured at Meadows.

Several species of metazoan parasites were observed in the lumen of the gastrointestinal tract of flounder collected at the two sites in the Humber Arm but were rare or absent in specimens from Norris Point (Table 2). Digenea (primarily *Hemiurus* spp., *Derogenes varicus*, and *Steringophorus furciger*), larval ascaridoid nematodes (*Phocascaris* and *Anisakis* spp.) and the acanthocephalan *Echinorhynchus gadi* were similar in prevalence and abundance in samples of fish taken from Birchy Cove and Summerside (Table 2). However, both the prevalence and abundance of the cestode *Bothriocephalus claviceps* were significantly greater in specimens taken from Birchy Cove than at Summerside whereas none was observed in flounder examined at Norris Point.

DISCUSSION

Results from the present study suggest that the health of winter flounder captured in the vicinity of the pulp and paper mill at Birchy Cove was impaired more often than in samples from Summerside and that fish health was generally not impaired at the reference site (Norris Point). It was also reported recently that flounder living near another pulp mill in Newfoundland that uses the sulfite bleaching process were affected more than samples from Birchy Cove (Khan et al. 1992, 1994). Skin ulcers, fin necrosis and condition factor were significantly different in fish taken at Port Harmon than at Birchy Cove (Khan et al. 1992). Our present study has also shown that, despite its up-current location, Summerside might not represent an appropriate effluent-free reference site since some of the flounder displayed macroscopic and microscopic lesions. A report on the Humber Arm refers to the Summerside area as a site that is slightly affected by outfall from the paper mill in contrast to the severe impact at Birchy Cove (Ledrew et al. 1989). Possibly fish migration across the inlet and/or drift of the outfall during periods of strong north easterly winds in the direction of Summerside might represent the underlying causes of the lesions. However, similarity of the Humber Arm to Bonne Bay as inlets from the Gulf of St. Lawrence and minimal to no macroscopic and microscopic lesions in flounder suggest that the site at Norris Point is not influenced by pollution and could conceivably be considered in the present study an adequate reference site. Meadows might also represent a potential reference site based on the absence of lesions in flounder and its designation as an area not influenced by outfall from the paper mill (Ledrew et al. 1989).

Table 2. Mean (± S.E.) abundance and prevalence (%) of intestinal helminths in subsamples of *P. americanus* collected from Birchy Cove and Summerside, Humber Arm and Norris Point, Bonne Bay in 1992-1993.

Parasite	Birchy Cove n = 32	Summerside n = 31	Norris Point n = 36
Digenea	1.09 ± 0.73	2.68 ± 1.01	1.21 ± 0.61
(%)	(15.6)	(25.8)	(14.4)
Cestoidea	1.97 ± 0.65**	0.10 ± 0.07	0
(%)	(46.9)***	(6.5)	(0)
Nematoda	0.53 ± 0.41	0.45 ± 0.19	0.11 ± 0.05**
(%)	(15.6)	(19.4)	(8.1)
Acanthocephala	1.25 ± 0.62	0.77 ± 0.36	0.21 ± 0.10**
(%)	(34.4)	(22.6)	(6.4)***

** $p < 0.05$
*** $p < 0.001$

The livers of flounder sampled at Birchy Cove displayed both non-proliferative and proliferative lesions. The presence of MMCs, lipidosis and pericholangial fibrosis has been noted previously in winter flounder captured at Port Harmon (Khan et al. 1994), Boston Harbor (Bodammer and Murchelano 1990; Murchelano and Wolke 1991; Moore 1991) and to some extent at Long Island Sound (Gronlund et al. 1991). We did not observe, however, a high prevalence of necrotic lesions or basophilic and eosinophilic foci as noted in samples from Port Harmon (Khan et al. 1994) or Boston Harbor (Murchelano and Wolke 1991; Moore 1991). However, proliferative lesions, primarily clear cell foci, were observed in flounder taken at Birchy Cove. Similar lesions were also seen in fish from Port Harmon (Khan et al. 1994), Boston Harbor (Murchelano and Wolke 1991; Moore 1991) and Long Island Sound (Gronlund et al. 1991). Both sites in New England are repositories for several environmental contaminants such as PAHs, PCBs, chlorinated pesticides and heavy metals which originate from urban and industrial discharges and shipping (Moore 1991; Gronlund et al. 1991). These lesions observed in species of flounder which inhabit estuaries and inshore waters where the sediment serves as a repository for anthropogenic compounds might be induced by a variety of contaminants acting singly or collectively. Several authors (Murchelano and Wolke 1991; Moore 1991; Gronlund et al. 1991) have identified these foci as degenerative hydropic cells but both a recent study (Khan et al. 1994) and the present study suggest that the cells contain lipid as determined by an oil red 0 positively stained material at the light microscopic level and osmiophilic inclusions at the fine structural level. Their occurrence appears to be more widespread in larger than smaller (<25 cm in length) fish (Khan et al. 1994). An in-depth study based on different ages of winter flounder would be essential to determine the occurrence and prevalence of the hepatic lesions at Birchy Cove.

The occurrence of multifocal hemosiderosis in the spleen and kidney of winter flounder captured at Birchy Cove coincided with low hemoglobin values. In contrast, hemosiderin concentration was lower and hemoglobin levels higher in samples from Summerside and Norris Point. These lesions occur in the tissues of fish after excessive destruction of erythrocytes, biodegradation and sequestration of the end-products which includes hemosiderin in macrophages (Khan and Nag 1993). Hemosiderosis has also been reported in winter flounder taken near another pulp and paper mill in Newfoundland but the concentration of the pigment in the spleen of fish of comparable length (21-25 cm) was considerably greater (~2x) than in samples from Birchy Cove (Khan et al. 1994) and in other species of fish exposed to pollutants (Haensly et al. 1982; Bowser et al. 1990; Santos et al. 1990; Pulsford et al. 1992). Both mills in Newfoundland release several resin acids in the effluent, primarily dehydroabietic acid, which can cause liver dysfunction and increased hemolysis (Mattsoff and Oikari 1987; Räbergh et al. 1992). The presence of high levels of bilirubin in the blood plasma and jaundice in fish exposed to resin acids (Mattsoff and Oikari 1987) suggests that these compounds might represent the underlying causes of anemia and multifocal hemosiderosis observed in winter flounder captured near the two pulp mills in Newfoundland.

Hyperplasia of the branchial epithelium occurred more often in flounder from Birchy Cove than samples from either Summerside or Norris Point. It was not, however, as severe as noted in fish taken at Port Harmon (Khan et al. 1994). Some of these observations have been reported previously in fish following exposure to BKME (Lehtinen et al. 1984; Couillard et al. 1988) and sublethal levels of a variety of pollutants (Mallatt 1985). In contrast to other studies of excessive gill hyperplasia and associated parasitism in fish exposed to or living near to pulp mills (Lehtinen et al. 1984; Barker et al. 1994), no protozoan parasites were observed infesting the gills of winter flounder in the present or a previous study (Khan et al. 1994). However, metacercariae of C. lingua appeared to be similar in prevalence and abundance in samples from Birchy Cove and Summerside but were absent at Norris Point and Meadows.

Differences in the parasite fauna between flounder caught at Birchy Cove and Summerside were minimal as only the prevalence and abundance of the cestode B. claviceps were significantly different. Previous studies have reported different densities of some parasites such as metacercariae of C. lingua and the gastrointestinal acanthocephalan E. gadi captured near and some distance from another pulp mill in Newfoundland (Khan et al. 1992; Barker et al. 1994). Barker (1993) noted that although the abundance of the two parasites was low in spring and higher in late summer, infestation of C. lingua was greater and E. gadi less in fish taken in the vicinity of the mill. It was suggested that changes in host physiology affected susceptibility to the ectoparasite and reduced feeding or ingestion of pulp mill waste culminated in voiding of the enteric parasites. Since several studies have reported a connection between exposure to pollutants and parasitic levels (Khan and Thulin 1991), it is likely that migration of flounder across the inlet from sulfite-laden sediment at Birchy Cove to the sandy bottom at Summerside might have transpired. Samples taken at sites more distant from the mill might clarify this aspect of the present study.

Delayed ovarian development was observed in winter flounder captured at Birchy Cove (present study) and at Port Harmon (Barker 1993) in Newfoundland. Prior to the onset of spawning, an increase in GSI is associated with ovarian hydration (Harmin and Crim 1992) and this was probably responsible for the elevated values observed in the sample from Birchy Cove. It is likely that the GSI values in flounder at Birchy Cove might have been lower than those at Summerside and Norris Point if the time of collection had been earlier. A number of studies have reported reduced gonadal sizes and presumably impaired development in perch (*Perca fluviatilis*) (Andersson et al. 1988), whitefish (*Coregonus clupeaformis*) (Munkittrick et al. 1992a) and white suckers (McMaster et al. 1991). In another report, the results on sexual maturation were inconclusive (Hodson et al. 1992). It has been shown that in BKME-exposed fish, decreased levels of plasma sex steroids occurred in both sexes (McMaster et al. 1991; Munkittrick et al. 1992a). Neither secondary treatment of the BKME effluent nor temporary shutdown of mill operations altered the levels of the plasma steroids in fish living in the exposure area (Munkittrick et al. 1992b). Reduced development of secondary sexual characteristics in male white suckers was also associated with BKME exposure (McMaster et al. 1991). Increased levels of mixed function oxygenase (MFO) enzymes appear to coincide with decreased plasma steroid hormonal levels (McMaster et al. 1991; Munkittrick et al. 1991). However, recent studies suggest no direct correlation exists between some sex steroids such as 17β-estradiol and MFO induction in female fish (Gagnon et al. 1994). Future studies on winter flounder inhabiting the Humber Arm should focus on plasma steroid levels in relation to ovarian development (Harmin 1991).

It is evident from the present study that although no significant differences in condition factor were apparent, epidermal anomalies and tissue lesions occurred more often in winter flounder collected near the pulp mill in the Humber Arm than at the two other sites. Moreover, the lesions observed in fish taken from the affected site were similar to those reported from samples captured in the vicinity of another pulp and paper mill which is also located in another inlet but is not under the influence of urban and industrial discharges (Khan et al. 1992, 1994; Barker 1993; Barker et al. 1994). Low lymphocyte levels noted in fish collected from Birchy Cove suggest impaired defense mechanisms and susceptibility to opportunistic organisms which subsequently could cause the epidermal lesions (Lindesjoo and Thulin 1990). Infection of organs, in addition to the gastrointestinal tract, by *G. stephani* might also be associated with immunosuppression. It is apparent, then, that an integrative histopathological approach to evaluate fish health, based on changes in specific organs and tissues, appears to be a reliable method to detect minimal alterations following exposure to xenobiotic agents (Adams 1990; Hinton and Lauren 1990). Future work on the assessment of fish health in the Humber Arm should focus on the examination of samples taken from sites more distant and upcurrent from the mill.

ACKNOWLEDGMENTS

This study was supported by a grant to R.A.K. from the Natural Sciences and Engineering Council of Canada. We are grateful to Mr. R. Ficken for photography, Ms. S. Kenny for typing, Mr. K. Nag for determination of hemosidcrin levels, Ms. S. Billiard for technical assistance and Dr. W. Threlfall for reviewing the manuscript.

REFERENCES

Adams, S.M. 1990. Status and use of biological indicators for evaluating the effects of stress on fish. In: S.M. Adams [ed.] Biological indicators of stress in fish. Amer. Fish. Soc. Symp. 8:1-8.

Adams, S.M., W.D. Crumby, M.S. Greeley Jr., L.R. Shugart and C.F. Saylor. 1992. Responses of fish populations and communities to pulp mill effluents: A holistic assessment. Ecotoxicol. Environ. Safety 24:347-360.

Andersson, T., B.-E. Bengtsson, L. Förlin, J. Härdig and A. Larsson. 1987. Long-term effects of bleached kraft mill effluents on carbohydrate metabolism and hepatic xenobiotic biotransformation enzymes in fish. Ecotoxiocol. Environ. Safety 13:53-60.

Andersson, T., L. Förlin, J. Härdig and A. Larsson. 1988. Physiological disturbances in fish living in coastal water polluted with bleached kraft pulp mill effluents. Can. J. Fish. Aquat. Sci. 45:1525-1536.

Axelsson, B. and L. Norrgren. 1991. Parasite frequency and liver anomalies in three-spined stickleback, *Gasterosteus aculeatus* (L.), after long-term exposure to pulp mill effluents in marine mesocosms. Arch. Environ. Contam. Toxicol. 21:505-513.

Barker, D.E. 1993. Evidence of chronic stress in winter flounder, *Pleuronectes* (=*Pseudopleuronectes*) *americanus* living adjacent to a pulp and paper mill in St. George's Bay, western Newfoundland. M.Sc. Thesis, Memorial University of Newfoundland, St. John's, Newfoundland, 101 pp.

Barker, D.E., R.A. Khan, E.M. Lee, R.G. Hooper and K. Ryan. 1994. Anomalies in sculpins (*Myoxocephalus* spp.) sampled near a pulp and paper mill. Arch. Environ. Contam. Toxicol. 26:491-496.

Bengtsson, B.-E., A. Bengtsson and V. Jjärnlund. 1988. Effects of pulp mill effluents on vertebrae of fourhorn sculpin *Myoxocephalus quadricornis*, bleak *Alburnus alburnus*, and perch *Perca fluviatilis*. Arch. Environ. Contam. Toxicol. 17:789-797.

Bodammer, J.E. and R.A. Murchelano. 1990. Cytological study of vacuolated cells and other aberrant hepatocytes in winter flounder from Boston Harbor. Cancer Res. 50:6744-6756.

Bowser, P.R., D. Martineau, R. Sloan, R. Brown and C. Carusone. 1990. Prevalence of liver lesions in brown bull heads from a polluted site and a nonpolluted reference site on the Hudson River, New York. J. Aquat. Animal Hlth. 2:177-181.

Bucher F., F. Hofer and W. Salvenmoser. 1992. Effects of treated pulp mill effluents on hepatic morphology in male bullhead (*Cottus gobio* L.). Arch. Environ. Contam. Toxicol. 23:410-419.

Couillard, C.M., R.A., Berman and J.C., Panisset. 1988. Histopathology of rainbow trout exposed to a bleached kraft pulp mill effluent. Arch. Environ. Contam. Toxicol. 17:319-323.

Gagnon, M.M., J.J. Dodson, P.V. Hodson, G. Van Der Kraak and J.H. Carey. 1994. Seasonal effects of bleached kraft mill effluent on reproductive parameters of white sucker (*Catostomus commersoni*) populations of St. Maurice River, Quebec, Canada. Can. J. Fish. Aquat. Sci. 51:337-347.

Gronlund, W.D., S.L. Chan, B.B. McCain, R.C. Clark Jr., M.S. Myers, J.E. Stein, D.W. Brown, J.T. Landahl, M.M. Krahn and U. Varanasi. 1991. Multidisciplinary assessment of pollution at three sites in Long Island Sound. Estuaries 14:299-305.

Haensly, W.E., J.M. Neff, J.R. Sharp, A.C. Morris, M.E. Bedgood and P.D. Boem. 1982. Histopathology of *Pleuronectes platessa* L. from Aber Wrac'h and Aber Benoit, Brittany, France: Long-term effects of the Amoco Cadiz crude oil spill. J. Fish Dis. 5:365-391.

Hall, T.J., R.K. Haley and L.E. LaFleur. 1991. Effects of biologically treated bleached kraft mill effluent on cold water stream productivity in experimental stream channels. Environ. Toxicol. Chem. 10:1051-1060.

Hall, T.J., R.K. Haley, D.L. Borton, A.H. Walsh and R.E. Wolke. 1992. Histopathology of rainbow trout (*Oncorhynchus mykiss*) after long-term exposure to biologically treated bleached kraft mill effluent in experimental stream channels. Can. J. Fish. Aquat. Sci. 49:939-944.

Härdig, J.T., T. Andersson, B.E. Bengtsson, J.L. Förlin and A. Larsson. 1988. Long-term effects of bleached kraft mill effluents on red and white blood cell status, ion balance, and vertebral structure in fish. Ecotoxicol. Environ. Safety 15:96-106.

Harmin, S.A. 1991. Studies of seasonal reproductive cycles and hormonal control of reproduction in winter flounder, *Pseudopleuronectes americanus* Walbaum. Ph.D. Thesis, Department of Biology, Memorial University of Newfoundland, St. John's, 314 pp.

Harmin, S.A. and L.W. Crim. 1992. Gonadotropic hormone-releasing hormone analog (GnRH-A) induced ovulation and spawning in female winter flounder, *Pseudopleuronectes americanus* (Walbaum). Aquaculture 104:375-390.

Hinton, D.E. and D.J. Lauren. 1990. Integrative histopathological approaches to detecting effects of environmental stressors on fish. Am. Fish. Soc. Symp. 8:51-66.

Hodson, P.V., R.K. McWhirter, B. Gray, D. Thiverge, J.H. Carey, G. Van Der Kraak, M. Whittle and M.C. Levesque. 1992. Effects of bleached kraft mill effluent on fish in the St. Maurice River, Quebec. Environ. Toxicol. Chem. 11:1635-1651.

Khan, R.A and K. Nag. 1993. Estimation of hemosiderosis in seabirds and fish exposed to petroleum. Bull. Environ. Contam. Toxicol. 50:125-131.

Khan, R.A. and J. Thulin. 1991. Influence of pollution on parasites of aquatic animals. Adv. Parasitol. 30:201-238.

Khan R.A., D. Barker, R. Hooper and E.M. Lee. 1992. Effect of pulp and paper effluent on a marine fish, *Pseudopleuronectes americanus*. Bull. Environ. Contam. Toxicol. 48:449-456.

Khan, R.A., D.E. Barker, R. Hooper, E.M. Lee, K. Ryan and K. Nag. 1994. Histopathology in winter flounder (*Pleuronectes americanus*) living adjacent to a pulp and paper mill. Arch. Environ. Contam. Toxicol. 26:95-102.

Ledrew, Fudge and Associates. 1989. Humber Arm monitoring program. 1988 Sampling. St. John's, Newfoundland, 86 pp.

Lehtinen, K-J., M. Notini and L. Landler. 1984. Tissue damage and parasite frequency in flounders *Platichthys flesus* chronically exposed to bleached kraft pulp mill effluents. Ann. Zool. Fenn. 21:23-28.

Lindesjoo, E. and J. Thulin. 1990. Fin erosion of perch (*Perca fluviatilis*) and ruffe (*Gymnocephalus cernua*) in a pulp mill effluent area. Dis. Aquat. Org. 8:119-126.

Lindstrom-Seppä, P. and A. Oikari. 1990. Biotransformation and other toxicological and physiological responses in rainbow trout (*Salmo gairdneri*) caged in a lake receiving effluents of pulp and paper industry. Aquat Toxicol. 26:187-204.

Mallatt, J. 1985. Fish gill structural changes induced by toxicants and other irritants: A statistical review. Can. J. Fish. Aquat. Sci. 42:630-648.

Mattsoff, L. and A. Oikari. 1987. Acute hyperbilirubinaemia in rainbow trout (*Salmo gairdneri*) caused by resin acids. Comp. Biochem. Physiol. 88C:263-268.

McLeay, D. and Associates Ltd. 1987. Aquatic toxicity of pulp and paper mill effluent: A review. Minister of Supply and Services, Ottawa, Canada, 191 pp.

McMaster, M.E., G.J. Van Der Kraak, C.B. Portt, K.R. Munkittrick, P.K. Sibley, I.R. Smith and D.G. Dixon. 1991. Changes in hepatic mixed-function oxygenase (MFO) activity, plasma sterol levels and age at maturity of a white sucker (*Catostomus commersoni*) population exposed to bleached kraft mill effluent. Aquat. Toxicol. 21:199-218.

Moore, M.J. 1991. Vacuolation, proliferation and neoplasia in the liver of Boston Harbor winter flounder (*Pseudopleuronectes americanus*). Ph.D. thesis, Woods Hole Oceanographic Institution, Massachusetts Institute of Technology, Boston, WHOI 91-28, 266 pp.

Munkittrick, K.R., C.B. Portt, G.J. Van Der Kraak, I.R. Smith, D.A. Rokosh. 1991. Impact of bleached kraft mill effluent on population characteristics, liver MFO activity, and serum steroid levels of a Lake Superior white sucker (*Catostomus commersoni*) population. Can. J. Fish. Aquat. Sci. 48:1371-1380.

Munkittrick, K.R., M.E. McMaster, C.B. Portt, G.J. Van Der Kraak, I.R. Smith and D.G. Dixon. 1992a. Changes in maturity, plasma, sex, steroid levels, hepatic mixed-function oxygenase activity, and the presence of external lesions in lake whitefish (*Coregonus clupeaformis*) exposed to bleached kraft mill effluent. Can. J. Fish. Aquat. Sci. 49:1560-1569.

Munkittrick, K.R., G.J. Van Der Kraak, M.E. McMaster and C.B. Portt. 1992b. Reproductive disfunction and MFO activity in three species of fish exposed to bleached kraft mill effluent at Jackfish Bay, Lake Superior. Water Poll. Res. J. Canada 27:439-446.

Murchelano, R.A. and R.E. Wolke. 1991. Neoplasms and nonneoplastic liver lesions in winter flounder, *Pseudopleuronectes americanus*, from Boston Harbor, Massachusetts. Environ. Health Perspect. 90:17-26.

Myers, M.S., L.D. Rhodes and B.B. McCain. 1987. Pathologic anatomy and patterns of occurrence of hepatic neoplasia, putative preneoplastic lesions, and other idiopathic hepatic conditions in English sole (*Parophrys vetulus*) from Puget Sound, Washington. J. Natl. Cancer Inst. 78:333-363.

Myllyvirta, T.P. and P.J. Vuorinen. 1989. Avoidance of bleached kraft mill effluent by pre-exposed *Coregonus albula*. L. Water. Res. 23:1219-1227.

Neuman, E. and P. Karås. 1988. Effects of pulp mill effluent on a Baltic coastal fish community. Water Sci. Technol. 20:95-106.

Oikari, A., B. Holmbom, E. Änäs, M. Miilunpalo, G. Kruzynski and M. Castren. 1985. Ecotoxicological aspects of pulp and paper mill effluents discharged to inland water system: Distribution in water, and toxicant residues and physiological effects in caged fish (*Salmo gairdneri*). Aquat. Toxicol. 6:219-239.

Pulsford, A.L., K.P. Ryan and J.A. Nott. 1992. Metals and melanomacrophages in flounder, *Platichthys flesus*, spleen and kidney. J. Mar. Biol. Assoc. U.K. 72:483-498.

Råbergh, C.M.I., B. Isomaa and J.E. Eriksson. 1992. The resin acids dehydroabietic acid and isopimaric acid inhibit bile acid uptake and perturb potassium transport in isolated hepatocytes from rainbow trout (*Onchorhynchus mykiss*). Aquat. Toxicol. 23:169-180.

Sandström, O. and G. Thoresson. 1988. Mortality in perch populations in a Baltic pulp mill effluent area. Mar. Pollut. Bull. 19:564-567.

Santos, M.A., F. Pires and A. Hall. 1990. Metabolic effects of kraft mill effluents on the eel, *Anguilla anguilla* L. Ecotoxicol. Environ. Safety 20:10-19.

Sokal, R.R. and F.J. Rohlf. 1987. Introduction to biostatistics. 2nd ed. W.H. Freeman, New York, 363 pp.

Takvorian, P.M. and A. Cali. 1984. Seasonal prevalence of the microsporidan, *Glugea stephani* (Hagenmuller), in winter flounder, *Pseudopleuronectes americanus* (Walbaum), from the New York-New Jersey Lower Bay Complex. J. Fish Biol. 24:655-663.

Waldichuk, M. 1990. Dioxin pollution near pulp mills. Mar. Pollut. Bull. 21:365-366.

Whittle, D.M. and K.W. Flood. 1977. Assessment of the acute toxicity, growth impairment, and flesh-tainting potential of bleached kraft mill effluent on rainbow trout (*Salmo gairdneri*). J. Fish. Res. Board Canada 34:869-878.

EFFECTS OF PULP MILL EFFLUENT ON BENTHIC FRESHWATER INVERTEBRATES: FOOD AVAILABILITY AND STIMULATION OF INCREASED GROWTH AND DEVELOPMENT

Richard B. Lowell, Joseph M. Culp, Frederick J. Wrona and Max L. Bothwell

National Hydrology Research Institute, Environment Canada, 11 Innovation Boulevard, Saskatoon, Saskatchewan, S7N 3H5 Canada

The potential short- and long-term responses of benthic river invertebrates to biotreated bleached kraft pulp mill effluent are illustrated in a review of two recent experimental and statistical analyses. To determine short-term responses, mayflies (*Baetis tricaudatus* Dodds) were exposed in artificial streams to control river water, 1% effluent, or 10% effluent while controlling periphyton food levels (low versus high food availability). After 2 wk, exposure to the effluent stimulated significant increases in growth and development of the experimental animals. This stimulation was greater than could be accounted for by differences in algal food availability. This suggests that the effluent may have (1) increased the nutritive value of the food, (2) stimulated an increase in mayfly feeding, and/or (3) directly stimulated increased mayfly growth by way of hormonal or other metabolic effects. Multivariate analysis (ordination) of the abundances of benthic invertebrates sampled over a 20-year period downriver of the pulp mill revealed that five families of stoneflies, caddisflies, and mayflies (including Baetidae) were more abundant in years when the mill output of suspended solids was greater. This positive correlation should be viewed with caution, however, until more direct evidence is available on the mechanism of the long-term effects of pulp mill effluent on river invertebrates.

INTRODUCTION

Biologically treated bleached kraft pulp mill effluent has the potential for both inhibitory (e.g., toxicity) and stimulatory (e.g., nutrient enhancement) effects on river invertebrates. These effluents contain a diverse array of compounds (e.g., metals, chlorophenolics, resin acids, polycyclic aromatic hydrocarbons), some of which have been shown to have deleterious sublethal or lethal effects (McLeay 1987). In contrast to these directly toxic compounds, however, pulp mill effluents often also contain high levels of phosphorus and nitrogen which act as algal nutrients. The enrichment effect of these nutrients can lead to enhanced productivity in the receiving water ecosystem (Hannson 1987; Feder and Pearson 1988; Hall *et al.* 1991). At present, little data is available to tease out the combined toxic and nutrient enhancement effects of pulp mill effluents (Solomon *et al.* 1993). We summarize and compare evidence from two studies on the Thompson River in British Columbia suggesting an overall stimulatory effect by biologically treated pulp mill effluent on certain benthic river invertebrates, such as mayflies.

Experimental Measures of Short-Term Growth and Development

In a pilot study designed to determine the relative contribution and nature of the contaminant versus nutrient enhancement effects, we conducted a factorial experiment measuring the survival, growth, and developmental responses of the mayfly *Baetis tricaudatus* Dodds (Ephemeroptera: Baetidae) to different concentrations of pulp mill effluent under two different feeding regimes. *B. tricaudatus* is one of the more abundant benthic macroinvertebrates in the Thompson River. It grazes on periphyton (Scrimgeour *et al.*

1991), making this mayfly a promising species for measuring the combined effects of nutrient loading (which affects food availability to the mayfly) and toxicity.

Full details of the methods and results of this study are given by Lowell et al. (1995a). Briefly, *B. tricaudatus* were exposed to three effluent concentrations (control river water, 1% effluent, 10% effluent) and two food levels (low, high) for 2 wk within artificial streams arranged in a 2x3 factorial design. The outdoor experimental stream enclosure was located on the Lower Thompson River beside the Weyerhaeuser bleached kraft pulp mill in Kamloops, British Columbia. The mill currently employs chlorine dioxide substitution in addition to biological treatment of the effluent. During periods of low flow (November-March), effluent concentrations in the Thompson River are typically 1% following complete mix. The 10% treatment was more typical of concentrations closer to the mill before complete mixing. Food was provided to the mayflies during the experiment by periphyton grown on ceramic blocks. The low versus high food treatment levels were controlled by how long the mayflies were allowed to feed on the periphyton-covered blocks. The experiments were run in arrays of circular Plexiglas artificial streams (diameter = 8.8 cm, stream bottom area = 50 cm^2) (Lowell et al. 1995b). Small water jets produced the current in each stream; these were driven by pumps drawing water from mixing reservoirs. Each of the six experimental treatments included 7 replicate streams (42 streams in all). The growth response of the mayflies to the effluent was determined by measuring dry body weights and number of molts at the end of the experiment. In addition, the relative length of the wing pads can be used as a measure of the degree of development of aquatic insect larvae as they mature (Clifford 1970; Clifford et al. 1979). This was measured as the ratio of length to spread of the wing pads.

The levels of most of those contaminants that were detected in the effluent were low. The measured contaminants included chlorophenolics, resin acids, polycyclic aromatic hydrocarbons, and several metals (Lowell et al. 1995a). Following dilution, the concentrations were at or below those federal or provincial water quality guideline levels that were available. In contrast, phosphorus, which is a limiting nutrient in the Thompson River (Bothwell et al. 1992), was at a great enough concentration in the effluent to stimulate increased algal growth in the streams. For example, the concentration of soluble reactive phosphorus in the full-strength effluent was 1116 $\mu g\ L^{-1}$ as compared to $\leq 6\ \mu g\ L^{-1}$ in the river water upstream of the mill outfall.

Although survival was not significantly affected (Fig. 1; $p > 0.1$ for comparisons of control versus experimental treatments using analysis of variance [ANOVA]), the effluent had a significant stimulatory effect on growth and development. For example, dry body weight was 20-50% greater for mayflies exposed to the effluent (Fig. 2; $p \leq 0.005$). Moreover, the initial growth and development trajectories suggested that mayflies that are exposed to effluent may emerge sooner and at a larger size than non-exposed individuals. This trend was investigated using analysis of covariance for plots of size (body weight) versus degree of development (wing length/spread). For example, within the high food treatment, the slope for the 1% effluent-exposed mayflies was significantly greater than for the control mayflies (Fig. 3; $p < 0.02$). The conclusion that effluent-exposed mayflies will emerge at a larger size is preliminary, however, due to the restricted time frame of the experiment and the amount of data overlap.

In addition, 1% effluent caused more frequent molting than 10% effluent (Fig. 4; $p < 0.05$ for ANOVA comparisons of 1% versus 10% experimental treatments). This suggests that as effluent concentration is increased from 1% to 10%, inhibitory effects may begin to mask the stimulatory effects observed at lower concentrations.

Long-Term Patterns in Benthic Invertebrate Community Structure

The growth stimulation study indicated that pulp mill effluent stimulated increased short-term growth and development of mayflies. This, in turn, raises the question of what the long-term effects of the effluent would be on benthic invertebrate populations in the Thompson River. This question can, in part, be addressed using the results of a multivariate analysis of benthic invertebrate samples taken yearly over a period of 20 yr at Walhachin, a site approximately 50 km downriver of the mill (more detailed methods and results in Lowell et al. in prep.). This was a period over which mill pulp production was gradually increased from 791 to 1056 air dry tonnes d^{-1}. The increased production was generally associated with increased mill total phosphorus load to the river, although this was partially offset by decreased sewage loading from the City of Kamloops (Bothwell et al. 1992).

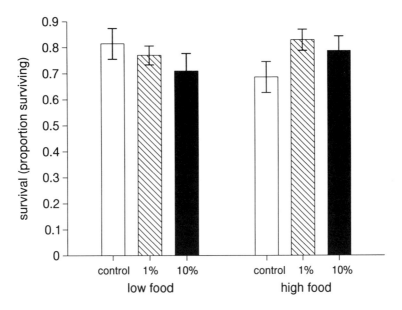

Figure 1. Mayfly survival at the end of the experiment at two food levels (low, high) and three concentrations (control river water, 1% and 10% effluent) (± 1 SE).

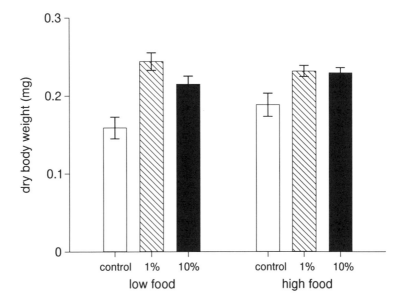

Figure 2. Dry body weight of mayflies at the end of the experiment at two food levels (low, high) and three concentrations (control river water, 1% and 10% effluent) (± 1 SE).

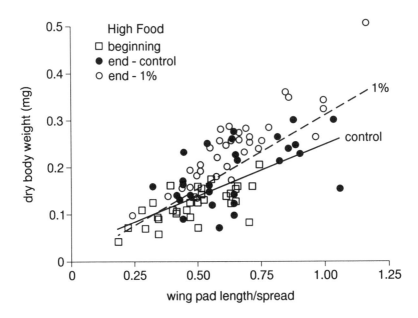

Figure 3. Body size (dry weight) versus degree of development (wing pad length/spread) for mayflies at the beginning (pretreatment) and end (control river water and 1% effluent treatments) of the experiment at the high food level. Solid line indicates regression for pooled beginning and end-control mayflies. Dashed line indicates regression for pooled beginning and end-1% mayflies.

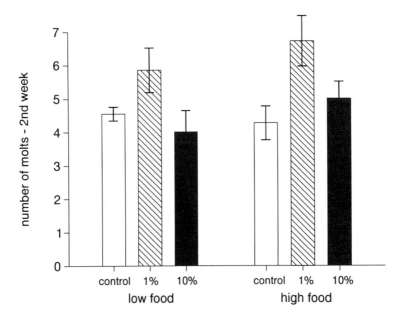

Figure 4. Number of molts produced by mayflies during the second week at two food levels (low, high) and three concentrations (control river water, 1% and 10% effluent) (± 1 SE).

To summarize, macroinvertebrates were sampled during February-March from 1973 to 1992 within 0.18-m^2 plots using a 0.2-mm mesh net to sample to a depth of 6-12 cm into the gravel substratum. For the purposes of this analysis, taxa were identified to the level of family for the insects and class, subclass, order, or suborder for the other taxa. Data were expressed as number of individuals per square meter present for each taxa, averaged over the three to six replicate samples obtained during each sampling period. The data set was then arranged into an object (year) by attribute (taxa) matrix and subjected to ordination analysis using the PATN multivariate statistical package (version 3.5, CSIRO, Lyneham, A.C.T., Australia). Belbin (1992) discusses more fully the rationale and advantages of the statistical approach described below.

The original data matrix contained 33 taxa. Before analysis, non-benthic taxa were removed from the matrix. As is often practiced in multivariate analyses, rare taxa not occupying unique roles in the community were also excluded to lower the ordination stress values to levels low enough to meaningfully summarize the multivariate data set in two dimensions. In addition, Harpacticoida, Ostracoda, and Acarina were removed because their small size may have resulted in inconsistent counts among years. Following this editing, 15 insect families and 5 other taxa (Hydroida, Turbellaria, Nematoda, Hirudinea, Oligochaeta) were included in the analysis.

Those extrinsic data that were available for most of the 20-year period were also arranged into a similarly constructed data matrix to incorporate into the analysis at a later stage of analysis (principal axis correlation below). These extrinsic variables included: temperature, total dissolved phosphorus, and single-wavelength color in the river at Walhachin; river discharge at the Spence's Bridge monitoring station approximately 50 km downriver of Walhachin; and total suspended solids and biochemical oxygen demand monitored in the Weyerhaeuser pulp mill effluent.

The multivariate data set was then reduced to two dimensions to simplify data interpretation. Initially, the invertebrate raw data matrix was transformed into a matrix of association values (Bray-Curtis dissimilarity coefficients). Then the matrix of association values was subjected to multidimensional scaling ordination, which finds the best fit of the objects (years) into a two-dimensional ordination space.

Principal axis correlation (PCC option of PATN) was performed on the resulting ordination to determine which taxa and extrinsic variables best accounted for the positioning of years within the ordination space. This is a multiple-linear regression technique that determines the linear relationship between the taxa or extrinsic data and the ordination space. It provides a vector within the ordination space giving the direction of best fit for a given taxon or extrinsic variable and the correlation coefficient (r) indicating the goodness of fit of that vector. This approach is similar to correlating individual taxa or extrinsic variables to ordination axes but is more powerful because it also indicates the direction of best fit and is not dependent upon the positioning of the axes.

The following one extrinsic variable and eight taxa best accounted for the positioning of years within the ordination space ($r > 0.65$, $p < 0.01$). The mill output of effluent suspended solids (measured in the same month that the invertebrate samples were taken) was greater in years on one end of the ordination space. These tended to be years after 1982, although the distribution of years within the ordination space was not always chronologically uniform. Five families of mayflies (Baetidae, Ephemerellidae, Heptageniidae), stoneflies (Perlodidae), and caddisflies (Hydropsychidae) were also more abundant in years on the same end of the ordination space, years when the output of effluent solids was higher. In contrast, Nematoda and Oligochaeta were more abundant in years on the opposite end of the ordination space (generally before 1983), years when the output of effluent solids was lower. Finally, the vector for midges (Chironomidae) was at right angles to the other significant vectors; that is, chironomid abundance was fairly independent of the output of effluent solids and the abundances for the other taxa.

The data suggest that the effluent could have caused increases in some of the mayfly, stonefly, and caddisfly families, but that the effluent impact may not have been great enough to lead to greater abundances of taxa often thought of as "polluted water" invertebrates, such as the nematodes and oligochaetes. These results should be interpreted with caution, however, because any conclusions are necessarily based on correlation rather than experimentally verified causation. For example, other causative extrinsic variables could have covaried with the mill output of effluent solids and these other variables may have been the actual driving force behind the changes in community structure observed over the 20-year period.

DISCUSSION

Interestingly, the experimentally measured stimulatory effects of the pulp mill effluent on *B. tricaudatus* occurred within both the low and high food treatments. This means that the effluent-exposed mayflies grew faster than observed even for the high food control animals, which already had access to more food than they could eat during the experiment. Thus, the stimulatory effects were greater than could be accounted for by the increase in the quantity of available food caused by phosphorus fertilization of benthic algae. This result suggests that the effluent may have (1) increased the nutritive value of the food, (2) stimulated an increase in mayfly feeding, and/or (3) directly stimulated increased mayfly growth by way of hormonal or other growth-related metabolic effects.

These three potential mechanisms, as well as the long-term patterns, can be partly evaluated in light of a few related studies. Regarding the first mechanism, the nutritive content of the food could be increased either in the algal cells themselves or in the microbial community and detritus intermingled with the algae. This raises the question of whether the mayflies (and other benthic invertebrates) were actually consuming significant quantities of material derived from the pulp mill effluent in addition to being exposed via the surrounding medium. This question has been addressed in a parallel pilot study examining stable isotope ratios in these invertebrates and the algal biofilm as compared to the ratios in the Thompson River up- and downstream of the outfall for the pulp mill. The relevant results of this study (Wassenaar and Culp 1996) can be summarized as follows. Naturally present stable isotopes of carbon and nitrogen sometimes occur in different proportions in anthropogenic pollutants as compared to unpolluted waters (Van Dover et al. 1992). When this happens, the isotopic signatures of ^{13}C and ^{15}N can potentially be used to trace the fate of these pollutants as they are incorporated into the food web. Background levels of $\delta^{13}C$ (using typical delta notation for isotopic composition given as ‰) in the Thompson River ranged from -7.1 to -8.1‰, whereas strong carbon isotopic depletion was observed just below the pulp mill outfall (-12.9‰). This suggests that 23-34% of the dissolved inorganic carbon (DIC) at this site was derived from mill effluent, assuming that the effluent originally had a signature close to that for terrestrially derived organic matter (-25 to -30‰) (Wassenaar and Culp 1995).

The algal biofilm 35-100 km downriver of the mill had an isotopic composition of -14 to -17‰, which would be expected following isotopic fractionation given that the primary algal carbon source was from background riverine DIC (LaZerte and Szalados 1982). The composition of the benthic invertebrates ranged from -18‰ for algal grazing chironomids to -27‰ for filter feeding trichopterans, with ephemeropterans and plecopterans falling in between; $\delta^{13}C$ for *B. tricaudatus* was -22‰. This depletion in the $\delta^{13}C$ values relative to the algal biofilm suggests that these insects consumed 51-76% terrestrially derived organic carbon. The $\delta^{15}N$ data also suggested that a large proportion (60-80%) of the insects' food was derived from a terrestrial source, thus supporting the calculations based on the carbon isotope data. Since the Lower Thompson is an open canopy river, a significant proportion of this terrestrially derived material could have been derived from pulp mill effluent. Thus, one possible interpretation of the experimental growth stimulation results and the long-term patterns in community structure is that the effluent increased the availability of high-quality detrital and/or microbial food to the mayflies and other grazers.

In reference to the possible stimulation of increased feeding, a pair of previous studies indirectly suggest that bleached kraft pulp mill effluent may stimulate an increase in the feeding rate of coho salmon (McLeay and Brown 1974, 1979). Salmon exposed to biotreated effluent and provided with an excess of food pellets exhibited an increased growth rate. But when the food pellets offered to the salmon were restricted to 70% of their satiation level, thereby preventing the salmon from exhibiting an increased feeding rate, the growth stimulation was reduced.

The wide variety of compounds found in pulp mill effluent also have the potential of directly stimulating increased growth or otherwise affecting the development of benthic invertebrates. For instance, many types of plants are known to contain insect hormones, antihormones, or their pharmacobiological mimics (Slama 1979). Woody plants, such as the trees used in pulp mills, commonly contain two key classes of compounds that affect insect growth: juvabione-type compounds (juvenile hormone) and ecdysone-type compounds (molting hormone). Some of the major components of the neutral fractions of effluents derived from pine, fir, and spruce are juvabione, juvabiol, and dehydrojuvabione (Leach et al. 1975). During our short-term growth experiments, the Weyerhaeuser mill used pulpwood derived from lodgepole pine (35%), douglas fir (25%), engelmann spruce (20%), cedar (10%), and smaller amounts of balsam fir and hemlock (W. Pehowich, personal communication). Other plant compounds that may have antihormonal effects include diterpenes that are related to the resin acid abietic acid (Slama 1979).

Growth stimulation may also occur due to hormesis, a widely observed response of organisms exposed to particularly low levels of chemicals that are normally toxic at high concentrations (Boxenbaum et al. 1988). Hormesis has been observed within a variety of taxa, such as bacteria, yeast, protists, algae, higher plants, nematodes, insects, and vertebrates; typical responses include increases in growth, development, reproductive success, disease resistance, and longevity. Some of the types of compounds that have been shown to cause hormesis are inorganic salts and acids, heavy metals, and a variety of organic compounds. For instance, crickets are known to exhibit increased growth when exposed to each of several kinds of pesticides applied in small doses ranging from 0.1 to 0.001 of the LD_{100} (Luckey 1968). Although the mechanism(s) causing hormesis are not yet well understood, an association has been observed between growth stimulation and protein turnover in some aquatic invertebrates (D.J. Baird, personal communication; Barber et al. 1990). Protein turnover may be increased, even at low toxicant concentrations, to repair the damage that would have occurred had the concentration of the toxicant been higher. This may shift more resources into structural materials and increased growth, possibly at the expense of energy storage and reproductive output.

The potential for interacting growth and reproductive effects should be considered before extrapolating the results of our short-term experiments to expected long-term patterns. The long-term patterns in community structure that we measured did indicate a positive correlation between the abundances of some families of mayflies, stoneflies, and caddisflies and the suspended solids output of the mill. At this point, however, any conclusions about the long-term effects of the effluent remain tentative until more direct evidence is available on the mechanism of these effects.

ACKNOWLEDGMENTS

These studies benefitted from the help of several people, and we would like to extend thanks to Monique Dube, Colin Gray, Don Holmes, Sue Burton, Nancy Glozier, Daryl Halliwell, Kevin Himbeault, Peter Jones, Steve Josefowich, Eric Marles, Wayne Pehowich, and Garnet Richards, in addition to Weyerhaeuser Canada for access to their facilities. Funding was provided by Environment Canada through the Fraser River Action Plan and the National Hydrology Research Institute, as well as by the British Columbia Ministry of Environment, Lands, and Parks.

REFERENCES

Barber, I., D.J. Baird and P. Calow. 1990. Clonal variation in general responses of *Daphnia magna* Straus to toxic stress. II. Physiological effects. Func. Ecol. 4:409-414.

Belbin, L. 1992. PATN pattern analysis package: technical reference. CSIRO, Lyneham, A.C.T., Australia, 235 p.

Bothwell, M.L., G. Derkson, R.N. Nordin and J.M. Culp. 1992. Nutrient and grazer control of algal biomass in the Thompson River, British Columbia: a case history of water quality management, pp. 253-266. *In* R.D. Robarts and M.L. Bothwell [eds.] Aquatic ecosystems in semi-arid regions: implications for resource management. National Hydrology Research Institute Symposium Series 7, Environment Canada, Saskatoon, SK.

Boxenbaum, H., P.J. Neafsey and D.J. Fournier. 1988. Hormesis, Gompertz functions, and risk assessment. Drug Metabolism Rev. 19:195-229.

Clifford, H.F. 1970. Analysis of a northern mayfly (Ephemeroptera) population, with special reference to allometry of size. Can. J. Zool. 48:305-316.

Clifford, H.F., H. Hamilton and B.A. Killins. 1979. Biology of the mayfly *Leptophlebia cupida* (Say) (Ephemeroptera: Leptophlebiidae). Can. J. Zool. 57:1026-1045.

Feder, H.M. and T.H. Pearson. 1988. The benthic ecology of Loch Linnhe and Loch Eil, a sea-loch system on the west coast of Scotland. V. Biology of the dominant soft-bottom epifauna and their interaction with the infauna. J. Exp. Mar. Biol. Ecol. 116:99-134.

Hall, T.J., R.K. Haley and L.E. LaFleur. 1991. Effects of biologically treated bleached kraft mill effluent on cold water stream productivity in experimental stream channels. Environ. Toxicol. Chem. 10:1051-1060.

Hansson, S. 1987. Effects of pulp and paper mill effluents on coastal fish communities in the Gulf of Bothnia, Baltic Sea. Ambio 16:344-348.

LaZerte, B. and J.E. Szalados. 1982. Stable carbon isotope ratio of submerged freshwater macrophytes. Limnol. Oceanogr. 27:413-418.

Leach, J.M., A.N. Thakore and J.F. Manville. 1975. Acute toxicity to juvenile rainbow trout (*Salmo gairdneri*) of naturally occurring insect juvenile hormone analogues. J. Fish. Res. Board Can. 32:2556-2559.

Lowell, R.B., J.M. Culp and M.L. Bothwell. (in preparation). Long-term patterns in benthic invertebrate community structure in the Lower Thompson River, British Columbia. NHRI Contribution, National Hydrology Research Institute, Environment Canada, Saskatoon, SK.

Lowell, R.B., J.M. Culp and F.J. Wrona. 1995a. Stimulation of increased short-term growth and development of mayflies by pulp mill effluent. Environ. Toxicol. Chem. 14: In press.

Lowell, R.B., J.M. Culp and F.J. Wrona. 1995b. Toxicity testing with artificial streams: effects of differences in current velocity. Environ. Toxicol. Chem. 14: In press.

Luckey, T.D. 1968. Insect hormoligosis. J. Econ. Entomol. 61:7-12.

McLeay, D.J. 1987. Aquatic toxicity of pulp and paper mill effluent: a review. EPS 4/PF/1, Report, Environment Canada, Ottawa, ON.

McLeay, D.J. and D.A. Brown. 1974. Growth stimulation and biochemical changes in juvenile coho salmon (*Oncorhynchus kisutch*) exposed to bleached kraft pulp mill effluent for 200 days. J. Fish. Res. Board Can. 31:1043-1049.

McLeay, D.J. and D.A. Brown. 1979. Stress and chronic effects of untreated and treated bleached kraft pulp mill effluent on the biochemistry and stamina of juvenile coho salmon (*Oncorhynchus kisutch*). J. Fish. Res. Board Can. 36:1049-1059.

Scrimgeour, G.J., J.M. Culp, M.L. Bothwell, F.J. Wrona and M.H. McKee. 1991. Mechanisms of algal patch depletion: importance of consumptive and non-consumptive losses in mayfly-diatom systems. Oecologia 85:343-348.

Slama, K. 1979. Insect hormones and antihormones in plants, pp. 683-700. *In* G.A. Rosenthal and D.H. Janzen [eds.] Herbivores: their interaction with secondary plant metabolites. Academic Press, Orlando, FL.

Solomon, K., H. Bergman, R. Hugget, D. Mackay and B. McKague. 1993. A review and assessment of the ecological risks associated with the use of chlorine dioxide for the bleaching of pulp. Report, Science Review Panel, Alliance for Environmental Technology, Erin, ON.

Van Dover, C.L., J.F. Grassle, B. Fry, R.H. Garritt and V.R. Starczak. 1992. Stable isotope evidence for entry of sewage derived organic material into a deep-sea food web. Nature 360:153-156.

Wassenaar, L.I., and J.M. Culp. 1996. The use of stable isotopic analyses to identify pulp mill effluent signatures in riverine food webs. *In* M.R. Servos, K.R. Munkittrick, J.H. Carey and G. Van Der Kraak (Eds.). Environmental fate and effects of pulp and paper mill effluents, St. Lucie Press, Delray Beach, FL.

COMPARATIVE EVALUATION OF MACROBENTHIC ASSEMBLAGES FROM THE SULPHUR RIVER ARKANSAS IN RELATION TO PULP MILL EFFLUENT

Chet F. Rakocinski[1], Michael R. Milligan[2], Richard W. Heard[1]
and Thomas L. Deardorff[3]

[1]Gulf Coast Research Laboratory, P.O. Box 7000, Ocean Springs, MS 39566 U.S.A.
[2]Center for Systematics and Taxonomy, P.O. Box 37534, Sarasota, FL 34278 U.S.A.
[3]International Paper - Erling Riis Research Laboratory, P.O. Box 2787, Mobile, AL 36652 U.S.A.

In October 1992 during low-flow conditions, we conducted an evaluation of spatial variation in macrobenthic assemblages from the lower Sulphur River system (Texas and Arkansas border) in relation to bleached kraft mill effluent (BKME). We adopted a multifaceted approach using detailed taxonomic data and various quantitative methods. A classification analysis based on assemblage similarity showed a close relationship between the effluent site and a neighboring site located 2.3 river miles (3.7 km) downstream of the effluent. Although the farthest upstream and downstream sites were separated by 14.0 river miles (22.5 km) as well as by intrinsic habitat differences, they were closely linked in the classification analysis. Complementary variation in relative abundances of three taxonomic core groups (tubificids, chironomids, and ceratopogonids) helped explain the classification analysis. Furthermore, the macrobenthic composition of the outfall canal resembled that of neighboring river sites. High total densities of macrobenthic organisms greater than 3400 m^{-2} occurred at both the effluent site and the neighboring downstream site. Parallel among-site variation in tubificid abundances implied that differences in this taxonomic group might be driving among-site variation in total densities. The predominant macrobenthic pattern suggested organic enrichment as a possible chronic effect of BKME on macrobenthic assemblages. Nonetheless, a relatively high species richness and typical taxonomic diversity occurred at the effluent site. This study illustrates the utility of taking a whole-community approach using detailed taxonomic data to identify chronic macrobenthic effects of pulp mill effluents.

INTRODUCTION

Macrobenthic assemblages provide excellent indicators of aquatic conditions (Wiederholm 1984; Gray *et al.* 1992; Reice and Wohlenberg 1993). Bottom-dwelling organisms occur in direct contact with sediments, which are the ultimate repositories of most pollutants (Reice and Wohlenberg 1993). Because macrobenthic organisms occur on a relatively small spatial scale (<1 m^2) and since many are relatively sessile, they cannot readily avoid pollution (Gray *et al.* 1992; Reice and Wohlenberg 1993). The temporal scale of macrobenthic life histories, ranging from several months to several years, appropriately reflects chronic environmental stress through population responses that can be detected for some time after an impact (Reice and Wohlenberg 1993). Population responses by different species can vary and may be either inhibitory or stimulatory. Resultant graded responses by macrobenthic assemblages allow assessments of impact severity, as macrobenthic abundance and diversity are greatly reduced from extreme stress (McNulty 1970; Wiederholm 1984). Although *in situ* biotic surveys cannot demonstrate mechanisms of stress, responses by macrobenthic assemblages integrate the effects of multiple causal agents associated with pollution sources (Wiederholm 1984). Of course, dynamic natural environmental variation must also be carefully considered (Ferraro *et al.* 1991), as well as any other potentially confounding factors.

In October 1992, we conducted a single macrobenthic survey of the lower Sulphur River near the Arkansas/Texas border, between the Wright Patman Dam and the confluence of the Sulphur and Red

Rivers. The purpose of this survey was to appraise spatial variation in macrobenthic assemblages among several sites along the lower Sulphur River. Of particular interest was the possible influence of bleached kraft mill effluent (BKME) from the International Paper (IP) Texarkana Mill outfall within the study reach. Although lotic systems continually change throughout their course, we expected mid-reach sites near the outfall to stand out against other gradients if they were impacted by BKME. Although this macrobenthic survey was limited to a one-time comparative evaluation of spatial variation, it was conducted during low-flow conditions, which were ideal for effective benthic sampling of this regulated lotic system. This study illustrates the utility of high taxonomic resolution, careful benthic sampling, and multifaceted data analyses for resolving clear macrobenthic spatial patterns in relation to a potential impact.

STUDY AREA

Since late 1972, IP has been operating a large 630 ton per day bleached kraft paper mill near Texarkana, Texas. This mill uses a chlorine bleaching sequence. Wastewater from the mill averages around 28 million gallons per day and is treated in an aeration basin and held in a 10,000 acre-foot pond for up to 1 yr before being released into the Sulphur River. Eventually, BKME is discharged through an outfall canal (IPOC) into the Sulphur River. BKME from kraft mills is typically discolored and carries considerable suspended solids (Scrimgeour 1989).

On the Sulphur River, the flow regime between Wright Patman Lake and the Red River is completely regulated by the Wright Patman dam (Fig. 1). Effluent discharge rates from the IP mill are typically restricted to a small fraction (<5%) of the discharge being released from the reservoir. Due to discharge fluctuations, water levels below the dam can vary by as much as 6 meters. Much of the surrounding watershed is made-up by extensive bottomland hardwood swamp. In upper and mid-reaches the substrate consists of silty mud with organic detritus, whereas in lower reaches it consists of scoured muddy clay.

For the quantitative macrobenthic survey, six sites on the Sulphur River and one site on Days Creek were selected, encompassing a wide area around the mill outfall (Table 1; Fig. 1). Locations of most sites were chosen to coincide with former studies of this area. These sites were located between 16.5 river miles (26.5 km) and 30.5 river miles (49.1 km) above the confluence of the Sulphur and Red Rivers. Henceforth, we will refer to these sites by their river mile distances either upstream or downstream of the IP outfall as follows: 5.2 miles (8.4 km) upstream (5.2U), 0.7 miles (1.1 km) upstream (0.7U), immediately adjacent to the IP outfall (0.0E), 2.3 miles (3.7 km) downstream (2.3D), 8.0 miles (12.9 km) downstream (8.0D), 8.5 miles (13.7 km) downstream on Days Creek (8.5D), and 8.8 miles (14.2 km) downstream (8.8D). The IPOC was sampled roughly one-quarter mile (0.4 km) above the Sulphur River; however, due to the abundant pulpy substrate this site was only qualitatively examined. Another potentially polluted small tributary to the Sulphur River, Days Creek (8.5D), also was sampled roughly one-quarter mile (0.4 km) above the Sulphur River as a counterpart of the outfall canal.

METHODS

Sampling Design

Field sampling was conducted between 19 October and 23 October 1992. River flow was low during the sampling period, with discharge from the Wright Patman dam averaging only 1000 cfs and mill effluent contributing about 3% of that amount. On 19 October, the sites were located and water quality measured at four of the sites which were being monitored for a complementary study. Water quality data included water temperature, dissolved oxygen, pH, conductivity, and photosynthetically active radiation (a measure of light attenuation). Benthic sampling proceeded between 20 October and 23 October 1992. In conjunction with benthic sampling, weather conditions and habitat characteristics were noted, including current velocity, depth, and sediment composition. Surface current velocities were measured by timing a semi-buoyant vial and sediments characterized in the field by visual inspection of dominant and subdominant constituents.

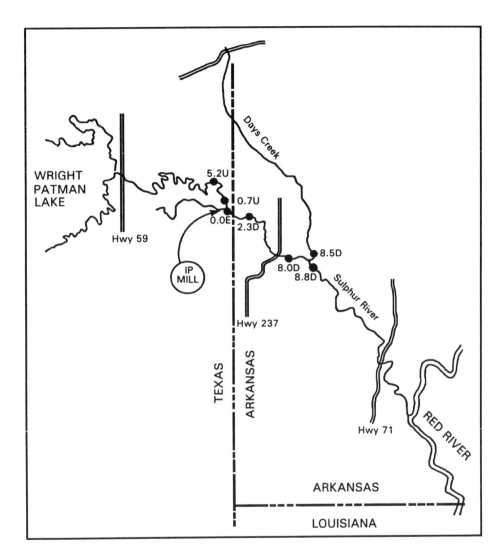

Figure 1. Map of Sulphur River area showing sites located between 49.1 and 26.6 km above the confluence of the Sulphur and Red Rivers. Sites are labeled as river mile distances upstream or downstream relative to the mill outfall. Map not drawn to scale.

A consistent benthic sampling protocol was employed at all sites which were always placed along the depositional bank of the river. Macrobenthos were sampled using a pole-mounted 6-inch-square (2.32×10^{-2} m^2) Ekman dredge. At each site, five stations were placed at 12.5-m intervals along the river bank. Stations were located away from the bank at a standard depth of 1.0 m. Three grabs (7×10^{-2} m^2) were combined to make up one sample for each station.

Table 1. Site characteristics for macrobenthic survey from the lower Sulphur River, 19-23 October 1992.

Site	Location	Cur Vel (m sec^{-1})	Dpth (m)	Temp (°C)	DO (mg L^{-1})	pH	Cond (μS)	PAR (Surf) (x 1000) Up	PAR (Surf) (x 1000) Down	PAR (1 m) (x 1000) Up	PAR (1 m) (x 1000) Down	Substrate	Comments
5.2U	Sulphur River - 5.2 river miles (8.4 km) upstream of IP outfall	0.25	1.0	20.5	8.4	7.7	150	2.50	0.15	0.15	0.01	silty mud with plant detritus and some fine sand	west depositional bank - just downstream of submerged branches along gradual bend in river
0.7U	Sulphur River - 0.7 river miles (1.1 km) upstream of IP outfall	0.40	1.0	NA	NA	NA	NA	NA	NA	NA	NA	silty mud with plant detritus and some fine sand	east depositional bank - along gradual bend in river - <1.2 km upstream of outfall
0.0E	Sulphur River - adjacent to IP outfall	0.14	1.0	20.5	6.0	7.7	365	1.00	0.50	0.50	<0.01	silty mud, very soft	southwest depositional bank - adjacent to outfall, between boat launch (20 m from closest station) and outfall (10 m from farthest station) - along gradual bend in river - slow clockwise eddy within the sample area and slow current caused by spit blocking the outfall - water colored red-brown
IPOC	IP outfall canal	<0.10	0.7	NA	NA	NA	NA	NA	NA	NA	NA	granular brown pulpy detritus with mud	sampled in mid-canal - water colored deep red-brown - low discharge rate from outfall (<1000 cfs) - abundant granular pulpy detritus necessitated qualitative sampling
2.3D	Sulphur River - 2.3 river miles (3.7 km) downstream of IP outfall	0.17	1.0	20.5	7.8	7.6	185	5.00	0.01	0.04	<0.01	fine silty mud with plant detritus	west depositional bank along gradual bend in river - first station downstream of railroad trestle, beyond old bridge pilings
8.0D	Sulphur River - 8.0 river miles (12.9 km) downstream of IP outfall	0.20	1.0	20.5	8.4	7.6	185	2.00	0.24	0.10	0.01	fine silty mud with plant detritus	west depositional bank - just downstream of Hwy 237
8.5D	Days Creek - 8.5 river miles (13.7 km) downstream of IP outfall	0.08	1.0	NA	NA	NA	NA	NA	NA	NA	NA	hard clay with some mud	north depositional bank - roughly 0.4 km up into Days Creek, along bend - hard clay substrate not conducive to sampling with Ekman grab
8.8D	Sulphur River - 8.8 river miles (14.2 km) downstream of IP outfall	0.20	1.0	NA	NA	NA	NA	NA	NA	NA	NA	hard red clay with some mud	west depositional bank - roughly 0.4 km below Days Creek - drop-off close to shore - river relatively scoured in this section, closer to the Red River confluence

Cur Vel = current velocity; Dpth = depth; Temp = water temperature; Cond = conductivity; PAR = photosynthetically active radiation; Surf = surface; Up = probe facing surface; Down = probe facing bottom.

All samples were transported to the IP boat ramp, and beginning with the furthermost upstream site, samples from each site were fully processed before proceeding to the next downstream site. Samples from each station were processed separately to allow estimates of within-site variability. Benthic samples were completely sifted through a 0.5-mm standard sieve to remove fine particulate material. Remaining material, including large particulate detritus and benthic organisms, was preserved in 10% formalin, labeled, and returned to the laboratory for benthic processing.

Laboratory Methods

At the laboratory, macrobenthic organisms were sorted from remaining material, placed into general taxonomic groups, and transferred to 70% ethanol. All samples were completely sorted, except those from the IPOC, which were sorted for the predetermined period of 4 h per sample. We followed a quality control protocol, whereby one station from each site was randomly chosen to check for at least 90% sorting accuracy. Once sorted, macrobenthic organisms were identified to the lowest possible taxonomic level, usually species, and enumerated for each station. All enumerated information was subsequently entered into DBase III+ files and checked using a standard procedure.

Data Analysis

Several common ecological indices were employed to describe variation in macrobenthic assemblage structure, including species richness (S), evenness (J') (natural log scale), diversity (H') (natural log scale), total density, and geometric mean density (natural log scale). Species richness is a direct count of the number of taxonomic categories identified from each site. A taxon is regarded as the lowest discernable taxonomic category (usually species) to which a specimen can be unambiguously attributed. Evenness expresses how equally the total number of organisms is distributed among the various recognized taxa. Diversity is a measure of the interaction of both species richness and evenness. Total density is the average number of macrobenthic organisms per unit area (m^{-2}), whereas the geometric mean density is the average of log-transformed densities.

For this study, the following macrobenthic indices were calculated:

Species richness (S) no. taxa × $0.35\ m^{-2}$
Total density Σ (no. organisms × m^{-2}) × no. stations^{-1}
Geometric mean density Σ (ln (N + 1) organisms × m^{-2}) × no. stations^{-1}

Along with accompanying 95% confidence intervals of two standard errors, geometric mean densities were compared among sites. Bartlett's test was used to test homogeneity of variance in log-transformed densities among sites. Because variances were heterogeneous, pairwise comparisons of geometric mean densities were made using the Games and Howell method (Sokal and Rohlf 1981). A Spearman rank correlation also was run between the geometric mean total density and its variance to examine dispersion.

Estimates of H' and J' were calculated following Pielou (1975). This method produces a curve of successive values for Brillouin's index (H_k), from which H', J', and their variances are estimated. Each estimate of H' was taken as the average of H_k values remaining after the H_k curve leveled off and fluctuated by no more than 0.2. Five sets of estimates were made for each site, and midpoint values for H', J', and their variances were used as final estimates.

To obtain an overview of assemblage similarities among the seven quantitative sites, a classification analysis was performed using detailed taxonomic data. Two steps included, (1) the calculation of a similarity matrix and (2) the classification analysis. To accomplish these steps, we used the software package Community Analysis System (CAS) (Bloom 1992). Three different classifications were run to ensure a robust interpretation. In the first classification, we took a direct approach by including all taxonomic categories and by using untransformed data for all seven sites. Total numbers of every taxon from each site were included in the first similarity matrix, and matrix elements were calculated using the proportional similarity index. The sorting algorithm Group Averaging was used to perform the classifications. The classifications generated dendrograms depicting groups of sites at various levels of assemblage similarity. A second classification was performed using log-transformed data (i.e., ln (N+1)) from which immature tubificids were excluded to downplay numerical dominance. Finally, a third classification was done exactly as the second, except it excluded the two lower river sites having distinctive clay substrate (i.e., 8.5D [Days Creek] and 8.8D).

RESULTS

Habitat Characteristics

Physical data were generally comparable among sites (Table 1). Of note was the comparatively low dissolved oxygen and high conductivity at the effluent site (0.0E). Surface current velocity along the Sulphur River was fastest (0.4 m × sec^{-1}) at site 0.7U and slowest (0.14 m × sec^{-1}) at site 0.0E. Velocities at other sites on the Sulphur River averaged 0.2 m × sec^{-1}. An especially slow surface current velocity of less than 0.1 m × sec^{-1} occurred at Days Creek (8.5D).

Most sites had a substrate consisting of silty mud and partly decomposed plant detritus (Table 1). The fine silty sediment at site 0.0E was exceptionally depositional. Unlike most sites, 8.8D and 8.5D had a characteristic substrate of hard red clay and mud. The substrate from the IPOC was characterized by an abundance of pulpy granular detritus.

Macrobenthic Indices

Macrobenthic assemblages in the Sulphur River study area were fairly diverse, constituting 75 taxa, including various aquatic insects, worms, crustaceans, and mollusks (Appendices 1 and 2). Two important taxonomic groups were the Diptera with 27 taxa and the Oligochaeta with 16 taxa. A total of 5038 macrobenthic organisms were enumerated from the seven quantitative sites. Three core groups included two dipteran families, Chironomidae (11.3%) and Ceratopogonidae (3.6%), as well as the oligochaete family Tubificidae (68.8%).

Species richness (S) was similar among most sites, ranging from 29 to 39 taxa, except at site 8.0D, where S only reached 19 taxa (Table 2; Fig. 2). The highest S value of 39 taxa occurred at site 0.0E, located on the Sulphur River adjacent to the IP outfall.

Table 2. Values for macrobenthic indices from quantitative sites sampled during October 1992 (geometric mean density values are presented on the natural log scale).

Site	Species Richness (S)	Avg Total Density (no. × m^{-2})	Geometric Mean Density	Standard Error Geom Mean	Estimated Diversity (H′)	Standard Deviation ln Est H′	Estimated Evenness (J′)	Standard Deviation ln Est J′
5.2U	37	2167	7.63	0.14	1.87	0.29	0.52	0.08
0.7U	30	1441	6.92	0.44	1.99	0.36	0.58	0.11
0.0O	39	3588	8.17	0.09	2.05	0.15	0.56	0.04
2.3D	30	3421	8.13	0.06	1.73	0.14	0.51	0.04
8.0D	19	1214	7.02	0.21	1.47	0.12	0.50	0.04
8.5D	33	620	6.02	0.46	2.95	0.18	0.84	0.05
8.8D	37	2049	7.52	0.23	2.23	0.50	0.62	0.14

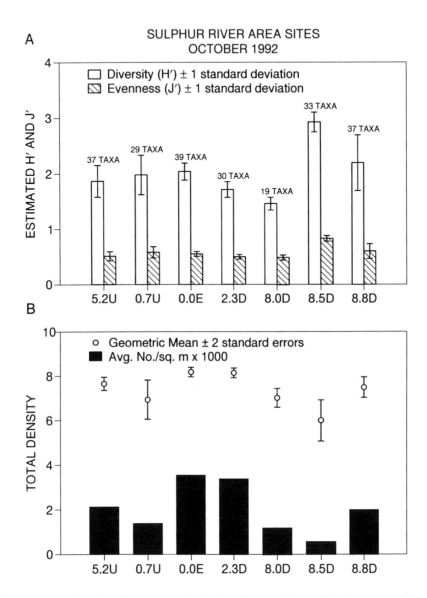

Figure 2. Graphs of macrobenthic indices for quantitative sites. Species richness (S) values occur above diversity bars in A. Geometric mean density values are plotted on the natural log scale in B.

Total densities varied widely among sites, from 620 to 3587 × m^{-2}. The highest total densities of macrobenthic organisms (>3400 × m^{-2}) occurred at sites 0.0E and 2.3D on the Sulphur River, while the lowest total density (620 × m^{-2}) occurred at site 8.5D on Days Creek. Tubificid abundances paralleled among-site variation in total densities, indicating that this core group was mainly driving variation in total densities (Fig. 3; Appendices 1 and 2). The magnitude of variation in log-transformed total densities was unequal among sites ($\chi^2 = 20$, $p < 0.005$) and inversely correlated with the geometric mean ($r_s = -0.93$, $p < 0.02$). Nevertheless, based on 95% confidence limits (i.e., 2 standard errors), geometric mean densities were significantly higher at sites 0.0E and 2.3D, because their lower confidence limits did not overlap with upper limits from any of the other sites (Allan 1984). Multiple comparisons of geometric means among all possible pairs of sites by the Games and Howell Method, however, failed to detect any other significant among-site differences in total densities ($p > 0.05$).

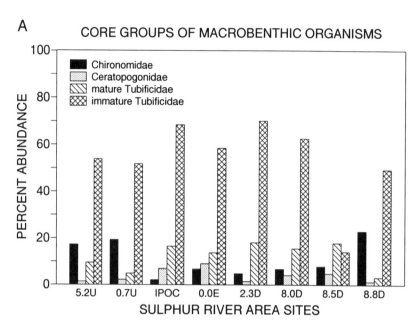

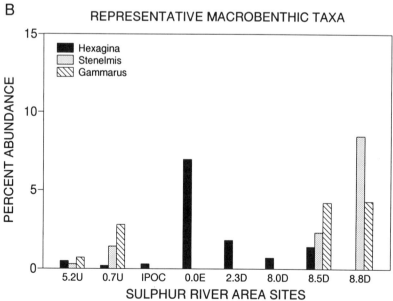

Figure 3. Among-site variation in relative abundances of core groups and selected macrobenthic taxa.

Estimated evenness (J') ranged from 0.50 at site 8.0D to 0.84 at site 8.5D (Days Creek), and estimated diversity (H') likewise spanned a twofold range, from 1.47 at site 8.0D to 2.95 at site 8.5D (Days Creek) (Fig. 2; Table 2). High diversity reflects corresponding high S and/or evenness, and high evenness often reflects low total density as well as low numerical dominance by any particular taxon. Notwithstanding relatively high densities at sites 0.0E and 2.3D, evenness and diversity estimates for these two sites were similar to those upstream of the IPOC. An interaction between high evenness and relatively low densities produced markedly high diversities at the two most downstream sites, 8.5D and 8.8D. The relatively low diversity at site 8.0D, however, was attended by both low S and low total density.

Classification Analysis

Results from the first classification paralleled differences in tubificid abundances and total densities. Three pairs of sites were linked at 70% assemblage similarity, including sites 0.0E and 2.3D, sites 0.7U and 8.0D, and sites 5.2U and 8.8D. Four of the sites from both above and below the outfall were similar at the 60% level, including sites 5.2U, 0.7U, 8.0D, and 8.8D. The 8.5D site at Days Creek was classified as uniquely distinct, showing less than 25% similarity with other sites.

The second classification based on log-transformed data from which immature tubificids were excluded produced a consistent picture of assemblage similarity among the same seven sites (Fig. 4). A mid-reach cluster of three sites still was separated from the other cluster of both upstream and downstream sites at the 40% level. The same two pairs of sites that were linked at 70% assemblage similarity in the first run now were closely linked near the 60% similarity level, including sites 0.0E and 2.3D, as well as sites 5.2U and 8.8D. In the second classification, however, site 8.0D was linked with the former pair of sites at the 51% level, while site 0.7U was linked with the latter pair of sites at the 58% level. The Days Creek site (8.5D) still was distinctly classified but now was weakly linked (47% similarity) with sites 0.7U, 5.2U, and 8.8D. Upon exclusion of the two lower sites (8.5D and 8.8D), the third classification retained former similarity relationships among the remaining sites. Sites 0.0E and 2.3D still were linked at the 61.7% level, while the two sites above the IPOC (5.2U and 0.7U) were now linked at the 57.1% level and separated from the other sites.

A close examination of relative abundances of the three core groups, Chironomidae, Ceratopogonidae, and Tubificidae, helped to interpret the pattern of assemblage similarity shown by the classification analysis (Fig. 3; Appendices 1 and 2). For example, high tubificid abundances at sites 0.0E and 2.3D occurred for all tubificid taxa, including several taxa that were not very abundant at other sites. Relative abundances of several other taxa, such as *Hexagenia*, *Stenelmis*, and *Gammarus*, also corresponded with the observed macrobenthic pattern. Extreme upstream (5.2U) and the downstream (8.8D) sites were most similar to each other in relative abundances of core group and other taxa, explaining their linkage in the classification dendrogram. Finally, relative abundances of core group and other taxa in the IPOC were most similar to those of neighboring sites, 0.0E and 2.3D.

DISCUSSION

Most macrobenthic organisms cannot readily move to avoid unfavorable conditions (Gray *et al.* 1992; Reice and Wohlenberg 1993). Thus, severe impacts to macrobenthic assemblages may reduce species richness and diversity by inhibiting taxa that have a low tolerance to stress. In this study, neither species richness nor diversity was found to be notably low near the IPOC. Instead, these macrobenthic indices were high where the IPOC joined the river, but markedly low at a site 12.9 km downstream (8.0D), which may have been stressed by sources other than mill effluent.

Macrobenthic spatial patterns among sites along the Sulphur River suggest that organic enrichment from BKME may have influenced macrobenthic assemblage structure. This interpretation is supported by the pattern of among-site -variation in tubificid abundances. High relative abundances of tubificids occurred within the IPOC and at neighboring sites on the Sulphur River. Altered relative abundances of macrobenthic taxa, including stimulatory responses by tubificids, are indicative of mildly polluted lotic conditions (Wiederholm 1984). Indeed, most of the common tubificids found in this study are opportunistic and favored by organic enrichment (Brinkhurst 1966). Increased tubificid abundances also often indicate organic enrichment effects from BKME (Scrimgeour 1989). It is still surprising, however, that a strong macrobenthic signal could be detected in such a complex, variable, and sedimentary lotic system as the lower Sulphur River. BKME is typically high in suspended solids, which can contribute to sedimentation of organic materials in receiving waters. Resultant stimulatory responses by tubificids can reflect both elevated food levels and tolerance to such conditions as high bio chemical oxygen demand.

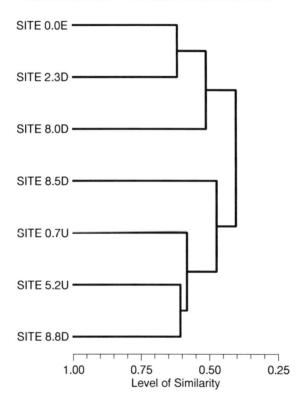

Figure 4. Classification dendrogram of quantitative sites based on macrobenthic assemblage similarities. This run included all quantitative sites and used log-transformed data (i.e., ln (N+1)) from which immature tubificids were excluded.

Some differences in macrobenthic assemblages were undoubtedly associated with natural habitat variation, a factor which always should be considered when assessing environmental impacts (Ferraro et al. 1991). For example, because *Hexagenia* nymphs burrow in fine depositional sediment, it is not surprising that this mayfly was so abundant at the sheltered river site near the mill effluent (0.0E), where such sediments accumulated. Lotic habitats change continually along a natural gradient (Vannote et al. 1980). In this study, characteristic clay/mud substrata at two lower reach sites indicative of a lotic gradient (8.5D [Days Creek] and 8.8D) corresponded with low total densities, high evenness, and high diversities of macrobenthic organisms. Accordingly, *Axarus* sp. was the dominant chironomid larva at the farthest downstream site (8.8D), while it was not prominent elsewhere. Extraneous anthropogenic variation also might obscure point-source effects of concern. For example, particularly low S and H' values suggestive of such extraneous stress occurred at one Sulphur River site (8.0D).

Notwithstanding such natural and extraneous variation, certain dissimilarities in macrobenthic assemblages did indicate effects from BKME. Despite the obvious habitat differences between the farthest upstream and downstream sites (5.2U and 8.5D), they still were closely linked in the classification analysis, indicating that their macrobenthic assemblages were relatively similar within the context of all quantitative sites. Parallel assemblage similarities between the two mid-reach sites near the mill effluent presumably reflected its distinctive influence. Consequently, the predominant macrobenthic pattern did not follow a natural gradient interpretation.

The pattern of macrobenthic assemblage structure expressed through detailed taxonomic data, careful sampling, and a multifaceted data analysis effectively revealed the influences of BKME. In this case, effects were mainly stimulatory, although macrobenthic community data certainly may reflect inhibitory effects as well, particularly when interpreted in light of experimental information. Indeed, the detection of such biotic patterns should accompany experimental studies of mechanisms. Moreover, such whole-community patterns must be related to lower level responses in order to validate their ecological significance.

ACKNOWLEDGMENTS

Dr. Robin M. Overstreet assisted in the coordination of this project and contributed valuable interpretive insights as well as editorial advice. Ms. Susan Powell also contributed valuable editorial assistance. Personnel of the Invertebrate Zoology and Parasitology Sections of GCRL provided needed technical assistance in the performance of this project. Dr. John H. Rodgers kindly let us use his physical data. This study was supported by International Paper.

REFERENCES

Allan, J.D. 1984. Hypothesis testing in ecological studies of aquatic insects, p. 484-507. *In* V.H. Resh and D.M. Rosenberg [ed.] The Ecology of Aquatic Insects. Praeger, New York, NY.
Bloom, S.A. 1992. The Community Analysis System 4.2. Ecological Data Consultants, Inc., Archer, FL. 134 p.
Brinkhurst, R.O. 1966. The Tubificidae (Oligochaeta) of polluted waters. Inter. Verein. Theor. Angew. Limnol. Verh. 16:854-859.
Ferraro, S.P., R.C. Swartz, F.A. Cole and D.W. Schults. 1991. Temporal changes in the benthos along a pollution gradient: Discriminating the effects of natural phenomena from sewage-industrial wastewater effects. Estuar. Coastal Shelf Sci. 33:383-407.
Gray, J.S., A.D. McIntyre and J. Štirn. 1992. Manual of methods in aquatic environment research. Part 11. Biological assessment of marine pollution with particular reference to benthos. FAO Fisheries Technical Paper. No. 324. Rome, FAO. 49 p.
McNulty, J.K. 1970. Effects of abatement of domestic sewage pollution on the benthos, volumes of zooplankton, and the fouling organisms of Biscayne Bay, Florida. Stud. Trop. Oceanogr. Miami 9:1-107.
Pielou, E.C. 1975. Ecological Diversity. John Wiley & Sons, New York, NY. 165 p.
Reice, S.R., and M. Wohlenberg. 1993. Monitoring freshwater benthic macroinvertebrates and benthic processes: Measures for assessment of ecosystem health, p. 287-305. *In* D.M. Rosenberg and V.H. Resh [ed.] Freshwater Biomonitoring and Benthic Macroinvertebrates. Chapman & Hall, New York, NY.
Scrimgeour, G.J. 1989. Effects of bleached kraft mill effluent on macroinvertebrate and fish populations in weedbeds in a New Zealand hydro-electric lake. New Zealand J. Mar. Fresh. Res. 23:373-379.
Sokal, R.R., and F.J. Rohlf. 1981. Biometry. The Principles and Practice of Statistics in Biological Research. Freeman & Co., San Francisco, CA. 859 p.
Vannote, R.L., G.W. Minshall, K.W. Cummins, J.R. Sedell and C.E. Cushing. 1980. The river continuum concept. Can. J. Fish. Aquat. Sci. 37:130-137.
Wiederholm, T. 1984. Responses of aquatic insects to environmental pollution, p. 508-557. *In* V.H. Resh and D.M. Rosenberg [ed.] The Ecology of Aquatic Insects. Praeger, New York, NY.

Appendix 1. Summary of macrobenthic organisms sampled from seven quantitative sites in October 1992. Entries are number per 0.35 m² (i.e., three 15.24-cm² Ekman dredge samples at five stations per site; total no. grabs = 15 per site).

Taxon	5.2U	0.7U	0.0O	2.3D	8.0D	8.5D	8.8D
PHYLUM Arthropoda							
CLASS Arachnoidea							
ORDER Hydracarina							
Hydracarina sp.	8	1	2	-	4	2	7
CLASS Insecta							
ORDER Diptera							
FAMILY Chironomidae							
Ablabesmyia sp.	-	-	16	4	-	2	-
Aeshium beckae	1	-	-	-	-	-	-
Axarus sp.	-	10	-	-	-	3	78
Chironomini sp. (PUPA)	10	7	2	-	-	-	2
Chironomus sp.	-	-	5	-	-	-	-
Coelotanypus sp.	-	-	5	-	-	-	-
Cryptochironomus sp.	79	24	17	8	26	3	47
Dicrotendipes sp.	-	1	-	-	-	-	-
Glyptotendipes sp.	17	16	-	1	-	-	16
Harnischia sp.	9	1	9	25	1	3	4
Microchironomus sp.	-	-	1	-	-	-	-
Nanocladius sp.	2	-	1	3	-	-	2
Natarsia sp.	-	-	1	-	-	-	-
Paralauterborniella sp.	6	-	3	7	1	2	5
Polypedilum convictum	1	1	1	1	-	-	-
Polypedilum halterale	2	13	1	1	-	2	-
Polypedilum scalaenum	-	4	-	-	-	-	4
Polypedilum sp.	-	-	-	-	-	1	-
Procladius sp.	-	-	2	-	-	-	-
Rheotanytarsus sp.	1	18	-	1	-	-	2
Stelechomyia sp.	-	-	-	1	-	-	-
Tanypodinae sp. (PUPA)	-	-	-	1	-	-	-
Tanytarsus sp.	1	-	18	2	-	-	2
Tribelos sp.	-	-	-	-	-	1	-
FAMILY Ceratopogonidae							
Ceratopogonidae sp.	11	10	111	16	17	10	7
FAMILY Chaoboridae							
Chaoborus sp.	4	1	-	1	-	-	1
FAMILY Empididae							
Empididae sp.	-	-	-	-	1	-	1
ORDER Ephemeroptera							
FAMILY Ephemeridae							
Hexagenia sp.	4	1	87	22	3	3	-
FAMILY Caenidae							
Caenis sp.	1	4	-	-	-	35	3
FAMILY Tricorythidae							
Tricorythodes sp.	1	-	-	-	-	-	2
FAMILY Heptageniidae							
Stenonema interpunctatum gp.	-	-	-	-	-	5	-
ORDER Trichoptera							

Taxon	5.2U	0.7U	0.0O	2.3D	8.0D	8.5D	8.8D
PHYLUM Arthropoda							
FAMILY Hydropsychidae							
Cheumatopsyche sp.	-	-	-	-	-	-	4
Hydropsychidae sp.	-	3	-	-	-	-	3
FAMILY Polycentropodidae							
Neureclipsis sp.	-	-	-	2	-	11	1
Polycentropodidae sp.	-	-	1	-	-	-	-
FAMILY Leptoceridae							
Nectopsyche sp.	2	7	-	-	-	-	-
Oecetis sp.	-	1	-	-	-	-	1
ORDER Odonata							
FAMILY Gomphidae							
Gomphidae sp.	9	1	1	1	5	-	2
FAMILY Macromiidae							
Macromia sp.	1	-	-	-	-	-	-
ORDER Coleoptera							
FAMILY Elmidae							
Dubiraphia sp.	-	-	1	-	-	-	-
Elmidae sp.	-	-	-	-	-	1	-
Stenelmis sp.	2	7	-	-	-	5	61
FAMILY Hydrophilidae							
Berosus sp.	-	-	1	-	-	-	-
ORDER Hemiptera							
Hemiptera sp. (JUV.)	-	-	1	-	-	-	-
CLASS Crustacea							
ORDER Amphipoda							
FAMILY Gammaridae							
Gammarus sp.	5	14	-	-	-	9	31
FAMILY Crangonyctidae							
Crangonyx sp.	1	-	-	-	-	1	4
PHYLUM Annelida							
CLASS Oligochaeta							
ORDER Haplotaxida							
FAMILY Naididae							
Dero digitata	-	-	6	-	-	1	-
Nais communis	-	1	-	-	-	-	-
Nais pardalis	2	-	-	-	-	-	-
Pristina leidyi	-	-	-	-	-	-	1
Pristinella acuminata	88	2	11	34	-	-	2
Pristinella jenkinae	2	1	1	3	-	6	3
FAMILY Tubificidae							
Aulodrilus pigueti	-	-	23	1	-	18	-
Bothrioneurum vejdovskyanum	5	1	-	-	-	7	7
Branchiura sowerbyi	-	-	22	32	7	9	9
Ilyodrilus templetoni	-	-	10	106	5	1	1
Limnodrilus cervix	3	-	17	13	5	-	-
Limnodrilus hoffmeisteri	61	22	87	60	48	2	1
Limnodrilus udekemeianus	1	-	-	-	-	1	1
Quistidrilus multisetosus	1	-	12	2	-	-	-
Tubificidae w/ (IMM.)	-	1	80	222	11	2	8
Tubificidae w/o (IMM.)	403	257	648	611	252	28	341

Taxon	5.2U	0.7U	0.0O	2.3D	8.0D	8.5D	8.8D
PHYLUM Annelida							
CLASS Hirudinea							
ORDER Rhynchobdellida							
FAMILY Glossiphoniidae							
Helobdella elongata	3	-	-	-	1	1	-
ORDER Pharyngobdellida							
FAMILY Erpobdellidae							
Mooreobdella microstoma	1	-	5	1	1	-	4
PHYLUM Entoprocta							
Urnatella gracilis	2	-	-	-	-	-	-
PHYLUM Mollusca							
CLASS Bivalvia							
Bivalvia sp.	2	58	-	-	-	23	2
SUPERFAMILY Corbiculacea							
FAMILY Corbiculidae							
Corbicula fluminea	3	-	1	9	33	15	44
FAMILY Sphaeriidae							
Sphaeriidae sp. (JUV.)	-	-	24	-	-	-	-
Sphaerium transversum	-	-	13	-	1	-	-
SUPERFAMILY Unionacea							
FAMILY Unionidae							
Unionidae sp.	-	-	-	1	-	-	-
CLASS Gastropoda							
Gastropoda sp.	-	-	1	-	-	-	-
ORDER Ctenobranchiata							
FAMILY Viviparidae							
Campeloma cf. *crassula*	-	-	2	-	1	-	-
ORDER Pulmonata							
FAMILY Physidae							
Physella sp.	-	-	-	-	-	1	-
FAMILY Planorbidae							
Menetus dilatutus	-	-	-	-	-	2	-
TOTAL NUMBER	755	488	1250	1192	423	216	714
TOTAL TAXA	37	29	39	30	19	33	37

Appendix 2. Summary of macrobenthic organisms qualitatively sampled from the IPOC to the Sulphur River. This site was sampled in a similar manner to the quantitative sites, allowing comparisons of relative abundances.

Taxon	Number	Percentage
PHYLUM Arthropoda		
CLASS Insecta		
ORDER Diptera		
FAMILY Chironomidae		
Ablabesmyia sp.	2	0.11
Chironomini sp. (PUPA)	4	0.22
Chironomus sp.	3	0.16
Clinotanypus sp.	2	0.11
Coelotanypus sp.	4	0.22
Cryptochironomus sp.	10	0.54
Harnischia sp.	3	0.16
Parachironomus sp.	1	0.05
Paralauterborniella sp.	1	0.05
Polypedilum halterale	2	0.11
Procladius sp.	7	0.38
Tanypus sp.	1	0.05
Tanytarsus sp.	2	0.11
Zavreliella sp.	1	0.05
FAMILY Ceratopogonidae		
Ceratopogonidae sp.	126	6.82
FAMILY Chaoboridae		
Chaoborus sp.	8	0.43
ORDER Ephemeroptera		
FAMILY Ephemeridae		
Hexagenia sp.	6	0.32
FAMILY Caenidae		
Caenis sp.	1	0.05
ORDER Odonata		
FAMILY Libellulidae		
Brachymesia sp.	1	0.05
ORDER Coleoptera		
FAMILY Hydrophilidae		
Berosus sp.	1	0.05
ORDER Hemiptera		
Hemiptera sp. (JUV.)	1	0.05
PHYLUM Annelida		
CLASS Oligochaeta		
ORDER Haplotaxida		
FAMILY Naididae		
Dero digitata	74	4.00
Pristinella acuminata	1	0.05
FAMILY Tubificidae		
Aulodrilus pigueti	76	4.11
Bothrioneurum vejdovskyanum	1	0.05
Branchiura sowerbyi	159	8.60
Ilyodrilus templetoni	19	1.03
Limnodrilus cervix	1	0.05
Limnodrilus hoffmeisteri	5	0.27
Quistidrilus multisetosus	43	2.33
Tubificidae w/ (IMM.)	92	4.98
Tubificidae w/o (IMM.)	1173	63.47
PHYLUM Mollusca		
CLASS Bivalvia		
Bivalvia sp.	7	0.38
SUPERFAMILY Corbiculacea		
FAMILY Sphaeriidae		
Sphaerium transversum	10	0.54
TOTAL	1848	99.95
TOTAL TAXA	34	

DESIGN AND APPLICATION OF A NOVEL STREAM MICROCOSM SYSTEM FOR ASSESSING EFFLUENT IMPACTS TO LARGE RIVERS

Joseph M. Culp and Cheryl L. Podemski

National Hydrology Research Institute, 11 Innovation Boulevard,
Saskatoon, Saskatchewan, S7N 3H5 Canada

The high degree of spatial heterogeneity and the challenge of obtaining true replicates make it difficult to predict or to quantify the impacts of complex effluents on riverine biota. We have developed an experimental stream system that simulates the riverine environment for the purpose of assessing the impacts of effluent discharges on large rivers. This transportable outdoor system is established beside the study river, providing ambient water temperature and light regimes. The system is comprised of 16 circular 0.9 m² streams, allowing for adequate replication. Current velocity is provided by an inexpensive belt-driven propeller that can produce mid-water velocities exceeding 20 cm sec^{-1}. The system is partially recirculating and the water depth and hydraulic residence times are easily adjusted. By increasing water residence time within the streams, the volume of toxic effluents or contaminants required during an experiment can be minimized. We have used this system to investigate the nutrient and contaminant effects of treated effluents from kraft pulp mills on complex food webs in the Athabasca River, Alberta, Canada. Multiple trophic-level effects are examined by seeding the tanks with natural substrata and biota (i.e., biofilm, invertebrates, fish) from the river.

INTRODUCTION

Artificial streams (i.e., stream microcosms) have been used as tools for investigating ecological interactions in running waters since the 1960s (see Lamberti and Steinman 1994 for a review). This approach has been used to investigate a variety of ecological phenomena including the effects of environmental factors such as irradiance, temperature, and nutrients on algae (McIntire 1966a, Bothwell 1988, 1989) and to examine specific trophic relationships among algae, insects, and fish (Lamberti et al. 1987, 1989; Schlosser 1988; Culp et al. 1991; Scrimgeour et al. 1991). Although field microcosms have been argued to be an important link between laboratory tests and field studies (Giddings and Eddlemon 1979), stream microcosms have received relatively little attention from researchers investigating the impacts of pulp mill effluents on riverine ecosystems.

In natural environments like the Athabasca River, Alberta, Canada, the high degree of spatial heterogeneity and the challenge of obtaining true replicates often makes it difficult to predict or verify quantitatively the impacts of complex effluents on riverine biota. Therefore, an important advantage of the microcosm approach lies in its capacity for investigating complex benthic food webs in model systems that simulate specific conditions in large rivers while allowing for true replication of treatments and controls. It is important to recognize that the microcosm approach simulates, rather than reproduces exactly, key components of the riverine environment. By locating the experimental system beside the study river, the stream microcosm can be supplied with natural river water under ambient water temperature and light regimes. Trophic mechanisms can then be examined by seeding the microcosms with natural substratum and biota (i.e., biofilm, invertebrates, and fish) from the river.

Our primary objective was to design and construct a transportable stream microcosm system for testing impacts of nutrients and contaminants from pulp mill effluents on complex food webs, including both primary (attached algae) and secondary producers (benthic invertebrates) found in the Athabasca River. Here we describe microcosm system design, specific hydraulic characteristics of the microcosms, important details and procedures of installation, and the results obtained during initial tests of the system at the Hinton, Alberta experimental site.

STREAM MICROCOSM DESIGN AND CONSTRUCTION

The experimental stream system consists of 16 circular 0.9 m^2 tanks placed in pairs on tables, requires a 9 x 5 m area of level ground for set-up, and is established beside the study river under ambient water temperature and light regimes (Fig. 1). Water from the study river is pumped into the head tank reservoir and delivered through a system of pipes to the stream microcosms. Water flow to individual streams is controlled and current in each stream is created by a belt-driven propeller system. Water depth in the tanks is maintained at 26.9 ± 0.1 cm ($\bar{x}$ ± 1 SE) by an overflow drain that returns all wastewater directly to the river. Availability of a water intake system (e.g., pulp mill water intake) and electrical power negates the need to provide generators and water pumps.

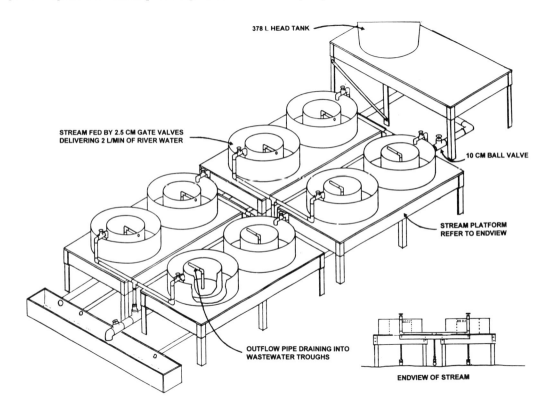

Figure 1. Oblique view of stream microcosm system showing circular streams, water delivery and wastewater systems (not drawn exactly to scale).

Water Delivery System

River water is pumped via a water intake and pumping system into a head tank, then gravity-fed to the streams. Water demand by the system depends upon the flow rate chosen by the experimenter. Generally, we deliver 2 L min^{-1} to each stream; therefore, water in excess of 32 L min^{-1} is required for the 16-stream system. Each stream microcosm contains 227-L; thus, hydraulic residence time is approximately 2 h. By increasing water residence time within the streams, the volume of effluents or contaminants required during an experiment can be minimized. The head tank is a 378 L polyethylene tank placed on a 1.2 m high platform. Gate valves control water input to each stream, allowing precise flow rate calibration for each stream. The head tank and all water delivery lines are wrapped with heat tape and insulated to prevent freezing. Note that the system is designed to withstand air and water temperatures near 0°C and has been operated at Hinton, Alberta, Canada during the autumn and early spring.

Stream Microcosms

The streams are 107 cm diameter tanks constructed of polyester fiberglass. Streams are constructed by cutting 38 cm long sections of 107 cm pipe and bonding a flat sheet of fiberglass to one end of the pipe. A 25 cm diameter section of pipe is then centered in the larger pipe and the bottom cut out to form a standpipe. Streams are placed on eight 74 cm high tables, two to a table (Fig. 1). When assembling this system, care must be taken to ensure that the tables are level. The water outflow pipe passes through the standpipe and drains into pipes beneath the tables which connect to a wooden trough (Fig. 1). These drain lines are wrapped with heat tape and insulated. Wastewater from the streams is returned directly to the study river.

Current in each stream is created by a belt-driven propeller system because this type of propulsive device is relatively gentle on organisms that might drift through it (Craig 1993). The motor assembly for each stream is comprised of a geared head motor (250 rpm, 1/40 amp) driving a 22.6 mm pulley. A belt-drive transmits power from the motor to a 49.3 mm pulley mounted on the propeller shaft. The motor and associated electronics are mounted on an aluminum frame in a weatherproof enclosure. This frame is clamped to the top edge (outside) of the stream microcosm. A 16 x 230 mm long copper strut extending downward from the aluminum frame holds the propeller shaft bushing and grease seal. A grease nipple at the top of this tube allows for lubrication of the propeller shaft and bushing. The propeller (one per stream) is a 23 cm (9 inch) diameter aluminum fan blade which rotates at a no-load speed of 115 rpm.

Contaminants are delivered independently and continuously to individual streams by peristaltic pumps (Masterflex®L/S Nema-type 13 wash down controllers and cartridge pump heads) and a series of insulated tubes for solution delivery. Peristaltic pumps are kept in insulated boxes to keep them within approved operating temperatures. Treatment solutions such as nutrients or effluent can be stored in insulated containers. Tubes carrying contaminant solutions are threaded through foam pipe insulation to the streams and then fed into the water delivery spout. Heating of the solutions and insulation of all supply lines is recommended to prevent the thin supply lines (<2 mm) from freezing. However, care must be taken with warming or heating solutions since solubility of some materials (e.g., surfactants) can be altered.

METHODS OF OPERATION

A variety of substrata types, including natural and artificial materials, can be used for stocking and/or sampling in the microcosms. During initial operation in autumn 1993, a standardized benthic environment was created in each stream tank to simulate typical riffle areas found along a reach of the Athabasca River, Alberta, upstream of a bleached kraft pulp mill operation. The bottom of each stream was covered with approximately 8 cm of thoroughly washed gravel (1-2 cm) upon which ten stones ($\bar{x}$ surface area = 535 cm^2) from the river were placed. The use of stones from the river provided a method of stocking the streams with a natural benthic community of invertebrates and periphyton. In addition, porcelain tiles (23.5 cm^2) were used to provide a standardized substratum with which to compare biofilm development and accumulation. During removal from the river each stone was surrounded with a U-net (Scrimgeour et al. 1993), carefully lifted from the stream bed, and placed into a container (two stones per container) of river water so that the periphyton and invertebrates associated with the stone were not dislodged. In addition, the substratum beneath the stones was gently disturbed to collect any invertebrates under and around the base of the stone. The stones and their biota were immediately transported to the streams and each pair of stones (and associated biota) randomly assigned to a stream for experimental tests that incorporated multiple trophic levels (i.e., periphyton and invertebrates).

Current Velocity, Contaminant Mixing, and Water Temperature

The distribution of water velocities in the streams was characterized in the laboratory using a Nixon Instruments® velocity meter. In this test the substratum consisted of gravel and stones as in the autumn 1993 experiments. Mean velocities recorded at 21 locations around the stream were similar for all three depths: 0.20 ± 0.03 m sec^{-1} 4 cm below the water surface, 0.20 ± 0.02 m sec^{-1} at the water column midpoint, and 0.23 ± 0.02 m sec^{-1} just above the highest point of each stone. During the autumn experiment overall mean velocity was similar to our laboratory tests ($\bar{x} = 0.26 \pm 0.01$ m sec^{-1}, n = 150). Observations of dye traces indicate that contaminants mix within the first quarter of the stream length (Culp et al. 1994).

Water temperature was monitored by placing a Ryan® thermograph in one of the streams and another in the head tank. Temperatures in the head tank reflected the temperature of incoming river water. In contrast, the 2 h hydraulic residence time in the streams resulted in slight heating or cooling of water in the streams depending upon ambient air temperatures. For example, over a 3 d period in autumn, the streams were cooler at night and warmer during the day as compared to the incoming river water (Fig. 2). For cold water river application, the degree of heating and cooling we have observed (<5°C) is not a substantial problem. Deployment in warmer climates, however, may require the addition of temperature control capability such as insulation.

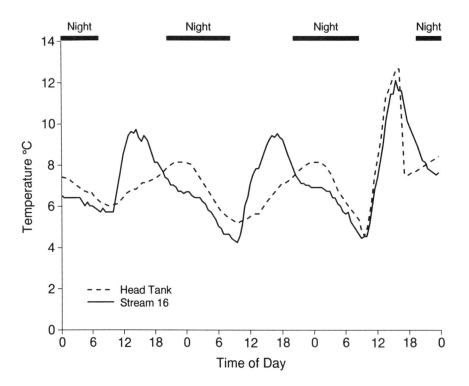

Figure 2. Diel water temperatures in experimental stream and head tank over a 72 h period in autumn 1993 at Hinton, Alberta, Canada. The head tank temperatures are equal to ambient values in the Athabasca River.

Maintenance of Stream Microcosm System

The stream system required a moderate amount of regular maintenance as the motors had to be inspected daily for loose or misaligned belts. Loose belts were commonplace, particularly in the first 3 wk of operation, but were easily fixed. In addition, the motors were lubricated with non-toxic, Permatex Superlube® (food grade USDA H1) every 3 d to prevent seizing of the propeller shafts. Debris falling into the streams, such as leaves during autumn, had to be removed daily as this material accumulated on the propellers, causing them to become unbalanced and rotate unevenly. Drain screens were brushed daily. On a weekly basis the 10 cm water delivery lines were flushed to remove any silt and sand deposits. Finally, delivery tubing had to be inspected for blockages and changed as required; this was particularly a problem in tubing carrying pulp mill effluent due to blockage by suspended solids and biofilm development.

Application to the Northern River Basins Study

Nutrients and contaminants in bleached kraft pulp mill effluents can affect water quality, periphyton abundance, and benthic invertebrate communities in receiving waters. Presently, it is difficult to predict their effects on large river ecosystems because the effluents contain (1) nutrients that stimulate the algal and microbial food supplies of invertebrates and (2) contaminant stressors that can reduce invertebrate growth and production. The importance of this nutrient and contaminant interaction in altering the production of key trophic linkages in receiving waters can only be revealed through experimentally based research designs.

In autumn 1993, we used the stream microcosm system to identify and measure the nutrient and contaminant effects of treated effluents from a bleached kraft pulp mill on complex food webs in the Athabasca River, Alberta, Canada. The experimental treatments consisted of (1) a control that received raw river water, (2) a 1% dilution of treated pulp mill and sewage effluent, and (3) a 1% dilution of the nitrogen and phosphorus contained in the concentrated effluent (i.e., nutrients in undiluted effluent: 218-307 µg L^{-1} P, 150-760 µg L^{-1} $N-NO_3$, 311-3390 µg L^{-1} $N-NH_4$). Five replicates of each treatment were randomly assigned to the streams and the nutrient (N/P) and pulp mill effluent (PME) treatments established by delivering predetermined quantities of the respective solution by peristaltic pumps and delivery tubes. The five control streams continued to receive only river water. Over the course of the 28 d experiment we measured water chemistry and temperature, current velocity, periphyton biomass, insect abundance and growth, and algal contaminant concentration in the streams. Note that the results of algal biomass, insect abundance, and growth are discussed fully by Podemski and Culp (1996). Our results indicated that during autumn both the nutrient treatments and the 1% pulp mill effluent treatments significantly increased final algal biomass relative to the control streams (Fig. 3). In fact, these increases in algal biomass followed a similar time course in treatments N/P and PME. Thus, by using the stream microcosm system, we were able to conclude that the combined municipal sewage and pulp mill effluent caused enhanced algal production because of P and N loading from the effluent. Furthermore, under the effluent concentrations tested, the periphytic response to nutrient enrichment appears to overwhelm any negative effects, such as sublethal toxicity, caused by effluent contaminants.

This example from the Athabasca River demonstrates the usefulness of stream microcosms as tools for establishing cause-effect relationships in investigations of the impact of effluents on large rivers. Our stream microcosm system overcomes the problems of spatial heterogeneity encountered in field studies and provides experimental control of the dose and exposure to effluents and strong inference from appropriate replication of treatments. If combined with appropriate field validation and comparison to river conditions at similar water depths, stream microcosms have the potential to increase greatly our understanding of the ecological effects of effluents on large rivers.

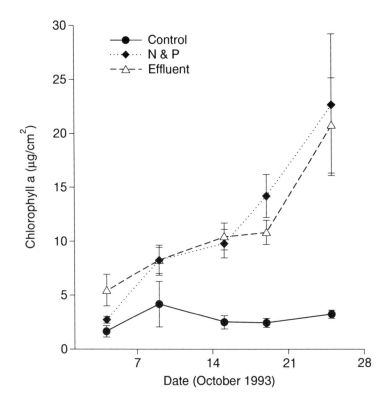

Figure 3. Mean (± 1 S.E.) chlorophyll a concentrations (µg/cm^2) on stones in the control (●), nutrient (♦), and pulp mill effluent (△) treatments over the 28 d experimental period.

ACKNOWLEDGMENTS

This work contributed to the Northern Rivers Basin Study, a joint initiative of the Province of Alberta and the Government of Canada.

We are indebted to John Mollison, Jim Banner, Bob Christie, Carol Casey, Daryl Halliwell, and Eric Marles whose technical assistance made the work possible. We also thank Kevin Cash, Patricia Chambers, Chris Perrin, Garry Scrimgeour, and Fred Wrona for their helpful comments during the development of the stream microcosm system and K. Munkittrick, C. Rakocinski, and J. Rodgers for constructive criticisms of an earlier draft. This study was supported by funds to J.M.C. from the National Hydrology Research Institute of Environment Canada and the Northern River Basins Study. C.L.P was supported by a Natural Sciences and Engineering Research Council of Canada post-graduate scholarship.

REFERENCES

Bothwell, M.L. 1988. Growth rate responses of lotic diatoms to experimental phosphorus enrichment: the influence of temperature and light. Canadian Journal of Fisheries and Aquatic Sciences 45:261-270.

Bothwell, M.L. 1989. Phosphorus-limited growth dynamics of lotic periphytic diatom communities: areal biomass and cellular growth rate responses. Canadian Journal of Fisheries and Aquatic Sciences 46:1293-1301.

Craig, D.A. 1993. Hydrodynamic considerations in artificial stream research. p. 324-327. *In* G.A. Lamberti and A.D. Steinman (ed.). Research in artificial streams: applications, uses and abuses. Journal of the North American Benthological Society 12:313-384.

Culp, J.M., N.E. Glozier and G.J. Scrimgeour. 1991. Reduction of predation risk under the cover of darkness: avoidance responses of mayfly larvae to a benthic fish. Oecologia 86:163-169.

Culp, J.M., C.L. Podemski and C. Casey. 1994. Design and application of a transportable experimental stream for assessing effluent impacts on riverine biota. Northern River Basins Study Contract No. 2611-C1.

Giddings, J.M. and G.K. Eddlemon. 1979. Some ecological and experimental properties of complex aquatic microcosms. International Journal of Environmental Studies 13:119-123.

Lamberti, G.A. and A.D. Steinman. 1994. Research in artificial streams: applications, uses and abuses. Journal of the North American Benthological Society 12:313-384.

Lamberti, G.A., L.R. Ashkenas, S.V. Gregory and A.D. Steinman. 1987. Effects of three herbivores on periphyton communities in laboratory streams. Journal of the North American Benthological Society 6:92-104.

Lamberti, G.A., S.V. Gregory, L.R. Ashkenas, A.D. Steinman and C.D. Mcintire. 1989. Productive capacity of the periphyton as a determinant of plant-herbivore interactions in streams. Ecology 70:1840-1856.

McIntire, C.D. 1966a. Some factors affecting respiration of periphyton communities in lotic environments. Ecology 47:918-930.

McIntire, C.D. 1966b. Some effects of current velocity on periphyton communities in laboratory streams. Hydrobiologia 27:559-570.

Podemski, C.L. and J.M. Culp. 1996. Nutrient and contaminant effects of bleached kraft mill effluent on benthic algae and insects of the Athabasca River. *In* Environmental Fate and Effects of Pulp and Paper Mill Effluents, M.R. Servos, K.R. Munkittrick, J.H. Carey and G. Van Der Kraak (ed.), St. Lucie Press, Delray Beach, FL.

Schlosser, I.J. 1988. Predation risk and habitat selection by two size classes of a stream cyprinid: experimental test of a hypothesis. Oikos 52:36-40.

Scrimgeour, G.J., J.M. Culp, M.L. Bothwell, F.J. Wrona and M.H. McKee. 1991. Mechanisms of algal patch depletion: importance of consumptive and non-consumptive losses in mayfly-diatom systems. Oecologia 85:343-348.

Scrimgeour, G.J., J.M. Culp and N.E. Glozier. 1993. An improved technique for sampling lotic invertebrates. Hydrobiologia 254:65-71.

WHOLE ORGANISM AND POPULATION RESPONSE OF SENTINEL FISH SPECIES IN A LARGE RIVER RECEIVING PULP MILL EFFLUENTS

Stella M. Swanson[1], Richard Schryer[2] and Barry Firth[3]

[1]Golder Associates Ltd., 1011 Sixth Ave. S.W., Calgary, Alberta, T2P 0W1 Canada
[2]Golder Associates Ltd., 2174 Airport Drive, Saskatoon, Saskatchewan, S7L 6M6 Canada
[3]Weyerhaeuser Technical Center, Tacoma, WA, 98477 U.S.A.

Fish populations of walleye and white sucker in the vicinity of a kraft pulp mill on the North Saskatchewan River, Saskatchewan, were compared to fish from a pristine reference river, the Cowan River in northwest Saskatchewan, on behalf of Weyerhaeuser Canada Ltd. in 1991 and 1992. The target sample size for the two sentinel species was 20 per sex per sample area. There were no indications of significantly slower or faster growing walleye in the North Saskatchewan River. Exposure indicators such as EROD activity and organochlorine body burdens in North Saskatchewan walleye were in the range found in the reference river. Livers were not enlarged. North Saskatchewan River white suckers were younger on average and matured earlier than Cowan River white suckers. Condition factors were nearly identical between the two rivers. The rate of growth between sites was equal or only marginally higher in the North Saskatchewan River. The main differences were a higher mean EROD activity (although the ranges were similar in both rivers), younger fish, smaller GSIs (because the fish were younger and some partially spent) and a higher incidence of pale livers and parasites in the North Saskatchewan. There were no detectable levels of organochlorine compounds in white suckers from the North Saskatchewan River. There were no correlations between indicators of exposure to bleached kraft mill effluent and population-level parameters in either species. A greater range of age classes observed in the walleye population of the North Saskatchewan River may have been a consequence of a difference in exploitation rates in the two rivers. Very low river flows affected the success of the main sampling year (1992) and thus the ability to interpret data on reproductive potential of walleye. Comparison of 1991 and 1992 data was difficult because of differences in spawning condition (pre-spawning versus spawning) in the two years. The cause of younger, faster maturing white suckers in the North Saskatchewan is not apparent, since the overall pattern observed is not "typical" of population responses observed in other studies. It appears that the usefulness of whole organism and population parameters as early indicators of stress from specific sources such as pulp mills depends on a strong understanding of the effects of other factors (e.g., exploitation and river flows). In these large rivers, detecting incremental population-level effects from one particular stressor in the absence of a strong contaminant gradient seems unlikely.

INTRODUCTION

One of the main environmental challenges facing industry and government is to ensure that industrial discharges do not have detrimental effects on populations. One of the programs that has been implemented with the protection of populations in mind is Environmental Effects Monitoring (EEM), enacted as a means of evaluating the effectiveness of the Canadian Fisheries Act in protecting fish and fish habitat. Pulp and paper is the first industry to have EEM added to monitoring requirements (Environment Canada and Department of Fisheries and Oceans 1992). An adult fish survey is one of the three main components of EEM. The aim of this paper is to communicate our early experiences with implementation of adult fish studies at pulp mill sites.

The adult fish survey of EEM focuses on indicators of growth and reproductive potential in two "sentinel species." The sentinel species are chosen on the basis of degree of potential exposure to the effluent, as well as on a number of practical considerations such as abundance and growth rate (Environment Canada and Department of Fisheries and Oceans 1993). The survey design must include a reference area with similar habitat and the same fish species but with no pulp mill discharges. Fish are collected during times of the year when exposure to the effluent is assumed to be high. Habitat characteristics from the study area and the reference area are mapped in detail during "pre-design" activities before actual sampling of the sentinel species takes place.

Weyerhaeuser Canada Ltd. operates a bleached kraft pulp mill and paper mill at Prince Albert, Saskatchewan, on the North Saskatchewan River. In 1992, fish population studies were conducted in anticipation of the EEM requirements in order to obtain a better understanding of the fish distribution and relative abundance. In addition, two of the more abundant species, white sucker (*Catostomus commersoni*) and walleye (*Stizostedion vitreum*), were collected to evaluate growth, age distribution and reproductive potential. Contaminant body burdens and mixed function oxidase levels in liver were also studied. Data from the North Saskatchewan River were compared to data from the Cowan River, a pristine system in northwest Saskatchewan. The overall objective of the 1992 study was to determine if any significant effects could be observed in walleye and white sucker populations as a result of exposure to bleached kraft mill effluent (BKME).

METHODS

Sampling Strategy

Sampling in the North Saskatchewan began in the spring of 1991, with additional sampling in the summer of 1991. Intensive sampling programs were carried out simultaneously on the North Saskatchewan River and the reference site during the 1992 spring spawning season. Fishing efforts on the North Saskatchewan were concentrated in an area between the mill discharge and 55 km downstream. Sampling at the reference site was done at one main site along the Cowan River. Efforts were made to sample a variety of habitat types including river run with either sand or rock substrates, riffles, rapids and backwaters. All fish from the North Saskatchewan River were captured using a boat-mounted electrofishing unit (Smith-Root Model VI-A powered by a 5-kW generator). Fish were placed in water-filled tubs on board the boat and allowed to recover before sampling. Fish from the Cowan River were collected in cooperation with personnel from Saskatchewan Fisheries Branch using a trap net. Fish were selected from the trap net catch and placed in separate holding pens until sampling. Supplemental collection of white suckers with gill nets was conducted on the Cowan River in order to obtain additional fully mature individuals. The target sample size for the two sentinel species was 20 per sex per sample area (North Saskatchewan and Cowan Rivers).

Fish Sampling

All fish sampled were subjected to a standardized external examination which made note of any abnormal external features (e.g., wounds, parasites, tumors, fin fraying). The sex and state of sexual maturity was also assessed during this examination and mature individuals of the two sentinel species (as indicated by the presence of secondary sexual characteristics, the size and firmness of the abdomen and/or the ability to extrude sexual products) were set aside for full-scale processing. Other standard measures included fork length and weight. Game fish (i.e., sauger, goldeye, northern pike, sturgeon) were quickly examined, tagged with plastic anchor tags and released.

All sentinel fish sampled were in spawning condition, i.e., sexual products could be easily extruded. The selected fish was first weighed. The next step was to strip the majority of sexual products (eggs, milt) from the fish. The eggs or milt were collected in a pre-weighed wide-mouthed container and put aside till the remainder of the sexual products and gonad could be removed. This procedure was necessary to avoid the loss of eggs/milt while blood was being taken from the fish. The length of the fish was then taken. Blood samples were drawn from the fish while they were still alive. Blood was kept cool on wet ice and centrifuged in the field. Following the drawing of blood, the fish was rinsed and sacrificed with a sharp blow to the head. Liver was collected quickly for mixed function oxidase (MFO) analysis and frozen in liquid nitrogen. Fillet samples were taken for organochlorine analysis using acetone/hexane-rinsed dissection tools and acetone/hexane-rinsed foil as a dissection surface. Samples were placed in rinsed foil and frozen on dry ice. The body cavity was then completely opened and bile withdrawn from the gall bladder, placed in a cryovial and frozen in liquid nitrogen. Remaining gonadal material was collected and placed with the previously collected material and weighed to the nearest 0.1 g. For males, a 1-g sample from the midsection of one of the testes was placed in a histology cassette and preserved in 10% neutral buffered formalin. For females, samples of ovarian tissue were collected for fecundity, histological staging and egg size. Exactly 1 g of eggs was placed in a histology cassette and preserved in 10% buffered formalin. These eggs were to be counted to calculate total fecundity and were also used for staging analysis. Egg diameter was measured by volumetric displacement. Exactly 200 eggs were placed in a 10-mL graduated cylinder and the volume of water displaced by the eggs was recorded. For suckers, the entire contents of their stomachs and intestines were collected and preserved in ethanol. For walleye, the contents of the stomachs were observed and recorded. Any diseased or abnormal tissues were collected and preserved in 10% buffered formalin. Ageing structures taken for the two sentinel species were pectoral fins for white suckers and pectoral and dorsal fins, otoliths and opercula for walleye.

Laboratory Methods

MFO function was determined through analysis of the activity of ethoxyresorufin-*o*-deethylase (EROD) using the method of Klotz *et al.* (1984) as described by Hodson *et al.* (1991). Quality control measures included the use of reference liver samples with known high EROD activity. In addition, analysis of cytochrome P450 and P420 content was conducted on selected samples; a large P420 absorption peak indicates sample degradation (Hodson *et al.* 1991).

Dioxin and furan congeners in fish fillets were determined according to the U.S. EPA method 8290 (1987) and 1613 revised (1990) with quality control as specified by ELAP (1988) and Environment Canada (1992a,b). Chlorinated phenolic analysis was performed following U.S. EPA Method 1653 and analyzed by GC/SIM-MS on a gas chromatograph equipped with a Mass Selective Detector. Diterpene resin acids and fatty acids were determined by GC/SIM-MS after methylation, concentration and gel permeation chromatography followed by alumina chromatography.

Total fecundity for the females of both species was calculated by counting the number of eggs in the 1-g sample. Total number of eggs was then calculated by multiplying the number of eggs in 1 g by the total weight of ovaries recorded in the field.

Contents of the sucker stomachs were sorted with a dissecting microscope. Food items were identified and recorded.

Suckers were aged using pectoral fins. The fins were cleaned and mounted in epoxy. Cross-sections were cut using a jeweler's saw. The cut sections were mounted and examined with a dissecting microscope. Walleye otoliths were placed in oil of wintergreen and examined under a dissecting microscope. Opercular bones were cleaned and backlit and the annuli counted by direct observation. In cases where a clear age could not be obtained from either the otolith or the opercular bone, the dorsal spines from walleye were also used. These were examined by the same method as sucker pectoral fins.

Data Analysis

All data were entered into a Lotus 123 spreadsheet. Functional length-weight regressions were performed using the Fisheries Science Applications Systems (FSAS) designed by Saila *et al.* (1988).

Analysis of variance and other linear regressions was performed using the Statistix statistical software. Data for length-weight analysis and weight-weight regressions (e.g., gonad-total weight) were log-transformed.

Condition factors were determined using the Fulton-type condition factor. Gonadal somatic indices (GSI) were calculated by dividing gonad weight by total body weight and multiplying by 100. Liver somatic indices (LSI) were calculated by dividing liver weight by total body weight and multiplying by 100.

RESULTS

Capture of spawning walleye from the North Saskatchewan River was hampered by a very dry spring with almost no spring run off. Therefore, the fish were forced to spawn within the main body of the river rather than moving up tributaries. Consequently, fish were dispersed throughout the spawning season and sample sizes for walleye were less than the desired 20 specimens per sex. Whenever necessary, data from a spring 1991 North Saskatchewan River study conducted for Weyerhaeuser were used to supplement information for walleye. It is recognized that use of data from different years is not the most appropriate method of comparison. Differences in growing seasons from one year to the next can affect population characteristics and reproductive potential. Thus, any interpretation based on use of data from both years was made with caution.

Condition factors for walleye and white sucker from the North Saskatchewan and Cowan Rivers varied little between sites and sampling seasons (Table 1). White sucker from the North Saskatchewan River had slightly higher condition factors in the spring of 1991. This is due to the proportionately higher number of mature females caught in this season.

Age class distribution of walleye was difficult to compare using 1992 data alone because of the small sample size from the North Saskatchewan River (n = 15). When capture data from both 1991 were used, both populations were dominated by individuals of ages 5 to 9 (Fig. 1). Males tended to be younger than females, e.g., almost all of the males in the Cowan River samples were either age 5 or 6 while females were in the 6- to 9-year groups. A greater overall range of age classes was observed in the walleye population of the North Saskatchewan River (Fig. 1).

Table 1. Condition factors for walleye and white sucker captured during all sampling programs.

		Walleye		White Sucker	
		North Sask. River	Cowan River	North Sask. River	Cowan River
Spring 1991	Mean Length (mm)	521.000		422.000	
	Mean Weight (g)	1811.000		1216.000	
	Condition Factor	1.281		1.615	
Summer 1991	Mean Length (mm)	419.000		399.000	
	Mean Weight (g)	935.000		880.000	
	Condition Factor	1.274		1.384	
Spring 1992	Mean Length (mm)	462.000	479.000	434.000	417.000
	Mean Weight (g)	1265.000	1360.000	1184.000	1053.000
	Condition Factor	1.286	1.236	1.446	1.452
Merkowsky 1987					
Prince Albert	Condition Factor	1.070		1.596	
Cecil Ferry	Condition Factor	1.080		1.571	
The Forks	Condition Factor	1.135		1.628	

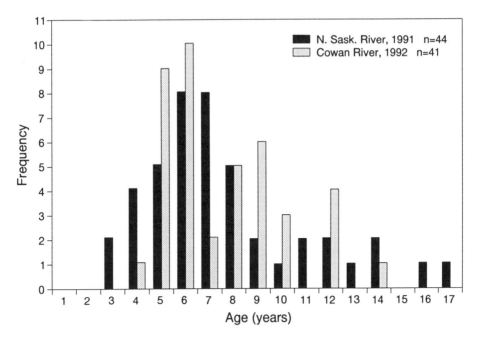

Figure 1. Age class distribution of walleye in the North Saskatchewan River 1991 and Cowan River 1992.

Age class distribution of white suckers showed a distinct difference in the major age classes. The North Saskatchewan white suckers were mostly found in the 5 to 7 age groups whereas the Cowan River fish were largely in the 7 to 10 age groups (Fig. 2). There were no consistent differences in dominant age classes between males and females in either river.

Length-weight regression analysis showed that the slope for the walleye from the Cowan River was slightly lower than that from the North Saskatchewan River (Table 2). This indicates that walleye from the Cowan River are slightly lighter for the same length class. The opposite was true for white suckers; the regression slope was slightly higher for Cowan River fish (Table 2).

Growth was examined by examining mean length at age for both species. Results for walleye were too variable to make any clear distinctions between North Saskatchewan and Cowan fish or between years of sampling (Table 3). Growth curves for the 1992 sample appear similar between the two rivers; however, sample size for walleye from the North Saskatchewan River was small and younger age classes were not adequately sampled for either species (because of the emphasis on obtaining mature specimens).

Age to maturity is traditionally stated as the age at which 50% of a particular sex within a population reaches sexual maturity. Both 1992 sampling programs examined mature fish only; thus determination of age to maturity was impossible. As an approximation, ages to maturity are stated as the youngest mature fish encountered in the sample. The youngest mature male and female walleye from the North Saskatchewan were 3 and 4 years, respectively. The youngest mature male and female walleye from the Cowan were 4 and 5 years, respectively. The youngest mature male and female white suckers were 5 years old for both sexes from both rivers.

Mean ages of mature walleye were generally similar between the two rivers, with the exception of mature males in 1991 from the North Saskatchewan (Table 4). White suckers from the Cowan River were older on average than those from the North Saskatchewan. These older fish were also larger.

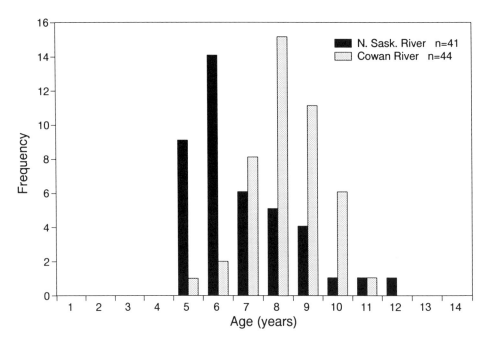

Figure 2. Age class distribution of white sucker in the North Saskatchewan and Cowan Rivers, spring 1992.

Table 2. Results of functional length-weight regressions (log-log transformation) for walleye and white sucker captured in the North Saskatchewan and Cowan Rivers in all sampling programs.

		Walleye		White Sucker	
		North Sask. River	Cowan River	North Sask. River	Cowan River
Spring 1991	n	45.000		354.000	
	Slope (b)	3.171		2.964	
	Intercept (a)	-5.426		-4.711	
	Correlation (r)	0.932		0.929	
Summer 1991	n	93.000		94.000	
	Slope (b)	3.219		2.675	
	Intercept (a)	-5.584		-4.020	
	Correlation (r)	0.981		0.790	
Spring 1992	n	15.000	41.000	41.000	44.000
	Slope (b)	3.171	3.049	2.796	3.597
	Intercept (a)	-5.424	-5.062	-4.309	-6.408
	Correlation (r)	0.997	0.985	0.924	0.906

Table 3. Mean length at age data for walleye and white sucker from the 1991 both spring and summer sampling programs, and from the 1992 spring sampling programs (sexes combined).

| | North Saskatchewan River | | | | | | | |
| | Walleye 1991 | | White Sucker 1991 | | Walleye 1992 | | White Sucker 1992 | |
Age	Spring (n = 45)	Summer (n = 90)	Spring (n = 105)	Summer (n = 89)	North Sask River (n = 15)	Cowan River (n = 41)	North Sask River (n = 41)	Cowan River (n = 44)
1	-	-	-	-	-	-	-	-
2	-	254.0	-	-	-	-	-	-
3	351.0	319.0	-	-	320.0	-	-	-
4	379.2	388.3	367.0	370.8	403.3	375.0	-	-
5	414.8	449.1	396.0	378.1	433.7	422.0	422.2	450.0
6	477.8	483.5	412.4	395.6	411.0	445.5	415.9	420.0
7	518.0	513.0	416.5	406.1	-	442.5	451.8	399.9
8	546.7	597.0	418.2	409.0	535.0	514.6	432.8	411.6
9	587.5	558.8	445.0	409.8	-	525.3	474.5	421.7
10	569.0	613.0	444.5	427.0	-	532.0	485.0	427.3
11	646.0	598.0	425.3	454.8	571.0	-	520.0	480.0
12	669.3	703.0	464.5	445.0	484.0	581.3	402.0	-
13	-	-	-	451.0	744.0	-	-	-
14	701.5	705.0	489.0	-	-	489.0	-	-
15	733.0	-	-	-	-	-	-	-
16	-	-	-	-	-	-	-	-
17	709.0	-	-	-	-	-	-	-

Table 4. Mean age of total sample of sexually mature fish by sex and site.

| | North Saskatchewan River | | | | | | Cowan River | | |
| | 1991 | | | 1992 | | | | | |
	Male	Female	Total	Male	Female	Total	Male	Female	Total
Walleye									
Mean Age	5.3	8.5	7.7	6.0	7.6	6.5	6.7	8.3	7.5
n	9	26	35	10	5	15	20	21	41
White Sucker									
Mean Age	7.1	7.2	7.2	6.8	6.9	6.8	8.0	8.5	8.2
n	44	60	104	21	20	41	23	21	44

Data on reproductive potential of walleye in the two rivers were difficult to compare because of the small sample size obtained from the North Saskatchewan River in 1992 (only two gravid females). When data from 1991 North Saskatchewan fish were included, mean fecundity appeared to be similar to Cowan River fish (42.8 vs. 46.5 eggs g^{-1}, respectively). However, the range in the number of eggs per female within an age class was very large. For example, 6-year-old females from the Cowan River had from 46,008 to 76,949 eggs (n = 5) while 6-year-old females from the North Saskatchewan had from 47,574 to 116,851 eggs (n = 2). Mean GSIs for the 1991 and 1992 seasons were 12.4 and 14.6, respectively, compared to the mean of 13.1 for walleye in the Cowan River. The 1991 fish were in pre-spawning condition whereas the Cowan River fish were all in spawning condition. Therefore, from this admittedly limited data, it appears that the GSIs of the North Saskatchewan River walleye were at least as high as those in the Cowan River. Regressions of total body weight with ovary weight showed similar slopes for North Saskatchewan and Cowan fish. Comparisons of body weight with testes weight were not possible because the males from the North Saskatchewan River had released a portion of their milt.

Data for reproductive potential of white sucker were far more complete. The only difficulty encountered was with a portion of the Cowan River females which did not achieve complete final maturation of their eggs. Consequently, egg diameters are not given for these individuals. The mean GSIs, total fecundities, mean fecundities and egg diameters were all somewhat smaller in the North Saskatchewan River sample (Table 5). GSIs were the only parameters found to be statistically different ($p = 0.03$). Regressions of total body weight and ovary weight showed similar slopes and correlations for the two rivers. However, regressions of total body weight and testes weight were quite different because of the much larger variability in North Saskatchewan males.

There were no major differences in the incidence of external and internal abnormalities in walleye from the two rivers. Occasional leeches on the anal fin were observed in the North Saskatchewan, while some walleye from the Cowan had recent abrasions to the body and fins. Cowan walleye also had a high incidence of small white cysts on their gills. Fish from both sites had abundant tapeworm infestations of the intestines and pyloric caecae.

White suckers from the Cowan River had a somewhat higher incidence of external abnormalities, particularly lesions on or splitting of fins (11.4 and 20.4%, respectively, compared to 0 and 2.4% in North Saskatchewan fish). This damage may have occurred during their spawning migration or during the period when they were kept penned in the trap net. White suckers from the North Saskatchewan River had a 19.5% incidence of pale livers whereas fish from the Cowan River had an incidence of 4.5% pale livers. North Saskatchewan white suckers also had a 34% incidence of intestinal tapeworms (34.1 versus 0%) and relatively high numbers of other cysts and nematodes (12.1 and 17.1%) compared to Cowan River white suckers (2.3 and 0%). The white sucker tapeworms probably indicate a different food/prey in the North Saskatchewan River.

In May 1992, 6 walleye (2 female and four male) and 6 white sucker (3 female and 3 male) from the North Saskatchewan River were analyzed for MFO activity. The data were compared with the MFO results from 20 walleye (10 female and 10 male) and 21 white sucker (10 male and 11 females) collected from the Cowan River in April 1992. The MFO activity in walleye as indicated by EROD values varied from <2.0 to 11 pmoles mg^{-1} min^{-1} in North Saskatchewan fish and from <1.0 to 20 pmoles mg^{-1} min^{-1} from the Cowan River. There was no significant difference in mean EROD activities between the two rivers. White sucker EROD activities varied from <4.0 to 39 pmoles mg^{-1} min^{-1} for North Saskatchewan fish and from <2.0 to 34 pmoles mg^{-1} min^{-1} for Cowan River fish. The mean value of 15.8 from the North Saskatchewan River was significantly higher than the mean value of 6.2 from the Cowan River.

Linear regressions between total body weight and liver weight showed little disparity in the slopes, intercepts or correlations for female walleye from the two rivers (Table 6). Log transformations removed a great deal of the variability from the male walleye data with little disparity in the slopes although the intercepts were somewhat different. Correlation was poor for female white sucker liver weight regressions in both sites. North Saskatchewan River females had a higher intercept and slope than their Cowan River counterparts. The reverse was true for white sucker males. The mean LSI for North Saskatchewan River female and male walleye was significantly smaller than the Cowan River female and male walleye ($p = 0.008$; $p = 0.000$, respectively). There were no significant differences among white sucker LSIs.

Table 5. Reproductive parameters calculated for female white sucker captured during the spring sampling program, North Saskatchewan and Cowan Rivers, 1992.

Species	Fish No.	Length (mm)	Weight (g)	Age	Ovary Weight (g)	GSI	Total Fecundity (# eggs per gram of fish)	Fecundity (# eggs per gram of fish)	Egg Diameter (mm)
White Sucker	2	470	1500	6	181.9	12.13	31903	21.27	2.12
North Sask River	5	450	1350	7	149.3	11.06	27621	20.46	2.25
	10	440	1425	9	168.8	11.85	36461	25.59	2.12
	14	431	1150	8	65.4	5.69	12961	11.27	2.12
	22	451	1400	7	189.3	13.52	38239	27.31	2.12
	23	422	1200	6	160.2	13.35	45103	37.59	1.79
	24	485	1625	10	213.8	13.16	57397	35.32	1.79
	29	436	1100	6	107.2	9.75	22999	20.91	2.12
	30	450	1150	8	175.3	15.24	41196	35.82	1.79
	31	509	1550	9	50.0	3.23	7688	4.96	
	32	456	1450	7	248.8	17.16	87613	60.42	1.79
	33	435	1225	5	153.3	12.51	47140	38.48	1.56
	34	405	1100	5	29.4	2.67	6076	5.52	2.12
	35	455	1250	5	121.2	9.70	23432	18.75	2.25
	36	520	1775	11	117.3	6.61	35731	20.13	1.97
	37	430	1200	5	161.2	13.43	30897	25.75	1.97
	38	467	1425	7	222.7	15.63	59273	41.59	1.79
	39	425	1200	5	133.2	11.10	29183	24.32	2.12
	40	405	950	6	105.2	11.07	24100	25.37	1.97
	41	395	925	6	20.6	2.23	3691	3.99	1.97
Mean						10.55		25.24	1.99
Standard Deviation						4.23		13.46	0.49
n						20.0		20.00	19.00
Cowan River	24	453	1550	10	295.0	19.03	55460	35.78	2.05
	25	440	1100	8	138.8	12.62	23735	21.58	2.12
	26	412	1100	9	180.5	16.41	48194	43.80	2.32
	27	450	1350	8	195.8	14.51	43468	32.20	2.12
	28	414	1125	9	154.6	13.74	33084	29.41	1.88
	29	435	1160	9	84.6	7.29	18358	15.83	1.97
	30	440	1300	10	156.6	12.05	21454	16.50	2.32
	31	428	1100	9	175.9	15.99	24098	21.91	2.25
	32	465	1500	9	230.3	15.35	47212	31.47	2.12
	33	430	1200	9	158.6	13.22	32672	27.23	
	34	435	1225	8	159.5	13.02	34452	28.12	
	35	420	1200	6	174.8	14.57	49818	41.52	
	36	425	1150	8	122.9	10.69	25809	22.44	
	37	414	1075	8	117.2	10.90	34457	32.05	
	38	455	1550	10	212.5	13.71	67150	43.32	
	39	480	1550	11	266.1	17.17	57744	37.25	
	40	405	975	9	133.8	13.72	33584	34.44	
	41	420	1150	6	143.3	12.46	43707	38.01	
	42	427	1250	8	154.3	12.34	42433	33.95	
	43	410	1000	10	47.8	4.78	5115	5.11	
	44	450	1100	5	142.4	12.95	36597	33.27	
Mean						13.17		29.77	2.13
Standard Deviation						3.07		9.55	0.14
n						21.0		21.00	9.00

Table 6. Means LSIs (sexes separate) and regression analysis results of liver weight versus total body weight for male and female fish from both study sites, 1992.

	LSI	LSI log	n	Intercept	Slope	Correlation
Walleye						
North Sask River						
Female	1.2046	0.1964	5	-2.3495	1.1295	0.9278
Male	1.0758	0.3921	10	-2.4456	1.1604	0.8177
Cowan River						
Female	1.4771	0.1817	21	-2.0920	1.0811	0.8942
Male	2.0107	0.4436	20	-1.9136	1.0690	0.7347
White Sucker						
North Sask River						
Female	1.2480	0.2224	20	-2.4576	1.1759	0.5240
Male	1.2943	0.2607	21	-1.9256	1.0094	0.4888
Cowan River						
Female	1.3191	0.2329	21	-1.3201	0.8166	0.3243
Male	1.2395	0.2078	23	-2.4669	1.1879	0.4908

As specified in the EEM regulations, a ten-fish composite sample of walleye and white suckers was analyzed. In addition, two additional walleye samples were analyzed in order to compare the contaminant burdens of gravid versus resting animals. None of the samples had any dioxin or furan congeners present above detection limits. Tetrachlorovertrole was found in the flesh of white suckers from the North Saskatchewan at a concentration of 3.0-3.8 ppb. No resin or fatty acids were found.

DISCUSSION

Because of the poor sampling success in 1992 in the North Saskatchewan River, 1991 data were used to supplement the analysis. This was not done until an analysis of the age class distributions (total and by sex) and growth rates showed that little difference existed between the two sampling years. Thus, use of the 1991 results (when necessary) was deemed to be justified. It is anticipated that this situation may arise frequently with EEM studies, especially those on large rivers or in other habitats with highly variable conditions year to year. The experience with this study serves as an example of how contingencies within EEM studies can be addressed. As our experience increases, a consensus may develop regarding how to deal with these inevitable difficulties.

Walleye

The greater range of age classes observed in the walleye population of the North Saskatchewan River may be a consequence of a difference in exploitation rates. Walleye in the North Saskatchewan River are subject to relatively heavy angling pressure, as demonstrated by the return of anchor tags by fishermen. The walleye in the Beaver River are subjected to heavy angling pressure as they return from their spawning run in the Cowan River. In addition, a commercial and native fishery exists on Lac Isle-a-la-Crosse where many of the walleye that participate in the Cowan River spawning overwinter. This gill net fishery is restricted to large mesh sizes which would eliminate many of the larger individuals of the population.

Both rivers had walleye populations with a greater number of males in the younger age classes. Older age classes were dominated by females. This distribution is typical of many walleye populations (Eschmeyer 1950; Rawson 1957; Payne 1964; Colby *et al.* 1979). Insufficient data on walleye reproductive potential were collected from the North Saskatchewan River in 1992 to draw any meaningful conclusions. The limited data from the 1991 North Saskatchewan River indicate that GSIs and fecundities are comparable to those observed in the Cowan River; however, differences in spawning condition between the 1991 North Saskatchewan fish (pre-spawning) and the 1992 Cowan fish (spawning) make any interpretation difficult.

Growth rates, condition factors and length-weight regressions were very similar in the two rivers, with the exception of a slight indication of heavier fish for length in the North Saskatchewan. Growth rates of females were slightly faster than males. This is typical of walleye populations (Colby *et al.* 1979). There were no significant differences in EROD induction in walleye from the two rivers. Regressions between total body weight and liver weight were similar (i.e., no evidence of enlarged livers in North Saskatchewan samples). There was also an absence of detectable organochlorines. A sample of walleye taken in 1988 did not have detectable levels of 2,3,7,8-tetrachlorodibenzo-*p*-dioxin or 2,3,7,8-tetrachlorodibenzofuran. The similarity in population characteristics and the absence of exposure indicators such as EROD induction and contaminant body burdens suggest that walleye from the North Saskatchewan River are not displaying a significant detectable response to exposure to BKME. We did not find comparable studies on the effects of BKME exposure on walleye populations. The focus has been on the members of the families Catostomidae and Salmonidae.

White Sucker

North Saskatchewan River white suckers were younger on average and matured earlier than Cowan River white suckers. Condition factors were nearly identical between the two rivers. The rate of growth between sites was equal or only marginally above that in the Cowan River. This difference in growth was most often only evident in the older age classes where sample sizes were smaller and the data more variable. Males dominated the younger age classes in both rivers whereas females were more numerous in the older age groups. This agrees with results obtained by Campbell (1935) and Gaboury (1985). Growth rates of females from both sites were also slightly higher than that observed for males.

Differences in reproductive indices were observed between North Saskatchewan and Cowan River fish; however, age differences and variability caused by partially spent individuals may account for most of the difference. Fecundities, GSIs and egg diameters were all lower in the North Saskatchewan River population; however, the only statistical difference was in GSIs. The differences are mainly attributable to the difference in age of the females sampled. Older, larger females are known to have higher GSIs, fecundities and egg diameters (Munkittrick *et al.* 1991). Thus, the older, larger females in the Cowan River had higher reproductive indices. In addition, 4 of the 20 females sampled from the North Saskatchewan River were partially spent. This substantially reduced their GSIs and fecundities although it did not affect egg diameter calculations. Removal of these four individuals from the data set brought the mean GSI up to 12.4, which is very similar to the Cowan River results when the difference in age is considered. The similar total body weight/ovary weight regressions in the two rivers also indicate little difference between the two populations. Total body weight/testes weight regressions were quite different because of the number of males that were already partially spawned in the North Saskatchewan sample.

The population characteristics, reproductive indices and exposure indicators of the North Saskatchewan white sucker sample captured in the BKME plume did not display the characteristics of other BKME-exposed white suckers in Jackfish Bay, Ontario (Munkittrick *et al.* 1991) or the St. Maurice

River, Quebec (Hodson et al. 1992; Gagnon et al. 1993). The Jackfish Bay fish had decreased growth, delayed age-to-maturity, increased condition factor, increased liver size, higher EROD activity, reduced survival, reduced fecundity with age, smaller gonads, lack of secondary sex characteristics in males and a lack of egg diameter increase with age in females. St. Maurice fish showed a response to nutrient enrichment by growing more rapidly and storing more lipid but also appeared to reflect BKME exposure by later maturation and smaller gonads. In contrast, white suckers from the North Saskatchewan had no significant differences in growth, age-to-maturity (as measured by recording youngest ages of mature fish), condition factors, liver sizes, gonad size:total body size regressions and egg diameter with age in females. There were normal secondary sex characteristics in males. The only differences were a higher mean EROD activity (although the ranges were similar in both rivers), younger fish, smaller GSIs (because the fish were younger and some partially spent) and a higher incidence of pale livers and parasites in the North Saskatchewan. The external and internal abnormalities in the North Saskatchewan fish were different from the lesions observed in lake whitefish in Jackfish Bay (Munkittrick et al. 1992). There were no detectable levels of organochlorine compounds in white suckers from the North Saskatchewan River.

Based upon these results, BKME-related effects as observed at Jackfish Bay or the St. Maurice River were not observed in walleye or white suckers in the North Saskatchewan River. Differences in age structure, gonad size and parasite infestation could be explained by differences in exploitation rates, natural differences in gonad size with age and increased numbers of intermediate hosts because of nutrient enrichment, respectively. BKME-related effects cannot be eliminated; however, indicators of exposure such as EROD and organochlorine body burden showed little or no response. Thus, comparisons between exposure and effect indicators were difficult. The focus of a second cycle of EEM studies may include the investigation of measures to ensure sampled fish are exposed and specific inclusion of habitat-related factors such as nutrient enrichment in the study design. It appears that the usefulness of whole organism and population parameters as early indicators of stress from specific sources such as pulp mills depends on a strong understanding of the effects of other factors (e.g., exploitation, river flows, nutrient enrichment). In these large rivers, detecting incremental population-level effects from one particular stressor in the absence of a strong contaminant gradient seems unlikely.

REFERENCES

Campbell, R.S. 1935. A study of the common sucker, *Catostomus commersoni* (Lacepede) of Waskesiu Lake. M.A. Thesis. Dept. of Biology, University of Saskatchewan, Saskatoon, Sask. 48 p.

Colby, P.J., R.E. McNichol and R.A. Ryder. 1979. Synopsis of biological data on the walleye *Stizostedion vitreum* (Mitchill 1818). FAO Fisheries Synopsis No. 119. 138 p.

ELAP (Environmental Laboratory Approval Program). 1988. Environmental Laboratory Approval Program Certification Manual. New York State Department of Health.

Environmental Canada. 1992a. Reference method for the determination of polychlorinated dibenzo-para-dioxins (PCDDs) and polychlorinated di-benzo furans (PCDFs) in pulp and paper mill effluents. Conservation and Protection, Ottawa, Ontario. Report EPS 1/RM/19.

Environment Canada. 1992b. Methods manual for the National Water Quality Laboratory. Environment Canada, Inland Waters Directorate, Burlington, Ontario.

Environment Canada and Department of Fisheries and Oceans. 1992. Aquatic Environmental Effects Monitoring Requirements. Annex 1: Aquatic Environmental Effects Monitoring Requirements at Pulp and Paper Mills and Off-Site Treatment Facilities under the Pulp and Paper Effluent Regulations of the Fisheries Act, May 20, 1992. Conservation and Protection, Ottawa, Ontario.

Environment Canada and Department of Fisheries and Oceans. 1993. Technical Guidance Document for Aquatic Environmental Effects Monitoring Related to Federal Fisheries Act Requirements. Department of Fisheries and Oceans, Ottawa, Ontario.

Eschmeyer, P.H. 1950. The life history of the walleye, *Stizostedion vitreum vitreum* (Mitchill) in Michigan. Bull. Mich. Dept. Conserv., Inst. Fish. Res. 3:99 p.

Gaboury, M.N. 1985. A fisheries survey of the Valley River, Manitoba, with particular reference to walleye (*Stizostedion vitreum*) reproductive success. Manitoba Nat. Res. Fisheries Rep. #85-02. 120 p.

Gagnon, M.M., J.J. Dodson and P.V. Hodson. 1993. BKME-induced and naturally occurring variations in growth and reproduction of white suckers (*Catostomus commersoni*). Abstracts, 14th Annual Meeting, Society of Environmental Toxicology and Chemistry. November 14-18, Houston, TX., p. 130.

Hodson, P.V., P.J. Kloepper-Sams, K.R. Munkittrick, W.L. Lockhart, D.A. Metner, P.L. Luxon, I.R. Smith, M.M. Gagnon, M. Servos and J.F. Payne. 1991. Protocols for measuring mixed function oxidases of fish liver. Canadian Technical Report of Fisheries and Aquatic Sciences. #1829, 51 p.

Hodson, P.V., M.McWhirter, K. Ralph, B. Gary, D. Thivierge, J.C. Carey, G. Van der Kraak, D.M. Whittle and M.-C. Levesque. 1992. Effects on bleached kraft mill effluent on fish in the St. Maurice River, Quebec. Environ. Toxicol. Chem. 11:1635-1651.

Klotz, A.V., J.J. Stegeman and C. Walsh. 1984. An alternative 7-ethoxyresorufin o-deethylase activity assay: a continuous visible spectrophometric method for measurement of cycteochrome P-450 monooxygenase activity. Anal. Biochem. 40:138-145.

Munkittrick, K.R., C.B. Portt, G.J. Van Der Kraak, I.R. Smith and D.A. Rokosh. 1991. Impact of bleached kraft mill effluent on population characteristics, liver MFO activity and serum steroid levels of a Lake Superior white sucker (*Catostomus commersoni*) population. Can. J. Fish. Aquat. Sci. 48:1371-1380.

Munkittrick, K.R., M.E. McMaster, C.B. Portt, G.J. Van der Kraak, I.R. Smith and D.G. Dixon. 1992. Changes in maturity, plasma sex steroid levels, hepatic mixed function oxygenase activity and the presence of external lesions in lake whitefish (*Coregonus clupeaformis*) exposed to bleached kraft mill effluent. Can. J. Fish. Aquat. Sci. 49:1560-1569.

Payne, N.R. 1964. The life history of the walleye, *Stizostedion vitreum vitreum* (Mitchill), in the Bay of Quinte. M.A. Thesis, University of Toronto, Toronto, Ontario.

Rawson, D.S. 1957. The life history and ecology of the yellow walleye, *Stizostedion vitreum*, in Lac La Ronge, Saskatchewan. Trans. Am. Fish. Soc. 86:15-37.

Saila, S.B., C.W. Recksiek and M.H. Prager. 1988. Basic fishery science programs. Developments in Aquaculture and Fisheries Science. 18. Elsevier Science Publishers, Amsterdam. 230 p.

NUTRIENT AND CONTAMINANT EFFECTS OF BLEACHED KRAFT MILL EFFLUENT ON BENTHIC ALGAE AND INSECTS OF THE ATHABASCA RIVER

Cheryl L. Podemski and Joseph M. Culp

National Hydrology Research Institute, 11 Innovation Boulevard, Saskatoon, Saskatchewan S7N 3H5 Canada

Pulp mill effluents may result in both inhibitory (chemical toxicity) and stimulatory (through nutrient enrichment) effects on aquatic ecosystems. In September 1993, we conducted an experiment to determine if there were differences in the response of benthic algae and insects to an addition of effluent versus the addition of similar levels of bioavailable phosphorus and nitrogen. This experiment was conducted in an artificial stream system constructed on-site at a bleached kraft pulp mill located in Hinton, Alberta. The stream system was stocked with a natural community of algae and invertebrates from the Athabasca River, collected just upstream of the mill's discharge. The 28 d experiment consisted of three treatments: (1) a control, (2) a 1% dilution of treated bleached kraft mill effluent (BKME) and (3) the addition of nitrogen and phosphorus (NP) to match levels of soluble reactive phosphorus and nitrogen in the forms of nitrate/nitrite and ammonium ($N-NO_3+NO_2$ and $N-NH_4$) in the effluent. The addition of effluent or nutrients resulted in a dramatic and significant increase in the chlorophyll a (Chl_a) content of periphyton on both natural and artificial substrates, compared to control streams. There was no difference in periphyton Chl_a content in effluent and nutrient addition steams. Composition of the invertebrate community was not affected by the addition of effluent, suggesting no effect on survival. Insect growth, as measured by differences in mean thorax length, increased with the addition of effluent. Growth of *Ameletus* (Ephemeroptera, Siphloneuridae) and capniid stoneflies (Plecoptera, Capniidae) was similarly increased in streams receiving a nutrient addition. Growth of baetid mayflies was not affected by the addition of effluent or nutrients. Our results suggest that the primary effect of 1% BKME addition to the Athabasca River is to stimulate primary productivity. If there were any inhibitory effects of the effluent on insect growth, these appear to have been mitigated by the concurrent increase in food resources.

INTRODUCTION

The complex nature of pulping effluent makes prediction of their impact on aquatic communities difficult. Mill effluents contain toxicants, such as chlorinated phenolics, and resin and fatty acids that may have inhibitory effects upon the growth and production of organisms exposed to these effluents. Pulp mill effluents also contain compounds that may stimulate the growth of the algal and microbial food supplies of invertebrates, and therefore effluent additions can enrich receiving waters (Byrd *et al.* 1986; Hall *et al.* 1991; Bothwell 1992). Thus, the net impact of the addition of effluent depends upon the nutrient-contaminant interaction, because inhibitory effects of toxicants may be mitigated by the increased availability of resources.

We tested the impacts of nutrients and contaminants from pulp mill effluents on a food web comprised of primary producers (attached algae) and secondary producers (benthic insects) found in the Athabasca River. This was accomplished through the use of a system of artificial streams (Culp and Podemski 1996) designed to simulate conditions such as substratum, temperature, water velocity and dissolved oxygen found in the Athabasca River. The use of artificial streams enabled us to design a replicated experiment that would allow an examination of the interaction of nutrients and contaminants. To accomplish this, we compared the effects of the addition of bleached kraft mill effluent (BKME) (to a 1% dilution) to our experimental community, with the effect of the addition of the nutrients (nitrogen and phosphorus) contained in the effluent.

METHODS AND MATERIALS

Study Site

Weldwood of Canada Ltd. operates a bleached kraft pulp mill in Hinton, Alberta (Canada). Weldwood uses softwood furnish (65% lodgepole pine, 30% black and white spruce, 5% alpine fir) and produces an average of 1100 metric tonnes per day of market pulp. The mill combines oxygen delignification (O) with chlorine dioxide (D) bleaching and peroxide (P) (bleach sequence $OD_{100}E_{OP}(DE_sD)$). Effluent from the mill is combined with domestic sewage from the town of Hinton, and the combined effluent receives primary (200' reactor clarifier) and secondary treatment (6.5 d aerated stabilization basin) before being released to the Athabasca River. Average water usage by the mill is 113,000 m^3 d^{-1} (approximately 110 m^3 ADT^{-1}); these flows include sewage from the town (approximately 5500 m^3 d^{-1}) and the mill's clearwater bypass (approximately 45,000 m^3 d^{-1}). The effluent treatment system achieves 91.8% BOD reduction (average for 1994, T. Andrews pers. comm.). Weldwood is the second point source of nutrient loading to the river, the first being sewage from the town of Jasper, Alberta. Upstream of the mill the river is phosphorus limited (Perrin *et al.* 1994). Higher periphyton biomass and invertebrate densities are observed for 22 km downstream of Weldwood's effluent release compared to upstream (Sentar Consultants Ltd. 1993).

Experimental Design

The experiment was conducted in an artificial stream facility located alongside the Athabasca River at the Weldwood pulp mill in Hinton, Alberta. The system is described in detail in Culp *et al.* (1994) and Culp and Podemski (1996). In brief, the system consisted of 16 circular streams each with a bottom surface area of 0.9 m^2. The streams were partially recirculating; they were supplied with water from the Athabasca River at a delivery rate of 2 L min^{-1} and had a residence time of approximately 2 h. Mean water depth was 26.9 ± 0.1 cm ($\bar{x}$ ± SE). Water movement was provided by a belt-driven propeller which created a mean water velocity of 0.26 ± 0.01 m sec^{-1} (n = 150) in each stream.

The bottom of each stream was covered with a layer of thoroughly washed crushed rock. Ten stones ($\bar{x}$ surface area = 535 cm^2) collected from the Athabasca River were placed on top of the gravel. These stones were used to stock the streams with a community of periphyton and invertebrates. In the river, stones were surrounded with a U-net (Scrimgeour *et al.* 1993) and carefully lifted into a container (two stones per container) of river water. The net captured any invertebrates that were dislodged during removal of the stones. In addition, the substratum underneath each stone was disturbed to collect any invertebrates under and around the base of the stone. The stones and their associated biota were immediately transported to the streams and each pair of stones (and associated biota) were randomly assigned to a stream (ten stones per stream). The surface area sampled by ten U-nets was equivalent to the bottom surface area of the microcosms. Unglazed porcelain tiles were placed on the gravel in each stream to provide a standardized substratum for comparison of periphyton development and accumulation.

The experiment included three treatments: (1) a control receiving raw river water, (2) a 1% dilution of treated mill effluent and (3) a 1% dilution of the nitrogen and phosphorus (NP treatment) found in the effluent (i.e., concentrations in undiluted effluent: 218-307 µg L^{-1} P, 150-760 µg L^{-1} $N-NO_3+NO_2$, 311-3390 µg L^{-1} $N-NH_4$). Five replicate streams were randomly assigned to each treatment. Continuous delivery of the treatment solutions was accomplished by peristaltic pumps (Masterflex® L/S Nema, type 13 wash down controllers and cartridge pump heads). Effluent was collected daily from the mill treatment system just prior to release to the river. Samples of the effluent were collected daily and analyzed for soluble reactive phosphorus (SRP), nitrogen in the form of nitrate/nitrite ($N-NO_3+NO_2$) and nitrogen in the form of ammonium (N-NH$_4$). Concentrations in the effluent were used to set the levels of P, N-NO$_3$, and N-NH$_4$ in the NP treatment. Nutrients were supplied in the form of a solution of KH$_2$PO$_4$, NaNO$_3$ and NH$_4$Cl. A solution with concentrations equivalent to the median values of the effluent samples was added to the NP streams for an 8 d period, followed by a 1 d nutrient spike application containing concentrations equivalent to the highest effluent value recorded during the previous 8 d period. This nutrient delivery schedule (8 d median, 1 d spike, 8 d median, etc.) was used because the laboratory processing time for the effluent analysis was 7-9 d. The median and spike effluent concentrations are listed in Table 1. A nutrient spike was used because effluent concentrations were highly variable (Fig. 1) and Bothwell (1992) showed that periphyton communities were able to utilize nutrient spikes to achieve higher long-term growth rates.

Table 1. Median and spike concentrations ($\mu g\ L^{-1}$) of SRP, N-NO$_3$ and N-NH$_4$ used over the 28 d experimental period. Note that these are the effluent concentrations used to set the nutrient solution concentrations and that streams received a 1% dilution of these concentrations.

Day	SRP ($\mu g\ L^{-1}$)	N-NO$_3$ ($\mu g\ L^{-1}$)	N-NH$_4$ ($\mu g\ L^{-1}$)
1-8	218.12	547.6	1502.4
9	240.1	760.5	1577.6
10-17	221.6	683.2	1310.6
18	218.5	251.5	3390.2
19-26	160.5	245.02	942.19
27	307.3	216.5	2762.5
28	256.8	150.6	1965.2

Periphyton samples were collected every 5 d from one randomly selected stone and one tile in each replicate stream. Rocks were sampled by using a scalpel to remove biofilm from within a 9.6 cm^2 template. The entire top surface (23 cm^2) of tiles was sampled. Periphyton samples were placed in vials and held on ice until frozen later the same day. In the laboratory, each sample was homogenized, split into two parts and the concentrations of chlorophyll a (Chl$_a$) and ash-free dry mass (AFDM) determined following Marker et al. (1980) and Aoli (1990).

At the end of the 28 d experiment, invertebrates were collected by washing the entire contents of each stream through a 250 μm sieve (an upper layer of 0.63 cm hardware cloth was used to remove gravel). Invertebrates were preserved immediately in 10% formalin. Samples were sorted under 12x magnification, identified to genus whenever possible and enumerated. Due to the large numbers of immature animals, many individuals were identified only to family. Growth of numerically dominant taxa, namely the mayflies *Ameletus* (Siphloneuridae) and Baetidae and the stonefly family Capniidae, was estimated by measuring thorax length with the aid of a *camera lucida* and a digitizing pad system.

The Chl$_a$ content and AFDM of periphyton on rocks and tiles were compared at day 25 with a one-way ANOVA after log transformation to normalize the distributions. Mean comparisons were done using Fisher's protected LSD with $\alpha = 0.05$. Insect community composition was compared with the use of β-flexible, group average fusion ($\beta = -0.25$). Taxa present in fewer than 10% of the streams and comprising less than 1% of the total number of individuals present were removed from the analysis. Thorax lengths of the insects in the three treatments were compared by one-way ANOVA, and mean comparisons were done using Fisher's protected LSD with $\alpha = 0.05$.

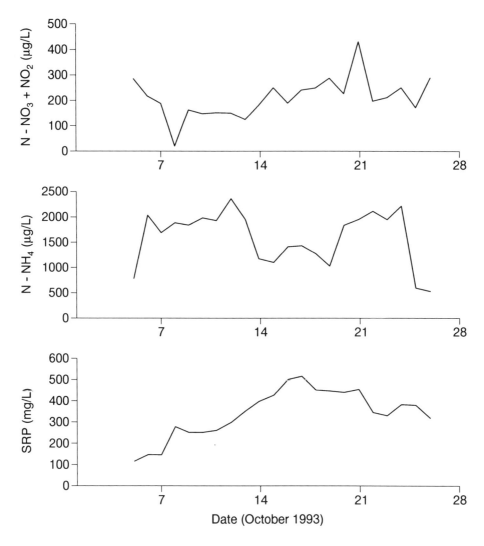

Figure 1. Concentrations of SRP, N-NO$_3$+NO$_2$ and N-NH$_4$ present in Weldwood effluent during the experiment.

RESULTS

The Chl$_a$ content of periphyton on rocks and tiles increased rapidly in streams receiving either effluent or NP additions compared to the control streams (Fig. 2). After 25 d, the Chl$_a$ concentration of periphyton on tiles in the effluent- and nutrient-treated streams was approximately 33 times higher than in the control streams and treatments were significantly different (ANOVA, $F = 242.77$, $p < 0.001$). There was no difference in the Chl$_a$ content of the periphyton on tiles in BKME and NP streams ($p = 0.38$), but BKME and NP streams were significantly different from control streams ($p = 0.05$). Significant differences were also found in the Chl$_a$ content of periphyton on rocks (ANOVA, $F = 7.14$, $p = 0.009$). Again, the Chl$_a$ content in control streams was found to be significantly lower than that of the effluent treatment ($p = 0.05$), and there was no difference in the Chl$_a$ content of periphyton in the effluent and NP streams ($p = 0.48$).

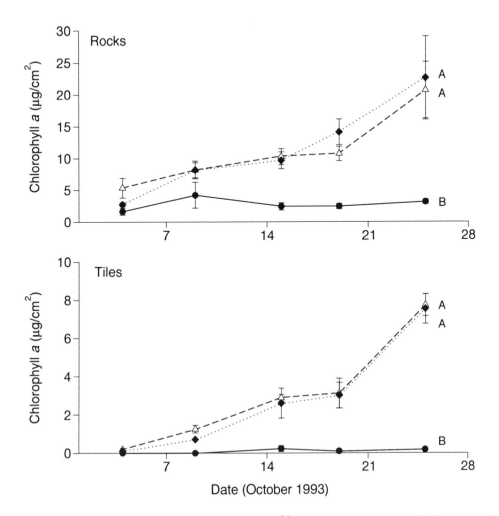

Figure 2. Mean (± 1 S.E.) chlorophyll a concentrations ($\mu g\ cm^{-2}$) of periphyton on rocks and tiles in control (●), nitrogen and phosphorus addition (NP) (♦) and BKME (△) streams; treatments with the same letter are not significantly different at $p = 0.05$.

The effect of the treatments on AFDM of periphyton was less clear (Fig. 3). During the first 2 wk of the experiment there was insufficient material on tiles to measure both Chl_a and AFDM. There was a significant effect of treatment (ANOVA, $F = 8.53$, $p = 0.05$) on the log-transformed AFDM of periphyton on tiles at day 25. Levels were significantly higher in the streams receiving effluent or nutrient additions as compared to the control ($p = 0.05$). There was no difference between the AFDM in the effluent and NP streams ($p = 0.12$). There was no significant effect of the treatment on the AFDM of periphyton on rocks (ANOVA, $F = 0.87$, $p = 0.445$).

The invertebrate community of the streams was largely comprised of mayflies and stoneflies. Numerically dominant taxa included the mayfly *Ameletus* (Siphloneuridae), mayflies belonging to the family Baetidae (most of which were too immature to identify past family) and the stonefly family Capniidae. Results of the cluster analysis on community composition data are shown in Fig. 4. The treatments did not cluster, indicating no difference in community composition between treatments.

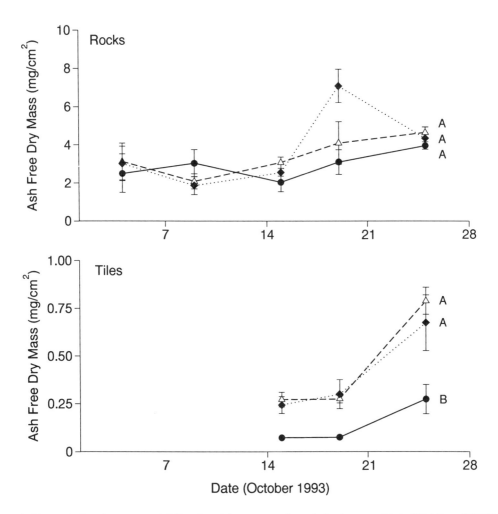

Figure 3. Mean ash-free dry mass (± 1 S.E.) of periphyton on rocks and tiles in control (●), NP (♦) and BKME (▲) streams; treatments with the same letter are not significantly different at p = 0.05.

There were significant differences in the growth of *Ameletus* nymphs, as measured by mean thorax length (ANOVA, F = 5.57, p = 0.021). Nymphs were significantly larger in streams receiving effluent or nutrient additions as compared to nymphs in the control streams (Fig. 5). There was no difference in mean thorax length of nymphs in the effluent and nutrient addition streams The same results were found for the capniid stoneflies; the effect of treatment was significant (ANOVA, F =7.5, p = 0.008). Again, nymphs in streams receiving either BKME or nutrient additions were significantly larger than those in the control, and there was no difference in mean thorax length of nymphs in the effluent and nutrient addition streams. There were no significant differences in the mean thorax length of baetid mayfly nymphs (ANOVA, F = 2.44, p = 0.129).

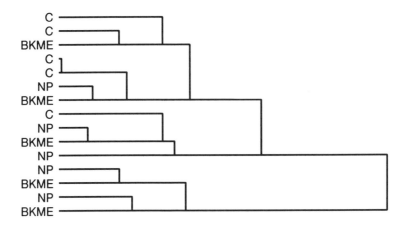

Figure 4. Dendrogram showing the results of a cluster analysis on the invertebrate community composition in the artificial streams.

DISCUSSION

The addition of a 1% dilution of BKME had a fertilizing effect on benthic algae, resulting in significant increases in periphyton Chl_a content. This result provides further evidence of nutrient limitation in the river upstream of the mill. After 25 d, Chl_a was approximately 33 times higher in streams receiving nutrient or effluent additions than in controls. This difference is much greater than the differences observed in the river; surveys in 1990-1992 showed increases of 2.2-7.5 times 0.8 km downstream of the mill's outfall (Sentar Consultants Ltd. 1993). Measurements undertaken at the same time as our experiment found mean Chl_a content to be 8.8 times higher downstream of the mill. In the river, however, grazer density is greatly increased downstream of the mill. This numerical response is not possible in our short-term experiment, and the difference in grazing pressure is probably responsible for most of the difference between the effect seen in the river and in the artificial streams. The increase in Chl_a content in response to effluent addition was not statistically different from the response to the addition of nutrients alone, suggesting that there were no inhibitory effects of the addition of 1% effluent. Biomass of periphyton on porcelain tiles showed a similar effect, while there was no effect of effluent on the biomass on rocks. This discrepancy was probably due to the larger variation in the AFMD from rocks, which was likely caused by variation in the amount of detritus present on the rocks when they were removed from the river.

The invertebrate community in the artificial streams showed no changes in composition in response to the addition of effluent or nutrients. This result suggests that there was no acute toxicity of the effluent. Changes in community composition consistent with nutrient enrichment, are often observed as a response to effluent additions (NCASI 1989a,b; Sentar Consultants Ltd. 1993). This type of community response is likely a long-term response that was not possible in our experiment given its relatively short duration and the closed nature of the system. Growth of baetid mayflies, as measured by differences in mean thorax length, was not significantly affected by the addition of either effluent or nutrients. The majority of baetids were quite immature (mean body length = 1.7 mm, n = 2718, SE = 0.0034). It is possible that insufficient growth occurred during the experiment to allow any response to be seen. These results differ from Lowell et al. (1994), who found increased growth in *Baetis* in response to exposure to a 1% dilution of pulp mill effluent over a period of only 2 wk. Growth of *Ameletus* and capniid stoneflies, however, was significantly increased by the addition of effluent. This increase was not different from the increase observed in response to the addition of nutrients. Our results suggest that either there were no sublethal effects of effluent on growth or that these effects were mitigated entirely by the increased food available.

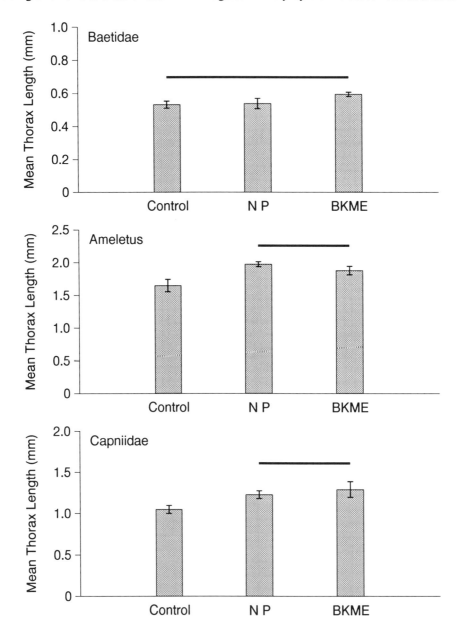

Figure 5. Mean thorax length (± 1 SE) of baetid and Ameletus (Siphloneuridae) mayflies, and capniid stoneflies; bars connect means that are not significantly different at $p = 0.05$.

The results of our experiment demonstrate that during autumn, a 1% dilution of the combined sewage and pulp mill effluent resulted in increased periphyton production. This response indicates that any sublethal effects on periphyton growth are overwhelmed by the nutritive effect of the effluent. The response to the effluent addition was not different from the response to the addition of nutrients alone, further demonstrating that the overwhelming effect of the effluent is one of nutrient enrichment. This conclusion refers, of course, only to a 1% dilution, as higher concentrations of effluent could produce chronic toxicity effects (Soniassay et al. 1977; Lowell et al. 1994) or decreased growth due to color-related reductions in light penetration (Soniassay et al. 1977; NCASI 1985). Effluent additions had no impact on insect community structure, suggesting no acute toxicity. The effect of effluent on insect growth was taxon-specific. For example, *Ameletus* and capniid nymphs grew to a larger size in the presence of effluent or nutrients, probably because the effect of effluent was to increase food resources and, thus, growth. Baetid mayflies, however, showed no response to either treatment, suggesting that effluent had no sublethal effects and that growth was not food-limited in any treatment, or simply that so little growth had occurred that no response could have been observed. In autumn 1994, we repeated this experiment and increased its duration to allow increased time for growth of *Baetis* to be measured. However, data from this second experiment were not available when this paper was submitted for publication.

ACKNOWLEDGMENTS

We are indebted to J. Mollison, J. Banner, C. Casey, C. Perrin and E. Marles for technical assistance in construction of the stream system. D. Kloeble, T. Luong, G. Hutchinson and P. Burgess provided valuable assistance in the laboratory and A. Dale and D. Halliwell assisted with manuscript preparation. P. Chambers, G. Scrimgeour, F. Wrona and K. Cash provided helpful comments during the development of this project. We also thank Weldwood of Canada (Hinton Division) and the staff of the Technical & Environmental Services Department for their cooperation during the project. T. Andrews (Weldwood) provided information on the mill's furnish, bleach sequence, water usage and treatment system. P. Chapman, D. Borton, and K. Munkittrick reviewed and improved the manuscript. This study was supported by funds to J.M.C. from the National Hydrology Research Institute (Environment Canada) and the Northern River Basins Study. C.L.P. was supported by a Natural Sciences and Engineering Research Council of Canada Post-graduate Scholarship.

REFERENCES

Aoli, J. 1990. A critical review of recent freshwater periphyton field methods. Can J. Fish. Aquat. Sci. 47:656-670.

Bothwell, M.L. 1992. Eutrophication of rivers by nutrients in treated kraft pulp mill effluent. Water Pollut. Res. J. Can. 27:447-472.

Byrd, J.F., E.J. Eysenbach and W.E. Bishop. 1986. The effect of treated pulping effluent on a river and lake ecosystem. TAPPI J. 69:94-98.

Culp, J.M., and C.L. Podemski. 1996. Design and application of a novel stream microcosm system for assessing effluent impacts to large rivers. *In* Environmental Fate and Effects of Pulp and Paper Mill Effluents. M.R. Servos, K.R. Munkittrick, J.H. Carey and G. Van Der Kraak (ed.). St. Lucie Press, Delray Beach, FL.

Culp, J.M., C.L. Podemski and C. Casey. 1994. Design and application of a transportable experimental stream system for assessing effluent impacts on riverine biota. Northern River Basins Study. Edmonton, AB.

Hall, T.J., R.K. Haley and L.E. Lafleur. 1991. Effects of biologically treated kraft mill effluent on cold water stream productivity in experimental stream channels. Environ. Toxicol. Chem. 10:1051-1060.

Lowell, R.B., J.M. Culp and F.J. Wrona. 1994. Stimulation of increased short-term growth and development of the mayfly *Baetis tricaudatis* from the Thompson River Basin following exposure to biologically treated pulp mill effluent. NHRI Contribution No. CS-94007. National Hydrology Research Institute, Saskatoon, SK.

Marker, A.F.H, E.A. Nusch, H. Rai and B. Reiman. 1980. The measurement of photosynthetic pigments in freshwaters and the standardization of methods: conclusions and recommendations. Arch. Hydrobiol. Bieh. Ergebn. Limnol. 14:91-106.

NCASI. 1985. Effects of biologically treated bleached kraft mill effluent on cold water stream productivity in experimental stream channels - fourth progress report. NCASI Technical Bulletin No. 474.

NCASI. 1989a. Pulping effluents in the aquatic environment - Part 1: A review of the published literature. NCASI Technical Bulletin No. 572.

NCASI. 1989b. Pulping effluents in the aquatic environment - Part 2: A review of the unpublished literature. NCASI Technical Bulletin No. 573.

Perrin, C.J., P.A. Chambers and M.L. Bothwell. 1994. Growth rate and biomass response of periphytic algae to nutrient enrichment of stable and unstable substrata in the Athabasca River, Alberta. Draft report prepared by Limnotek Research and Development Inc. for Northern River Basins Study, Edmonton, AB. 23 p.

Scrimgeour, G.J., J.M. Culp and N.E. Glozier. 1993. An improved technique for sampling lotic invertebrates. Hydrobiologia 254:65-71.

Sentar Consultants Ltd. 1993. Overview of the long-term changes in benthic macroinvertebrates in the Athabasca River near Hinton. Report prepared for Weldwood of Canada, Hinton Division. 36 p.

Soniassay, R.N., J.C. Mueller and C.C. Walden. 1977. The effects of BKME colour and toxic constituents on algal growth. Pulp Paper Can. 78(8):T179.

FIELD AND LABORATORY RESPONSES OF WHOLE ORGANISMS ASSOCIATED WITH PULP MILLS

Stella Swanson

Golder Associates Ltd., 1011 Sixth Avenue S.W., Calgary, Alberta, T2P 0W1 Canada

The measurement and assessment of the response of whole organisms to pulp mill effluent covers a wide variety of approaches, organisms and types of receiving environments. In addition, the type of pulp mill (wood furnish, bleaching sequence, type of effluent treatment, volume of discharge) has a direct bearing on the nature and extent of expected impacts. Recent studies summarized in this paper reflect the diversity of pulp mill types and receiving environments.

INTRODUCTION

This review summarizes the major advances in investigating and understanding the effects of pulp mill effluents on whole organisms. Many of the papers described approaches that combined whole organism endpoints with population or community level endpoints. Thus, there was a continuum from whole organism to higher level effects. This continuum formed the basis for much of the discussion surrounding the applicability of whole organism responses as indicators of population- or community-level effects.

ADVANCES IN THE INVESTIGATION AND UNDERSTANDING OF WHOLE ORGANISM EFFECTS

Freshwater Fish (Temperate/Subtropical)

A wide range of investigations on the responses of fish to pulp mill effluents were recently presented at the 2nd International Conference on Fate and Effects of Bleached Pulp Mill effluents. Whole organism responses reported included histopathology and parasite load, size at age, condition, relative gonad size, relative liver size and behavior. Investigations of whole organism responses were often combined with biomarkers such as mixed function oxidase (MFO) activity and blood cell counts. Less frequently, multi-level studies combined biomarker, whole organism and population responses; population endpoints included size/age distribution, growth, production and survival. Approaches included laboratory toxicity testings, field microcosms, field mesocosms and full-scale field studies.

Overstreet and Hawkins (1996) suggested that the mosquitofish (*Gambusia affinis*) is a good sentinel species because of its abundance, restricted home range, position in food webs, susceptibility to parasitic infections and a tendency towards rapid development of neoplasms. It can also be bred in captivity. Effects of effluents from a combined softwood/hardwood pulp mill utilizing oxygen delignification and chlorine dioxide substitution were investigated. Fish in a canal receiving direct effluent as well as those downstream were "healthy" in terms of species richness of parasites as well as the absence of pathogenic parasites known to be associated with stress. No cancerous or other lesions were present. The most stressed site was upstream from the effluent canal (as measured by low species richness of parasites, invasion into host tissues by a ciliate, a high prevalence of macrophage aggregates in the spleen and high prevalence of vacuolated hepatocytes). Parasite infection was greatest in fish from the effluent canal, indicating a biologically rich area.

Gibbons et al. (1994) reported on the use of a small forage fish (peamouth chub *Mylocheilus caurinus*) as a possible sentinel species in the Fraser River. Peamouth chub are an alternative to more mobile large fish species. Whole organism measures (size at age, mean age, condition, gonad size, liver weight) were not different upstream and downstream of the mill. There was a 2 to 2.5-fold ethoxyresorufin-*o*-deethylase (EROD) induction in exposed fish. Final research results will indicate whether MFO activity and steroid levels in wild peamouth chub show similar responses to laboratory-exposed goldfish and rainbow trout.

Krozynski et al. (1994) reported that fish behavior can influence the degree of exposure. Chinook salmon (*Oncorhynchus tschawytcha*) behavior in the Fraser River brought them into close proximity to an effluent plume because of selection of nearshore habitats.

Simultaneous biomarker/whole organism/population-level endpoints in controlled experiments were part of a sophisticated investigation of pulp mill effects in NCASI (National Council for Air and Stream Improvement) experimental streams. These experiments are examples of the rare studies that combine several levels of response. Therefore, they provide valuable information about the predictive capability of physiological and whole organism measurements with respect to population-level effects. In the NCASI experiments described by Borton et al. (1996), effluent from a bleached kraft pulp mill utilizing high chlorine dioxide substitution and secondary treatment was used to produce 4 and 8% treatments. No effects on survival, growth or production of four test species, largemouth bass (*Micropterus salmoides*), channel catfish (*Ictalurus punctatus*), bluegill sunfish (*Lepomis macrochirus*) and golden shiners (*Notemogonus chrysoleucas*), were observed at 4% effluent. At 8% effluent, bluegill sunfish growth was less but production and survival were not adversely affected. Survival and production of channel catfish and largemouth bass were greatest in 8% effluent. Total production of all fish in 8% effluent was nearly double the other treatments (0 and 4%). There were no effects on parasite infestation in either treatment. Condition factors of largemouth bass were less than controls in both effluents and spleen somatic index was greater; however there were significant differences in control fish in these parameters as well. Channel catfish from the two control streams had significantly different condition factors, liver somatic indices (LSIs), hematocrits and leucocrits, and largemouth bass from controls had significantly different LSIs. These differences among control fish cast doubt on interpretation of differences between exposed and control fish for these whole organism parameters. EROD activity was elevated in both effluent treatments. However, since the survival, growth and production of fish were not adversely affected, there was no correspondence between biomarker and population-level responses. It was concluded that we need more knowledge of normal ranges of biomarkers.

Results of an Environmental Effects Monitoring (EEM) style study by Swanson et al. (1996) on North Saskatchewan River white sucker (*Catostomus catostomus*) and walleye (*Stizostedion* vitreum) populations exposed to bleached kraft mill effluent (high chlorine dioxide substitution, secondary treatment) showed that although EROD activity was higher in exposed white suckers, the pattern of responses to pulp mill effluent seen in other Canadian studies was not observed. This pattern has included decreased growth, delayed age to maturity, increased condition factor, increased liver size, reduced survival, reduced fecundity with age and smaller gonads. In contrast, white suckers from the North Saskatchewan had no significant differences in growth, age to maturity, condition factors, liver sizes, gonad size, total body size regressions and egg diameter with age. The only differences were younger and faster maturing white suckers with smaller gonadal somatic indices (GSIs) and a higher incidence of pale livers and parasites. There was no apparent cause of the younger, faster growing fish. The higher incidence of parasites may be due to a greater abundance of intermediate invertebrate hosts in the North Saskatchewan. Differences observed in walleye populations were also not apparently related to effluent exposure. Age structure differences in walleye may have reflected differences in commercial exploitation rates. River flows affected sample success and spawning success. It was concluded that detecting incremental population-level effects from one particular stressor in the absence of a strong contaminant gradient seems unlikely. The results of other EEM studies (due out in 1996) will show whether this conclusion is correct.

Sandström (1994) presented a review of Scandinavian and North American studies of *in situ* effects. Most were kraft mills (some with high chorine dioxide substitution). There were two sulfite mills. Treatment was not specified. MFO induction was the most consistent response. Some increases in growth were noted (especially in Swedish studies of perch) as well as some decreases, but the majority showed no effect. Several studies reported increases in condition factor (11 out of 18); there was one decrease and several no effects. An increased age to maturity was noted in 9 of 12 studies. Reduced GSIs were observed in 14 of 24 studies; there was 1 increased GSI. Decreased fecundity occurred in 2 of 6 studies and decreased egg or sperm viability was noted in 5 of 5 studies. Sandström (1996) recommended that more effort should be devoted to a more precise application of life history modeling, i.e., age, size at first maturity and recruitment. Larger samples are required for growth comparisons and important parameters such as temperature are rarely considered in growth comparisons. Sandström (1994) observed that secondary treatment could not eliminate effects on growth and condition factor. Follow-up studies at Norrsundett after chlorine dioxide substitution showed a significant recovery in sexual maturation and gonad growth in perch. The effects of natural wood sterols require investigation. The implementation of EEM requirements at Canadian pulp mills was seen as a positive step.

Benthic Algae/Benthic Invertebrates

Advances in the understanding of pulp mill effects have focused on the overall effect of nutrients combined with potentially toxic constituents. Lowell *et al.* (1996) observed mayfly response to biologically treated kraft effluent (1 and 10%) in artificial streams in an outdoor enclosure beside a pulp mill on the Thompson River. Effects were stimulatory; effluent-exposed mayflies grew faster than high-food controls, which already had access to more food than they could eat. The authors suggested that there may be increases in nutritive value of food, a stimulatory effect on mayfly feeding or hormonal or other metabolic effects. Stable isotope work showed a link to a terrestrial source of carbon (pulp wood); thus, effluent may have increased availability of high-quality detrital and/or microbial food. The authors also suggested that juvabione-type (juvenile hormones) compounds and ecdysone-type compounds (molting hormone) may play a role. Juvabione may act by inhibiting morphogenesis and differentiation of reproductive organs, allowing continued somatic growth. Diterpenes could interfere with hormonal control of growth and development. An increase in growth and molting could be due to combined response to several of these compounds. Alternatively, hormesis may be involved, i.e., low levels of stressors having a stimulatory effect. Long-term patterns in community structure indicated a positive correlation between abundances of some families of mayflies, stoneflies and caddisflies and mill output. This very interesting study illustrates the complexity of pulp-mill-related effects and also points to some additional hypotheses to be tested.

Classification analysis can be an effective tool for examining faunal assemblages in exposed and reference areas. Rakocinski *et al.* (1996) studied benthic communities in the Sulphur River on the Texas/Arkansas border which receives treated effluent from a paper mill. Classification analysis showed close macrobenthic similarity between the effluent site and a downstream site. The macrobenthic composition of the outfall canal resembled that of neighboring river sites with high total densities at both. The predominant macrobenthic pattern suggested organic enrichment; nonetheless, high species richness and typical taxonomic diversity occurred at the effluent site.

Artificial stream systems were a common theme among researchers studying benthic organisms, especially among those interested in the effects of nutrients. Podemski and Culp (1996) described an artificial stream system in the Athabasca River which was used for a 28-d experiment involving addition of 1% bleached kraft mill effluent (BKME) or nutrients at levels found in the effluent. No difference in periphyton chlorophyll *a* between effluent and nutrient-addition streams was observed. Composition of the invertebrate community was not affected, suggesting no effect on survival. Insect growth was increased by the addition of effluent, although not uniformly among species. Growth of *Ameletus* was also increased by nutrient addition but baetid mayfly growth was not affected by effluent. The authors concluded that the primary effect of BKME in the Athabasca River was to stimulate primary productivity.

Another interesting *in situ* experimental technique was described by Scrimgeour *et al.* (1994) who used diffusion substrata containing various combinations of nutrients, nutrients plus a herbivore-inhibiting compound and whole mill effluent plus phosphorus in the Athabasca River. Epilithic biomass and invertebrate density were highest in P and N+P treatments. Epilithic biomass was twofold higher when diffusing substrata contained nutrients and a herbivore-inhibiting compound. Algal biomass and invertebrate density were affected by whole effluent concentration and typically decreased with increasing effluent concentration. When combined, results suggested that P is the nutrient limiting primary production and that the magnitude of multiple trophic-level responses may be best understood as arising from combined effects on nutrients and contaminants. The technique used in these experiments appears to be a very useful tool for examining the effects of various combinations of parameters on *in situ* biota.

Marine/Estuarine Biota

Khan *et al.* (1996) documented effects from untreated thermomechanical/sulfite effluent in a marine inlet (Newfoundland). Effects included higher liver and ovarian somatic indices (related to delayed spawning in exposed fish), lower hemoglobin and lymphocyte levels, higher incidence of liver and gill lesions and elevated levels of hemosiderin in spleen and kidney (associated with sulfite-laden sediment). These results clearly indicate the obvious impacts of untreated effluent.

The meiofauna community was suggested as an alternate and often overlooked monitoring tool by Sundelin and Eriksson (1996). They reported effects on meiofauna in the Baltic Sea when no effects on macrofauna were evident several kilometers outside an area reported as affected in earlier studies. The differences between macrofauna and meiofauna may be because macrofauna respond to enriched organic load while meifoauna are sensitive to toxicants. The ostracod *Paracyprideis fennica* was particularly sensitive, followed by nematodes. Kinorynchs and harpacticoid copepods had a weaker response and turbellarians were resistant. Comparison of before and after change in bleaching from chlorine to oxygen and chlorine dioxide showed increased biomass though no increase in total abundance; ostracods returned.

Deavin (1994) described a toxicity test using the Tasmanian blenny as a organism for indicating early life-stage effects from primary-treated eucalypt effluent. Significant effects were seen on hatch rate, mortality at hatch and length of the larvae at effluents concentrations above 7%. Craniofacial deformities and spinal curvatures were also observed in the 2.5, 5 and 10% treatments. The blenny test appears to be a sufficiently sensitive and practical test for effects on marine fish.

METHODS AND APPROACHES

In their review of laboratory responses of whole organisms to pulp mill effluents, Kovacs and Megraw (1996) noted that increased chlorine dioxide substitution reduces toxicity of bleaching effluents but not combined effluents (pulping and bleaching). The benefits of oxygen delignification and totally chlorine free (TCF) versus elemental chlorine free (ECF) are inconclusive. Other factors that appear to be important determinants of toxicity potential are wood furnish, effectiveness of in-plant control measures and efficiency of secondary treatment. Kovacs and Megraw (1996) emphasized that comparisons of before and after changes in process or treatment technology can only be done at one specific mill. He also concluded that pulping contributes to toxicity, and changes to pulping in addition to changes in bleach plant are required to achieve further improvements.

Kovacs and Megraw (1996) also concluded that because field surveys reflect historical process operations plus current conditions, we should use predictive tools such as toxicity tests and model ecosystem studies (such as those used by Lehtinen [1996]). He cautioned that extrapolation to the field is very difficult because of highly variable field conditions, the presence of other stressors and the uncertainty attached to extrapolation to other species. He stressed that laboratory tests should make use of a "battery approach," using several species. He also stressed that effluent concentrations in laboratory tests should be representative of concentrations seen in receiving environments.

New methods and approaches described during the conference included transportable stream microcosms, a basin-wide study of fate and effects of contaminants and, risk assessment. Ecological versus statistical significance was discussed. The findings of laboratory toxicological tests were reviewed, including discussion of the effects of various process and treatment combinations on effluent toxicity and the ability of laboratory tests to predict field effects at individual and population levels.

Culp and Podemski (1996) described a transportable stream microcosm system consisting of 16 circular 0.9-m^2 streams. Multiple trophic-level effects are examined by seeding the tanks with natural substrate and biota from the river. The system is partially recirculating and water depth and hydraulic residence times are easily adjusted. The system has been used to study the nutrient and contaminant effects of treated effluents from a kraft pulp mill on complex food webs in the Athabasca River and appears to be a useful design for other river studies.

Model ecosystems using functional and structural endpoints were described by Lehtinen (1996) Functional endpoints included carbon dioxide consumption, organic carbon turnover and net system oxygen production. Structural endpoints included growth of bladder-wrack, biomass of filamentous algae, sediment invertebrates and fish production. Responses in both functional and structural levels were observed regardless of bleaching levels. External treatment reduced the responses; responses were relatively weak on a structural level. There were no strong differences between ECF and TCF effluents. It can be argued that these functional and structural endpoints are more ecologically meaningful than single, whole organism responses.

Risk assessment approaches have the potential to assist in the screening and prioritization of issues surrounding the effects of bleached pulp mill effluents. Solomon (1994) presented a risk assessment of organochlorine compounds produced by the use of chlorine dioxide. They found that if dilution factors are higher than ten, risk quotients for the 95th percentile concentrations of chlorinated phenols will have margins of safety for aquatic organisms of more than 1000-fold (based upon laboratory toxicity test data). Significant reductions in effects from mills with high chlorine dioxide substitution have been observed in microcosm and field studies. The authors concluded that mills bleaching with chlorine dioxide and employing secondary treatment and with dilution typical of North American mills present an insignificant risk to the environment from organochlorine compounds. This effectively sums up one of the main overall themes of the conference: the decline in significance of organochlorine compounds and a switch to interest in nutrient enrichment and possible effects from natural wood compounds.

Ecological versus statistical significance is an important issue with respect to the design and interpretation of both laboratory and field studies. Kilgour and Sommers (1994) suggested that if the difference between two means is greater than two standard deviations, then differences are likely to be ecologically significant. Two times the standard deviation means that the observed difference provides a signal that is four times stronger than background variation; to use this approach, sampling design must adequately quantify variation in reference areas. This implies considerable effort in reference areas, a prominent theme in EEM studies.

Issues

A number of issues were raised during a very lively discussion period. These included:
- What is an "effect?"
- Do whole organism effects predict population, community or ecosystem effects?
- Laboratory versus mesocosm versus field approaches: what is the best mix?
- The importance of secondary treatment in eliminating whole organism effects
- ECF versus TCF effluent effects
- The influence of confounding variables such as habitat gradients and other discharges
- The influence of the type of ecosystem on observed response (relative resilience, robustness)
- Stimulatory versus inhibitory responses
- What are the best effects indicators?

- What is "significant" at the population level?
- How well do we need to understand the ecosystem to extrapolate from whole organism responses?
- The need to normalize results to type of mill (wood furnish, process, treatment)
- The role of adaptation; is it "bad" to have slight MFO induction?
- Risk assessment (predictive) versus impact assessment (retrospective)
- The importance of understanding recruitment and mortality in populations
- Need better measures of reproduction (e.g., fry survival)
- The need for better sample design, more replication, *in situ* mesocosm approaches, a "suite" of indicators
- Ecological relevance and the role of common sense - very sensitive bioassays are available yet effects in the field are not always observed
- Is it worth trying to look for population-level effects? Need connection between chemical/physiological data and populations
- What does "no observed effect" at the population level mean? Have we looked hard enough? Do we know enough to say this?
- Risk assessment shows that we appear to be close to the threshold of effect/no effect

The overall conclusion stated by Kovacs and Megraw (1996) deserves special emphasis: there needs to be a greater effort towards understanding the sources and agents responsible for residual toxicity of BKME, especially with respect to effects on reproduction and on the relationship between laboratory and field responses. This will lead to more defensible conclusions regarding future strategies for effluent management.

REFERENCES

Borton, D.L., P.A. Van Veld, R.E Wolke and L. Lafleur. 1996. Survival, growth, production and biomarker responses of fish exposed to high substitution bleached kraft pulp mill effluent in experimental streams. *In* Environmental Fate and Effects of Pulp and Paper Mill Effluents, M.R. Servos, K.R. Munkittrick, J.H. Carey and G. Van Der Kraak (ed.), St. Lucie Press, Delray Beach, FL.

Culp, J.M., and C.L. Podemski. 1996. Design and application of a novel stream microcosm system for assessing effluent impacts to large rivers. *In* Environmental Fate and Effects of Pulp and Paper Mill Effluents, M.R. Servos, K.R. Munkittrick, J.H. Carey and G. Van Der Kraak (ed.), St. Lucie Press, Delray Beach, FL.

Deavin, J.G. 1994. Development of an early life-stage bioassay to determine sublethal effects of primary-treated eucalypt-based pulp mill effluent using a marine species, the Tasmanian blenny (*Parablennius tasmanianus*). Presented at the Second International Conference on Environmental Fate and Effects of Bleached Pulp Mill Effluents, November 6-10, 1994, Vancouver, B.C., Canada.

Gibbons, W.N., K.R. Munkittrick and W.D. Taylor. 1994. Peamouth chub (*Mylocheilus caurinus*) response to BKME exposure to effluent in the Upper Fraser River, B.C. Presented at the Second International Conference on Environmental Fate and Effects of Bleached Pulp Mill Effluents, November 6-10, 1994, Vancouver, B.C., Canada.

Khan, R.A., D.E. Barker, K. Ryan, B. Murphy and R.G. Hooper. 1996. Abnormalities in winter flounder (*Pleuronectes americanus*) living near a paper mill in the Humber Arm, Newfoundland. *In* Environmental Fate and Effects of Pulp and Paper Mill Effluents, M.R. Servos, K.R. Munkittrick, J.H. Carey and G. Van Der Kraak (ed.), St. Lucie Press, Delray Beach, FL.

Kilgour, B.W and K.M. Sommers. 1994. Resolving ecologically versus statistically significant effect sizes: invoking a 95% rule using ± 2 SDs. Presented at the Second International Conference on Environmental Fate and Effects of Bleached Pulp Mill Effluents, November 6-10, 1994, Vancouver, B.C., Canada.

Kovacs, T.G., and S.R. Megraw. 1996. Laboratory responses of whole organisms exposed to pulp and paper mill effluents: 1991-1994. *In* Environmental Fate and Effects of Pulp and Paper Mill Effluents, M.R. Servos, K.R. Munkittrick, J.H. Carey and G. Van Der Kraak (ed.), St. Lucie Press, Delray Beach, FL.

Kruzynski, G.M., I.K. Birtwell and B. Emmett. 1994. Behavioral considerations in the design of a study of the effects of BKME on juvenile chinook salmon (*Oncorhynchus tschawytcha*) in the upper Fraser River. Presented at the Second International Conference on Environmental Fate and Effects of Bleached Pulp Mill Effluents, November 6-10, 1994, Vancouver, B.C., Canada.

Lehtinen, K-J. 1996. Biochemical responses in organisms exposed to effluents from pulp production - are they related to bleaching? *In* Environmental Fate and Effects of Pulp and Paper Mill Effluents, M.R. Servos, K.R. Munkittrick, J.H. Carey and G. Van Der Kraak (ed.), St. Lucie Press, Delray Beach, FL.

Lowell, R.B., J.M. Culp and F.J. Wrona. 1996. Effects of pulp mill effluent on benthic freshwater invertebrates: food availability and stimulation of increased growth and development. *In* Environmental Fate and Effects of Pulp and Paper Mill Effluents, M.R. Servos, K.R. Munkittrick, J.H. Carey and G. Van Der Kraak (ed.), St. Lucie Press, Delray Beach, FL.

Overstreet, R.M. and W.E. Hawkins. 1996. The western mosquitofish as an environmental sentinel: parasites and histological lesions. *In* Environmental Fate and Effects of Pulp and Paper Mill Effluents, M.R. Servos, K.R. Munkittrick, J.H. Carey and G. Van Der Kraak (ed.), St. Lucie Press, Delray Beach, FL.

Podemski, C.L. and J.M. Culp. 1996. Nutrient and contaminant effects of bleached kraft mill effluent on benthic algae and insects of the Athabasca River (Alberta, Canada). *In* Environmental Fate and Effects of Pulp and Paper Mill Effluents, M.R. Servos, K.R. Munkittrick, J.H. Carey and G. Van Der Kraak (ed.), St. Lucie Press, Delray Beach, FL.

Rakocinski, C.F., M.R. Milligan and R.W. Heard. 1996. Comparative evaluation of macrobenthic assemblages from the Sulphur River Arkansas in relation to pulp mill effluent. *In* Environmental Fate and Effects of Pulp and Paper Mill Effluents, M.R. Servos, K.R. Munkittrick, J.H. Carey and G. Van Der Kraak (ed.), St. Lucie Press, Delray Beach, FL.

Sandström, O. 1994. *In situ* assessments of pulp mill effluent impact on life-history variables in fish. National Board of Fisheries, Institute of Coastal Research, Gamla Slipvägen 19, S-740 71, Öregrund, Sweden.

Sandström, O. 1996. *In situ* assessment of the impact of pulp mill effluent on life-history variables in fish. *In* Environmental Fate and Effects of Pulp and Paper Mill Effluents, M.R. Servos, K.R. Munkittrick, J.H. Carey and G. Van Der Kraak (ed.), St. Lucie Press, Delray Beach, FL.

Scrimgeour, G.J., P.A. Chambers, J.M. Culp and C.L. Podemski. 1994. Effects of bleached pulp mill effluent on algal and benthic communities: separating between nutrient and contaminant impacts. Presented at the Second International Conference on Environmental Fate and Effects of Bleached Pulp Mill Effluents, November 6-10, 1994, Vancouver, B.C., Canada.

Solomon, K., H. Bergman, D. Mackay, B. McKauge and D. Pryke. 1994. Ecotoxicological risks from organochlorine compounds produced by the use of chlorine dioxide for the bleaching of pulp. Presented at the Second International Conference on Environmental Fate and Effects of Bleached Pulp Mill Effluents, November 6-10, 1994, Vancouver, B.C., Canada.

Sundelin, B. and A-K. Eriksson. 1996. Effect monitoring in pulp mill areas - response of the meiofauna community to altered process technique. *In* Environmental Fate and Effects of Pulp and Paper Mill Effluents, M.R. Servos, K.R. Munkittrick, J.H. Carey and G. Van Der Kraak (ed.), St. Lucie Press, Delray Beach, FL.

Swanson, S.M., R. Schryer and B. Firth. 1996. Whole organism and population response of sentinel fish species in a large river receiving pulp mill effluents. *In* Environmental Fate and Effects of Pulp and Paper Mill Effluents, M.R. Servos, K.R. Munkittrick, J.H. Carey and G. Van Der Kraak (ed.), St. Lucie Press, Delray Beach, FL.

SECTION VI

INTEGRATED MONITORING

The shift of monitoring programs away from simple chemical "end-of-the-pipe" measurements to fully integrated monitoring programs is necessary to ensure the protection of complex ecosystems. The chemical composition of effluents, environmental fate and bioavailability interact to result in toxicity, or lack of toxicity, in different species at different levels of organization. Simply protecting a single sensitive species or component within the environment does not ensure the integrity of aN ecosystem in the receiving environment. Integrated monitoring increases the validity, confidence and interpretation of the monitoring data. The success of the implementation of new technologies can be best documented using a well-designed, ecologically based monitoring approach.

THE ROLE OF BIOLOGICAL MEASUREMENTS IN MONITORING PULP AND PAPER MILLS IN AUSTRALASIA AND S.E. ASIA

Bruce M. Allender[1], John S. Gifford[2] and Paul N. McFarlane[2]

[1]Amcor Research & Technology Centre, PO Box 1, Fairfield, Victoria, 3078, Australia
[2]New Zealand Forest Research Institute, Private Bag 3020, Rotorua, New Zealand

The approach to biological monitoring in environmental regulation is reviewed for Australia, Indonesia, Malaysia, New Zealand, Philippines and Thailand. The European and North American influences are considered in relation to local needs. Local developments and procedures for biomonitoring of pulp and paper mill effluents are described.

INTRODUCTION

Over the past 20 years there has been substantial research on the biological impacts of pulp and paper mill effluents into both fresh and marine waters in North America and Scandinavia (McLeay 1987; Priha 1991). This has led to a number of different environmental monitoring procedures which are now built into the regulations to control discharges. Should these same procedures be applied directly in other countries? Can prescriptive regulations be applied to all discharges? Should site-specific or country-specific environmental assessment approaches be developed?

This paper considers the role of biological monitoring in the pulp and paper industry in Australasia and S.E. Asia with regard to the local industry and regulatory environment. Biomonitoring developments in Australia, New Zealand, Indonesia, Malaysia, Thailand and the Philippines are reviewed because they are the countries with the most significant pulp and paper developments in the region.

Water quality issues are of concern to a greater or lesser extent in all countries. Although the focus here is on pulp and paper effluents, the trends discussed also apply to other industries and their effluents.

A particular concern with the pulp and paper industry has been the discharge of chlorinated organic compounds associated with the bleaching of chemical pulp. Only examples from the chemical pulp sector of the industry are discussed, but it is acknowledged that effluents from other pulp and paper operations such as mechanical pulping and recycled fiber paper making can also have biological impacts and that biological monitoring is relevant for the whole industry.

REGIONAL REGULATORY FRAMEWORKS

Each of the countries in the region has environmental requirements for discharges and ambient water quality. In this discussion, biological monitoring for ecosystem protection is placed in the context of these regulatory frameworks. Bacteriological monitoring for the protection of human health will not be reviewed.

The environmental policies and legislation in individual countries influence the extent and nature of effluent monitoring (including any biological monitoring) programs. For example, Thailand and Indonesia currently set different end-of-pipe standards for different industrial wastewaters, while in the Philippines, Malaysia, Australia and New Zealand water usage of the recipient water is the main factor in deciding effluent standards (ANZECC 1992; Mino 1993).

BIOLOGICAL MONITORING REGULATORY ENVIRONMENTS IN S.E. ASIA

In neither Indonesia, Malaysia, the Philippines nor Thailand is there any biological monitoring or toxicity testing in the licensing schedules for the pulp and paper industry. Nevertheless regulatory authorities are very much aware of such requirements in other countries, and all of these countries are making enormous efforts to improve their environmental regulation framework, often based on models from other countries. For example, Thailand has a joint project with the United States Environmental Protection Agency (U.S. EPA) to develop environmental policy and procedures. Indonesia is using off-shore consulting groups for the same purpose and now has the Environmental Impact Assessment Board (BAPEDAL) to implement pollution control strategies (Pratomo 1994).

It is not clear how imported concepts will be adapted in each country to meet national and local environmental goals as distinct from global goals. Metha (1994), in discussing the impact of the U.S. EPA's proposed cluster rule on Asia and the Pacific, pointed out that as the pulp and paper industry is globalized, there will be international pressures to control a large number of pollutants. He emphasized the importance of each country coming up with standards that are reasonable, rational and appropriate for the region. But only local modifications to the U.S. EPA approach are suggested. There may be more appropriate ways for achieving this objective based on water quality objectives.

Probably different national approaches will be developed. Indonesia, for example, has substantially increased end-of-pipe monitoring in the pulp and paper industry. There are more stringent effluent standards for conventional parameters (flow, BOD, COD and TSS) and additional limits (including AOX, phenol, phosphoros and sulfide) at different levels for different types of pulp and paper processes (Metha 1994). These are applicable to new or expanded mills from 1997 and are causing some concern to the industry in terms of compliance (Pratomo 1994).

Overall, there is an internationalization of environmental policy and infrastructure. This comes from using international consultants and international funding, as well as market expectations (Odendahl and Holloran 1992).

Currently, direct biological assessment is only required when undertaking an Environmental Impact Assessment (EIA) for new projects. All the countries have legislated EIA conditions, but the exact detail of biological assessment requirements is determined on a case-by-case basis. The biological survey which is usually part of an EIA could be a baseline for ongoing biological monitoring of effluent impacts.

So far, the countries have developed environmental polices and established conventional end-of-pipe discharge limits. Compliance with these alone will be costly for industry (e.g. pulp and paper executive interviews in *Paper Asia* Mar./Apr. and Nov./Dec. 1992), but will improve the receiving water quality enormously. However, direct biological monitoring as part of the end-of-pipe controls (e.g., bioassays) or environmental impact (field monitoring) has yet to be formally established in any of these countries.

There is a view that biological monitoring requirements may come sooner than expected, due to increased competition for common water resources (T & T Konsult pers. comm.). If that same water is also required for high-value food industries such as aquaculture, then the regulation of the water quality from a biological perspective becomes increasingly important. This is expected to lead to the introduction of suites of local bioassay tests.

There are several impediments to implementing regulatory bioassays and biomonitoring. The limited amount of local bioassay development work does not appear to be leading to the development of standard test methods. A further impediment is the lack of test laboratories and technical staff for routine contract work to a regulatory standard. Verification and enforcement require independent technical resources which do not appear to exist at this stage.

BIOLOGICAL MONITORING REGULATORY ENVIRONMENTS IN AUSTRALIA AND NEW ZEALAND

Australia

Environmental legislation in Australia is the responsibility of the individual states. There is usually no direct commonwealth control, although there is now an Environment Protection Authority with commonwealth environmental responsibilities. The National Environmental Protection Council sets national environmental quality guidelines which will result in a greater level of national harmonization. The council will assume some of the activities of Australian and New Zealand Environment and Conservation Council (ANZECC) described below.

ANZECC has an impact on harmonizing standards through the National Water Quality Guidelines (1992), which contain criteria for protecting the environmental values of natural waters in a variety of use categories from potable water to water for industrial uses. The levels set are largely based on the water quality criteria developed by the U.S. EPA using North American toxicity studies, in the absence of local data. A 0.02 safety factor on the U.S. EPA or Environment Canada Lowest Observable Effect Level values is considered sufficient protection for the Australian environment. The guidelines do recognize that toxicant levels should be set on "chronic sublethal tests on a range of sensitive local species representing different taxa and trophic levels," but with the present state of bioassay development this is not always achievable.

Environmental legislation differs between states. Victoria and Tasmania are two states where there are chemical pulp mills.

Under the Victorian Environment Protection Act 1970, State Environment Protection Policies (SEPP) are required to "establish the basis for maintaining environmental quality sufficient to protect existing and anticipated beneficial uses." This is a broad management framework and is the responsibility of the Victorian Environment Protection Authority (EPA). The principal SEPP for managing Victoria's fresh waters is "Waters of Victoria." Detailed ambient requirements are then set for regional watershed SEPPs within the state, taking into account local water use and quality values, through extensive public consultation. This includes assessments of the status and health of aquatic biological communities and ensuring that there is "no acute toxicity to aquatic life."

All effluents and waters entering the waterways are managed to ensure that the ambient water quality criteria are met and then improved. This can include ongoing biological surveys as part of discharge licences. Results of bioassay testing, in agreement with the EPA, are used for monitoring rather than enforcement. For most industrial discharges to marine environments, licenses also include bioassay monitoring.

In Tasmania, the Environmental Management and Pollution Control Bill 1993 provides for sustainable development and protection of the state's environment. The concept is to prevent "environmental harm." The Division of Environmental Management establishes ambient water quality requirements based largely on the ANZECC (1992) water quality criteria and sets license conditions. Where an effluent discharge causes non-compliance, Environment Improvement Plans are required, with specific objectives and time-to-achieve commitments. Regulation is achieved by a system of mandatory and voluntary environmental audits.

In all the Australian states, EIA policies and procedures are required for new industrial developments or major expansions. These vary in detail, but assessment of possible biological impacts based on field surveys is an integral part of each case.

Arising out of major national concerns about a possible pulp mill in Tasmania, there is also a specific set of environmental guidelines for new bleached eucalypt kraft mills (NBEK) (CoA 1989). The guidelines are comprehensive (Sprague 1991), covering all emissions from the mill, site suitability criteria, extensive baseline studies, and end-of-pipe and environmental monitoring.

The guidelines were deficient in several areas, particularly bioassay and biological monitoring. A 5 year, A$7.8 million National Pulp Mills Research Program (NPMRP) was initiated to understand kraft eucalypt pulping and bleaching processes and to develop tools for monitoring the potential impact of the resultant effluents. This new knowledge will be incorporated into the revision of the guidelines which will be released in 1995.

New Zealand

The key environmental legislation in New Zealand is the Resource Management Act (1991). The purpose of the act is to "promote the sustainable management of natural and physical resources." The act is intended to balance the needs of people and communities with the maintenance of ecological values.

The act emphasizes the prevention of adverse effects on the environment and, in the case of aquatic environments, effects in receiving waters. A schedule to the act presents descriptive guidelines for effects which are not permitted in different classes of waters. For example, in water being managed for aquatic ecosystem protection, "there shall be no undesirable biological growths as a result of any discharge of a contaminant into the water." For all water bodies, rules governing discharges require that there be no "significant adverse effect on aquatic life."

There has been further work to develop quantitative guidelines which enable enforcement agencies to implement discharge permits. To date, guideline documents have been prepared to define:
- The maximum incremental concentrations of organic and nitrogenous compounds which are permissible to prevent heterotrophic or phototrophic growths in rivers (Quinn 1992).
- The maximum incremental effluent quantities which are permissible to prevent conspicuous changes in color or clarity (Davie-Colley 1994).

In addition, there is a draft document on the assessment and definition of "significant adverse effect on aquatic life."

BIOLOGICAL MONITORING IN AUSTRALIA AND NEW ZEALAND

Activity in this area is illustrated by the following examples from chemical pulping operations in both countries.

Australia

The Maryvale Pulp Mill in Victoria is an integrated site with a kraft and neutral sulfite semi-chemical pulping component of 400,000 tonnes per annum (tpa), of which about 125,000 tonnes is bleached (EOP,DC)(EOP)D eucalypt. After extensive secondary treatment, paper machine water is discharged into a river. The pulping and bleaching effluent is mixed with an equal amount of domestic sewage, followed by extended lagoon treatment, and then discharge to the marine environment.

For many years the benthic invertebrate populations in the river sediment have been monitored up- and downstream of the outfall, with no discernible differences. The sewer discharge has recently been diverted from a freshwater lake system to a marine outfall. This was preceded by a 3 yr baseline and control site study of intertidal and subtidal sediment invertebrate populations. This survey has continued, supplemented by studies of local fish biochemical biomarkers ethoxyresorufin-o-deethylase (EROD) and bioaccumulation in local black mussels. Bioassay tests have included an algal growth test and Microtox®, all showing non-toxic responses.

The Burnie Mill in Northern Tasmania is also an integrated mill site manufacturing about 92,000 tpa of soda-AQ and semichemical pulp. Most chemical pulp is bleached using a CE(DH) sequence. There have been independent intertidal and subtidal sediment invertebrate surveys around the mill outfall, fish bile chemistry surveys for chlorophenolic compounds, and acute fish bioassay studies, using local species, all showing minimal effects. Now population monitoring as well as bioassays are required as part of new license conditions.

The Millicent Mill in South Australia is an integrated tissue manufacturer with present bisulfite pulping capacity of 130,000 tpa. A mill expansion, including a change to hydrogen peroxide bleaching with secondary effluent treatment, involved an extensive study on the biology of the freshwater interdunal lake into which the effluent was discharged. Ongoing monitoring as part of license conditions shows an increase in ecosystem diversity in the lake. An acute growth bioassay with a local *Daphnia* species, as agreed with the state regulatory authority, now shows no toxicity.

New Zealand

In New Zealand, biomonitoring has a significant role in the development of the Tarawera River regional plan. The Tarawera River is a major water resource in the Bay of Plenty Region, North Island.

Two pulp mills discharge wastewater into the river between 21 and 27 km from the sea. Caxton Paper Mill is an integrated mill producing sulfonated chemithermomechanical unbleached and peroxide bleached pulp at a rated capacity of 75,000 tpa, with three paper machines (55,000 tpa). The second mill (Tasman Pulp and Paper Company) is a fully integrated operation producing 225,000, 125,0000 and 280,0000 tpa of stone groundwood, refiner mechanical pulp and kraft pulp, respectively; 205,000 tpa of the pulp is bleached using HH or OD/CeoDED sequences.

Both mills discharge into the Tarawera River. At Caxton, there is anaerobic treatment into rapid infiltration ponds and clarification. For Tasman, discharge follows extensive aerated lagoon treatment and is equivalent to approximately 9% of the river flow.

Information on the toxicity of pulp and paper effluents and other sources has been an important input to the development of the Tarawera River Management Plan. The biological monitoring program has consisted of the following activities:
- Assessing the chemistry of all effluent discharges, river water, sediments and biota;
- Studying the river and coastal ecology (including river invertebrates and drift of coastal plankton);
- Biomonitoring of freshwater mussels (health and bioaccumulation of contaminants);
- Assessment of fish health (histopathology and biochemical markers, blood chemistry and short term toxicity tests).

DEVELOPMENT OF BIOLOGICAL ASSESSMENT PROCEDURES

As monitoring of the biological effects of effluent discharges from all sources becomes incorporated into legislation and performance monitoring programs, test procedures need to be developed and used in a consistent manner. Existing biotests include lethal and sublethal short term laboratory toxicity, health assessments, biomarkers, bioaccumulation assays and population studies. Although some of these are already used in Australia and New Zealand (and to a small extent in SE Asian countries), many are insufficiently developed for reliable use within a regulatory framework (i.e., part of a discharge license or consent performance limits). Some progress has been made as described below, but more methods for both fresh and marine waters are required to incorporate:
- Standardized test protocols
- Selection of appropriate test species (local and introduced)
- Quality assurance and control procedures
- Comparison of the tests with other international standard procedures

Single Species Bioassays

The process of developing several standard single-species toxicity assessment procedures is well advanced in Australia (Stauber *et al.* 1994). As a result of the NPMRP, specific marine test protocols are now available for:
- *Nitzschia* growth Test: This is a 96 h growth test using the marine planktonic diatom *Nitzschia closterium*. There are several unique features of the test. The alga is euryhaline and is grown in un-enriched seawater medium, and it is sensitive to chlorate, particularly under low nutrient conditions.
- *Hormosira* fertilization test: Uses the local fucoid, *Hormosira banksii*, similar to fertilization tests developed for North American macroalgae.

- Doughboy scallop (*Chlamys asperrimus*) larval abnormality test: A local adaptation of similar tests with North American bivalves.
- Sea urchin (*Heliocidaris tuberculata*) fertilization inhibition: A local adaptation of similar tests with North American echinoderms.
- *Allorchestes* growth inhibition test: The amphipod *Allorchestes compressa* has been widely used in S.E. Australia, but a formal protocol is now developed.
- Blenny larvae survival test: This is an acute toxicity test on young fish of local rock pool species, *Parablennius tasmanianus*.

Biomarkers and Field Monitoring

The future of biomarkers in regulations is uncertain. While they have been developed for local fish species and used in both Australia and New Zealand, it remains unclear whether they can be an early warning routine monitoring tool (Johannes 1994). There are currently no proposals to use biomarkers for regulatory compliance of pulp and paper discharges in either Australia or New Zealand.

Field monitoring to determine biological baseline conditions and impacts is an essential part of EIA programs and may be a discharge permit requirement. Yet considerable effort and cost can be expended for equivocal results. One of the NPMRP projects was to develop a statistically reliable experimental design of field monitoring programs. This has resulted in an comprehensive report to be published as a technical paper of the program and will be referred to in the NBEK guidelines revision.

CONCLUSIONS

- Australia, New Zealand and the countries in S.E. Asia have developed different environmental regulations depending upon their historical legislative developments, social and political systems and the nature of environmental issues in individual countries.
- Local modifications and new developments are required to meet these national differences, rather than directly transferring specific biomonitoring tests and standards used in other countries. This is happening in Australia and New Zealand and needs to happen in the S.E. Asian countries as well.
- In S.E. Asia, monitoring of pulp and paper and other discharges is largely based on conventional parameters. However it is envisaged that requirements for toxicity testing of effluents will become increasingly important in the future.
- Using international consultants and aid agencies, the S.E. Asian countries are rapidly putting into place environmental management polices and procedures similar to those in Western countries. With limited resources and technology bases, local biological monitoring procedures are developing slowly. There will need to be considerable effort, funding and training to regulate based on local biomonitoring.
- In both Australia and New Zealand, where descriptive water quality guidelines exist in current legislation, there are moves to develop quantitative guidelines and tests which will enable enforcement agencies to implement meaningful discharge permits.
- In Australia, the federal government has developed specific numerical guidelines for different discharge parameters from NBEK mills discharging to marine waters. As part of the process a suite of specific toxicity testing protocols has been developed. These could be used more widely for other discharges.
- In New Zealand, requirements for biological testing and environmental monitoring are emerging. As result, toxicity tests for regulatory compliance monitoring will be needed.
- In Australia and New Zealand, there is a high level of understanding of environmental issues, well established environmental agencies and sufficient technical resources to provide a level of local environmental responsibility equal to the best international standards.

ACKNOWLEDGMENTS

The advice from T & T Konsult, Jakko Poyryi (Thailand) consulting companies, and Messrs. Suwat Kamolpanus, Hanafi Pratomo and Edgar Furatero is very much appreciated.

REFERENCES

Australian and New Zealand Environment and Conservation Council (ANZECC). 1992. Australian water quality guidelines for fresh and marine waters.
Commonwealth of Australia (CoA). 1989. Environmental guidelines for new bleached eucalypt pulp mills. 29 pp.
Davie-Colley, D. 1994. Guidelines for the management of water colour and clarity. Water Quality Guidelines No. 2. New Zealand Ministry for the Environment.
Johannes, R.E. 1994. Evaluating the impacts of pulp mill effluents on the marine environment: monitoring mesocosms, biomarkers and future trends. National Forum on Bleached Eucalypt Kraft Mills in Australia, Melbourne. 4-6 October. pp. 145-153.
McLeay, D. 1987. Aquatic toxicity of pulp and paper mill effluent: a review. Environment Canada Report EPS 4/pf/1. 191 pp.
Metha, Y. 1994. U.S. EPA: impact on Asia Pacific. Asian Pacific Papermaker 4(2):51-55.
Mino, T. 1993. Comparison of legislation for water pollution control in Asian countries. Wat. Sci. Tech. 28(7):251-255.
Odendahl, S. and M. Holloran. 1992. International approaches to pulp and paper regulations: how do we compare? Beak Consultants Paper. 25 p.
Pratomo, H. 1994. Impact of new government regulations on future existence of pulp mill. Proc. Appita Annual Conference, Melbourne. pp. 69-74.
Priha, M. 1991. Toxic characteristics of effluents from bleached kraft pulp mills using modern technologies. Proc. International Conference on Bleached Kraft Mills, Melbourne, 4-7 February. pp. 203-218.
Quinn, J.M. 1992. Guidelines for the control of undesirable biological growths in water. Water Quality Guidelines No. 1. New Zealand Ministry for the Environment.
Sprague, J.B. 1991. Environmentally desirable approaches for regulating effluents from pulp mills. Wat. Sci. Tech. 24(3/4):361-371.
Stauber, J.L., L. Gunthorpe, J.G. Deavin, B.L. Munday, R. Krassoi, J. Simon and M. Ahsananullah. 1994. Single species Australian bioassays for monitoring new bleached eucalypt kraft pulp mills. National Forum on Bleached Eucalypt Kraft Mills in Australia, Melbourne. 4-6 October. pp. 127-143.

A REVIEW OF RECENT STUDIES IN NORTH AMERICA USING INTEGRATED MONITORING APPROACHES TO ASSESSING BLEACHED KRAFT MILL EFFLUENT EFFECTS ON RECEIVING WATERS

Timothy J. Hall

National Council for Air and Stream Improvement (NCASI), 1900 Shannon Point Road, Anacortes, Washington, 98221 U.S.A.

A review is provided of recent studies carried out in North America, where an integrated monitoring approach has been used in assessing possible bleached pulp mill effluent effects. Integrated monitoring has been here defined as an approach in which environmental impact assessment is based on multiple ecosystem measurements including population structure, standing crop, or production at several trophic levels. Examples of these types of studies include those carried out in mill receiving waters or experimental streams in which fish population, macroinvertebrate, and/or periphyton community characteristics have been described. In some cases it has been possible to compare and contrast the findings from these studies with the results from laboratory bioassay, biomarker, or biochemical measurements carried out at organism or suborganism levels with the same effluent. Although integrated monitoring studies can be expensive to conduct, they are considered of important value in addressing questions of ecological relevance with respect to assessing the environmental significance of mill effluents. This review summarizes the types of studies that were not previously reported, have been completed, or that are in progress in North America since the 1991 review by Owens and the SEPA "Environmental Fate and Effects of Bleached Pulp Mill Effluents" symposium (Södergren 1991).

INTRODUCTION

Owens (1991) provided a recent review of the hazard assessment of pulp and paper effluents on the aquatic environment, including those related to bleached pulp mill effluents. Included in that review was a general discussion of the various approaches which have been used, including those that are based on traditional structure and function measurements of aquatic communities as well as those that are based on sublethal measurements within single species. Relatively few of the studies reviewed included multiple component in-stream receiving monitoring of biota at the population or community levels. The importance of issues related to bleached pulp mill effluents was highlighted by an international conference on the fate and effects of bleached mill effluents held in Stockholm in 1991. The conference proceedings also provide an additional source of information on integrated monitoring studies (Södergren 1991). The current review provides an update on integrated monitoring studies carried out with bleached pulp mill effluents that have taken place or have appeared in the gray or published literature since these earlier reviews.

Integrated monitoring has been considered in various ways. In some cases emphasis has been on integrating chemical monitoring data with biological data, and in other cases with integrating laboratory bioassay data with in-stream community health data. For this review the emphasis has been placed on biological monitoring with bleached kraft mill effluent effects assessment based on the integration of data from in-stream measurements of multiple components of the aquatic community. Although the emphasis has been placed on biological data, a brief review is also provided of other study components, including those from laboratory bioassays, chemical characterization, and those from biomarker and biochemical measurements.

Eysenbach *et al.* (1990) and NCASI (1989) reported receiving water monitoring data compilation which included 200 studies representing 45 different mills and 40 receiving waters in North America. A number of mills in North American continue to carry out these types of surveys on a periodic basis. Although monitoring of this type may be based on regulatory requirements, it is also done by some mills voluntarily in an effort to address receiving water concerns and ecological relevance questions related to laboratory bioassays which are more often used in regulatory programs. No attempt was made in the current review to provide an update on the status of these routine surveys.

This review has also largely excluded regulatory agency related monitoring programs, primarily due to space considerations and the lack of specificity in terms of addressing bleached kraft mill effluent issues. Some good examples of integrated monitoring exist within this sector. For example, the Environmental Effects Monitoring Program, which has been implemented for pulp and paper mills in Canada, includes a range of in-stream monitoring parameters in addition to laboratory bioassays (Environment Canada and Department of Fisheries and Oceans 1992). In the U.S., under EPA guidelines, states are in the process of adopting "biological criteria" for detecting chronic pollution effects which might not be otherwise detected with measurements of conventional chemical and physical parameters (U.S. EPA 1990). Various approaches have been incorporated by the states, but most are based on fish and macroinvertebrate survey work.

Eight integrated monitoring studies were selected for this review, including six of which took place in North American rivers and two of which took place in experimental streams facilities constructed to mimic natural streams. The study locations (Fig. 1) approximate the most southern and northern mill locations of bleached kraft mills in North America as well as corresponding extreme distribution from east to west coast. Consequently, a broad range of ecological conditions are represented, including those representative of both warm and cold water fish species. For each study, a brief itemization of major study components is included as well as a brief review of the approach used and progress or findings to date. A synopsis of study highlights is provided following each individual study section.

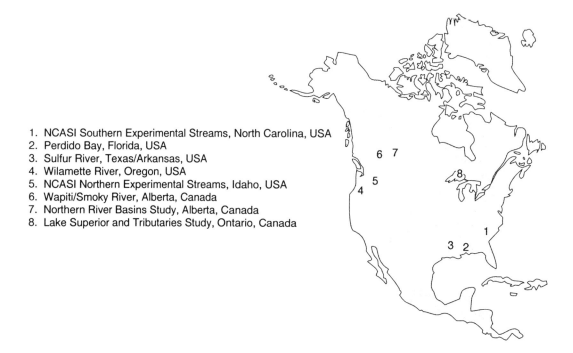

1. NCASI Southern Experimental Streams, North Carolina, USA
2. Perdido Bay, Florida, USA
3. Sulfur River, Texas/Arkansas, USA
4. Wilamette River, Oregon, USA
5. NCASI Northern Experimental Streams, Idaho, USA
6. Wapiti/Smoky River, Alberta, Canada
7. Northern River Basins Study, Alberta, Canada
8. Lake Superior and Tributaries Study, Ontario, Canada

Figure 1. Geographic locations of integrated monitoring studies with bleached kraft mill effluents.

SUMMARY OF INTEGRATED MONITORING STUDIES WITH BLEACHED KRAFT MILL EFFLUENTS

Perdido Bay, Florida

A bleached kraft mill is located on Elevenmile Creek, a tributary to Perdido Bay and the Gulf of Mexico (Brim 1993). This is a study of point and non-point discharges into the bay. While the bay is generally considered free of toxic compounds, concerns remain about overall water quality. The study indicated that anthropogenic sources of nutrients into Perdido Bay are dominated by input from Elevenmile Creek. A summary of study parameters is provided in Table 1.

Table 1. Study parameters for Perdido Bay, Florida (U.S.), integrated monitoring study. Bioassays are short-term chronic tests unless otherwise indicated.

Laboratory bioassay	Periphyton	Macroinvertebrate	Fish	
			Population	Health indices
Bacteria-bioluminescent		Species	Species	Histopathology
Daphnid		Species diversity		
Amphipod				
Gambusia-masculinization				

Laboratory daphnid bioassays indicated an absence of acute effluent toxicity, although some chronic responses occurred. Longer term 10-d bioassays with the amphipod *Hyalella azteca* were carried out with sediments and water from a number of sites, and indicated reduced feeding activity for both Elevenmile Creek and the Perdido River. Pore water bioassays carried out using the Microtox® bacterial bioluminescence test indicated responses at all sites except the Perdido River; however, some responses appeared to be anomalous and related to pore water extract storage time.

Sampling of benthic macroinvertebrates from Elevenmile Creek indicated fewer species and lower species diversity compared to other streams in the region. Although effluent secondary treatment operated at 95% BOD_5 removal efficiency, seasonal dissolved oxygen level depressions occurred in Elevenmile Creek below the 5 mg L^{-1} U.S. EPA water quality criteria. A synopsis of a 4-yr ecosystem monitoring study provided by Livingston (1992) addressed the question of whether the mill was having adverse effects on the Perdido Bay ecosystem. Mill effects included organic enrichment and elevations in levels of ammonia, nutrients, color, and turbidity. Mill effluent did not appear to be responsible for degradation downstream in Perdido Bay. Areas near the inflow of Elevenmile Creek had increased primary and secondary productivity and healthy populations of commercially important species, including blue crabs, penaeid shrimp, and several finfish.

The number of fish species in Elevenmile Creek was substantially lower (9 species) than from the adjoining Perdido River (41 species). The absence of pollution-intolerant species and the presence of pollution-tolerant species suggested habitat degradation in Elevenmile Creek. Field-collected fish in Elevenmile Creek indicated a 14% tissue abnormality rate based on external and internal examination, a rate which was considered natural for wild fish populations. An additional survey of fish livers also indicated similar liver histopathology to wild fish. There were indications of masculinization of female mosquitofish (*Gambusia affinis*) on Elevenmile Creek, a finding later confirmed in laboratory exposure studies (Bortone et al. 1989).

Perdido Bay Study Highlights

- No acute toxicity, some reduced feeding activity in amphipod sediment tests
- Fewer macroinvertebrate species/lower diversity near mill discharge
- Effluent increased primary/secondary productivity due to organic enrichment
- Fewer total fish species with shift toward pollution-tolerant species
- Absence of fish histopathological effects
- Indication of *Gambusia* masculinization

Sulfur River, Texas/Arkansas

The Sulfur River, located in the southeast U.S. along the Texas/Arkansas border, is the location of a bleached kraft mill which has undergone mill process conversions to increased ClO_2 substitution. Studies carried out over the past 10 yr have addressed effluent and receiving water changes resulting from the mill process changes. A summary of study parameters is provided in Table 2.

Table 2. Study parameters for Sulfur River, Texas/Florida (U.S.), integrated monitoring study. Bioassays are short-term chronic tests unless otherwise indicated.

Laboratory bioassay	Periphyton	Macroinvertebrate	Fish	
			Population	Health indices
Bacteria-bioluminescent		Species	Species	Gross pathology
Selenastrum		Species richness		Histopathology
Daphnid		Species diversity		
Menidia				

This study included a determination of effluent color effects on algal photosynthesis using the *Selenastrum capricornutum* laboratory bioassay (Rodgers et al. 1994). Additional laboratory tests included sediment bioassays, 7-d tests with embryo/larvae *Menidia beryllina*, the Microtox® bioluminescence test, and *Ceriodaphnia dubia* 7-d chronic tests (Middaugh et al. 1994). Bioassay results indicated that color-based light effects were limited to within a few meters of the effluent outfall and an absence of effects measured with the other toxicity tests.

Macroinvertebrate surveys carried out above and below the effluent discharge and in the outflow canal indicated similar faunal composition (Rakocinski et al. 1996). Although some faunal enrichment was attributed to effluent, there was relatively high species richness and typical taxonomic diversity at the sample site closest to the discharge point.

Histopathology monitoring of *G. affinis* above and below the effluent discharge canal indicated an absence of neoplasia (Overstreet et al. 1996). Melanomacrophage aggregates were greater upstream of the effluent discharge and elevated numbers of mitotic figures in hepatocytes, although prevalent in the fish sampled, did not appear to be related to mill discharge. Mosquitofish collected near the effluent canal had a higher incidence of parasite infections than those upstream, which was attributed to a possibly greater richness of invertebrate and vertebrate hosts associated with the effluent canal (Overstreet et al. 1996). Many parasites involve a sequence of multiple hosts and the absence of parasitic infections upstream of the effluent canal may have reflected the absence of some of these intermediate hosts. The mill associated with this study is undergoing further process changes to 100% ClO_2 substitution and some elements of this monitoring program will continue following this conversion.

Sulfur River Study Highlights

- Before/after increased ClO_2 substitution comparisons
- Absence of acute/chronic toxicity
- Algal bioassays predicted limited effects from effluent color
- Limited benthic effects agreed with bioassays
- Absence of *Gambusia* histopathological effects
- Increased *Gambusia* parasite incidence used as indication of healthy invertebrate/vertebrate intermediate hosts populations

Willamette River, Oregon

A legislative mandate was responsible for a multi-year study of the Willamette River which began in 1992 (Tetra Tech *et al.* 1993; Tetra Tech 1993, 1994). This river basin covers approximately 12% of the state of Oregon and receives the discharge from several pulp and paper mills and municipalities in addition to being influenced by agricultural non-point source effects). A bleached kraft mill is located on the river, the secondary treatment system of which also receives the effluent of an adjoining de-ink mill. The study included nine elements with broad goals of:
1. Developing predictive water quality, hydrology, and sediment transport models
2. Calibrating these models using field-collected data
3. Developing and implementing standard biological assessment protocols for use in future water quality monitoring

The ecological system component included measurements of benthic macroinvertebrates, fish, and periphyton. A summary of study parameters is provided in Table 3.

Table 3. Study parameters for Willamette River, Oregon (U.S.), integrated monitoring study. Bioassays are short-term chronic tests unless otherwise indicated.

Laboratory bioassay	Periphyton	Macroinvertebrate	Fish	
			Population	Health indices
	Chlorophyll	Species	Species	Gross pathology
	Biomass	Functional feeding groups		Histopathology
	Productivity			Biomarkers

Water quality measurements and comparisons of concentrations with toxicity data for indigenous species were used in lieu of laboratory bioassays to assess the potential for direct toxicity. Periphyton biomass and chlorophyll *a* content decreased downstream along the river irrespective of a variety of point discharges. Highest concentrations of periphyton occurred below the mill discharge. Gross primary production and respiration rates indicated a net positive production of oxygen possibly important to the river during late summer periods of lower dissolved oxygen.

Benthic macroinvertebrates were assessed in both shallow riffle areas and deeper pools using kicknet and grab sample techniques. The U.S. EPA Rapid Bioassessment Protocol was used to determine whether benthic community responses related to habitat or water quality effects. Riffle sample sites downstream of the mill discharge, along with sites downstream of several municipal wastewater treatment plants, were identified as having moderate biological impairment compared to reference sites. Similar degradation was not indicated for the deeper pool stations.

Fish community health measurements included community structure, diversity, abundance, and surveys for external or skeletal deformities. There were a number of fish community changes measured over the 300 km of river surveyed. None of these changes appeared to relate to the discharge point of the bleached kraft mill. Changes primarily related to distinct river reach characteristics according to upstream/downstream position along the river. Fish health measurements included adult northern squawfish (*Ptychocheilus oregonensis*) and largescale sucker (*Catostomus macrocheilus*). No fish health conclusions were drawn due to the absence of salmonid fish in the survey and lack of historical data to define normal conditions for the two species examined. A survey of juvenile squawfish for skeletal deformities showed no change in frequency of anomalies above or below the mill discharge. Related studies used the cytochrome P4501A biomarker with cutthroat trout (*Salmo clarki*) from above and below the mill discharge. Although there was a 2-fold increase in fish 2,3,7,8-tetrachlorodibenzo-*p*-dioxin (TCDD) below the outfall during an early fall sample, there was no difference in cytochrome P4501A induction (Curtis et al. 1992, 1993). Elevated TCDD levels below the mill were not present in an early summer sample and the fall TCDD levels were substantially lower than reported in fish from the southeastern U.S.

Willamette River Study Highlights

- Water quality analysis indicated concentrations below levels of concern and used in lieu of bioassays
- Periphyton increased downstream over river course; no indication of effluent effects
- Riffle macroinvertebrate habitat but not pools indicated moderate biological impairment
- Population/health indices measurements indicate no adverse fish effects
- Absence of cytochrome P4501A induction

Wapiti/Smoky River, Alberta

Secondary treated effluent from a bleached kraft mill enters the Wapiti River at Grand Prairie. Recent modifications to the mill have included conversions to 70 and 100 ClO_2 substitution. Studies initiated in 1990 addressed questions related to:
1. The fate of chlorinated organics in river food chains
2. The overall health of river fish populations
3. Possible habitat degradation which might be affecting fish populations (Swanson et al. 1993)

The study included sampling above and below the effluent discharge and before and after the mill process changes to increased ClO_2 substitution. A summary of study parameters is provided in Table 4.

Table 4. Study parameters for Wapiti/Smoky River, Alberta (Canada), integrated. Bioassays are short-term chronic tests unless otherwise indicated.

Laboratory bioassay	Periphyton	Macroinvertebrate	Fish	
			Population	Health indices
		Species	Species	Gross pathology
		Species diversity	Age structure	Histopathology
				Condition factor
				Reproductive condition
				Somatic indices
				Biomarkers

Benthic monitoring carried out between 1979 (pre-mill) and 1992 has indicated some enrichment effects downstream of the mill that may be in part due to the city wastewater treatment plant. Benthic responses included increased abundance with some reduction in community diversity and a shift in species composition. Water quality monitoring indicated increases in color, salts, and ammonia-nitrogen immediately below the mill discharge as well as increases in AOX. There did not appear to be a relationship between AOX and concentrations of dioxins, furans, or chlorinated phenolics in the water. Dioxin and furan concentration declined to below detection limits after the conversion to 70% ClO_2 substitution. The only consistently detectable mill contaminants found in river sediments were resin and fatty acids. Chlorinated phenolics were not detected in bottom sediments. Comparisons of water and sediment chemical concentrations with literature values indicated little potential for toxic responses of the mill effluent to stream biota.

The North Saskatchewan River was used as a reference site for the fisheries portions of the study since there were no pulp mills located there. Mountain whitefish (*Prosopium williamsoni*) and longnose sucker (*Catostomus catostomus*) were the primary fish species located on these rivers. Dioxin/furan concentrations were greater in mountain whitefish than in longnose suckers or several other native fish. Higher concentrations in mountain whitefish appeared to be related to their feeding on filter feeding insects and the chemical burden associated with suspended solids. Extensive analysis was performed to evaluate fish health. Both longnose sucker and mountain whitefish indicated an absence of neoplastic lesions or incidence of gross pathology or histopathology associated with exposure to mill effluent (Kloepper-Sams *et al.* 1994; Kloepper-Sams and Benton 1994). Other measurements, including blood parameters, liver somatic indices, sex steroid levels, and reproductive capacity provided no consistent evidence of an effluent effect. Both species of fish indicated cytochrome P4501A induction but this was not associated with effects as measured by other fish health parameters. This supports the use of ethyresorufin-*o*-deethylase (EROD) induction as an indicator of effluent exposure rather than effect. Fish population studies indicated those exposed to mill effluent were similar to those from the reference river (Swanson *et al.* 1994). Differences in fish distribution and abundance were linked to natural events, including floods and low flow, as well as to habitat quality variables. There was no measurable effluent effect on population age structure that would indicate age class loss or reproductive failure.

Wapiti/Smoky River Study Highlights

- Before/after increased ClO_2 substitution comparisons
- Macroinvertebrate responses partly due to other dischargers
- Water/sediment chemical concentrations indicative of no toxicity
- No adverse fish effects fish based on population/health indices measurements
- Biomarker responses (EROD) were indicators of exposure, not effect
- Natural events were a major factor affecting fish populations and their habitat

Lake Superior and Tributaries, Ontario

A series of studies have been conducted on tributary rivers to Lake Superior and in the adjacent area of Ontario to determine environmental response from effluent for a variety of mill process and treatment types. Monitoring carried out above and below effluent discharges included eight bleached kraft mills, four of which had secondary treatment. A summary of study findings is provided in Table 5.

Chronic bioassays were carried out with receiving waters using *Ceriodaphnia dubia* and fathead minnows *(Pimephales promelas)* (Robinson *et al.* 1994). No significant effects were indicated for any of the test endpoints with samples downstream of mills with secondary treatment. Bioassays with samples downstream of mills without secondary treatment or non-bleached mills provided a range of responses and indicated a lack of correlation with AOX or chlorinated organic content of the effluents. Bioassay results were found to be predictive of benthic macroinvertebrate responses based on comparisons with historical benthic macroinvertebrate surveys made on some of these same rivers.

Table 5. Study parameters for Lake Superior and tributaries, Ontario (Canada), integrated monitoring study. Bioassays are short-term chronic tests unless otherwise indicated.

Laboratory bioassay	Periphyton	Macroinvertebrate	Fish	
			Population	Health indices
Ceriodaphnia		Species	Species	Gross pathology
Pimephales		Species diversity	Age structure	Condition factor
		Species abundance	Size structure	Organ weights
		Pollution-tolerant species	Community structure	Biomarkers

Fish components to these studies included organ weights, mixed function oxygenase (MFO) induction, plasma steroids, and liver dioxins (Munkittrick et al. 1994). Whitefish (*Coregonus catostomus*) collected near bleached kraft mills indicated elevated EROD induction but similar elevations were observed adjacent to unbleached mills. The incidence of EROD induction, and decreases in blood steroid, and gonadal or liver size appeared to be unrelated to the degree of effluent treatment and also occurred adjacent to unbleached mills. These results were also inconsistent with results from chronic bioassays and macroinvertebrate surveys where effects were not indicated.

Concentrations of polychlorinated dibenzo p-dioxins (PCDDs) and dibenzofurans (PCDFs) were measured in white sucker (*Catostomus commersoni*) downstream of five of the bleached kraft mills and compared to reference sites (Servos et al. 1994). There was no relationship indicated for PCDD/PCDF toxicity equivalent concentrations (TEQs) and fish condition factor, gonadosomatic index, liver somatic index, or plasma 11-ketotestosterone. Lower TEQs were associated with mills that used higher levels of ClO_2 substitution and that were practicing secondary treatment. TEQs for the bleached kraft mills ranged over an order of magnitude but this appeared to have no correlation with the level of EROD activity. Additional analysis of white sucker included use of the H4IIE rat hepatoma bioassay to establish toxic equivalent concentrations (TECs) for 2,3,7,8-tetrachlorodibenzo-*p*-dioxin downstream of the mills (van den Heuvel et al. 1994). The liver TECs measured using the H4IIE assay were accounted for largely by the measured concentrations of dioxin and furan congeners. Higher TECs were generally associated with mills practicing chlorine bleaching, which similarly correlated with higher EROD induction levels and lower plasma testosterone concentrations. Two mills using little or no chlorine bleaching also generated high liver TECs. This could not be accounted for by dioxin/furan concentrations and may be another indication of the possible presence of EROD-inducing compounds unrelated to chlorine bleaching.

Lake Superior and Tributaries Study Highlights

- Absence of chronic bioassay effects for mills with secondary treatment
- Macroinvertebrate responses agree with bioassay findings
- EROD and other fish health responses unrelated to chlorine bleaching
- EROD induction not predicted by bioassay or macroinvertebrates

Northern River Basins Study, Alberta

The northern river basins study was initiated in 1991 to address concerns about development effects on the Peace, Athabasca, and Slave River basins (Northern River Basins Study Status Report 1994; Northern River Basins Study Annual Report 1994). The Athabasca River is the location of two bleached kraft mills. This is a 4.5-yr study with $5 million allocated to scientific research. Much of the work completed to date has involved methods development and baseline data collection. A summary of study parameters is provided in Table 6. The summary table does not reflect all study parameters since some elements of the experimental design are still under development.

Table 6. Study parameters for northern river basins study, Alberta (Canada), integrated monitoring study. Bioassays are short-term chronic tests unless otherwise indicated. Matrix is incomplete since some elements of monitoring program are still being developed.

Laboratory bioassay	Periphyton	Macroinvertebrate	Fish	
			Population	Health indices
Periphyton-outdoor flumes			Species	Gross pathology
Macroinvertebrate-outdoor flumes			Age structure	Histopathology
				Reproductive condition
				Gonad structure
				Somatic indices
				Biomarkers

A substantial portion of the research is directed at contaminant and nutrient effects on the aquatic food web. Contaminant measurements have included dioxins and furans in fish, sediment, and selected biota. Nutrient monitoring has been included and interfaces with in-stream monitoring of bacteria, periphyton, and macroinvertebrate populations. These monitoring data are being used in conjunction with outdoor flume studies in developing computer models used to predict nutrient and dissolved oxygen effects. Additional flume work with effluent from the two mills focuses on macroinvertebrate and periphyton effects. Another concern addressed in the study relates to nutrient-enhanced productivity and possible reduced intergravel dissolved oxygen depressions which might result from subsequent decomposition. Sediment oxygen demand investigations were consequently included as part of the study.

Slight elevations in dioxins and furans were indicated for the monitored fish, which included mountain whitefish (*Prosopium williamsoni*), longnose suckers (*Catostomus Catostomus*), northern pike (*Esox lucius*), goldeye (*Hiodon alosoides*), and walleye (*Stizostedion vitreum vitreum*). Field studies using stable isotopes were used to distinguish food chain relationships for various fish, including white sucker (*Catostomus commersoni*), longnose suckers (*C. catostomus*), and mountain whitefish (*P. williamsoni*). Whitefish were found to obtain much of their food from tributaries rather than the main river. Although autopsies and steroid measurements have been completed on a large number of fish, data evaluations and conclusions have not been reached regarding these fish health parameters. Liver MFO measurements were discontinued after initial measurements indicated difficulties with the analysis procedure and data interpretation.

Northern River Basin Study Highlights

- Study in progress, some monitoring elements not yet implemented
- Outdoor flume studies used for bacteria, periphyton, and macroinvertebrates
- Computer model development directed at predicting nutrient effects
- Fish population/health measurements are in progress

NCASI Experimental Streams, Idaho and North Carolina

There are two experimental stream facilities in operation in the U.S. which address the question of effluent effects on receiving waters. Each of these facilities consists of four replicate outdoor experimental streams. Comparisons made between replicate control and effluent exposure streams allow for a statistically based assessment of effluent effects under precisely defined exposure conditions. Both stream facilities operate in conjunction with secondary treated effluent from a bleached kraft mill located adjacent to each site. Each mill has undergone process changes in recent years including conversions to ClO_2 substitution and oxygen delignification. Previous studies carried out while these mills practiced conventional bleaching have allowed a comparison of before/after conversion changes in effluent quality. A summary of study parameters is provided in Table 7.

Table 7. *Study parameters for NCASI experimental streams, Idaho and North Carolina (U.S.), integrated monitoring studies. Bioassays are short-term chronic tests unless otherwise indicated.*

Laboratory bioassay	Periphyton	Macroinvertebrate	Fish	
			Population	Health indices
Champia	Biomass	Species	Growth	Histopathology
Ceriodaphnia	Chlorophyll	Species diversity	Survival	Condition factor
Echinoderm		Functional feeding groups	Production	Somatic indices
Mysid				Biomarkers
Menidia				
Cyprinodon				
Pimephales				

Effluent exposure studies at the northern experimental streams have included measurements of fish production and health with rainbow trout (*Oncorhynchus mykiss*) as well as measurements of the supporting food web. Comparisons have been made for 6-9 mo exposures at 1.5 and 5.0% v/v effluent before and after mill conversion to 70% ClO_2 substitution (Haley *et al.* 1994). Mill conversion resulted in decreased effluent chlorinated organics, AOX, and color. Neither macroinvertebrate community characteristics nor periphyton chlorophyll *a* nor biomass indicated significant differences between control and effluent exposure streams before or after the mill process change. Rainbow trout production as well as other fish health parameters indicated no significant difference between fish exposed to effluent and those that were not. Fish histopathology indicated an absence of neoplasia or significant differences in liver somatic indices, sexual maturity, sex ratios, blood hematology, or condition factor. Significant cytochrome P4501A and EROD induction occurred at the highest effluent concentration, but the absence of any corresponding measurable fish health impact emphasizes MFO induction as an indication of exposure rather than effect.

Corresponding laboratory chronic bioassays indicated that both before and after mill conversion many responses were near the upper limit of detection (Hall *et al.* 1993). There was consequently no significant change in bioassay response following mill conversion. Similar studies and comparisons have recently been completed after mill conversion to oxygen delignification.

A similar study with a variety of warm water fish species, including largemouth bass (*Micropterus salmoides*), channel catfish (*Ictalurus punctatus*), golden shiner (*Notemigonus crysoleucas*) and bluegill sunfish (*Lepomis macrochirus*), has been carried at the southern experimental streams with effluent during a 7-mo period before and a 9-mo period after mill conversion to 70% ClO_2 substitution (NCASI 1994a,b). Mill conversion resulted in significant decreases in effluent chlorinated phenolics, dioxins, and furans, while resulting in increased concentrations of chlorovanillins.

Response endpoints for bioassays with *Ceriodaphnia dubia* and fathead minnow (*Pimephales promelas*) were not significantly different before or after conversion. Although there were species differences in growth and survival, overall fish production was greatest during pre-conversion at 8% v/v effluent, the highest concentration tested. Survival, growth, and production of all four species were not significantly different from control stream fish at either 4 or 8% v/v effluent following mill conversion. Benthic macroinvertebrate sampling carried out during the 70% ClO_2 substitution period indicated no adverse effects on species diversity, numbers, or biomass. Fish exposed in the experimental streams during the high substitution period were also examined for the EROD and P4501A biomarkers, histopathology, condition factor, liver somatic index, spleen somatic index, and blood hematocrit and leucocrit.

For many of the fish health parameters, significant differences were identified between the two control streams which were used as a basis for comparing effluent treatment stream responses. This limited the detection of effluent-related responses and highlighted the need for a better understanding of normal ranges and variability for fish health parameters. The EROD and P4501A biomarkers did indicate a clear dose-response relationship to effluent exposure; however, as noted, this did not correspond with adverse fish health responses. Additional studies with this effluent have been completed after mill conversion to 100% ClO_2 substitution and oxygen delignification.

NCASI Experimental Stream Study Highlights

- Outdoor experimental streams (two locations) used as cold and warm water streams simulations
- Studies carried out before/after host mills increased ClO_2 substitution
- Chronic bioassays indicated few differences before/after mill process changes
- Periphyton and macroinvertebrates not significantly affected by effluent before/after process changes
- No adverse effects on rainbow trout at highest effluent test concentration (5.0% v/v) based on production and fish health indices
- No adverse effects on warm water fish species at highest test concentration (8.0% v/v) based on production and fish health indices
- Studies recently completed after mill conversions to oxygen delignification

SUMMARY

Six receiving water and two experimental stream studies were reviewed in which bleach kraft mill effluent effects were addressed. Fish and macroinvertebrate populations were the common emphasis for all of the studies, although some also addressed algal or periphyton populations. Effluent effects when measured were generally restricted to the area immediately adjacent to the effluent discharge. The review of North American mill receiving water studies (Eysenbach *et al.* 1990; NCASI 1989) submitted to an independent scientific panel led to their conclusion that "secondary treated bleached kraft mills had little or no adverse impact on aquatic life in receiving waters." In addition, there were no marked differences in toxicity to aquatic life with bleached vs. unbleached pulp mill effluents. Five out of the eight studies integrated laboratory bioassays with in-stream measurements. Bioassays generally confirmed in-stream biological measurements with secondary treated effluents, indicating a lack of acute or chronic toxicity. Most effluent effects appeared to be related to organic enrichment rather than to direct toxicity.

Several of the studies in this review were initiated to determine whether mill process changes to increased ClO_2 substitution and oxygen delignification resulted in improved effluent quality relative to in-stream biota or laboratory bioassay responses. Although these process changes have resulted in decreased effluent AOX and chlorinated organic content, there is little evidence to date that this results in a direct biological benefit based on either in-stream measurements or laboratory bioassays.

The majority of these studies included MFO biomarker measurements. Several of the studies provided evidence of a poor correlation between EROD or cytochrome P4501A and the AOX, dioxin/furan, or chlorinated organic content of effluent or receiving water. There was also evidence presented for a poor correlation between MFO induction and fish health effects as measured by histopathology, hematology, sex steroids, or reproductive capacity. There continues to be a major focus on fish reproduction, with reproductive steroid measurements included in the majority of studies. Evidence has been provided in at least one study that steroid-related reproductive effects are not specific to bleach plant effluents. Although several of the studies included population structure analysis, there was no indication of population level reproductive effects as would be indicated by age class loss. The significance of biomarker responses has not been verified by in-stream measurements of fish at the population or community level or by effects on the supporting food web as determined by in-stream measurements or laboratory bioassays. The validation of biomarker responses for their ecological relevance based on in-stream measurements at the population and community levels provides an example of the importance of integrated monitoring approaches.

Although there is a significant cost to conducting integrated receiving water studies, there is also commensurate significant value in the data generated. A problem common to all receiving water studies is the interpretation of the influence of a single point source bleached kraft mill effluent from other point and non-point source discharges in the broader spectrum of normal seasonal and temporal variability. This is a primary advantage of deferring the determination and interpretation of receiving water effects to laboratory bioassays and procedures such as biomarker and steroid measurements. It is crucial, however, to establish the relevance of these measurements in predicting receiving water effects. There remains, in some cases, a discrepancy between conclusions being drawn with some of these measurements and measures of fish and aquatic community health measured in receiving waters. The integrated in-stream monitoring studies in this review and the earlier reviews by Owens (1991), Eysenbach (1990), and NCASI (1989) continue to indicate an absence of effects in bleached kraft mill effluent receiving waters at the population or community level and provide further evidence of the importance of studies which emphasize a broad base of in-stream measurements.

REFERENCES

Bortone, S.A., W.P. Davis and C.M. Charles. 1989. Morphological and behavioral characters in mosquitofish as potential bioindication of exposure to kraft mill effluent. Bull. Env. Contam. Toxicol. 43:370-377.

Brimm, M.S. 1993. Toxics characterization report for Perdido Bay, Alabama and Florida. U.S. Fish and Wildlife Service Publication PCFO-EC-93-04, U.S. Fish and Wildlife Service, Atlanta, GA.

Curtis, L.R. H.M. Carpenter, R.R. Donohoe, M.L. Deinzer, M.A. Beilstein, D.E. Williams and O.R. Hedstrom. 1992. Toxicity and longitudinal distribution of persistent organochlorines in the Willamette River. Phase 1 Final Report, Oregon Department of Environmental Quality, Portland, OR.

Curtis, L.R. H.M. Carpenter, R.M. Donohoe, D.E. Williams, O.R. Hedstrom, M.L. Deinzer, M.A. Beilstein, E. Foster and R. Gates. 1993. Sensitivity of cytochrome P450-1A1 induction in fish as a biomarker for distribution of TCDD and TCDF in the Willamette River, Oregon. Environ. Sci. Techn. 27:2149-2157.

Environment Canada and Department of Fisheries and Oceans. 1992. Aquatic environmental effects monitoring requirements. Annex I: Aquatic environmental effects monitoring requirements at pulp and paper mills and off-site treatment facilities regulated under the Pulp and Paper Effluent Regulations of the Fisheries Act. EPS 1/RM/18, Ottawa, Ontario.

Eysenbach, E.J., L.W. Neal and J.W. Owens. 1990. Pulping effluents in the aquatic environment. Data compilation and scientific panel report. TAPPI J. pp. 104-106.

Haley, R.K., T.J. Hall and T.M. Bousquet. 1994. Effects of biologically treated bleached kraft mill effluent before and after mill conversion to increased chlorine dioxide substitution: Results of an experimental streams study. Environ. Toxicol. Chem.: In press.

Hall, T.J., R.K. Haley, R. Miner, D. Borton and L. Lafleur. 1993. An assessment of the ecological impacts of chlorine-based bleaching and its variants. International Environmental Symposium, April 27-29, 1993, Paris, France.

Kloepper-Sams, P.J. and E. Benton. 1994. Exposure of fish to biologically treated bleached-kraft effluent. 2. Induction of hepatic cytochrome P450A in mountain whitefish (*Prosopium williamsoni*) and other species. Environ. Toxicol. Chem. 13:1483-1496.

Kloepper-Sams, P.J., S.M. Swanson, T. Marchant, R. Schryer and J.W. Owens. 1994. Exposure of fish to biologically treated bleached-kraft effluent. 1. Biochemical, physiological and pathological assessment of Rocky Mountain whitefish (*Prosopium williamsoni*) and longnose sucker (*Catostomus catostomus*). Environ. Toxicol. Chem. 13:1469-1482.

Livingston, R.J. 1992. Ecological study of the Perdido drainage system. I. Summary of results. Environmental Planning & Analysis, Inc., Tallahassee, FL.

Middaugh, D.P., N. Beckham, T.L. Deardorf and J.W. Fournie. 1994. Responses of fish embryos to bleached kraft mill process water. Presented at the Second International Conference on Environmental Fate and Effects of Bleached Pulp Mill Effluents, November 6-10, 1994, Vancouver, B.C., Canada.

Munkittrick, K.R., G.J. Van Der Kraak, M.E. McMaster, C.B. Portt, M.R. van den Heuvel and M.R. Servos. 1994. Survey of receiving-water environmental impacts associated with discharges from pulp mills. 2. Gonad size, liver size, hepatic EROD activity and plasma sex steroid levels in white sucker. Environ. Toxicol. Chem. 13:1089-1101.

NCASI. 1989. Pulping effluents in the aquatic environment - Part II: A review of unpublished studies of in-stream aquatic biota in the vicinity of pulp mill discharges. Tech. Bull. 573, New York, NY.

NCASI. 1994a. Characterization of a bleach kraft mill effluent with high substitution of chlorine dioxide for chlorine using laboratory bioassays and experimental streams. Part 1. Comparison of effects of a bleached kraft pulp mill effluent on laboratory bioassay responses and fish growth and production in experimental streams before and after high substitution with ClO_2. Tech. Bull. National Council of the Paper Industry for Air and Stream Improvement, New York, NY.

NCASI. 1994b. Characterization of a bleach kraft mill effluent with high substitution of chlorine dioxide for chlorine using laboratory bioassays and experimental streams. Part 2. Comparison of biochemical, physiological, and histopathological biomarkers to the survival, growth and production of fish exposed to high substitution bleached kraft mill effluent in experimental streams. Tech. Bull. National Council of the Paper Industry for Air and Stream Improvement, New York, NY.

Northern River Basins Study. 1994. Status Report, January, 1994. Northern River Basins Study, Edmonton, Alberta.

Northern River Basins Study. 1994. Annual Report, 1993-1994. Northern River Basins Study, Edmonton, Alberta.

Overstreet, R.M., W.E. Hawkins and T.L. Deardorff. 1996. The western mosquitofish as an environmental sentinel: Parasites and histological lesions. *In* Environmental Fate and Effects of Pulp and Paper Mill Effluents, M.R. Servos, K.R. Munkittrick, J.H. Carey and G. Van Der Kraak (ed.), St. Lucie Press, Delray Beach, FL.

Owens, W.J. 1991. The hazard assessment of pulp and paper effluents in the aquatic environment: A review. Environ. Toxicol. Chem. 10:1511-1540.

Rakocinski, C.F., M.R. Milligan, R.W. Heard and T.L. Deardorff. 1996. Comparative evaluation of macrobenthic assemblages from the Sulfur River Arkansas in relation to pulp mill effluent. *In* Environmental Fate and Effects of Pulp and Paper Mill Effluents, M.R. Servos, K.R. Munkittrick, J.H. Carey and G. Van Der Kraak (ed.), St. Lucie Press, Delray Beach, FL.

Robinson, R.D., J.H. Carey, K.R. Solomon, I.R. Smith, M.R. Servos and K.R. Munkittrick. 1994. Survey of receiving-water environmental impacts associated with discharges from pulp mills. 1. Mill characteristics, receiving water chemical profiles and lab toxicity tests. Environ. Toxicol. Chem. 13:1075-1088.

Rodgers, J.H., W.B. Hawkins, A.W. Dunn and T.L. Deardorff. 1994. An evaluation of potential impacts of bleached kraft mill effluents on the lower Sulfur River, Texas-Arkansas. Presented at the Second International Conference on Environmental Fate and Effects of Bleached Pulp Mill Effluents, November 6-10, 1994, Vancouver, B.C., Canada.

Servos, M.R., S.Y. Huestis, D.M. Whittle, G.J. Van Der Kraak and K.R. Munkittrick. 1994. Survey of receiving-water environmental impacts associated with discharges from pulp mills. 3. Polychlorinated dioxins and furans in muscle and liver of white sucker (*Catostomus commersoni*). Environ. Toxicol. Chem. 13:1103-1115.

Swanson, S.M., R. Schryer, R. Shelast, K. Holley, I. Berbekar, P. Kloepper-Sams, J.W. Owens, L. Steeves, D. Birkholz and T. Marchant. 1993. Wapiti/Smoky River Ecosystem Study. Weyerhaeuser Canada, Grand Prairie, Alberta.

Swanson, S.M., R. Schryer, R. Shelast, P.J. Kloepper-Sams and J.W. Owens. 1994. Exposure of fish to biologically treated bleached-kraft mill effluent. 3. Fish habitat and population assessment. Environ. Toxicol. Chem. 13:1497-1507.

Södergren, A. (Ed.). 1991. Environmental fate and effects of bleached pulp mill effluents. Proceedings, SEPA Conference, Nov. 19-21, 1991, Stockholm, Sweden.

Tetra Tech. 1993. Willamette River Basin Water Quality Study. Willamette River Ecological Systems Investigation Component Report. Final Report TC 8983-09, Oregon Department of Environmental Quality, Portland, OR.

Tetra Tech. 1994. Willamette River Basin Water Quality Study. Phase II. Ecological Monitoring Component: Benthic Metric Selection and Data Evaluation. Oregon Department of Environmental Quality, Portland, OR.

Tetra Tech, E&S Environmental Chemistry, Limno-Tech, Inc. and Taxon Aquatic Monitoring Co. 1993. Willamette River Basin Water Quality Study, Summary Report. Final Report TC 8983-10, Oregon Department of Environmental Quality, Portland, OR.

U.S. EPA (United States Environmental Protection Agency). 1990. Update on the biological criteria program. Newsletter, Criteria and Standards Division Office of Water Regulations and Standards, U.S. EPA, Washington, D.C. 3:1-7.

van den Heuvel, M.R., K.R. Munkittrick, G.J. Van Der Kraak, M.E. McMaster, C.B. Portt, M.R. Servos and D.G. Dixon. 1994. Survey of receiving-water environmental impacts associated with discharges from pulp mills. 4. Bioassay-derived 2,3,7,8-tetrachlorodibenzo-p-dioxin toxic equivalent concentrations in white sucker (*Catostomus commersoni*) in relation to biochemical indicators of impact. Environ. Toxicol. Chem. 13:1117-1126.

ECOTOXICOLOGICAL EFFECTS OF PROCESS CHANGES IMPLEMENTED IN A PULP AND PAPER MILL: A NORDIC CASE STUDY

Aimo Oikari[1,3] and Bjarne Holmbom[2]

[1]Helsinki University of Technology, Laboratory of Environmental Protection, FIN-02150 Espoo, Finland
[2]Åbo Akademi University, Department of Forest Products Chemistry, FIN-20500 Turku, Finland
[3]Present address: University of Jyväskylä, Department of Biology and Environmental Sciences, Box 35, FIN-40351 Jyväskylä, Finland

An integrated mill producing kraft pulp (ca. 420,000 t a^{-1}) and printing paper from hardwood and softwood was investigated for the ecotoxicological advantages due to complete substitution of chlorine by chlorine dioxide in the bleaching and replacement of aerated lagoons by an activated sludge plant in the treatment of effluents. A comparative study on Lake Saimaa (SE Finland) was performed before (1990/91) and after (1993) the process changes, and included the following sections: discharges of chlorinated organics and selected wood extractives from the mill, distribution and persistence of xenobiotic chemicals in the lake, their association with sedimenting particles and bioavailability, and biomarker responses in caged whitefish and feral fish (roach, bream, perch) exposed in waters polluted by bleached kraft mill effluent. The data revealed dramatic reductions in organochlorines, like chlorinated phenolics, both in the lake and fish. Based on the responses in fish, the novel environmental technology in the mill was reflected as much less exposure and effects in the aquatic biota. As a consequence, the lake area affected toxicologically was substantially reduced. Laboratory data on exposed whitefish supported the conclusions drawn from the field observations.

INTRODUCTION

Lake ecosystems seem to be capable of recovering, to a large extent, within a few years following substantial reduction of organochlorines, other bioactive compounds and nutrients discharged from a pulp and paper mill. For example, closing a sulfite mill and development of new processes in pulp production and effluent treatment lead towards substantial recovery of fish community within years (Hakkari 1992). However, due to large annual fluctuation from natural reasons, a long-term survey may be necessary to associate the community recovery to process alterations. On the other hand, by combining information on ambient xenobiotic chemicals with their availability and effects in individuals, direct evidence on advantages of large environmental investments in ecosystem may be achieved. This approach was used in the present study.

The Kaukas Inc. (Lappeenranta, Finland) pulp mill of the Kymmene Group produced during the years of the present study, 1991-93, about 1050-1200 t d^{-1} of hardwood (mainly birch) and softwood (mainly pine) kraft pulp. Hardwood and softwood pulp was processed on separate production lines in the approximate ratio of 60:40, respectively. The integrated paper mill at Kaukas produced in 1991-93 about 1060-1180 t d^{-1} of LWC and MFC magazine papers on two paper machines, using kraft pulp from the pulp mill and spruce groundwood pulp yielded in the paper mill as fiber furnish. The Kaukas mill complex also includes a large sawmill and a mill producing crude and refined tall oil as well as sitosterols. In 1991-93, the combined flow through the effluent treatment system and discharged to the lake was about 120-130 m^{-3}d^{-1}.

During the first stage of this study, in 1990-91, the mill used traditional chlorine and chlorine dioxide bleaching in a D/C(EO)DED sequence. In bleaching of hardwood and softwood pulp the approximate proportion of C was 5 and 45%, respectively. In February 1992, the mill practically stopped using molecular chlorine (<<1%) and adopted the bleaching sequence OD(EO)D(EP)D for hardwood pulp and D(EOP)DED for softwood pulp.

In 1990-91, the combined pulp and paper mill effluent was treated in an aerated lagoon system with a hydraulic retention of 3-4 days. Since April 1992, the combined mill effluent has been treated in a new, modern activated sludge treatment plant. This is a super low-loaded activated sludge process, incorporating an aerobic selector (Simpura and Pakarinen 1993).

Aims of the study (ECOBALANCE project) were:
1. To compare, in 1991 and 1993, the emissions from Kaukas Mills, and concentrations of selected chlorinated organics and unchlorinated wood extractives in the recipient Lake Saimaa system, before and after process changes were implemented at the mill in 1992.
2. To evaluate the ecotoxicological effects of fundamental bleaching and effluent treatment process changes in a Nordic pulp and paper mill discharging its effluents to a lake ecosystem.

MATERIALS AND METHODS

The Study Area and Emissions to Lake Saimaa

The mill discharges its effluents (about 1.5 m^3 s^{-1}) to southern Lake Saimaa. From upstream of the effluent outlet, the lake water is passing the mill with a flow of about 40 m^3 s^{-1}. Thus in the mixing zone, within 1 km of the outfall, a theoretical concentration of approximately 4% of effluent is achieved.

Collection and Analysis of Effluent and Water Column Samples

Water samples were taken of the treated mill effluent and of recipient waters at one site upstream from the mill and eight sites below the mill, during 4-5 weeks in May-June 1991 and in May-June 1993 (Fig. 1).

Mill effluent samples were taken at the outlet to the lake with an automatic sampler. The daily composite samples were stored separately in a freezer until analysis. Samples were analyzed daily at the mill laboratory for common wastewater parameters such as pH, suspended solids, sodium, COD_{Cr}, BOD_7 and AOX. Both internal and external QC/QA are applied in the laboratory.

Recipient waters were collected twice a week as composite samples of the whole water column and later combined as 30-day samples for analysis. Sites farther away from the mill were sampled with a time delay to take into consideration the hydrological data for the recipient lake. For some general water quality parameters (pH, conductivity, sodium and chloride), individual samples were analyzed on the day of collection. Samples were stored in glass bottles in a freezer (-20°C) until further analysis for individual organic components.

Analysis of waters for organic components was by GC after relevant extraction and derivatization, as described in Hemming and Holmbom (1992). Free chlorinated phenolics (CPs) were analyzed by GC with EC detector after *in situ* acetylation and hexane extraction. For the determination of total phenolics, the samples were hydrolyzed for 16 h at 70°C in a 2.5 M KOH solution of ethanol-water (1:1 by vol.). Lipophilic wood extractives were analyzed after filtration (GF/A glass fiber filter, approx. 1-10 μm) separately on the solid matter and the filtrates. Solid matter was extracted with a 0.5 M ethanol-water solution at 70°C for 3 h. Filtrates were eluted through XAD-8/XAD-2 columns (75:25 ratio) and adsorbed lipophilic material was eluted by diethyl ether. The extracts were methylated by diazomethane and furthermore silylated before analysis on GC with FI detector.

Collection and Analysis of Sedimenting Material

Sedimenting particulate material from the study area was collected only in 1991 (Kukkonen *et al.* 1992). At two reference sites upstream and at four sites downstream from the mill, sediment traps (in 3 glass tubes: diameter 5 cm, height 50 cm) were lowered to depths of 10 m and 2 m above the bottom. Sedimentation rates (as g(dw) m^{-2} d^{-1}, 30-d periods) were determined in May-June, and samples collected were analyzed for organic carbon, total organically bound halogen and chlorophenolics. Sedimentation rate calculations of chlorophenolics assume that no resuspension occurs in the lake (i.e., maximal values are presented).

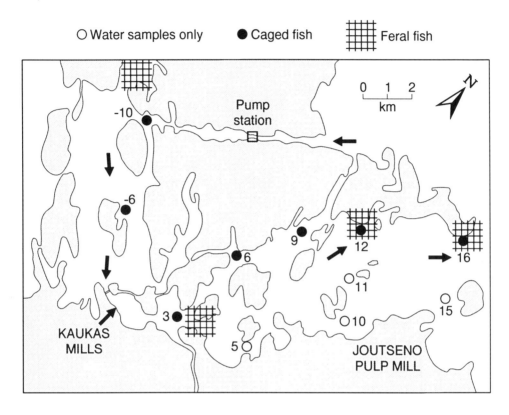

Figure 1. The study area of Lake Saimaa in 1991 and 1993. Sites of sampling water (all circles), caging whitefish (black circles) and collecting feral fish (shaded) are shown. The approximate distances in kilometers upstream (reference) and downstream to Kaukas Mills are indicated.

Assessment of Bioavailability of Organochlorines

The bioavailable part of chlorophenolics present in water was assessed in 1991 and 1993 using the total concentration in the bile of whitefish exposed for 30 days at five areas 3-16 km from the mill (Fig. 1). This assumes that the sum of free and conjugated CPs secreted into the bile is proportional to their available concentration in ambient water.

In order to investigate in more detail the role of dissolved organic carbon (i.e., natural humics and discharged chlorolignins) in the bioavailability of hydrophobic xenobiotic chemicals, lake water samples were collected in October 1990 from four stations 3-16 km downstream from the mill and compared to an upstream control (Kukkonen 1992). Bioconcentration factor (BCF) achieved in *Daphnia magna* within 24 h was used as an indicator of biological availability of 2,3,7,8-tetrachlorodibenzo-*p*-dioxin (TCDD), as modified by the quantity and quality of DOC in the lake water. As revealed by Kukkonen (1992), BCF is proportional to the fraction of TCDD freely dissolved in water.

Experimental Exposure of Whitefish in the Field and Analysis of Biomarkers

Immature 1⁺-year-old whitefish (*Coregonus lavaretus* L. s.l.), a particulate feeder on plankton and seston, were exposed by caging in May-June 1991 and 1993 for 30 days at two reference areas 6 and 10 km upstream and five areas 3-16 km downstream from the mill sewer (Soimasuo *et al.* 1994). Oval-shaped 250-liter cages, where juvenile whitefish can be kept unstressed for several weeks, were used. The cages, with 6-8 fish (less than 1.5 kg m³) in each, were placed on the lake bottom at a depth of ca. 5 m. After exposure, whitefish were sampled on the research vessel "Muikku" as described (Soimasuo *et al.* 1994). Blood and blood plasma, bile and liver samples were stored in liquid nitrogen for later analyses.

The sum of conjugated and free CPs in the bile was analyzed by GC (Lindström-Seppä and Oikari 1989, 1990; Hemming and Holmbom 1992). Samples from two to four fish were combined for one analysis. Conjugates of CPs were hydrolyzed with HCl (37%) at 70°C for 2.5 h, and CPs extracted with two 2-mL lots of *n*-hexane, acetylated and measured by GC. The following CPs were monitored by GC: 2,4,5- and 2,4,6-trichlorophenols, 2,3,4,6-tetrachlorophenol, pentachlorophenol, 3,4,5- and 4,5,6-trichloroguaiacols, and tetrachloroguaiacol.

Microsomal and cytosolic fractions were separated from homogenized liver by ultracentrifugation. Ethoxyresorufin-*o*-deethylase (EROD), uridine-5′-diphosphate glucuronosyltransferase (UDP-GT, substrate: *p*-nitrophenol) and glutathione-*S*-transferase (GST, substrate: 1-chloro-2,4-dinitrobenzene) activities were determined (Lindström-Seppä and Oikari 1989). In order to yield enough tissue material for analysis, liver samples from two to four fish were combined in 1991 (3-5 assays from 10-14 fish per site); individual animals were analyzed in 1993 (14-18 fish per site). Concentration of immunoglobulin M (IgM) in whitefish plasma was measured by enzyme-linked immunosorbent assay (Aaltonen *et al.* 1994). Antibodies against chromatographically purified plasma IgM of trout (*Salmo trutta* L.) and whitefish in 1991 and 1993, respectively, were raised in rabbits and used as the trapping and detecting agents, applying the double antibody sandwich method. For both species, the cross-reactivity with whitefish IgM was strong. Plasma lactate dehydrogenase (LDH) and aspartate aminotransferase (ASAT) activities in 1991 and 1993, respectively, were assayed using Boehringer diagnostic kits (191353, 191337).

Collection of Feral Fish Population Samples

Biotransformation enzyme activity in fish populations living in the study area at southern Lake Saimaa was determined in 1987 (September; Lindström-Seppä and Oikari 1990). This was repeated in 1993 (June-July). Perch (*Perca fluviatilis* L.), roach (*Rutilus rutilus* L.) and bream (*Abramis brama* L.), all rather stationary species spawning in the spring, were caught by seine, weir trap or hook. Animals were allowed to recover in cages kept in local water, and livers were sampled within 5 h after the fish were caught. Freezing, storage and analyses of livers for enzyme biomarkers were conducted as described above for whitefish.

Statistical comparison of fish data was carried out using one-way analysis of variance, followed by Tukey's test ($P < 0.05$ denoting significance).

RESULTS

Effluent and Recipient Water Quality

The process changes in 1992 effectively reduced the effluent load (Table 1). The average reductions over the effluent treatment system are impressive, and the activated sludge plant was extremely efficient.

The improvement in effluent quality also was clearly seen in the lake receiving wastewaters (Table 2). In contrast to 1991, water TOC values were only slightly elevated downstream to the pulp mill sewer in 1993. Water color in the study area below the mill was approximately halved in 1993 compared to 1991. The AOX concentration of the mill effluent decreased, proportionally, even more: from 16.3 mg L⁻¹

in 1991 to 2.1 mg L^{-1} in 1993 (averages of daily samples during the 4-5 week study periods). Consequently, the AOX values in the recipient water columns dropped from about 500 µg L^{-1} to about 100 µg L^{-1} at 3-6 km downstream to the mill, and at 10-15 km dropped from about 200 µg L^{-1} in 1991 to about 50 µg L^{-1} in 1993. In reality, the background levels (30-40 µg L^{-1}) were approached in 1993.

Table 1. Average values for standard effluent load parameters for the Kaukas Mills 1991-93. Reductions are for 1993 when the activated sludge plant operation was stabilized (E. Simpura 1994, personal communication).

Parameter	1991	1992	1993	Reduction in Treatment Plant % (1993)
BOD, t d^{-1}	14.3	8.4	1.7	98
COD, t d^{-1}	99.2	55.5	38.6	76
Susp. solids, t d^{-1}	9.6	4.6	3.4	93
AOX, t d^{-1}	1.7	0.4	0.4	62
Phosphorus, kg d^{-1}	111	59	34	75
Nitrogen, kg d^{-1}	1272	598	404	43

Table 2. Effects of process changes in pulp bleaching and effluent treatment, implemented in 1992, on water quality at southern Lake Saimaa. Values are depicted for 30-day periods in May-June 1991 and 1993. Distances upstream (-6) and downstream (3-16) from the mill are in kilometers.

Sampling Site	TOC (mg L^{-1})		Absorbance x 10^3 (at 460 nm)		AOX (µg L^{-1})		Na^+ (mg L^{-1})	
km from Mill	1991	1993	1991	1993	1991	1993	1991	1993
-6	6.9	7.1	10	7	40	30	2.9	2.9
Mill effluent, after dilution[a]	8.8	3.1			720	75	10.2	9.0
3	22.9	9.3	66	13	520	90	13.1	11.9
5	12.2	9.1	23	13	540	80	12.7	10.7
6	11.5	9.1	24	13	540	80	12.9	10.8
10	10.5	8.8	24	12	420	90	11.3	9.6
11	8.8	8.3	20	11	250	110	8.6	8.0
12	9.4	8.6	20	11	190	180	8.9	8.0
15	8.3	8.5	20	11	210	60	6.9	6.6
16	8.1	8.0	18	11	680	50	6.8	6.1

[a] Adjusted for theoretical dilution in the mixing zone outside the mill, based on determinated sodium concentration.

Replacing elementary chlorine with chlorine dioxide in bleaching and introduction of efficient secondary treatment resulted in a dramatic decrease in CPs in the mill effluent. Overall, total CPs, including CPs released by alkaline treatment, decreased 98%, from 350 µg L^{-1} in 1991 to a mere 5.6 µg

L^{-1} in 1993 (average of samples during the 4-5 week study periods). Free CPs decreased correspondingly from 101 µg L^{-1} to only 0.7 µg L^{-1}, i.e., by more than 99%. Among the CPs still present in the mill effluent in 1993, the dichlorinated phenolics dominated (monochlorinated phenolics were not analyzed).

The sharp decreases in effluent CPs were also seen in the study area downstream from the mill (Table 3, Fig. 2). Comparable to the sum of all CPs, total chloroguaiacols (data not shown), which are specific to chlorine bleaching effluents, decreased by 97-98%. In 1991, a clear gradient was noticed over the study area, but this was not seen in 1993, possibly because the concentrations of CPs in water were approaching the analytical detection limit. Somewhat unexpectedly, the reference site upstream to the pulp mill also was cleaned (Table 3, Fig. 2), apparently due to hydrological reasons.

Table 3. Total and free chlorinated phenolics in water at southern Lake Saimaa in 1991 and 1993. Sum of chlorinated phenols, guaiacols and catechols with two or more chlorine atoms are shown. Total chlorophenolics include phenolics released by alkaline treatment in 2.5 M KOH solution for 16 h at 70°C.

Sampling Site km from Mill	Total Cps (µg L^{-1})		Free CPs (µg L^{-1})	
	1991	1993	1991	1993
-6	3.2	0.25	0.13	0.07
Mill effluent[a]	15.5	0.20	4.45	0.003
3	14.6	0.38	2.95	0.09
5	n.d.[b]	0.48	1.87	0.10
6	15.5	0.25	2.34	0.08
10	11.5	0.30	1.09	0.08
11	7.6	0.34	0.93	0.07
12	8.1	0.55	2.19	0.09
15	6.0	0.32	0.44	0.10
16	6.2	0.22	2.22	0.09

[a] Adjusted for theoretical dilution in the mixing zone outside the mill, based on determinated sodium concentration.
[b] Not determinated.

Wood extractives also exhibited a significant decrease from 1991 to 1993 (Table 4). Total fatty acids in treated mill effluent decreased from ca. 2.1 to 0.3 mg L^{-1} and resin acids from 0.33 to 0.02 mg L^{-1}. On the other hand, the concentrations of sitosterol and betulinol (originating from birch bark) remained unaltered. In lake waters, although less fatty and resin acids were discharged in 1993 than in 1991, the concentrations were rather similar. The essential role of adsorption to particulates for the transport of wood extractives in lake, noted earlier (Holmbom *et al.* 1992), must be assumed in the 1993 study.

Sedimentation of Particulates and the Fate of Chlorophenolics

A continuously decreasing gradient over the study area in rates of sedimenting particles and organic carbon, total organically bound halogen (SOX) and the sum of CPs (including chlorocatechols) was revealed (Kukkonen *et al.* 1992). Concentrations of CPs were 2-8 times higher in the particles collected downstream from the mill than in those upstream. In all, when compared to the upstream situation, sedimentation rates of chloroguaiacols and chlorocatechols were substantially higher at downstream locations, revealing continuous decrease down to 16 km from pulp mill (Fig. 3). About 60% of chloroguaiacols and about 90% of chlorocatechols were removed by sedimentation or degradation processes on the recipient area. On the other hand, while comparing the changes with an inert tracer Na, it was evident that chlorophenols mostly remain in the water and flow through the lake system (Kukkonen et al. 1992).

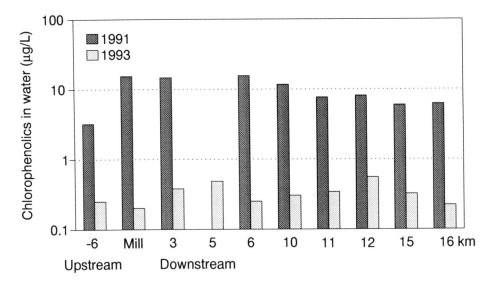

Figure 2. Effects of process changes in pulp bleaching and effluent treatment, implemented in 1992, on the total chlorinated phenolics in water at southern Lake Saimaa. Sum of chlorinated phenols, guaiacols and catechols with two or more chlorine atoms are shown. The effluent value "Mill" has been adjusted to the concentration after dilution in the mixing zone outside the mill. Total chlorophenolics include phenolics released by alkaline treatment prior to extraction and GC analysis.

Table 4. Concentrations of wood extractives in mill effluent (1991 and 1993) and in lake water at a site 3 km downstream from the outfall of the mill.

	Mill Effluent, 1991 ($\mu g\ L^{-1}$)		Mill Effluent, 1993 ($\mu g\ L^{-1}$)		3 km from Mill, 1993 ($\mu g\ L^{-1}$)	
	In Water[a]	At Solids[b]	In Water	At Solids	In Water	At Solids
Fatty acids	520	1710	170	170	14.4	29.4
Resin acids	220	110	5.3	12.5	0.8	1.69
Sitosterol	14	1	12.6	1.8	0.9	3.5
Betulinol	22	35	4.2	120	0.9	3.2

[a] Determinated on the filtrate passing through GF/A glass fiber filter (Whatman).
[b] Determined on the filter residue.

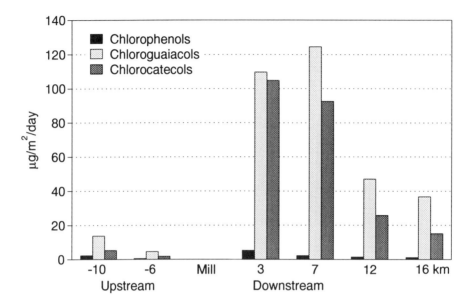

Figure 3. Maximum sedimentation rates (i.e., assuming no resuspension) of chlorophenolics at different distances upstream and downstream from the sewer of Kaukas pulp and paper mill at southern Lake Saimaa in May-June 1991. The period corresponds with the time when whitefish were exposed in cages for biomarker responses.

No comparative study was done on influences of process changes implemented in the Kaukas Mills on sedimenting CPs. However, the baseline situation in 1991 indicates that particulate feeders like whitefish and lake mussel must have been significantly loaded by chloroguaiacols, other chlorophenolics and related compounds as well as sorbed in particulate material.

Organochlorines and Biomarkers in Whitefish

When compared to 1991, the concentrations of CPs and the sum of metabolic conjugates and unmetabolized CPs (chlorophenols and -guaiacols) in the bile of whitefish were decreased dramatically in 1993 (Fig. 4). While the levels of CPs in water downstream to the mill decreased by about 98%, the reduction in concentration of bile CPs over the whole study area was proportional or even more. For instance, at 3 km from the mill, where the bile content of CPs was highest (ca. 540 µg mL^{-1}) in 1991, the reduction was over 99.5% (conc. 1.6 µg mL^{-1}). Based on CPs in fish body, the changes indicate that the exposure intensity of whitefish to chlorophenolics and related compounds in 1993 was only 1-3% of what whitefish experienced in 1991.

In whitefish held at upstream locations from the mill, the low level of "clean" background in 1991 got even "cleaner" in 1993 (bile CP reduction: 95.7%). Like CPs in water (Fig. 2), the hydrology of the study area explains this, as a leak of a minor portion of the principal water mass, generally transferring towards the Vuoksi River in the east (not shown in Fig. 1), will flow around the pump station to upstream locations relative to the Kaukas Mills.

Decreased induction, by 80% on an average, of liver EROD activity (Fig. 5), the most characteristic biomarker of conventional chlorine-bleached kraft pulp mill effluents, was revealed. It is noteworthy, however, that a slight but statistically significant induction of EROD still prevailed near the mill (3-6 km) in 1993. In addition, the lower amounts of IsM, indicative of possible immunosuppression, in whitefish plasma near the mill outfall in 1991 disappeared in 1993 (Fig. 6). The decreasing impact of Kaukas Mills down to 12 km from the outfall is then reversed by the influence of effluents discharged from Joutseno Pulp mills, another producer located about 4 km south from the farthest experimental area (16-km site) on Lake Saimaa (Fig. 1).

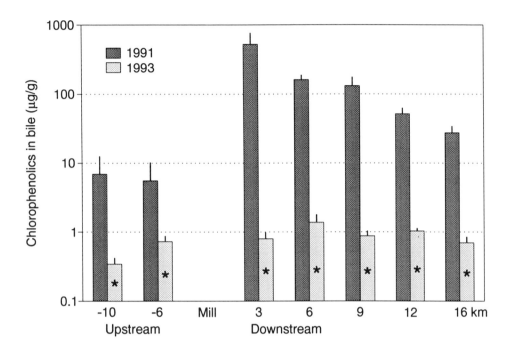

Figure 4. Accumulation of conjugated and free chlorophenolics in the bile of whitefish exposed to bleached kraft pulp and paper mill effluents at southern Lake Saimaa for 30 days (May-June) in 1991 and 1993. The columns represent averages for 10-14 and 14-18 fish in 1991 and 1993, respectively. Bars indicate SDs, and asterisks statistical significance ($p < 0.05$, ANOVA) between the years.

Other biomarkers measured (UDP-GT, GST, LDH and ASAT, see methods), which potentially might be indicative of effects of BKME in fish, revealed no responses which are related to a pollution gradient on southern Lake Saimaa in 1991 or 1993.

Biomarkers in Feral Fish Populations

Induction of liver EROD activity, when compared with fish populations (roach, bream and perch) living downstream to Kaukas Mills in 1987, was diminished in 1993. In 1987, EROD activity in roach caught 1-7 km from the mill was, on an average, 6.7 times higher and at 13 km around twice that in fish living in upstream reference locations. In 1993, instead, the average increase was less than double and not statistically significant at any fishing areas (4, 11 and 16 km from mill).

Conclusions on bream are rather similar with roach. Bream caught 4 km downstream from the mill still revealed higher (three times, $p < 0.01$) liver EROD activity when compared to upstream locations. In 1987, however, at the same distance (1-7 km) the relative average induction was much stronger (7 times) and still significantly elevated (4.8 times) at 13 km from the mill. In 1993, on the other hand, liver EROD in bream caught 11 km from the mill showed only a slight and statistically insignificant sign of induction.

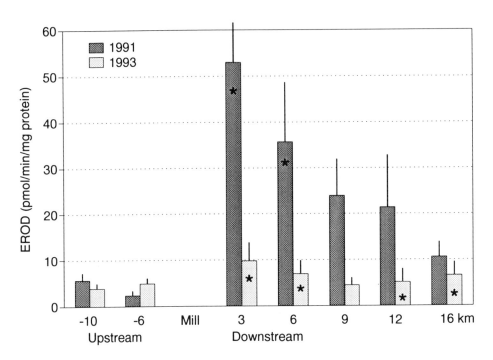

Figure 5. Activity of EROD in the liver of whitefish exposed in cages to bleached kraft pulp and paper mill effluents at southern Lake Saimaa for 30 days (May-June) in 1991 and 1993. For each site, 3-5 pooled samples of 10-14 fish (1991) or 14-18 individual fish (1993) were assayed. Means and SDs are depicted; asterisks denote significance (Tukey) when compared to upstream references.

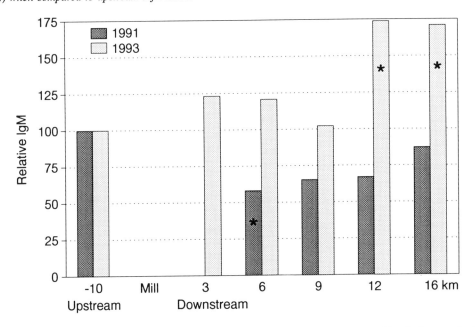

Figure 6. Relative concentrations of immunoglobulin M (upstream IgM = 100) in the blood plasma of whitefish exposed in cages to bleached kraft pulp and paper mill effluents at southern Lake Saimaa for 30 days (May-June) in 1991 and 1993. For further details, see legend to Fig. 5.

DISCUSSION AND CONCLUSIONS

The Process Changes Improved Water Quality in Lake Saimaa

The process changes in pulp bleaching and wastewater treatment implemented in Kaukas Mills in 1992 resulted in a considerable decrease in mill effluent load. This was seen in all measured parameters and concentrations of potentially harmful organic compounds analyzed in this study. The discharge of AOX decreased to about 10% of the value before the process alterations. CPs decreased even more dramatically, down to 2-3% of the amounts prevailing in the lake in 1991, before the change.

Elemental chlorine-free bleaching results in very little formation of polychlorinated compounds, noticed both for high molar mass lignin and phenolics (Dahlman *et al.* 1994; O'Connor *et al.* 1994). Evidently, the large reduction of AOX and the nearby complete elimination of CPs in effluent was due mainly to the omission of elemental chlorine in bleaching. The effluent treatment plant further decreased the amounts of CPs by about 60%. Moreover, in a wastewater treatment plant incorporating an effective activated sludge secondary stage, AOX emissions can consistently be more than halved over the treatment plant (E. Simpura, personal communication).

In conclusion, at Kaukas Mills the nearly complete replacement of chlorine (C << 1%) by chlorine dioxide in pulp bleaching in combination with efficient secondary treatment results in essentially complete elimination of highly chlorinated phenolics in the mill effluents. In treated effluent a nanogram per liter level of CPs was reached. Emission of AOX is significantly decreased as well, by 80-90%, and the color is approximately halved. Wood extractives are also degraded in modernized secondary treatment process, but are not totally eliminated. For example, sitosterol seems to be rather stable in secondary treatment.

Factors Changing the Bioavailability of Organochlorines

Both particulate organic material (POM) and dissolved organic carbon (DOC) may modify the availability of lipophilic contaminants by animals (Kukkonen 1992). Because pulp mills discharge suspended solids as well as AOX and other DOC, forming gradients like that seen at southern Lake Saimaa, we regarded as important to assess the possible roles of these factors in the present study. In upstream (6-10 km)-downstream (3-16 km) comparison, the impact of Kaukas Mills on the amount of sedimenting material was distinct (Kukkonen *et al.* 1992). For whitefish (the type, or race, we used), a plankton feeder by its nutrition, ingestion of particulates may have markedly contributed to amounts of chlorinated guaiacols and catechols absorbed by fish. On the other hand, uptake of chlorinated phenols *via* food seems to be less important than, for instance, chloroguaiacols. Additionally, lipophilic wood extractives (e.g., fatty acids, sitosterol and betulinol) were adsorbed to particulates. In all, along the study area, loading the whitefish depended on the variable bioavailability due to POM.

Laboratory studies revealed that chlorolignins and kraft lignins in pulp mill effluents have a considerable high capacity to bind hydrophobic xenobiotics, like TCDD (Kukkonen 1992). The measured partition coefficients (K_p) of the lignins actually were somewhat higher than K_p of natural lake humics. As water samples were collected from southern Lake Saimaa in 1990, the binding of TCDD (and two other hydrophobic xenobiotics studied) was higher when water DOC was composed of increasing amounts of chlorolignin material. Eventually, the tendency of chlorolignin to bind TCDD resulted in decreased accumulation of TCDD in *Daphnia*. Maximally (3-km sample downstream from mill), the decrease was more than 80%. Therefore, based on decreased concentrations of TOC and AOX in Lake Saimaa water due to mill process changes (Table 2), a lesser significance of ambient DOC for the bioavailability of hydrophobic organics near the mill may be suggested in 1993 when compared to 1991.

Importantly, however, the binding of TCDD to chlorolignins, studied quantitatively by using equilibrium dialysis (Kukkonen 1992), was more or less completely reversible during a 4-d dissociation period. Thus, although the accumulation of TCDD in a planktonic crustacean may decrease in chloroligninous water near the mill, this is counteracted by the reversibility of binding at locations farther away. This may lead to a wider impact area through bioaccumulation. In reality, on the other hand, binding to sedimenting particulates and decreased emissions of hydrophobic organochlorines from the mill may have counteracted this ecotoxicological disadvantage at southern Lake Saimaa in 1993.

Decreased Expression of Ecotoxicity Endpoints

In whitefish experimentally exposed to contaminated waters at southern Lake Saimaa, the improved water quality led to decreased responses in the three toxicological variables (bile CPs, liver EROD and plasma IgM), which were significantly altered in 1991. With increasing distance from the mill, the previously consistent distance-related decreases in concentrations of bile CPs were nearly lost in 1993. Immunosuppression in exposed whitefish (Soimasuo et al. 1994, 1995), distinct in 1991 at receiving waters near the mill, was deleted in 1993. An increase in plasma IgM concentration at the farthest sites (12-16 km) at Lake Saimaa may be explained by the impact of another mill in Joutseno. Thus far, no good explanation for biological significance of this response exists.

Earlier, in 1986 and 1990, an apparent inhibitory zone for the induction of liver EROD in caged whitefish was revealed in locations near the pulp mill (Lindström-Seppä and Oikari 1989). This kind of suppressed inducibility of EROD was not seen in 1991. The relationship between exposure distance and the dose received by fish made the direct comparison of 1991 and 1993 possible. Although the decreases in EROD induction were substantial, and most distinct in waters nearest to the mill, a lesser induction of liver EROD in whitefish exposed downstream from the mill outfall was still shown in 1993. Observations on feral fish, particularly bream (1987 vs. 1993), were in accordance with this. It is noteworthy that in a laboratory experiment, the use of a low-chlorine (C 1-5%) bleaching sequence followed by aerated lagoons as the secondary treatment (Kaukas mills) did not decrease the inducibility of liver EROD (Soimasuo et al. 1995) when compared to the sequence used in 1991 (C 5-45%).

A nearly complete replacement of elemental chlorine (C << 1%) with chlorine dioxide with an efficient secondary treatment of effluents significantly decreased but did not fully eliminate the presence of inducing potential present in wastewaters. Therefore, when considering the process alternatives in pulp production, besides the possible role of chlorine dioxide in production of EROD inducers, more emphasis ought to be given to unchlorinated organic components (Martel et al. 1994). There remains a need to understand the induction of EROD as an toxicity indicator in fish and fish populations and its physiological role in animal health and fitness.

Reduced Impact Area Due to Pulp Mill Effluents in Lake Saimaa

The recovery of the Lake Saimaa ecosystem after process alterations implemented in the spring of 1992 has been noticeable. This conclusion is based on reduced exposure and toxicity endpoint responses measured experimentally and epidemiologically, but it cannot yet be extended to fish community structure. Return of several species demanding improved water quality, however, can be expected according to earlier experiences in Finland (Hakkari 1992).

Reduced concentrations of organochlorines in water corresponded with lesser responses of biomarkers in fish downstream from the pulp and paper mill. In particular, correlation between reductions of CPs in ambient water and, the dose received by fish in the bile, was distinct. It is not known whether or not CPs in fish can be used as an exposure index for other bioactive organochlorines accumulated by fish from pulp and paper mill effluents as well. The reduced exposure, indicated by bile CPs in this study, is expected to mean less ecotoxicity. This is supported by responses of liver EROD and plasma IgM in the contaminated lake area investigated. If cytochrome P-450-related mono-oxygenation and immunological systems in fish are biologically meaningful indicators, both the intensity and the area of impact were more than halved at southern Lake Saimaa in 1993 when compared to 1991.

The investments in water protection and environmental technologies at Kaukas mills in 1992 have resulted in very significant reductions exposure to bioactive chemicals and effects in fish living in southern Lake Saimaa.

ACKNOWLEDGMENTS

This investigation received grants from the Maj and Tor Nessling Foundation, Kymmene/Kaukas Corp. and The Academy of Finland. The technical support provided by Mr. Esa Simpura and his staff at Kaukas Mills is greatly appreciated. We also thank all our colleagues and collaborators who participated in this ECOBALANCE project.

REFERENCES

Aaltonen, T., I. Jokinen and E.T. Valtonen. 1994. Antibody synthesis in roach (*Rutilus rutilus*); analysis of antibody secreting cells in lymphoid organs with ELISPOT-assay. Fish Shellfish Immunol. 4:129-140.

Dahlman, O., A. Reiman, L. Strömberg and R. Mörck. 1994. On the nature of high molecular weight effluent materials from modern ECF- and TCF-bleaching. Proc. Int. Pulp Bleaching Conf., Tech. Sec. CPPA, Montreal, 1994, pp. 123-132.

Hakkari, L. 1992. Effects of pulp and paper mill effluents on fish populations in Finland. Finnish Fish. Res. 13:93-106.

Hemming, J. and B. Holmbom. 1992. Analytical methods used at the Department of Forest Products Chemistry; Environmental chemistry. Report C1/92, Åbo Akademi University, Turku/Åbo, Finland, pp. 1-5.

Holmbom, B., J. Hemming and P. Mäki-Arvela. 1992. Environmental fate of effluent components from the Kaukas pulp and paper mill in south Lake Saimaa system. *In* M. Viljanen, and S. Ollikainen (ed.). Proc. Lake Saimaa Seminar, Univ. Joensuu, Karelian Inst., 103, pp. 39-52.

Kukkonen, J. 1992. Effects of lignin and chlorolignin in pulp mill effluents on the binding and bioavailability of hydrophobic organic pollutants. Water Res. 11:1523-1532.

Kukkonen, J., J. Pellinen, J. Hemming, B. Holmbom and A. Oikari. 1992. Effects of sedimenting particulate matter on the fate of some chlorinated compounds in pulp mill effluent. Abstract and poster at SETAC 13th Annual Meeting, Cincinnati, Ohio, pp. 1-6.

Lindström-Seppä, P. and A. Oikari. 1989. Biotransformation and other physiological responses in whitefish caged in a lake receiving pulp and paper mill effluents. Ecotoxicol. Environ. Safety 18:191-203.

Lindström-Seppä, P. and A. Oikari. 1990. Biotransformation activities of feral fish in waters receiving bleached pulp mill effluents. Environ. Toxicol. Chem. 9:1415-1424.

Martel, P.H., T.G. Kovacs, B.I. O'Connor and R.H. Voss. 1994. A survey of pulp and paper mill effluents for their potential to induce mixed function oxidase enzyme activity in fish. Water Res. 28:1835-1844.

O'Connor, B.I., T.G. Kovacs, R.H. Voss, P.H. Martel and B. van Lierop. 1994. A laboratory assessment of the environmental quality of alternative pulp bleaching effluents. Pulp Pap. Can. 95:47-56.

Simpura, E. and K. Pakarinen. 1993. Super low loaded activated sludge process incorporating an aerobic selector for the treatment of pulp and paper waste water. TAPPI Proc., Environ. Conf., Atlanta, pp. 865-877.

Soimasuo, R., T. Aaltonen, M. Nikinmaa, J. Pellinen, T. Ristola and A. Oikari. 1985. Physiological toxicity of low-chlorine bleached pulp and paper mill effluent on whitefish (*Coregonus lavaretus* L. s.l.): A laboratory exposure simulating lake pollution. Ecotoxicol. Environ. Safety: In press.

Soimasuo, R., I. Jokinen, J. Kukkonen, T. Petänen, T. Ristola and A. Oikari. 1994. Biomarker responses in a pollution gradient: Effects of pulp and paper mill effluents on whitefish studied by caging technique. Aquatic Toxicol.: In press.

MONITORING ENVIRONMENTAL EFFECTS AND REGULATING PULP AND PAPER DISCHARGES: BAY OF PLENTY, NEW ZEALAND

Paul Dell, Fergus Power, Robert Donald, John McIntosh, Stephen Park and Liping Pang

Environment BOP, P.O. Box 364, Whakatane, New Zealand

In October 1991, New Zealand introduced the Resource Management Act (RMA). This legislation represents a comprehensive and integrated approach to "promote the sustainable management of natural and physical resources." Under the RMA the "effect" of an activity is paramount in considering its approval. The preparation of the Tarawera River Regional Plan was initiated because the Tarawera River receives discharges from two pulp and paper mills and a geothermal bore field. A comprehensive monitoring program was designed to assess the impact of the discharges. This included detailed investigations of ecology, toxicity and chemistry within the river and marine environment. While the instream concentrations of pulp mill contaminants are relatively low, chronic impacts on biota consistent with bleached kraft mill effluent exposure are present. Ecological surveys have revealed significant impacts on macroinvertebrate and aquatic plant communities. Large inputs of color degrade the river visually while high effluent oxygen demand occasionally reduces dissolved oxygen concentrations to critical levels. Despite concerns with toxicity, it is the authors' view that the dissolved oxygen and color are the most significant long-term issues facing the management of the river.

INTRODUCTION

The Tarawera River flows from oligotrophic Lake Tarawera to the coast at Matata, a distance of 55 km (Fig. 1). The flow regime is stable due to the lake reservoir and the sponge-like nature of the pumice soils. Flows are relatively small with a mean flow of 26 m^3 s^{-1}. Long- term changes in vegetation cover (particularly to production forestry) and a regional rainfall decline have resulted in a reduction of flows over the last 30 years (Fig. 2; Pang 1993).

Two pulp and paper mills discharge bleached pulp effluent to the lower Tarawera River at Kawerau. Tasman Pulp and Paper Company produces bleached kraft and mechanical pulp (270,000 and 310,000 t yr^{-1}, respectively). Treatment of Tasman bleached kraft mill effluent occurs in an aerated oxidation pond system with a residence time of 5-6 d. The final discharge rate is 2 m^3 s^{-1}. Caxton Paper Mills operates a bleached sulfonated-chemithermomechanical pulp mill with a rated capacity of 50,000 t yr^{-1} paper and 75,000 t yr^{-1} pulp. Effluent from the pulp mill and sewage from the Kawerau township is treated in an anaerobic system with discharge (ca. 0.25 m^3 s^{-1}) to the river and/or rapid infiltration basins. In addition, Works Geothermal operates a bore field supplying steam to the Tasman plant. The discharge of waste geothermal fluid occurs at two points, giving a combined river discharge of 0.24 m^3 s^{-1}.

Between 1990 and 1994 Environment BOP carried out detailed investigations of the water quality and quantity in the Tarawera River catchment. This integrated study included four modules designed to identify the impacts of the pulp and paper industries (Fig. 3). A synthesis of the results of these investigations and a discussion of the regulatory approach that Environment BOP intends to adopt is presented in this paper.

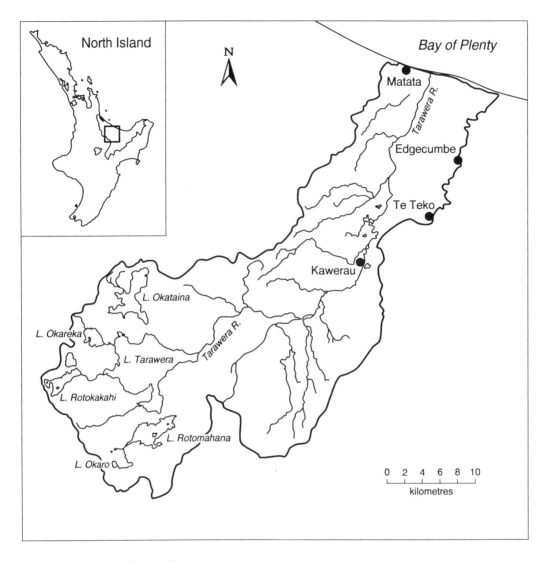

Figure 1. The Tarawera River catchment.

CHEMISTRY

Extensive areas of geothermal activity contribute inorganic elements to Lake Tarawera, the source of the Tarawera River. Compared to other rivers in New Zealand, the concentrations of dissolved salts are high (Smith and Maasdam 1994). Many geothermal constituents are potentially toxic to aquatic life. While discharges of geothermal fluid elevate the concentrations of several metals (e.g., mercury, lithium, arsenic, boron), these do not exceed established guidelines for the protection of aquatic life. Hydrogen sulfide may exceed aquatic life criteria under worst case conditions (river at 1 in 10-year low flow and maximum permitted effluent flow).

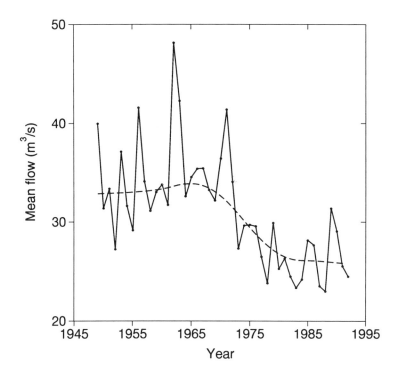

Figure 2. Mean annual flows recorded in the Tarawera River.

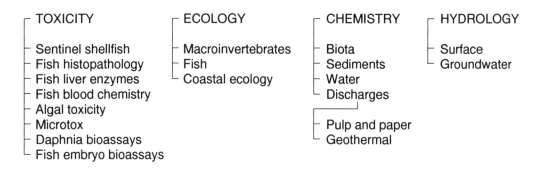

Figure 3. Technical investigations carried out for the Tarawera River Regional Plan.

Extractable organic substances in effluent, water and sediment samples have been investigated using a combination of capillary column GC-FID, GC-ECD and GC-MS procedures (Singh-Thandi 1993). A unique aspect of the chemistry of the final Tasman discharge is the presence of some saturated resin acids (mainly abietan-18-oic acid) and a series of degraded resin hydrocarbons. Fichtelite, dehydroabietin, tetrahydroretene and retene are the most prevalent of these degraded hydrocarbons. These compounds are indicative of the anaerobic modification of abietic acid. They have not been detected in untreated effluent but are thought to come from a sludge lagoon which is dewatered to the oxidation pond treatment system (Wilkins and Panadam 1986).

The organic composition of the Tasman and Caxton effluent differs markedly. Organic compounds in the Tasman effluent include degraded resin acids, fatty acids and chlorophenolic substances (mean levels of 382, 52 and 8 µg L^{-1}, respectively). Appreciably higher concentrations of fatty acids and resin acids are present in the Caxton discharge (mean levels of 654 and 7204 µg L^{-1}, respectively). The concentrations of kinleithic acid (a hydroxylated, process-modified resin acid; Wilkins et al. 1989) in the Caxton discharge (0 - 1351 µg L^{-1}) appear to correlate with the operation of the anaerobic digester. Despite a 10-fold difference in effluent volumes, each mill has a similar mass discharge of resin and fatty acids (Table 1).

Table 1. Fatty and resin acid mass discharges to the Tarawera River January-February 1992.

Compound (kg d^{-1})	Caxton	Tasman
Myristic acid	0.4	
Palmitoleic acid	0.6	
Palmitic acid	2.5	3.1
Oleic acid	2.0	1.5
Stearic acid	1.8	1.5
Other fatty acids	1.1	5.0
Total fatty acids	8.3	11.1
Secodehydroabietic acid 1	1.6	3.1
Secodehydroabietic acid 2	1.0	1.6
Pimaric acid	7.1	1.8
Sandaracopimaric acid	1.2	
I/P/L group	13.1	
Abietan-18-oic acid		35.7
Dehydroabietic acid	41.6	16.9
Abietic acid	4.2	
Kinleithic acid	10.7	
Other resin acids	20.8	21.5
Total resin acids	90.5	80.8

The instream concentrations of these organic compounds following discharge may differ from that calculated from dilution rates. Downstream of the Tasman discharge, contaminant concentrations agree with the expected river dilution factor (10 - 12 fold). In contrast, a greater than expected attenuation of extractable organic compounds was found downstream of the Caxton discharge. While the exact mechanism is unclear, this rapid assimilation of organic contaminants may be due to bacterial metabolism, complexation with sulfur, flocculation or adsorption to sediments (Singh-Thandi 1993).

High levels of organic contaminants are present in river bank sediments. Resin acid levels in sediments downstream of the Tasman discharge range from 600-1400 mg L^{-1}. Sediment and seepage water samples collected upstream of the Caxton discharge contain an array of fatty acids (56-110 µg L^{-1}), resin acids (166-266 µg L^{-1}) and degraded resin hydrocarbons (2-12 µg L^{-1}). These contaminants travel by groundwater flow and are derived from the Caxton infiltration basins. They include a novel degraded resin acid and a degraded resin hydrocarbon, designated caxtonic acid (30 µg L^{-1}) and caxtonellite (283 µg L^{-1}), respectively. The structure of caxtonellite has been established using one- and two-dimensional techniques (Singh-Thandi 1993).

Bioaccumulation of pulp and paper contaminants occurs in freshwater and marine mussels (Hickey et al. 1993). GC/MS techniques have revealed that degraded resin hydrocarbons bioaccumulate to a much greater extent than is the case for resin acids (ca 20-fold). Apart from pentachlorophenol, which is found at all river sites, the accumulation of chlorinated organic compounds predominantly occurs downstream of the Tasman discharge. Resin acids are detectable in mussels downstream of the Caxton discharge but are present in higher concentrations downstream of the Tasman discharge. Little degradation of organic contaminants occurs with increasing distance down the Tarawera River.

During the last decade, process and effluent treatment modifications at the Tasman mill have resulted in reductions in the quantities of contaminants discharged. These measures have included the introduction of oxygen delignification and efforts to further reduce elemental chlorine usage by chlorine dioxide substitution. Since mill modernization, the resin acid levels in freshwater mussel tissue have declined by 14-75%. Chlorinated compounds showed a more marked reduction (54-97%), with the exception of PCP which increased by 38%. Chlorinated resin acids, which are the major chlorinated species, showed a high reduction (>92%).

Dioxin and furan analyses have been carried out on river water, effluents and biota (freshwater and marine). Concentrations in the pulp and paper effluents are relatively low by world standards 2,3,7,8-tetrachlorodibenzo-p-dioxin (TCDD) toxic equivalents (NATO basis): Tasman 7.5 pg L^{-1}, Caxton 20 pg L^{-1}). Correspondingly low levels (reported on a fresh weight basis) are present in freshwater eels (0.2-6 pg g^{-1} toxic equivalents [TEQ]), coastal shellfish (0.07-0.13 pg g^{-1} TEQ) and finfish (0.03-0.23 pg g^{-1} TEQ). Dioxin and furan levels have decreased considerably since the modernization and process alterations at the Tasman mill discussed above.

Dissolved oxygen concentrations are of major concern to the biological health of the lower Tarawera River. On the bed of the river, moving sediment dunes, composed of porous pumice, provide ideal habitat for oxygen-consuming microorganisms. The efficiency of this phenomenon is such that benthic metabolism accounts for at least 90% of the present deoxygenation in the Tarawera River (Rutherford et al. 1991). Because of the high rate of dissolved oxygen depletion, the ability of the river to assimilate effluents with high biochemical oxygen demand is limited. In recent years dissolved oxygen concentrations have frequently fallen below the minimum standard for the lower river of 5 g m^{-3} (Fig. 4).

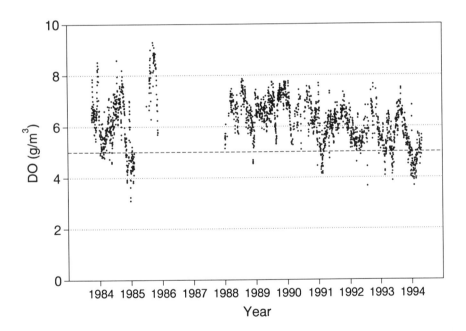

Figure 4. Daily mean dissolved oxygen concentrations recorded in the lower Tarawera River. Dashed line indicates 5 g m^{-3} limit.

The discharge of kKraft effluent from the Tasman mill degrades the appearance and visual appeal of the river. Kraft mill effluent is intensely light absorbing and weakly light scattering such that a small amount of effluent produces a disproportionately large increase in color (Davies-Colley et al. 1993). Munsell hue declines by 15 units (from green to yellow) downstream of the Tasman discharge, giving the river a brown-black appearance. Associated with this change are reductions in visual clarity, as measured by black disc (Davies-Colley 1988), and increases in the absorbance of the water (Table 2).

Table 2. Color and clarity measurements on the Tarawera River 1992-94.

	Upstream Caxton	Downstream Caxton	Upstream Tasman	Downstream Tasman	Caxton Effluent	Tasman Effluent
Munsell hue	45 (green)	42 (green)	42 (green)	27 (yellow)	ND	ND
Absorbance[a]	0.7 (0.3)	1.2 (0.4)	1.2 (0.4)	4.6 (0.7)	33.9 (27.4)	29.9 (5.5)
Black disc (m)	2.5	1.8	1.8	0.6	ND	ND

[a] At 440 nm, unfiltered (mean ± SD).
ND = not determined.

TOXICITY

The toxicities of the pulp and paper and geothermal effluents have been investigated using several standard test protocols. Hickey (1994) tested the response of *Daphnia magna* (14-d chronic test), *Selenastrum capricornutum* (algal acute 96-h test) and Microtox™ (bacterium acute test). On a whole effluent basis, toxicity was defined as moderate for the geothermal effluent, slight for the Caxton CTMP effluent and slight to threshold for the Tasman oxidation pond effluent. High concentrations of arsenic and hydrogen sulfide are thought to contribute significantly to the toxicity of the geothermal effluent. Beresford (1993) assessed the hatching success and larval mortality of zebrafish (*Brachydanio rerio*) exposed to various concentrations of Tasman and Caxton effluent. Hatching and mortality of eggs and larvae were severely affected in 20% Caxton effluent, suggesting high toxicity, while the Tasman oxidation pond effluent had low toxicity. None of the above tests have demonstrated toxicity in river water collected from below the effluent discharges.

Several techniques have shown chronic effects in feral eels (predominantly *Anguilla australis*) exposed to the effluent streams. These eels generally show no external evidence of stress and are most abundant downstream of the Tasman discharge. Analysis of the liver, an organ that has a major role in detoxification, has revealed fatty changes, tissue necrosis and enlargement in eels exposed to the pulp mill effluents. Changes in blood plasma enzyme levels also suggest liver disease, while urea and potassium levels indicate altered kidney function (Beresford 1993).

Preliminary work has shown that biomarkers have potential in assessing contaminant exposure in the Tarawera River. This research has focused on ethoxyresorufin-*o*-deethylase (EROD). EROD is induced in fish livers in response to dioxin-like chemicals (although see Munkittrick et al. 1992). Because of this response EROD has been widely used in assessing exposure to pulp and paper contaminants. Significant increases in EROD activity have been recorded in both caged and feral eels at sites downstream of the Tasman effluent outfall. Using this data, a relationship has been defined between liver EROD activity and muscle burdens of 2,3,7,8-TCDD in feral eels (Dr. Paul Jones, ESR Environmental, P.O. Box 30-547, Lower Hutt, New Zealand, pers. comm.).

ECOLOGY

Ecological investigations have focused on the status of fish communities in the catchment and on the impacts of the industrial discharges on macroinvertebrate and aquatic plant communities (Donald 1994).

Inputs of color from the Tasman effluent discharge reduce the extent of submerged plant communities. In 1984, a diverse community (14 species) covered 35% of the river bed upstream of the Tasman effluent discharge. Downstream of the discharge, diversity declined to five species covering just 3% of the river bed. These differences are due to reductions in light penetration, as substrate composition and water velocity are similar at both sites. Further downstream, channelization and straightening of the river have increased the mobility of the bed. It is unclear whether these relatively unstable areas would support submerged plant communities if light penetration were adequate.

During summer low flows, extensive biological growths occur on the bed of the river downstream of the Caxton outfall. These green filamentous algae (*Ulothrix zonata*) are associated with filamentous sulfur bacteria (*Thiothrix* sp.) forming visible slimes. In 1993, areal chlorophyll a and ash free dry weight (AFDW) were measured to give an assessment of biomass in relation to an upstream control site. Chlorophyll a increased downstream of the outfall from 243 to 509 mg m^{-3} (ANOVA, $p<0.05$) and AFDW increased from 28 to 79 mg m^{-3} (ANOVA, $p < 0.01$). These increases result from inputs of nutrients and readily degradable organic compounds from the Caxton discharge.

The Caxton anaerobic treatment plant discharge has marked impacts on macroinvertebrate communities. These impacts have been assessed using artificial substrate methodology. Upstream of the discharge the river supported a community composed of similar proportions of mayflies, caddisflies and chironomid larvae. Downstream of the discharge the community consisted largely of chironomid larvae and oligochaetes, with a halving of caddisfly abundance and the near total loss of the sensitive mayfly taxa.

Several measures (metrics) derived from the macroinvertebrate data allow these community changes to be described. Most of the metrics changed significantly in response to the Caxton discharge. In particular, the EPT index (the number of mayfly, caddisfly and stonefly taxa) and the QMCI index (a measure of organic pollution specific to New Zealand streams) declined significantly. The bioassessment protocols of the U.S. EPA (1989) have been used to integrate this information and give an assessment of the level of impairment or ecological impact. Using these protocols, the level of impact downstream of the mixing zone of the Caxton discharge is defined as moderate to severe.

The status of fish communities in the Tarawera River and tributary streams has been assessed using netting and electro-fishing techniques. In the tributaries of the lower catchment, indigenous fish are diverse, with a tendency towards markedly lower diversity in the upper catchment. This pattern is not due to poor habitat quality. Most of the indigenous fish in New Zealand are diadromous, requiring free access to the sea to complete their life cycle. It is possible that poor water quality in the lower river is influencing juvenile fish migration and ultimately distribution. Unfortunately, there is little information on the tolerance of these species during migration periods.

Common smelt (*Retropinna retropinna*) are abundant in the lower Tarawera River. These populations have been compared with those in the nearby Rangitaiki River, a system which does not receive pulp and paper effluent. Length-weight and diet analyses suggest that smelt production is enhanced by the discharge of "food" from the pulp and paper industries. Littoral Cladocera are the dominant smelt food in the Tarawera River. These filter and detrital feeders feed on the high concentrations of bacteria derived from and promoted by the major discharges. Hickey *et al.* (1993) found that the condition of freshwater mussels and the survival, growth and reproduction of fingernail clams are enhanced downstream of the Tasman discharge. The authors suggested that high food levels present in the Tasman effluent favor these filter feeders.

The impact of Tarawera River contaminants on colonization in the marine environment has been investigated using artificial substrates (Environment BOP 1994). These samplers were deployed within the surface plume approximately 500 m out from the river mouth. No significant difference in mean species richness (predominantly invertebrates) occurred between the Tarawera River and a Rangitaiki River control site (means of 12.7 and 11, respectively). Almost identical lists of species were present at both sites, with little evidence that fauna avoided settling outside the Tarawera River mouth. Some significant variations in individual species abundance did occur between the two river sites. This was minor and probably reflected natural recruitment variation.

REGULATION AND MONITORING

The regulation of the discharge of bleached pulp mill effluents must be flexible enough to evolve along with advances in monitoring technology and environmental research. This is especially so for the Tarawera River, a relatively small waterway receiving large volumes of effluent. As environmental managers, it is the authors' view that dissolved oxygen and color are the most important long-term issues in the lower Tarawera River. Numerical standards for these parameters will be set in the Tarawera River Regional Plan.

Under current legislation, a single standard based on a minimum dissolved oxygen concentration of 5 g m^{-3} applies in the lower Tarawera River. Environment BOP considers this approach to be inflexible and has adopted the research-based standards advocated by the U.S. EPA (1986). These provide protection for cold water salmonid fisheries and are consistent with the aim of reinstating a trout fishery in the lower river. The standards adopted impose limits on the 30-d mean (6.5 g m^{-3}), 7-d mean-minimum (5 g m^{-3}) and absolute minimum oxygen concentrations (4.5 g m^{-3}).

A simplified computer model has provided guidance on the BOD load limits needed to consistently meet these standards (Rutherford 1991, 1993). This Gross Oxygen Uptake Model (GUM) shows that the 30-d mean standard is the most restrictive on BOD loading. To meet this standard, industry will need to achieve a large reduction in their combined BOD load (from 6-7 to 2-3 t d^{-1}). The revised oxygen standards will be introduced in a staged manner to allow industry to progressively reduce BOD loads.

The public perception of the lower Tarawera River is mainly influenced by the poor visual appearance of the water. Perception studies of New Zealand waters have clearly defined the color and clarity thresholds for bathing and aesthetic purposes (Smith 1994). Waters with a Munsell hue of at least 30 (green-yellow) and visual clarity of at least 1.2 m (black disc) are generally considered acceptable for use. In certain cases, discoloration and low clarity may be acceptable if the source is perceived to be natural. Examples include tannin-stained forest streams and high country lakes of poor clarity.

The achievement of an "acceptable" green hue in the lower Tarawera River is dependent on a 95% reduction in the color of the Tasman effluent (Davies-Colley *et al.* 1994). Suitable technology to economically reduce kraft effluent color on this scale is not yet available (Dell 1993). From a management perspective, it is likely that reductions in effluent color will be achieved by gradual improvements in plant operation and process modifications. Discharges of large amounts of color will become a non-complying activity after notification of the Tarawera River Regional Plan. In this case industry will have recourse to continue discharging under Section 107 of the Resource Management Act. This provision requires that the applicant demonstrate exceptional circumstances or that the discharge is of a temporary nature.

The issue of toxicity in the Tarawera River has been intensively investigated with conflicting results. While chronic and acute testing methods have suggested that there is no toxicity in river water, analysis of macroinvertebrate and eel communities has revealed impacts which have been attributed to the pulp and paper industries. There is considerable debate over the ability of standard chronic and acute toxicity testing to predict instream biological impacts (U.S. EPA 1992). Despite the uncertainty, it is considered necessary to specify a standard based on toxicity measurement. This is a move away from specifying numerical contaminant standards, recognizing that it is not feasible to specify allowable concentrations of every potentially toxic compound. In the Tarawera River it will not be permissible for a discharge to produce a toxic response as measured by standard acute (salmonid) and chronic (*Daphnia*) tests.

International debate over the environmental impact of bleached pulp mills has focused on the production and discharge of organochlorine compounds. New bleaching technology recently introduced at the Tasman mill has substantially reduced the production of chlorinated organics. Since 1988 the production of adsorbable organic halide (AOX) by Tasman has declined from 3.0 to 0.9 kg ADt^{-1} of pulp. The present AOX levels are below limits set by a number of Canadian provincial regulators (Dell 1993) and further reductions are expected in the future. Concentrations of 2,3,7,8-TCDD and 2,3,7,8-tetrachlorodibenzofuran in both Tasman and Caxton effluent are substantially below Canadian federal limits (set at 15 and 50 ppq, respectively; Dell 1993).

Monitoring and plan review are essential to the long-term management of the river. During the ten-year life span of the Tarawera River Regional Plan two intensive monitoring periods will be initiated. These will include the integrated monitoring techniques and parameters covered in this paper. The results of this monitoring will show the success of certain policies promulgated in the plan. In particular, it is expected that the quantities of contaminants discharged into the Tarawera River will decline over the next ten years.

REFERENCES

Beresford, D. 1993. Progress report on the investigations into the effects of geothermal and pulp and paper mill effluent on the health of fish living in the Tarawera River. Doctoral progress report to Environment BOP (December 1991 to December 1992). 24 p.

Davies-Colley, R.J. 1988. Measuring water clarity with a black disc. Limnol. Oceanog. 33:616-623.

Davies-Colley, R.J., W.N. Vant and D.G. Smith. 1993. Colour and clarity of natural waters: Science and management of optical water quality. Ellis Horwood, Great Britain. 310 p.

Davies-Colley, R.J., D.G. Smith, D. Speed and J.W. Nagels. 1994. Water discolouration of the Tarawera River. NIWA Consultancy Report No. BPR 007 (August 1994) prepared for Environment BOP. 26 p.

Dell, P.M. 1993. Pulp and paper industry study tour, July 1993. Environment BOP Environmental Report 93/4.

Donald, R.C. 1994. Tarawera River Regional Plan technical investigations. Freshwater ecology component. Environment BOP Environmental Publication 94/1. 122 p.

Environment BOP. 1994. Proceedings of a toxicity workshop held in Whakatane, 17-18 May 1994. Environment BOP Environmental Report 94/15. 123 p.

Hickey, C.W. 1994. An assessment of the toxicity of Tarawera River: Effluent discharges and receiving waters and assessment of geothermal discharges. NIWA Consultancy Report No. SCJ008/2161 (May 1994) prepared for Environment BOP.

Hickey, C.W., D.S. Roper and M.L. Martin. 1993. Mussel biomonitoring in the Tarawera River: 1993. NIWA Consultancy Report BPR 215 (October 1993) prepared for Environment BOP. 30 p.

Munkittrick, K.R., M.R. van den Heuvel, D.A. Metner, W.L. Lockhart and J.J. Stegman. 1992. Interlaboratory comparison and optimisation of hepatic ethoxyresorufin o-deethylase activity in white sucker (*Catostomus commersoni*) exposed to bleached kraft pulp mill effluent. Environ. Toxicol. Chem. 12:1273-1282.

Pang, L. 1993. Tarawera River flow analysis. Environment BOP Environmental Publication 93/2. 102 p.

Rutherford, J.C. 1991. Deoxygenation in a mobile-bed river - II. Model calibration and post-audit. Water Res. 25:1499-1508.

Rutherford, J.C. 1993. Revised BOD load limits for the Tarawera River. NIWA Consultancy Report No. BPR045/3 (September 1993) prepared for Environment BOP. 6 p.

Rutherford, J.C., R.J. Wilcock and C.W. Hickey. 1991. Deoxygenation in a mobile-bed river - I. Field studies. Water Res. 25:1487-1497.

Singh-Thandi, M. 1993. Some aspects of the organic chemistry of the Tarawera River. M.Sc. thesis, University of Waikato, Hamilton, New Zealand. 154 p.

Smith, D.G. 1994. Some management implications of human clarity and colour perception of fresh waters. Paper for the 5th International Symposium on Society and Resource Management, Colorado State University, Fort Collins, June 1994.

Smith, D.G., and R. Maasdam. 1994. New Zealand's National River Water Quality Network 1. Design and physico-chemical characterisation. N.Z. J. Mar. Freshwater Res. 28:19-35.

United States Environmental Protection Agency. 1986. Quality criteria for water 1986. U.S. EPA Publication EPA/440/5-86-001. Washington D.C.

United States Environmental Protection Agency. 1989. Rapid Bioassessment Protocols for use in streams and rivers. Benthic macroinvertebrates and fish. U.S. EPA Publication EPA/444/4-89-001. Washington D.C.

United States Environmental Protection Agency. 1992. Water quality standards for the 21st century. Proceedings of the third national conference, Las Vegas, Nevada, August 31-September 3, 1992. U.S. EPA Publication 823-R-92-009. Washington D.C. 320 p.

Wilkins, A.L., and S. Panadam. 1986. Extractable organic substances from the discharges of a New Zealand pulp and paper mill. Appita 40:208-212.

Wilkins, A.L., A. Langdon and S. Panadam. 1989. Kinleithic acid: A new hydroxylated resin acid from the biological treatment system of a New Zealand kraft pulp and paper mill. Aust. J. Chem. 42:983-986.

ECOTOXICOLOGICAL IMPACTS OF PULP MILL EFFLUENTS IN FINLAND

Maarit H. Priha

The Finnish Pulp and Paper Research Institute, Box 70, Tekniikantie 2, FIN-02151 Espoo, Finland

The total effluent from all 15 Finnish pulp mills was biotested with five laboratory-scale sublethal and lethal toxicity tests. The biotests were chosen so as to represent different types of organisms and impact mechanisms: immobilization of the water flea *Daphnia magna* (24-h EC50), inhibition of light emission of the luminescent bacterium *Vibrio fischeri* (15-min EC50), growth inhibition/stimulation of the green alga *Selenastrum capricornutum* (96-h EC50), early life stage development of the zebra fish *Brachydanio rerio* and MFO activity (EROD) in isolated hepatocytes from rainbow trout (*Oncorhynchus mykiss*). The effluent samples were collected during normal production periods (1 to 30-d pooled samples) and frozen daily. Thirteen mills had biological effluent treatment (activated sludge or aerated lagoon) and two mechanical treatment systems.

In general the biologically treated bleached kraft mill effluents showed little or no toxic impact on test organisms regardless of mill technology. None of the effluents was toxic to *Daphnia* even at a 100% concentration. Three effluents were clearly toxic to bacteria and two showed some toxic effects which may have been due to the interfering effect of effluent color. All the effluents stimulated algal growth in some of the test concentrations, and three also caused growth inhibition at rather high concentrations. None of the secondary treated effluents at 100% concentration affected the survival of zebra fish embryos and larvae, whereas the mechanically treated effluents were toxic. Six of the secondary treated effluents had a slight effect on zebra fish hatching by shortening the hatching time, while the mechanically treated effluents prolonged the hatching time. Only two of the effluents exhibited a slight induction in EROD activity, but seven effluents inhibited the activity.

INTRODUCTION

Extensive research has been performed during the last ten years on the biological impacts of pulp and paper industry effluents. A wide range of organisms and impact mechanisms have been studied both in natural environments and in laboratory conditions. The progress in process and effluent treatment technology has reduced the acute lethal effects in receiving waters and focused the attention on sublethal and long-term effects.

Axegård *et al.* (1993) reviewed a variety of toxicity studies concerning bleached pulp mill effluents conducted with laboratory tests, model ecosystems and field studies. It was concluded that effluents from modern kraft pulp technology (i.e., oxygen delignification, high chlorine dioxide substitution, external treatment) exhibit no or low acute toxicity in single-species tests and considerably lower responses in subacute toxicity when compared with the effluents from older bleaching and pulping technologies. Model ecosystem/experimental stream studies have shown that it is impossible to predict the environmental impact of an effluent on the basis of its AOX (chlorinated organics) content at the current low AOX levels. Based on field studies performed during the last ten years, it was concluded that the development in pulping and bleaching technology during the 1980s most probably resulted in a decrease in environmental impact on receiving waters. The acute and subacute toxicity of the total effluents before secondary treatment from elemental chlorine free (ECF) and totally chlorine free (TCF) pulping was studied by Lövblad and Malmström (1994). Both TCF/ECF softwood and TCF hardwood effluents showed no or very low acute and subacute toxicity. Although at a low toxicity level, the TCF effluents in all tests showed lower toxicity emission factor values as compared to the ECF effluent. Five secondary treated kraft mill effluents, of which four were from bleached and one from unbleached kraft production, were tested by Martel et al. (1993) for hepatic MFO (mixed-function oxidase) induction by exposing rainbow trout in laboratory conditions. MFO activity was significantly induced as a result of exposure to bleached as well as unbleached kraft mill effluents. The data suggested that the replacement of chlorine in bleaching by chlorine dioxide or nonchlorine compounds did not significantly alter the ability of the effluent to induce MFO activity, but instead MFO induction was observed in fish exposed to black liquor.

In a field study on downstream sites of 11 Canadian pulp and paper mills, Robinson et al. (1994) discovered toxic effects on *Ceriodaphnia* and fathead minnow with recipient water in laboratory tests, the toxicity being generally associated with the low dilution discharge of primary treated effluent. The same sites were studied for physiological changes in wild fish (Munkittrick et al. 1994), and induction of ethoxyresorufin-*o*-deethylase (EROD) enzymes and depression of plasma sex steroid levels were found downstream of several pulp mills, including some mills not using chlorine in bleaching or mills that have secondary treatment. The laboratory toxicity tests could not predict these responses in wild fish. In another field study of a riverine ecosystem receiving treated bleached kraft effluent, it was shown that the effluent did not have discernible impacts on either fish populations or individual fish health. Elevated EROD induction was observed, but it was not predictive of biological impacts (Swanson et al. 1992). In a Swedish field study on five bleached kraft pulp mill sites (one mill with aerated lagoon, the others with sedimentation basins or effluent treatment not indicated), several types of population or individual level impact were observed, of which the biochemical/physiological disturbances in wild fish showed the largest scale distribution (Södergren et al. 1993).

It is obvious that each pulp mill and receiving water system is unique, as well as the individual organisms, populations and response patterns. Thus generalizations on biological impacts based on case studies are difficult to make. However, the recent trend on process modifications in pulp industry (chlorine dioxide bleaching, chlorine-free bleaching, modified cooking, biological treatment of effluents, etc.) has clearly been shown to reduce the environmental impacts in receiving waters.

The aim of this study was to create a "state-of-the-art" picture of the biological characteristics of Finnish pulp mill effluents, when almost every mill that applies biological treatment for effluents and ECF (and TCF) bleaching has replaced conventional bleaching. The whole-mill effluents were collected after treatment to represent the true effluent discharge into receiving waters. The sampling time was usually from several days up to one month to obtain a good average discharge. The biotesting methods were chosen so as to represent a wide scale of organisms and impact mechanisms, but also to be short-term laboratory tests with low effluent volume requirements. The profound causal relationships with the chemical characteristics of effluents were not considered here, but the results were compared with conventional effluent monitoring parameters.

MATERIALS AND METHODS

Effluent Samples

The whole-mill effluent samples were collected from all 15 Finnish pulp mills. All the mills, with one exception, produced bleached kraft pulp and two of them also produced unbleached kraft pulp. One mill produced unbleached kraft and semialkaline pulp. Seven of the mills were integrated with paper production, such that both pulp mill and paper mill had a joint effluent treatment, resulting in a combined effluent from pulp and paper production. The annual mean bleached pulp production of the mills studied was 178,000-491,000 t year^{-1} including both softwood and hardwood pulp. In the three mills which also produced unbleached pulp the proportion of such pulp was 16-45% of total pulp production. The annual mean paper production in the integrated mills was 138,000-551,000 t year^{-1}.

Thirteen mills carry out biological effluent treatment (10 activated sludge plants, 3 aerated lagoons) and two have mechanical treatment (clarifier, settling basin). The samples were taken from treated effluents discharged into the receiving waters. The mills have variable amounts of clean water (cooling water, etc.) in effluent treatment systems, which affects the final concentration rate of the effluent. Flows of the discharged effluents into receiving waters were 22,000-142,000 m^3 d^{-1}. Nine of the mills are located inland and the rest on the Baltic coast.

All the whole-mill effluent samples were taken during normal production periods in connection with regular effluent monitoring. The effluents were collected as composite samples from a longer period of time so as to represent an average discharge of a mill. Hence preservation was seen as necessary to stop the biological activity in the effluents and minimize the changing of the samples. Freezing to -20°C was regarded as the best storage method and is an accepted preservation method for toxicity test samples; ISO Standard 5667-3:1994(E) provided guidance on the preservation and handling of the samples. Preservation was not expected to cause any significant changes in the biologically treated whole-mill effluents as they are stabilized after the vigorous aeration and biological degradation in the treatment process and are not likely to contain any highly volatile or easily degradable compounds.

The sampling took place in the period September 1993-May 1994 and was performed by mill staff. The effluent samples were taken from automatic sample collectors and frozen daily to form a composite sample period of 5-30 d (two samples were collected during one day and one was a 3-d pooled sample). The frozen samples were sent to the testing laboratory, thawed, divided into smaller amounts to fit one single toxicity test volume, refrozen and stored at -20°C until use.

Chemical Characteristics

The mills reported the effluent monitoring characteristics and the production and effluent treatment specifications for the sampling period. The effluent monitoring by the mills is controlled by the local District of Waters and the Environment. The chemical analyses were conducted in each mill according to standard methods approved by the local authorities. No additional chemical analysis was done in connection with the biotesting of the samples except for pH and conductivity measurements. The effluent characteristics are listed in Table 1.

Toxicity Testing

All the effluent samples were biotested with five different laboratory-scale lethal and sublethal tests. The biotests were chosen so as to represent different types of organisms and impact mechanisms. Methods 1 and 2 are national standards in Finland and are similar to the corresponding ISO standards. Method 3 is based on the ISO standard draft and method 4 is a national standard based on a corresponding Swedish standard. Method 5 is not a standardized method and was adapted from several sources. As a whole, it is a method under development.

1. Inhibition of the mobility of the water flea Daphnia magna (Cladocera, Crustacea)
Principle: determination of the 24-h EC50 value, i.e., the concentration of the effluent that causes a 50% immobilization of *Daphnia* in 24 h.
Test method: SFS Standard 5062 (1984). Test organism: *Daphnia magna*, age < 24 h. Dilution water: according to ISO Standard 6341 (1989). Dilution series: five sample concentrations (5-100%) and a control, three replicates each, five *Daphnia* neonates/10 mL. Test conditions: temperature 21 ± 2°C, laboratory illumination.

Table 1. Process and effluent characteristics of the pulp mills studied. The process characteristics indicate the mill operation during the sampling period. The effluent characteristics are mean values of the sampling period.

Mill/ Sample No.	Bleaching Type	Wood Furniture	Effluent Treatment	Production	Chemical Characteristics of the Whole-Mill Effluents (mg L⁻¹)					
					BOD7	COD Cr	SS	Total P	Total N	AOX
1	ECF	SW+HW	AL	BK (int.)	54	452	92	1.1	8.6	3.4
2	ECF	SW+HW	AL	BK	56	586	52	0.7	8.1	3.3
3	ECF	SW+HW	AL	BK	83	500	25	0.6	4.9	0.8
4	ECF	SW+HW	AS	BK	32	596	32	0.4	2.8	7.4
5	ECF	SW+HW	AS	BK	35	470	39	0.9	4.0	4.5
6	ECF	SW+HW	AS	BK	6	316	6	0.1	0.6	2.7
7	ECF	SW+HW	AS	BK (int.)	8	460	16	0.5	1.8	5.7
8	ECF	SW+HW	AS	BK	58	802	61	1.3	4.5	10.2
9	ECF+TCF	SW	AS	BK (int.)	12	233	34	0.2	3.7	2.1
10	ECF+TCF	SW+HW	AS	BK+UBK (int.)	16	250	11	0.3	1.8	1.2
11	ECF	SW+HW	AS	BK+UBK	5	410	8	0.7	6.1	6.7
12	ECF+CONV	SW+HW	AS	BK (int.)	10	464	66	0.3	3.0	10.4
13	CONV	SW+HW	AS	BK (int.)	16	738	51	1.3	13.1	12.0
14	ECF	SW	MECH	BK	155	568	22	0.3	2.6	-
15	-	-	MECH	UBK+SAP (int.)	335	783	72	0.9	5.7	-

Abbreviations: ECF (elemental chlorine free), TCF (totally chlorine free), CONV (D/C stage in bleaching), SW (softwood), HW (hardwood), AL (aerated lagoon), AS (activated sludge treatment), MECH (clarifier and/or settling basin), BK (bleached kraft), UBK (unbleached kraft), SAP (semialkaline pulp), int. (integrated with paper and/or board production, indicated if joint effluent treatment).

2. Growth inhibition/stimulation of the green alga Selenastrum capricornutum (Chlorophyta, Chlorophyceae)

Principle: determination of the 72-h EC50 value, i.e., the concentration of the effluent that causes a 50% decrease in algal growth with respect to the control in 72 h; determination of the most stimulative concentration (MSC), i.e., the concentration of effluent that causes the highest increase in algal growth with respect to the control.

Test method: SFS Standard 5072 (1986). Test strain: *Selenastrum capricornutum* Niva oh11 280384. Dilution water: sterile nutrient solution (constant concentration in each test solution, total P 275 g L⁻¹, total N 4 200 g L⁻¹). Dilution series: five to six sample concentrations (1-50%) and a control, four replicates each, of which one served as a color reference without algal inoculum), algal inoculum from an exponentially growing preculture, 20,000-30,000 cells/mL test solution. Culturing conditions: temperature 23 ± 1°C, illumination 5,000 ± 500 lux and continuous agitation. Algal density measurements: spectrophotometric absorbance measurement, wavelength 685 nm, test solutions without algal inoculum as reference to eliminate the possible interference of effluent color. Cell density (microscope counting) and absorbance at 685 nm in algae solution were linearly related (correlation coefficient 0.99).

3. Inhibition of the light emission of the luminescent bacterium Vibrio fischeri

Principle: determination of the 15-min EC50 value, i.e., the concentration of effluent that causes, in a 15-min contact time, a 50% decrease in luminescent light intensity with respect to the initial light emission of the test bacteria.

Test method: applied from the ISO standard draft ISO/TC147/SC5 N107 (1993). Test strain: *Vibrio fischeri* NRRL B-11177. Dilution water: 2% NaCl solution. Dilution series: five sample concentrations (6.25 - 87%) and controls. Test conditions: temperature 15 ± 1°C. Test bacteria: freshly prepared from a preculture (2 d at 20°C on nutrient agar on a petri dish). Measurements: luminometric measurement of the initial light intensity of the bacterial suspensions in 2% NaCl solution, light intensity measurements immediately after combining the sample dilutions with the bacterial suspensions and after a 15- and 30-min contact time.

4. Embryo-larval toxicity in zebra fish (Brachydanio rerio)
Principle: determination of the impact of effluents on the median time from fertilization to hatching and the median survival time of embryos and larvae with respect to the control exposures.
Test method: SFS Standard 5501 (1991). Test fish: parent generation, age 18 months, maintained in glass aquaria in activated carbon-filtered tap water at 22 ± 1°C, acclimatized to the dilution water at 23 ± 1°C at least two weeks before spawning for the test. Test concentrations: only 100% effluent and a control were tested when the *Daphnia* test showed a nontoxic result and the bacteria test a nontoxic or close to nontoxic result; otherwise lower concentrations were used, two replicates per effluent concentration, four replicates per control, 20 viable eggs per 50 mL (time after fertilization <4 h). Spawning: in dilution water at 26 ± 1°C, ratio of females:males 1:2 (reunited after 5-7 d separation), spawning induced with light after overnight dark period. Test conditions: photoperiod 12 h light/12 h dark, temperature 26 ± 1°C, no feeding, daily replacement of test solutions with freshly prepared solutions. Measurements: daily counting of viable eggs, hatched embryos and viable larvae (until >90% of the larvae in the control solutions were dead), pH and oxygen control.

5. Hepatic MFO enzyme activity induction/inhibition in rainbow trout (Oncorhynchus mykiss) isolated liver cells (hepatocytes)
Principle: determination of the lowest effluent concentration that causes induction/inhibition in the MFO enzyme system of exposed isolated liver cells with respect to the control liver cell cultures, based on the catalytic activity of EROD.
Test method: isolation and culturing of hepatocytes (Moon *et al.* 1985; Klaunig *et al.* 1985); EROD activity (Burke and Mayer 1974). Test fish: female rainbow trout from a local hatchery, age 1+ years, weight 150-250 g, maintained in aerated stainless steel aquaria in activated carbon-filtered tap water at 7-8°C, pH 7.8, acclimation time at least one week, feeding once weekly with commercial trout chow pellets at a rate of 0.2-0.3% of body weight (feeding was stopped 5 d before test). Test concentrations: 0.008, 0.04, 0.2, 1, 5 and 10% and a control, three replicates each, six control replicates. Isolation of hepatocytes: two-step perfusion (I with Ca-free buffer solution, II with buffer solution + collagenase) of the liver of an anaesthetized (MS 222) fish via cannula in the hepatic portal vein at 20°C, dissociation of the liver cells with a stainless steel comb, separation and collection of the hepatocytes by filtering through sterile cotton gauze and by washing with cold culture medium (centrifugation at 4°C), testing for viability by Trypan Blue exclusion. Culture medium: Medium 199 supplemented with 10 mL L^{-1} of antibiotic-antimycotic solution, 5 mL L^{-1} of L-glutamine, 0.15 g L^{-1} of $NaHCO_3$ and 0.5 g L^{-1} of Na_2HPO_4, gassing with 5% CO_2/95% O_2, pH adjustment to 7.6, sterile filtration with 0.2-m membrane filter. Test cultures: distribution of freshly prepared hepatocytes on petri dishes in precooled (10°C) culture medium (5-10·10^6 cells per plate), culturing for 24-h at 10°C before exposure, after 24 h culture medium replaced with fresh medium and the test substance dilutions, incubation for 48 h at 10°C, collection of the hepatocytes by gentle pipetting, centrifugation 3,000 rpm for 4 min at 4°C, storage in 0.25 M saccharose solution with 20% of glycerol in liquid nitrogen. EROD activity measurement: a kinetic spectrofluorometric measurement of the formation of resorufin at 18°C, continuous stirring, excitation wavelength 530 nm, emission wavelength 585 nm. Reaction solution: 100 L hepatocyte suspension in 0.1 M Tris-HCl buffer with 1 M of ethoxyresorufin as the substrate and with 10 M NADPH, calibration with standard resorufin. Chemicals: purchased from Sigma Chemical Co. except for Collagenase H and NADPH (Boehringer Mannheim GmbH) and antibiotic-antimycotic solution (Gibco BRL). Protein assay of the cell suspension: bicinchoninic acid protein assay kit (Sigma Procedure No. TPRO-562). Results: expressed as pmol resorufin · mg protein^{-1} · min^{-1}.

Preparation of Samples for Biotesting

pH of the samples was adjusted to 7 (*Daphnia* test, algae test), unless the original pH fell to 6-8.5 (luminescent bacteria test) and to 7.8 (zebra fish embryo/larvae test). pH adjustment for EROD test was not required because of the low test concentrations and high buffering capacity of the culture medium. Salinity was adjusted for luminescent bacteria test by adding 20 g NaCl L^{-1} sample. The interference of suspended solids on the luminescence measurement in bacteria test and respiration of rainbow trout hepatocytes was eliminated by centrifugation of the effluents (3000 rpm, 10-15 min). Microbial activity in effluents may suppress algal growth by competition and extracellular products. Thus to eliminate the interference of bacteria and other microbes the samples were filtered before testing (membrane filtration 0.45 or 0.2 µm).

RESULTS

None of the whole-mill effluents, regardless of bleaching type or effluent treatment system, caused *Daphnia* toxicity, even when tested at 100% concentration (Table 2). The chemical effluent characteristics are listed in Table 1, showing the concentration ranges of parameters related to the content of organic matter (BOD, COD) and chlorinated organics (AOX). The sensitivity of the test organism was tested with potassium dichromate (mean 24-h EC50 1.5 mg L^{-1}, SD 0.2 mg L^{-1}) and was in accordance with the mean 24-h EC50 value of 1.47 mg L^{-1} for potassium dichromate given in the ISO Standard 6341 (1989).

Table 2. The effect of whole-mill effluents on Daphnia mobility, bacterium luminescence and algal growth.

Mill/ Sample No.	*Daphnia*, Immobilization 24-h EC50 %	Bacterium, Light Inhibition 15-min EC50[1] %	Alga, Growth Inhibition 72-h EC50 %	Alga, Growth Stimulation 72-h MSC %
1	NE	NE	NE	10
2	NE	NE	28	10
3	NE	NE	NE	20
4	NE	79 (NE)	50	20
5	NE	38 (47)	NE	30
6	NE	NE	NE	3
7	NE	NE	NE	30
8	NE	NE	NE	30
9	NE	NE	NE	10
10	NE	NE	NE	50
11	NE	NE	NE	20
12	NE	NE	NE	20
13	NE	79 (NE)	NE	10
14	NE	9.1	NE	30
15	NE	2.7	66[(2]	10

NE = no effect.
[1)] Color-corrected values in parentheses.
[2)] Extrapolated result.

The effect of effluent on algal growth is a combination of two opposite but interrelated reactions. The toxic components of the effluents may inhibit the algal growth, and the nutrient content of the effluents may stimulate it with respect to the initial growth achieved with the standard nutrient solution. The latter, stimulating effect was more dominant with the tested effluents. All the effluents showed a degree of stimulation (10-60% in the MSC) in some or all of the test concentrations (Table 2). Only three effluents (2, 4 and 15) inhibited algal growth, with this occurring at rather high concentrations: 72-h EC50 28, 50 and 66% (Table 2). The nutrient content of the effluents gave a total P concentration of 50 up to 650 µg L^{-1} in the most concentrated test solution (50% effluent) and a total N concentration of 300 up to 6,500 µg L^{-1}, suggesting in some cases a high increase in macronutrient concentration with respect to the initial nutrient concentration in the medium (total P 275 µg L^{-1} and total N 4, 200 µg L^{-1}). However, no direct relationship occurred between the amount of nutrient addition and the percentage stimulation, but the final growth rate was probably affected by other simultaneous factors, such as the proportion of bioavailable P and N in the effluents.

Filtration of the effluents to remove bacteria and other microbes is in most cases necessary as their competition may strongly suppress algal growth in test conditions. Hence some losses in chemical content of the samples may occur through adsorption on suspended solids and biomass.

In general, a typical inhibition/stimulation curve of algal growth showed an increasing stimulation at lower test concentrations and after attaining MSC, a turn towards decreasing stimulation and/or increasing inhibition at higher concentrations. The sensitivity of the test alga to potassium dichromate was high (mean 72-h EC50 23 µg L^{-1}, SD 5 µg L^{-1}) when compared to the corresponding 72-h EC50 range of 200-750 µg L^{-1} given in ISO Standard 8692 (1989). This might be due to a somewhat different nutrient medium in SFS and ISO standards.

Inhibition of bacterial luminescence was detected with five whole-mill effluents (Table 2). However, as the method was based on direct measurement of the decrease in light intensity, the brownish color of the effluent may have had an interfering effect, causing an intensity decrease in itself without any toxic input. To evaluate the color effect, the luminescence of the test solutions was also measured immediately after combination of the bacterial suspensions and the sample dilutions. It was assumed that most of the intensity decrease from the initial intensity of the bacterial suspension seen immediately after contact of the sample and the bacteria would be due to the color effect. This kind of measurement could not exclude the possibility of effluent containing very fast-acting toxicants, but it gave an estimate of the highest possible color effect. In addition, if no further decrease in light intensity occurred during a 30-min contact time, the interference of effluent color was very likely. The test results of effluents 4, 5 and 13 were corrected for the "highest possible" color effect using the relationship of initial light intensity and light intensity at 0-min contact time (the results indicated in parentheses in Table 2). Effluents 4 and 13 became nontoxic, showing no further decrease after the immediate reduction in light intensity, during the 30-min contact time. The color effect of Effluent 5 was only slight and it was truly inhibitive. Effluents 14 and 15 were strongly inhibitive and they did not show any color effect in the concentration levels of the EC50 values. The sensitivity of the test bacteria strain to the positive control 3,5-dichlorophenol was somewhat higher (mean 15-min EC50 3.6 mg L^{-1}, SD 0.6 mg L^{-1}) than indicated in the ISO standard draft (mean EC50 6.1 mg L^{-1}, SD 1.7 mg L^{-1}) for freshly prepared bacteria.

The embryo-larvae test with zebra fish was conducted as a screening method, i.e., only 100% effluent was tested, except for two effluents (14 and 15) which were clearly toxic to bacteria. For these, 50% effluent was tested for effluents 14 and 15 and 25% for effluent 15. The mean hatching time of eggs and the mean survival time of larvae were calculated for each concentration (Table 3). Comparisons of the treatments with the controls were made using Dunnett's two-tailed t-test at a confidence level of 0.95. The significant differences are indicated with an asterisk in Table 3.

Five of the effluents had no effect on hatching time, while seven effluents accelerated and two retarded hatching. The 100% concentration of effluent 15 had a strong toxic effect and killed all the eggs in <24 h, and effluent 14 also caused 50% mortality in the eggs with a 100% concentration. Oxygen concentration in these test solutions remained on a 70-80% level of saturation, and hence oxygen depletion was not the cause for the mortality. The mortality of eggs in other samples and concentrations did not exceed the rate in the control samples. Only two effluents (14 at 100 and 50% concentrations and 15 at

50% concentration) had a significant effect on the survival of larvae. The 50% concentration was the lowest tested with effluent 14, and thus the nontoxic concentration could not be estimated. The 25% concentration of effluent 15 had no significant effect on larval survival. Effluent 12 was not tested with embryo-larvae test in this study, but the same sample was tested using the same method in National Board of Waters and the Environment and showed no effect (T. Nakari, P.O.Box 250, FIN-00101 Helsinki, pers. comm.).

Table 3. *The effect of whole-mill effluents on mean hatching and survival time in zebra fish embryo-larvae test. The statistically significant difference (p = 0.95) with respect to the control is indicated by an asterisk (*). SD = standard deviation of the replicates.*

Mill/ Sample No.	Sample Concentration %	Mean Hatching Time				Hatching Time (difference from control)	Mean Survival Time				Survival Time (difference from control)
		Sample d	SD %	Control d	SD %		Sample d	SD %	Control d	SD %	
1	100	2.8	1.1	2.6	5.0	NE	11.3	10.6	12.1	1.9	NE
2	100	1.8	5.7	2.6	5.0	*[2]	11.6	2.7	12.1	1.9	NE
3	100	2.2	0.1	2.7	11.9	NE	9.8	23.6	10.6	4.1	NE
4	100	2.8	1.8	2.7	1.8	NE	11.3	0.9	10.8	11.7	NE
5	100	2.8	1.8	2.7	1.8	NE	12.6	0.4	10.8	11.7	NE
6	100	1.9	11.3	2.6	5.0	*[2]	11.9	6.4	12.1	1.9	NE
7	100	2.0	7.5	2.6	5.0	*[2]	12.4	0.3	12.1	1.9	NE
8	100	2.0	13.6	2.6	5.0	*[2]	11.7	0.3	12.1	1.9	NE
9	100	2.2	0.5	2.6	5.0	*[2]	12.0	0.3	12.1	1.9	NE
10	100	2.2	2.7	2.6	5.0	*[2]	11.9	0.2	12.1	1.9	NE
11	100	2.2	2.3	2.6	5.0	*[2]	11.3	2.9	12.1	1.9	NE
12	NT	-	-	-	-		-	-	-	-	
13	100	2.9	0.7	2.7	1.8	NE	10.8	0.1	10.8	11.7	NE
14	100	3.8	4.7	2.7	1.8	*[3]	5.7	3.5	10.8	11.7	*
14	50	2.7	0.4	2.7	1.8	NE	5.9	7.3	10.8	11.7	*
15	100	[1]					-				
15	50	2.7	2.9	2.7	1.8	NE	6.5	2.6	10.8	11.7	*
15	25	3.4	0.0	2.7	1.8	*[3]	9.0	2.0	10.8	11.7	NE

NT = not tested.
NE = no effect.
[1] All the eggs were dead in <24 h.
[2] Shortened hatching time.
[3] Prolonged hatching time.

The MFO activity measured as the catalytic activity of EROD in isolated rainbow trout hepatocyte cultures showed two types of response in the exposures, both induction and/or inhibition. Comparisons of the treatments with the controls were made using Dunnett's two-tailed t-test at a confidence level of 0.95. Seven of the effluents had no significant effect on EROD activity, six had an inhibitive effect on enzyme activity, one caused enzyme induction and one caused both induction (in lower concentrations) and inhibition (in higher concentrations). Table 4 indicates the lowest sample concentrations that had a significant effect on EROD activity with respect to the control.

Table 4. The potential of whole-mill effluents to induce or inhibit EROD activity in isolated liver cell cultures of rainbow trout. The lowest effluent exposure concentrations causing statistically significant (p = 0.95) change in activity with respect to the control are indicated as well as the maximal detected induction and/or inhibition per effluent.

Mill / Sample No.	EROD Activity			
	Lowest Inducing Test Concentration, %	Max. Induction, % of Control	Lowest Inhibitive Test Concentration, %	Max. Inhibition, % of Control
1	NE		NE	
2	NE		0.2	17.1
3	NE		5.0	72.5
4	NE		NE	
5	NE		NE	
6	NE		5.0	53.2
7	NE		1.0	32.8
8	NE		NE	
9	1.0	30.3	NE	
10	0.008	19.3	1.0	15.9
11	NE		NE	
12	NE		NE	
13	NE		NE	
14	NE		0.008	52.4
15	NE		10.0	42.8

NE = no effect.

The mean EROD activities in the control samples (hepatocytes in culture medium) of separate test series varied from 81-172 pmol mg prot^{-1} min^{-1}, with one exceptionally high value of 286 pmol mg prot^{-1} min^{-1}. The mean standard deviation of the control replicates per test was 17.3% (2.2-29.5%) and of the sample replicates per concentration 7.8% (0.3-28.0%). All the fish used for liver cell isolation were females of the same size and age class, but were from four different lots from the same hatchery. The time fish were kept in laboratory tanks also varied from 1-6 weeks. These differences are probably the reason for the variation of the control level between different fish. However, as each sample exposure was compared to the control of the same test series, the variation could be ignored. Whether these differences in the basic activity level also indicate a change in the sensitivity of response to the exposure is not known.

With effluents having an effect on the MFO system, the maximal increase/decrease in EROD activity in the exposed hepatocytes vs. control cells was moderately low, although statistically significant. In cases of enzyme induction (effluents 9 and 10) the highest increases in EROD activity were 30 and 19%, respectively. The effluent concentrations causing significant induction were 1 and 5% with effluent 9 and

0.008 and 0.04% with effluent 10. With inhibitive samples the decrease in EROD activity varied from 15.9-72.5%. In general, both induction and inhibition showed a dose-response relationship, except effluents 2 and 10, for which no clear dose-response relationship for enzyme inhibition was detected.

DISCUSSION

The biological responses of the effluents are summarized in a frequency block chart (Fig. 1). The figure does not show the rate of impact, but only the different types of responses on a yes/no scale. The common characteristics for the effluents were that none caused *Daphnia* toxicity and all stimulated algal growth at some concentration. The mechanically treated effluents 14 and 15 showed the most frequent responses. The stimulation of algal growth was the only response for effluents 1, 12 and 13. Nine effluents affected the hatching of zebra fish, but it must be noted that seven samples shortened the hatching time, while the two mechanically treated effluents prolonged it. In general, the response profiles of the effluent samples are quite variable, which indicates the inconsistency of the sensitivities of the biological tests to different effluents. In our experience, this anomaly is typical in effluent testing and has also been shown by O'Connor *et al.* (1994).

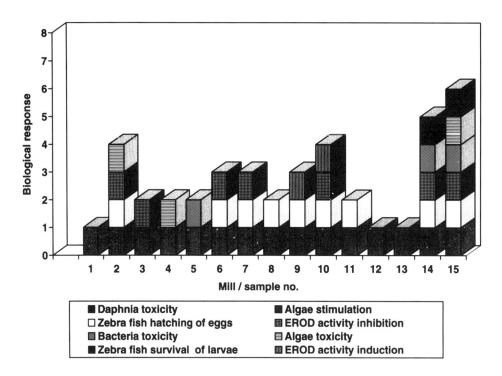

Figure 1. Frequency block chart of different biological responses to whole-mill effluent exposures in laboratory tests. A missing block indicates no observed effect. The height of the block does not indicate the strength of the observed effect.

Comparison of the biological responses and effluent monitoring parameters (Figs. 2 and 3) shows that biological impacts cannot be predicted from either the content of organic matter, measured as BOD and COD, or the AOX content. For instance, effluents 12 and 13 had the highest AOX concentrations and the lowest response frequencies. The lack of relationship between the current AOX levels and environmental effects has been shown in several studies, as reviewed by Axegård *et al.* (1993). Likewise, a high COD level did not indicate a high biological response. The highest BOD levels were in the mechanically treated effluents, which also showed high biological responses. The BOD level depends on the effluent treatment system and most likely does not indicate a causal relationship between BOD and biological activity, but rather the effectiveness of concurrent degradation processes in effluent treatment. In contrast, the algal growth stimulation detected with all effluents is most probably due to the nutrient content of the effluents, although not in a dose-response manner with the total amounts of phosphorus and nitrogen, as previously discussed.

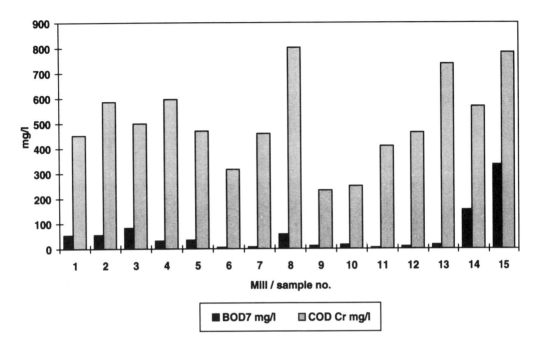

Figure 2. The mean concentration of BOD and COD in whole-mill effluents during the sampling period.

No systematic differences in biological activity between effluents treated in aerated lagoons or activated sludge plants could be detected. Likewise, the bleaching combination did not explain the variations in the biological responses except for one similarity in EROD induction, which was caused by two effluents both also running TCF bleaching during effluent sampling. This, however, cannot be interpreted as a causal relationship as it is based on only two measurements, and the effluent samples also contained many other types of process streams.

The *Daphnia* test (usually *Ceriodaphnia* in North America) has been widely used in effluent testing and has many advantages, being rapid, fairly easy to conduct and repeatable. However, the need is for more sensitive tests measuring long-term rather than acute lethal effects. In this study, both bacteria and algae tests turned out to be somewhat more sensitive to the effluents than the *Daphnia* test. The disadvantage in using bacteria test for effluent testing is that the effluent color may interfere with the results. A proper method for color correction has not yet been standardized. The algal growth potential test gives information on the potential of the effluent for causing toxicity and growth stimulation (i.e., eutrophication) impact mechanisms which are both of interest when estimating the effects on the natural environment. The embryo-larvae test with zebra fish determines the true effects on fish reproduction although in laboratory conditions and with a tropical fish species. Several of the biologically treated effluents in this study slightly shortened the hatching time, but none affected the survival of embryo and larvae. The shortening may be interpreted as a stimulative effect. The mechanism for the stimulation is not known.

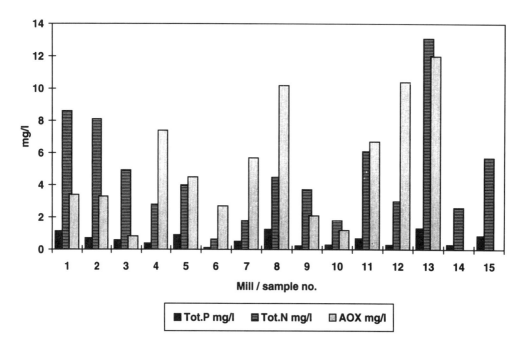

Figure 3. Mean concentration of total phosphorus, total nitrogen and AOX in whole-mill effluents during the sampling period.

The EROD enzyme activity is regulated by cytochrome P4501A1, which belongs to polyaromatic hydrocarbons inducible enzyme proteins, and is thus of major toxicological significance because of its role in the metabolism of aromatic hydrocarbons (Andersson and Förlin 1992). Numerous experiments both in laboratory and field conditions have shown that exposure to bleached kraft mill effluents can cause induction of EROD activity in many species of fish (Martel et al. 1993). However, EROD induction has not been found to be predictive for other biological impacts (Swanson et al. 1992), but serves rather as a sensitive indicator of effluent exposure (Munkittrick et al. 1994). In this study both induction and inhibition of EROD activity were discovered in isolated rainbow trout liver cells. The inhibition of activity was a more frequent response of effluent exposure. Pesonen and Andersson (1992) also found the same type of response patterns when exposing isolated hepatocyte cultures to extracts from unbleached and bleached pulp mill effluents. It was concluded that the inhibition of activity may be a nonspecific toxic effect or partly due to the presence of resin acids.

Biotesting of composite effluent samples with a battery of test methods necessitates preservation of the samples in order to minimize the changing of the samples and to be able to test "the same sample" in each test. In this study, freezing was considered the best method for the circumstances, as the microbial activity in the effluents after the biological treatment would probably have caused lots of changes in the effluents if stored at +4°C. On the other hand, freezing the samples was not likely to cause significant changes in samples, as the biologically treated effluents are quite stabilized after the treatment process (BOD reduction 85-99%) and do not contain highly volatile or easily degradable compounds.

In general, the biological effects observed were low level, if any. The toxicities discovered for algae and bacteria were few and the EC50 values were at effluent concentrations not likely to occur in receiving waters, except for the bacterial toxicities of the mechanically treated effluents. Likewise, the 100%

effluents after biological treatment showed little if any effect on zebra fish hatching and no effect on embryo and larval survival. The induction and inhibition of EROD activity, although at low level, were detected in concentration ranges that may occur in receiving waters.

CONCLUSIONS

Finnish kraft pulp mills have gone through a significant change in pulping and bleaching technology and effluent treatment during the late 1980s and early 1990s. The whole-mill effluents of the modern technology mills show little biological activity in various test organisms and response mechanisms in laboratory tests.

The results indicate that no serious effects are likely to occur in the effluent dilutions present in receiving waters of modern technology mills, yet the laboratory test results as such are not directly applicable to predicting biological effects in natural environments. No causal relationship was found between biological effects and effluent monitoring parameters BOD, COD and AOX. The common feature in all the effluents is their potential for increasing eutrophication.

ACKNOWLEDGMENTS

The technical assistance of Veli Kovanen, Irina Toiviainen and Peter Vihra is greatly appreciated. Thanks are due to Ari Langi for discussions and helpful comments throughout the study. The good cooperation of the Finnish pulping companies made this study possible.

REFERENCES

Andersson, T., and L. Förlin. 1992. Regulation of the cytochrome P450 enzyme system in fish. Aquat. Toxicol. 24:1-20.

Axegård, P., O. Dahlman, I. Haglind, B. Jacobson, R. Mörck and L. Strömberg. 1993. Pulp bleaching and the environment - the situation 1993. Nordic Pulp & Paper Res. J. 8(4):365-378.

Burke, D.M., and R.T. Mayer. 1974. Ethoxyresorufin: Direct fluorimetric assay of a microsomal o-dealkylation which is preferentially inducible by 3-methylcholanthrene. Drug. Metab. Dispos. 2:583-588.

Klaunig, J.E., J.R. Randall and P.J. Goldblatt. 1985. Trout hepatocyte culture: Isolation and primary culture. In Vitro Cell. Dev. Biol. 21:221-228.

Lövblad, R., and J. Malmström. 1994. Biological effects of kraft pulp mill effluents - A comparision between ECF and TCF pulp production. International Non-chlorine Bleaching Conference, March 6-10, 1994, Amelia Island, Florida, Proceedings 11-2. 18 p.

Martel, P., T. Kovacs, B. O'Connor and R. Voss. 1993. A survey of pulp and paper mill effluents for their potential to induce mixed function oxidase enzyme activity in fish. 79th Annual Meeting, Technical Section, CPPA 26.-27.1.1993, A165-A177.

Moon, T.W., P.J. Walsh and T.P. Mommsen. 1985. Fish hepatocytes: A model metabolic system. Can. J. Fish. Aquat. Sci. 42:1772-1782.

Munkittrick, K., G. Van Der Kraak, M. Mcmaster, C. Portt, M. van den Heuvel and M. Servos. 1994. Survey of receiving-water environmental impacts associated with discharges from pulp mills. 2. Gonad size, liver size, hepatic EROD activity and plasma sex steroid levels in white sucker. Env. Tox. Chem. 13(7):1089-1101.

O'Connor, B., T. Kovacs, R. Voss, P. Martel and B. Van Lierop. 1994. A laboratory assessment of the environmental quality of alternative pulp bleaching effluents. Pulp Paper Can. 95(3):47-56.

Pesonen, M., and T. Andersson. 1992. Toxic effects of bleached and unbleached paper mill effluents in primary cultures of rainbow trout hepatocytes. Ecotoxicol. Environ. Safety 24(1):63-71.

Robinson, R., J. Carey, K. Solomon, I. Smith, M. Servos and K. Munkittrick. 1994. Survey of receiving-water environmental impacts associated with discharges from pulp mills. 1. Mill characteristics, receiving-water chemical profiles and lab toxicity tests. Env. Tox. Chem. 13(7):1075-1088.

Swanson, S., R. Shelast, R. Schryer, P. Kloepper-Sams, T. Marchant, K. Kroeker, J. Bernstein and J.W. Owens. 1992. Fish populations and biomarker responses at a Canadian bleached kraft mill site. TAPPI J. 75(12):139-149.

Södergren, A., M. Adolfsson-Erici, B-E. Bengtsson, P. Jonsson, S. Lagergren, L. Rahm and F. Wulff. 1993. Environmental impact of bleached pulp mill effluents. *In*: Södergren, A. (ed.). Bleached pulp mill effluents. Composition, fate and effects in the Baltic Sea. Swedish Environmental Protection Agency, Report 4047:8-126.

RECENT PERSPECTIVES IN INTEGRATED MONITORING

Peter M. Chapman

EVS Environment Consultants, 195 Pemberton Avenue, North Vancouver, B.C., V7P 2R4 Canada

Integrated monitoring goes beyond, for instance, chemical analyses, biomarkers or whole-organism measures and tests. It is defined as a combination of chemical analyses, experimentation (e.g., toxicity tests, mesocosms) and observation (e.g., measures of resident community structure, biomarkers), with overall conclusions derived from a preponderance (or "weight") of evidence approach. Critical factors for the success of integrated monitoring include choice of suitable, appropriate reactive and proactive components and reference area comparisons. Both of these are more likely to be site- and situation-specific than generic. Environmental effects monitoring (EEM) programs comprise integrated monitoring and, although still in development, are a hopeful trend away from limited, single-component monitoring. Major future improvements to integrated monitoring programs are expected to and should include new, less subjective methods for deriving overall conclusions and appropriate use of new, "trendy" tools whose utility is still arguable (e.g., biomarkers) without excluding established, proven components (e.g., laboratory toxicity tests). Perhaps the greatest expectation (and need, if facts are to triumph over perceptions and politics) is that the results of integrated monitoring programs will eventually drive the regulations rather than the reverse.

INTRODUCTION

Integrated monitoring, per Chapman *et al.* (1992), combines more than one measure of environmental quality to provide an ongoing assessment of the status of the system under study. More specifically, chemical analyses together with both experimentation (e.g., laboratory toxicity tests, caged organism exposures, mesocosms) and observation (e.g., measures of resident communities) are required to define an integrated monitoring program (cf. Chapman 1995a). Neither experimentation nor observation alone will provide an adequate answer to the ultimate question of the status of the system under study: prediction provided by experimentation is as necessary as the understanding provided by observation. Conclusions are drawn from individual components and considered relative to one another. Then, a preponderance (or "weight") of evidence approach is used to derive overall conclusions. In an ideal world, integrated monitoring would be both efficient and unambiguous. However, in the ambiguous real world, accuracy is a more attainable goal.

Hall's (1996) definition is similar to the above. Accordingly, his review of integrated monitoring approaches in North America contains mainly studies which include chemistry, experimentation and observation. However, he did include a few studies with only chemistry and observation, which thus could not be (see below) predictive of effects before they occurred, nor could they provide information for elucidating mechanisms of cause and effect. The majority of examples provided by Allender *et al.* (1996) for Australasia and S.E. Asia also comprised chemistry, experimentation and observation, as did the Nordic case study provided by Oikari and Holmbom (1996). However, of nine recent presentations on integrated monitoring (i.e., Dell *et al.* 1996; Deardoff 1994; Friesen *et al.* 1994; Ikonomou *et al.* 1994; Oanh *et al.* 1994; Peddle and Robertson 1994; Priha 1996; Haglind *et al.* 1994; Rodgers *et al.* 1994), only three included chemistry, observation and experimentation. Two each of the remaining six were concerned with observation and chemistry, experimentation and chemistry, and chemistry alone. In contrast, several papers on field and laboratory responses (e.g., Cash and Wrona 1995; Scrimgeour *et al.* 1995) were integrative per the above definition. There appears to be some confusion over what constitutes integrated monitoring. The definition proposed in the paragraph above, if generally accepted, would provide needed clarification.

There is no disagreement that truly integrated monitoring provides more information than the sum of its individual components. For instance, Robinson et al. (1994) conducted the first attempt at pulp mills to integrate traditional predictors of effluent toxicity (specifically laboratory whole-organism toxicity tests) with responses in wild fish. They found that the toxicity tests were generally predictive of impact to the benthic macroinvertebrate community, but were not predictive of physiological changes which may occur in wild fish.

The environmental significance of such physiological changes is presently uncertain. It is argued that they may act as an early warning signal for whole-organism effects (Munkittrick et al. 1994). It is also argued that they may simply comprise biochemical markers of exposure not associated with any discernable adverse effects on, for instance, individual fish health (Kloepper-Sams et al. 1994). However, the only way to determine the significance of physiological changes (or of various other individual measures of exposure, stress or effects) in field populations is to conduct integrated monitoring which includes experimentation.

Integrated monitoring serves to answer the following individual questions: (1) What are the contaminants of concern? (2) What is their environmental fate? (3) What are their effects? The last question, effects, includes bioaccumulation, biochemical and whole-organism toxic responses. As noted above, the overall question is holistic: What is the status of the system under study? Answering this question is not without financial cost, but failing to answer this question or achieving incorrect answer(s) is far more costly both financially and environmentally. Specifically, appropriate integrated monitoring provides the most powerful evidence for concluding that either there is a need for management action (e.g., process changes, remediation) or there is a need for continued monitoring, with change(s) dictated by previous findings or enough is enough.

The answer to the above ultimate question also depends heavily on comparison to at least one reference area. Thus, the choice of suitable reference area(s) is critical to a successful integrated monitoring program. Reference area comparisons must be appropriate and, in particular, must adequately account for any habitat differences in drawing conclusions at the population level of organization (cf. Swanson et al. 1994; Dell et al. 1996). Further, we must decide (as scientists and as members of society) what ecological (and other) endpoints comprise a reference area in specific and generic terms. For instance, biodiversity can be the same upstream and downstream of a mill despite species differences. Do we want to maintain biodiversity or individual species?

NEW APPROACHES (1991-1994)

The pulp and paper industry has a relatively long tradition of environmental fate and effects studies. An initial focus on water pollution has now expanded in scope and extent, including biota (bacteria, protozoans, invertebrates, fish, birds, mammals), sediments and a variety of sublethal responses and physiological indicators of exposure and/or stress. Integrated monitoring is a logical development given the broad scope and extent of many environmental studies at pulp mills but one that is still relatively new, though it is receiving increasing attention since the 1991 Conference on Environmental Fate and Effects of Bleached Kraft Mill Effluents (Södergren 1991).

Integrated monitoring has advanced to the point that it is now a major focus because of the interrelationship of technology (process/treatment), research (scientists and regulators) and political input. For instance, in New Zealand (Dell et al. 1996), legislation deals with the effects of activities rather than with the activity itself. Thus case-by-case assessment is a reality which requires integrated monitoring, which also occurs in Sweden (Haglind et al. 1994). Other reasons for the increasing emphasis on integrated monitoring identified during discussion include the realities that: (1) despite the tremendous research effort into identifying techniques to assess effect and response (e.g., biomarkers are the focus of much current effort), no single test will allow informed decision-making; (2) contamination rarely results from a single point source, and multiple point and non-point sources require multiple assessment tools; (3) major mill process changes have occurred and are occurring, and each mill is unique, as is its discharge environment, requiring detailed case-by-case monitoring; (4) high natural variability makes it difficult to detect meaningful change, and even more difficult to ascribe such to anthropogenic activities, without an integrated approach; (5) we are only beginning to understand the structure and function of populations under very specialized circumstances, and being able to predict the results of anthropogenic insults to ecosystems requires full integration of different disciplines (e.g., ecological literature on population- and community-level consequences of environmental stressors indicate that multiple outcomes are possible, that major changes can occur at higher levels of organization without significant effects at the level of the individual organism or the reverse and that subtle changes and the timing of these changes combine to produce major responses); and (6) the use of data selectively to support preconceived ideas rather than for illumination is far less likely to be an issue in an integrated, holistic monitoring program.

Recent, regulated integrated monitoring approaches in many countries (e.g., Canada, Sweden) have been termed environmental effects monitoring (EEM). EEM typically involves: (1) a framework for implementation; (2) criteria for evaluating effects and defining acceptable environmental quality, including possible remedial measures where needed; and (3) mechanisms for updating or modifying the monitoring, particularly with regard to new research information and/or tools. Typically, EEM uses ecological characteristics of the receiving environment (e.g., community composition) together with laboratory studies (e.g., toxicity tests) and chemical analyses. Thus EEM comprises integrated monitoring. Specific EEM components vary by ecoregion, recognizing wide regional variations in receiving water characteristics and water use requirements. As noted by Priha (1996), studying 15 different Finnish pulp mills and commenting that generic predictions are impossible "It is obvious that each pulp mill and receiving water system is unique as well as the individual organisms, populations and response patterns."

EEM is the reverse of earlier "technology-based" approaches. Rather than relying on effluent testing (chemical and biological), EEM places emphasis on the environmental characteristics of the receiving environment, coupled with early warning measures using sensitive laboratory organisms and biomarkers.

PREDICTIONS FOR THE FUTURE (1994-1997)

The integrated monitoring approach typified by EEM is expected to expand in use. Improvements which are needed, and which are expected to occur, include: (1) large-scale, long-term field verification (but not necessarily validation; see Chapman 1995b); (2) regulatory refinement such that simplistic and unrealistic pass/fail criteria are replaced by more complex but environmentally realistic criteria (i.e., integrated monitoring obviates the need for numerical environmental quality criteria except possibly in an initial screening function); (3) the concurrent development of technical and other (e.g., social, economic) criteria for defining "acceptable" environmental quality for different sites and situations; (4) increasing emphasis on scientific considerations for inclusion of specific parameters (i.e., just because we can conduct a certain test or measurement, should we?); (5) improvements in experimental (e.g., the development of toxicity identification evaluation procedures for wastewaters provided essential information related to cause and effect, and these procedures are now being extended to sediment and tissues) and observational tools (e.g., the abundance of parasites as indicators of stress to their hosts; Overstreet *et al.* 1996); and (6) improvements in data analysis methods, in particular the integration of different monitoring programs and differential weighting of measures reliably shown to provide early warning (proactive components, presently restricted to certain whole-organism laboratory toxicity tests) compared to those which simply indicate that an effect has occurred (reactive components, e.g., changes in community structure). This last improvement basically reflects the integration of risk assessment (which is predictive and thus proactive) with impact assessment (which is reactive, assessing impact(s) which have already occurred). All of these improvements require different disciplines (e.g., toxicologists, chemists, ecologists, modellers, statisticians, etc.) working together during all phases of the monitoring.

The most critical future need is to improve interpretative methods for determining the significance of environmental effects, such that regulatory or management controls and limits can be properly applied to allow industry to operate and develop without environmental degradation. Note that the term "significance" is not used in the statistical sense; statistical and ecological significance are generally very different and so-called "effects" or "no effects" can be statistical artifacts, especially since generally used statistical tests do not consider intermediate states of Nature.

However, if this critical need is to be met, science must play a dominant role in determining site-specific integrated monitoring components. This is not presently the case, as too often a "shopping list" approach is taken which includes such non-scientific factors as socio-political perceptions and regulatory requirements (many of which are badly dated). And the magnitude of the monitoring program must be commensurate with the severity of the issue. The ideal situation would involve: using scientific criteria to determine components *a priori* both generically and on a site- and situation-specific basis; basing the study objectives on hypothesis testing rather than on what are too often "fishing expeditions"; using a tiered approach where, as cost and complexity increase with increasing tiers, uncertainty decreases (and where increasing cost is associated with increasing severity); and focusing the monitoring to determine whether environmental effects are occurring or will occur (core monitoring - combining risk assessment and environmental impact assessment [including assessing how well we have set assessment values]) and, if so, on deriving the causal factor(s) (investigative studies, often primarily experimental, triggered by core monitoring results).

As noted previously, this last point involves both reactive approaches (e.g., measures of community structure which reflect whether or not an effect has occurred) and proactive approaches (e.g., laboratory studies capable of predicting that an effect will occur before such is manifest or detectable in the receiving environment). In this regard, a great deal of work is presently being undertaken using biomarkers to predict effects. Biomarkers can be defined as biochemical, physiological and histological endpoints which reflect an organism's attempt to compensate for, or tolerate, stressors in the environment (definition adapted from Cormier and Daniel 1994). At their present stage of development, the greatest utility of biomarkers appears to be in screening (i.e., documenting exposure), rather than in decision-making. This will hopefully change in the future as the chemical(s) responsible for triggering various biomarker responses are identified and the relevance (or lack thereof) of specific biomarkers for predicting receiving environment effects is determined. However, until such occurs, sublethal toxicity measures provide a more immediately useful focus, particularly when such measures (e.g., mortality, growth, reproduction) provide population-level information on the species most severely impacted (which may not necessarily, despite the focus of many recent papers, be fish).

Interpretative methods are predicted to and must improve. Present interpretative methods are an arbitrary mix of comparisons with guidelines or criteria (for contaminant concentrations in various media), various types of statistical analyses (often varying between investigators) and "best professional judgment" (which is highly variable depending on, for instance, individual biases and which therefore is always arguable). Overall there is a need for consistency in interpretation. Without this, there will be no corresponding consistency in regulatory and management decisions and, hence, no "level playing field" for either industry or the environment.

Of particular importance is the basis upon which integrated monitoring is implemented. All pulp mills, regardless of where they are constructed, require some form of initial environmental assessment. Too often this initial assessment is not fully related to subsequent monitoring. Enlightened self-interest by industry (i.e., the most effective use of monies for environmental monitoring) and hopefully also regulatory initiatives are expected to result in not only better linkages between the initial environmental assessment and subsequent monitoring (operational and post-operational), but also in integrated (as defined previously) initial environmental assessments. However, to do this most effectively, researchers need to obtain and use basic information on mill processes as well as characterizing the receiving environment. Basic information required includes source, type and quantity of wood pulped; technology used for pulping and bleaching; and wastewater/sludge treatment used.

Finally, it is important to note that integrated monitoring programs are often driven by regulations which are not always based on good science. In Canada, the primary purpose of EEM for the pulp and paper industry is to assess the adequacy of existing regulations in protecting the environment (Environment Canada and Department of Fisheries and Oceans 1992). If science is truly a part of the process, then within a few years, there should be examples of how the results of EEM and/or other integrated monitoring programs have actually changed existing regulations. There should also be examples of how science has interacted with all stakeholders to educate them, and to be educated by them. Specifically, science strives to determine what level of stress to the environment is sustainable, whereas society determines what level of response is acceptable (i.e., the reference comparison which is so critical, as noted previously, is determined by combined scientific and societal considerations). If this does occur, then facts triumph over public perceptions and consequent political reactions. Otherwise, a great deal of time and money will have been expended without adequate returns, particularly for environmental protection. There was a hopeful sign in this regard on the last day of the Conference on Environmental Fate and Effects of Bleached Pulp Mill Effluents, November 10, 1994. On that day the Canadian national newspaper, *The Globe and Mail*, had an article on the conference in which British Columbia's Environment Minister was quoted relative to a ban on chlorine from pulp mills which he had imposed some years previously: "If the science on this issue crystallizes, one is never adverse to reconsideration."

ACKNOWLEDGMENTS

I thank all presenters and participants at the Second International Conference on Environmental Fate and Effects of Bleached Pulp Mill Effluents, as my thought processes and hence this paper greatly benefited from listening to them. In particular I thank the members of my Poster Panel Discussion group, Kevin Cash, Fred Wrona, Paul Dell, Tom Deardorff, and John Rodgers, who ignored my suggestions but (because of this?) provided what I feel were some of the best comments and discussion of the conference.

REFERENCES

Allender, B.M., J.S. Gifford and P.N. McFarlane. 1996. The role of biological measurements in monitoring pulp and paper mills in Australasia and S.E. Asia. *In* Environmental Fate and Effects of Pulp and Paper Mill Effluents, M.R. Servos, K.R. Munkittrick, J.H. Carey and G. Van Der Kraak (ed.), St. Lucie Press, Delray Beach, FL.

Cash, K.J., and F.J. Wrona. 1994. Assessing the impact of pulp-mill effluent and other contaminants on large river systems - the Northern River Basins framework. Presented at the Second International Conference on Environmental Fate and Effects of Bleached Pulp Mill Effluents, November 6-10, 1994, Vancouver, B.C., Canada.

Chapman, P.M. 1995a. Extrapolating laboratory toxicity results to the field - lessons to learn from. Environ. Toxicol. Chem.: Iin press.

Chapman, P.M. 1995b. Do laboratory toxicity tests require field validation? Environ. Toxicol. Chem.: In Press.

Chapman, P.M., E.A. Power and G.A. Burton, Jr. 1992. Integrative assessments in aquatic ecosystems. pp. 313-340. *In* Sediment Toxicity Assessment, G.A. Burton, Jr. (ed.). Lewis Publishers, Boca Raton, FL.

Cormier, S.M., and F.B. Daniel. 1994. Biomarkers: taking the science forward. Environ. Toxicol. Chem. 13:1011-1012.

Deardorff, T.L. 1994. Generation of polychlorinated dioxins (PCDD) and furans (PCDF) by boat motor operation, and impact on aquatic sediments. Presented at the Second International Conference on Environmental Fate and Effects of Bleached Pulp Mill Effluents, November 6-10, 1994, Vancouver, B.C., Canada.

Dell, P., F. Power, R. Donald, J. McIntosh, S. Park and L. Pang. 1996. Monitoring environmental effects and regulating pulp and paper discharges: Bay of Plenty, New Zealand. *In* Environmental Fate and Effects of Pulp and Paper Mill Effluents, M.R. Servos, K.R. Munkittrick, J.H. Carey and G. Van Der Kraak (ed.), St. Lucie Press, Delray Beach, FL.

Environment Canada and Department of Fisheries and Oceans. 1992. Aquatic environmental effects monitoring requirements. Annex I: Aquatic environmental effects monitoring requirements at pulp and paper mills and off-site treatment facilities regulated under the Pulp and Paper Effluent Regulations of the Fisheries Act. Ottawa, Ontario. EPS 1/RM/18.

Friesen, C., P.L. Wong, S.B. Brown and W.L. Lockhart. 1994. Changes in fish and aquatic habitat of the Winnipeg River downstream from the Pine Falls pulp mill. Presented at the Second International Conference on Environmental Fate and Effects of Bleached Pulp Mill Effluents, November 6-10, 1994, Vancouver, B.C., Canada.

Haglind, I., B. Hultman and L. Strömberg. 1994. Results from an environmental research program conducted by the Swedish pulp and paper industry. Presented at the Second International Conference on Environmental Fate and Effects of Bleached Pulp Mill Effluents, November 6-10, 1994, Vancouver, B.C., Canada.

Hall, T.J. 1996. A review of recent studies in North America using integrated monitoring approaches to assessing bleached kraft mill effluent effects on receiving waters. *In* Environmental Fate and Effects of Pulp and Paper Mill Effluents, M.R. Servos, K.R. Munkittrick, J.H. Carey and G. Van Der Kraak (ed.), St. Lucie Press, Delray Beach, FL.

Ikonomou, M.G., N. Crewe, T. Fraser, I.H. Rogers and T.G. Smith. 1994. PPT detection of dioxins, furans and toxic mono-ortho and non-ortho substituted PCBs in marine mammals by HRGC/HRMS using ultra small sample sizes, 100 mg. Presented at the Second International Conference on Environmental Fate and Effects of Bleached Pulp Mill Effluents, November 6-10, 1994, Vancouver, B.C., Canada.

Kloepper-Sams, P.J., S.M. Swanson, T. Marchant, R. Schryer and J.W. Owens. 1994. Exposure of fish to biologically treated bleached-kraft effluent. 1. Biochemical, physiological and pathological assessment of Rocky Mountain whitefish (*Prosopium williamsoni*) and longnose sucker (*Catostomus catostomus*). Environ. Toxicol. Chem. 13:1469-1482.

Munkittrick, K.R., G.J. Van Der Kraak, M.E. McMaster, C.B. Portt, M.R. van den Heuvel and M.R. Servos. 1994. Survey of receiving-water environmental impacts associated with discharges from pulp mills. 2. Gonad size, liver size, hepatic EROD activity and plasma sex steroid levels in white sucker. Environ. Toxicol. Chem. 13:1089-1101.

Oanh, K., N. Thi and B-E. Bengtsson. 1994. Contamination and toxicity of effluents from a Vietnamese kraft pulp and paper mill. Presented at the Second International Conference on Environmental Fate and Effects of Bleached Pulp Mill Effluents, November 6-10, 1994, Vancouver, B.C., Canada.

Oikari, A., and B. Holmbom. 1996. Ecotoxicological effects of process changes implemented in a pulp and paper mill: a Nordic case study. *In* Environmental Fate and Effects of Pulp and Paper Mill Effluents, M.R. Servos, K.R. Munkittrick, J.H. Carey and G. Van Der Kraak (ed.), St. Lucie Press, Delray Beach, FL.

Overstreet, R.M., W.E. Hawkins and T.L. Deardorff. 1996. The western mosquitofish as an environmental sentinel: parasites and histological lesions. *In* Environmental Fate and Effects of Pulp and Paper Mill Effluents, M.R. Servos, K.R. Munkittrick, J.H. Carey and G. Van Der Kraak (ed.), St. Lucie Press, Delray Beach, FL.

Peddle, J., and K. Robertson. 1994. Slave River environmental quality monitoring program. Presented at the Second International Conference on Environmental Fate and Effects of Bleached Pulp Mill Effluents, November 6-10, 1994, Vancouver, B.C., Canada.

Priha, M. 1996. Ecotoxicological impacts of pulp mill effluents in Finland. *In* Environmental Fate and Effects of Pulp and Paper Mill Effluents, M.R. Servos, K.R. Munkittrick, J.H. Carey and G. Van Der Kraak (ed.), St. Lucie Press, Delray Beach, FL.

Robinson, D.D., J.H. Carey, K.R. Solomon, I.R. Smith, M.R. Servos and K.R. Munkittrick. 1994. Survey of receiving-water environmental impacts associated with discharges from pulp mills. 1. Mill characteristics, receiving-water chemical profiles and lab toxicity tests. Environ. Toxicol. Chem. 13:1075-1088.

Rodgers, J.H. Jr., W.B. Hawkins, A.W. Dunn and T.L. Deardorff. 1994. An evaluation of potential impacts of bleached kraft mill effluents on the lower Sulphur River, Texas–Arkansas. Presented at the Second International Conference on Environmental Fate and Effects of Bleached Pulp Mill Effluents, November 6-10, 1994, Vancouver, B.C., Canada.

Scrimgeour, G.J., P.A. Chambers, J.M. Culp and C.L. Podemski. 1995. Effects of bleached pulp mill effluent on algal and benthic communities: separating between nutrient and contaminant impacts. Presented at the Second International Conference on Environmental Fate and Effects of Bleached Pulp Mill Effluents, November 6-10, 1994, Vancouver, B.C., Canada.

Södergren, A. (ed.). 1991. Environmental fate and effects of bleached Pulp mill effluents. Swedish Environmental Protection Agency Report 4031. 394 pp.

Swanson, S.M., R. Schryer, R. Shelast, P.J. Kloepper-Sams and J.W. Owens. 1994. Exposure of fish to biologically treated bleached-kraft mill effluent. 3. Fish habitat and population assessment. Environ. Toxicol. Chem. 13:1497-1507.

SECTION VII

FUTURE DIRECTIONS

The pulp and paper industry has responded quickly to implement new process and treatment technologies which directly reduce the release of a wide variety of chemicals into the environment. The move toward elemental chlorine and totally chlorine-free processes has been rapid. The modernization of process and bleaching technologies including extended delignification, increased use of chlorine dioxide substitution, ozone, peroxide and biological (enzymes) bleaching has resulted in dramatic changes and reductions in the toxicity and chemical composition of the effluents. Although enormous progress has been made, the industry is in a state of transition and many scientific challenges remain. Environmental remediation and any possible regulatory developments need to be based on a sound scientific understanding of the production and control of contaminants within the mills as well as the movement, fate, exposure and toxicology of the chemicals released into the environment.

REGULATION OF PULP MILL AQUATIC DISCHARGES: CURRENT STATUS AND NEEDS FROM AN INTERNATIONAL PERSPECTIVE

J.W. Owens

Environmental Sciences, Procter & Gamble, 5299 Spring Grove Avenue, Cincinnati, OH, 45217 U.S.A.

The co-evolution of pulp mill process engineering, environmental science, and regulatory discharge requirements which began in the early 1970s continues today. There is a general international convergence in both regulatory parameters and current regulatory issues. Organic/nutrient enrichment and acute toxicity issues are largely resolved. Similarly, the discovery of persistent and bioaccumulating substances in the bleach plant has resulted in the virtual replacement of chlorine gas. Currently, the primary issues are identification of the sources of potential sublethal toxicity in aquatic organisms and the further minimization of aquatic discharges. In Canada, efforts focus on the development of environmental surveillance and laboratory techniques to address sublethal alterations in reproductive hormones. Unmet needs in this approach include the interpretive experience to understand site and ecosystem variability in surveillance data. In the U.S., efforts focus on a revision of effluent command and control regulations. In Nordic geographies, efforts focus on process development to minimize discharge loads and volumes to low levels. Unmet needs for these approaches include the capability to identify, quantify, and then balance environmental benefits against the environmental and social costs of further effluent regulation and minimization.

INTRODUCTION

Emissions from pulp mills are controlled by government regulation as these operations are potentially large dischargers to both air and water. These national regulations are determined by several factors: (1) the country's basic water and air pollution legislation which provides both a national goal/purpose and a legal framework, (2) the relative state of scientific knowledge on environmental impacts associated with pulp mill, and (3) the relative production, abatement, and treatment engineering technologies available to control emissions.

These factors have an intertwined history that has led to different regulatory specifications in each country, but with a focus on similar issues. They have also led to a progressive change in regulatory approaches over time to combine end-of-pipe abatement with newer pollution prevention strategies. This overview focuses on four major producing nations: Canada, the United States, Sweden, and Finland, as most significant regulatory and engineering activity has originated in these nations. This is not to ignore the regulations in Japan (where all mills have oxygen delignification), Brazil (where regulations are as stringent as in "industrialized" nations), Australia (which has a well-designed environmental framework to establish pre-operational baselines and to continue monitoring after mill start-up), or other nations.

NATIONAL LEGISLATION

Enabling legislation or laws establish regulatory authority, set specific national goals or objectives, and establish a regulatory framework (e.g., discharge permits). These laws and especially the resulting regulations and guidelines establish the basic chemical and biological discharge criteria. In Canada and the U.S., federal environmental agencies issue national guidelines (minimum requirements), and provincial or state authorities may also impose more, but not less, stringent limits, depending upon their local/regional interests. In contrast, Nordic countries set basic national goals and guidelines, but actual emission limits or criteria are set by local authorities on a case-by-case basis. Almost universally, discharge permits are revised after a certain time period. Limits normally become more stringent when the permits are renewed or reissued and are more stringent for new mills than existing mills (inherently recognizing the potentially unfavorable economics of retrofitting old mills). Discharge limits can be stated in several ways:

1. A maximum concentration not to be exceeded (usually per day or per tonne of pulp produced);
2. An average load per unit of time, such as pounds or kilograms BOD per day or month, or per unit of production, such as kilograms absorbable organic halides (AOX) per air-dried tonne (ADt);
3. Biological test criteria such as acute lethality (survival for 96 h in 100% effluent) or chronic tests such as fathead minnow survival and growth or *Ceriodaphnia* reproduction; and
4. In some cases, criteria for receiving waters such as dissolved oxygen levels or metal concentrations are set.

Normally, mills are obliged to monitor regulated parameters and to report these to the authorities. It is also common for authorities to inspect mills and to analyze effluent samples and conduct sampling in receiving waters. When discharge limits are exceeded, plants may be fined and, in some cases, forced to halt production until the situation is remedied.

ENVIRONMENTAL IMPACTS

Environmental impacts at pulp mills have been recognized in three historical stages. At each stage, national agencies have responded with specific regulations for observed or perceived environmental effects. First, pulp mill regulations (beginning in the 1950s and 1960s) limited local environmental disturbances with specific conventional parameter surrogates such as for oxygen deficiency (BOD_x; x = 5 d in North America and 7 d in Scandinavia), fiber deposition (total suspended solids, TSS), or acute lethality (LD50). Second, (beginning in the mid-1980s), the desire to control environmental impacts believed to be associated with chemical toxicity led to new chemical parameters such as resin acids or AOX and new biological tests such as chronic bioassays (fathead minnow and *Ceriodaphnia*). In a nearly concurrent third stage that continues today, the focus has been the emission of persistent, toxic, bioaccumulative compounds into the environment, such as 2,3,7,8-tetrachlorodibenzo-*p*-dioxin (TCDD) and 2,3,7,8-tetrachlorodibenzo-furan (TCDF). The effort to remove TCDD/TCDF led to the misperception that most chlorinated organics discharged might be as toxic and bioaccumulative. However, most AOX is very water soluble with a log K_{ow} of -2 versus a log K_{ow} for TCDD of 6. This means that the difference in bioaccumulative potential is different between AOX and TCDD by 100,000,000-fold. Most recently, the environmental focus has shifted to sometimes subtle sublethal biological effects (biomarkers) on fish reproduction (reproductive steroids) and enzyme induction (P450IA). However, these issues have not yet been translated into regulations anywhere in the world. Increasingly, as part of an integrated multi-media regulatory effort, atmospheric emissions and solid wastes are considered in conjunction with the regulation of aquatic discharges..

Since the mid-1980s, three studies or findings have had significant influence on the formation of recent effluent regulations: the Environment Cellulose I study conducted at Norrsundet in Sweden in 1983-85 the discovery of TCDD and TCDF in bleached kraft effluents, and studies on fish reproduction in Canadian waters downstream of mills from 1988 to the present. A short review of each is useful.

Environment Cellulose I

The Environment Cellulose I project was one of the first integrated environmental studies and has been interpreted to implicate chlorinated organics having environmental impacts (despite the caution of original investigators that conventional pollutants were also disrupting the site). The Norrsundet mill was being rebuilt to include oxygen delignification and was thus considered a current state-of-the-art mill. However, the site had a history of major impacts including an anoxic mat of fiber and bark (Landner *et al.* 1977), and the mill start-up during the study was not smooth, i.e., with apparent black liquor spills (Landner 1990). Significant disruption of the benthic community (Kautsky *et al.* 1988), significant changes in fish population abundance and distribution (Neuman and Karås 1988), fin erosion, potential decrements in adult reproductive capacity (Sandström *et al.* 1988), disruption of recruitment for several species (Karås *et al.* 1991), and significant changes in a large suite of biomarkers (Andersson *et al.* 1988) were all observed. As a small unbleached mill was used as a reference site and far lesser effects were observed there, the Norrsundet results were thought to implicate the bleach discharge of chlorinated organics or AOX. Since the studies the 1980s, continued studies at the Norrsundet site have seen a gradual improvement in several measurements in the fish populations, fish reproduction, and fish physiology/biochemistry. These have occurred in parallel with improved process control, removal of chlorine gas, and the addition of external treatment to reduce BOD by ~50% (Förlin *et al.* 1992). However, for reasons that are as yet unidentified, there continues to be evidence for recruitment failure of perch fry and larvae at this site (Sandström 1994).

Discovery of TCDD/TCDF in Bleach Plant Effluents

In the late 1980s, the discharge of TCDD and TCDF from pulp mills employing chlorine gas bleaching was simultaneously discovered in the United States and Sweden (see for example Swanson *et al.* 1988; Amendola *et al.* 1989). This led to the systematic investigation of mill discharges in Sweden, Canada, and the United States and the analysis of fish and organisms in receiving waters below pulp mill discharges, which led to a number of fisheries closures. Within a short period of time, the role of chlorine gas in TCDD and TCDF formation was confirmed, and substitution of chlorine gas with chlorine dioxide began throughout the industry. Today, analytical results from Sweden, Canada, the United States, and other countries show that TCDD/TCDF have been virtually eliminated from the industry's discharges where chlorine gas has been removed.

Fish Reproduction in Canadian Receiving Waters

Recently, studies in Canadian receiving waters have been conducted at several sites with different ecosystems and mill operations (McMaster *et al.* 1991; Munkittrick *et al.* 1991, 1992a,b,c, 1994; Hodson *et al.* 1992; Servos *et al.* 1992; Kloepper-Sams and Benton 1994; Kloepper-Sams *et al.* 1994; Swanson *et al.* 1994). From a scientific and regulatory standpoint, the results are significant in several major respects:

(1) Certain effects found at Norrsundet, particularly a potential decrement in adult reproductive capacity and induction of hepatic ethoxyresorufin-*o*-deethylase (EROD) activity (or P450IA), were found at many Canadian sites.

(2) Reproductive effects, when they were observed, were associated with a decrease in circulating levels of sex steroids in both sexes.

(3) There was a wide range of these effects from mill to mill. For example, there was no decrease in gonad size in two fish species and no decrease in sex steroids at Grande Prairie, versus significant decreases in the gonad size of three species and their sex steroid levels at Terrace Bay, regardless of the fact that both mills use softwoods in similar bleached kraft processes. More importantly, the effects were also observed at unbleached mills and did not correlate with AOX levels in receiving waters or body burdens of chlorinated chemicals.

In an overall regulatory context, the Environment Cellulose I findings in conjunction with the discovery of TCDD/TCDF let to an almost exclusive focus on chlorinated organics and bleaching. Several nations rapidly implemented or proposed AOX regulations, even though the scientific basis for using AOX as a regulatory parameter was questioned (Folke *et al.* 1991). Then, the Canadian data dispelled the association between effects and chlorinated organics. As a result, Canada has declined to use AOX as a

Table 1. Current and proposed U.S. effluent regulations: bleached kraft softwood market pulp.

	Current (existing sources)	Proposed (existing sources)	Proposed (new sources)[a]
BOD_5 (kg tonne^{-1})	8.06	2.19	0.365
TSS (kg tonne^{-1})	16.4	8.75	0.383
COD (kg tonne^{-1})	Unregulated	25.4	Same as existing
AOX (kg tonne^{-1})	Unregulated	0.16	Same as existing
TCDD	Unregulated	ND[b]	Same as existing
TCDF	Unregulated	359 ng t^{-1}[c]	Same as existing
Chlorophenols[d]	Biocides only	ND[c]	Same as existing
Color	Some states regulate	76	Same as existing

[a] Only differences from existing sources are shown.
[b] ND = non-detectable.
[c] Daily maximum, other values are monthly average.
[d] The proposed regulations list specific chlorophenolic compounds.

FINLAND

The Finnish Water Act was first issued in 1962, with subsequent amendments. Permits are granted by independent courts of justice, Water Courts. The Water Court sets effluent targets after considering the reduction or elimination of detrimental environmental effects, economic costs, and the available technology. Usually, a specific technology is not required by the Water Court. Final appeals can be made to the national Superior Water Court. The National Board of Waters and the Environment proposes permit criteria to the Water Court and other parties may present their arguments to the Water Court.

The primary historical considerations have been on reducing BOD, suspended solids, and phosphorus emissions to the receiving waters. Permits are monitored via routine chemical, physical, and biological studies of the receiving waters followed by water quality modeling studies. Mills may be required to conduct studies. Finland's effluent criteria (Table 2) were patterned on the recommendations and proposals within the Helsinki Convention regarding the discharges from pulp and paper industry (HELCOM 1988, 1989), but may be upgraded to the suggested parameters negotiated by the Nordic environmental ministers in 1993.

Table 2. Finnish bleached kraft pulp parameters.

	HELCOM parameters (kg ADt^{-1})	Nordic ministers parameters kg ADt^{-1}	
		Existing	New
AOX	2	0.4	0.2
COD	65	30	15
Total P	0.06	0.04	0.02
Total N	Not applicable	0.2	0.15

SWEDEN

Discharges are regulated by the Environment Protection Act of 1969 with revisions in 1981 and 1988. Individual (case-by-case) permits are given by the local licensing board, which considers both available internal and external technologies and economical feasibility with the objective of minimizing environmental impacts. The national Swedish Environmental Protection Agency (SEPA) proposes the permit terms based on overall national goals. An appeal against a local permit decision may be filed by SEPA, the company, the local trade union, or the community. The government is the final court of appeal. Permits may contain upper limit values for discharges (BOD, COD, TSS, chlorate, etc.) and specific process steps, such as oxygen delignification and biological treatment.

Beginning in the mid-1980s, Sweden has emphasized pollution prevention. In contrast to the U.S. evaluation in the mid-1970s, oxygen delignification had emerged as a proven technology in the early 1980s. Therefore, oxygen delignification and, more recently, extended delignification have been emphasized. Target brownstock kappa numbers of 12-14 prior to the bleach plant have been discussed. Best management practices have been encouraged, such as limits on washing losses to 10 kg Na_2SO_4 ADt^{-1}, along with internal technologies such as condensate stripping. The SEPA has also promoted the testing of new technologies ranging from membrane filtration to ozone delignification. External treatment, however, exists at only 9 of 15 bleached kraft mills. In 1991, Sweden modified its environmental legislation to include the objective of limiting discharge to such a level that "no harm" would be done to the environment, to encourage BAT, and to give the industry a technical and economic priority in reviewing permits.

Technologically, emphasis has been on reduced chlorine gas use for bleaching, substituting with chlorine dioxide or other bleaching agents, and on reduced bleaching charges (maximum equivalent Cl_2 charge of 18-20 kg ADt^{-1} for softwood). As of 1993, chlorine gas was no longer used in Sweden, with 87% of all bleached pulp produced by ECF and 13% produced by TCF (Langerhan 1994). Average 1993 AOX and COD data for all Swedish bleached kraft mills, the six best mills in 1993, and year 2000 goals for leading mills emphasizing further efforts to minimize discharges are given in Table 3.

Table 3. Current Swedish mill performance and year 2000 goals.

	1993 - Average	1993 - 6 Best	2000 Goal
AOX (kg tonne^{-1})	0.39	0.18	<0.1
COD (kg tonne^{-1})	39	27	20-25

DISCUSSION

All national regulatory schemes for the pulp and paper industry have in common a desire to protect the environment, to use new and better process technology, and to balance the economic and social costs of implementing regulatory changes. These regulatory schemes on are on the verge of facing an unprecedented challenge. The environmental trends in the pulp and paper industry are very positive: evidence of adverse effects at the community and the population levels is diminishing rapidly, the evidence is strong that persistent and bioaccumulative chlorinated materials have been eliminated by process changes, evidence for subtle effects at the physiological and biochemical level is also diminishing, and in the case of reproductive effects investigators appear close to identifying the causative agents. Simultaneously, undiminished engineering innovations across the pulp and paper industry, such as enhanced delignification, chlorine dioxide substitution, and biological wastewater treatment, have made possible a significant decline in overall emissions as well as the virtual elimination of specific chemicals such as TCDD. Combined with new developments-in-progress which are moving rapidly towards mill trials, it has been suggested that a "minimal impact" mill with very limited liquid discharges can be engineered and that even a closed mill is possible. These advances may largely supersede current regulatory expectations and establish a new level of BAT, as has occurred in the past. However, engineering designs are directed at a very few greenfield mills likely to be built in the next few decades in North America and Scandinavia. There has been the widespread, and potentially mistaken, perception that existing mills could be efficiently and economically retrofitted with these developing technologies.

The regulatory questions are:
1. Would closed or minimal impact mills actually offer significant environmental improvements given the current data trends?
2. Would the economic and societal costs of mill closures and the disruption of whole communities be appropriate to achieve this improvement?

The fundamental issue is that BAT may be moving beyond the demands of actual environmental protection so that complete installation of these closure technologies in all mills may not be a requirement. This would not preclude implementation of appropriate equipment and retrofits where discharges of specific parameters must be reduced. In summary, the BAT demand inherent in several regulatory frameworks is not constructed to balance the environmental benefits and the economic costs.

There are additional challenges to the current regulatory framework. First, existing mills have a very large and diverse array of wood furnishes, existing equipment, and market products. Therefore, retrofit engineering must be implemented on a case-by-case basis. This demonstrates the advantage inherent in allowing site-specific assessments of need to enter national BAT and economic considerations. Second, by focusing strictly on BAT, more subtle trade-offs may be ignored. For example, there is a significant potential for the closed mill to require external energy sources to evaporate future bleachery wastewaters where current mills are largely energy self-sufficient. Is the burning of fossil fuels for this energy requirement an acceptable trade-off when combined with the societal costs and the environmental benefits? This may depend upon the engineering capabilities of decreasing water usage to the range of 8-10 m^3 per tonne versus the current best performance range of 25-40 m^3.

On the environmental side, there is a very significant question to address: Are the *scientific methods* to quantify the environmental benefits from "minimal impact" changes in place? The answer is yes for many receiving water and fish population endpoints and a qualified no for others such as subtle physiological and biochemical changes. First, chemical analyses and site-specific models can be put into place for dissolved oxygen, metals, and suspended solids, but those issues are largely resolved, namely a properly designed, sited, and operated pulp mill is not today expected to have any measurable acute or chronic adverse environmental effects. Second, chemical analysis of effluents and species at or near the apex of the food chain can also be monitored for persistent and bioaccumulative chemicals. The change to chlorine dioxide as a bleaching chemical has resulted in effluents which are non-detectable for TCDD and for most other polychlorinated chemicals. With some time lag, the body burdens of biota in the receiving waters are now in sharp decline. Third, biological measurements are available for fish populations and benthic communities. These data agree with chronic effluent toxicity tests: the previous observed effects at some sites are declining. However, these tests are time and resource consuming to conduct.

At a more subtle biological level, decrements in gonad size, decreases in circulating sex steroids, and induction of P450IA at intensely studied sites (e.g., Canada) appear to be declining. However, these tests are not widely and routinely available at this time and standardization of protocols is incomplete. Simultaneously, the experimental systems to test species in *controlled settings*, away from the biological and chemical "noise," complexity, and dynamic variations inherent in real environments are not fully developed. However, it should be recognized that there are promising efforts well underway such as the mesocosm studies in Scandinavia, the experimental streams studies in the U.S., and the recent laboratory exposures of fathead minnows and other fish species in Canada and the U.S. Therefore, regulatory policies, which continue to emphasize a general reduction in discharges or BAT and find it increasingly difficult to assess costs and environmental benefits, should consider fostering the development of these systems and incorporating the results into regulatory implementation.

Again, on the environmental side: Are the *field monitoring systems* in place to assess the benefits across the diversity of ecosystems and mill designs? The answer at this time is no. Only Environment Canada has laid down in principle an environmental effects monitoring system for the pulp and paper industry (Environment Canada 1992). This system is still in its infancy as the overall design is still under discussion and only a few individual endpoint protocols have been finalized, such as the P450IA assay (Hodson *et al.* 1991). No other nation or regulatory authority has come forward to implement a similar systematic concept. This should not fail to recognize that significant pieces of such a system are moving forward, such as the fisheries monitoring program and criteria in Sweden (Thorresson 1993) and significant efforts in individual programs within the U.S. EPA at several levels of biological organization. Instead, the concentration of effort outside of Canada has been on effluent parameters and estimates of engineering capabilities.

In the final analysis, as we look to the future, this brings us to a final question. The scientific methods, controlled exposures, and field monitoring systems will be needed to address societal questions for regulatory policymakers: Where is the balance between a sole focus on effluent parameters and the actual ecological status of receiving environments? Which should be the future technological basis for pulp mill regulations? Therefore, much hinges on the so-called cost-benefit analysis. The mechanism for this analysis has not currently been explored and tested. In the U.S., the Clean Water Act is based on technically driven standards with proposed legislative amendments to add cost-benefit analysis. By presidential executive order, regulatory agencies must theoretically demonstrate that benefits to human and environmental health are sufficient to justify the industry's anticipated capital expenditures. As might be expected, there are now substantial differences over both the calculation of benefits and the estimates of the capital expenses. In theory, when the rules are finalized, the option exists for the industry or individual companies to challenge the calculations in court. In the same vein, there is strong and growing sentiment that the cost-benefit and risk assessment mechanisms within the regulatory framework should be strengthened. Therefore, the implication to scientists is that the importance of analyzing the environmental benefits will continue to increase. As a result, all parties will be directed more and more towards a long-term interest in accurately assessing environmental benefits, rather than a simple focus on "end-of-pipe" performance. Intuitively, this calls for an increasing emphasis on the development of tools for measuring the meaningful impact of technological progress.

As noted, one alternative to regulatory action is the marketplace. In Europe, ECF pulps have grown to comprise 80% of the bleached market pulp volume and TCF has grown to approximately 12%. However, TCF penetration has slowed due to pricing pressures despite significant marketing campaigns for TCF pulp. Given the high capital and increased operating costs for a minimal discharge mill, a significant portion of the market does not appear willing to support a premium price without identifiable benefits. Therefore, beyond a few "market niche" minimal effluent mills, regulatory action is likely to remain the driving force and environmental performance floor for the industry. Left to its own, the installation of "minimal impact" technology at most mills would be limited, at best, over the next one to two decades.

In closing, there is a significant need from regulatory, industry, and market perspectives to understand and to accurately dimension the meaningful and measurable adverse effects of pulp mill discharges, where they exist. Today, that task is becoming increasingly difficult, suggesting that pulp mill environmental controls may have nearly reached the point of diminishing return, where the cost of "improvement" cannot be easily justified. To move this debate forward, it is worthwhile to reiterate that there are three critical areas of environmental science which must be improved in regards to pulp and paper discharges:
1. The ability to identify and to quantitate with measurement systems those chemical and biological endpoints which are linked to environmental protection,

2. The ability to measure sometimes subtle response in controlled environments so that causative agents can be identified and effluents reliably tested, and
3. The ability to monitor real ecosystems for these subtle effects in diverse and fluctuating environments to ensure that the interplay of effluent, habitat, and other factors is included in environmental protection systems.

Only in this way, with the proper science, can the industry implement new technologies that lead to meaningful environmental improvements and remain economically competitive. From a regulatory standpoint, these developments must be coupled with assessments of environmental risk, as well as a cost-:benefit analysis that measures the real effectiveness of different regulatory proposals. This risk should be evaluated not only at the national level, but on a site-specific basis where special needs may arise.

REFERENCES

Amendola, G.A., D. Barana, R. Blosser, L. LaFleur, A. McBride, F. Thomas, T. Tiernan and R. Whittemore. 1989. The occurrence and fate of PCDDs and PCDFs in five bleached kraft pulp and paper mills. Chemosphere 18:1181-1189.

Andersson, T., L. Förlin, J. Härdig and Å. Larsson. 1988. Physiological disturbances in fish living in the receiving body of water of a kraft mill industry. Can. J. Fish. Aquat. Sci. 45:1525-1536.

Environment Canada. 1992. Pulp and paper effluent regulations. Dept. of Fisheries and Oceans. Can. Gaz. Part II 126:1967-1997.

Folke, J., H. Edde and K.-J. Lehtinen. 1991. The scientific foundation of adsorbable organohalogens (AOX) as a regulatory parameter for control of organochlorine compounds. Proceedings of the 1991 TAPPI Environmental Conference. San Antonio, TX. April 7-10. pp. 517-527.

Förlin, L., T. Andersson, L. Balk and Å. Larsson. 1992. Biochemical and physiological effects of pulp mill effluents on fish. In: Environmental Fate and Effects of Bleached Pulp Mill Effluents. Report 4031. Swedish Environmental Protection Agency, Solna, Sweden. pp. 235-247.

HELCOM. 1988. Recommendations concerning restrictions of discharges from the pulp and paper industry. Recommendation 9/6, Helsinki Convention.

HELCOM. 1989. Finnish comments on draft recommendations concerning restriction of discharges from the sulphite pulp industry. Helsinki Convention.

Hodson, P.V., P.J. Kloepper-Sams, K.R. Munkittrick, W.L. Lockhart, D.A. Metner, P.L. Luxon, I.R. Smith, M.M. Gagnon, M. Servos and J. Payne. 1991. Protocols for measuring mixed function oxygenases of fish liver. Can. Tech. Rep. Fish. Aquat. Sci. 1829.

Hodson, P.V., M. McWhirter, K. Ralph, B. Gray, D. Thivuerge, J.C. Carey, G. Van Der Kraak, D.M. Whittle and M.-C. Levesque. 1992. Effects of bleached kraft mill effluent on fish in the St. Maurice River, Quebec. Environ. Toxicol. Chem. 11:1635-1651.

Kautsky, H., U. Kautsky and S. Nellbring. 1988. Distribution of flora and fauna in an area receiving pulp mill effluents in the Baltic Sea. Ophelia 28:139-155.

Kloepper-Sams, P.J., and E. Benton. 1994. Exposure of fish to biologically treated bleached kraft effluent. 2. Induction of hepatic cytochrome P450IA in mountain whitefish *(Prosopium williamsoni)* and other species. Environ. Toxicol. Chem. 13:1483-1496.

Kloepper-Sams, P.J., S.M. Swanson, T. Marchant, R. Schryer and J.W. Owens. 1994. Exposure of fish to biologically treated bleached kraft effluent. 1. Biochemical, physiological, and pathological assessment of Rocky Mountain whitefish *(Prosopium williamsoni)* and longnose suckers *(Catostomus catostomus)*. Environ. Toxicol. Chem. 13:1469-1482.

Kåras, P., E. Neuman and O. Sandström. 1991. Effects of a pulp mill effluent on the population dynamics of perch *Perca fluviatilis* L. Can. J. Fish. Aquat. Sci. 48:28-34

Landner, L. 1990. Sammanställning och utvärdering av miljötillståndet i Norrsundet-recipienten under senare delen 1980-talet. MiljöForskar Gruppen Report F90/075:3. Stockholm, Sweden. 46 pp.

Landner, L., K. Nilsson and R. Rosenberg. 1977. Assessment of industrial pollution by means of benthic macrofauna surveys along the Swedish Baltic coast. Vatten 3:324-379.

Langerhan, S. 1994. The status of regulatory parameters for the pulp and paper industry in Sweden. International Conference on Non-Chlorine Bleaching. Amelia Island, FL., March 5-9.

McMaster, M.E., G.J. Van Der Kraak, C.B. Portt, K.R. Munkittrick, P.K. Sibley, I.R. Smith and D.G. Dixon. 1991. Changes in hepatic mixed-function oxygenase (MFO) activity, plasma steroid levels and age at maturity of a white sucker (*Catostomus commersoni*) population exposed to a bleached kraft pulp mill effluent. Aquat. Toxicol. 21:199-218.

Munkittrick, K.R., C. Portt, G.J. Van Der Kraak, I.R. Smith and D. Rokosh. 1991. Impact of bleached kraft mill effluent on liver MFO activity, serum steroids, and population characteristics of a Lake Superior white sucker population. Can. J. Fish. Aquat. Sci. 48:1371-1380.

Munkittrick, K.R., G.J. Van Der Kraak, M.E. McMaster and C.B. Portt. 1992a. Response of hepatic mixed function oxygenase (MFO) activity and plasma sex steroids to secondary treatment and mill shutdown. Environ. Toxicol. Chem. 11:1427-1439.

Munkittrick, K.R., M.E. McMaster, C.B. Portt, G.J. Van Der Kraak, I.R. Smith and D.G. Dixon. 1992b. Changes in maturity, plasma sex steroid levels, hepatic mixed-function oxidase levels, and the presence of external lesions in lake whitefish (*Coregonus clupeaformis*) exposed to bleached kraft mill effluent. Can. J. Fish. Aquat. Sci. 49:1560-1569.

Munkittrick, K.R., G.J. Van Der Kraak, M.E. McMaster and C.B. Portt. 1992c. Response of hepatic mixed function oxidase (MFO) activity and plasma sex steroids to secondary treatment and mill shutdown. Environ. Toxicol. Chem. 11:1427-1439.

Munkittrick, K.R., G.J. Van Der Kraak, M.E. McMaster, C.B. Portt, M.R. van den Heuvel and M.R. Servos. 1994. Survey of receiving water environmental impacts associated with discharges from pulp mills. 2. Gonad size, liver size, hepatic EROD activity, and plasma sex steroid levels in white sucker. Environ. Toxicol. Chem. 13:1089-1301.

Neuman, E., and P. Karås. 1988. Effects of pulp mill effluent on a Baltic coastal fish community. Water Sci. Technol. 20:95-106.

Sandström, O. 1994. Incomplete recovery in a coastal fish community exposed to effluent from a modernized Swedish bleached kraft mill. Can. J. Fish. Aquat. Sci. 51:2195-2202.

Sandström, O., E. Neuman and P. Karås. 1988. Effects of a bleached pulp mill effluent on growth and gonad function in Baltic coastal fish. Water Sci. Technol. 20:107-118.

Servos, M., J. Carey, M. Ferguson, G.J. Van Der Kraak, H. Ferguson, J. Parrott, K. Gorman and R. Cowling. 1992. Impact of a modern bleached kraft mill with secondary treatment on white suckers. Water PolluT. Res. J. Can. 27:423-427.

Swanson, S.E., C. Rappe, K. Kringstad and J. Malstrom. 1988. Emissions of PCDDs and PCDFs from the pulp industry. Chemosphere 17:681-691.

Swanson, S.M., R. Schryer, R. Shelast, P.J. Kloepper-Sams and J.W. Owens. 1994. Exposure to biologically treated bleached kraft effluent. III. Habitat and population assessment. Environ. Toxicol. Chem. 13:1497-1507.

Thoresson, G. 1993. Guidelines for coastal monitoring: Fishery biology. National Board of Fisheries, Institute of Coastal Research, Öregrund, Sweden.

U.S. EPA. 1993. Proposed rules for the pulp and paper industry. Fed. Reg. 58(241):66100-66500. Dec. 17, 1993.

SWEDISH ENVIRONMENTAL REGULATIONS FOR BLEACHED KRAFT PULP MILLS

Staffan Lagergren

Swedish Environmental Protection Agency, S-171 85 Solna, Sweden

In Sweden, the discharge of chlorinated organic substances from pulp mills has received considerable attention since the late 1970s. The large research project Environment/Cellulose demonstrated detrimental effects in the receiving waters outside pulp mills. Several of these effects were suspected to be related to exposure to chlorinated organics. Furthermore, chlorinated organic substances have been widely distributed throughout the Baltic Sea. The first steps to reduce bleach plant effluents were taken in the early 1970s when oxygen delignification stages began to be installed. Discharges of AOX at Swedish mills at that time were estimated to have been 8-10 kg AOX tonne pulp^{-1}. Recent implementation of process modifications including the elimination of molecular Cl has reduced discharges of AOX to <1 kg tonne pulp^{-1}. Additional process changes (best available technology) could reduce AOX levels at Swedish mills to less than 0.1-0.2 kg tonne pulp^{-1}.

SWEDISH GOVERNMENT BILLS

The 1988 Swedish Government Bill on Environmental Policy stated that all Swedish mills should have started to take steps no later than 1992 in order to reduce the discharges of chlorinated organic substances to a level of 1.5 kg total organically bound chlorine (TOCl) per tonne pulp. A long term goal was set requiring that the discharges of chlorinated organic substances be eliminated. It is important to note that in Sweden emission limit values are set individually for each mill in the licensing process. This meant that the conditions of the operating permit for each mill had to be reviewed to reduce the AOX, TOCl limit values to the levels set in the Government Bill.

In the 1991 Government Bill, the Swedish EPA was asked to draw up a ten-year program for a review of the conditions of the operating permits for industrial discharges. The pulp and paper industry was given priority in this review program. Discharges from pulp and paper mills should be reduced by the beginning of the next century to such a level that the environment is not noticeably affected. The implication in terms of discharge levels is suggested to be clarified through continued research and environmental monitoring. In the process of reviewing the conditions of the operating permits, discharges of environmentally harmful substances other than chlorinated organic compounds from the manufacture of both bleached and unbleached pulp are to be dealt with. This means that focus should not merely be on AOX, but also on specific chlorinated and unchlorinated substances, as well as on nutrients, metals and on chemicals used in the manufacturing process.

REVIEW OF EMISSION LIMIT VALUES

In December 1992, the Swedish EPA presented a program for review of emission conditions in the operating permits as asked for in the 1991 Government Bill. This program included a presentation of the environmental problems in different industries and remedial measures for dealing with them as well as the costs of taking such measures.

Consideration of an application for an operating permit under the Environment Protection Act is based on best available technology (BAT, i.e., the best achieved technology to date at production scale in Sweden or any other country). The environmental effects and cost are then weighed together when the emission limits are being set. Table 1 shows the emission levels for kraft mills after internal and external treatment measures have been implemented.

Table 1. Emission levels for Swedish kraft mills after internal and external treatment measures.

	AOX	COD	P	N
Bleached pulp before external treatment				
Current normal level	0.3-1	35-70	0.05-0.1	0.2-0.4
Current BAT	0*-0.3	35-40	0.05	0.2
Development potential	0*-0.1	10-15	0.02	0.2
Bleached pulp after external treatment, BAT	0*-0.2	10-15	0.01-0.02	0.1-0.2

* TCF-pulp.

The BAT levels are based on a combination of in-plant remedial measures and external treatment including oxygen delignification, biological treatment and chemical precipitation. However, the Swedish EPA favors in-plant measures like extended delignification, efficient washing, bleaching without chlorine gas and recycling of bleach plant effluents. Systems for spillage collection and condensate treatment are also important steps to reduce the discharges of COD and nutrients.

From Table 1, it can be seen that there is a potential to reach the BAT levels without external treatment. In this case, bleach plant effluents must be recycled to the recovery boiler or be incinerated separately and discharges from other parts of the process reduced to very low levels (Table 2).

Table 2. Possible COD discharge from different parts of a mill.

	Normal	BAT	Future
Debarking	1-10	<1	<1
Condensate, SW	3-6	3	2
Condensate, HW	4-7	4	3
Spillage	2-7	2	<1
Washing	6-10	6	4
	8-12	8	6
Total bleached, SW	12-30	12	8
Total unbleached, HW	15-25	15	10
Bleaching plant	20-40	25	5
Bleaching plant	20-30	20	3
Total, SW	32-70	37	13
Total, HW	35-65	35	13

SW = softwood; HW = hardwood.

There is a major environmental gain to be achieved from changing over to chlorine-free chemicals, since the prospects of attaining a very high degree of closure in the bleach plant improve substantially. This in turn facilitates a reduction in emissions not only of organic substances, but also of metals and nutrients. The aim should be to increase the degree of recycling in the bleach plant gradually and to effectively treat the wastewater from processes. Factors affecting the scope for recycling are, for example:
- The use of chemicals containing chlorine.
- The ability to exclude chlorides from the system.
- The occurrence of other substances causing disturbance, such as calcium, potassium, aluminium and silicon.
- The size of the flow of acid and alkaline effluent.

CURRENT SITUATION

The emissions of AOX and COD from the pulp industry are shown in Fig. 1. The reduction of AOX emissions has lately been very pronounced due to market demands for totally chlorine-free (TCF) and elemental chlorine-free (ECF) pulps (Fig. 2). The combination of extended delignification and oxygen delignification is becoming increasingly common in Swedish mills. All mills today have systems for condensate stripping and collection of spillage. Eight of the fifteen bleached kraft mills have biological treatment plants (Table 3). Today no mill in Sweden uses chlorine gas (Fig. 3) and all mills in Sweden have installed an oxygen delignification stage. Three mills use ozone and peroxide for the production of TCF pulps. Several mills use peroxide for a part of the production only. The consumption of bleaching chemicals (Cl_2, $NaClO_3$, O_2, H_2O_2) since 1986 can be seen in Fig. 4.

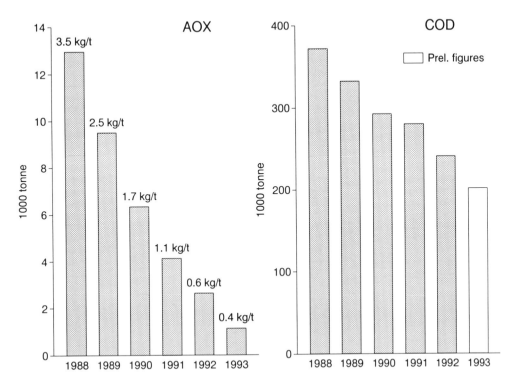

Figure 1. AOX and COD discharges from Swedish bleached kraft mills.

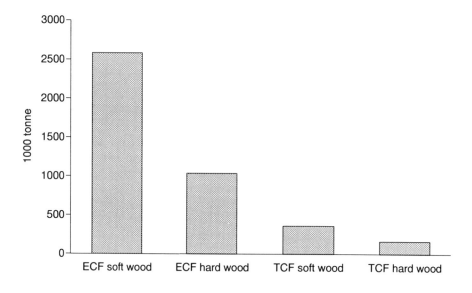

Figure 2. Bleached kraft pulp production, Sweden 1993.

Table 3. Process equipment in the 15 Swedish bleached kraft mills in 1994.

Extended delignification	11
Oxygen delignification	15
100% chlorine dioxide	15
Hydrogen peroxide	11
Ozone	3
Secondary treatment	8

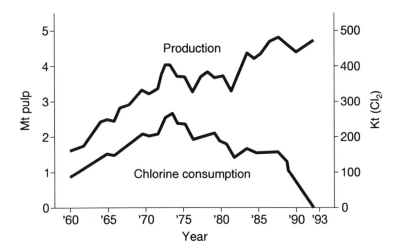

Figure 3. Chlorine consumption and pulp production in Swedish pulp mills.

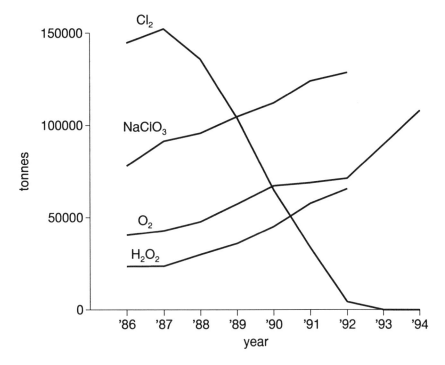

Figure 4. Chemical consumption in Swedish pulp mills.

FUTURE DEVELOPMENT

A significant question to be answered is whether the reduction in AOX also has resulted in improvements in the environment. No doubt the risk of effects of persistent chlorinated compounds like dioxins and phenolic compounds has been greatly reduced (Figs. 5 and 6). Yet we do not fully understand the biological effects of wastewaters from the production of ECF and TCF pulps. Some effects remain even after the reduction or elimination of chlorinated substances. Nonchlorinated substances may very well have contributed to the effects observed earlier in many receiving waters outside pulp mills using chlorine bleaching. One advantage is however apparent with the replacement of chemicals containing chlorine; the possibilities increase to recycle wastewater from the bleach plant to the recovery boiler. The final goal of a closed bleach plant seems to be within reach. The development of cooking and bleaching processes and discharges is seen in Table 4. It is my conviction that before the turn of the century several mills will have reached the BAT level indicated.

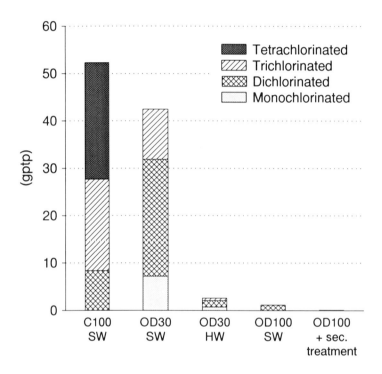

Figure 5. Chlorinated phenolic substances in effluents produced using different production or treatment strategies.

INTERNATIONAL WORK

In November 1993, the Nordic Council of Ministers issued a statement on the impact on the environment of the pulp and paper industry and actions to limit this impact. The statement was based on a report from a working group with representatives from the Nordic countries. The ministers emphasized the importance of continued work to develop new closed processes. They also supported the proposal for general levels (see Tables 5 and 6) for maximum permissible emissions. Exceptions may however be motivated in certain special cases due to local conditions and the different policies of the Nordic countries in the field of environmental protection. The ministers especially stressed the need for continued research on the effects of pulp and paper industry emissions.

Sweden also works in the Helsinki Commission (HELCOM) and Paris Commission (PARCOM) to reach agreements on the reduction of pollutants to the Baltic Sea and the Northeast Atlantic, respectively. According to a PARCOM decision, the discharges of chlorinated organic substances should not, as an annual average for each mill, exceed 1 kg of AOX tonne^{-1} air-dried bleached kraft or sulfite pulp.

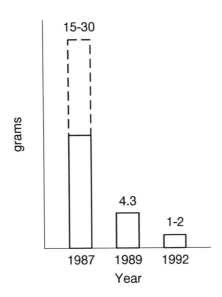

Figure 6. Recent reductions in polychlorinated dibenzo-p-dioxins and dibenzofurans (g yr^{-1}) discharges at Swedish pulp mills.

Table 4. Development of process and discharge parameter history in Sweden.

Year	Process	Secondary Treatment	AOX kg t^{-1}	COD kg t^{-1}
1970	CEHDED	No	6-8	80-100
1980	O(C90+D10)EDED	No	3-4	50-60
1990	O(D50+C50)(EO)D(EP)D	Yes	1.5-2	25-35
1993	Extended delig. OD(EO)D(EP)D, OQPP, OQZPP	Yes	<0.2	20-25
1998	Further system closure	Yes	<0.1	10-15

O = oxygen delignification, P = peroxide, D = chlorine dioxide, Q = chelating agent, C = chlorine (molecular), Z = ozone, E = extraction stage, H = hypochlorite.

Table 5. Levels not to be exceeded by the end of this century by any mill as annual averages (kg tonne pulp^{-1}).

Type of Mill	AOX	COD	Tot-P	Tot-N
Bleached kraft pulp	0.4	30	0.04	0.2
Unbleached kraft pulp	-	15	0.02	0.2

Table 6. Levels not to be exceeded by any new or considerably (on the order of 30%) enlarged mills (kg tonne pulp^{-1}).

Type of Mill	AOX	COD	Tot-P	Tot-N
Bleached kraft pulp	0.2	15	0.02	0.15
Unbleached kraft pulp	-	8	0.01	0.15

For new installations or existing installations subject to a substantial increase in production capacity or subject to retrofitting, the formation of polychlorinated organic substances should be reduced by using a molecular chlorine multiple below 0.05 (i.e., per tonne of pulp less than 0.5 kg of molecular chlorine, Cl_2, per kappa number of pulp entering the bleach plant). For existing installations not subject to a substantial increase in production capacity or not subject to retrofitting, this value should preferably be below 0.05 and should in any case not exceed 0.15. These objectives apply to all mills starting to operate on or after January 1, 1993. For mills in operation before January 1, 1993, the objectives should be attained by January 1, 1996.

ENVIRONMENTAL ASPECTS OF ECF VS. TCF PULP BLEACHING

Jens Folke[1], Lars Renberg[2] and Neil McCubbin[3]

[1]MFG—European Environmental Research Group, Østergade 16, 3250 Gilleleje, Denmark
[2]Akzo-Nobel, Eka Nobel AB, 44580 Bohus, Sweden
[3]N. McCubbin Consultants, 140 Fisher's Point, Foster, Quebec, J0E 1R0 Canada

This paper provides an update on the environmental aspects of elemental chlorine free (ECF) vs. totally chlorine free (TCF) pulp bleaching with an emphasis on literature published during the last two years. In general, there is no evidence of any adverse effects related to the use of chlorine dioxide, i.e., the trace amounts of chlorinated organics formed during ECF bleaching have no environmental effects and resemble naturally formed organochlorines. TCF pulp bleaching is less well documented, but it seems that the environmental effects of a pulp mill operation cannot be deduced from knowing the bleach sequence of the mill. One will have to know about the pulping operation, spill control, brownstock washer loss, etc. Biological data are mostly logarithmic by nature. Therefore, when comparing toxicity emission factors (TEF) between mills we suggest that $\log TEF = Q \times \log TU$ is used (where Q is the amount of effluent per tonne of pulp produced and TU is the toxic units, EC_{xx}^{-1} instead of the hitherto used $TEF = Q \times \log TU$. Bleaching sequences based on peroxide generally result in lower energy demand than those based on ozone, which in turn are more energy efficient than those based on chlorine dioxide. However, mill operating conditions, product specifications and operator skill can have just as much of an effect on energy consumption. The agenda is thus not whether to use ECF or TCF as the choice of bleaching sequence, but rather to modernize the pulping operation itself.

INTRODUCTION

Organochlorines in pulp mill effluents first became of interest in the mid-1980s. Since then, there has been widespread discussion and controversy concerning their environmental significance (Folke 1989; Folke et al. 1991; O'Connor et al. 1992). Apparently, a number of environmentalist groups and regulatory agencies, including several European countries as well as some Canadian provinces, have failed to make a clear distinction between chloride (which is very abundant in the environment), the generally harmless organochlorines discharged by modern chlorine dioxide based bleach plants, the harmful organochlorines discharged by old-fashioned chlorine-based bleached plants and the really harmful types of organochlorines, such as PCB, DDT and others. The latter have very different characteristics from the ones related to the pulp and paper mill discharges and have nothing to do with the pulp and paper industry (Amato 1993). These groups also failed to consider the abundance of naturally occurring chloroorganics in the environment (Grimvall 1993) in their crusade against chlorine.

The result has been that the pulp and paper industry in many countries has been forced into investments in technologies that have reduced AOX discharges from bleach plant effluents significantly. Initially, the industry lowered the molecular chlorine charge, which is environmentally sound (Kringstad et al. 1989; Folke, et al. 1991, 1992). A move towards total elimination of chlorine (elemental chlorine-free [ECF]) then followed. Recently, several world-scale mills have replaced chlorine dioxide with hydrogen peroxide, ozone and peroxyacetic acid (totally chlorine-free [TCF]), bleaching), and competent authors have questioned whether or not these latter investments were driven more by environmental scaremongering and commercial protectionism than by science or environmental protection (Reeve 1993; Clark 1994). Fortunately, these investments were generally accompanied by improved operating practices, upgraded brownstock washing systems and installation of spill recovery systems, which also reduced the discharges of unchlorinated organics from mills. Scientific evidence suggests that the reduction in total amount of organics discharged, which frequently accompanies the implementation of AOX control measures, is the real reason for the reduction of toxicity observed when effluents from mills with low, but differing, AOX discharge rates are compared. This paper uses as the base case the OD, C/DEopDEpD bleaching sequence used at Mönsterås before ozone bleaching was introduced. The environmental impact in the receiving waters of this mill's effluent is known to be minimal (Landner *et al.* 1994). The paper compares the effluent from that mill at that time with a number of ECF and TCF processes in terms of toxicity, mill energy consumption, costs and energy balance.

ENVIRONMENTAL QUALITY

We have previously published a literature study on the technical aspects and environmental effects of chlorine dioxide in pulp bleaching (Folke and Männistä 1993; Folke *et al.* 1993a). This review provides an update with emphasis on literature published since the last review.

The rapidity of acceptance of ultra low-chlorine and TCF bleaching in Scandinavia, without recovery of bleach plant effluents, to accommodate customer and regulatory demands to eliminate AOX discharges, demonstrates the weakness of a zero AOX regulation. It is ironic that the most environmentally effective recovery of a bleach plant effluent currently being practiced is probably at a U.S. mill which is not required to do so to comply with any regulation (Nutt *et al.* 1992). The TCF operations in Scandinavia have eliminated AOX discharges, probably with little or no environmental benefit relative to the prior low AOX operations.

It is technically feasible to recycle most effluents from TCF bleaching to the recovery system, where it is reasonably safe to assume that all organics will be effectively destroyed. However, few of the mills practicing TCF bleaching have recycled these effluents, although they report a long term objective of doing so and some are practicing partial recycle.

Substances other than the chlorinated organic ones have to be taken into account when the goal is to decrease the environmental impact of bleached pulp mill wastewaters (Folke *et al.* 1993b; McCubbin and Folke 1993). A realistic medium term goal for the elimination of the effects of kraft mill effluents on receiving waters should be a "near effluent-free mill" concept, where most of the spent bleaching liquors could be included in the recovery cycle. There would be no planned discharge of unbleached pulping liquors and accidental discharges of the latter would be minimized.

EXPOSURE

Chlorophenolics

PAPRICAN has reported laboratory investigations of seven different bleaching sequences: $C_{90}+D_{10}E$, $C_{50}+D_{50}E$, $D_{100}E$, $D_{100}E$, $OD_{100}E$, OQP and OZEP (O'Connor *et al.* 1994). Chemical analysis showed that polychlorinated phenolics (PoCPs, i.e., phenolics with three or more chlorine atoms) were not formed in sequences producing ECF pulp (O'Connor *et al.* 1993b; 1994). The same conclusion was reached in laboratory investigations by Renberg (1992). Renard and Philips and others reported on formation of PoCPs during high chlorine dioxide substitution bleaching in three papers (Deardorff *et al.* 1994; Tsai *et al.* 1994; Yin *et al.* 1994). Tsai *et al.* (1994) confirmed that PoCPs were not formed and also found that oxygen delignification increased the concentration of unchlorinated phenolics in the internal recycle stream. Mill trial results confirmed these results (Deardorff *et al.* 1994).

Chelating Agents

TCF sequences involving peroxide necessarily lead to the use of EDTA, DTPA or similar metal chelating agents (Table 1). Generally, these substances are discharged with the effluent. Literature on the biodegradability of EDTA under aerobic conditions generally supports the conclusion that EDTA is persistent or at least very slowly degradable (Alder et al. 1990). However, we are not aware of any studies of the effects of these rather large discharges of chelating agents that result from current TCF bleaching. At the very least, one must assume that the metal balances of the receiving waters could be affected as pointed out by Sandström and Larsson (1993). Also, any degradation of nitrogen-containing chelating compounds such as EDTA or DTPA could lead to increased eutrophication, an effect that could beneficially be studied in mesocosm experiments.

Table 1. Summary of mill data for the Swedish "SödraCell."

Mill sample	Flow m^3 ADt^{-1}	AOX kg ADt^{-1}	COD kg ADt^{-1}	ClO_2 kg ADt^{-1}	EDTA kg ADt^{-1}	H_2O_2 kg ADt^{-1}	O_3 kg ADt^{-1}
Värö (ODEo, D, E, D)							
SW ECF-1	111	0.23	42	27			
SW ECF-2	95	0.26	34	26			
SW ECF-3	133	0.5	40	30			
Värö (OQP)							
SW TCF-1	95	<0.05	32		1.8	36	
SW TCF-2	93	<0.05	37		1.7	32	
SW TCF-3	89	<0.05	40		1.8	44	
Mönsterås (OZQP)							
HW TCF-1	45	<0.05	24		2.1	35	3
HW TCF-2	50	<0.05	32		2	34	5.5
HW TCF-3	46	<0.05	33		2	31	3

SW = softwood; HW = hardwood.

AOX

Dahlman et al. (1993) and O'Connor et al. (1993b, 1994) have both characterized the high molecular mass material of effluents from ECF pulping. They reported C/Cl ratios of 89-272, which is only slightly lower than the C/Cl ratio of naturally found humic substances. It is also in the same order as the C/N ratio of the high molecular mass material and considerably higher than the C/S ratio, i.e., this material would be more precisely characterized as organosulfur materials. O'Connor et al. (1993a) conducted tests on a variety of bleached and unbleached kraft mill effluents, sampled before and after biological treatment systems, and showed that there was no correlation between AOX and acute or sublethal toxicity to *Ceriodaphnia* or fathead minnows. The AOX content of most of O'Connor's biologically treated samples corresponded to discharges of under 2.5 kg t^{-1} pulp, with the highest value being 4.8 kg t^{-1}. Yin et al. (1994) found that the AOX formed in sequences producing ECF pulp is more easily removed in biological treatment plants than AOX formed in sequences with low chlorine dioxide substitution. Once more, this confirms that an AOX regulation to a uniform level (e.g., 1 kg ADt^{-1} as in the European PARCOM and HELCOM conventions) is an inappropriate regulation due to the great differences in the chemical composition of AOX. Even worse in this regard is the recent Ecolabel criteria of kitchen and toilet papers decided by the European Union member countries (CEU 1994). The criteria include an AOX scale from

0.1-0.5, i.e. mills that discharge 0.3-0.5 kg ADt^{-1} are penalized with a maximum of loads points for AOX, while mills above 0.5 kg ADt^{-1} meet a hurdle and cannot obtain an Ecolabel at all. There are no literature data that support any differentiation in environmental effects between mills discharging 0.05 and 0.6 kg ADt^{-1} of AOX. The proposed Canadian EcoLogo guidelines rejected AOX as a useful ecolabel criterion and Environment Canada rejected AOX as a regulatory parameter after a detailed analysis.

EFFECTS

Acute Lethal Toxicity

There are no examples in the literature of any acute lethal toxicity to fish or crustacea of effluents from ECF or TCF sequences even without biotreatment.

Sublethal and Chronic Toxicity

O'Conner et al. (1994) showed in lab-scale experiments that *Ceriodaphnia* reproduction was more severely affected by the TCF (OZEP and OQP) sequences than sequences using ECF bleaching. In fact the TCF sequences were more toxic than the chlorine-based ones. The least toxic one for *Ceriodaphnia* reproduction was the OD$_{100}$ sequence. The effect on fathead minnow growth had the ranking D$_{100}$ > OZEP > OQP > OD$_{100}$. In fact the OD$_{100}$ sequence was non-toxic to fathead minnow. These results are all from laboratory experiments. As the authors point out, biological treatment typically reduces toxicity by a factor of 10 or more, so the data cannot be used to predict the toxicity of total mill effluents after biotreatment. In fact some of the chlorine-containing bleaching sequences may give rise to insignificant effects after biotreatment (Landner et al. 1994).

Toxicity Emission Factors

Toxicity emission factors (TEFs) are frequently used to compare the toxicity of two different effluents: TEF = TU × Q, where toxic units (TU) is defined as 100 EC50^{-1}, and Q is the amount of process water discharged per tonne of pulp produced. When two pulp mill effluents are compared, clearly the one with the highest TEF is the most toxic. The question is, however, whether a mill discharging a TEF of 400 is twice as toxic as one discharging a TEF of 200. Biological data are essentially logarithmic, e.g., decibel and toxicity data are normally attributed to an exponential distribution. In a LOGIT or PROBIT analysis the log EC50 value is used. Therefore, it may be more correct to use logTEF = Q × log (TU) as the basis for comparing two effluents. In this way logTEF = 0 if the effluent is non-toxic. Figure 1 shows how our previously published data would plot using the logTEF.

The figure shows even more clearly now that efforts to lower AOX discharges below ≈1.2 kg ADt^{-1} do not reduce toxicity any further. However, it does not show whether the relationship between AOX and toxicity is an independent one or whether it depends on the reduction of other compounds in the effluent. To analyze this we have re-analyzed previously published data (Lövblad and Malmström 1994). The data are summarized in Table 1.

Table 2 presents the corresponding toxicity data calculated as logTEF. As mentioned, non-toxic effluents have a logTEF value of zero, as opposed to the traditional TEF value which for a non-toxic effluent would be equal to the specific water flow per tonne of pulp. Also, the analysis indicates that a slight toxicity may be found in some bioassays for both ECF and TCF pulp, i.e., as long as the spent bleach liquors are discharged to the environment from both the ECF and TCF processes, it cannot be concluded on the basis of those results that one or the other bleaching technology is superior to the other in terms of effluent toxicity.

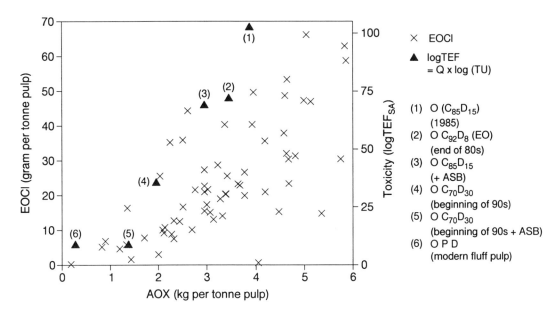

Figure 1. EOCl and toxicity relationship to AOX. (Copyright MFG; reprinted with permission. Source: Folke et al. 1993a; Folke and Männistö 1993.)

Table 2. Summary of logTEF data for the Swedish "SödraCell."

Mill sample	Microtox® bacteria sublethal tox.	*Selenastrum* growth inhibition	Zebra fish reproduction	*Ceriodaphnia* acute toxicity & reproduction
Värö				
SW ECF-1	15	0		
SW ECF-2			24	
SW ECF-3				0
Värö				
SW TCF-1		0		
SW TCF-2	5.3		3.7	
SW TCF-3				0
Mönsterås				
HW TCF-1	12	9.5		
HW TCF-2			2.5	
HW TCF-3				0

SW = softwood; HW = hardwood.
Source of raw data: Lövblad and Malmström 1994.

Biomarkers

Since it was discovered that fish exposed to chlorine bleached kraft pulp mill effluent in Jackfish Bay, Lake Superior, exhibited a wide variety of physiological responses, including reproduction effects and induction of cytochrome P450 enzymes (ethoxyresorufin-o-deethylase [EROD] activity), there have been intensive discussions about the source and the significance of those findings. It is now believed that at least two effects are expressed: steroid problems related to a breakdown in the control and production of gonodal sex steroid, and induction of liver enzymes (Munkittrick et al. 1994). These studies indicate that the same responses are found outside mills that do not use chlorine and that biological treatment or large dilutions do not always eliminate the problem(s).

Detailed studies at PAPRICAN showed that increased EROD activity was seen with chlorine-based sequences and one $D_{100}E$ sequence, that the sequence $OD_{100}E$ suppressed EROD activity and that OQP, OZEP and another $D_{100}E$ sequence did not significantly alter hepatic mixed function oxygenase (MFO) activity (O'Connor et al. 1994).

CLOSING UP THE MILL

Obviously, if a mill succeeds in making high quality bleached pulp without an effluent, this mill will cause no adverse effects in the aqueous environment. However, as the total yield of pulp on wood would not increase in an effluent-free mill using current technology, obviously, the waste has to go to solid waste or be incinerated. The resulting atmospheric emissions may be trivial, if suitable technology is used. While it is true in principle that recycling bleaching effluents to recovery does increase air emissions, the effect may be quite minor. The increase in CO_2 discharge is offset by a reduction in fossil fuel consumption somewhere. The increases in discharges of other pollutants are trivial and SO_2 discharges will drop, simply because the bleach liquor has less S than fuel oil or coal. In mills burning natural gas or hog fuel, this would be irrelevant.

The driving force for mills that need to invest in new technology should be towards technology that will reduce operating costs or improve the yield or energy balance of the mill, i.e., technology that will survive even after tomorrow. These technologies are truly cleaner technologies and include systems for spill prevention and recovery, closing the screen room process, equipment for extended cooking and/or oxygen delignification, and change in bleach plant operations to more closed operation (e.g., chlorine dioxide or ozone bleaching) if the effluents can be recovered. Investments in increased chlorine dioxide capacity solely for the purpose of lowering the AOX content of the effluent lack environmental justification, unless one has already installed equipment to reduce the kappa number of the brownstock pulp entering the bleach plant, e.g., extended cooking or oxygen delignification. Last but not least, any technology performs no better than the people operating it. We believe that resources set aside for the training of mill personnel can result in great improvements in the overall environmental performance of the mill.

We consider that the priority of process changes to limit environmental effects from effluent discharges should be:
(1) Upgrade the pulping and recovery operation to prevent black liquor spills from entering the sewage treatment system.
(2) Change the bleach plant to use a molecular chlorine multiple of less than ≈ 0.15.
(3) Upgrade the brownstock washing system to reduce wash loss carryover to the bleach plant to less than 10 COD kg ADt^{-1}.
(4) Install extended delignification systems such as extended continuous cooking (e.g., Kamyr), extended batch cooking (e.g., Sunds) and/or oxygen delignification.
(5) Change the bleach plant to use a large fraction of chlorine dioxide or oxygen based chemicals. Ozone bleaching with filtrate recycle is a further extension of the delignification. The same can be said if the chlorine dioxide spent bleach liquor is recycled.

Using large amounts of hydrogen peroxide as a substitute for chlorine dioxide will not benefit the environment unless most or all of the P-stage effluents are recycled to the recovery system. The use of chelating agents to prevent trace metals from reducing hydrogen peroxide may prevent this recycling and may have negative effects on the receiving water.

It is often stated that TCF pulping is on the track to the closed cycle bleach plant (Pulp and Paper Canada Report 1993). Several competent authors have suggested that ECF pulping can achieve the same (Strömqvist 1994). In both cases non-process elements brought in with the wood are dissolved from the pulp. These elements include silica, aluminium, manganese and potassium. These elements must each be balanced by the installation of additional "kidney" functions—the spent bleach liquor cannot just be recirculated to the recovery system. For ECF pulp an additional chloride "kidney" must be installed.

Chloride removal is a major problem, and the only full-scale process yet installed (at Thunder Bay, Ontario in the 1970s) failed, despite considerable research and investment. However, modern ECF bleach sequences such as the increasingly widely used ODEODED discharge only about 20% of the chloride common in the traditional CEDED and similar bleach sequences. Maples et al. (1994) describe a bleach filtrate recovery (BFR) process that Champion intends to install full-scale in its Canton mill. This recovers the first D and Eo filtrates which represent about 90% of the organic loading of the ODEODED sequence. The BFR process is based on using the recovered filtrates to wash the pulp counter currently to the recovery boiler and includes a metal removal process for the D filtrate and a crystallization/washing system for removal of chlorides and potassium from the dust collected in the recovery boiler's electrostatic precipitator. The calculated increase in evaporator load is only 10%.

Presuming it performs as predicted, the BFR process will be simpler, more energy efficient and probably more economical than most other systems that have been proposed for the recovery of the effluent from ECF bleach plants. Champion does not intend to recover the filtrates from the later bleaching stages in this system, but expects to be able to do so eventually.

Once the chloride is removed the ECF mill will have to face the same problems with trace metals, scaling, and maintaining the hydraulic balance in day-to-day operation as a TCF mill attempting to operate with a similar degree of closure of the water system. At present, both ECF and TCF appear to be viable routes toward substantial reduction of the environmental effects of mills. The degree of recovery of the organic substances achieved will be more important than the bleaching sequence used.

ENERGY CONSUMPTION OF BLEACHING PROCESSES

The principal energy requirement for bleaching pulp is the manufacture of the bleaching chemicals. This energy may be consumed either on the mill site or at a remote location. Mill site manufacturing is common for oxygen and essential for chlorine dioxide and ozone. A very few mill site systems exist for manufacturing the sodium chlorate required for mill site manufacture of chlorine dioxide, but most mills purchase sodium chlorate from external sources. Whether manufactured on-site or off-site, the energy consumption is quite similar. From an environmental perspective, the location where the energy is used is inconsequential.

The energy required for the manufacture of the principal bleaching chemicals is summarized in Table 3. These data assume modern equipment with good but not exceptional operating practices and efficiencies. Actual values vary somewhat in practice, but since they are mostly quite dependent on electrochemistry, this variation is normally relatively minor. Most of the energy required for the production of chlorine dioxide is for the production of sodium chlorate, which is normally carried out off the mill site. Energy used to mix oxygen, ozone and chlorine dioxide with the pulp is not trivial, and allowances are included in the data shown.

Energy used for the transportation of chemicals is relatively minor, and the differences between different bleaching processes in this respect are negligible. The differences in the quantities of energy used to transport pulp within the mill are considered to be negligible.

Table 4 presents the energy consumption calculated by the authors for bleaching sequences in several mills. Except for Mönsterås, the data are based on feeding of softwood, and the end point is marketable bleached pulp. In the table, the bleach sequence symbols have the usual meanings. Lower case "p" and "o" refer to hydrogen peroxide and/or oxygen reinforced extraction stages, while upper case "P" and "O" refer to an independent hydrogen peroxide bleach stage, generally operating at over 70°C with a peroxide

Table 3. Specific energy consumption for manufacture of bleaching chemicals (includes energy to mix chemical with pulp).

0.00	kWh kg^{-1}		kWh kg^{-1}
Chlorine	1.6	Hypo (available Cl$_2$)	3.5
Caustic (ECU)	1.8	Hydrogen peroxide	3.5
Caustic (non-ECU)	1.8	Chlorine dioxide	11
Oxygen by truck	2.8	Ozone	14
On-site oxygen	2.8		

Source of raw data: N. McCubbin Consultants, Inc.

Table 4. Summary of energy demand and bleaching costs for actual mills.

Mill #	Bleach sequence	Kapp no. into bleach plant	Energy demand kWh t^{-1} pulp	Bleaching cost $U.S. t^{-1}
A	OWP (Värö)	12	192	38
B	OWPPP	14	213	43
C	OWPPP	14	213	43
D	OZEopD	6	216	26
B	OZZEopPP	3	227	40
C	OZZEopPP	3	227	40
E	ZQP (Mönsterås)	10	234	43
D	CdEoHD	31	256	27
E	OD,C/DEopDEpD	12	265	30
D	ODEoDED	17	301	28
F	OCdEoHD	15	304	39
F	ODEoDED	10	309	26
C	OZZEopDED	6	321	42
D	CdEoHD	31	323	35
B	OZZEopDED	6	325	43
B	CdEoHDED	31	349	35
D	DEoDED	31	363	33
B	CdEoHDED	31	385	42
B	ODEoDED	17	404	38
F	ODEoDED	13	419	36
A	ODEoDED (Värö)	15	425	37
C	DcEoHDED	33	437	50
C	DcEoHDED	33	442	46
B	DEoDED	31	462	42
C	ODEoDED	18	466	41
C	DEoDED	33	503	46

Source of raw data: N. McCubbin Consultants, Inc.

charge well over 10 kg t^{-1} pulp or an oxygen delignification stage. Oxygen and ozone stages are all medium consistency.

The consumption of each chemical is based on either data reported in literature or operating bleach plants known to the authors. In the case of Mönsterås and Värö, the data in Table 1 were used, and the consumption of the other bleach chemicals estimated by the authors. The brightness in various mills ranges from approximately 80 ISO to 90 ISO, which partially explains the scatter in the data. The table shows the kappa number of the pulp, at the entry to the first bleaching stage which discharges effluent to sewer. It was assumed that effluent from the Z and any immediately subsequent E_{op} stages would be recovered, but that the filtrate from later P stages would not. This corresponds to the best industrial practice in 1994, but as mills develop technology and operating techniques to recover bleach process effluents from peroxide and chlorine dioxide stages, the effective "kappa into bleach plant" will drop.

The progression from the highest to the lowest energy-consuming sequences is not smooth, reflecting the variations from mill to mill. In practice there is considerable variation between bleach plants using identical bleaching sequences. The authors have seen data where the consumption of bleaching chemicals differed by well over 25% in substantially identical bleach plants producing competitive products from similar wood, using identical bleaching sequences. The reasons are primarily differences in operating skills, which in turn depend heavily on management commitment, the quality of training and the supervision of process operators, and the skills of the maintenance personnel. In many cases these differences outweigh the advantage of one bleach sequence over the other with respect to energy efficiency and effluent quality.

Notice that the sequences using a mixture of chlorine dioxide and chlorine in the first stage are often more energy efficient than more modern sequences. Also, the sequences using ozone and peroxide are generally less energy intensive than the others, but some mills relying on chlorine dioxide have lower specific energy consumption than those with ozone and/or peroxide. It is trivial to say that there is no direct correlation between the kappa numbers of the pulp entering the bleach plant and the toxicity of the bleach plant effluent. It depends, on among other things, the wash loss, i.e., the amount of low molecular mass materials in the pulp prior to bleaching.

The costs of chemicals for different bleaching sequences are shown in Table 4 for typical 1994 U.S. prices. They are not closely correlated with the energy consumption, reflecting the fact that the costs of some bleaching chemicals are more dependent on electricity costs than others.

CONCLUSIONS

Conclusions of our review two years ago (Folke *et al.* 1993a) were that there was no evidence of any adverse effects related to the use of chlorine dioxide, i.e., the chlorinated organics formed had no effects in the environment and resembled naturally formed organochlorines (Grimvall 1993). Since then it has become even more clear that the environmental effects cannot be deduced from knowing the bleach sequence of a mill. One will have to know about the pulping operation, spill control, wash loss and so on. Comparison between mills must consider all substances discharged or the environmental effects of the whole effluent. Particularly in the case of today's low AOX bleaching sequences, comparison of bleaching technology is environmentally irrelevant from a toxicity point of view.

Bleaching sequences based on peroxide generally result in lower energy demand than those based on ozone, which in turn are more energy efficient than those based on chlorine dioxide. However, mill operating conditions, product specifications and operator skills can have just as much of an effect on energy consumption. The agenda is thus not whether to use ECF or TCF as the choice of bleaching sequence, but rather to modernize the pulping operation itself.

REFERENCES

Alder, A.C., H. Siegrist, W. Gujer and W. Giger. 1990. Behaviour of NTA and EDTA in biological wastewater treatment. Water Res. 24:733-742.

Amato, I. 1993. The crusade against chlorine. Science 261(9 July):152-154.

CEU. 1994. Commission decision of 1994 establishing the ecological criteria for the award of the community eco-label to toilet paper and kitchen rolls. Commission of the European Communities. *Off. J. Europ. Comm.:* In press.

Clark, D. 1994. A European perspective. International Non-Chlorine Bleaching Conference Proceedings, 6-10 March, Amelia Island, FL. Miller Freeman Inc. Distribution Center, Gilroy, CA.

Dahlman, O., I. Haglind, R. Mörck, F.d. Sousa and L. Strömberg. 1993. Chemical composition of effluents from chlorine dioxide bleaching of kraft pulps before and after secondary effluent treatment. EU CE PA International Environmental Symposium, Paris, France. Centre Technique du Papier, Grenoble, France, Vol. 1, pp. 193-215.

Deardorff, T.L., R.R. Willhelm, A.J. Nonni, J.J. Renard and R.B. Phillips. 1994. Formation of polychlorinated phenolic compounds during high chlorine dioxide substitution bleaching. Part 3: Mill trial results. TAPPI J. 77(8):163-168.

Folke, J., COWIconsult. 1989. The technical and economic aspects of measures to reduce water pollution caused by the discharges from the pulp and paper industry. Final Report. B6612-551-88 (Study Contract), Commission of the European Communities, DG XI, Brussels, Belgium.

Folke, J., and E. Männistö. 1993. Chlorine dioxide in pulp bleaching—environmental influence on future use. EU CE PA International Environmental Symposium, Paris, France. Centre Technique du Papier, Grenoble, France, Vol. 1, pp. 217-232.

Folke, J., H. Edde and K.-J. Lehtinen. 1991. The scientific foundation of adsorbable organohalogens (AOX) as a regulatory parameter for control of organochlorine compounds. TAPPI Environmental Conference, San Antonio, TX. TAPPI Press, Atlanta, GA, pp. 517-527.

Folke, J., L. Landner and N. McCubbin. 1992. Is AOX removal by biological effluent treatment consistent with environmental protection objectives? TAPPI Environmental Conference, Richmond, VA. TAPPI Press, Atlanta, GA, pp. 849-857.

Folke, J., L. Landner, K.-J. Lehtinen, E. Männistö, H. Männistö and N. McCubbin, European Environmental Research Group Ltd. and Duoplan Oy. 1993a. Chlorine dioxide in pulp bleaching—technical aspects and environmental effects. Literature study. CEFIC, Chlorate Sector Group, Brussels, Belgium.

Folke, J., L. Landner, K.-J. Lehtinen and N. McCubbin. 1993b. Simplified bioassays and chemical analysis to be used for regulatory purposes in the pulp industry. TAPPI Press, Atlanta, GA, pp. 413-425.

Grimvall, A. 1993. Naturally produced chloro-organics in the environment. First Global Conference on Paper and the Environment, 6-8 June, Brussels, Belgium. Papercast and Economie Papetière, Brussels, Belgium, pp. 122-132.

Kringstad, K.P., L. Johannesson, M.-C. Kolar, S. Swanson, B. Glas and C. Rappe. 1989. Influence of chlorine ratio and oxygen bleaching on the formation of PCDFs and PCDDs in pulp bleaching. Part 2: A full mill study. TAPPI J. 72:163-170.

Landner, L., O. Grahn, J. Härdig, K.-J. Lehtinen, C. Monfelt and J. Tana. 1994. A field study of environmental impacts at a bleached kraft pulp mill site on the Baltic Sea coast. Ecotoxicol. Environ. Safety 27:128-157.

Lövblad, R. and J. Malmström. 1994. Biological effects of kraft mill effluents. A comparison between ECF and TCF pulp production. International Non/Chlorine Bleaching Conference, March 6-10, Amelia Island, FL. Miller Freeman Inc. Distribution Center, Gilroy, CA.

Maples, G., R. Ambady, J.R. Caron, S. Stratton and R.V. Canovas. 1994. BFR. A new process toward bleach plant closure. Proc. Intl. Pulp Bleaching Conf., Vancouver, BC, pp. 253-262.

McCubbin, N. and J. Folke. 1993. Significance of AOX vs unchlorinated organics. CPPA Annual Meeting, Thunder Bay, Ontario, Canada.

Munkittrick, K.R. and G.J. Van Der Kraak. 1994. Receiving water environmental effects associated with discharges from Ontario pulp mills. Pulp and Paper Canada 95(5):57-59.

Nutt, W.E., B.F. Griggs, S.W. Eachus and M.A. Pikulin. 1992. Development of an ozone bleaching process. Proc. TAPPI Pulping Conference, Nov. 1-5, Boston, MA, pp. 1109.

O'Connor, B.I., R.G. Kovacs, R.H. Voss and P.H. Martel. 1992. A study of the relationship between laboratory bioassay response and AOX content for pulp mill effluents. 78th Annual Meeting, Technical Section CPPA, Montreal, Quebec, p. A223.

O'Connor, B.I., T.G. Kovacs, R.H. Voss and P.H. Martel. 1993a. A study of the relationship between laboratory bioassay response and AOX content for pulp mill effluents. Proceedings of the Annual Meeting of the CPPA Technical Section, Montreal, Canada, p. 223.

O'Connor, B.I., T.G. Kovacs, R.H. Voss, P.H. Martel and B.V. Lierop. 1993b. A laboratory assessment of the environmental quality of alternative pulp bleaching effluents. EU CE PA International Environmental Symposium, Paris, France. Centre Technique du Papier, Grenoble, France, Vol. 1, pp. 273-297.

O'Connor, B.I., T.G. Kovacs, R.H. Voss, P.H. Martel and B.V. Lierop. 1994. A laboratory assessment of the environmental quality of alternative pulp bleaching effluents. Pulp and Paper Canada 95(3):47-56.

Pulp and Paper Canada Report. 1993. Non-chlorine bleaching aims at closing the process loop. Pulp and Paper Canada 94(5):8-12.

Reeve, D. 1993. Banning AOX unjustified. Pulp and Paper Canada 94(6):7.

Renberg, L. 1992. The use of cost-effective chemical and biological tests for the estimation of the environmental impact of bleaching plant effluents. TAPPI Environmental Conference. TAPPI Press, Richmond, VA, pp. 317-329.

Sandström, O. and Å. Larsson. 1993. Environmental impact of the pulp and paper mill effluents. Summary of results from Nordic investigations, in Swedish. *In:* Environmental Questions in Relation to the Nordic Pulp and Paper Industry. Nordic Council of Ministers, Copenhagen, Denmark.

Strömqvist, D. 1994. New technology development for the closed cycle bleach plant. Second Global Conference on Paper and the Environment, 6-8 June, Frankfurt, Germany. Papercast and Economie Papetière, Brussels, Belgium.

Tsai, T.Y., J.J. Renard and R.B. Phillips. 1994. Formation of polychlorinated phenolic compounds during high chlorine dioxide substitution bleaching. Part 1: Laboratory investigation. TAPPI J. 77(8):149-157.

Yin, C., J.J. Renard and R.B. Phillips. 1994. Formation of polychlorinated phenolic compounds during high chlorine dioxide substitution bleaching. Part 2: Biotreatment of combined bleach plant effluents. TAPPI J. 77(8):158-162.

FUTURE DIRECTIONS FOR ENVIRONMENTAL HARMONIZATION OF PULP MILLS

Jens Folke

MFG—European Environmental Research Group, Østergade 16, 3250 Gilleleje, Denmark

This paper summarizes three areas of discussion related to future directions for environmental harmonization of pulp mills: (1) the future direction of scientific research, (2) regulatory needs of the future and (3) the technological implications of the scientific findings. The conference made it clear that no more research is needed in the fields of organochlorines such as AOX, chlorophenols and polychlorinated dibenzo-p-dioxins (PCDDs). Instead more research is needed to investigate the environmental exposure and effects of natural wood products and their transformation products. An optimal regulatory strategy was to combine three different tactics: (1) discharge limits, at the end of the effluent pipe, for amounts of pollutants emitted, to attain a reasonable level of industrial practice; (2) site-specific limits, strict enough to eliminate sublethal effects within the aquatic community beyond a mixing zone; and (3) periodic ecological surveys to check the effectiveness of the first two tactics. While it is scientifically unfounded to regulate AOX, the inclusion of COD or DOC under tactic 1 is more controversial. On the one hand there is no direct link between COD/DOC and environmental effects. On the other hand a tight COD/DOC control improves the mill energy balance and may eliminate the compounds in this fraction that cause the remaining weak toxic effects observed from modern bleach plant operations. The reason for the controversy is that many North American pulp mills have invested in chlorine dioxide bleaching instead of oxygen delignification and other techniques that would close up the water streams of the mill. Pollutants measured as COD are lost resources and a tight COD regulation will improve the energy balance and substitute fossil fuel for biofuels. But what are the social costs for society of forcing the development too quickly?

INTRODUCTION

This paper summarizes the papers and the panel discussion of the session on future directions for pulp mills. Three areas were discussed in particular:
- The future direction of scientific research.
- Regulatory needs of the future.
- The technological implications of the scientific findings.

Table 1 is an attempt to briefly summarize the discussion on those three subjects. The table shows how science has developed from assessing acute lethal toxicity in the 1970s and beginning of the 1980s, how the regulatory authorities responded by BOD/TSS and toxicity controls and how the mills responded somewhat differently in North America and Scandinavia. In the mid-1980s tetrachlorodibenzo-p-dioxin (TCDD) became a big issue, particularly in North America, while Scandinavia took the whole world on an "AOX ride" by misinterpreting the results of the Norrsundet study and basically overlooking the effects of black liquor spills as a source of toxicity (refer to the paper by Owens 1991). In Germany, the sulfite mills at that time had financial problems due to a new requirement for SO_2 emission control, and they were quick to grab the opportunity to market totally chlorine free (TCF) bleached pulp at a supreme price with the assistance of NGOs. Scandinavian mills responded to this market change by introducing elemental chlorine free (ECF) or TCF papers for all qualities of bleached pulp, while the majority of North American went to ECF to eliminate the TCDD problem.

Table 1. The development of environmental science, regulations and technologies for pulp and paper mill effluent control. Copyright MFG; reproduced with permission.

Science:		
• Effluents acutely, lethal toxic	• fish contamination by dioxins	• Search for substance 'X' and control options
• Oxygen depletion	• dioxins in paper products	• Evaluation of metabolic impairment, immunosuppression, invertebrate reproduction
• Habitat degradation	• AOX wrongly linked to effects (Norrsundet study)	
	• Effluents sub-lethal toxicity	
	• Biomarkers (induction/ suppression of liver enzymes)	
Regulation:		
• BOD/TSS controls	• Dioxin control to prevent formation through changes to bleach plants	• Search for a science based regulation or use The Principle of Precautionary Action?
• Toxicity controls	• AOX control (in Europe)	• COD limitation? and/or control of substance 'Y', 'Z',...?
Technology:		
• Secondary effluent treatment (in North America)	• ECF bleaching from Kappa No. 32 (in North America)	• Extended cooking or O-stage delignification + ECF/TCF bleaching
• O-stage delignification + primary treatment (in Sweden)	• Extended cooking; low Cl$_2$-multiple + secondary treatment (in Europe)	• Progressive systems closure
		• Minimum impact mills? or total effluent free mills?
1980	1990	2000

Meanwhile, as it became clear to the scientific community, particularly after the 1991 Saltsjöbaden conference, that AOX was a non-issue, Canadian and Scandinavian environmental research programs started to focus on sublethal toxicity (e.g., effects on reproduction, metabolic effects) and biomarkers (e.g., MFO induction, UDP-GT suppression/induction, etc.). Substances responsible for these effects were also identified in effluents that did not use chlorine for bleaching or did not use bleaching at all, and a search for the actual compound(s) began. The regulators have not responded uniformly to this new challenge. It is probably fair to say that the Canadian federal regulators are in search of a new science-based regulation (Carey et al. 1994), while several provinces along with some European regulators are pleading for "the Principle of Precautionary Action," meaning that they promote zero AOX and zero discharge merely on the speculation that this will be of benefit to the environment, but without any scientific back-up (Lagergren 1994). The U.S. EPA is promoting a "Best Available Technology" philosophy based on a requirement to do so from the Congress, meaning that if a certain level of discharge of a particular contaminant can be achieved, then that level will be required by all the mills independently of whether or not that level will be of benefit to the environment. Therefore, AOX along with a number of other parameters and specific compounds are proposed to be regulated in the cluster rules. In the following, the three issues shall be examined in more detail.

SCIENTIFIC RESEARCH

As mentioned, in the course of the last five years, the focus of the scientific research on adverse effects on the aquatic environment from pulp and paper mills has changed from a focus on bleaching towards a focus on the whole process and in particular a focus on the pulping process. The conclusion crystallizes that the remaining weak, toxic effects of pulp and paper mill effluents from modern operations cannot be explained from the discharge of organochlorine compounds, be that AOX, chlorinated phenolics or any other specific chlorinated compound. To explain the observed chronic sublethal effects one has to focus on the entire pulp and paper mill operation. Clearly, non-chlorinated compounds discharged from the pulping operation is the major factor.

We think it is fair to say that there is a general consensus among scientists that EROD, UDP-GT and similar biomarker responses should be interpreted as only an indicator of exposure. Several scientists mentioned that future research should focus more on actual bioindicators and that plots of dose-response curves against time should be established, so that both inhibitory and stimulatory responses can be assessed. Lehtinen (1994) referred to results on the effects of 2,4,6-trichlorophenol and dehydroabietic acid on UDP-GT response in exposed rainbow trout. While 2,4,6-trichlorophenol induced UDP-GT response and dehydroabietic acid inhibited UDP-GT response, the combined effect of both substances resulted in a fluctuation of inhibitory and stimulatory responses with time. Had UDP-GT only been measured, either an inhibition, a stimulation or no response could have been achieved. This is the concern that research be more focused on impacts or long-term effects and less on undefined impacts of discharges, and that there be more clear communication of results from scientists to regulators and the general public.

Folke *et al.* (1996) emphasized the need that biological field studies as well as studies on whole mill effluents describe the particular production method in greater detail than just referring to ECF or TCF bleaching. Also, to compare the toxicity among different effluents they suggested using $logTEF = Q \times logTU$ (where Q is the amount of effluent per tonne of pulp produced and TU is the toxic units, EC_{xx}^{-1} instead of the hitherto used $TEF = Q \times TU$, the rational being that biological data are essentially logarithmic and in this way $logTEF = 0$, if the effluent is non-toxic.

A weakness of some studies is that they refer to the concentration of the effluent and do not give enough details of the pulping and bleaching sequences to compare to other technologies. Another weakness of experiments is improper or poor experimental designs which can lead to accepting a false hypothesis.

There is a feeling among some scientists that no more research is needed in the field of organochlorines such as AOX, chlorophenols and dioxins. Instead they would like to see more research on natural wood products and their transformation products. The COD or DOC of a biologically treated effluent contains exactly that fraction of the organic substances, and the natural ecosystems cannot readily degrade this fraction, as it has already been biotreated. However, this provides no indication of whether or not these substances are harmful to the environment.

Figure 1 shows the relation between BOD and COD in a conceptual way. (This does not imply that BOD can be calculated directly from a COD analysis. The actual COD/BOD ratio is quite variable.) DOC is an instrumental analysis in which a filtered water sample is burned and the evolved CO_2 is quantified by IR detection, or it is catalytically reduced to methane, which is subsequently burned and quantified in a flame ionization detector. Thus, as opposed to COD, it is independent of the oxidation stage of organic carbon and does not include other elements such as hydrogen, sulfur and nitrogen. The shaded areas of Figure 1 depict the slowly degradable substances among which we will find those compounds that are responsible for the remaining weak toxic effects of pulp mill effluents. We expect to see many contributions addressing the exposure and impact of those compounds in the years to come.

REGULATIONS

Environmental protection philosophies have changed from a virtual non-concern in the 1960s, to the "solution to the pollution is dilution" approach of the 1970s, the "end-of-pipe" treatment of the 1980s on to the "cleaner technology" option of the 1990s. The underlying philosophy of the 1970s and 1980s can best be characterized as the "environmental protection" era (Fig. 2). The environment was regarded as an externality and the industry was required to set aside money to pay the cost for avoiding detrimental effects on what was outside the fence (Landner 1994). The philosophy of the 1990s has been driven by the Bruntland Report's "sustainable development" concept (World Commission on Environment and Development 1987) into a "resource management" philosophy, by which the industry has to save resources and energy and take responsibility for its products during the whole life cycle. The "polluter pays principle" has been introduced as a means of internalizing the social costs of pollution rather than requiring a particular "end-of-pipe" treatment. While this philosophy has many advantages compared to the simpler "environmental protection" philosophy, it is predicted that a new era of "eco-development" will take over, in which the environment is no longer regarded as an externality that can be more or less internalized (ecology being economized), but rather will be regarded as an integrated part of the whole production cycle (economy being ecologized) (Landner 1994).

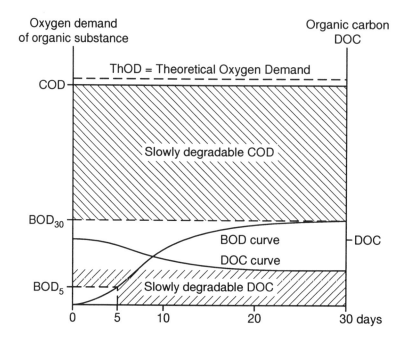

Figure 1. DOC, COD and BOD_5 relations. Copyright MFG; reproduced with permission.

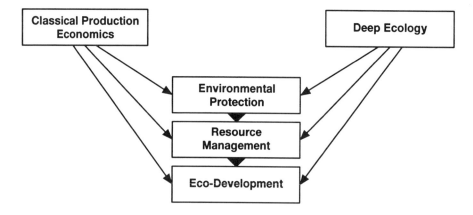

Figure 2. The evolution of environmental protection philosophies. Copyright MFG; reproduced with permission.

The successively evolving philosophy is reflected in the pulp and paper industry effluent limitations set by regulatory authorities worldwide. In the 1970s, effluent regulations were based on the same parameters as those used as design parameters for biological wastewater treatment facilities, i.e., BOD_5 and TSS. As mentioned, AOX gained popularity among some regulators in the late 1980s, partly on the basis of a misinterpretation of scientific results (Folke *et al.* 1991). None of the papers presented in this book supports the use of AOX as a meaningful regulatory parameter. In the 1990s a tendency to select criteria that force better resource management on the industry has emerged. In a discussion of various approaches for regulating effluents from pulp mills, Sprague (1991) suggested as an optimal strategy combining three different tactics:

(1) Discharge limits, at the end of the effluent pipe, for amounts of pollutants emitted, to attain a reasonable level of industrial practice;
(2) Site-specific limits, strict enough to eliminate sublethal effects within the aquatic community beyond a mixing zone;
(3) Periodic ecological surveys to check the effectiveness of the first two tactics.

Tactic 1 limits the discharges exclusively at the end of the effluent pipe and is applicable to all mills to attain an acceptable level of industrial practice. Tactic 2 is site-specific and is used to eliminate sublethal effects within the aquatic community beyond a mixing zone. The variables that might be regulated under tactic 1 are various physical and chemical analyses such as COD, TSS, Tot-P and acute toxicity, all expressed as amounts per tonne of pulp (or paper) produced. Also, acute and chronic toxicity should be expressed as an amount, i.e., logTEF. The objective of tactic 2 should be to eliminate all ecosystem damage, beyond the edge of a specified mixing zone. Chemical analyses could include BOD_5, chlorate and color, as well as analyses of specific substances known to cause defined ecological effects. Effect analyses should be based on sensitive bioassays for acute and chronic toxicity using species that are characteristic for the particular receiving environment such as applied by Canadian and Australian regulatory authorities or species that have a large data base, e.g., the crustacean *Ceriodaphnia dubia*, the zebra fish (*Brachydanio rerio*) and the alga *Selenastrum capricornutum*. The most efficient way to apply tactic 2 is to analyze the effluent itself and then calculate whether conditions will be satisfactory at the edge of the mixing zone. The need for tactic 3 has been stressed by several participants and it is important to look for a range of ecosystem effects when monitoring the receiving waters.

The package of regulations for discharges from bleached kraft mills issued by Australia in 1989 is an application of the three tactics outlined that discuss the parameters to be included under tactics 1 and 2 (Gifford and McFarlane 1991). The outcome of a five-year research program was presented by Abel (1994). According to Abel (1994) AOX is to be included under tactic 1 for "political reasons" only.

The inclusion of COD or DOC under tactic 1 is somewhat controversial. With bleach plant effluents, DOC has the advantage over COD that the ratio between AOX and DOC is a direct measure of the degree of chlorination, i.e., the number of carbon atoms per chlorine atom. Thus, the DOC:AOX ratio could be regulated to avoid discharging highly chlorinated materials. The cluster rules avoid the formation of highly chlorinated materials by regulating polychlorinated phenolics directly. COD has been regulated in Europe and has been proposed to be regulated in the new U.S. EPA cluster rules (U.S. EPA 1993). It appears that control of accidental spills, closing up the screen room and reducing the wash loss would reduce the remaining, non-lethal toxicity from pulping operations. However, since it is not feasible to regulate technology per se, regulation of COD and DOC could be a substitute at least until the scientific data base is sufficiently strong to regulate the wood extractives causing effects directly. COD or DOC in the effluent of a pulp mill can be regarded as a biofuel that has not been utilized. Tight regulation under tactic 1 will force technology in places that will utilize this resource and thus give a better energy balance of the mill and possibly diminish the remaining weak toxic effects. Tight COD control may be sufficient to meet the concerns of some NGOs and a totally effluent-free mill may not be needed after all. Tightening up the mill too quickly may increase air pollution and solid waste due to the use of inappropriate technologies.

TECHNOLOGY

Technology has been developed and will continue to be developed at a much faster pace than biological science can provide the answers as to what effects will be associated with a particular technology. Lagergren (1994) from the Swedish EPA referred to two different actions that could be taken as a result of scientific research:
- Require "hard evidence" before regulatory action is taken.
- Regulatory action should be taken on the basis of "early warnings," "before it's too late."

Lagergren (1994) pointed as an example to the difficulties in predicting effects at the population level and quite rightly, as pointed out by Carey (1994), regulators would have done a poor job if large-scale effects on populations were seen. The problem arises when the use of the precautionary principle is extended to taking action despite scientific evidence showing that this action will not correct the "problem" (refer to the AOX discussion above). Of course this is less of a problem if there is no social or other costs associated with the regulation. But in case of the North American industry the "chlorophobia" has led to investments in much too large chlorine dioxide capacity instead of technology that would have reduced the overall organic waste load. If there are social or other costs associated with the regulation, Owens (1994) and Vrooman and Wearing (1994) both stressed the need to conduct a cost-benefit analysis. It was suggested that the industry and regulatory authorities work together in voluntary programs such as the Canadian ARET (Accelerated Reduction of Environmental Toxicants). The technology is there to close up a green field mill entirely (total effluent free), but at what cost in terms of environmental trade-offs? For many existing mills a progressing system closure may be the only alternative to closing down the mill entirely. Mills that have already invested heavily in chlorine dioxide capacity to bleach, for example, softwood from kappa no. 32 and extensive wastewater treatment to handle the discharge may be unable to invest any further for the first couple of years to write off the already large investment, and this may not be needed if the mill is situated at a good site where the receiving waters have a large capacity for dealing with the weak toxic effects associated with these operations.

For a mill that has not yet invested in additional chlorine dioxide capacity a more viable route for upgrading an "old" North American mill to limit environmental effects from effluent discharges would be (Folke *et al.* 1994):
(1) Upgrade the pulping and recovery operation to prevent black liquor spills from entering the sewage treatment system.
(2) Change the bleach plant to use a molecular chlorine multiple of less than ≈ 0.15.
(3) Upgrade the brownstock washing system to reduce wash loss carryover to the bleach plant to less than 10 COD kg ADt^{-1}.
(4) Install extended delignification systems such as extended continuous cooking (e.g., Kamyr), extended batch cooking (e.g., Sunds) and/or oxygen delignification.
(5) Change the bleach plant to use a large fraction of chlorine dioxide or oxygen based chemicals. Ozone bleaching with filtrate recycle is a further extension of the delignification. The same can be said if the chlorine dioxide spent bleach liquor is recycled.

Using large amounts of hydrogen peroxide as a substitute for chlorine dioxide will not be of benefit to the environment unless most or all of the P-stage effluents are recycled to the recovery system. The use of chelating agents to prevent trace metals from reducing hydrogen peroxide may prevent this and may have negative effects on the receiving water.

Swedish mills have to a large extent followed that route. According to Lagergren (1996), 11 Swedish mills apply extended delignification, 15 mills apply oxygen delignification, 15 mills use ECF bleaching, 11 mills use peroxide and 3 mills use ozone.

The big question is: How far do we need to go? To a totally effluent-free mill or to a minimum environmental impact mill? What are the environmental benefits and trade-offs in the two cases? Pollutants measured as COD are lost resources and tight COD regulation will improve the energy balance and substitute fossil fuel for biofuels. But what are the social costs of forcing the development too quickly?

There are already examples of effluent-free mills (e.g., the BCTMP mill at Meadow Lake (Canada) and the sulfite mill at Domsjö, Sweden), but this does not mean that all existing mills can be upgraded or even that it is beneficial for all greenfield mills to be constructed totally effluent free.

It is fair to say that this conference completed the scientific research on the bleach plant priorities. The remaining areas are pulping, air, solid waste and forestry/fiber supply issues. Vrooman (1994) pointed out that research in these fields needs to be precise and clear, with quantifiable goals set. What do we want to protect? What effects are present? Better public communication of scientific results would greatly benefit the industry.

ACKNOWLEDGMENTS

Thanks to Ms. Lauren Blum, Environmental Defense Fund in New York; Dr. John Carey, National Water Research Institute at the Canada Centre for Inland Waters in Burlington, Ontario; Professor Bjarne Holmbom, Åbo University, Finland; Dr. J. William Owens, The Proctor & Gamble Company, Cincinnati, Ohio; and Mr. Wally M. Vrooman, Avenor Inc., Thunder Bay, Ontario, for participating in the panel discussion that formed the basis of this paper.

REFERENCES

Abel, K. 1994. Research supporting environmental guidelines for eucalypt kraft pulp mills in Australia. Presented at the Second International Conference on Environmental Fate and Effects of Bleached Pulp Mill Effluents, November 6-10, 1994, Vancouver, B.C., Canada.

Carey, J., A. Chung, H. Cook and D. Halliburton. 1994. Overview of current Canadian federal regulatory regime for the bleached pulp and paper industry and future regulatory developments. Presented at the Second International Conference on Environmental Fate and Effects of Bleached Pulp Mill Effluents, November 6-10, 1994, Vancouver, B.C., Canada.

Folke, J., H. Edde and K.-J. Lehtinen. 1991. The scientific foundation of adsorbable organohalogens (AOX) as a regulatory parameter for control of organochlorine compounds. Tappi Environmental Conference, San Antonio, TX. Tappi Press, Atlanta, GA, pp. 517-527.

Folke, J., L. Renberg and N. McCubbin. 1994. Environmental aspects of ECF vs. TCF pulping. Presented at the Second International Conference on Environmental Fate and Effects of Bleached Pulp Mill Effluents, November 6-10, 1994, Vancouver, B.C., Canada.

Folke, J., L. Renberg and N. McCubbin. 1996. Environmental aspects of ECF vs. TCF pulp bleaching. *In* Environmental Fate and Effects of Pulp and Paper Mill Effluents, M.R. Servos, K.R. Munkittrick, J.H. Carey and G. Van Der Kraak (ed.), St. Lucie Press, Delray Beach, FL.

Gifford, J.S., and P.N. McFarlane. 1991. The development of environmental control legislation and effluent standards for Australasian wood processing industries. Wat. Sci. Technol. 24(3/4):37-44.

Lagergren, S. 1994. A regulatory perspective on future direction for pulp mills. Presented at the Second International Conference on Environmental Fate and Effects of Bleached Pulp Mill Effluents, November 6-10, 1994, Vancouver, B.C., Canada.

Lagergren, S. 1996. Swedish environmental regulations for bleached kraft pulp mills. *In* Environmental Fate and Effects of Pulp and Paper Mill Effluents, M.R. Servos, K.R. Munkittrick, J.H. Carey and G. Van Der Kraak (ed.), St. Lucie Press, Delray Beach, FL.

Landner, L. 1994. How do we know when we have done enough to protect the environment? Mar. Poll. Bull.: Submitted.

Lehtinen, K.-J. 1994. Biochemical responses in organisms exposed to effluents from pulp production—are they related to bleaching? resented at the Second International Conference on Environmental Fate and Effects of Bleached Pulp Mill Effluents, November 6-10, 1994, Vancouver, B.C., Canada.

Owens, J.W. 1991. A critical review of scandinavian studies on the aquatic impacts of pulp and paper effluents. Tappi Environmental Conference, San Antonio, TX. Tappi Press, Atlanta, GA, pp. 271-281.

Owens, J.W. 1994. Regulation of pulp mill aquatic discharges: current status and needs from an international perspective. Presented at the Second International Conference on Environmental Fate and Effects of Bleached Pulp Mill Effluents, November 6-10, 1994, Vancouver, B.C., Canada.

Sprague, J. 1991. Environmentally desirable approaches for regulating effluents from pulp mills. Water Sci. Technol. 24(3/4):361-371.

U.S. EPA. 1993. Proposed technical development document for the pulp, paper and paperboard category effluent limitations guidelines, pretreatment standards, and new source performance standards. Engineering and Analysis Division, Office of Science and Technology, U.S. Environmental Protection Agency, Washington D.C.

Vrooman, W.M., and J. Wearing, 1994. Future directions for Canadian pulp mills. Presented at the Second International Conference on Environmental Fate and Effects of Bleached Pulp Mill Effluents, November 6-10, 1994, Vancouver, B.C., Canada.

World Commission on Environment and Development. 1987. Our common future. Gro Harlem Brundtland (ed.), Oxford University Press, Oxford, UK.

Keyword Index

A

Adsorbable organohalide (AOX), 3, 41, 53, 83, 129, 151, 160, 170, 179, 182, 203, 204, 229, 239, 335, 379, 673, 683, 693
Adsorption kinetics, 139
Adult fish survey, 558
Algae growth, 33, 41
Alkylaromatic structures, 341
Aquatic birds, 303
Aquatic mammals and reptiles, 307
Artificial streams, 525, 549, 571

B

Benthic algae, 571, 583
Benthic macroinvertebrates, 525, 571, 583
Bioaccumulation, 171, 281, 283, 335, 341
 invertebrates, 315, 335, 341
 fish, 283, 327
 wildlife, 297
Bioavailability, 283, 613
Biochemical responses, 347, 359, 391, 439
 biomarkers, 369, 473, 621, 685
 EROD, 79, 350, 360, 425, 473, 557, 620, 637
 MFO, 79, 162, 349, 359, 369, 379, 360, 401
Biodegradation, 179, 241
Biofilms, 239
Biological monitoring, 591
Biological responses, 69, 161, 307
Biological treatment, 27, 42, 107, 352, 401, 526, 637
Biomarkers, 369, 473, 621, 685
Biota-sediment accumulation factors, 283
Biotransformation, 171, 273
Black liquor, 195
Bleaching process, 401
Bleaching technologies, 275

C

Chemical composition, 3, 21, 159, 169
Chemical characterization, 42, 85, 271
Chlorinated acetic acids, 8
Chlorinated etherified lignin structures, 129
Chlorinated degradation products, 129
Chlorinated organics, 203, 613, 673, 681
Chlorinated phenolic compounds, 7, 8, 24, 25, 41, 56, 79, 107, 170, 335, 369, 391
Chlorine bleaching, 351, 359

Chlorine dioxide substitution, 3, 24, 53, 107, 253, 369, 463, 473, 526, 613, 617, 681
Chloroform, 9
Chlorolignin, 171, 180
Chronic toxicity, 33, 53, 69, 95, 391, 459, 684
Contaminant loads, 229
Contaminant profiles, 151, 315
Contamination sources, 315
Cycling of chloroorganic substances, 203

D

Degradation processes, 239, 275
Detoxification enzymes, 360, 369
Disease challenge, 393
Dispersal, 170
Dissolved organic matter, 273
Dynamic surface tension, 139

E

Elemental chlorine free (ECF), 3, 21, 24, 33, 41, 69, 95, 151
Ecotoxicological effects, 613, 637
Effluent characterization
 toxicity, 69, 79
 chemistry, 129, 151, 154
Effluent fractionation, 79, 95, 360
Effluent quality, 151, 254, 468
Effluent toxicity, 462
Elemental chlorine free (ECF), 3, 41, 69, 95, 159, 359, 369, 401, 460, 483, 637, 675, 681, 693
Energy consumption, 681
Environmental effects, 662, 681
Environmental fate, 169, 179, 219, 271, 618
Environmental regulation, 591, 627, 661, 673, 693, 695
Etherified lignin structures, 129
Ethoxyresorufin-o-deethylase (EROD), 79, 350, 360, 425, 473, 557, 620, 637
Experimental streams, 54, 473
Exposure, 169
Extended delignification, 3
Extractable organic chlorine (EOCl), 172, 203, 483
Extractives, 3, 22, 25

Keyword Index

F
Fate, 167, 169, 179, 203, 219, 229, 239, 253, 261, 271, 618
Fathead minnow, 53
Fatty and resin acid, 9, 41, 630
Fecundity, 454
Fish population response, 449, 557, 581
Fish production, 473
Fish larval growth tests, 53
Fish reproduction, 425, 449, 563, 663
Food chain models, 283
Fugacity, 220
Future directions, 659, 661, 673, 681, 693

G
Growth inhibition, 33, 41
Growth rate, 452, 473, 563

H
High molecular weight material, 9, 25, 129, 160, 239
Histological lesions, 495, 511
HPLC-fractionated extracts, 69
Hydrogen peroxide, 3
Hydrophilic fraction, 95

I
Identity of MFO inducers, 353
Integrated monitoring, 589, 591, 599, 613, 627, 651
Internal process changes, 3
Invertebrate growth and development, 525

K
Kappa number, 14

L
Laboratory whole-organism tests, 401, 459
Larval growth, 53
Life cycle experiments, 460
Life-history variables, 449
Lignin, 22
Luminescence bacteria test, 33, 41

M
Macrobenthic assemblages, 533
Marine mammals, 303
Marine organisms, 95
Marine and estuarine birds, 298
Meiofauna community, 483
Mesocosm, 179, 261, 373
Mixed function oxygenase (MFO), 79, 162, 349, 359, 369, 379, 360, 401
Microcosm, 70
Mill processes, 468
Mobility inhibition, 33
Modernization, 174
Modified cooking, 3
Molecular size distribution, 41, 179, 180, 242, 261
Monitoring environmental effects, 627

N
Nanofiltration, 79
National legislation, 661
Nutrient enrichment, 525, 549, 571

O
Organic enrichment, 533
Organic compounds, 629
Organochlorines, 341, 557, 693
Organosulfur compounds, 195
Origins of chemicals, 1, 21, 159
Origins of toxicity, 1, 33, 41, 53, 69, 159
Oxygen delignification, 3, 53, 107
Ozone, 3, 5

P
Parasites, 495, 511
Particle dynamics, 207
Partitioning and fate, 288
Pathways, 204
PCDD/PCDF, 9, 276, 297, 315, 327, 350, 631
Peroxide, 681
Photodegradation, 239
Pollutants in wildlife, 297
Polychlorinated anisoles, 327
Polychlorinated veratroles, 327
Population responses, 449, 557
Principal components analysis, 315
Process change, 483, 613
Process modification, 3, 160, 613, 637
Pulping technologies, 274

Keyword Index

Q
Quantifying resin acids, 119
QWASI model, 220, 221

R
Red algae reproduction, 95
Regulation, 591, 627, 661, 673, 693, 695
Resin acids, 9, 41, 119, 139, 173, 391

S
Secondary treatment, 11, 33, 41, 79, 95, 179, 183, 360, 425, 617
Sediment extracts, 69
Sediment formation, 261
Sedimentation, 207
Sediments, 172
Sexual maturation, 452, 563
Shutdown, 79
Size exclusion chromatography, 180
Sources of chemicals and toxicity, 21, 26, 352
Sources of pulping chemicals, 21
Sources of bleaching derived chemicals, 21
Species richness, 538
Stable isotopes, 413
Steroid hormones, 425
Sterols, 9, 41, 359
Surface tension, 139
Swimming performance, 393

T
Tainting, 327
Terpenes, 41
Tetrahydrofuran-soluble, 266
Thioacidolysis, 131

Totally chlorine free (TCF), 3, 21, 26, 33, 41, 95, 159, 359, 401, 460, 637, 675, 681, 693
Toxicity, 95, 161, 632
Toxicity emission factors, 681
Toxicity tests, 33, 41, 637
Toxicological assessment, 461

U
Ultrafiltration, 7, 130, 240, 180
Unbleached process, 95

W
Water quality, 107, 616
  model, 219
  protection values, 107, 108
  screening value, 108
Whole organism responses, 447, 449, 459, 557, 581
  algae, 571
  benthic invertebrates, 525, 533, 549, 571
  effect monitoring meiofauna community, 483
  experimental streams, 473, 549
  lab responses, 459
  life-history variables, 449
  wild fish, 557
  parasites, histological lesions, 495, 511
Whole organism field studies, 449, 473
Wildlife, 297
Wood extractives, 613
Wood chemistry, 22

Z
Zebra fish hatching and survival, 41
Zebra fish hatching and survival test, 33